T0181519

Lecture Notes in Computer Science 13270

Advanced Research in Computing and Software Science

Subline of Lecture Notes in Computer Science

More information about this series at https://link.springer.com/bookseries/558

Cristina Bazgan · Henning Fernau (Eds.)

Combinatorial Algorithms

33rd International Workshop, IWOCA 2022
Trier, Germany, June 7–9, 2022
Proceedings

Editors
Cristina Bazgan ⓘ
Université Paris-Dauphine
Paris, France

Henning Fernau ⓘ
Universität Trier
Trier, Germany

ISSN 0302-9743   ISSN 1611-3349   (electronic)
Lecture Notes in Computer Science
ISBN 978-3-031-06677-1   ISBN 978-3-031-06678-8   (eBook)
https://doi.org/10.1007/978-3-031-06678-8

This Springer imprint is published by the registered company Springer Nature Switzerland AG
The registered company address is: Gewerbestrasse 11, 6330 Cham, Switzerland

Preface

The 33rd International Workshop on Combinatorial Algorithms (IWOCA 2022) was planned as a hybrid event, with the on-site activities taking place at the University of Trier, Germany. Due to the COVID-19 pandemic and also in order to lower the carbon footprint of the conference, it was decided that the conference could be also attended online. The conference was scheduled during June 7–9, 2022, followed by the Graphmaster and Stringmaster Workshops.

IWOCA is an annual conference series that started in 1989 as AWOCA (Australasian Workshop on Combinatorial Algorithms), and became an international conference in 2007, having been held in Australia, Canada, the Czech Republic, Finland, France, Indonesia, India, Italy, Japan, Singapore, South Korea, the UK, and the USA. Now, Germany can be added to this list of IWOCA countries. The conference brings together researchers on diverse topics related to combinatorial algorithms, such as algorithms and data structures; algorithmic and combinatorial aspects of cryptography and information security; algorithmic game theory and complexity of games; approximation algorithms; complexity theory; combinatorics and graph theory; combinatorial generation, enumeration, and counting; combinatorial optimization; combinatorics of words; computational biology; computational geometry; decompositions and combinatorial designs; distributed and network algorithms; experimental combinatorics; fine-grained complexity; graph algorithms and modeling with graphs; graph drawing and graph labeling; network theory and temporal graphs; quantum computing and algorithms for quantum computers; online algorithms; parameterized and exact algorithms; probabilistic and randomized algorithms; and streaming algorithms.

The Program Committee (PC) of IWOCA 2022 received 96 abstract submissions; finally, 86 full papers were submitted. Each submission was reviewed by at least three PC members and some trusted external referees, and evaluated on its quality, originality, and relevance to the conference. The PC selected 35 papers for presentation at the conference and inclusion in the proceedings. Three invited talks were scheduled for IWOCA 2022, given by Akanksha Agrawal (Indian Institute of Technology Madras, India) Erik Demaine (MIT, USA), and Bhaskar Ray Chaudhury (University of Illinois Urbana-Champaign, USA), as also testified by these proceedings.

The Program Committee also selected two papers to receive the Best Paper Award and the Best Student Paper Award, respectively. These awards were sponsored by Springer. The awardees are

- Best Paper Award: Hideo Bannai, Tomohiro I, Tomasz Kociumaka, Dominik Köppl, Simon Puglisi. Computing Longest (Common) Lyndon Subsequence.
- Best Student Paper Award: Kanae Yoshiwatari, Hironori Kiya, Tesshu Hanaka, Hirotaka Ono. Winner Determination Algorithms for Graph Games with Matching Structures.

We would like to thank all invited speakers for accepting to give a talk at the conference, the Program Committee members who graciously gave their time and energy, and the 122 external reviewers for their expertise.

The organization of IWOCA 2022 started in the middle of the pandemic crisis. To cope with the related uncertainties, we decided to run IWOCA as a hybrid event. We still hope that, in particular, a number of younger participants were able to enjoy IWOCA as an on-site event, possibly as their first-ever conference.

We also thank Springer for publishing the proceedings of IWOCA 2022 in their ARCoSS/LNCS series and for their financial support towards the best paper awards. Also, we were one of the last conferences collecting experiences with OCS, a Springer system we used for managing the collection and editing of the papers.

Finally, we thank the Steering Committee for giving us the opportunity to serve as program chairs of IWOCA 2022 and for their continuous support.

April 2022

Cristina Bazgan
Henning Fernau

Organization

Program Chairs

Henning Fernau   Universität Trier, Germany
Cristina Bazgan   Université Paris-Dauphine, France

Program Committee

Faisal Abu-Khzam   Lebanese American University, Lebanon
Jérémy Barbay   Universidad de Chile, Chile
Cristina Bazgan   Université Paris-Dauphine, France
Ljiljana Brankovic   University of New England, Australia
Katrin Casel   Hasso-Plattner-Institut, Germany
Katarina Cechlárová   Pavol Jozef Šafarik University in Košice, Slovakia
Charles Colbourn   Arizona State University, USA
Thomas Erlebach   Durham University, UK
Henning Fernau   Universität Trier, Germany
Florent Foucaud   Université Clermont Auvergne, France
Kshitij Gajjar   National University of Singapore, Singapore
Adriana Hansberg   Universidad Nacional Autónoma de México, Mexico
Toru Hasunuma   Kyoto University, Japan
Sun-Yuan Hsieh   National Cheng Kung University, Taiwan
Philipp Kindermann   Universität Trier, Germany
Gunnar Klau   Heinrich-Heine-Universität Düsseldorf, Germany
Yasuaki Kobayashi   Kyoto University, Japan
Veli Mäkinen   University of Helsinki, Finland
Florin Manea   Universität Göttingen, Germany
David Manlove   University of Glasgow, UK
André Nichterlein   TU Berlin, Germany
Mateus de Oliveira Oliveira   University of Bergen, Norway
Jakub Radoszewski   University of Warsaw, Poland
Giovanna Rosone   University of Pisa, Italy
Irena Rusu   University of Nantes, France
Jeffrey Shallit   University of Waterloo, Canada
Blerina Sinaimeri   Luiss Guido Carli, Italy
Wing-Kin Sung   National University of Singapore, Singapore
Sue Whitesides   University of Victoria, Canada
Meirav Zehavi   Ben-Gurion University, Israel

Steering Committee

Maria Chudnovsky   Princeton University, USA
Costas Iliopoulos   King's College London, UK
Ralf Klasing   CNRS and University of Bordeaux, France
Wing-Kin Sung   National University of Singapore, Singapore

Additional Reviewers

Emmanuel Arrighi
Max Bannach
Alexey Barsukov
Matthias Bentert
Sergey Bereg
Sebastian Berndt
William Billingsley
Niclas Boehmer
Benjamin Merlin Bumpus
Onur Çağırıcı
Tiziana Calamoneri
Armando Castañeda
Luca Castelli Aleardi
Dibyayan Chakraborty
Prerona Chatterjee
Yong Chen
Yu Han Chen
Guan Hao Chen
Dun Wei Cheng
Yu Ting Chou
Antoine Dailly
Colin de la Higuera
Gabriele Di Stefano
Changyu Dong
Guillaume Ducoffe
Maël Dumas
Lech Duraj
Jan Ekstein
Edith Elkind
Carl Feghali
Silvia Fernandez-Merchant
Jiří Fiala
Rudolf Fleischer
Travis Gagie
Harmender Gahlawat
Arnab Ganguly

Philip Geevarghese
Petr Golovach
Veronica Guerrini
Sushmita Gupta
Lianna Hambardzumyan
Tesshu Hanaka
Tim A. Hartmann
David Hartvigsen
Klaus Heeger
Davis Issac
Wojciech Janczewski
Paul Jungeblut
Leon Kellerhals
Ralf Klasing
Jonathan Klawitter
Ton Kloks
Tomohiro Koana
Christian Komusiewicz
Maria Kosche
Tore Koß
Pascal Kunz
Abhiruk Lahiri
Stefan Lendl
Vincent Limouzy
Cheng Yi Liu
Felipe Louza
Mike Luby
Soumen Maity
Andrea Marino
Arnaud Mary
Terry A. Mckee
Nikolaos Melissinos
Harshil Mittal
Shuichi Miyazaki
Bojan Mohar
Doost Ali Mojdeh

Tulasimohan Molli
Trung Thanh Nguyen
Pavlos Nikolopoulos
Pascal Ochem
Eunjin Oh
Andres Olivares
Yota Otachi
Arti Pandey
Nina Pardal
David Paul
Christophe Picouleau
Théo Pierron
Varun Ramanathan
Srinivas Rao
Felix Reidl
Malte Renken
Cléophée Robin
Edmund Sadgrove
Sanjib Sadhu
Vibha Sahlot
Abhishek Sahu
Chris Schewiegelshohn
Simon Schierreich
Markus Schmid
Christiane Schmidt

Melanie Schmidt
Henning Schnoor
Marinella Sciortino
Rik Sengupta
Ashutosh Shankar
Suhail Sherif
Stefan Siemer
Juliusz Straszyński
Bernardo Subercaseaux
Ondřej Suchý
Sylwester Swat
Yuma Tamura
Meng Shiou Tsai
Jianhua Tu
Ugo Vaccaro
Rohit Vaish
Agastya Vibhuti Jha
Tomasz Waleń
Kunihiro Wasa
Hsu Chun Yen
Yu Yokoi
Ryo Yoshinaka
Janez Zerovnik
Johannes Zink
Wiktor Zuba

Graphs as Algorithms: Characterizing Motion-Planning Gadgets through Simulation and Complexity (Abstract of Invited Talk)

Erik D. Demaine

Computer Science and Artificial Intelligence Laboratory, Massachusetts Institute
of Technology, Cambridge, MA 02139, USA
edemaine@mit.edu

Abstract. Most motion planning problems—designing the route for one or more agents (robots, humans, cars, drones, etc.) through a changeable environment—are computationally difficult: NP-hard, PSPACE-hard, or worse. Such hardness proofs usually consist of several *gadgets*—local pieces of environment with limited agent interactions/traversals, some of which change local state, which in turn change available interactions/traversals—that can be pasted together into the overall reduction. Such gadgets essentially act like finite automata, where the transitions are controlled by one or more agents traversing the environment.

In this talk, I'll describe our quest to characterize exactly which such gadgets suffice to prove different kinds of hardness, in our *motion-planning-through-gadgets framework* that has developed over the past few years [1–9]. This framework enables many hardness proofs, old and new, to be distilled down to a single diagram of a single gadget.

Even stronger, we aim to characterize which motion-planning gadgets can *simulate* which others. Gadget simulations are given by a graph describing how to connect together the simulating gadgets (along with their initial states) in a way that acts like the simulated gadget, essentially representing a reduction algorithm as a graph. See Fig. 1.

Keywords: Gadgets · Motion planning · Computational complexity

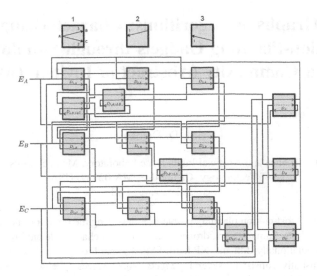

Fig. 1. Illustrative example of a proof that a "door" gadget is *universal*, meaning that it can simulate all other gadgets, from [2]. This graph of connections between door gadgets (drawn as shaded squares) describes an algorithmic reduction from motion planning on doors to motion planning on the desired gadget (with state diagram at top).

References

1. Ani, J., et al.: PSPACE-completeness of pulling blocks to reach a goal. J. Inf. Process. **28**, 929–941 (2020)
2. Ani, J., Bosboom, J., Demaine, E.D., Diomidov, Y., Hendrickson, D., Lynch, J.: Walking through doors is hard, even without staircases: Proving PSPACE-hardness via planar assemblies of door gadgets. In: Proceedings of the 10th International Conference on Fun with Algorithms (FUN 2020), pp. 3:1–3:23, La Maddalena, Italy, September 2020. Full paper available at https://arXiv.org/abs/2006.01256
3. Ani, J., Chung, L., Demaine, E.D., Diomidov, Y., Hendrickson, D., Lynch, J.: Pushing blocks via checkable gadgets: PSPACE-completeness of Push-1F and block/box dude. In: Proceedings of the 11th International Conference on Fun with Algorithms (FUN 2022), pp. 2:1–2:30, Island of Favignana, Sicily, Italy, May 30 – June 3 2022
4. Ani, J., Demaine, E.D., Diomidov, Y., Hendrickson, D., Lynch, J. (2022). Traversability, reconfiguration, and reachability in the gadget framework. In: Mutzel, P., Rahman, M.S., Slamin (eds.) WALCOM 2022. LNCS, vol 13174, pp. 47–58. Springer, Cham. https://doi.org/10.1007/978-3-030-96731-4_5
5. Ani, J., Demaine, E.D., Hendrickson, D., Lynch, J.: Trains, games, and complexity: 0/1/2-player motion planning through input/output gadgets. In: Mutzel, P., Rahman, M.S., Slamin (eds.) WALCOM 2022. LNCS, vol. 13174, pp. 187–198. Springer, Cham. xhttps://doi.org/10.1007/978-3-030-96731-4_5
6. Demaine, E.D., Grosof, I., Lynch, J., Rudoy, M.: Computational complexity of motion planning of a robot through simple gadgets. In: Proceedings of the 9th International Conference on Fun with Algorithms (FUN 2018), pp. 18:1–18:21, La Maddalena, Italy, June 2018

7. Demaine, E.D., Hendrickson, D., Lynch, J.: Toward a general theory of motion planning complexity: Characterizing which gadgets make games hard. In: Proceedings of the 11th Conference on Innovations in Theoretical Computer Science (ITCS 2020), pp. 62:1–62:42, Seattle, Washington, January 2020

8. Hendrickson, D.: Gadgets and gizmos: A formal model of simulation in the gadget framework for motion planning. Master's thesis, Massachusetts Institute of Technology, June 2021

9. Lynch, J.: A Framework for Proving the Computational Intractability of Motion Planning Problems. PhD thesis, Massachusetts Institute of Technology, September 2020

Contents

Invited Papers

Distance from Triviality 2.0: Hybrid Parameterizations

Akanksha Agrawal[1](✉) and M. S. Ramanujan[2]

[1] Indian Institute of Technology Madras, Chennai, India
akanksha@cse.iitm.ac.in
[2] University of Warwick, Coventry, England
R.Maadapuzhi-Sridharan@warwick.ac.uk

Abstract. Vertex deletion problems have been at the heart of numerous major advances in Algorithms and Combinatorial Optimization, and especially so in the area of Parameterized Complexity. For a family of graphs $\mathcal{H}$, the input to VERTEX DELETION TO $\mathcal{H}$ is a graph G and an integer k, and the objective is to decide whether there is a vertex-subset, called a *modulator*, whose removal from G results in a graph contained in the family $\mathcal{H}$, and such that $|S| \leq k$. Traditionally, the majority of the study of VERTEX DELETION TO $\mathcal{H}$ problems in Parameterized Complexity has been limited to parameterization by modulator size and structural graph width measures of the input graph such as treewidth. Recent years have seen systematic efforts at: i) quantifying the complexity of modulators in ways other than their size, and ii) studying the complexity landscape of various graph problems under parameterizations that are simultaneously better than both the modulator size and certain width measures of the graph. In this talk we will look at some exciting developments in this direction in relation to two such parameters that are "hybridizations" of the modulator size, and the well-explored graph parameters – treewidth and treedepth.

Keywords: Parameterized Complexity · Vertex deletion · Elimination distance · $\mathcal{H}$-treewidth

1 Introduction to Parameterized Algorithmics

The central goal in Algorithm Design and Analysis is to obtain "provably" the fastest algorithm for various problems. Classical Complexity Theory focuses on understanding the complexity of a problem based on the input size, and the holy grail of efficient solvability is polynomial-time computability. Arguably, most of the real world problems that we encounter turn out to be NP-hard, and thus we cannot hope to obtain a polynomial-time algorithm for them. Often the complexity of a problem is not just governed by its input size, but also certain structural properties of the input or output. The framework of Parameterized Complexity was originally developed by Downey and Fellows to cope with NP-hardness. Intuitively, it is a two-dimensional generalization of "P vs. NP", where

© Springer Nature Switzerland AG 2022
C. Bazgan and H. Fernau (Eds.): IWOCA 2022, LNCS 13270, pp. 3–20, 2022.
https://doi.org/10.1007/978-3-031-06678-8_1

in addition to the input size, one studies how relevant a secondary measure affects the computational complexity of problem. Here, the secondary measures are some quantification over the input/output (or some combination of them).

Parameterized Complexity. A *parameterized problem* Π is a subset of $\Sigma^* \times \mathbb{N}$, where Σ is a finite alphabet set. An instance of a parameterized problem is a tuple (I, κ), where I is a classical problem instance and κ is a natural number, which is called the *parameter*. One of the central concepts in the Parameterized Complexity is *fixed-parameter tractability (FPT)*, which intuitively speaking is solvability of instances of a parameterized problem in time $f(\kappa) \cdot |I|^{\mathcal{O}(1)}$, where (I, κ) is the given instance, $|I|$ is the encoding length of the instance and f is a function on κ. When we talk about solvability, it means that there are some algorithm(s) that resolve the given instance of the parameterized problem. Either there can be a single algorithm that works for all instances of the problem, or there maybe a particular algorithm that we wish to invoke based on the value of the parameter. This may leads cause some non-uniformity, and thus we have the following two notions of fixed-parameter tractability.

Definition 1 (Uniform and non-uniform FPT, Definition 2.2.1 [17]). Let Π be a parameterized problem.

  (i) We say that Π is *uniformly* FPT if there is an algorithm $\mathcal{A}$, a constant c, and an arbitrary function $f : \mathbb{N} \to \mathbb{N}$ such that: the running time of $\mathcal{A}(I, \kappa)$ is at most $f(\kappa) \cdot |I|^c$ and $(I, \kappa) \in \Pi$ if and only if $\mathcal{A}(I, \kappa) = 1$.
 (ii) We say that Π is *non-uniformly* FPT if there is collection of algorithms $\{\mathcal{A}_\kappa \mid \kappa \in \mathbb{N}\}$, a constant c, and an arbitrary function $f : \mathbb{N} \to \mathbb{N}$, such that: for each $\kappa \in \mathbb{N}$, the running time of $\mathcal{A}_k(I, \kappa)$ is $f(\kappa) \cdot |I|^c$ and $(I, \kappa) \in \Pi$ if and only if $\mathcal{A}_\kappa(I, \kappa) = 1$.

We say that a parameterized problem is in the complexity class FPT if it is uniformly FPT or non-uniformly FPT. Moreover, the algorithms involved in the above two items are called FPT and non-uniform FPT algorithms, respectively.

Not all parameterized problems are FPT under reasonable complexity-theoretic assumptions. Similar to the notion of NP-hardness and NP-hard reductions in the Classical Complexity Theory, we have the notion of W[t]-hardness, where $t \in \mathbb{N}$ and parameterized reductions in the Parameterized Complexity. Given that not all parameterized problems can be (non-uniformly) FPT, one of the quests in the field of Parameterized Complexity is to classify problems for which an FPT algorithm can exist. This leads us to the notion of FPT-equivalence. We say that two parameterized problems are (non-uniformly) FPT-*equivalent* if, given an FPT algorithm for any one of the two problems we can obtain a (non-uniform) FPT algorithm for the other problem.

  Another central notion in Parameterized Complexity is *kernelization*, which mathematically captures the efficiency of a pre-processing/data reduction routine. A *kernelization algorithm* or a *kernel* for a parameterized problem Q takes as input an instance (I, κ) of Q and in time polynomial in $|I| + \kappa$ returns an instance (I', κ') such that $(I, \kappa) \in Q$ if and only if $(I', \kappa') \in Q$. Furthermore, $|I'| + \kappa' \leq g(\kappa)$, where $g(\cdot)$ is some function on κ.

For more details on the topic, please see the textbooks of Downey and Fellows [17], Flum and Grohe [20], Niedermeier [42], and Cygan et al. [16].

2 Vertex Deletion Problems and Hybrid Parameters

Given the modeling power of graphs, graph problems have been extensively studied in the literature. Many of the classical graph problems can be defined as follows. For a family of graphs $\mathcal{H}$, in the VERTEX DELETION TO $\mathcal{H}$ problem we are given a graph and the goal is to compute a minimum sized set of vertices to delete from the graph in order to obtain a graph contained in $\mathcal{H}$. For a graph G, $\mathbf{mod}_{\mathcal{H}}(G)$ denotes the size of a smallest vertex set S such that $G - S \in \mathcal{H}$. If $G - S \in \mathcal{H}$, then S is called a *modulator* to $\mathcal{H}$. Lewis and Yannakakis [36] obtained that for all non-trivial families of graphs, the VERTEX DELETION TO $\mathcal{H}$ problems are NP-complete. Thus, the parameterized complexity of these problems have been extensively explored and arguably, their study has led to the development of several important tools and techniques in the field. The problems have been studied for numerous choices of structured $\mathcal{H}$, e.g., when $\mathcal{H}$ is planar, bipartite, chordal, interval, acyclic or edgeless, respectively, we get the classical PLANAR VERTEX DELETION, ODD CYCLE TRANSVERSAL, CHORDAL VERTEX DELETION, INTERVAL VERTEX DELETION, FEEDBACK VERTEX SET, and VERTEX COVER problems (see, for example, [16]). Moreover, Parameterized Complexity offers the perfect tools to design and analyze algorithms that *utilize* small vertex modulators to various graph classes. This is because it is often the case that inputs that have vertex modulators of small size to $\mathcal{H}$ turn out to be tractable for many problems that are NP-complete in general while being polynomial-time solvable on graphs in $\mathcal{H}$. In fact, using the size of the smallest vertex modulator of a graph into tractable graph classes as a parameter that captures "distance from triviality" was proposed by Guo, Hüffner and Niedermeier [28] as a methodology for systematically studying the parameterized complexity of a problem. In the last two decades, this methodology has become a rich source of interesting and useful parameters for graph problems.

Over the past decades much of the study of the vertex-deletion problems in parameterized complexity have mainly focused on parameters like solution size and various graph-width measures , and thus, their power and complexity has been well understood. In particular, numerous vertex-deletion problems are known to be fixed-parameter tractable (i.e., can be solved in time $f(k) \cdot n^{\mathcal{O}(1)}$, where k is the parameter and n is the input size) under these parameterizations. In light of this state of the art, recent efforts have shifted to the goal of identifying and exploring "hybrid" parameters such that:

(a) they are upper bounded by the solution size as well as certain graph-width measures,

(b) they can be arbitrarily (and simultaneously) smaller than both the solution size and graph-width measures, and

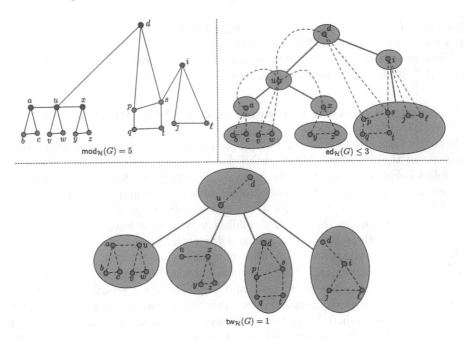

Fig. 1. Let $\mathcal{H}$ be the family of triangle-free graphs. The figure shows a graph G with a modulator to $\mathcal{H}$ of size 5 (blue vertices), an $\mathcal{H}$-elimination decomposition of depth 3 and an $\mathcal{H}$-tree decomposition of width 1. The vertices in L for the $\mathcal{H}$-elimination decomposition and $\mathcal{H}$-tree decomposition are the red vertices, and the edges of G in these decompositions are shown using dotted lines. The yellow ovals and the blue edges between them represents the associated trees in these decompositions. (Color figure online)

(c) a significant number of the problems that are in the class FPT when parameterized by solution size or graph-width measures, can be shown to be in the class FPT also when parameterized by these new parameters.

Two such recently introduced parameters are: (a) $\mathcal{H}$-elimination distance and (b) $\mathcal{H}$-treewidth of a graph, where $\mathcal{H}$ is a family of graphs. Bulian and Dawar [7] introduced the notion of $\mathcal{H}$-elimination distance of a graph G (denoted $\mathbf{ed}_{\mathcal{H}}(G)$), which intuitively speaking, is the number of rounds one needs to obtain a graph in $\mathcal{H}$ where in each round we are allowed to remove one vertex from each of the connected components. We next formally define $\mathcal{H}$-elimination distance (see Fig. 1 for an illustration). We remark that our definition is stated differently than the one given by Bulian and Dawar [7], but it is equivalent and (almost) in-line with the definition used by Jansen et al. [33].

Definition 2. For a graph class $\mathcal{H}$, an $\mathcal{H}$-*elimination decomposition* of graph G is a pair (T, χ, L), where T is a rooted forest, $\chi \colon V(T) \to 2^{V(G)}$ and $L \subseteq V(G)$, such that:

1. For each internal node t of T we have $|\chi(t)| \leq 1$ and $\chi(t) \subseteq V(G) \setminus L$.
2. The sets $(\chi(t))_{t \in V(T)}$ form a partition of $V(G)$.

3. For each edge $uv \in E(G)$, if $u \in \chi(t_1)$ and $v \in \chi(t_2)$ then t_1, t_2 are in ancestor-descendant relation in T.
4. For each leaf t of T, we have $\chi(t) \subseteq L$ and the graph $G[\chi(t)]$, called a base component, belongs to $\mathcal{H}$.

The *depth* of T is the maximum number of **edges** on a root-to-leaf path (see Fig. 1). We refer to the union of base components as the set of base vertices. The $\mathcal{H}$-elimination distance of G, denoted $\mathbf{ed}_{\mathcal{H}}(G)$, is the minimum depth of an $\mathcal{H}$-elimination forest for G.

It is straight-forward to verify that for any G and $\mathcal{H}$, the minimum depth of an $\mathcal{H}$-elimination forest of G is equal to the $\mathcal{H}$-elimination distance given by the (intuitve) recursive definition stated previously. (We remark that, due to the above, we have defined the depth of an $\mathcal{H}$-elimination forest in terms of the number of edges, while the traditional definition of treedepth counts vertices on root-to-leaf paths.) Readers familiar with the notion of treedepth [41] will be able to see that if $\mathcal{H}$ is the class of empty graphs, then the $\mathcal{H}$-elimination distance of G is nothing but the treedepth of G. In fact, if $\mathcal{H}$ is union-closed (*as will be the case for all graph classes we consider*), then one gets the following equivalent perspective on this notion. The $\mathcal{H}$-elimination distance of G is defined as the minimum possible *treedepth* of the torso of a modulator of G to $\mathcal{H}$. Here, the torso of a vertex set S in a graph G is the graph with vertex set S and an edge between two vertices $u, v \in S$ if there is a path between u and v in G whose internal vertices all lie outside S. The treedepth of G, denoted $\mathbf{td}(G)$, is the minimum depth of a standard elimination forest.

The second parameter, $\mathcal{H}$-treewidth, introduced by Eiben et al. [18], "generalizes" treewidth and solution size, the same way that elimination distance generalizes treedepth and solution size. This notion builds on a similar parameterization that was first defined in the context CSPs [26], which also found applications in algorithms for SAT [25] and Mixed ILPs [24]. Intuitively speaking, a $\mathcal{H}$-tree decomposition of a graph G of width ℓ is simply a tree decomposition of G paired with a vertex subset $L \subseteq V(G)$ that already induces in a graph in $\mathcal{H}$ and thus are ignored in the width computation. Additionally, in order for this subset to not interfere with many vertices, we required that each vertex in L is contained in only one bag in the tree decomposition. Notice that a bag in the tree decomposition can contain at most $\ell + 1$ vertices from $V(G) \setminus L$, but can contain any number of vertices from L. The $\mathcal{H}$-treewidth of G (denoted $\mathbf{tw}_{\mathcal{H}}(G)$) is the minimum width taken over all $\mathcal{H}$-tree decompositions of G. We will now formally define $\mathcal{H}$-treewidth (see Fig. 1 for an illustration).

Definition 3 ([33]). For a graph class $\mathcal{H}$, an $\mathcal{H}$-*tree decomposition* of graph G is a triplet (T, χ, L) where $L \subseteq V(G)$, T is a rooted tree and $\chi \colon V(T) \to 2^{V(G)}$, such that:

1. For each $v \in V(G)$ the nodes $\{t \mid v \in \chi(t)\}$ form a non-empty connected subtree of T.
2. For each edge $\{u, v\} \in E(G)$ there is a node $t \in V(G)$ with $\{u, v\} \subseteq \chi(t)$.

3. For each vertex $v \in L$, there is a unique $t \in V(T)$ for which $v \in \chi(t)$, with t being a leaf of T.
4. For each node $t \in V(T)$, the graph $G[\chi(t) \cap L]$ belongs to $\mathcal{H}$.

The *width* of a $\mathcal{H}$-tree decomposition is defined as $\max(0, \max_{t \in V(T)} |\chi(t) \backslash L| - 1)$. The $\mathcal{H}$-treewidth of a graph G, denoted $\mathbf{tw}_{\mathcal{H}}(G)$, is the minimum width of a $\mathcal{H}$-tree decomposition of G. The connected components of $G[L]$ are called base components and the vertices in L are called base vertices.

Note that a pair (T, χ) is a (standard) *tree decomposition* if $(T, \chi, \emptyset)$ satisfies all conditions of an $\mathcal{H}$-decomposition; the choice of $\mathcal{H}$ is irrelevant. In the definition of width, we subtract one from the size of a largest bag to mimic treewidth. The maximum with zero is taken to prevent $G \in \mathcal{H}$ from having $\mathbf{tw}_{\mathcal{H}}(G) = -1$.

From the definition it is easy to see that $\mathcal{H}$-treewidth of G is upper bounded by its treewidth, as one could simply take a tree-decomposition of G and the set $L = \emptyset$. Furthermore, $\mathcal{H}$-treewidth can always be upper bounded by $\mathbf{mod}_{\mathcal{H}}(G)$.

From our definitions, it is easy to see that $\mathbf{ed}_{\mathcal{H}}(G)$ (resp., $\mathbf{tw}_{\mathcal{H}}(G)$) can be arbitrarily smaller than both $\mathbf{mod}_{\mathcal{H}}(G)$ and the treedepth of G (respectively, the treewidth of G).[1] We note that, since treewidth of a graph can be upper bounded by its treedepth, $\mathbf{tw}_{\mathcal{H}}(G)$ can be arbitrarily smaller than $\mathbf{ed}_{\mathcal{H}}(G)$. As both $\mathbf{ed}_{\mathcal{H}}(G)$ and $\mathbf{tw}_{\mathcal{H}}(G)$ satisfy Properties (a) and (b) stated previously, in the recent times there have been efforts in understanding the extent to which Property (c) is satisfied by these parameters. This leads us to the following two fundamental and challenging questions that will be the focus of this talk:

Question 1: For which families $\mathcal{H}$ of graphs is ELIMINATION DISTANCE TO $\mathcal{H}$ (resp., TREEWIDTH DECOMPOSITION TO $\mathcal{H}$) FPT when parameterized by $\mathbf{ed}_{\mathcal{H}}$ (resp., $\mathbf{tw}_{\mathcal{H}}$)?

Question 2: For which families $\mathcal{H}$ of graphs is VERTEX DELETION TO $\mathcal{H}$ parameterized by $\mathbf{ed}_{\mathcal{H}}(G)$ (or $\mathbf{tw}_{\mathcal{H}}(G)$) FPT?

The ELIMINATION DISTANCE TO $\mathcal{H}$ (resp., TREEWIDTH DECOMPOSITION TO $\mathcal{H}$), the input is a graph G and integer k and the goal is to decide whether $\mathbf{ed}_{\mathcal{H}}(G) \leq k$ (resp., $\mathbf{tw}_{\mathcal{H}}(G) \leq k$). Note that for any of VERTEX DELETION TO $\mathcal{H}$, ELIMINATION DISTANCE TO $\mathcal{H}$ or TREEWIDTH DECOMPOSITION TO $\mathcal{H}$, one may study its parameterized complexity for any choice of the parameter $\mathbf{mod}_{\mathcal{H}}(G)$, $\mathbf{ed}_{\mathcal{H}}(G)$ or $\mathbf{tw}_{\mathcal{H}}(G)$ (see Fig. 2). Whenever we skip mentioning the parameter for these problems, then the parameter will be the number k.

We remark that, although for most families $\mathcal{H}$,[2] VERTEX DELETION TO $\mathcal{H}$ has a trivial XP-time algorithm , i.e., an $n^{\mathcal{O}(k)}$-time algorithm, it is not obvious how we can obtain an XP algorithm even for structured $\mathcal{H}$ for the problems ELIMINATION DISTANCE TO $\mathcal{H}$ or TREEWIDTH DECOMPOSITION TO $\mathcal{H}$, when

---

[1] For this, we always assume that $\mathcal{H}$ contains the empty graph and so $V(G)$ is a trivial modulator to $\mathcal{H}$.

[2] Here we make a mild assumption that, checking whether a graph is in $\mathcal{H}$ can be done in polynomial time.

(Classical) Problems   Parameters

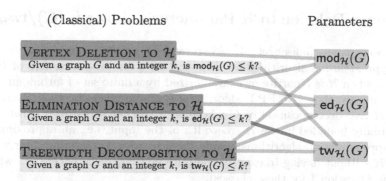

Fig. 2. Different parameterized problem that are of interest to us. Note one can obtain a parameterized problem by picking a (classical) problem from the left and a parameter from the right (which is illustrated by a connecting line between them). All the parameterized problems illustrated using the green connecting lines are FPT equivalent when $\mathcal{H}$ satisfies a mild condition [2].

parameterized by k. In the absence of a resolution to Question 1, Question 2 then brings with it the challenge of solving VERTEX DELETION TO $\mathcal{H}$ (or indeed, any problem) without necessarily being able to efficiently compute $\mathsf{ed}_{\mathcal{H}}$ or $\mathsf{tw}_{\mathcal{H}}$.

3   Elimination Distance/Tree Decomposition to $\mathcal{H}$

Bulian and Dawar [7] showed that ELIMINATION DISTANCE TO $\mathcal{H}$ is FPT, when $\mathcal{H}$ is a minor-closed class of graphs and asked whether it is FPT, when $\mathcal{H}$ is the family of graphs of degree at most d. In a partial resolution to this question, Lindermayr et al. [37] obtained that ELIMINATION DISTANCE TO $\mathcal{H}$ is FPT when we restrict the input graphs to be planar. After this, we (along with a few others) [3] resolved this question completely by showing that the problem is (non-uniformly) FPT. In fact, our result extends for all the families of graphs $\mathcal{H}$ that can be characterized by a finite family of induced subgraphs.

Jansen and de Kroon [32] extended the aforementioned result, and showed that TREEWIDTH DECOMPOSITION TO $\mathcal{H}$ is also (non-uniformly) FPT when $\mathcal{H}$ is characterized by a finite family of excluded induced subgraphs. They also showed that TREEWIDTH DECOMPOSITION TO $\mathcal{H}$ and ELIMINATION DISTANCE TO $\mathcal{H}$ are non-uniformly FPT when $\mathcal{H}$ is the family of bipartite graphs. Recently, Fomin et al. [21] showed that for every graph family $\mathcal{H}$ expressible by a particular fragment of first order-logic, ELIMINATION DISTANCE TO $\mathcal{H}$ is (non-uniformly) FPT. Since a family of graphs characterized by a finite set of forbidden induced subgraphs is expressible in this fragment of logic, this result also generalizes the result from [3].

4   Vertex Deletion to $\mathcal{H}$ Parameterized by $\mathbf{ed}_{\mathcal{H}}(G)/\mathbf{tw}_{\mathcal{H}}(G)$

In a recent paper, Jansen et al. [33] provide a general framework to design FPT-*approximation* algorithms for $\mathbf{ed}_{\mathcal{H}}$ and $\mathbf{tw}_{\mathcal{H}}$ for various choices of $\mathcal{H}$. For instance, when $\mathcal{H}$ is bipartite or characterized by a finite set of forbidden (topological) minors, they give FPT algorithms (parameterized by $\mathbf{tw}_{\mathcal{H}}$) that compute a $\mathcal{H}$-tree decomposition of G whose width is *not necessarily optimal*, but polynomially bounded in the $\mathcal{H}$-treewidth of the input, i.e., an approximation. These approximation algorithms enable them to address Question 2 for various classes $\mathcal{H}$ without having to exactly compute $\mathbf{ed}_{\mathcal{H}}(G)$ or $\mathbf{tw}_{\mathcal{H}}(G)$ (i.e., without resolving Question 1 for these classes).

Towards answering Question 2, Jansen et al. [33] give the following FPT algorithms for VERTEX DELETION TO $\mathcal{H}$ parameterized by $\mathbf{tw}_{\mathcal{H}}$. Let $\mathcal{H}$ be a hereditary class of graphs that is defined by a finite number of forbidden connected (a) minors, or (b) induced subgraphs, or (c) $\mathcal{H}$ is the family of bipartite graphs or chordal graphs.[3] There is an algorithm that, given an n-vertex graph G, computes a minimum vertex set X such that $G - X \in \mathcal{H}$ in time $f(\mathbf{tw}_{\mathcal{H}}(G)) \cdot n^{\mathcal{O}(1)}$. We note that all of these FPT algorithms are uniform.

5   Equivalence of Elimination/Decomposition/Deletion

A closer look at the results in [3,21,33] reveals an interesting property: these algorithms for ELIMINATION DISTANCE TO $\mathcal{H}$ and TREEWIDTH DECOMPOSITION TO $\mathcal{H}$ utilize the corresponding (known) algorithms for VERTEX DELETION TO $\mathcal{H}$ in a non-trivial manner. This raises the following natural questions.

> **Question 3:** When (if at all) is the parameterized complexity of ELIMINATION DISTANCE TO $\mathcal{H}$ or TREEWIDTH DECOMPOSITION TO $\mathcal{H}$ different from that of VERTEX DELETION TO $\mathcal{H}$?
>
> **Question 4:** When (if at all) is the parameterized complexity of VERTEX DELETION TO $\mathcal{H}$ different from that of VERTEX DELETION TO $\mathcal{H}$ parameterized by $\mathbf{ed}_{\mathcal{H}}(G)$ or $\mathbf{tw}_{\mathcal{H}}(G)$?

Intrigued by the above questions, we (along with others) very recently obtained that, when $\mathcal{H}$ is a hereditary family of graphs that is CMSO[4] definable and closed under disjoint union, then all the parameterized problems illustrated by green connecting lines in Fig. 2 are (non-uniformly) FPT-equivalent [2]. We formally state this result below.

---

[3] A family of graphs $\mathcal{H}$ is hereditary if for each graph $G \in \mathcal{H}$, every induced subgraph of G belongs to $\mathcal{H}$.

[4] When we say CMSO, we refer to the fragment that is sometimes referred to as $CMSO_2$ in the literature. We will only be using a meta-result regarding CMSO definable problems, and thus, we refer to [5,11,12] for a an introduction to this topics.

Theorem 1. $\mathcal{H}$ *be a hereditary family of graphs that is* CMSO *definable and closed under disjoint union. Then the following problems are (non-uniformly)* FPT*-equivalent.*

1. VERTEX DELETION TO $\mathcal{H}$ *parameterized by* $\mathbf{mod}_{\mathcal{H}}(G)$
2. VERTEX DELETION TO $\mathcal{H}$ *parameterized by* $\mathbf{ed}_{\mathcal{H}}(G)$
3. VERTEX DELETION TO $\mathcal{H}$ *parameterized by* $\mathbf{tw}_{\mathcal{H}}(G)$
4. ELIMINATION DISTANCE TO $\mathcal{H}$ *parameterized by* $\mathbf{mod}_{\mathcal{H}}(G)$
5. ELIMINATION DISTANCE TO $\mathcal{H}$ *parameterized by* $\mathbf{ed}_{\mathcal{H}}(G)$
6. TREEWIDTH DECOMPOSITION TO $\mathcal{H}$ *parameterized by* $\mathbf{mod}_{\mathcal{H}}(G)$
7. TREEWIDTH DECOMPOSITION TO $\mathcal{H}$ *parameterized by* $\mathbf{ed}_{\mathcal{H}}(G)$
8. TREEWIDTH DECOMPOSITION TO $\mathcal{H}$ *parameterized by* $\mathbf{tw}_{\mathcal{H}}(G)$

We will look at one of the implications from the above theorem, which we summarize in the following lemma.

Lemma 1. *Consider a family $\mathcal{H}$ of graphs that is* CMSO *definable and is closed under disjoint union and induced subgraphs. If* ELIMINATION DISTANCE TO $\mathcal{H}$ *(resp.* TREEWIDTH DECOMPOSITION TO $\mathcal{H}$*) parameterized by* $\mathbf{mod}_{\mathcal{H}}(G)$ *is* FPT*, then* VERTEX DELETION TO $\mathcal{H}$ *parameterized by* $\mathbf{mod}_{\mathcal{H}}(G)$ *is also* FPT*.*

Before proving the above lemma, we introduce some notations and state some useful results regarding them.

Preliminaries. Consider a graph G. A pair (X, Y) where $X \cup Y = V(G)$ is a *separation* if there is no edge $\{u, v\} \in E(G)$ such that $u \in X \setminus Y$ and $v \in Y \setminus X$. The order of (X, Y) is $|X \cap Y|$. For $s, c \in \mathbb{N}$, we say that G is (s, c)-*breakable* if there exists a separation (X, Y) of order at most c such that $|X \setminus Y| \geq s$ and $|Y \setminus X| \geq s$. Moreover, if such a separation does not exists, then G is (s, c)-*breakable.*

We will crucially use the following result of Lokshtanov et al. [39] that allows one to obtain a (non-uniform) FPT algorithm for CMSO-expressible graph problems by designing an FPT algorithm for the problem on unbreakable graphs.

Proposition 1 (Theorem 1, [39]). *Let ψ be a CMSO sentence and let $d > 4$ be a positive integer. There exists a function $\alpha : \mathbb{N} \to \mathbb{N}$, such that for every $c \in \mathbb{N}$ there is an $\alpha(c) \in \mathbb{N}$, if there exists an algorithm that solves* CMSO$[\psi]$ *on $(\alpha(c), c)$-unbreakable graphs in time $\mathcal{O}(n^d)$, then there exists an algorithm that solves* CMSO$[\psi]$ *on general graphs in time $\mathcal{O}(n^d)$.*

Hereafter, α will denote the function obtained using the above proposition. We next state a result which immediately follows from the definition of (s, c)-unbreakable graphs.

Proposition 2. *Consider a graph G, an integer k and a set $S \subseteq V(G)$ of size at most k, where G is an $(\alpha(k), k)$-unbreakable graph with $|V(G)| > 2\alpha(k) + k$. Then, there is exactly one connected component C^* in $G - S$ that has at least $\alpha(k)$ vertices and $|V(G) \setminus V(C^*)| < \alpha(k) + k$.*

Proof of Lemma. 1. Fix any family $\mathcal{H}$ of graphs that is CMSO definable, hereditary and closed under disjoint union, such that ELIMINATION DISTANCE TO $\mathcal{H}$ (resp. TREEWIDTH DECOMPOSITION TO $\mathcal{H}$) admits an FPT algorithm, say, $\mathscr{X}_{mod}$ running in time $f(\ell) \cdot n^{\mathcal{O}(1)}$, where n is the number of vertices in the given graph G and $\ell = \mathbf{mod}_{\mathcal{H}}(G)$.

Let (G, k) be an instance of VERTEX DELETION TO $\mathcal{H}$. To prove Lemma 1, from Proposition 1 it is enough to design an algorithm for $(\alpha(k), k)$-unbreakable graphs, and thus, we assume that G is $(\alpha(k), k)$-unbreakable. We begin with the following simple sanity checks.

Base Case 1. If $G \in \mathcal{H}$ and $k \geq 0$, then return that (G, k) is a yes-instance of the problem.[5] Moreover, if $k < 0$, then return that the instance is a no-instance.

Base Case 2. If $|V(G)| \leq 2\alpha(k) + k$, then for each $S \subseteq V(G)$ of size at most k, check if $G - S \in \mathcal{H}$. If for any such S we obtain that $G - S \in \mathcal{H}$, return that (G, k) is a yes-instance, and otherwise return that it is a no instance.

Base Case 3. If G does not admit an $\mathcal{H}$-elimination decomposition (resp. $\mathcal{H}$-tree decomposition) of depth (resp. width) at most k, then return that (G, k) is a no-instance of the problem.

The correctness of the Base Case 1 and 2 is immediate from their descriptions. Note that $\mathbf{tw}_{\mathcal{H}}(G) \leq \mathbf{ed}_{\mathcal{H}}(G) \leq \mathbf{mod}_{\mathcal{H}}(G)$. Thus, if (G, k) is a yes-instance of VERTEX DELETION TO $\mathcal{H}$, then it must admit an $\mathcal{H}$-elimination decomposition (resp. $\mathcal{H}$-tree decomposition) of depth (resp. width) at most k. The above implies the correctness of Base Case 3. Note that using $\mathscr{X}_{mod}$, we can test/apply all the base cases in time bounded by $\max\{2^{\alpha(k)+k}, f(k)\} \cdot n^{\mathcal{O}(1)}$.

Hereafter we assume that the base cases are not applicable. We compute an $\mathcal{H}$-elimination decomposition (resp. $\mathcal{H}$-tree decomposition), say, (T, χ, L), by using the algorithm $\mathscr{X}_{mod}$, of depth (resp. width) at most k.[6] Note that the above decomposition can be computed as Base Case 3 is not applicable.

Let C^* be a connected component in $G[L]$ with maximum number of vertices, and let $S^* = N_G(C^*)$ and $Z^* = V(G) \setminus N_G[C^*]$. Note that as (T, χ, L) is an $\mathcal{H}$-elimination decomposition (resp. $\mathcal{H}$-tree decomposition) of depth (resp. width) at most k, we can obtain that $|S^*| \leq k$. As G is $(\alpha(k), k)$-unbreakable, the above together with Observation 2 implies that $|Z^* \cup S^*| \leq \alpha(k) + k$. We have the following property which can be immediately derived from the fact that (T, χ, L) is an $\mathcal{H}$-elimination decomposition (resp. $\mathcal{H}$-tree decomposition) for G of depth (resp. width) at most k.

Proposition 3. *Either $G - S^* \in \mathcal{H}$, or for every $S \subseteq V(G)$ of size at most k such that $G - S \in \mathcal{H}$, we have $Z^* \cap S \neq \emptyset$.*

The above result leads us to the following base case and our branching rule.

---

[5] We can check if $G \in \mathcal{H}$ by calling $\mathscr{X}_{mod}$ for the instance $(G, 0)$. We recall that $\mathcal{H}$ is closed under disjoint union.

[6] Even if $\mathscr{X}_{mod}$ is a decision algorithm, using the self-reducibility like property, we can compute the decomposition itself, see Lemma 3.5 and 3.6 in [1] or [2] for details.

Base Case 4. $G - S^* \in \mathcal{H}$, then return that (G, k) is a yes-instance.

As $|S^*| \leq k$, the correctness of the above base case immediately follows. Moreover, using $\mathcal{X}_{\text{mod}}$, we can apply Base Case 3 in time bounded by $f(k) \cdot n^{\mathcal{O}(1)}$.

Branching Rule. For each $z \in Z^*$, (recursively) solve the instance $(G - \{z\}, k - 1)$. Return that (G, k) is a yes-instance if and only if one of these instances is a yes-instance.

The correctness of the branching rule follows from Proposition 3 and non-applicability of Base Case 4. Moreover, we can create instances in the branching rule in polynomial time, given the decomposition (T, χ, L).

Note that the depth of the recursion tree is bounded by $k+1$. Also, each of the steps can be applied in time bounded by $\max\{2^{\alpha(k)+k}, f(k)\} \cdot n^{\mathcal{O}(1)}$. Thus we can bound the running time of our algorithm by $k^{\mathcal{O}(k)} \cdot \max\{2^{\alpha(k)+k}, f(k)\} \cdot n^{\mathcal{O}(1)}$. The correctness of the algorithm is immediate from its description and Proposition 3. This concludes the proof of Lemma 1. □

6   Implication/Applications of the Equivalence Result

Unification/Extension of Known Results. Theorem 1 gives us a powerful classification tool which states that as far as the (non-uniform) fixed-parameter tractability of computing any of the parameters $\mathbf{mod}_{\mathcal{H}}(G)$, $\mathbf{ed}_{\mathcal{H}}(G)$ and $\mathbf{tw}_{\mathcal{H}}(G)$ is concerned, they are essentially the "same parameter" for many frequently considered graph classes $\mathcal{H}$. In other words, to obtain an FPT algorithm for any of the problems stated in Theorem 1, it is sufficient to design an FPT algorithm for the standard vertex-deletion problem, namely, VERTEX DELETION TO $\mathcal{H}$. This implication unifies several known results in the literature. For example, let $\mathcal{H}$ be the family of graphs of degree at most d and recall that it was only recently in [4, 32] it was shown that ELIMINATION DISTANCE TO $\mathcal{H}$ and TREEWIDTH DECOMPOSITION TO $\mathcal{H}$ are (non-uniformly) FPT, respectively. However, using equivalence result, the fixed-parameter tractability of these two problems and in fact, even the fixed-parameter tractability of VERTEX DELETION TO $\mathcal{H}$ parameterized by $\mathbf{ed}_{\mathcal{H}}(G)$ (or $\mathbf{tw}_{\mathcal{H}}(G)$), is implied by the straightforward $d^k n^{\mathcal{O}(1)}$-time branching algorithm for VERTEX DELETION TO $\mathcal{H}$ (i.e., the problem of deleting at most k vertices to get a graph of degree at most d).

Moreover, for various well-studied families of $\mathcal{H}$, we immediately derive FPT algorithms for all combinations of VERTEX DELETION TO $\mathcal{H}$, ELIMINATION DISTANCE TO $\mathcal{H}$, TREEWIDTH DECOMPOSITION TO $\mathcal{H}$ parameterized by any of $\mathbf{mod}_{\mathcal{H}}(G)$ $\mathbf{ed}_{\mathcal{H}}(G)$ and $\mathbf{tw}_{\mathcal{H}}(G)$, which are covered by the equivalence. For instance, we can invoke it using well-known FPT algorithms for VERTEX DELETION TO $\mathcal{H}$ for several families of graphs that are CMSO definable and closed under disjoint union, such as families defined by a finite number of forbidden connected (a) minors, or (b) topological minors, or (c) induced subgraphs, or (d) $\mathcal{H}$ being bipartite, chordal, proper-interval, interval, or distance-hereditary; to name a few [6, 8–10, 19, 22, 23, 34, 35, 38, 40, 48–50]. Thus, it provides a unified understanding of many recent results and resolves the parameterized complexity of several questions that were open prior to this result. Of particular significance among the new results is the case where $\mathcal{H}$ is a class defined by a finite number of forbidden connected topological minors, as this gives the *first* FPT algorithms for computing $\mathbf{ed}_{\mathcal{H}}$ and $\mathbf{tw}_{\mathcal{H}}$, resolving an open problem posed by Jansen et al. [33].

Deletion to Families of Bounded Rankwidth. We observe that Theorem 1 can be invoked by taking $\mathcal{H}$ as the class of graphs of bounded *rankwidth*, extending a result of Eiben et al. [18].

Rankwidth is a graph parameter introduced by Oum and Seymour [47] to approximate yet another graph parameter called Cliquewidth. The notion of cliquewidth was defined by Courcelle and Olariu [14] as a measure of how "clique-like" the input graph is. One of the main motivations was that several NP-complete problems become tractable on the family of cliques (complete graphs), the assumption was that these algorithmic properties extend to "clique-like" graphs [13]. However, computing cliquewidth and the corresponding cliquewidth decomposition seems to be computationally intractable. This then motivated the notion of rankwidth, which is a graph parameter that approximates cliquewidth well while also being algorithmically tractable [44,47]. For more information on cliquewidth and rankwidth, we refer to the surveys by Hlinený et al. [29] and Oum [46].

For a graph G, we use $\mathbf{rwd}(G)$ to denote the rankwidth of G. Let $\eta \geq 1$ be a fixed integer and let $\mathcal{H}_\eta$ denote the class of graphs of rankwidth at most η. It is known that VERTEX DELETION TO $\mathcal{H}_\eta$ is FPT [15]. The algorithm is based on the fact that for every integer η, there is a finite set $\mathcal{C}_\eta$ of graphs such that for every graph G, $\mathbf{rwd}(G) \leq \eta$ if and only if no vertex-minors of G are isomorphic to a graph in $\mathcal{C}_\eta$ [43,45]. Further, it is known that vertex-minors can be expressed in CMSO, this together with the fact that we can test whether a graph H is a vertex-minor of G or not in $f(|H|)n^{\mathcal{O}(1)}$ time on graphs of bounded rankwidth leads to the desired algorithm [15, Theorem 6.11]. It is also important to mention that for VERTEX DELETION TO $\mathcal{H}_1$, also known as the DISTANCE-HEREDITARY VERTEX-DELETION problem, there is a dedicated algorithm running in time $2^{\mathcal{O}(k)}n^{\mathcal{O}(1)}$ [19]. For us, two properties of $\mathcal{H}_\eta$ are important: (a) expressibility in CMSO and (b) being closed under disjoint union. These two properties, together with the result in [15] imply that our result

Theorem 1 is also applicable to $\mathcal{H}_\eta$. Thus, we are able to generalize and extend the result of Eiben et al. [18], who showed that for every η, computing $\mathbf{tw}_{\mathcal{H}_\eta}$ is FPT. For different notions and definitions related to rankwidth and vertex-minors we refer the reader to [29,46].

Cut Problems. Notice that in the same spirit as we have seen so far, one could also consider the parameterized complexity of other classical problems such as cut problems (e.g., MULTIWAY CUT), as long as the parameter is smaller than the standard parameter studied so far. Note that at a first look, problems like, MULTIWAY CUT, SUBSET FVS and SUBSET OCT do not seem to fit under the umbrella of vertex deletion problems to a particular graph class.

Ihe MULTIWAY CUT problem, where one is given a graph G and a set of vertices S (called terminals) and an integer ℓ and the goal is to decide whether there is a set of at most ℓ vertices whose deletion separates every pair of these terminals. The standard parameterization for this problem is the solution size ℓ. Jansen et al. [33] propose to consider annotated graphs (i.e., undirected graphs with a distinguished set of terminal vertices) and study the parameterized complexity of MULTIWAY CUT parameterized by the elimination distance to a graph where each component has at most one terminal. Notice that this new parameter is always upper bounded by the size of a minimum solution.

We can obtain an FPT algorithm for MULTIWAY CUT with the above stated parameter by using Theorem 1. We note that although Theorem 1 is defined only when $\mathcal{H}$ is a family of graphs, in order to capture such problems, we can express them in terms of appropriate notions of structures and then give a reduction to a pure graph problem on which Theorem 1 can be invoked. These results make concrete advances in the direction proposed by Jansen et al. [33] to develop FPT algorithms for MULTIWAY CUT parameterized by the elimination distance to a graph where each component has at most one terminal. We remark that such results can also be obtained for problems like SUBSET FVS and SUBSET OCT.

Modulators to Scattered Families. Recent years have seen another new direction of research on VERTEX DELETION TO $\mathcal{H}$ – instead of studying the computation of a modulator to a single family of graphs $\mathcal{H}$, one can focus to compute small vertex sets whose deletion leaves a graph where each connected component comes from a particular pre-specified graph class [27,30,31]. For example, given a graph G and a number k, find a vertex set S of size at most k (or decide whether one exists) such that in $G - S$, each connected component is either chordal or bipartite. Let us call such an S, a *scattered modulator*. Such scattered modulators (if small) can be used to design new FPT algorithms for certain problems by taking separate FPT algorithms for the problems on each of the pre-specified graph classes and then combining them in a non-trivial way "through" the scattered modulator. The quality of the modulators considered in this line of research has mainly been in terms of the size. We can extend the idea of studying graph problems with hybrid parameters like scattered elimination distance or scattered tree decompositions as well.

The first study of scattered modulators was undertaken by Ganian et al. [27], who introduced this notion in their work on constraint satisfaction problems. Recently, Jacob et al. [30,31] initiated the study of scattered modulators explicitly for "scattered" families of graphs. In particular, let $\mathcal{H}_1, \ldots, \mathcal{H}_d$ be families of graphs. Then, the scattered family of graphs $\otimes(\mathcal{H}_1, \ldots, \mathcal{H}_d)$ is defined as the set of all graphs G such that every connected component of G belongs to $\bigcup_{i=1}^{d} \mathcal{H}_i$. That is, each connected component of G belongs to some $\mathcal{H}_i$. As their main result, Jacob et al. [30] showed that VERTEX DELETION TO $\mathcal{H}$ is FPT whenever VERTEX DELETION TO $\mathcal{H}_i$, $i \in \{1, \ldots, d\}$, is FPT, and each of $\mathcal{H}_i$ is CMSO expressible. Here, $\mathcal{H}$ is the scattered family $\otimes(\mathcal{H}_1, \ldots, \mathcal{H}_d)$. Notice that if each of $\mathcal{H}_i$ is CMSO expressible then so is $\mathcal{H}$. Further, it is easy to observe that if each of $\mathcal{H}_i$ is closed under disjoint union then so is $\mathcal{H}$. The last two properties together with the result of Jacob et al. [30] enable us to invoke Theorem 1 even when $\mathcal{H}$ is a scattered graph family.

Cross Parameterizations. Another direction of research in Parameterized Complexity is cross parameterizations: parameterization of one problem with respect to alternate parameters. For example, consider ODD CYCLE TRANSVERSAL (OCT) on chordal graphs. Let $\mathcal{H}$ denote the family of chordal graphs. It is well known that OCT is polynomial-time solvable on chordal graphs. Further, given a graph G and a modulator to chordal graphs of size $\mathbf{mod}_{\mathcal{H}}(G)$, OCT admits an algorithm with running time $2^{\mathcal{O}(\mathbf{mod}_{\mathcal{H}}(G))}n^{\mathcal{O}(1)}$. It is therefore natural to ask whether OCT admits an algorithm with running time $f(\mathbf{ed}_{\mathcal{H}}(G))n^{\mathcal{O}(1)}$ or $f(\mathbf{tw}_{\mathcal{H}}(G))n^{\mathcal{O}(1)}$, given an $\mathcal{H}$-elimination forest of G of depth $\mathbf{ed}_{\mathcal{H}}(G)$ or an $\mathcal{H}$-tree decomposition of G of width $\mathbf{tw}_{\mathcal{H}}(G)$, respectively. The question is also relevant, in fact more challenging, when an $\mathcal{H}$-elimination forest of G of depth $\mathbf{ed}_{\mathcal{H}}(G)$ or an $\mathcal{H}$-decomposition of G of width $\mathbf{tw}_{\mathcal{H}}(G)$, respectively, is *not* given. Jansen et al. [33] specifically mentioned this research direction in their paper.

A step in this direction of research can be seen in the work of Eiben et al. [18, Thm. 4]. They present a meta-theorem that yields non-uniform FPT algorithms when Π satisfies several conditions, which require a technical generalization of an FPT algorithm for Π parameterized by deletion distance to $\mathcal{H}$. If the problem satisfies certain requirements (see [2], for more details), then in fact it enables existence of FPT algorithms for vertex-deletion problems parameterized by $\mathbf{ed}_{\mathcal{H}}(G)$ (or $\mathbf{tw}_{\mathcal{H}}(G)$) when given an $\mathcal{H}$-elimination forest of G of depth $\mathbf{ed}_{\mathcal{H}}(G)$ (resp., an $\mathcal{H}$-decomposition of G of width $\mathbf{tw}_{\mathcal{H}}(G)$).

Towards Uniform FPT Algorithms. Recall that the FPT algorithms obtained via Theorems 1 and the extension to families of structures are non-uniform. In fact, all of the current known FPT algorithms for ELIMINATION DISTANCE TO $\mathcal{H}$ or TREEWIDTH DECOMPOSITION TO $\mathcal{H}$, are non-uniform; except for ELIMINATION DISTANCE TO $\mathcal{H}$, when $\mathcal{H}$ is the family of empty graphs (which, as discussed earlier, is simply the problem of computing treedepth). However, we note that the FPT-approximation algorithms in Jansen et al. [33] (in fact, all the algorithms obtained in [33]) are uniform.

The paper [2], presents a general set of requirements that, when satisfied, shows that ELIMINATION DISTANCE TO $\mathcal{H}$ parameterized by $\mathbf{ed}_{\mathcal{H}}(G)$ is uniformly FPT. Like before, we need $\mathcal{H}$ to be hereditary, CMSO definable, closed under disjoint union and VERTEX DELETION TO $\mathcal{H}$ is FPT, where the additional requirement is a strengthening of the last two demands. These strengthening is based on certain equivalence classes obtained over boundaried graphs, for more details on this, we refer the reader to [2].

We remark that these strengthened requirements in [2] are "simple" because many of the known algorithms for the several problems already implicitly yields them as part of their analysis. So, the satisfaction of these conditions do not seem (in various cases) to require much "extra" work compared to the design of an FPT algorithm (or a kernel) to the problem at hand. Using these sufficient conditions, *uniform* FPT algorithms for computing $\mathbf{ed}_{\mathcal{H}}$ can be obtained, when $\mathcal{H}$ is defined by excluding a finite number of connected (a) minors, or (b) topological minors, or (c) induced subgraphs, or when $\mathcal{H}$ is any of bipartite, chordal or interval graphs. For most of these problems, the existence of a uniform (and even non-uniform) FPT algorithm previously was open in the literature.

7 Conclusion and Open Problems

We looked at the studies pushing the boundaries of tractability in Parameterized Complexity for the two recently introduced hybrid parameters that combine solution size and width measures ($\mathcal{H}$-elimination distance and $\mathcal{H}$-treewidth). We saw a surprising result that these parameters are effectively only as powerful as the standard parameterization by the size of the modulator to $\mathcal{H}$ for a host of commonly studied graph classes $\mathcal{H}$. This unifies several recent results in the literature. We also looked at implications of this equivalence result for problems like MULTIWAY CUT and how similar results can be obtained for cross parameterizations of problems. Furthermore, as many of the algorithms are non-uniform algorithms, we have saw that there is a framework to design uniform FPT algorithms to compute elimination distance to $\mathcal{H}$ when $\mathcal{H}$ has certain properties.

Hybrid parameterizations have been the subject of a flurry of interesting results in the last half-a-decade and it would be interesting to identify new, algorithmically useful, hybrid parameterizations that are provably stronger than existing ones. We restate some of the interesting future research directions from [2].

- Is ELIMINATION DISTANCE TO $\mathcal{H}$ parameterized by $\mathbf{tw}_{\mathcal{H}}(G)$ equivalent to the eight problems in Theorem 1 for hereditary, union-closed and CMSO definable $\mathcal{H}$? We conjecture that the answer is yes.
- Can we design uniform FPT algorithms for TREEWIDTH DECOMPOSITION TO $\mathcal{H}$ in the same spirit as the framework to design uniform FPT algorithms for ELIMINATION DISTANCE TO $\mathcal{H}$?

Finally, the fact still remains that $\mathbf{tw}_{\mathcal{H}}(G) \leq \mathbf{ed}_{\mathcal{H}}(G) \leq \mathbf{mod}_{\mathcal{H}}(G)$ and that each parameter could be arbitrarily smaller than the parameters to its right. Thus, it is an interesting direction of research to study VERTEX DELETION TO $\mathcal{H}$ for specific classes $\mathcal{H}$ parameterized by $\mathbf{ed}_{\mathcal{H}}(G)$ and $\mathbf{tw}_{\mathcal{H}}(G)$ and aim to optimize the running time, in the spirit of Jansen et al. [33].

References

1. Agrawal, A., et al.: Deleting, eliminating and decomposing to hereditary classes are all FPT-equivalent (2021). https://doi.org/10.48550/ARXIV.2104.09950, https://arxiv.org/abs/2104.09950
2. Agrawal, A., et al.: Deleting, eliminating and decomposing to hereditary classes are all FPT-equivalent. In: ACM-SIAM Symposium on Discrete Algorithms (SODA) SIAM, pp. 1726–1736 (2022)

3. Agrawal, A., Kanesh, L., Panolan, F., Ramanujan, M.S., Saurabh, S.: An FPT algorithm for elimination distance to bounded degree graphs. In: 38th International Symposium on Theoretical Aspects of Computer Science (STACS), vol. 187, pp. 5:1–5:11 (2021). https://doi.org/10.4230/LIPIcs.STACS.2021.5
4. Agrawal, A., Ramanujan, M.S.: On the parameterized complexity of clique elimination distance. In: Cao, Y., Pilipczuk, M. (eds.) 15th International Symposium on Parameterized and Exact Computation, (IPEC), vol. 180, pp. 1:1–1:13 (2020). https://doi.org/10.4230/LIPIcs.IPEC.2020.1
5. Arnborg, S., Lagergren, J., Seese, D.: Easy problems for tree-decomposable graphs. J. Algorithms **12**(2), 308–340 (1991). https://doi.org/10.1016/0196-6774(91)90006-K
6. van Bevern, R., Komusiewicz, C., Moser, H., Niedermeier, R.: Measuring indifference: unit interval vertex deletion. In: Thilikos, D.M. (ed.) WG 2010. LNCS, vol. 6410, pp. 232–243. Springer, Heidelberg (2010). https://doi.org/10.1007/978-3-642-16926-7_22
7. Bulian, J., Dawar, A.: Graph isomorphism parameterized by elimination distance to bounded degree. Algorithmica **75**(2), 363–382 (2015). https://doi.org/10.1007/s00453-015-0045-3
8. Cai, L.: Fixed-parameter tractability of graph modification problems for hereditary properties. Inf. Process. Lett. **58**(4), 171–176 (1996). https://doi.org/10.1016/0020-0190(96)00050-6
9. Cao, Y., Marx, D.: Interval deletion is fixed-parameter tractable. ACM Trans. Algorithms **11**(3), 21:1-21:35 (2015). https://doi.org/10.1145/2629595
10. Cao, Y., Marx, D.: Chordal editing is fixed-parameter tractable. Algorithmica **75**(1), 118–137 (2015). https://doi.org/10.1007/s00453-015-0014-x
11. Courcelle, B.: The monadic second-order logic of graphs. I. Recognizable sets of finite graphs. Inf. Comput. **85**(1), 12–75 (1990). https://doi.org/10.1016/0890-5401(90)90043-H
12. Courcelle, B.: The expression of graph properties and graph transformations in monadic second-order logic. In: Handbook of Graph Grammars and Computing by Graph Transformations, vol. 1: Foundations, pp. 313–400. World Scientific (1997)
13. Courcelle, B., Makowsky, J.A., Rotics, U.: Linear time solvable optimization problems on graphs of bounded clique-width. Theory Comput. Syst. **33**(2), 125–150 (2000). https://doi.org/10.1007/s002249910009
14. Courcelle, B., Olariu, S.: Upper bounds to the clique width of graphs. Discret. Appl. Math. **101**(1–3), 77–114 (2000). https://doi.org/10.1016/S0166-218X(99)00184-5
15. Courcelle, B., Oum, S.: Vertex-minors, monadic second-order logic, and a conjecture by seese. J. Comb. Theory Ser. B **97**(1), 91–126 (2007). https://doi.org/10.1016/j.jctb.2006.04.003
16. Cygan, M., et al.: Parameterized Algorithms. Springer, Cham (2015). https://doi.org/10.1007/978-3-319-21275-3
17. Downey, R.G., Fellows, M.R.: Fundamentals of Parameterized Complexity. TCS, Springer, London (2013). https://doi.org/10.1007/978-1-4471-5559-1
18. Eiben, E., Ganian, R., Hamm, T., Kwon, O.: Measuring what matters: a hybrid approach to dynamic programming with treewidth. J. Comput. Syst. Sci. **121**, 57–75 (2021). https://doi.org/10.1016/j.jcss.2021.04.005
19. Eiben, E., Ganian, R., Kwon, O.: A single-exponential fixed-parameter algorithm for distance-hereditary vertex deletion. J. Comput. Syst. Sci. **97**, 121–146 (2018). https://doi.org/10.1016/j.jcss.2018.05.005
20. Flum, J., Grohe, M.: Parameterized Complexity Theory. TTCSAES, Springer, Heidelberg (2006). https://doi.org/10.1007/3-540-29953-X

21. Fomin, F.V., Golovach, P.A., Thilikos, D.M.: Parameterized complexity of elimination distance to first-order logic properties. In: 36th Annual ACM/IEEE Symposium on Logic in Computer Science (LICS), pp. 1–13 (2021). https://doi.org/10.1109/LICS52264.2021.9470540

22. Fomin, F.V., Lokshtanov, D., Misra, N., Saurabh, S.: Planar F-deletion: approximation, kernelization and optimal FPT algorithms. In: 53rd Annual IEEE Symposium on Foundations of Computer Science (FOCS), pp. 470–479 (2012). https://doi.org/10.1109/FOCS.2012.62

23. Fomin, F.V., Lokshtanov, D., Panolan, F., Saurabh, S., Zehavi, M.: Hitting topological minors is FPT. In: Proccedings of the 52nd Annual ACM-SIGACT Symposium on Theory of Computing (STOC), pp. 1317–1326 (2020). https://doi.org/10.1145/3357713.3384318

24. Ganian, R., Ordyniak, S., Ramanujan, M.S.: Going beyond primal treewidth for (M)ILP. In: Singh, S.P., Markovitch, S. (eds.) Proceedings of the Thirty-First AAAI Conference on Artificial Intelligence, pp. 815–821. AAAI Press (2017)

25. Ganian, R., Ramanujan, M.S., Szeider, S.: Backdoor treewidth for SAT. In: Gaspers, S., Walsh, T. (eds.) SAT 2017. LNCS, vol. 10491, pp. 20–37. Springer, Cham (2017). https://doi.org/10.1007/978-3-319-66263-3_2

26. Ganian, R., Ramanujan, M.S., Szeider, S.: Combining treewidth and backdoors for CSP. In: 34th Symposium on Theoretical Aspects of Computer Science (STACS), vol. 66, pp. 36:1–36:17 (2017). https://doi.org/10.4230/LIPIcs.STACS.2017.36

27. Ganian, R., Ramanujan, M.S., Szeider, S.: Discovering archipelagos of tractability for constraint satisfaction and counting. ACM Trans. Algorithms **13**(2), 291–2932 (2017). https://doi.org/10.1145/3014587

28. Guo, J., Hüffner, F., Niedermeier, R.: A structural view on parameterizing problems: distance from triviality. In: Parameterized and Exact Computation, First International Workshop, (IWPEC), vol. 3162, pp. 162–173 (2004). https://doi.org/10.1007/978-3-540-28639-4_15

29. Hlineny, P., Oum, S., Seese, D., Gottlob, G.: Width parameters beyond tree-width and their applications. Comput. J. **51**(3), 326–362 (2008). https://doi.org/10.1093/comjnl/bxm052

30. Jacob, A., de Kroon, J.J.H., Majumdar, D., Raman, V.: Parameterized complexity of deletion to scattered graph classes. CoRR abs/2105.04660 (2021). https://arxiv.org/abs/2105.04660

31. Jacob, A., Majumdar, D., Raman, V.: Parameterized complexity of deletion to scattered graph classes. In: 15th International Symposium on Parameterized and Exact Computation, (IPEC), vol. 180, pp. 18:1–18:17 (2020). https://doi.org/10.4230/LIPIcs.IPEC.2020.18

32. Jansen, B.M.P., de Kroon, J.J.H.: FPT algorithms to compute the elimination distance to bipartite graphs and more. In: Kowalik, Łukasz, Pilipczuk, Michał, Rzążewski, Paweł (eds.) WG 2021. LNCS, vol. 12911, pp. 80–93. Springer, Cham (2021). https://doi.org/10.1007/978-3-030-86838-3_6

33. Jansen, B.M.P., de Kroon, J.J.H., Wlodarczyk, M.: Vertex deletion parameterized by elimination distance and even less. In: Proceedings of the 53rd Annual ACM-SIGACT Symposium on Theory of Computing (STOC), pp. 1757–1769 (2021). https://doi.org/10.1145/3406325.3451068

34. Jansen, B.M.P., Lokshtanov, D., Saurabh, S.: A near-optimal planarization algorithm. In: Proceedings of the Twenty-Fifth Annual ACM-SIAM Symposium on Discrete Algorithms, pp. 1802–1811 (2014). https://doi.org/10.1137/1.9781611973402.130

35. Kim, E.J., et al.: Linear kernels and single-exponential algorithms via protrusion decompositions. ACM Trans. Algorithms **12**(2), 21:1-21:41 (2016). https://doi.org/10.1145/2797140

36. Lewis, J.M., Yannakakis, M.: The node-deletion problem for hereditary properties is NP-complete. J. Comput. Syst. Sci. **20**(2), 219–230 (1980). https://doi.org/10.1016/0022-0000(80)90060-4

37. Lindermayr, A., Siebertz, S., Vigny, A.: Elimination distance to bounded degree on planar graphs. In: 45th International Symposium on Mathematical Foundations of Computer Science, (MFCS), vol. 170, pp. 65:1–65:12 (2020). https://doi.org/10.4230/LIPIcs.MFCS.2020.65

38. Lokshtanov, D., Narayanaswamy, N.S., Raman, V., Ramanujan, M.S., Saurabh, S.: Faster parameterized algorithms using linear programming. ACM Trans. Algorithms **11**(2), 15:1–15:31 (2014). https://doi.org/10.1145/2566616. https://doi.org/10.1145/2566616

39. Lokshtanov, D., Ramanujan, M.S., Saurabh, S., Zehavi, M.: Reducing CMSO model checking to highly connected graphs. In: 45th International Colloquium on Automata, Languages, and Programming, (ICALP), vol. 107, pp. 135:1–135:14 (2018). https://doi.org/10.4230/LIPIcs.ICALP.2018.135

40. Marx, D.: Chordal deletion is fixed-parameter tractable. Algorithmica **57**(4), 747–768 (2010). https://doi.org/10.1007/s00453-008-9233-8

41. Nesetril, J., de Mendez, P.O.: Tree-depth, subgraph coloring and homomorphism bounds. Eur. J. Comb. **27**(6), 1022–1041 (2006). https://doi.org/10.1016/j.ejc.2005.01.010

42. Niedermeier, R.: Invitation to Fixed-Parameter Algorithms. Oxford University Press, Oxford (2006). https://doi.org/10.1093/acprof:oso/9780198566076.001.0001

43. Oum, S.: Rank-width and vertex-minors. J. Comb. Theory Ser. B **95**(1), 79–100 (2005). https://doi.org/10.1016/j.jctb.2005.03.003

44. Oum, S.: Approximating rank-width and clique-width quickly. ACM Trans. Algorithms **5**(1), 101–1020 (2008). https://doi.org/10.1145/1435375.1435385

45. Oum, S.: Rank-width and well-quasi-ordering. SIAM J. Discret. Math. **22**(2), 666–682 (2008). https://doi.org/10.1137/050629616

46. Oum, S.: Rank-width: algorithmic and structural results. Discret. Appl. Math. **231**, 15–24 (2017). https://doi.org/10.1016/j.dam.2016.08.006

47. Oum, S., Seymour, P.D.: Approximating clique-width and branch-width. J. Comb. Theory Ser. B **96**(4), 514–528 (2006). https://doi.org/10.1016/j.jctb.2005.10.006

48. Reed, B.A., Smith, K., Vetta, A.: Finding odd cycle transversals. Oper. Res. Lett. **32**(4), 299–301 (2004). https://doi.org/10.1016/j.orl.2003.10.009

49. Robertson, N., Seymour, P.D.: Graph minors. XIII. The disjoint paths problem. J. Comb. Theory Ser. B **63**(1), 65–110 (1995). https://doi.org/10.1006/jctb.1995.1006

50. Sau, I., Stamoulis, G., Thilikos, D.M.: An FPT-algorithm for recognizing k-apices of minor-closed graph classes. In: 47th International Colloquium on Automata, Languages, and Programming, (ICALP), vol. 168, pp. 95:1–95:20 (2020). https://doi.org/10.4230/LIPIcs.ICALP.2020.95

On the Existence of EFX Allocations

Bhaskar Ray Chaudhury[✉]

University of Illinois at Urbana Champaign, Champaign, USA
braycha@illinois.edu

Abstract. We focus on a fundamental problem in discrete fair division, where the goal is to divide indivisible goods among a set of agents "fairly". Ideally, one would aim to divide the goods such that no agent envies another agent. However, since the goods are indivisible, such allocations may not always exist (a simple scenario involving two agents and a single good). Therefore, relaxations of envy-freeness have been proposed and extensively studied. We focus on one of the most fundamental and sought out relaxations – envy-freeness up to any good (EFX), where no agent envies another, following the removal of any single good from the other's bundle. Despite substantial effort from the community, the existence of EFX allocations has not been settled. In this paper, we sketch the proof of existence of "almost" EFX allocations and the existence of EFX allocations when there are only three agents. In the end, we reduce the problem of finding improved guarantees on EFX allocations to a problem in zero sum extremal combinatorics.

Keywords: Fair division · EFX allocations · Extremal graph theory

Fair division is a fundamental branch of mathematical economics over the last seven decades (since the seminal work of Hugo Steinhaus in the 1940s [19]). In a classical fair division problem, the goal is to "fairly" allocate a set of items among a set of agents. Such problems even find a mention in ancient Greek mythology and the Bible. Even today, several real-life scenarios are paradigmatic of the problems in this domain, e.g., division of family inheritance [17], divorce settlements [6], spectrum allocation [13], air traffic management [20], course allocation [4] and many more[1]. For the past two decades, the computer science community has developed concrete formulations and tractable solutions to fair division problems and thus contributing substantially to the development in the field. With the advent of the Internet and the rise of centralized electronic platforms that intend to impose fairness constraints on their decisions (e.g., Airbnb would like to fairly matching hosts and guests, and Uber would like to fairly match drivers and riders etc.), there has been an increasing demand for computationally tractable protocols to solve fair division problems.

In an instance of a fair division problem, we have a set of *agents* and a set of *items*, and the goal is to determine an allocation of the items among the agents

---

[1] Check [1] and [2] for more detailed explanation of fair division protocols used in day to day problems.

© Springer Nature Switzerland AG 2022
C. Bazgan and H. Fernau (Eds.): IWOCA 2022, LNCS 13270, pp. 21–27, 2022.
https://doi.org/10.1007/978-3-031-06678-8_2

that makes every agent content i.e., is "fair" and achieves high welfare, i.e., is "efficient". The items to be divided can be *divisible* or *indivisible*, and they can be *desirable (goods)* or *undesirable (bads or chores)*. Motivated by applications, there are several notions of *fairness* and *efficiency*, which lead to several distinct problems (see Fig. 1).

The most extensively studied setting is that of divisible goods. An allocation through a *Competitive Equilibrium with Equal Incomes (CEEI)* is often considered a canonical way to achieve fairness and efficiency. In a CEEI, one creates a virtual market with the agents and the goods and equips each agent with the same amount of money, say 1 dollar. At a CE, one determines a price for each item and an allocation of the items to the agents such that (i) each agent gets her most preferred bundle of items[2] in exchange for her initial budget of 1 dollar and (ii) all items are completely allocated. Such an allocation is fair and efficient. To be precise, it is *envy-free*, i.e., no agent strictly prefers any other agent's bundle to her own and *Pareto-optimal*, i.e., there exists no allocation where some agent can be made happier without making any other agent worse. At first glance, it is not clear why such prices and allocations exist. However, there is an extensive line of work on *competitive equilibrium* (also referred to as market equilibrium) that not only shows the existence of such prices and allocations but also describes several fast algorithms to compute them. In fact, competitive equilibrium theory has a long history going back to the works of Léon Walras in 1874 [21]. However, the emphasis always, was on determining prices at which demand equals supply and it is not until quite recently the techniques and concepts from competitive equilibrium theory have been leveraged to find fair and efficient allocation of items that go beyond divisible goods. Most notably, fair and efficient division of *indivisible goods* has got some recent attention from the CS and the economics community in the past decade. This setting is also practically relevant: For instance, jewellery, artworks, estates, and electronics are indivisible goods that frequently require allocation. Despite the similarity in the nature of the problems, both settings pose far more challenges than the setting with divisible goods. In this paper, we highlight the aforementioned challenges and answer some fundamental questions in this setting. Our main contributions are elaborated in the upcoming section.

1   Fair and Efficient Allocation of Indivisible Goods

Fair division problems involving indivisible goods have been relatively understudied, primarily because classic fairness notions such as envy-freeness cannot be guaranteed even in trivial instances, such as a setting with two agents and a single indivisible good that both agents find valuable. However, over the last decade, several relaxations of envy-freeness and proportionality have been proposed and studied. In this paper, we consider one of the most important relaxations of envy-freeness: envy-freeness up to any good (EFX).

---

[2] Bundle that minimizes disutility in the case of bads or the bundle that maximizes utility in the case of goods.

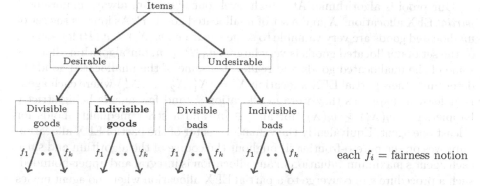

each f_i = fairness notion

Fig. 1. Spectrum of problems in fair division. The setting with divisible goods has been extensively studied. In this paper, we elaborate answers to some fundamental questions in fair division of indivisible goods.

Envy-freeness up to any good (EFX). "The closest analogue of envy-freeness" in the context of indivisible goods is that of *envy-freeness up to any good* (EFX) [7]. An allocation is said to be EFX if no agent envies another agent following the removal of *any* single good from the other agent's bundle. Until now, it is not known whether EFX allocations exist even when agents have *additive valuations*, despite "significant effort" by the research community [7,15]. Ariel Procaccia, in an editorial note in Communications of the ACM [18], refers to the question as

"fair division's biggest open problem".

1.1 EFX with Bounded Charity [11]

As the foundational result in this paper, we show that even when agents have much more general valuations than additive valuations[3], an EFX allocation always exists if we allow a small number of goods to remain unallocated [11]. Formally, there exists a partition $\langle X_1, X_2, \ldots, X_n, P \rangle$ of the good set M such that

- $X = \langle X_1, X_2, \ldots, X_n \rangle$ is EFX,
- $v_i(X_i) \geq v_i(P)$ for all $i \in [n]$, and
- $|P| \leq n - 1$[4],

where $v_i : 2^M \to \mathbb{R}_{\geq 0}$ denotes the valuation function of agent i. We remark that prior to this result, the only settings in which EFX allocations were known to exist for general valuations were the setting with only two agents or the setting in which all agents have the same valuation [16].

---

[3] A valuation v is additive if $v(S) = \sum_{s \in S} v(\{s\})$ for all S.

[4] Note that n is the number of agents which is typically much smaller than the number of goods.

Our proof is algorithmic. At a high-level, our algorithm always maintains a *partial* EFX allocation[5] X and a set of unallocated goods P. As long as the set of unallocated goods are very valuable to some agent i.e., $v_i(X_i) < v_i(P)$ for some i, or the set of unallocated goods is very large, i.e., $|P| \geq n$, our algorithm allocates some of the unallocated goods and reallocates some of the unallocated goods to determine a new partial EFX allocation $X' = \langle X_1', X_2', \ldots, X_n' \rangle$ where each agent is at least as happy as they were in allocation X and few agents being strictly happier, i.e., $v_i(X_i') \geq v_i(X_i)$ for all $i \in [n]$ and a strict inequality holds for atleast one agent. Equivalently the valuation vector of the partial EFX allocation improves on the pareto-frontier throughout the course of the algorithm and since each agent's maximum valuation in any allocation is integral and upper-bounded, such a procedure will converge to a partial EFX allocation where no agent envies the number of unallocated goods is at most $n - 1$.

1.2 Complete EFX Allocations for Three Agents [8]

Despite the above results on EFX with bounded charity, the existence of complete EFX allocation (where no good is unallocated) remained a hard problem, even with only three agents with additive valuations ("highly non trivial" problem according to [16]). *As the main result of this paper, we show that EFX allocations always exist when there are three agents with additive valuations* [8].

Note that the existence of EFX allocations with bounded charity already implies the existence of a partial EFX allocation with at most two unallocated goods when there are only three agents. However, allocating these remaining two goods may require some massive reallocation of the already allocated goods and can also render some agents worse off in the final allocation. In particular, we construct an instance with three agents, seven goods and an EFX allocation on six goods such that no complete EFX allocation pareto-dominates the existing partial EFX allocation, i.e., in any complete EFX allocation, there will always be some agent whose valuation will strictly decrease. This rules out the possibility of generalizing the approach used for proving the existence of EFX allocations with bounded charity. We circumvent this issue by defining a new potential function: Let a, b and c be the three agents and X_a, X_b and X_c be any partial EFX allocation. We define $\phi(X) = \langle v_a(X_a), v_b(X_b), v_c(X_c) \rangle$. We show that given any partial EFX allocation X and an unallocated good, we can always find another partial EFX allocation X' where either $v_a(X_a') > v_a(X_a)$ or $v_a(X_a') = v_a(X_a)$ and $v_b(X_b') > v_b(X_b)$ or $v_a(X_a') = v_a(X_a)$ and $v_b(X_b') = v_b(X_b)$ and $v_c(X_c') > v_c(X_c)$, i.e., $\phi(X') \succ_{lex} \phi(X)$. Since the valuation function are upper-bounded and integral, such an update procedure will finally converge to a complete EFX allocation. The update rules that determine X' from X involve significantly more reallocation than the update rules used in the context of EFX with bounded charity.

---

[5] An allocation where not all the goods are allocated.

1.3   EFX with Bounded Charity and High Nash Welfare [10]

As mentioned earlier, alongside fairness, another desirable property of an allocation is "efficiency" i.e., a measure of the overall welfare that the allocation achieves. One of the most common measures of economic efficiency is *Nash welfare*[6]– defined as the geometric mean of the valuations of the agents. It is intuitive that an allocation having high Nash welfare will have less skew in the valuation functions of the agents. At a high-level, Nash welfare captures the natural balance between fairness and efficiency and therefore is widely regarded as a direct indicator of the fairness and efficiency of an allocation. As a result, the problem of maximizing Nash welfare has independently received a great deal of attention from the research community [3,5,12,14]. Therefore, our goal now is to get all the aforementioned fairness guarantees, with high Nash welfare.

We show that in polynomial-time, we can determine an EFX allocation with bounded charity that also has high Nash welfare. In particular, our allocation achieves a $2e^{1/e}$ approximation of the maximum Nash welfare when agents have additive valuation functions and an $\mathcal{O}(n)$ approximation when agents have subadditive valuation functions [10]. The approximation achieved when agents have subadditive valuation functions improves upon the previous best approximation guarantees (even for the further restricted class of submodular valuation functions) and is *tight* under value queries[7].

1.4   Reduction to a Problem in Extremal Graph Theory [9]

Finding better relaxations of EFX allocations (improving the approximation factor or reducing the number of unallocated goods in a partial EFX allocation) is a systematic way to approach the problem, when there are arbitrary number of agents. To this end, we show the existence of $(1 - \varepsilon)$-EFX allocations[8] with sublinear charity. In particular, we establish a connection between the number of unallocated goods and a problem in extremal graph theory. Formally, given any integer $d > 0$, we define the rainbow cycle number $R(d)$ as the largest k such that there exists a k-partite graph $G = (V_1 \cup V_2 \cup \cdots \cup V_k, E)$ and

– each part has at most d vertices, i.e., $|V_\ell| \leq d$ for all $\ell \in [k]$,
– for all parts V_ℓ and $V_{\ell'}$, each vertex in V_ℓ has an incoming edge from some vertex in $V_{\ell'}$ and vice-versa, and
– there exists no cycle in G that visits each part at most once.

   We show that any finite upper-bound on $R(d)$ will already give us a polynomial time algorithm that finds $(1 - \varepsilon)$-EFX allocations with sublinear charity.

---

[6] An allocation with maximum Nash welfare is also Pareto-optimal (an alternate measure of efficiency).
[7] One would need exponentially many value queries to get any sublinear approximation of Nash welfare when agents have subadditive valuation functions.
[8] An allocation $X = \langle X_1, X_2, \ldots, X_n \rangle$ is a $(1 - \varepsilon)$-EFX allocation if and only if for all pairs of agents i and i', we have $v_i(X_i) \geq (1 - \varepsilon) \cdot v_i(X_{i'} \setminus \{g\})$ for all $g \in X_{i'}$.

In particular, there is a polynomial time algorithm that finds $(1 - \varepsilon)$-EFX allocations with $n/(\varepsilon h^{-1}(n/\varepsilon))$ charity, where $h^{-1}(d)$ is the smallest k such that $h(k) = k \cdot R(k) > d$ [9]. We show that $R(d) \in \mathcal{O}(d^4)$, implying the existence of $(1 - \varepsilon)$-EFX with $\mathcal{O}((n/\varepsilon)^{4/5})$ charity. We suspect that $R(d)$ is linear, which would then show the existence of $(1 - \varepsilon)$-EFX with $\mathcal{O}(\sqrt{n/\varepsilon})$ charity. However, we leave finding tighter upper-bounds on the rainbow cycle number as an interesting question for future research. Lastly, we believe that the idea of reducing a fair division problem (which involves cardinal preferences of agents) to a pure graph theoretic problem (without any numerical quantities) might find broader applicability in the future.

References

1. www.spliddit.org
2. www.fairoutcomes.com
3. Anari, N., Mai, T., Gharan, S.O., Vazirani, V.V.: Nash social welfare for indivisible items under separable, piecewise-linear concave utilities. In: Proceedings of 29th Symposium Discrete Algorithms (SODA), pp. 2274–2290 (2018)
4. Budish, E., Cantillon, E.: The multi-unit assignment problem: theory and evidence from course allocation at Harvard. Am. Econ. Rev. **102** (2010)
5. Barman, S., Krishnamurthy, S.K., Vaish, R.: Finding fair and efficient allocations. In: Proceedings of the 19th ACM Conference on Economics and Computation (EC), pp. 557–574 (2018)
6. Brams, S.J., Taylor, A.D.: Fair division - from Cake-Cutting to Dispute Resolution. Cambridge University Press, Cambridge (1996)
7. Caragiannis, I., Kurokawa, D., Moulin, H., Procaccia, A.D., Shah, N., Wang, J.: The unreasonable fairness of maximum Nash welfare. In: Proceedings of the 17th ACM Conference on Economics and Computation (EC), pp. 305–322 (2016)
8. Chaudhury, B.R., Garg, J., Mehlhorn, K.: EFX exists for three agents. In: Proceedings of the 21st ACM Conference on Economics and Computation (EC), pp. 1–19. ACM (2020)
9. Chaudhury, B.R., Garg, J., Mehlhorn, K., Mehta, R., Misra, P.: Improving EFX guarantees through rainbow cycle number. In: Proceedings of the 22nd ACM Conference on Economics and Computation (EC), pp. 310–311. ACM (2021)
10. Chaudhury, B.R., Garg, J., Mehta, R.: Fair and efficient allocations under subadditive valuations. In: Proceedings of the 35th AAAI Conference on Artificial Intelligence (AAAI) (2021)
11. Chaudhury, B.R., Kavitha, T., Mehlhorn, K., Sgouritsa, A.: A little charity guarantees almost envy-freeness. SIAM J. Comput. **50**(4), 1336–1358 (2021)
12. Cole, R., Gkatzelis, V.: Approximating the Nash social welfare with indivisible items. SIAM J. Comput. **47**(3), 1211–1236 (2018)
13. Etkin, R., Parekh, A., Tse, D.: Spectrum sharing for unlicensed bands. In: Proceedings of the First IEEE Symposium on New Frontiers in Dynamic Spectrum Access Networks (2005)
14. Garg, J., Kulkarni, P., Kulkarni, R.: Approximating Nash social welfare under submodular valuations through (un)matchings. In: Proceedings of the 31st Symposium on Discrete Algorithms (SODA 2020) (2020)
15. Moulin, H.: Fair division in the internet age. Ann. Rev. Econ. **11**, 407–441 (2019)

16. Plaut, B., Roughgarden, T.: Almost envy-freeness with general valuations. SIAM J. Discret. Math. **34**(2), 1039–1068 (2020)
17. Pratt, J.W., Zeckhauser, R.J.: The fair and efficient division of the Winsor family silver. Manag. Sci. **36**(11), 1293–1301 (1990)
18. Procaccia, A.D.: Technical perspective: an answer to fair division's most enigmatic question. Commun. ACM **63**(4), 118 (2020). https://doi.org/10.1145/3382131, https://doi.org/10.1145/3382131
19. Steinhaus, H.: The problem of fair division. Econometrica **16**, 101–104 (1948)
20. Vossen, T.W.: Fair allocation concepts in air traffic management. Ph.D. thesis. University of Maryland, College Park (2002)
21. Walras, L.: Éléments d'économie politique pure, ou théorie de la richesse sociale (Elements of Pure Economics, or the theory of social wealth). Lausanne, Paris (1874), (1899, 4th ed.; 1926, rev ed., 1954, Engl. transl.)

Contributed Papers

Lower Bounds for Restricted Schemes in the Two-Adaptive Bitprobe Model

Sreshth Aggarwal, Deepanjan Kesh, and Divyam Singal[(✉)]

Indian Institute of Technology Guwahati, Guwahati, Assam, India
dsingal@iitg.ac.in

Abstract. In the adaptive bitprobe model answering membership queries in two bitprobes, we consider the class of restricted schemes as introduced by Kesh and Sharma [1]. In that paper, the authors showed that such restricted schemes storing subsets of size 2 require $\Omega(m^{\frac{2}{3}})$ space. In this paper, we generalise the result to arbitrary subsets of size n, and prove that the space required for such restricted schemes will be $\Omega\left(\left(\frac{m}{n}\right)^{1-\frac{1}{\lfloor n/4\rfloor+2}}\right)$.

1   Introduction

In the bitprobe model, we store subsets $\mathcal{S}$ of size n from an universe $\mathcal{U}$ of size m in a data structure taking s amount of space, and answer membership queries by reading t bits of the data structure. The conventional notation to denote such schemes is as an (n, m, s, t)-scheme. Each such scheme has two components – the *storage scheme* sets the bits of the data structure according to the given subset $\mathcal{S}$, and the *query scheme* probes at most t bits of the data structure to answer membership queries. Schemes are categorised as *adaptive* if the location of each bitprobe in their query scheme depends on the answers obtained in the previous bitprobes. If the location of the bitprobes in the query scheme is independent of the answers obtained in the earlier bitprobes, the corresponding scheme is called *non-adaptive*. For further reading about bitprobe and other related models and their associated results, Nicholson *et al.* [2] has quite a detailed survey of the area.

In this paper, we restrict ourselves to those bitprobe schemes that answer membership queries using two adaptive bitprobes, i.e. $t = 2$. The data structure of such schemes can be thought of as having 3 tables, namely $\mathcal{A}, \mathcal{B}$, and $\mathcal{C}$. The first bitprobe is made in table $\mathcal{A}$, and if the bit probed in table $\mathcal{A}$ has been set to 0 the next bitprobe is made in table $\mathcal{B}$. On the other hand, the second bitprobe is made in table $\mathcal{C}$ if the bit queried in table $\mathcal{A}$ has been set to 1. The answer to the membership query is "Yes" if the second bitprobe returns 1, "No" otherwise.

The best known scheme for storing subsets of size two and answering membership queries using two adaptive bitprobes is due to Radhakrishnan *et al.* [3] which takes $O(m^{\frac{2}{3}})$ amount of space; the best known lower bound for the problem is $\Omega(m^{\frac{4}{7}})$ [4]. Though the problem is yet to be settled for subsets of size

© Springer Nature Switzerland AG 2022
C. Bazgan and H. Fernau (Eds.): IWOCA 2022, LNCS 13270, pp. 31–45, 2022.
https://doi.org/10.1007/978-3-031-06678-8_3

two, it has recently been shown for subsets of size three that the space required is $\Theta(m^{\frac{2}{3}})$ [5,6]. Garg and Radhakrishnan [7] proved that for arbitrary sized subsets, the space bounds for two adaptive bitprobe schemes are $\Omega(m^{1-\frac{1}{\lfloor n/4 \rfloor}})$ and $O(m^{1-\frac{1}{4n+1}})$, where $n \leq c \cdot \log m$.

Due to space constraints, some of the proofs and figures have been omitted in this paper. The full version of the paper has been uploaded to Arxiv [8] and has been referenced to wherever the proof has been omitted.

2 Restricted Schemes

Kesh and Sharma [1] proved that $\Omega(m^{\frac{2}{3}})$ is indeed the lower bound for two adaptive bitprobe schemes storing subsets of size two, albeit for a restricted class of schemes. We now introduce the restriction that characterises this class of schemes.

In the literature, elements of the universe $\mathcal{U}$ that query, or equivalently map to, the same bit in table $\mathcal{A}$ are said to form a *block*. We label the elements of a block uniquely as 1, 2, 3, ..., which we will refer to as the *index* of the element within a block. The element with index i of a block $\mathbf{a}$ will be denoted as $\mathbf{a}_i$. Elements of $\mathcal{U}$ that query, or map to, the same bit in tables $\mathcal{B}$ or $\mathcal{C}$ form a *set*. This departure in labels is made to distinguish the collections of elements in tables $\mathcal{B}$ and $\mathcal{C}$ from those of table $\mathcal{A}$, which will prove useful henceforth. The set to which the element $\mathbf{a}_i$ belongs to in table $\mathcal{B}$ will be denoted as $S_{\mathcal{B}}(\mathbf{a}_i)$; similarly for sets of table $\mathcal{C}$.

We impose the following restriction on the schemes designed to store subsets $\mathcal{S}$ and answer membersip queries using two bitprobes.

Restriction 2.1. *If two elements belong to the same set either in table $\mathcal{B}$ or in table $\mathcal{C}$, then their indices are the same.*

To take an example, in the schemes that we consider if $\mathbf{a}_i \in S_{\mathcal{C}}(\mathbf{b}_j)$, then it must be the case that $i = j$. We further simplify our premise by imposing the following restrictions on the schemes we are addressing. They are being made for the sake of simplicity and do not affect the final result.

Restriction 2.2. *Our class of schemes satisfy the following constraints.*

1. *The three tables $\mathcal{A}, B$, and $\mathcal{C}$ do not share any bit.*
2. *All the three tables are of the same size.*
3. *All the blocks in table $\mathcal{A}$ are of equal size. Let that size be b.*
4. *There are no singleton sets in tables $\mathcal{B}$ and $\mathcal{C}$.*
5. *All of the sets in the tables $\mathcal{B}$ and $\mathcal{C}$ are clean [6], i.e. no two elements of a block belong to the same set.*

As discussed in Sect. 1.5 of [1], the motivation for these kind of restrictions is from the schemes presented in such works as by Radhakrishnan *et al.* [3], Lewenstein *et al.* [9], and Radhakrishnan *et al.* [4]. The final restriction is motivated from Sect. 3 of Kesh [6], where it is shown that any scheme can be converted to a scheme with only clean sets with no asymptotic increase in the size of the data structure.

The main result of the paper (Theorem 8.3) is as follows.

Theorem. *Two adaptive bitprobe schemes for storing subsets of size at most* n *and satisfying Restriction 2.1 require* $\Omega\left(\left(\frac{m}{n}\right)^{1-\frac{1}{\lfloor n/4 \rfloor + 2}}\right)$ *space.*

In this restricted setting, as one might expect, the lower bound of Theorem 8.3 improves upon the bound proposed by Garg and Jaikumar [7] for all schemes, which is $\Omega(m^{1-\frac{1}{\lfloor n/4 \rfloor}})$ for $n \leq c \cdot \sqrt{\frac{\log m}{\log n}}$, and the comparison can be found in Sect. 8. To generalise the proof presented, Lemma 5.3, which shows that indices increase as the i in i-Universe (Definition 5.1) increases, and Lemma 6.2, which works because the subsets $\mathcal{S}$ and $\mathcal{X}$ (defined in Sect. 3) are disjoint, need to be proven for the generalised setting. Other lemmas, including those of Sect. 8, lend itself to generalisation without much effort.

3   Premise

As mentioned earlier, the subset of the universe $\mathcal{U}$ that we want to store in our data structure will be referred to as $\mathcal{S}$. In the subsequent discussion, it will be necessary to build certain subsets that we would like to store in the data structure of the restricted schemes and, consequently, arrive at certain contradictions – such subsets will be denoted at various places as $\mathcal{S}, \mathcal{S}', \mathcal{S}'', \mathcal{S}_1, \mathcal{S}_2$, etc. As we build the subsets to store, it will also be required to keep track of certain elements that cannot be part of $\mathcal{S}$ – such subsets will be denoted as $\mathcal{X}, \mathcal{X}', \mathcal{X}'', \mathcal{X}_1, \mathcal{X}_2$, etc.

As Restriction 2.1 forces sets in tables $\mathcal{B}$ and $\mathcal{C}$ to contain elements of only a certain index, it will prove helpful to refer to the various structures in the two tables by their indices. To start with, $\mathcal{U}_i$ will denote those elements of $\mathcal{U}$ which have index i. $\mathcal{B}_i$ will refer to the collection of all sets of table $\mathcal{B}$ comprised of elements of $\mathcal{U}_i$; similarly $\mathcal{C}_i$. Sometimes, we will also use $\mathcal{T}$ and $\mathcal{T}'$ to refer to either of the tables $\mathcal{B}$ or $\mathcal{C}$. Hence, if we have two distinct tables $\mathcal{T}$ and $\mathcal{T}'$, then one of them will be $\mathcal{B}$ and the other $\mathcal{C}$, which is not important. On the other hand, table $\mathcal{A}$ will always be referred to by its name.

In the literature the size of the data structure has always been denoted by s. As the sizes of the three tables are equal (Restriction 2.2), we would instead use s to denote the size of any particular table; this would alleviate the need of using the fraction $\frac{s}{3}$ whenever we refer to the table sizes. So, the schemes will henceforth be referred to as an $(n, m, 3 \cdot s, 2)$-schemes.

In the following two results, we present some self-evident and one essential property of the notations as defined above. They will be referenced to later, as needed.

Observation 3.1. *The size s of a table and the elements of index i are related as follows.*

1. $|\mathcal{A}| = |\mathcal{B}| = |\mathcal{C}| = |\mathcal{U}_i| = s$.
2. $\mathcal{U}_i = \bigcup\limits_{S \in \mathcal{B}_i} S$.

Proof. The first of the two observations follows from the fact that each block of table $\mathcal{A}$ has exactly one element of any particular index. The second observation follows from the definition of $\mathcal{B}_i$.   □

Lemma 3.2. *The correctness of a scheme remains unaffected under a permutation of the indices.*

Proof. Consider an $(n, m, 3s, t)$-scheme that satisfies Restrictions 2.1 and 2.2. Suppose π is some permutation on the indices of the blocks of table $\mathcal{A}$. We observe that a permutation of the indices do not affect the membership of a block – two elements which belonged to block **a** before, still belongs to **a** but with their indices changed according to π. The same is true for any set of table $\mathcal{B}$ or $\mathcal{C}$ – elements of a set all had the same index, say i, to start with, and they will now have the index $\pi(i)$. So, the data structure of a scheme remains unaffected under the permutation, only the labels of the sets have changed. Thus, if a scheme was correct to begin with, it will remain so after a permutation of the indices.   □

We end the section with a final notational convenience. We would, in the discussion to follow, require to perform some arithmetic on index i, like $i + 1$ or $2i + 2$. As the range of indices lie between 1 and b, inclusive, all such expressions should be considered $(\mathrm{mod}\ b) + 1$. This would help us to keep the expressions simple and avoid repetition.

4   Nodes and Paths

In this section we define *nodes*, *edges*, and *paths*, structures that are defined on top of the elements belonging to a set.

Definition 4.1. A *node* of table $\mathcal{T}$, denoted as $(\mathbf{e}_k, \mathbf{f}_k)_{\mathcal{T}}$, is an ordered pair of distinct elements $\mathbf{e}_k$ and $\mathbf{f}_k$ such that they belong to the same set in $\mathcal{T}$.

Each of the components of a node are called its *terms*, the first being referred to as the *antecedent* and the second as the *consequent*.

We say that a block **a** is *stored* in table $\mathcal{B}$ if the bit corresponding to the block **a** in table $\mathcal{A}$ is set to 0. Then any query for any element of block **a** will be made in table $\mathcal{B}$, and the sets corresponding to those elements in table $\mathcal{B}$ should be set to 1 or 0 according as the elements are in $\mathcal{S}$ or not. We can, hence, say that the elements of block **a** are being stored in table $\mathcal{B}$. Storing a block or an element in table in $\mathcal{C}$ can similarly be defined as when the the bit in table $\mathcal{A}$ corresponding to the block **a** is set to 1.

Observation 4.2. *Suppose a node be such that one of its terms is in the subset S and the other in $\mathcal{X}$. Then, if the antecedent of the node is stored in its own table, the consequent of the node cannot be stored in its table.*

Proof. Consider the node $(\mathbf{e}_k, \mathbf{f}_k)_{\mathcal{T}}$. If we store the antecedent in its table, namely $\mathcal{T}$, then there is way to ensure that the consequent cannot be stored in its table. To that end, we put $\mathbf{e}_k$ in S and $\mathbf{f}_k$ in $\mathcal{X}$. Then as we are storing $\mathbf{e}_k$ in table $\mathcal{T}$, the set corresponding to $\mathbf{e}_k$ in the table must be set to 1. The element $\mathbf{f}_k$ belongs to the same set in $\mathcal{T}$ yet it is not part of S. So, $\mathbf{f}_k$, and consequently its block $\mathbf{f}$, cannot be stored in table $\mathcal{T}$, because if we do the query for element $\mathbf{f}_k$ will incorrectly return "Yes".

An equally good choice to force the antecedent and the consequent to separate tables is to have $\mathbf{e}_k \in \mathcal{X}$ and $\mathbf{f}_k \in S$. □

Definition 4.3. There is said to be an *edge* from the node $(\mathbf{e}_k, \mathbf{f}_k)_{\mathcal{T}_1}$ to the node $(\mathbf{g}_l, \mathbf{h}_l)_{\mathcal{T}_2}$ if the following holds.

1. The nodes belong to distinct tables, i.e. $\mathcal{T}_1 \neq \mathcal{T}_2$.
2. $l = k + 1$.
3. The consequent of the first node and the antecedent of the second node belong to the same block, i.e. $\mathbf{f} = \mathbf{g}$.

The second node above can be rewritten as $(\mathbf{f}_{k+1}, \mathbf{h}_{k+1})_{\mathcal{T}_2}$. The nodes with the edge between them are connected via the common block $\mathbf{f}$, and hence will be will be shown as

$$(\mathbf{e}_k, \mathbf{f}_k)_{\mathcal{T}_1} \xrightarrow{\ \mathbf{f}\ } (\mathbf{f}_{k+1}, \mathbf{h}_{k+1})_{\mathcal{T}_2}.$$

Definition 4.4. A sequence of nodes is said to be a *path* if between every pair of adjacent nodes there is an edge from the former to the latter. The *length* of a path is the number of edges it contains.

A path will be denoted as

$$(\mathbf{e}_k, \mathbf{f}_k)_{\mathcal{T}_1} \xrightarrow{\ \mathbf{f}\ } (\mathbf{f}_{k+1}, \mathbf{g}_{k+1})_{\mathcal{T}_2} \xrightarrow{\ \mathbf{g}\ } (\mathbf{g}_{k+2}, \mathbf{h}_{k+2})_{\mathcal{T}_1} \xrightarrow{\ \mathbf{h}\ } \ldots$$

For our discussion, we will only consider paths of length at most $\lfloor \frac{b}{2} \rfloor - 1$; we will see in Sect. 7 as to the reason why.

Observation 4.5. *Any element occurs at most once in a path.*

Proof. It follows from the definition of a path which dictates that indices increase (mod b) $+ 1$ from the first node onwards, and from our upper bound on the length of a path. On the other hand, it should be noted that a block may occur multiple times along a path. □

Lemma 4.6. *Suppose for every node in a path one of the terms of the node is in S and the other is in $\mathcal{X}$. Then, if the antecedent of the first node is stored in its own table, antecedents of all the nodes will have to be stored in their respective tables and the consequents of the nodes cannot be stored in their respective tables.*

Proof. The lemma is a direct consequence of Observation 4.2, and the lemma and its proof appears as Lemma A.1 in [8]. □

5   Universe of Elements

In this section, we define the universe of an element of $\mathcal{U}$ recursively, and establish its relation with nodes and paths.

Definition 5.1. The i-*Universe* of an element e_k w.r.t. table $\mathcal{T}$, denoted as $\mathcal{U}_{\mathcal{T}}^i(e_k)$, is defined as follows.

$$\mathcal{U}_{\mathcal{T}}^i(e_k) = \begin{cases} \{\; u_{k+1} \mid u_k \in S_{\mathcal{T}}(e_k) \setminus \{e_k\} \;\}, & \text{for } i = 1; \\[2mm] \bigcup_{u_l \in \mathcal{U}_{\mathcal{T}}^{i-1}(e_k)} \mathcal{U}_{\mathcal{T}'}^1(u_l), & \text{for } i > 1. \end{cases}$$

The table $\mathcal{T}'$ is defined as follows.

$$\mathcal{T}' = \mathcal{T}, \text{ if } i \text{ is odd};$$
$$\mathcal{T}' \neq \mathcal{T}, \text{ otherwise.}$$

Similar to the upper bound on paths, we will consider i-universes for $1 \leq i \leq \lfloor \frac{b}{2} \rfloor - 1$, and, as stated before, we will see in Sect. 7 as to the reason why.

The i-Universe of an element is the union of the 1-Universes of all the elements in its $(i-1)$-Universe. So, as i increases so does the size of the universe. We will show that the elements of the i-Universe must necessarily belong to distinct sets in a table, so the larger the i the larger has to be the size of the data structure to accomodate the i-Universe. We start with a few properties of the elements belonging to the i-Universe of an element.

Observation 5.2. $\left| \mathcal{U}_{\mathcal{T}}^1(e_k) \right| = |S_{\mathcal{T}}(e_k) \setminus \{e_k\}|$.

Lemma 5.3. *If an element* x_l *belongs to the* i-*Universe of* e_k, *then* $l = k + i$.

Proof. The statement of the lemma can be established by induction on i, and the lemma and its proof appears as Lemma A.2 in [8]. □

Lemma 5.4. *If the element* x_{k+i} *belongs to the* i-*Universe of* e_k *w.r.t. table* $\mathcal{T}$, *then there is a path such that*

1. *The first node is in table* $\mathcal{T}$ *with its antecedent being* e_k.
2. *The last node is in table* $\mathcal{T}'$ *with its antecedent being* x_{k+i}. *The table* $\mathcal{T}'$ *is defined as follows.*

$$\mathcal{T} = \mathcal{T}', \text{ if } i \text{ is even}$$
$$\mathcal{T} \neq \mathcal{T}', \text{ otherwise.}$$

3. *The length of the path is* i.

It is important to observe that the nature of the table $\mathcal{T}'$ in the lemma above is contrary to that in Definition 5.1 in that $\mathcal{T}'$ is the same as $\mathcal{T}$ when i is even in the lemma above, whereas they are equal when i is odd in the definition of the i-Universe.

Proof. We will prove the aforementioned statement by induction on i, and the lemma and its proof appears as Lemma A.3 in [8]. □

6 Bad Elements

In this section, we show that large universes of elements give rise to *bad elements*, which put constraints on how and what subsets can be stored.

Definition 6.1. An element $\mathbf{e}_k$ is said to be *i-bad* w.r.t. table $\mathcal{T}$ if for any j between 1 and i, inclusive, there exist distinct elements $\mathbf{u}_{k+j}$ and $\mathbf{v}_{k+j}$ in $\mathcal{U}_{\mathcal{T}}^{j}(\mathbf{e}_k)$ s.t.

$$\mathbf{v}_{k+j} \in S_{\mathcal{T}'}(\mathbf{u}_{k+j}).$$

The table $\mathcal{T}'$ is defined as follows.

$$\mathcal{T} = \mathcal{T}', \text{ if } j \text{ is even}$$
$$\mathcal{T} \neq \mathcal{T}', \text{ otherwise.}$$

Elements which are not *i-bad* are said to be *i-good*.

The above definition suggests that if l is some constant less than or equal to i and the element $\mathbf{e}_k$ is l-bad, then it is also i-bad.

Lemma 6.2. *If an element $\mathbf{e}_k$ is i-bad w.r.t. table $\mathcal{T}$, then there exists a choice of the sets S and $\mathcal{X}$, each of size at most $2i$, s.t. the block $\mathbf{e}$ cannot be stored in table $\mathcal{T}$.*

Proof. This lemma and its proof appears as Lemma A.4 in [8]. □

7 Modified Schemes

Consider any restricted adaptive $(n, m, 3s, 2)$-scheme, the last component 2 denoting the number of bitprobes allowed. Let some element $\mathbf{e}_1$ of its universe $\mathcal{U}$ be i-bad w.r.t. table $\mathcal{B}$. Lemma 6.2 states that there exist sets S_1 and $\mathcal{X}_1$, each of size at most $2i$, s.t. the block $\mathbf{e}$ cannot be stored in table $\mathcal{B}$. Also, as an element becomes i-bad due to the elements of its i-Universe, the indices of the elements in the either of the sets S_1 and $\mathcal{X}_1$ lie between 1 and $i + 1$. Consider the element $\mathbf{e}_{i+2}$. If this element is i-bad w.r.t. table $\mathcal{C}$ there will exist sets S_2 and $\mathcal{X}_2$, again of size at most $2i$ each, s.t. the block $\mathbf{e}$ cannot be stored in table $\mathcal{C}$. The range of the indices in the two sets in this case would be from $i + 2$ to $2i + 2$.

We already know that the sets S_1 and $\mathcal{X}_1$ are disjoint, as are the sets S_2 and $\mathcal{X}_2$. Furthermore, as the range of indices in the two pairs of sets do not overlap, we can deduce that all the four sets are disjoint. Let us then consider the sets

$$S = S_1 \cup S_2 \text{ and } \mathcal{X} = \mathcal{X}_1 \cup \mathcal{X}_2,$$

each of their sizes being at most $4i$. As discussed above, this pair of sets imply that the block $\mathbf{e}$ cannot be stored in either of the tables $\mathcal{B}$ or $\mathcal{C}$, which is absurd as the scheme is deemed to be correct. So, we may conclude the following.

Lemma 7.1. *For any block of table* $\mathcal{A}$*, say* **e***, if the element* $\mathbf{e}_1$ *is* i*-bad w.r.t. table* $\mathcal{B}$*, then the element* $\mathbf{e}_{i+2}$ *cannot be* i*-bad w.r.t. table* $\mathcal{C}$*.*

Let us partition the universe $\mathcal{U}$ based on good and bad elements w.r.t. table $\mathcal{B}$. One part will be the union of all those blocks whose index 1 elements are good. The other part will be union of the remaining blocks.

$$\mathcal{U}' = \bigcup_{\substack{\mathbf{a}_1 \text{ is } i\text{-good} \\ \text{w.r.t. } \mathcal{B}}} \mathbf{a}; \quad \mathcal{U}'' = \bigcup_{\substack{\mathbf{a}_1 \text{ is } i\text{-bad} \\ \text{w.r.t. } \mathcal{B}}} \mathbf{a}.$$

According to Lemma 7.1, we know that though the index 1 elements of the blocks of $\mathcal{U}''$ are bad w.r.t. table $\mathcal{B}$, the index $i+2$ elements must necessarily be good w.r.t. table $\mathcal{C}$.

We now split our data structure in the following way. For any set X in either of table $\mathcal{B}$ or $\mathcal{C}$, we split it into two sets, one containing the elements of $\mathcal{U}'$ and the other containing the elements of $\mathcal{U}''$. More formally,

$$X = X' \cup X''; \quad X' \subset \mathcal{U}', \; X'' \subset \mathcal{U}''.$$

It is important to note that the indices of the elements in the two sets X' and X'' are the same as that of X. Consequently, the table $\mathcal{B}$ has been split into two parts, namely $\mathcal{B}'$ containing the sets with elements from $\mathcal{U}'$, and $\mathcal{B}''$ containing sets with elements from $\mathcal{U}''$. Thus, the collection of all sets in table $\mathcal{B}$ containing elements with index k, namely $\mathcal{B}_k$, is now $\mathcal{B}'_k \cup \mathcal{B}''_k$. The table $\mathcal{C}$ have similarly been split into two parts - $\mathcal{C}'$ and $\mathcal{C}''$.

Observation 7.2. *The size of* $\mathcal{B}_k$ *has at most doubled due to the above modification.*

The table $\mathcal{A}$ is also split into two tables, namely $\mathcal{A}'$ and $\mathcal{A}''$, containing elements of $\mathcal{U}'$ and $\mathcal{U}''$, respectively. As per our definition of $\mathcal{U}'$ and $\mathcal{U}''$, either a block belongs entirely in $\mathcal{U}'$ or entirely in $\mathcal{U}''$, and thus individual blocks are not split.

We now have two sets of data structures, one corresponding to the elements of $\mathcal{U}'$ and the other corresponding to $\mathcal{U}''$ –

$$(\mathcal{A}', \mathcal{B}', \mathcal{C}') \text{ and } (\mathcal{A}'', \mathcal{B}'', \mathcal{C}'').$$

This also means that within the original scheme, we have two independent schemes, one for the elements of $\mathcal{U}'$ and the other for the elements of $\mathcal{U}''$. Any subset $\mathcal{S}$ that is to be stored can now be split into $\mathcal{S}' \subset \mathcal{U}'$ and stored in the data structure corresponding to $\mathcal{U}'$, and $\mathcal{S}'' \subset \mathcal{U}''$ which can be stored in the data structure corresponding to $\mathcal{U}''$. The storage and query schemes remain as before for each of the parts of the data structure. So, to store a subset, if any block was earlier set to 0, in the new data structure it will still be set to 0. If any set X was being set to 1, now both X' and X'' will be set to 1; and so on.

We further modify the new data structure as follows. For the part $(\mathcal{A}'', \mathcal{B}'', \mathcal{C}'')$, we interchange the parts $\mathcal{B}''$ and $\mathcal{C}''$ in the tables $\mathcal{B}$ and $\mathcal{C}$ so that

$\mathcal{B}''$ will now be part of table $\mathcal{C}$ and $\mathcal{C}''$ will now be part of table $\mathcal{B}$. With this modification, for any index k the tables will be as follows.

$$\mathcal{B}_k = \mathcal{B}'_k \cup \mathcal{C}''_k; \quad \mathcal{C}_k = \mathcal{C}'_k \cup \mathcal{B}''_k$$

As the part pertaining to $\mathcal{U}'$, i.e. $(\mathcal{A}', \mathcal{B}', \mathcal{C}')$, is unaffected, the query scheme and storage scheme for it remains unchanged. For the part $(\mathcal{A}'', \mathcal{B}'', \mathcal{C}'')$, if a block was earlier set to 0 and thus sent to table $\mathcal{B}$, it should now be set to 1 and sent to table $\mathcal{C}$. Similarly, a block which was earlier set to 1 will now have to be set to 0.

Let us consider the sizes of the tables. For lack of a better notation, we will use $\mathcal{T}^{(0)}, \mathcal{T}^{(1)}, \mathcal{T}^{(2)}$ to refer to the original table, the table after the first modification, and after the second modification, respectively. Observation 7.2 tells us that

$$|\mathcal{T}_k^{(1)}| \le 2 \cdot |\mathcal{T}_k^{(0)}|.$$

After the second modification, we note that

$$|\mathcal{B}_k^{(2)}| + |\mathcal{C}_k^{(2)}| = |\mathcal{B}_k^{(1)}| + |\mathcal{C}_k^{(1)}|.$$

We make the third and final modification to our scheme. Before the second modification, all the elements of index 1 in $\mathcal{B}'$ were good w.r.t. table $\mathcal{B}$. After the second modification, all the elements of index $i+2$ in $\mathcal{C}''$, which were earlier good w.r.t. table $\mathcal{C}$, are now good w.r.t. table $\mathcal{B}$ because $\mathcal{C}''$ is now part of table $\mathcal{B}$. In Lemma 3.2, we have seen that the correctness of a scheme remains unaffected under a permutation of its indices. We now apply the following permutation over the indices of the data structure corresponding to $\mathcal{U}''$ – the labels k and $k+i+1$ are interchanged, where $1 \le k \le i+1$, whereas the rest of the indices remain unchanged. With this further modification, all the indices labelled from $i+2$ to $2i+2$ will now be labelled 1 to $i+1$ in that order, whereas the previously labelled indices 1 and $i+1$ will now be labelled $i+2$ to $2i+2$, in that order.

With this final modification, we now have a scheme where all the elements of index 1 are good w.r.t. table $\mathcal{B}$.

Lemma 7.3. *Any given restricted $(n, m, 3s, 2)$-scheme can be modified into a $(n, m, 6s, 2)$-scheme such that in the modified scheme all the elements of index 1 are i-good w.r.t. table $\mathcal{B}$.*

As the third and final modification does not affect indices larger than $2i+2$, we can say that

$$|\mathcal{B}_k^{(3)}| + |\mathcal{C}_k^{(3)}| = |\mathcal{B}_k^{(2)}| + |\mathcal{C}_k^{(2)}| \le 2 \cdot (|\mathcal{B}_k^{(0)}| + |\mathcal{C}_k^{(0)}|),$$

for $k > 2i+2$. As for the indices 1 to $2i+2$, the sets have been relabelled but not created, and as a result the total number of sets remain unchanged, i.e.

$$\sum_{k=1}^{2i+2} \left(|\mathcal{B}_k^{(3)}| + |\mathcal{C}_k^{(3)}| \right) = \sum_{k=1}^{2i+2} \left(|\mathcal{B}_k^{(2)}| + |\mathcal{C}_k^{(2)}| \right) \le 2 \cdot \sum_{k=1}^{2i+2} \left(|\mathcal{B}_k^{(0)}| + |\mathcal{C}_k^{(0)}| \right). \quad (1)$$

Finally, all of this can only be proven for subsets $\mathcal{S}$ and $\mathcal{X}$ whose sizes are at least $4i$, and for the range of indices 1 to $2i+2$. So, for the first condition we can set $n = 4i$. As for the range of indices, the size of a block b has to be larger than $2i+2$, which implies that universes and path lengths are bounded by $\lfloor \frac{b}{2} \rfloor - 1$.

8   Lower Bound

In this section, we will present our theorem on the space lower bound on restricted schemes. We start by presenting an estimate of the total sizes of all the t-universes of good elements.

Lemma 8.1. *Suppose all the elements with index 1 are t-good w.r.t. table $\mathcal{B}$. Then the sizes of their t-Universes satisfy the following inequality.*

$$\sum_{e \in \mathcal{A}} |\mathcal{U}_{\mathcal{B}}^t(e_1)| \geq c \cdot \frac{s^{t+1}}{\left(\sum_{i=1}^t (|\mathcal{B}_i| + |\mathcal{C}_i|)\right)^t},$$

for some constant c.

Proof. In this lemma, for the sake of convenience, we introduce two new notations. If $\mathbf{h}$ is a block then $\mathbf{h}_i$ was meant to denote that element of $\mathbf{h}$ which has index i. We now abuse the notation and use $\mathbf{h}_{i,j}$ to denote the element with index j in the block $\mathbf{h}_i$. We also introduce $P_{\mathcal{T}}(e_k)$ to denote the set $S_{\mathcal{T}}(e_k) \setminus \{e_k\}$. These notations will help us keep the expressions to follow succint.

Let us assume that t is odd. The sum of the sizes of the t-universes of all index 1 elements can be expressed as follows.

$$\sum_{e \in \mathcal{A}} |\mathcal{U}_{\mathcal{B}}^t(e_1)| = \sum_{e \in \mathcal{A}} \left| \left(\bigcup_{\mathbf{h}_{t-1,t} \in \mathcal{U}_{\mathcal{B}}^{t-1}(e_1)} \mathcal{U}_{\mathcal{B}}^1(\mathbf{h}_{t-1,t}) \right) \right|$$

$$= \sum_{e \in \mathcal{A}} \left(\sum_{\mathbf{h}_{t-1,t} \in \mathcal{U}_{\mathcal{B}}^{t-1}(e_1)} |\mathcal{U}_{\mathcal{B}}^1(\mathbf{h}_{t-1,t})| \right)$$

$$= \sum_{e \in \mathcal{A}} \sum_{\mathbf{h}_{t-1,t} \in \mathcal{U}_{\mathcal{B}}^{t-1}(e_1)} |P_{\mathcal{B}}(\mathbf{h}_{t-1,t})|$$

The above derivation follows from the definition of t-universe, the fact that all elements of index 1 are t-good, and Observation 5.2 about the size of 1-universes. We have now arrived at a summation indexed by the elements of $\mathcal{U}_{\mathcal{B}}^{t-1}(e_1)$, and applying Lemma A.5 in [8], we get –

$$\sum_{e \in \mathcal{A}} |\mathcal{U}_{\mathcal{B}}^t(e_1)| = \sum_{e \in \mathcal{A}} \sum_{\mathbf{h}_{1,1} \in P_{\mathcal{B}}(e_1)} \sum_{\mathbf{h}_{2,2} \in P_{\mathcal{C}}(\mathbf{h}_{1,2})} \sum_{\mathbf{h}_{3,3} \in P_{\mathcal{B}}(\mathbf{h}_{2,3})}$$

$$\cdots \sum_{\mathbf{h}_{t-1,t-1} \in P_{\mathcal{C}}(\mathbf{h}_{t-2,t-1})} |P_{\mathcal{B}}(\mathbf{h}_{t-1,t})| \cdot$$

The summation $\sum_{e \in \mathcal{A}}$ can be equivalently expressed as $\sum_{S \in \mathcal{B}_1} \sum_{e_1 \in S}$. As the elements e_1 and $h_{1,1}$ both belong to the set S and are distinct from each other, we can now reorder the first three indices of the summation as

$$\sum_{S \in \mathcal{B}_1} \sum_{h_{1,1} \in S} \sum_{e_1 \in P_{\mathcal{B}}(h_{1,1})} .$$

By pushing the summation indexed by e_1 inside, we can finally rewrite down the summation as –

$$\sum_{e \in \mathcal{A}} |\mathcal{U}_{\mathcal{B}}^t(e_1)| = \sum_{h_1 \in \mathcal{A}} \sum_{h_{2,2} \in P_{\mathcal{C}}(h_{1,2})} \sum_{h_{3,3} \in P_{\mathcal{B}}(h_{2,3})} \cdots$$
$$\sum_{h_{t-1,t-1} \in P_{\mathcal{C}}(h_{t-2,t-1})} |P_{\mathcal{B}}(h_{1,1})| \cdot |P_{\mathcal{B}}(h_{t-1,t})| . \quad (2)$$

Each term of this summation is determined by a tuple such as

$$(h_1 \in \mathcal{A}, \ h_{2,2} \in P_{\mathcal{C}}(h_{1,2}), \ h_{3,3} \in P_{\mathcal{B}}(h_{2,3}), \ \ldots, \ h_{t-1,t-1} \in P_{\mathcal{C}}(h_{t-2,t-1}))$$

where each block, except for the first, is dependent on the previous blocks. On the other hand, if any of the sets in the tuple is fixed, then the other blocks and the terms of the summation they index are fixed by the set. With this insight, we are going to put a lower bound on the sum of all t-universes.

Suppose X_1 be the smallest set that occurs in the summation above (Eq. 2), either as one of its terms or as one of its indices. We first consider the case when X_1 occurs as the index under the ith summation. Let us also consider that i is odd, which would imply that X_1 belongs to table $\mathcal{B}$. Thus the terms of the summation in which X_1 participates is determined as follows – the indices under the ith summation and beyond is determined as

$$(\ h_{i,i} \in X_1, \ h_{i+1,i+1} \in P_{\mathcal{C}}(h_{i,i+1}), \ h_{i+2,i+2} \in P_{\mathcal{B}}(h_{i+1,i+2}), \ \ldots) ,$$

and the indices prior to that is determined as

$$(h_{i-1,i} \in P_{\mathcal{B}}(h_{i,i}), h_{i-2,i-1} \in P_{\mathcal{C}}(h_{i-1,i-1}), \ldots, h_{2,3} \in P_{\mathcal{B}}(h_{3,3}), h_{1,2} \in P_{\mathcal{C}}(h_{2,2})) .$$

It is important to note that in the latter of the two tuples, $P_{\mathcal{B}}(h_{i,i})$ is the set $X_1 \setminus \{h_{i,i}\}$.

The sum of all the terms in which the set X_1 participates is as follows.

$$\sum_{h_{i-1,i} \in P_{\mathcal{B}}(h_{i,i})} \sum_{h_{i-2,i-1} \in P_{\mathcal{C}}(h_{i-1,i-1})} \sum_{h_{i-3,i-2} \in P_{\mathcal{B}}(h_{i-2,i-2})} \cdots \sum_{h_{2,3} \in P_{\mathcal{B}}(h_{3,3})} \sum_{h_{1,2} \in P_{\mathcal{C}}(h_{2,2})}$$
$$\left(\sum_{h_{i,i} \in X_1} \sum_{h_{i+1,i+1} \in P_{\mathcal{C}}(h_{i,i+1})} \cdots \sum_{h_{t-1,t-1} \in P_{\mathcal{C}}(h_{t-1,t-2})} (|P_{\mathcal{B}}(h_{1,1})| \cdot |P_{\mathcal{B}}(h_{t-1,t})|) \right)$$

As all of the sets involved have sizes $\geq |X_1|$, the above sum is at least $c_1 \cdot |X_1|^{t+1}$, for some constant c_1. From the remaining terms and index sets of the summation in Eq. 2, we remove all the blocks that belong to set X_1. So, if the initial sum in Eq. 2 is denoted by $\mathbb{S}_0$, and the remaining sum after the above procedure is $\mathbb{S}_1$, we have

$$\mathbb{S}_0 \geq \mathbb{S}_1 + c_1 |X_1|^{t+1}.$$

We next identify the smallest set, say X_2, in the summation $\mathbb{S}_1$ and repeat the above procedure which ends up in an estimation of all the terms associated with X_2, the estimation being $\geq c_2 \cdot |X_2|^{t+1}$, and removing the terms and blocks associated with X_2 from the remainder. We repeat this until all the blocks have thus been removed, upon which we will have a family of sets labelled X_is and they partition the blocks of table $\mathcal{A}$. The number of sets would, in the worst case, be the total number of sets in the tables $\mathcal{B}$ and $\mathcal{C}$ with index at most t. Consequently, we have

$$\sum_{e \in \mathcal{A}} |\mathcal{U}^t_{\mathcal{B}}(\mathbf{e}_1)| \geq c \cdot \sum_i |X_i|^{t+1} \geq c \cdot \sum_i \left(\frac{\sum_i |X_i|}{\sum_i 1} \right)^{t+1} \geq c \cdot \frac{s^{t+1}}{\left(\sum_{i=1}^t (|\mathcal{B}_i| + |\mathcal{C}_i|) \right)^t},$$

where c is some suitable constant. The final bound arises using the Cauchy-Schwarz inequality.

All the other scenarios including the case where t is presumed to be even, can be similarly argued. $\qquad\square$

Lemma 8.2. *If all elements of $\mathcal{U}_1$ are t-good w.r.t. table $\mathcal{B}$, then*

$$\sum_{j=1}^{t+1} (|\mathcal{B}_j| + |\mathcal{C}_j|) \geq c \cdot s^{\frac{t}{t+1}},$$

for some constant c.

Proof. As before, we will establish the statement of the lemma assuming that t is odd. The case when t is even will follow similarly. According to the definition of bad elements (Definition 6.1), a necessary property for an element to be t-good w.r.t. table $\mathcal{B}$ is that the elements of its t-Universe belong to distinct sets in $\mathcal{C}_{t+1}$, t being odd. Consequently, we have

$$\sum_{\mathbf{e}_1 \in \mathcal{U}_1} |\mathcal{U}^t_{\mathcal{B}}(\mathbf{e}_1)| \leq \sum_{\mathbf{e}_1 \in \mathcal{U}_1} |\mathcal{C}_{t+1}| = s \cdot |\mathcal{C}_{t+1}| \quad \text{(Observation 3.1)}$$

From Lemma 8.1, the inequalities follows.

$$s \cdot |\mathcal{C}_{t+1}| \geq \sum_{\mathbf{e}_1 \in \mathcal{U}_1} |\mathcal{U}^t_{\mathcal{B}}(\mathbf{e}_1)| \geq c \cdot \frac{s^{t+1}}{\left(\sum_{i=1}^t (|\mathcal{B}_i| + |\mathcal{C}_i|) \right)^t}$$

$$\implies \left(\sum_{i=1}^{t+1} (|\mathcal{B}_i| + |\mathcal{C}_i|) \right)^{t+1} \geq c \cdot s^t,$$

and the lemma follows. $\qquad\square$

Lemma 7.3 states that given a restricted $(n, m, 3s, 2)$-scheme, it can be modified into a $(n, m, 6s, 2)$-scheme such that in the modified scheme all the elements of $\mathcal{U}_1$ are i-good for some constant i. Furthermore, in that case we require the subset size, n, should be at least $4i$. So, from Eq. 1 and Lemma 8.2, we can deduce the following.

$$\sum_{k=1}^{2i+2} \left(|\mathcal{B}_k^{(0)}| + |\mathcal{C}_k^{(0)}| \right) \geq \frac{1}{2} \sum_{k=1}^{2i+2} \left(|\mathcal{B}_k^{(2)}| + |\mathcal{C}_k^{(2)}| \right) = \frac{1}{2} \sum_{k=1}^{2i+2} \left(|\mathcal{B}_k^{(3)}| + |\mathcal{C}_k^{(3)}| \right)$$
$$\geq c \cdot (2s)^{\frac{i}{i+1}}, \tag{3}$$

where $2s$ comes from the fact that the first modification splits the sets of tables $\mathcal{B}$ and $\mathcal{C}$ (Observation 7.2).

Let the indices in the original scheme be so chosen that the sum on the first $2i + 2$ indices in Eq. 3 is the minimum among all choices. We can then derive the following.

$$\sum_{k=1}^{b} \left(|\mathcal{B}_k^{(0)}| + |\mathcal{C}_k^{(0)}| \right) \geq \frac{b}{2i+2} \sum_{k=1}^{2i+2} \left(|\mathcal{B}_k^{(0)}| + |\mathcal{C}_k^{(0)}| \right)$$
$$\implies 2 \cdot s \geq \frac{1}{2i+2} \frac{m}{s} \cdot c \cdot (2s)^{\frac{i}{i+1}},$$

which upon simplification gives us

$$s \geq c' \cdot \left(\frac{m}{n} \right)^{1 - \frac{1}{\lfloor n/4 \rfloor + 2}},$$

for some suitable constant c'. Hence, the main result of the paper is as follows.

Theorem 8.3. *Two adaptive bitprobe schemes for storing subsets of size at most n and satisfying Restriction 2.1 require $\Omega(\left(\frac{m}{n} \right)^{1 - \frac{1}{\lfloor n/4 \rfloor + 2}})$ space.*

Comparing our result in this restricted setting with the bound proposed by Garg and Jaikumar [7] for all schemes, we see that our result improves on [7] for $n \leq c \cdot \sqrt{\frac{\log m}{\log n}}$.

Our lower bound is better if the following holds –

$$c_1 \cdot \left(\frac{m}{n} \right)^{1 - \frac{1}{\lfloor n/4 \rfloor + 2}} \geq c_2 \cdot m^{1 - \frac{1}{\lfloor n/4 \rfloor}}$$

Taking logarithm on both sides, we have

$$\left(\frac{1}{\frac{n}{4}} - \frac{1}{(\frac{n}{4} + 2)} \right) \log m \geq c' \cdot \left(1 - \frac{1}{(\frac{n}{4} + 2)} \right) \log n$$

for some constant c'.

Upon further simplification,

$$\log m \geq c' \cdot n^2 \log n$$

$$n \leq c \cdot \sqrt{\frac{\log m}{\log n}}$$

for some constant c, and thus our claim holds.

9   Conclusion

In this paper, we addressed a class of schemes, as devised by Kesh and Sharma [1], in the two adaptive bitprobe model and provided a space lower bound on such schemes for subsets of arbitrary sizes, thereby generalising the lower bound presented in that paper. As discussed earlier, one of the key lemmas that our lower bound proof hinges upon is Lemma 6.2, which demonstrates the generation of bad elements, and establishing this lemma is crucial in generalising the proof to arbitrary schemes. We hope that this issue can be resolved and the structure of our proof could serve as a template to provide bounds stronger that those presented by Garg and Jaikumar [7].

References

1. Kesh, D., Sharma, V.S.: On the bitprobe complexity of two probe adaptive schemes. Discrete Applied Mathematics 2021 (in press)
2. Nicholson, P.K., Raman, V., Rao, S.S.: A survey of data structures in the bitprobe model. In: Brodnik, A., López-Ortiz, A., Raman, V., Viola, A. (eds.) Space-Efficient Data Structures, Streams, and Algorithms. LNCS, vol. 8066, pp. 303–318. Springer, Heidelberg (2013). https://doi.org/10.1007/978-3-642-40273-9_19
3. Radhakrishnan, J., Raman, V., Srinivasa Rao, S.: Explicit deterministic constructions for membership in the bitprobe model. In: auf der Heide, F.M. (ed.) ESA 2001. LNCS, vol. 2161, pp. 290–299. Springer, Heidelberg (2001). https://doi.org/10.1007/3-540-44676-1_24
4. Radhakrishnan, J., Shah, S., Shannigrahi, S.: Data structures for storing small sets in the bitprobe model. In: de Berg, M., Meyer, U. (eds.) ESA 2010. LNCS, vol. 6347, pp. 159–170. Springer, Heidelberg (2010). https://doi.org/10.1007/978-3-642-15781-3_14
5. Baig, M.G.A.H., Kesh, D.: Two new schemes in the bitprobe model. In: Rahman, M.S., Sung, W.-K., Uehara, R. (eds.) WALCOM 2018. LNCS, vol. 10755, pp. 68–79. Springer, Cham (2018). https://doi.org/10.1007/978-3-319-75172-6_7
6. Kesh, D.: Space complexity of two adaptive bitprobe schemes storing three elements. In: 38th IARCS Annual Conference on Foundations of Software Technology and Theoretical Computer Science (FSTTCS) 2018, Ahmedabad, India. Proceedings, pp. 12:1–12:12 (2018)
7. Garg, M., Radhakrishnan, J.: Set membership with a few bit probes. In: Proceedings of the Twenty-Sixth Annual ACM-SIAM Symposium on Discrete Algorithms (SODA)2015, San Diego, CA, USA, pp. 776–784 (2015)

8. Aggarwal, S., Kesh, D., Singal, D.: Lower Bounds for Restricted Schemes in the Two-Adaptive Bitprobe Model. https://arxiv.org/abs/2204.03266
9. Lewenstein, M., Munro, J.I., Nicholson, P.K., Raman, V.: Improved explicit data structures in the bitprobe model. In: Schulz, A.S., Wagner, D. (eds.) ESA 2014. LNCS, vol. 8737, pp. 630–641. Springer, Heidelberg (2014). https://doi.org/10.1007/978-3-662-44777-2_52
10. Baig, M.G.A.H., Kesh, D.: Improved bounds for two query adaptive bitprobe schemes storing five elements. In: Li, Y., Cardei, M., Huang, Y. (eds.) COCOA 2019. LNCS, vol. 11949, pp. 13–25. Springer, Cham (2019). https://doi.org/10.1007/978-3-030-36412-0_2

Perfect Matchings with Crossings

Oswin Aichholzer[1] (iD), Ruy Fabila-Monroy[2] (iD), Philipp Kindermann[3] (iD),
Irene Parada[4] (iD), Rosna Paul[1(✉)] (iD), Daniel Perz[1] (iD), Patrick Schnider[5] (iD),
and Birgit Vogtenhuber[1] (iD)

[1] Institute of Software Technology, Graz University of Technology, Graz, Austria
{oaich,ropaul,daperz,bvogt}@ist.tugraz.at
[2] Departamento de Matemáticas, Cinvestav, Ciudad de México, Mexico
ruyfabila@math.cinvestav.edu.mx
[3] Fachbereich IV – Informatikwissenschaften, Universität Trier, Trier, Germany
kindermann@informatik.uni-trier.de
[4] Department of Applied Mathematics and Computer Science,
Technical University of Denmark, Kongens Lyngby, Denmark
irmde@dtu.dk
[5] Department of Mathematical Sciences, University of Copenhagen,
Copenhagen, Denmark
ps@math.ku.dk

Abstract. For sets of $n = 2m$ points in general position in the plane
we consider straight-line drawings of perfect matchings on them. It is
well known that such sets admit at least C_m different plane perfect
matchings, where C_m is the m-th Catalan number. Generalizing this
result we are interested in the number of drawings of perfect matchings which have k crossings. We show the following results. (1) For every
$k \leq \frac{1}{64}n^2 - O(n\sqrt{n})$, any set of n points, n sufficiently large, admits a perfect matching with exactly k crossings. (2) There exist sets of n points
where every perfect matching has fewer than $\frac{5}{72}n^2$ crossings. (3) The
number of perfect matchings with at most k crossings is superexponential in n if k is superlinear in n. (4) Point sets in convex position minimize
the number of perfect matchings with at most k crossings for $k = 0, 1, 2$,
and maximize the number of perfect matchings with $\binom{n/2}{2}$ crossings and
with $\binom{n/2}{2} - 1$ crossings.

Keywords: Perfect matchings · Crossings · Combinatorial geometry ·
Order types

Research on this work has been initiated at the 16th European Research Week on Geo-
metric Graphs which was held from November 18 to 22, 2019, near Strobl (Austria).
We thank all participants for the good atmosphere as well as for discussions on the
topic. Further, we thank Clemens Huemer for bringing this problem to our attention
in the course of a meeting of the H2020-MSCA-RISE project 73499 - CONNECT.
O. A. and R. P. supported by FWF grant W1230. R. M. partially supported by CONA-
CYT(Mexico) grant 253261. P. S. supported by ERC Grant ERC StG 716424-CASe.
D. P., I. P., and B. V. supported by FWF Project I 3340-N35. *Some results of this work
have been presented at the "Computational Geometry: Young Researchers Forum" in
2021* [2].

C. Bazgan and H. Fernau (Eds.): IWOCA 2022, LNCS 13270, pp. 46–59, 2022.
https://doi.org/10.1007/978-3-031-06678-8_4

1   Introduction

The question of how many different plane (that is, crossing-free) straight-line perfect matchings can be drawn on a point set P in general position (that is, no three points are colinear) has been extensively studied; see for example [4,5,9,13, 14]. It is known that $n = 2m$ points in general position admit at least C_m plane perfect matchings, where $C_m = \frac{1}{m+1}\binom{2m}{m} \in 2^{\Theta(n)}$ is the mth *Catalan number*. This bound is tight, as point sets of size n in convex position (for short, convex point sets) allow exactly $C_{n/2} = C_m$ plane perfect matchings, and (almost) all other point sets allow strictly more [4,9]. On the other hand, for general point sets the number of plane perfect matchings is at most $O(10.05^n)$ [13]. Finally, there exist point sets which allow $\Omega(3.09^n)$ many plane perfect matchings [5].

If we allow crossings, then we can draw every possible perfect matching. On n vertices there exist $(n-1)!! \in 2^{\Theta(n \log n)}$ such drawings, each having at most $\binom{n/2}{2} \in O(n^2)$ crossings. However, not much is known about the existence or number of straight-line perfect matchings with k crossings. For convex point sets there are several results on the distribution of crossings over all perfect matchings [8,11,12]. Considering general point sets, Pach and Solymosi [10] gave a complete characterization of point sets admitting perfect matchings with the maximum of $\mu := \binom{n/2}{2}$ crossings.

In this work, we analyze the number of straight-line perfect matchings with exactly or at most k crossings that a point set can admit. All considered point sets are in general position and have an even number of points. Further, *k-crossing matchings* and *($\leq k$)-crossing matchings* refer to perfect matchings with exactly k and at most k crossings, respectively.

We denote by $\mathrm{pm}_k(P)$ the number of k-crossing matchings on a point set P, by $\mathrm{pm}_k^{\max}(n)$ the maximum of $\mathrm{pm}_k(P)$, taken over all sets of n points P, and by $\mathrm{pm}_k^{\min}(n)$ the minimum of $\mathrm{pm}_k(P)$, also taken over all sets of n points P. Similarly, we denote with $\mathrm{pm}_{\leq k}(P)$ the number of $(\leq k)$-crossing matchings on a point set P and let $\mathrm{pm}_{\leq k}^{\max}(n)$ and $\mathrm{pm}_{\leq k}^{\min}(n)$ be defined analogously as before. Finally, $\mathrm{pm}_k^{\mathrm{conv}}(n)$ is the number of k-crossing matchings on a set of n points in convex position.

We start by investigating matchings with exactly k crossings in Sect. 2. There we prove that for every $k \leq \frac{1}{64}n^2 - O(n\sqrt{n})$, any set of n points, n sufficiently large, admits a perfect matching with exactly k crossings (Theorem 1) and that there exist sets of n points where every perfect matching has fewer than $\frac{5}{72}n^2$ crossings (Theorem 2). We also investigate point sets where the values of numbers of crossings in matchings are not consecutive. In Sect. 3 we then consider matchings with at most k crossings. We show that the number of perfect matchings with at most k crossings is superexponential in n if k is superlinear in n (Theorem 3), but only exponential if k is in $O(\frac{n}{\log n})$ (Corollary 1). Finally, in Sect. 4, we show that convex position is an extremal configuration in several cases. More specifically, we show that point sets in convex position minimize the number of perfect matchings with at most k crossings for $k = 0, 1, 2$ (Theorem 7), and maximize the number of perfect matchings with $\binom{n/2}{2}$ crossings and with

$\binom{n/2}{2}-1$ crossings (Theorem 6). Due to space restrictions, we are sometimes only able to sketch our proofs. The interested reader can find complete proofs in the full paper.

2 Exactly k Crossings

In this section, we show that for every set P of n points (sufficiently large) and every $k \in \{0, \ldots, \frac{1}{64}n^2 - O(n\sqrt{n})\}$, P admits a k-crossing matching, while this is not the case for $k \geq \frac{5n^2}{72}$.

For even values of $n \leq 10$, we have computed the numbers of perfect matchings with k crossings for all combinatorially different sets of n points using the order type data base; see [1] for details on order types. Table 1 lists the obtained numbers for sets of $n = 6, 8$, and 10 points, and for 0 up to the maximum number of $\binom{n/2}{2}$ crossings. We obtain the following observation.

Table 1. Number of k-crossing matchings for $n = 6, 8, 10$ points. Given is the number k of crossings, the minimum number of matchings, the number of matchings for the convex set, and the maximum number of matchings.

$n = 6$				$n = 8$				$n = 10$			
k	Min.	Conv.	Max.	k	Min.	Conv.	Max.	k	Min.	Conv.	Max.
0	5	5	12	0	14	14	56	0	42	42	311
1	2	6	10	1	20	28	60	1	120	120	442
2	0	3	3	2	4	28	33	2	135	180	350
3	0	1	1	3	0	20	28	3	39	195	308
				4	0	10	10	4	0	165	165
				5	0	4	4	5	0	117	117
				6	0	1	1	6	0	70	72
								7	0	35	35
								8	0	15	15
								9	0	5	5
								10	0	1	1

Observation 1. *For any $k \leq 3$, every set of 10 points admits a k-crossing matching.*

Proposition 1. *For a sufficiently large value of n every set P of n points admits a (rectilinear drawing of a) perfect matching with at least $\frac{1}{64}n^2$ crossings.*

Sketch of proof. We use a probabilistic approach to calculate the expected number of crossings in a random matching. Then we use the bound for the *rectilinear crossing number* to reach the lower bound of $\frac{1}{64}n^2$.   □

We next show two technical lemmas which we afterwards use to show in Theorem 1 that for sufficiently large n and any $0 \le k \le \frac{1}{64}n^2 - O(n\sqrt{n})$ there always exists a perfect matching with k crossings.

Lemma 1. *Let P be a point set with n points and a matching M with $\mathrm{cr}(M) = m$ crossings. Let $0 < k \le m$. Then P has a matching M' with $k - n + 3 \le \mathrm{cr}(M') \le k$ crossings.*

Proof. Let $M_0 := M$ and let $p_0, \ldots, p_{n-1}$ be the points of P ordered from top to bottom. We obtain a matching M_{i+1} from matching M_i as follows. If p_{2i} is matched to p_{2i+1}, then $M_{i+1} = M_i$. Otherwise, let q_{2i}, q_{2i+1} be the points of P matched to p_{2i} and p_{2i+1}, respectively, in M_i. We replace the edges (p_{2i}, q_{2i}) and (p_{2i+1}, q_{2i+1}) by the edges (p_{2i}, p_{2i+1}) and (q_{2i}, q_{2i+1}). Note that the edges $(p_0, p_1), \ldots,$ (p_{2i}, p_{2i+1}) have no crossing in M_{i+1}. Furthermore, the number of crossings of the edges (p_{2i}, q_{2i}) and (p_{2i+1}, q_{2i+1}) is at most $n - 2i - 3$ in M_i: in the worst case, each edge crosses all $((n - 2i)/2) - 1$ other edges, but the crossing between (p_{2i+1}, q_{2i+1}) and (p_{2i}, q_{2i}) is counted twice. Hence, we have $\mathrm{cr}(M_{i+1}) \ge \mathrm{cr}(M_i) - n + 2i + 3 \ge \mathrm{cr}(M_i) - n + 3$ and $\mathrm{cr}(M_{n/2}) = 0$. Then, the bound follows from choosing $M' = M_{i+1}$ such that $k \ge \mathrm{cr}(M_{i+1}) \ge \mathrm{cr}(M_i) - n + 3 \ge k - n + 3$. $\square$

Lemma 2. *For sufficiently large even n, and $0 \le k \le \frac{9}{169}\frac{n^2}{64}$ crossings, we have $\mathrm{pm}_k^{\min}(n) \ge 1$.*

Proof. Let $n_2 := 10\lfloor \frac{1}{13}n \rfloor$, and let $n_1 := \lceil \frac{3}{13}n \rceil$ if $\lceil \frac{3}{13}n \rceil$ is even, or $n_1 := \lceil \frac{3}{13}n \rceil + 1$ otherwise. Note that $n_1 + n_2 \le n$, since n is even. We linearly separate the point set P into a point set P_1 of the leftmost n_1 points and a point set P_2 of the rightmost n_2 points.

Let M_1 be the matching of P_1 with the largest number of crossings; then M_1 has at least $\frac{n_1^2}{64} \ge \frac{9}{169}\frac{n^2}{64}$ crossings by Proposition 1.

By Lemma 1, P_1 has a matching M_1' with $k - n_1 + 3 \le \mathrm{cr}(M_1') \le k$ crossings. Let $\ell = k - \mathrm{cr}(M_1') \le \lceil \frac{3}{13}n \rceil + 1 - 3 < \frac{3}{13}n - 1$. As ℓ is an integer, we get $\ell \le \frac{3}{13}n - 2$.

By Observation 1, every set of 10 points can be matched such that the matching has 0, 1, 2 or 3 crossings. We linearly separate P_2 into $\lfloor \frac{1}{13}n \rfloor$ sets of 10 points each. This way, we can find a matching of P_2 with exactly x crossings for every $0 \le x \le 3\lfloor \frac{1}{13}n \rfloor$. Since $3\lfloor \frac{1}{13}n \rfloor \ge \frac{3}{13}n - 2 \ge \ell$, we can find a matching M_2 of P_2 with exactly ℓ crossings. Thus, there is a matching $M = M_1' \cup M_2$ of P with exactly $\mathrm{cr}(M) = \mathrm{cr}(M_1') + \mathrm{cr}(M_2) = k - \ell + \ell = k$ crossings. If $n_1 + n_2 < n$ we match the remaining points (which lie between P_1 and P_2) without additional crossings. $\square$

Theorem 1. *For sufficiently large even n and $0 \le k \le \frac{1}{64}n^2 - O(n\sqrt{n})$ crossings, $\mathrm{pm}_k^{\min}(n) \ge 1$.*

Proof. We linearly separate the point set P into a point set P_1 of the leftmost $n_1 = n - 2 \cdot \lfloor \frac{52}{3} \sqrt{n} \rfloor$ points and a point set P_2 of the rightmost $n_2 = 2 \cdot \lfloor \frac{52}{3} \sqrt{n} \rfloor$ points. Here note that n_1 is even since n is even. Let M_1 be the matching of P_1 with the largest number of crossings; then M_1 has at least

$$\frac{1}{64} n_1^2 = \frac{1}{64}(n - 2 \cdot \lfloor \frac{52}{3} \sqrt{n} \rfloor)^2 \overset{(n > 1225)}{\geq} \frac{1}{64}(n - \frac{105}{3} \sqrt{n})^2$$

$$\geq \frac{1}{64} n^2 - \frac{35}{32} n \sqrt{n} + \frac{1225}{64} n \geq \frac{1}{64} n^2 - O(n\sqrt{n})$$

crossings by Proposition 1.

By Lemma 1, P_1 has a matching M_1' with $k - (n - 2 \cdot \lfloor \frac{52}{3} \sqrt{n} \rfloor) + 3 = k - n_1 + 3 \leq \mathrm{cr}(M_1') \leq k$ crossings. Let $\ell = k - \mathrm{cr}(M_1') \leq n - 2 \cdot \lfloor \frac{52}{3} \sqrt{n} \rfloor - 3$.

By Lemma 2, P_2 has a matching with exactly x crossings for every $0 \leq x \leq \frac{9}{169} \frac{n_2^2}{64}$. Note that

$$\frac{9}{169} \frac{n_2^2}{64} = \frac{9}{169 \cdot 64}(2 \cdot \lfloor \frac{52}{3} \sqrt{n} \rfloor)^2$$

$$\geq \frac{9}{169 \cdot 16} \left(\frac{52}{3} \sqrt{n} - 1 \right)^2$$

$$= n - \frac{9}{169 \cdot 16} \left(2 \cdot \frac{52}{3} \sqrt{n} - 1 \right)$$

$$\geq n - \frac{9}{169 \cdot 16} \left(2 \cdot \lfloor \frac{52}{3} \sqrt{n} \rfloor + 1 \right)$$

$$\geq n - 2 \cdot \lfloor \frac{52}{3} \sqrt{n} \rfloor - 3.$$

Hence, there is a matching M_2 of P_2 with exactly ℓ crossings. Thus, there is a matching $M = M_1' \cup M_2$ of P with exactly $\mathrm{cr}(M) = \mathrm{cr}(M_1') + \mathrm{cr}(M_2) = k - \ell + \ell = k$ crossings. $\square$

Theorem 2. *For $n \equiv (0 \bmod 6)$ and $k \geq \frac{5n^2}{72}$ crossings, $\mathrm{pm}_k^{\min}(n) = 0$.*

Sketch of proof. We show that the number of crossings that can be reached by a set of n points in a form of a windmill (see Fig. 1) is less than $\frac{5n^2}{72}$. This result is obtained by carefully analysing the maximum possible number of crossings of internal (I_i) and outgoing (O_i) matching edges of each wing. $\square$

2.1   Gaps in the Number of Crossings

In the above two theorems we considered all sets of n points. Now we are focusing on a fixed set P of n points and analyze for which values of k there exists a perfect matching with k crossings. By Theorem 1, it holds for all $k \leq \frac{1}{64} n^2 - O(n\sqrt{n})$. For some point sets P, for example points in convex position, there exist perfect matchings with all possible numbers of crossings (see Lemma 3). On the other

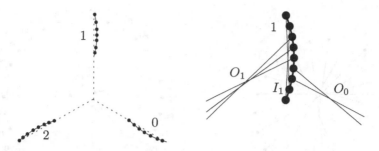

Fig. 1. Illustration for the proof of Theorem 2: A point set P for which every perfect matching has $< \frac{5n^2}{72}$ crossings (left). Interior (I_1) and outgoing (O_0, O_1) matching edges for *wing* 1 of P (right).

hand, Fig. 3 gives an example of a set of n points with exactly $n/2$ halving edges which does not have a perfect matching with $\binom{n/2}{2} - 1$ crossings but has a perfect matching with $\binom{n/2}{2}$ crossings. For any two integers r, t with $r < t$, if a set P of n points has perfect matchings with r and t crossings and does not have a perfect matching with s crossings for any $r < s < t$, then we say that the point set P has a *gap* between r and t (or, equivalently, for $k \in \{r+1, \ldots, t-1\}$). For example, the point set in Fig. 3 has a gap between 102 and 105, as we will show in the proof of Proposition 2.

Lemma 3. *Every set of n points in convex position admits a perfect k-crossing matching for every $0 \le k \le \mu = \binom{n/2}{2}$.*

Proof. Let P be a set of n points in convex position and label them in cyclic order from 1 to n. A crossing-family of size c is a set of c edges which all pairwise cross and thus has $\frac{c(c-1)}{2}$ crossings. Let c' be the largest integer such that $\frac{c'(c'-1)}{2} \le k$. To obtain k crossings in total observe that the number of crossings we still need to add to a crossing family of size c' is $x = k - \frac{c'(c'-1)}{2} < c'$. We first construct the crossing-family of size c' by connecting the points i and $(c'+i+1)$, for $i \in \{1, \ldots, x\}$, and by connecting the points i and $(c'+i+2)$, for $i \in \{x+1, \ldots, c'\}$. See Fig. 2 for an example. Next we connect points $c'+1$ and $c'+x+2$ which were not yet connected within the crossing family and contribute the remaining x crossings. Finally, we match all remaining points (if any, with indices $2c'+3$ to n) without crossings. Then M has in total k crossings as desired. □

We next construct point sets with a large gap. Let $n = 4g^2 + 6g + 2$ where $g \ge 1$ is an integer. Consider a regular n-gon with vertices $p_0, \ldots, p_{n-1}$ and circumcenter c. For the description we consider all indices modulo n. Let C_ϵ be a circle with center c and small enough such that no line spanned by p_i and $p_{\frac{n}{2}+i+1}$ for $1 \le i \le \frac{n}{2}$ intersect C_ϵ. We push the points $p_{((2g+2)\cdot i)+1}$ onto C_ϵ along the line spanned by $p_{((2g+2)\cdot i)+1}$ and $p_{((2g+2)\cdot i)+1\pm\frac{n}{2}}$ (see Fig. 3). Note that the points $p_{((2g+2)\cdot i)+1}$ form a regular $2g+1$-gon and are on C_ϵ. We call this set P_g.

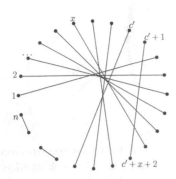

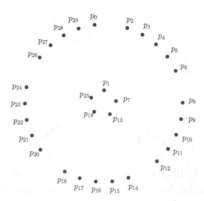

Fig. 2. Proof of Lemma 3: Matching with 42 crossings obtained from a crossing family of size $c' = 9$ plus $x = 6$ extra crossings.

Fig. 3. A set of 30 points with exactly 15 halving edges and a perfect matching with $\binom{15}{2} = 105$ crossings, but with no perfect matching with $k \in \{103, 104\}$ crossings.

Obviously, P_g spans a crossing family of size $\frac{n}{2}$. In the following proposition we prove that P_g, although we can draw a maximal crossing family on it, does not admit a matching with $\binom{n/2}{2} - i$ crossings for any $1 \le i \le g$.

Proposition 2. *For infinitely many values of n, there exists a set of n points which admits a matching with $\binom{n/2}{2}$ crossings, but does not admit a matching with $\binom{n/2}{2} - i$ crossings for any $1 \le i \le \frac{\sqrt{n}}{2} - O(1)$.*

Sketch of proof. We take the point set P_g. We show first that if our matching contains an edge of the form (p_x, p_y) with $p_y \ne p_{x+\frac{n}{2}}$ where p_x and p_y are on the boundary of the convex hull, then the matching has at most $\binom{n/2}{2} - g + 1$ crossings. So some edges are forced if there exists a matching with more than $\binom{n/2}{2} - g + 1$ crossings. We then show if our matching also contains an edge of the form (p_x, p_y) with $p_y \ne p_{x+\frac{n}{2}}$ (p_x and p_y not neccessarily on the boundary of the convex hull), then the matching has again at most $\mu - g + 1$ crossings. □

3 At Most k Crossings

We next show that if k is superlinear in n, then the number of $(\le k)$-crossing matchings is superexponential for every set of n points.

Theorem 3. *For $k \in \omega(n)$ crossings, $\mathrm{pm}_{\le k}^{\min}(n) \in 2^{\Omega(n \log(\frac{k}{n}))}$.*

Proof. Let P be any set of n points. Process the points from left to right and partition them into $\frac{n^2}{k}$ groups. Since we have n points, each group is of size $\frac{k}{n}$. Consider some group P_i. As we have mentioned in the introduction, we can draw

$2^{\Theta(\frac{k}{n}\log(\frac{k}{n}))}$ perfect matchings on P_i, and each of them has at most $\binom{k/n}{2} < \frac{k^2}{n^2}$ crossings. Thus, the number of perfect matchings where all edges are within some P_i is in

$$(2^{\Theta(\frac{k}{n}\log(\frac{k}{n}))})^{(\frac{n^2}{k})} = 2^{\Theta(n\log(\frac{k}{n}))}.$$

Further, each such matching has at most $\frac{n^2}{k} \cdot \frac{k^2}{n^2} = k$ crossings.   □

Note that for a $(\leq k)$-crossing matching, at most $4k$ points can be incident to crossing edges. Hence, the next theorem implies the upper bound on $\mathrm{pm}_{\leq k}^{\max}(n)$ stated in Corollary 1.

Theorem 4. *For a set P of n points and $0 \leq x \leq n$, let $\mathrm{pm}^x(n)$ be the number of perfect matchings whose crossing edges are incident to at most x points. Then $\mathrm{pm}^x(n) \in 2^{O(n+x\log x)}$.*

Proof. Consider some subset $P' \subset P$ of size x. We want to count the number of perfect matchings on P whose crossing edges are incident to points in P'. There are $2^{\Theta(x\log x)}$ perfect matchings on P'. We extend the matching in P' to P by adding matching edges of $P\backslash P'$ such that the matching on $P\backslash P'$ is plane. It is known that on any point set in general position with n points, the number of plane perfect matching is in $2^{\Theta(n)}$. Thus for each choice of matchings on P', there are at most $2^{O(n-x)}$ plane perfect matchings on $P\backslash P'$. Note, however, that edges of the matching on $P\backslash P'$ might intersect with edges from the matching on P'. But as we are only interested in an upper bound of the number of matchings where all edges with intersections are incident to points in P' it is sufficient that these matchings are a subset of the matchings we consider in our construction. Finally, the choices for P' are $\binom{n}{x} \leq 2^n$. Hence, we get $\mathrm{pm}^x(n) \leq 2^n \cdot 2^{\Theta(x\log x)} \cdot 2^{O(n-x)} \leq 2^{O(n+x\log x)}$.   □

Corollary 1. $\mathrm{pm}_{\leq k}^{\max}(n) \in 2^{O(n+k\log k)}$.

For $k \in \Omega(n)$, this bound is worse than the trivial upper bound from the number of all perfect matchings. For $k \in O(\frac{n}{\log n})$ we get a bound of $2^{O(n)}$, which is asymptotically tight.

4 Convex Position

In this section, we study the number $\mathrm{pm}_k^{\mathrm{conv}}(n)$ of k-crossing matchings on a set of n points in convex position. Obviously, $\mathrm{pm}_k^{\min}(n) \leq \mathrm{pm}_k^{\mathrm{conv}}(n) \leq \mathrm{pm}_k^{\max}(n)$. It is well known that convex sets minimize the number of plane perfect matchings; see for example [3,9]. Hence, we have $\mathrm{pm}_0^{\min}(n) = \mathrm{pm}_0^{\mathrm{conv}}(n)$. On the other hand, considering the maximum number $\mu = \binom{n/2}{2}$ of crossings, we can show that for $k \in \{\mu, \mu-1\}$, convex sets maximize the number of different k-crossing matchings, and that all sets of n points achieving these maximum numbers have exactly $\frac{n}{2}$ *halving edges* (edges that have $\frac{n-2}{2}$ points of the set on each side of their supporting line). The result for $k = \mu$ is a direct consequence of Theorems 1 and 2 by Pach and Solymosi [10].

Theorem 5. *For* $\mu = \binom{n/2}{2}$ *crossings,* $\mathrm{pm}_\mu^{\mathrm{conv}}(n) = \mathrm{pm}_\mu^{\max}(n) = 1$.

Proof. Let P be a set of n points in convex position and label them in cyclic order from 1 to n. By Pach and Solymosi [10, Theorem 1], a point set P admits a μ-crossing matching if and only if P has exactly $\frac{n}{2}$ halving edges. Thus, we construct our matching M to entirely consist of halving edges by connecting the points i and $(\frac{n}{2}+i)$, for $i \in \{1, \ldots, \frac{n}{2}\}$. Then any two edges of M cross and hence M has μ crossings. This is the only possible μ-crossing matching as [10, Theorem 2] shows that every set P of n points has at most one μ-crossing matching. $\square$

Theorem 6. *For* $n \geq 6$ *and* $\mu - 1 = \binom{n/2}{2} - 1$ *crossings,*

(1) $\mathrm{pm}_{\mu-1}^{\mathrm{conv}}(n) = \mathrm{pm}_{\mu-1}^{\max}(n) = \frac{n}{2}$, *obtained by any set of n points with exactly $\frac{n}{2}$ halving edges.*
(2) *Any set P of n points with more than $\frac{n}{2}$ halving edges has* $\mathrm{pm}_{\mu-1}(P) \leq 2$.

Proof. Consider a perfect matching M on n points with $\mu - 1$ crossings and let P denote its underlying point set. Then there is exactly one pair of non-crossing edges, say e and f, in M. There are two cases on how e and f can be positioned: they can be *parallel* (their endpoints are in convex position) or *stabbing* (their endpoints are not in convex position). In the stabbing case, we call the endpoint that is inside the convex hull of e and f as the *stabbing vertex*.

Case 1: The edges e and f are parallel. In this case, we remove e and f and add the diagonals of the quadrilateral defined by their endpoints to find a matching M' with μ crossings. Since before the change all other edges crossed e and f, they also cross the diagonals of this quadrilateral and thus the new edges. So we do not lose any crossing, but gain a crossing between e and f. Thus, by [10, Theorem 1], the underlying point set has exactly $n/2$ different halving edges.

Case 2: The edges e and f are stabbing. We assume w.l.o.g. that e is horizontal and incident to the stabbing vertex, that f is vertical, and that no other edges are horizontal or vertical; see Fig. 4 for an illustration.

There are two types of edges other than e and f in the matching: a family A that cross e from above (edges with negative slope) and a family B that cross e from below (edges with positive slope). Consider the lines through the stabbing vertex and the endpoints of f. As both e and f cross all the other edges of M, so does each of these lines. In particular, the stabbing vertex is incident to at least 3 halving edges, and thus the underlying point set has more than $n/2$ different halving edges. By the result of Pach and Solymosi [10], this implies that P does not admit any matching with μ crossings. Further, by the reasoning in Case 1, in all matchings of P with $\mu - 1$ crossings, the two non-crossing edges are stabbing.

Claim. The stabbing vertex is incident to exactly 3 halving edges of P and all other points are incident to exactly one halving edge of P.

Proof of Claim. Clearly, every edge of M except f and the two edges between the stabbing vertex and the endpoints of f are halving edges of P. Assume for

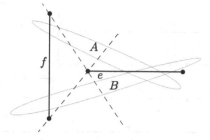

Fig. 4. The case where the non-crossing edges e and f are stabbing.

a contradiction that some other edge g spanned by P is a halving edge as well. Let $M_e = M\backslash\{e\}$, $M_f = M\backslash\{f\}$, and $M_{e,f} = M\backslash\{e,f\}$. Similarly, let P_e, P_f, and $P_{e,f}$ be the point sets obtained from P by removing the endpoints of e, f, or e and f, respectively. Note that M_e, M_f, and $M_{e,f}$ are pairwise crossing perfect matchings of P_e, P_f, and $P_{e,f}$, respectively, and hence contain exactly all halving edges for their underlying point sets. Thus, g cannot cross e (or f or both), as otherwise g would be a halving edge of P_e (or P_f or $P_{e,f}$) and hence an edge of M_e (or M_f or $M_{e,f}$) contained in M. If e and f lie in the same closed halfspace of g then each edge of $M_{e,f}$ must have at least one endpoint in that halfspace as well, which gives a count of at least $2 + \frac{n-4}{2} > \frac{n-2}{2}$ points of P in the open halfspace, a contradiction to g being a halving edge. If e and f lie in opposite closed halfspaces of g, then at most one endpoint of g is incident to e or f (as otherwise, g would be one of the halving edges incident to the stabbing vertex). Let p be the endpoint of g that is not incident to e and f. The edge in M incident to p lies in exactly one closed halfspace of g and hence crosses at most one of e and f, again a contradiction. ∎

From our claim, it follows that the stabbing vertex is the same for every matching of P with $\mu - 1$ crossings. Further, in any such matching, the stabbing vertex must be incident to a halving edge. Thus, for matchings other than M, it can only be matched with one of the endpoints of f. In particular, to obtain a matching with $\mu - 1$ crossings, there are at most three choices for an edge incident to the stabbing vertex, each choice uniquely determines the whole matching, and the edges in A and B must appear in any such matching.

As the edge incident to the stabbing vertex must cross the edges in both A and B, a different matching exists exactly if A or B are empty. If both A and B are empty, there are three possible matchings, but the underlying point set has only 4 points. We have thus shown that if $n \geq 6$ and there are more than $n/2$ different halving edges, there can be at most two matchings with $\mu - 1$ crossings.

We still have to show that if there are exactly $n/2$ different halving edges, then there are $n/2$ matchings with $\mu - 1$ crossings. For this, note that a *perfect crossing family* (a perfect matching of pairwise crossing edges) defines a natural rotational order on its edges by sorting them by slopes. For any matching with $\mu - 1$ crossings as in the Case 1, that is, with the non-crossing edges parallel,

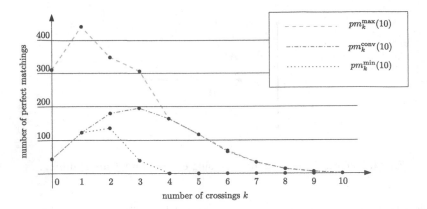

Fig. 5. Comparison of $\mathrm{pm}_k^{\min}(n)$, $\mathrm{pm}_k^{\max}(n)$, and $\mathrm{pm}_k^{\mathrm{conv}}(n)$ for $n = 10$ and $0 \leq k \leq 10$.

the construction above gives a mapping from matchings with $\mu - 1$ crossings to perfect crossing families with two neighboring edges marked. Clearly, this map is a bijection, showing that there are exactly $n/2$ matchings of the type of Case 1. On the other hand, as we have shown that in Case 2 there are more than $n/2$ different halving edges, we get no additional matchings from this case. □

It is natural to ask for which values of k and n it holds that $\mathrm{pm}_k^{\mathrm{conv}}(n) \in \{\mathrm{pm}_k^{\min}(n), \mathrm{pm}_k^{\max}(n)\}$; exhaustive computations for all point sets of small size indicate that this might be true for more than just $k \in \{0, \mu-1, \mu\}$; Fig. 5 shows the results for $n = 10$, c.f. Table 1.

As a variant of the above question, we consider for which values of k the convex set minimizes the number of matchings with at most k crossings, that is, $\mathrm{pm}_{\leq k}^{\mathrm{conv}}(n) = \mathrm{pm}_{\leq k}^{\min}(n)$? In the following, we prove the statement for any n and $k \leq 2$. We start by showing a useful technical lemma.

Lemma 4. *Let S_1 and S_2 be two point sets in general position which are separated by a line h. W.l.o.g. let h be horizontal, S_1 above h and S_2 below h. Then for any $a, b, c, d \in \mathbb{N}_0$, $a + b = |S_1| - 1$ and $c + d = |S_2| - 1$, there is a unique pair of points $p_1 \in S_1$ and $p_2 \in S_2$ such that the supporting line ℓ spanned by p_1 and p_2 splits S_1 such that there are a points to the left of ℓ and b points to the right of ℓ, and at the same time ℓ splits S_2 so that there are c points to the left of ℓ and d points to the right of ℓ.*

Proof. W.l.o.g. assume that no two points of $S_1 \cup S_2$ have the same x-coordinate and no three points of $S_1 \cup S_2$ are collinear. The existence of line ℓ follows from variants of the ham sandwich theorem. For self-containment, we give a short argumentation here. Consider the points of S_1 ordered from left to right and let p be the point with index $a + 1$ in this order. Let ℓ be the vertical line through p and note that ℓ splits S_1 in the required way. If ℓ has c points of S_2 to its left, we are done. W.l.o.g. assume that ℓ has more than c points of S_2 to its left, and start rotating ℓ clockwise around p. (The case that ℓ has less than c points of S_2 to its left can be handled analogously by rotating ℓ counter clockwise around p.)

Whenever ℓ touches a point $q \in S_1$, we set $p = q$ and continue the rotation around the new point p. Note that the splitting of S_1 remains $a : b$. If ℓ touches a point $q \in S_2$, then the number of points of S_2 to the left of ℓ is reduced by 1. If this number is c, then we stop the process and set $p_1 = p$ and $p_2 = q$. Otherwise, we continue the rotation. Before ℓ becomes horizontal, there are no points of S_2 to the left of ℓ, so the process terminates.

We now show that there is only one pair of points p_1 and p_2 that spans a line ℓ which splits the sets in the required way. Assume for the sake of contradiction that there are two such lines ℓ' and ℓ''. Then ℓ' and ℓ'' intersect at most once, w.l.o.g. above h and below h line ℓ' is to the left of ℓ''. If ℓ' has, as required, c points of S_2 to its left, then ℓ'' has at least $c + 1$ points to its left, as also the point $q \in S_2$ which spans ℓ' is to the left of ℓ''; a contradiction.   □

Theorem 7. *For a set of n points, the number of perfect matchings with at most k crossings, for $k = 0, 1, 2$, is minimized by convex point sets.*

Sketch of proof. Let S_C and P be sets of n points with leftmost points v and v', respectively, where S_C is in convex position. For each matching M with $k \leq 2$ crossings on S_C, we construct as follows a unique matching M' on P with at most k crossings. Consider the edge vw in M. The line through v and w splits S_C into an upper half U and a lower half L. There is a unique edge $v'w'$ in P whose supporting line splits P into parts of the same size as U and L. We add this edge to M'. If vw is not crossed by any edge in M, we iterate the construction on the smaller parts. Otherwise, consider the leftmost crossing, where we denote the endpoints of the crossing edge $l \in L$ and $u \in U$. By Lemma 4, there are unique points l' and u' in P such that the sizes of the induced parts agree in S_C and P. We add the edge $l'u'$ to M'. Note that this edge does not necessarily cross $v'w'$. If vw is crossed by a second edge, we repeat this step. Now, we can again iterate on the smaller parts.   □

5   Conclusion

We have given bounds for the number of perfect matchings with k crossings on sets of n points. As with many other counting problems in discrete geometry and some decision problems on point sets in the plane, the computational complexity of deciding the existence of and counting the number of perfect matchings with k crossings is in general unknown.

We have shown that if $k \leq \frac{1}{64}n^2 - O(n\sqrt{n})$, every set of n points, n sufficiently large, admits a perfect matching with k crossings. For those values of k the existential question is therefore settled but, as stated in [7], it is not even know whether counting the number of plane perfect matchings on a set of n points is hard (#P-complete).

Given a set P of n points, the problem of deciding whether it admits a perfect matching with k crossings can be reformulated in terms of the intersection graph G of line segments connecting points in P. This graph contains a vertex for each segment connecting two points in P and edges connect intersecting segments, where an intersection is either a proper crossing or a common endpoint. We consider a 2-edge-coloring of this graph where edges corresponding to segments sharing an endpoint are colored blue and edges corresponding to crossing segments are colored red. The point set P admits a perfect matching with k crossings if and only if G has an induced subgraph with $n/2$ vertices, k red edges, and no blue edge. This graph problem is in general NP-complete for segment intersection graphs by a reduction from the clique problem [6] However, the input parameters and the subset of segment intersection graphs that we are interested in, are very specific (though not so well understood) and it is therefore possible that the problem is polynomial-time solvable.

References

1. Aichholzer, O., Aurenhammer, F., Krasser, H.: Enumerating order types for small point sets with applications. Order **19**, 265–281 (2002). https://doi.org/10.1023/A:1021231927255
2. Aichholzer, O., et al.: Perfect matchings with crossings. In: Abstracts of the Computational Geometry: Young Researchers Forum, pp. 24–27 (2021). https://cse.buffalo.edu/socg21/files/YRF-Booklet.pdf#page=24
3. Aichholzer, O., Hackl, T., Huemer, C., Hurtado, F., Krasser, H., Vogtenhuber, B.: On the number of plane geometric graphs. Graphs Comb. **23**(1), 67–84 (2007). https://doi.org/10.1007/s00373-007-0704-5
4. Asinowski, A.: The number of non-crossing perfect plane matchings is minimized (almost) only by point sets in convex position (2015). arXiv preprint arXiv:1502.05332
5. Asinowski, A., Rote, G.: Point sets with many non-crossing perfect matchings. Comput. Geom. **68**, 7–33 (2018). https://doi.org/10.1016/j.comgeo.2017.05.006
6. Cabello, S., Cardinal, J., Langerman, S.: The clique problem in ray intersection graphs. Discrete Comput. Geom. **50**(3), 771–783 (2013). https://doi.org/10.1007/s00454-013-9538-5
7. Eppstein, D.: Counting polygon triangulations is hard. Discrete Comput. Geom. **64**(4), 1210–1234 (2020). https://doi.org/10.1007/s00454-020-00251-7
8. Flajolet, P., Noy, M.: Analytic combinatorics of chord diagrams. In: Krob, D., Mikhalev, A.A., Mikhalev, A.V. (eds.) Formal Power Series and Algebraic Combinatorics, pp. 191–201. Springer, Heidelberg. (2000). https://doi.org/10.1007/978-3-662-04166-6_17
9. García, A., Noy, M., Tejel, J.: Lower bounds on the number of crossing-free subgraphs of K_n. Comput. Geom. **16**(4), 211–221 (2000). https://doi.org/10.1016/S0925-7721(00)00010-9
10. Pach, J., Solymosi, J.: Halving lines and perfect cross-matchings. Adv. Discrete Comput. Geom. **223**, 245–249 (1999)
11. Pilaud, V., Rue, J.: Analytic combinatorics of chord and hyperchord diagrams with k crossings. Adv. Appl. Math. **57**, 60–100 (2014). https://doi.org/10.1016/j.aam.2014.04.001
12. Riordan, J.: The distribution of crossings of chords joining pairs of $2n$ points on a circle. Math. Comput. **29**(129), 215–222 (1975). https://doi.org/10.1090/S0025-5718-1975-0366686-9

13. Sharir, M., Welzl, E.: On the number of crossing-free matchings, cycles, and partitions. SIAM J. Comput. **36**(3), 695–720 (2006). https://doi.org/10.1137/050636036
14. You, C.: Improving Sharir and Welzl's bound on crossing-free matchings through solving a stronger recurrence (2017). arXiv preprint arXiv:1701.05909

Graph Parameters, Implicit Representations and Factorial Properties

Bogdan Alecu[1], Vladimir E. Alekseev[2], Aistis Atminas[3], Vadim Lozin[4(✉)],
and Viktor Zamaraev[5]

[1] School of Computing, University of Leeds, Leeds LS2 9JT, UK
B.Alecu@leeds.ac.uk
[2] Lobachevsky University of Nizhny Novgorod, 23 Gagarina Avenue,
603950 Nizhny Novgorod, Russia
[3] Department of Mathematical Sciences, Xi'an Jiaotong-Liverpool University,
111 Ren'ai Road, Suzhou 215123, China
Aistis.Atminas@xjtlu.edu.cn
[4] Mathematics Institute, University of Warwick, Coventry CV4 7AL, UK
V.Lozin@warwick.ac.uk
[5] Department of Computer Science, University of Liverpool, Liverpool L69 3BX, UK
Viktor.Zamaraev@liverpool.ac.uk

Abstract. A representation of an n-vertex graph G is implicit if it
assigns to each vertex of G a binary code of length $O(\log n)$ so that
the adjacency of two vertices is a function of their codes. A necessary
condition for a hereditary class $\mathcal{X}$ of graphs to admit an implicit representation is that $\mathcal{X}$ has at most factorial speed of growth. This condition,
however, is not sufficient, as was recently shown in [10]. Several sufficient
conditions for the existence of implicit representations deal with boundedness of some parameters, such as degeneracy or clique-width. In the
present paper, we analyse more graph parameters and prove a number
of new results related to implicit representation and factorial properties.

Keywords: Graph parameter · Hereditary class · Implicit
representation · Factorial property

1 Introduction

A representation of an n-vertex graph G is implicit if it assigns to each vertex of G
a binary code of length $O(\log n)$ so that the adjacency of two vertices is a function
of their codes. The idea of implicit representation was introduced in [11]. Its
importance is due to various reasons. First, it is order-optimal. Second, it allows
one to store information about graphs locally, which is crucial in distributed
computing. Finally, it is applicable to graphs in various classes of practical or
theoretical importance, such as graphs of bounded vertex degree, of bounded
clique-width, planar graphs, interval graphs, permutation graphs, line graphs,
etc.

This work was supported by the Russian Science Foundation Grant No. 21-11-00194.

C. Bazgan and H. Fernau (Eds.): IWOCA 2022, LNCS 13270, pp. 60–72, 2022.
https://doi.org/10.1007/978-3-031-06678-8_5

To better describe the area of applicability of implicit representations, let us observe that if graphs in a class $\mathcal{X}$ admit an implicit representation, then the number of n-vertex labelled graphs in $\mathcal{X}$, also known as the *speed* of $\mathcal{X}$, must be $2^{O(n \log n)}$, since the number of graphs cannot be larger than the number of binary words representing them. In the terminology of [5], hereditary classes containing $2^{\Theta(n \log n)}$ n-vertex labelled graphs have *factorial* speed of growth. The family of factorial classes, i.e. hereditary classes with a factorial speed of growth, is rich and diverse. In particular, it contains all classes mentioned earlier and a variety of other classes, such as unit disk graphs, classes of graphs of bounded arboricity, of bounded functionality [1], etc. The authors of [11], who introduced the notion of implicit representation, ask whether *every* hereditary class of speed $2^{O(n \log n)}$ admits such a representation.

Recently, Hatami and Hatami [10] answered this question negatively by proving the existence of a factorial class of bipartite graphs that does not admit an implicit representation. This negative result raises the following question: if the speed is not responsible for implicit representation, then what is responsible for it?

Looking for an answer to this question, we observe that most positive results on implicit representations deal with classes where certain graph parameters are bounded. In an attempt to produce more positive results, in Sect. 3 we analyse more graph parameters and in Sect. 4 we reveal new classes of graphs that admit an implicit representation.

In spite of the negative result in [10], factorial speed remains a necessary condition for an implicit representation in a hereditary class $\mathcal{X}$, and determining the speed of $\mathcal{X}$ is the first natural step towards deciding whether such a representation exists. A new result on this topic is presented in Sect. 5. All relevant preliminary information can be found in Sect. 2.

2   Preliminaries

All graphs in this paper are simple, i.e. undirected, without loops or multiple edges. The vertex set and the edge set of a graph G are denoted $V(G)$ and $E(G)$, respectively. The neighbourhood of a vertex $x \in V(G)$, denoted $N(x)$, is the set of vertices adjacent to x, and the degree of x, denoted $\deg(x)$, is the size of its neighbourhood. The codegree of x is the number of vertices non-adjacent to x.

As usual, K_n, P_n and C_n denote a complete graph, a chordless path and a chordless cycle on n vertices, respectively. By nG we denote the disjoint union of n copies of G.

The subgraph of G induced by a set $U \subseteq V(G)$ is denoted $G[U]$. If G does not contain an induced subgraph isomorphic to a graph H, we say that G is H-free and that H is a forbidden induced subgraph for G. A *homogeneous set* is a subset U of $V(G)$ such that $G[U]$ is either complete or edgeless.

A graph $G = (V, E)$ is *bipartite* if its vertex set can be partitioned into two independent sets. A bipartite graph given together with a bipartition of its vertex set into two independent sets A and B will be denoted $G = (A, B, E)$, where

$E \subseteq A \times B$. The *bipartite complement* of a bipartite graph $G = (A, B, E)$ is the bipartite graph $\tilde{G} := (A, B, (A \times B) - E)$. By $K_{n,m}$ we denote a complete bipartite graph with parts of size n and m. The graph $K_{1,n}$ is called a *star*. The *bi-codegree* of a vertex x in a bipartite graph $G = (A, B, E)$ is the degree of x in $\tilde{G}$.

Given two bipartite graphs $G_1 = (A_1, B_1, E_1)$ and $G_2 = (A_2, B_2, E_2)$, we say that G_1 does not contain a one-sided copy of G_2 if all induced occurrences of G_2 in G_1 have the vertices of A_2 in the same part of G_1.

2.1 Graph Classes

A class of graphs is *hereditary* if it is closed under taking induced subgraphs. It is well known that a class $\mathcal{X}$ is hereditary if and only if $\mathcal{X}$ can be described by a set of minimal forbidden induced subgraphs. In this section, we introduce a few hereditary classes that play an important role in this paper.

Motivated by the negative result in [10], which proves the existence of a factorial class of *bipartite* graphs that does not admit an implicit representation, we focus on hereditary subclasses of bipartite graphs. In particular, we study *monogenic* classes of bipartite graphs, i.e. classes defined by a single forbidden induced bipartite subgraph. The results in [2] and [13] provide a complete dichotomy for monogenic classes of bipartite graphs with respect to their speed. This dichotomy is presented in Theorem 1 below, where $S_{1,2,3}$ and $F_{t,p}$ are the graphs represented in Fig. 1.

   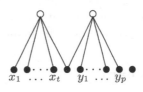

Fig. 1. The graphs $S_{1,2,3}$ (left) and $F_{t,p}$ (right)

Theorem 1. [2,13] *For a bipartite graph H, the class of H-free bipartite graphs has at most factorial speed of growth if and only if H is an induced subgraph of one of the following graphs: P_7, $S_{1,2,3}$ and $F_{t,p}$.*

2.2 Tools

Several useful tools to produce an implicit representation have been introduced in [3]. In this section, we mention two such tools, and generalise one of them.

The first result deals with the notion of locally bounded coverings, which can be defined as follows. Let G be a graph. A set of graphs $H_1, \ldots, H_k$ is called a *covering* of G if the union of $H_1, \ldots, H_k$ coincides with G, i.e. if $V(G) = \bigcup_{i=1}^{k} V(H_i)$ and $E(G) = \bigcup_{i=1}^{k} E(H_i)$.

Theorem 2. [3] *Let $\mathcal{X}$ be a class of graphs and c a constant. If every graph $G \in \mathcal{X}$ can be covered by graphs from a class $\mathcal{Y}$ admitting an implicit representation in such a way that every vertex of G is covered by at most c graphs, then $\mathcal{X}$ also admits an implicit representation.*

The second result deals with the notion of partial coverings and can be stated as follows.

Theorem 3. [3] *Let $\mathcal{X}$ be a hereditary class. Suppose there is a constant d and a hereditary class $\mathcal{Y}$ which admits an implicit representation such that every graph $G \in \mathcal{X}$ contains a non-empty subset $A \subseteq V(G)$ with the properties that $G[A] \in \mathcal{Y}$ and each vertex of A has at most d neighbours or at most d non-neighbours in $V(G) - A$. Then $\mathcal{X}$ admits an implicit representation.*

Next we provide a generalisation of Theorem 3 that will be useful later.

Theorem 4. *Let $\mathcal{X}$ be a hereditary class. Suppose there is a constant d and a hereditary class $\mathcal{Y}$ which admits an implicit representation such that every graph $G \in \mathcal{X}$ contains a non-empty subset $A \subseteq V(G)$ with the following properties:*

(1) $G[A] \in \mathcal{Y}$,
(2) $V(G) - A$ can be split into two non-empty subsets B_1 and B_2 with no edges between them, and
(3) every vertex of A has at most d neighbours or at most d non-neighbours in B_1 and at most d neighbours or at most d non-neighbours in B_2.

Then $\mathcal{X}$ admits an implicit representation.

Proof. Let G be an n-vertex graph in $\mathcal{X}$. We assign to the vertices of G pairwise distinct *indices* recursively as follows. Let $\{1, 2, \ldots, n\}$ be the *index range* of G, and let A, B_1, and B_2 be the partition of $V(G)$ satisfying the conditions (1)-(3) of the theorem. We assign to the vertices in A indices from the interval $\{|B_1| + 1, |B_1| + 2, \ldots, n - |B_2|\}$ bijectively in an arbitrary way. We define the indices of the vertices in B_1 recursively by decomposing $G[B_1]$ and using the interval $\{1, 2, \ldots, |B_1|\}$ as its index range. Similarly, we define the indices of the vertices in B_2 by decomposing $G[B_2]$ and using the interval $\{n - |B_2| + 1, n - |B_2| + 2, \ldots, n\}$ as its index range.

Now, for every vertex $v \in A$ its label consists of six components:

1. the label of v in the implicit representation of $G[A] \in \mathcal{Y}$;
2. the index of v;
3. the index range of B_1, which we call the *left index range* of v;
4. the index range of B_2, which we call the *right index range* of v;
5. a boolean flag indicating whether v has at most d neighbours or at most d non-neighbours in B_1 and the indices of those at most d vertices;
6. a boolean flag indicating whether v has at most d neighbours or at most d non-neighbours in B_2 and the indices of those at most d vertices.

For the third and the fourth component we store only the first and the last elements of the ranges, and therefore the total label size is $O(\log n)$. The labels of the vertices in B_1 and B_2 are defined recursively.

Note that two vertices can only be adjacent if either they have the same left and right index ranges or the index of one of the vertices is contained in the left or right index range of the other vertex. In the former case, the adjacency of the vertices is determined by the labels in the first components of their labels. In the latter case, the adjacency is determined using the information stored in the components 5 and 6 of the labels.   □

In the context of bipartite graphs, Theorem 4 can be restated as follows.

Theorem 5. *Let $\mathcal{X}$ be a hereditary class of bipartite graphs. Suppose there is a constant d and a hereditary class $\mathcal{Y}$ which admits an implicit representation such that every graph $G \in \mathcal{X}$ contains a non-empty subset $A \subseteq V(G)$ with the following properties:*

(1) $G[A] \in \mathcal{Y}$,
(2) $V(G) - A$ can be split into two non-empty subsets B_1 and B_2 with no edges between them, and
(3) every vertex v of A has at most d neighbours or at most d non-neighbours in the part of B_1, which is opposite to the part of A containing v, and at most d neighbours or at most d non-neighbours in the part of B_2, which is opposite to the part of A containing v.

Then $\mathcal{X}$ admits an implicit representation.

3   Graph Parameters

It is easy to see that classes of bounded vertex degree admit an implicit representation. More generally, bounded degeneracy in a class provides us with an implicit representation, where the *degeneracy* of a graph G is the minimum k such that every induced subgraph of G contains a vertex of degree at most k.

Spinrad showed in [14] that bounded clique-width also yields an implicit representation. The recently introduced parameter *twin-width* generalises clique-width in the sense that bounded clique-width implies bounded twin-width, but not vice versa. It was shown in [6] that bounded twin-width also implies the existence of an implicit representation.

The notion of graph functionality, introduced in [1], generalizes both degeneracy and twin-width in the sense that bounded degeneracy or bounded twin-width implies bounded functionality, but not vice versa. As we mentioned earlier, the graphs of bounded functionality have at most factorial speed of growth. However, whether they admit an implicit representation is wide-open. To approach this question, in Sect. 3.1 we analyse a parameter intermediate between twin-width and functionality. Then in Sect. 3.2, we introduce one more parameter.

3.1 Symmetric Difference

Let G be a graph. Given two vertices x, y, we define the *symmetric difference* of x and y in G as the number of vertices in $G - \{x, y\}$ adjacent to exactly one of x and y, and we denote it by $\mathrm{sd}(x, y)$. We define the symmetric difference $\mathrm{sd}(G)$ of G as the smallest number such that any induced subgraph of G has a pair of vertices with symmetric difference at most $\mathrm{sd}(G)$.

This parameter was introduced in [1], where it was shown that bounded clique-width implies bounded symmetric difference. Paper [1] also identifies a number of classes of bounded symmetric difference. Below we show that symmetric difference is bounded for $F_{t,p}$-free bipartite graphs (see Fig. 1 for an illustration of $F_{t,p}$). These classes have unbounded clique-width for all $t, p \geq 2$. To show that they have bounded symmetric difference, we assume without loss of generality that $t = p$.

Theorem 6. *For each $t \geq 2$, every $F_{t,t}$-free bipartite graph $G = (B, W, E)$ has symmetric difference at most $2t$.*

Proof. It is sufficient to show that G has a pair of vertices with symmetric difference at most $2t$. For two vertices x, y, we denote by $\mathrm{dd}(x, y)$ the degree difference $|\deg(x) - \deg(y)|$ and for a subset $U \subseteq V(G)$, we write $\mathrm{dd}(U) := \max\{\mathrm{dd}(x, y) : x, y \in U\}$. Assume without loss of generality that $\mathrm{dd}(W) \leq \mathrm{dd}(B)$ and let x, y be two vertices in B with $\mathrm{dd}(x, y) = \mathrm{dd}(B)$, $\deg(x) \geq \deg(y)$.

Write $X := N(x) - N(y)$. Clearly, $\mathrm{dd}(B) \leq |X|$. If $|X| \leq 2$, then $\mathrm{sd}(x, y) \leq 4 \leq 2t$ and we are done.

Now assume $|X| \geq 3$. Since $\mathrm{dd}(X) \leq \mathrm{dd}(W) \leq \mathrm{dd}(B) \leq |X|$, the set X contains two vertices p and q with $\mathrm{dd}(p, q) \leq 1$. Then $\mathrm{sd}(p, q) \leq 2t$, since otherwise both $P := N(p) - N(q)$ and $Q := N(q) - N(p)$ have size at least t, in which case x, y, p, q together with t vertices from P and t vertices from Q induce the forbidden graph $F_{t,t}$. □

Symmetric difference is also bounded in the class of $S_{1,2,3}$-free bipartite graphs, since these graphs have bounded clique-width [12]. For the remaining class from Theorem 1, i.e. the class of P_7-free bipartite graphs, the boundedness of symmetric difference is an open question.

Conjecture 1. The symmetric difference is bounded in the class of P_7-free bipartite graphs.

We also conjecture that every class of graphs of bounded symmetric difference admits an implicit representation and verify this conjecture for the classes of $F_{t,p}$-free bipartite graphs in Sect. 4.

Conjecture 2. Every class of graphs of bounded symmetric difference admits an implicit representation.

3.2   Double-Star Partition Number

Let us say that a class $\mathcal{X}$ of bipartite graphs is *double-star-free* if there is a constant p such that no graph G in $\mathcal{X}$ contains an unbalanced copy of $2K_{1,p}$, i.e. an induced copy of $2K_{1,p}$ in which the centres of both stars belong to the same part of the bipartition of G. In particular, every class of *double-star-free* graphs is $F_{t,p}$-free for some t, p.

We will say that a class $\mathcal{X}$ of graphs is of bounded *double-star partition number* if there are constants k and p such that the vertices of every graph in $\mathcal{X}$ can be partitioned into at most k homogeneous subsets so that the edges between any pair of subsets form a bipartite graph that does not contain an unbalanced copy of $2K_{1,p}$.

The classes of bounded double-star partition number have been defined in the previous paragraph through two constants, k and p. By taking the maximum of the two, we can talk about a single constant, which can be viewed as a graph parameter defining the family of classes of bounded double-star partition number. This parameter has never been formally defined in the literature. Our motivation is based on the results in [4], where the author identifies ten minimal hereditary classes of graphs, which, in our terminology, have unbounded double-star partition number. One of them is the class $\mathcal{S}$ of star forests in which the centers of all stars belong to the same part of the bipartition. One more class is the class of bipartite complements of graphs in $\mathcal{S}$. Moreover, $\mathcal{S}$ and the class of bipartite complements of graphs in $\mathcal{S}$ are the only two minimal hereditary classes of *bipartite* graphs of unbounded double-star partition number.

Theorem 7. [4] *A hereditary class $\mathcal{X}$ of bipartite graphs is of bounded double-star partition number if and only if $\mathcal{X}$ excludes a graph from $\mathcal{S}$ and the bipartite complement of a graph from $\mathcal{S}$.*

Our interest to this parameter is due to the fact that any class of bounded double-star partition number admits an implicit representation, as we show in Sect. 4.

4   Implicit Representations

In this section, we identify a number of new hereditary classes of graphs that admit an implicit representation.

4.1 $F_{t,p}$-Free Bipartite Graphs

In this section we show that $F_{t,p}$-free bipartite graphs admit an implicit representation for any t and p. Together with Theorem 6 this verifies Conjecture 2 for these classes.

Without loss of generality we assume that $t = p$ and split the analysis into several intermediate steps. The first step deals with the case of double-star-free bipartite graphs.

Theorem 8. *Let $G = (A, B, E)$ be a bipartite graph that does not contain an unbalanced induced copy of $2K_{1,t}$. Then G has a vertex of degree at most $t - 1$ or bi-codegree at most $(t - 1)(t^2 - 4t + 5)$.*

Proof. Let $x \in A$ be a vertex of maximum degree. Write Y for the set of neighbours of x, and Z for its set of non-neighbours in B (so $B = Y \cup Z$). We may assume $|Y| \geq t$ and $|Z| \geq (t - 1)(t^2 - 4t + 5) + 1$, since otherwise we are done.

Note that any vertex $w \in A$ is adjacent to fewer than t vertices in Z. Indeed, if $w \in A$ has t neighbours in Z, then it must be adjacent to all but at most $t - 1$ vertices in Y (since otherwise a $2K_{1,t}$ appears), so its degree is greater than that of x, a contradiction.

We now show that Z has a vertex of degree at most $t - 1$. Pick members $z_1, \ldots, z_{t-1} \in Z$ in a non-increasing order of their degrees, and write W_i for the neighbourhood of z_i. Since G is $2K_{1,t}$-free and $\deg(z_{i+1}) \leq \deg(z_i)$, for all $1 \leq i \leq t - 2$, $|W_{i+1} - W_i| \leq t - 1$. It is not difficult to see that in fact, $|W_{i+1} - \bigcap_{s=1}^{i} W_s| \leq (t - 1)i$, and in particular, $|W_{t-1} - \bigcap_{i=1}^{t-2} W_i| \leq (t - 1)(t - 2)$.

With this, we can compute an upper bound on the number of vertices in Z which have neighbours in W_{t-1}: by the degree condition given above, each vertex in $W_{t-1} \cap \bigcap_{i=1}^{t-2} W_i$ is adjacent to no vertices in Z other than $z_1, \ldots, z_{t-1}$. Each of the at most $(t-1)(t-2)$ vertices in $W_{t-1} - \bigcap_{i=1}^{t-2} W_i$ has at most $t - 2$ neighbours in Z other than z_{t-1}. This accounts for a total of at most $(t - 1) + (t - 1)(t - 2)^2 = (t - 1)(t^2 - 4t + 5)$ vertices which have neighbours in W_{t-1}, including z_{t-1} itself. By assumption on the size of Z, there must be a vertex $z \in Z$ which has no common neighbours with z_{t-1}. Since $2K_{1,t}$ is forbidden, one of z and z_{t-1} has degree at most $t - 1$, as claimed.   □

An immediate implication of this result, combined with Theorem 5, is that double-star-free bipartite graphs admit an implicit representation.

Corollary 1. *The class of $2K_{1,t}$-free bipartite graphs admits an implicit representation for any fixed t.*

Together with Theorem 2, this yields yet another interesting conclusion.

Corollary 2. *The classes of graphs of bounded double-star partition number admit an implicit representation.*

In the context of bipartite graphs, this corollary together with Theorem 7 implies the following generalization of Corollary 1.

Corollary 3. *Every class of bipartite graphs excluding a star forest and the bipartite complement of a star forest admits an implicit representation.*

Our next step towards implicit representations of $F_{t,t}$-free bipartite graphs deals with the case of $F_{t,t}^1$-free bipartite graphs, where $F_{t,t}^1$ is the graph obtained from $F_{t,t}$ by deleting the isolated vertex.

Theorem 9. *The class of $F_{t,t}^1$-free bipartite graphs admits an implicit representation.*

Proof. It suffices to prove the result for connected graphs (this follows for instance from Theorem 2). Let G be a connected $F_{t,t}^1$-free bipartite graph and let v be a vertex of maximum degree in G. We denote by V_i the set of vertices at distance i from v.

First, we show that the subgraph $G[V_1 \cup V_2]$ admits an implicit representation. To this end, we denote by u a vertex of maximum degree in V_1, by U the neighbourhood of u in V_2, $W := V_2 - U$, and $V_1' := V_1 - \{u\}$.

Let x be a vertex in V_1' and assume it has t neighbours in W. Then x has at least t non-neighbours in U (due to maximality of u), in which case the t neighbours of x in W, the t non-neighbours of x in U together with x, u and v induce an $F_{t,t}^1$. This contradiction shows that every vertex of V_1' has at most $t - 1$ neighbours in W, and hence the graph $G[V_1' \cup W]$ admits an implicit representation by Theorem 5.

To prove that $G[V_1' \cup U]$ admits an implicit representation, we observe that this graph is $2K_{1,t}$-free. Indeed, if the centers of the two stars belong to V_1', then they induce an $F_{t,t}^1$ together with vertex v, and if the centers of the two stars belong to U, then they induce an $F_{t,t}^1$ together with vertex u. Therefore, the graph $G[V_1 \cup V_2]$ can be covered by at most three graphs (one of them being the star centered at u), each of which admits an implicit representation, and hence by Theorem 2 this graph admits an implicit representation.

To complete the proof, we observe that every vertex of V_2 has at most $t - 1$ neighbours in V_3. Indeed, if a vertex $x \in V_2$ has t neighbours in V_3, then x has at least t non-neighbours in V_1 (due to maximality of v), in which case the t neighbours of x in V_3, the t non-neighbours of x in V_1 together with x, v, and any neighbour of x in V_1 (which must exist by definition) induce an $F_{t,t}^1$.

Now we apply Theorem 3 with $A = \{v\} \cup V_1 \cup V_2$ to conclude that G admits an implicit representation, because every vertex of A has at most $t-1$ neighbours outside of A. $\qquad\square$

The last step towards implicit representations of $F_{t,t}$-free bipartite graphs is similar to Theorem 9 with some modifications.

Theorem 10. *The class of $F_{t,t}$-free bipartite graphs admits an implicit representation.*

Proof. By analogy with Theorem 9 we consider a *connected* $F_{t,t}$-free bipartite graph G, denote by v a vertex of maximum degree in G and by V_i the set of vertices at distance i from v. Also, we denote by u a vertex of maximum degree in V_1, by U the neighbourhood of u in V_2, $W := V_2 - U$, and $V_1' := V_1 - \{u\}$.

Let x be a vertex in V_1' and assume it has t neighbours and one non-neighbour y in W. Then x has at least t non-neighbours in U (due to maximality of u), in which case the t neighbours of x in W, the t non-neighbours of x in U together with x, y, u and v induce an $F_{t,t}$. This contradiction shows that every vertex of

V_1' has either at most $t-1$ neighbours or at most 0 non-neighbours in W, and hence the graph $G[V_1' \cup W]$ admits an implicit representation by Theorem 5.

To prove that $G[V_1' \cup U]$ admits an implicit representation, we show that this graphs is $\widetilde{F}_{t,t}^1$-free. Indeed, if the centers of the two stars of $\widetilde{F}_{t,t}^1$ belong to V_1', then $\widetilde{F}_{t,t}^1$ together with vertex v induce an $F_{t,t}$, and if the centers of the two stars of $\widetilde{F}_{t,t}^1$ belong to U, then $\widetilde{F}_{t,t}^1$ together with vertex u induce an $F_{t,t}$. Therefore, the graph $G[V_1 \cup V_2]$ can be covered by at most three graphs, each of which admits an implicit representation, and hence by Theorem 2 this graph admits an implicit representation.

To complete the proof, we observe that every vertex of V_2 has either at most $t-1$ neighbours or 0 non-neighbours in V_3. Indeed, if a vertex $x \in V_2$ has t neighbours and one non-neighbour y in V_3, then x has at least t non-neighbours in V_1 (due to maximality of v), in which case the t neighbours of x in V_3, the t non-neighbours of x in V_1 together with x, y, v, and any neighbour of x in V_1 induce an $F_{t,t}$.

Finally, we observe that if a vertex $x \in V_2$ has t neighbours in V_3, then V_5 (and hence V_i for any $i \geq 5$) is empty, because otherwise an induced $F_{t,t}$ arises similarly as in the previous paragraph, where vertex y can be taken from V_5. Now we apply Theorem 5 with $A = \{v\} \cup V_1 \cup V_2$ to conclude that G admits an implicit representation. Indeed, if each vertex of V_2 has at most $t-1$ neighbours in V_3, then each vertex of A has at most $t-1$ neighbours outside of A, and if a vertex of V_2 has at least t neighbours in V_3, then $V_i = \emptyset$ for $i \geq 5$ and hence every vertex of A has at most $t-1$ neighbours or at most 0 non-neighbours in the *opposite* part outside of A. $\qquad\square$

4.2 One-Sided Forbidden Induced Bipartite Subgraphs

In the context of bipartite graphs, some hereditary classes are defined by forbidding one-sided copies of bipartite graphs. Consider, for instance, the class of star forests, whose vertices are partitioned into an independent set of black vertices and an independent set of white vertices. If the centers of all stars have the same colour, say black, then this class is defined by forbidding a P_3 with a white center. Very little is known about implicit representations for classes defined by one-sided forbidden induced bipartite subgraphs. It is known, for instance, that bipartite graphs without a one-sided P_5 admit an implicit representation. This is not difficult to show and also follows from the fact P_6-free bipartite graphs have bounded clique-width and hence admit an implicit representation (note that P_6 is symmetric with respect to swapping the bipartition). Below we strengthen the result for one-sided forbidden P_5 to one-sided forbidden $F_{t,1}$, where again $F_{t,1}^1$ is the graph obtained from $F_{t,1}$ by deleting the isolated vertex.

Lemma 1. *The class of bipartite graphs containing no one-sided copy of $F_{t,1}^1$ admits an implicit representation.*

Proof. Let $G = (U, V, E)$ be a bipartite graph containing no copy of $F_{t,1}^1$ with the vertex of largest degree in U. Let u be a vertex of maximum degree in U.

We split the vertices of V into the set V_1 of neighbours and the set V_0 of non-neighbours of u. Consider a vertex $x \in U$ such that x has a neighbour in V_1 and a neighbour in V_0 and denote by V_{10} the set of non-neighbours of x in V_1 and by V_{01} the set of neighbours of x in V_0. We note that $|V_{01}| \leq |V_{10}|$, since $\deg(x) \leq \deg(u)$. Besides, $|V_{10}| < t$, since otherwise t vertices in V_{10}, a vertex in V_{01} and a common neighbour of u and x (these vertices exist by assumption) together with u and x induce a forbidden copy of $F_{t,1}^1$. Therefore, x has at most $k - 1$ non-neighbours in V_1 and at most $k - 1$ neighbours in V_0.

Now we define three subsets A, B_1, B_2 as follows:

A consists of vertex u and the set of vertices of U that have neighbours both in V_1 and in V_0,

B_1 consists of V_1 and the vertices of U that have neighbours only in V_1,

B_2 consists of V_0 and the vertices of U that have neighbours only in V_0.

With this notation, the result follows from Theorem 5.   □

Theorem 11. *The class of bipartite graphs containing no one-sided copy of $F_{t,1}$ admits an implicit representation.*

Proof. Let $G = (U, V, E)$ be a connected bipartite graph containing no one-sided copy of $F_{t,1}$ with the vertex of largest degree in U. Let v be a vertex in V and let V_i the set of vertices at distance i from v. Then the graph $G_1 := G[V_1 \cup V_2]$ does not contain a one-sided copy of $\widetilde{F}_{t,1}^1$ with the vertex of largest degree in V_1, since otherwise together with v this copy would induce a one-sided copy of $F_{t,1}$ with the vertex of largest degree in U. Therefore, by Lemma 1 the graph G_1 admits an implicit representation.

For any $i > 1$, the $G_i := G[V_i \cup V_{i+1}]$ does not contain a one-sided copy of $F_{t,1}^1$ with the vertex of largest degree in V_i (for odd i) or with the vertex of largest degree in V_{i+1} (for even i), since otherwise together with v this copy would induce a one-sided copy of $F_{t,1}$ with the vertex of largest degree in U. Therefore, by Lemma 1 the graph G_i admits an implicit representation for all $i > 1$. Together with Theorem 2 this implies an implicit representation for G. □

For larger indices of one-sided forbidden $F_{t,p}$ the question remains open. Moreover, it remains open even for one-sided forbidden $2P_3$. It is interesting to note that if we forbid $2P_3$ with black centers and all black vertices have incomparable neighbourhoods, then the graph has bounded clique-width [7] and hence admits an implicit representation. However, in general the clique-width of $2P_3$-free bipartite graphs is unbounded and the question of implicit representation for one-sided forbidden $2P_3$ remains open.

Problem 1. Determine whether the class of bipartite graphs containing no one-sided induced $2P_3$ admits an implicit representation.

5 Factorial Properties

We repeat that bounded functionality implies at most factorial speed of growth. Whether the reverse implication is also valid was left as an open question in [1]. It turns out that the answer to this question is negative. This is witnessed by the class $\mathcal{Q}$ of induced subgraphs of hypercubes. Indeed, in [1] it was shown that $\mathcal{Q}$ has unbounded functionality. On the other hand, it was proved in [8] that there exists an implicit representation for $\mathcal{Q}$ and, in particular, the class is factorial; in fact, a result from recent work [9] implies that, more generally, the hereditary closure of Cartesian products of any finite set of graphs admits an implicit representation. These results however are non-constructive and they provide neither explicit labeling schemes, nor specific factorial bounds on the number of graphs. Below we give a concrete bound on the speed of $\mathcal{Q}$.

Theorem 12. *There are at most n^{2n} n-vertex graphs in $\mathcal{Q}$.*

Proof. Let Q_n denote the n-dimensional hypercube, i.e. the graph with vertex set $\{0,1\}^n$, in which two vertices are adjacent if and only if they differ in exactly one coordinate. To obtain the desired bound, we will produce, for each labelled n-vertex graph in $\mathcal{Q}$, a sequence of $2n$ numbers between 1 and n which allows us to retrieve the graph uniquely.

As a preliminary, let $G \in \mathcal{Q}$ be a connected graph on n vertices. By definition of $\mathcal{Q}$, G embeds into Q_m for some m. We claim that, in fact, G embeds into Q_{n-1}. If $m < n$, this is clear. Otherwise, using an embedding into Q_m, each vertex of G corresponds to an m-digit binary sequence. For two adjacent vertices, the sequences differ in exactly one position. From this, it follows inductively that the n vertices of G all agree in at least $m - (n-1)$ positions. The coordinates on which they agree can simply be removed; this produces an embedding of G into Q_{n-1}. Additionally, by symmetry, if G has a distinguished vertex r, we remark that we may find an embedding sending r to $(0, 0, \ldots, 0)$.

We are now ready to describe our encoding. Let $G \in \mathcal{Q}$ be any labelled graph with vertex set $\{x_1, \ldots, x_n\}$. We start by choosing, for each connected component C of G:

– a spanning tree T_C of C;
– a root r_C of T_C;
– an embedding φ_C of T_C into Q_{n-1} sending r_C to $(0, 0, \ldots, 0)$.

Write C^i for the component of x_i. We define two functions $p, d : V(G) \to \{1, \ldots, n\}$ as follows:

$$p(x_i) = \begin{cases} i, & \text{if } x_i = r_{C^i}; \\ j, & \text{if } x_i \neq r_{C^i}, \text{and } x_j \text{ is the parent of } x_i \text{ in } T_{C^i}. \end{cases}$$

$$d(x_i) = \begin{cases} 1, & \text{if } x_i = r_{C^i}; \\ j, & \text{if } x_i \neq r_{C^i}, \text{and } \varphi(x_i) \text{ and } \varphi(p(x_i)) \text{ differ in coordinate } j. \end{cases}$$

One easily checks that the above maps are well-defined; in particular, when x_i is not a root, the embeddings of x_i and of its parent do, indeed, differ in exactly one coordinate. The reader should also know that the value of d on the roots is, in practice, irrelevant – setting it to 1 is an arbitrary choice.

We now claim that G can be restored from the sequence

$$p(x_1), d(x_1), \ldots, p(x_n), d(x_n).$$

To do so, we first note that this sequence allows us to easily determine the partition of G into connected components. Moreover, for each connected component, we may then determine its embedding φ_C into Q_{n-1}: $\varphi_C(r_C)$ is by assumption $(0, 0, \ldots, 0)$; we may then identify its children using p, then compute their embeddings using d; we may then proceed inductively. This information allows us to determine the adjacency in G as claimed, and the encoding uses $2n$ integers between 1 and n as required. □

Problem 2. Find specific implicit representation for the class $\mathcal{Q}$ of induced subgraphs of hypercubes.

References

1. Alecu, B., Atminas, A., Lozin, V.: Graph functionality. J. Combin. Theory Ser. B **147**, 139–158 (2021). https://doi.org/10.1007/978-3-030-30786-8_11
2. Allen, P.: Forbidden induced bipartite graphs. J. Graph Theory **60**, 219–241 (2009)
3. Atminas, A., Collins, A., Lozin, V., Zamaraev, V.: Implicit representations and factorial properties of graphs. Discrete Math. **338**, 164–179 (2015)
4. Atminas, A.: Classes of graphs without star forests and related graphs. arXiv preprint arXiv:1711.01483
5. Balogh, J., Bollobás, B., Weinreich, D.: The speed of hereditary properties of graphs. J. Combin. Theory Ser. B **79**, 131–156 (2000)
6. Bonnet, É., Geniet, C., Kim, E.J., Thomassé, S., Watrigant, R.: Twin-width II: small classes. SODA 2021: 1977–1996
7. Boros, E., Gurvich, V., Milanic, M.: Characterizing and decomposing classes of threshold, split, and bipartite graphs via 1-Sperner hypergraphs. J. Graph Theory **94**, 364–397 (2020)
8. Harms, N.: Universal communication, universal graphs, and graph labeling, ITCS (2020)
9. Harms, N., Wild, S., Zamaraev, V.: Randomized communication and implicit graph representations. STOC (2022)
10. Hatami, H., Hatami, P.: The implicit graph conjecture is false. arXiv preprint arXiv:2111.13198
11. Kannan, S., Naor, M., Rudich, S.: Implicit representation of graphs. SIAM J. Discrete Math. **5**, 596–603 (1992)
12. Lozin, V.: Bipartite graphs without a skew star. Discrete Math. **257**, 83–100 (2002)
13. Lozin, V., Zamaraev, V.: The structure and the number of P_7-free bipartite graphs. Eur. J. Comb. **65**, 143–153 (2017)
14. Spinrad, J.P.: Efficient graph representations. Fields Institute Monographs, 19. American Mathematical Society, Providence, RI, pp. xiii+342 (2003)

Approximating Subset Sum Ratio
via Subset Sum Computations

Giannis Alonistiotis[1], Antonis Antonopoulos[1], Nikolaos Melissinos[2],
Aris Pagourtzis[1], Stavros Petsalakis[1], and Manolis Vasilakis[1]($\boxtimes$)

[1] School of Electrical and Computer Engineering, National Technical University
of Athens, Polytechnioupoli, 15780 Zografou, Athens, Greece
{ialonistiotis,aanton,spetsalakis,mvasilakis}@corelab.ntua.gr,
pagour@cs.ntua.gr
[2] Université Paris-Dauphine, PSL University, CNRS, LAMSADE,
75016 Paris, France
nikolaos.melissinos@dauphine.eu

Abstract. We present a new FPTAS for the SUBSET SUM RATIO problem, which, given a set of integers, asks for two disjoint subsets such that the ratio of their sums is as close to 1 as possible. Our scheme makes use of exact and approximate algorithms for the closely related SUBSET SUM problem, hence any progress over those—such as the recent improvement due to Bringmann and Nakos [SODA 2021]—carries over to our FPTAS. Depending on the relationship between the size of the input set n and the error margin ε, we improve upon the best currently known algorithm of Melissinos and Pagourtzis [COCOON 2018] of complexity $\mathcal{O}(n^4/\varepsilon)$. In particular, the exponent of n in our proposed scheme may decrease down to 2, depending on the SUBSET SUM algorithm used. Furthermore, while the aforementioned state of the art complexity, expressed in the form $\mathcal{O}((n+1/\varepsilon)^c)$, has constant $c = 5$, our results establish that $c < 5$.

Keywords: Approximation scheme · Combinatorial optimization ·
Knapsack problems · SUBSET SUM · SUBSET SUM RATIO

1 Introduction

One of Karp's 21 NP-complete problems [18], SUBSET SUM has seen astounding progress over the last few years. Koiliaris and Xu [21], Bringmann [7] and Jin and Wu [17] have presented pseudopolynomial algorithms resulting in substantial improvements over the long-standing standard approach of Bellman [6], and the improvement by Pisinger [29]. Moreover, the latter two algorithms [7,17] match the SETH-based lower bounds proved in [1]. Additionally, recently there has been progress in the approximation scheme of SUBSET SUM, the first such improvement in over 20 years, with a new algorithm introduced by Bringmann

Aris Pagourtzis and Stavros Petsalakis were supported in part by the PEVE 2020 basic research support program of the National Technical University of Athens.

C. Bazgan and H. Fernau (Eds.): IWOCA 2022, LNCS 13270, pp. 73–85, 2022.
https://doi.org/10.1007/978-3-031-06678-8_6

and Nakos [9], as well as corresponding lower bounds obtained through the lens of fine-grained complexity.

The EQUAL SUBSET SUM problem, which, given an input set, asks for two disjoint subsets of equal sum, is closely related to SUBSET SUM. It finds applications in multiple different fields, ranging from computational biology [10,13] and computational social choice [22], to cryptography [30], to name a few. In addition, it is related to important theoretical concepts such as the complexity of search problems in the class TFNP [28].

The centerpiece of this paper is the SUBSET SUM RATIO problem, the optimization version of EQUAL SUBSET SUM, which asks, given an input set $S \subseteq \mathbb{N}$, for two disjoint subsets $S_1, S_2 \subseteq S$, such that the following ratio is minimized

$$\frac{\max\left\{\sum_{s_i \in S_1} s_i, \sum_{s_j \in S_2} s_j\right\}}{\min\left\{\sum_{s_i \in S_1} s_i, \sum_{s_j \in S_2} s_j\right\}}$$

We present a new approximation scheme for this problem, highlighting its close relationship with the classical SUBSET SUM problem. Our proposed algorithm is the first to associate these closely related problems and, depending on the relationship of the cardinality of the input set n and the value of the error margin ε, achieves better asymptotic bounds than the current state of the art [23]. Moreover, while the complexity of the current state of the art approximation scheme expressed in the form $\mathcal{O}((n + 1/\varepsilon)^c)$ has an exponent $c = 5$, we present an FPTAS with constant $c < 5$.

1.1 Related Work

EQUAL SUBSET SUM as well as its optimization version called SUBSET SUM RATIO [5] are closely related to problems appearing in many scientific areas. Some examples include the PARTIAL DIGEST problem, which comes from computational biology [10,13], the allocation of individual goods [22], tournament construction [20], and a variation of SUBSET SUM, called Multiple Integrated Sets SSP, which finds applications in the field of cryptography [30]. Furthermore, it is related to important concepts in theoretical computer science; for example, a restricted version of EQUAL SUBSET SUM lies in a subclass of the complexity class TFNP, namely in PPP [28], a class consisting of search problems that always have a solution due to some pigeonhole argument, and no polynomial time algorithm is known for this restricted version.

EQUAL SUBSET SUM has been proven NP-hard by Woeginger and Yu [31] (see also the full version of [25] for an alternative proof) and several variations have been proven NP-hard by Cieliebak et al. in [11,12]. A 1.324-approximation algorithm has been proposed for SUBSET SUM RATIO in [31] and several FPTASs appeared in [5,23,27], the fastest so far being the one in [23] of complexity $\mathcal{O}(n^4/\varepsilon)$, the complexity of which seems to also apply to various meaningful special cases, as shown in [24].

As far as exact algorithms are concerned, recent progress has shown that EQUAL SUBSET SUM can be solved probabilistically in[1] $\mathcal{O}^*(1.7088^n)$ time [25], faster than a standard "meet-in-the-middle" approach yielding an $\mathcal{O}^*(3^{n/2}) \leq \mathcal{O}^*(1.7321^n)$ time algorithm.

These problems are tightly connected to SUBSET SUM, which has seen impressive advances recently, due to Koiliaris and Xu [21] who gave a deterministic $\tilde{\mathcal{O}}(\sqrt{n}t)$ algorithm, where n is the number of input elements and t is the target, and by Bringmann [7] who gave a $\tilde{\mathcal{O}}(n+t)$ randomized algorithm, which is essentially optimal under SETH [1]. See also [2] for an extension of these algorithms to a more general setting. Jin and Wu subsequently proposed a simpler randomized algorithm [17] achieving the same bounds as [7], which however seems to only solve the decision version of the problem. Recently, Bringmann and Nakos [8] have presented an $\mathcal{O}(|\mathcal{S}_t(Z)|^{4/3}\text{poly}(\log t))$ algorithm, where $\mathcal{S}_t(Z)$ is the set of all subset sums of the input set Z that are smaller than t, based on top-k convolution.

Regarding approximation schemes for SUBSET SUM, recently Bringmann and Nakos [9] presented the first improvement in over 20 years, since the scheme of [19] had remained the state of the art. Making use of modern techniques, they additionally provide lower bounds based on the popular *min-plus convolution* conjecture [14]. Furthermore, they present a new FPTAS for the closely related PARTITION problem, by observing that their techniques can be used to approximate a slightly more *relaxed* version of the SUBSET SUM problem, firstly studied in [25].

1.2 Our Contribution

We present a novel approximation scheme for the SUBSET SUM RATIO problem. Our algorithm makes use of exact and approximation algorithms for SUBSET SUM, thus, any improvement over those carries over to our proposed scheme. Additionally, depending on the relationship between n and ε, our algorithm improves upon the best existing approximation scheme of [23].

We start by presenting some necessary background in Sect. 2. Afterwards, in Sect. 3 we introduce an FPTAS for a restricted version of the problem. In the following Sect. 4, we explain how to make use of the algorithm presented in the previous section, in order to obtain an approximation scheme for the SUBSET SUM RATIO problem. The complexity of the final scheme is thoroughly analyzed in Sect. 5, followed by some possible directions for future research in Sect. 6.

2 Preliminaries

Let, for $x \in \mathbb{N}$, $[x] = \{0, \ldots, x\}$ denote the set of integers in the interval $[0, x]$. Given a set $S \subseteq \mathbb{N}$, denote its *largest element* by $\max(S)$ and the sum of its

---

[1] Standard $\mathcal{O}^*$ notation is used to hide polynomial and $\tilde{\mathcal{O}}$ to hide polylogarithmic factors.

elements by $\Sigma(S) = \sum_{s \in S} s$. If we are additionally given a value $\varepsilon \in (0,1)$, define the following *partition* of its elements:

- The set of its *large* elements as $L(S,\varepsilon) = \{s \in S \mid s \geq \varepsilon \cdot \max(S)\}$. Note that $\max(S) \in L(S,\varepsilon)$, for any $\varepsilon \in (0,1)$.
- The set of its *small* elements as $M(S,\varepsilon) = \{s \in S \mid s < \varepsilon \cdot \max(S)\}$.

In the following, since the values of the associated parameters will be clear from the context, they will be omitted and we will refer to these sets simply as L and M.

The following definitions will be useful for the rest of this paper.

Definition 1 (Closest Set and pair). *Given a set $S_i \subseteq A$, we define as its closest set a set $S_{i,\mathrm{opt}}$ such that $S_{i,\mathrm{opt}} \subseteq A \backslash S_i$ and $\Sigma(S_i) \geq \Sigma(S_{i,\mathrm{opt}}) \geq \Sigma(S')$ for all $S' \subseteq A \backslash S_i$. The pair $(S_i, S_{i,\mathrm{opt}})$ is called* closest pair.

Definition 2 (ε-close set and pair). *Given a set $S_i \subseteq A$, we define as its ε-close set a set $S_{i,\varepsilon}$ such that $S_{i,\varepsilon} \subseteq A \backslash S_i$ and $\Sigma(S_i) \geq \Sigma(S_{i,\varepsilon}) \geq (1-\varepsilon) \cdot \Sigma(S_{i,\mathrm{opt}})$. The pair $(S_i, S_{i,\varepsilon})$ is called ε-close pair.*

Remark 1. Note that $S_{i,\mathrm{opt}}$ is also an ε-close set of S_i for any $\varepsilon \in (0,1)$.

Definition 3 (Subset Sum). *Given a set X and target t, compute a subset $Y \subseteq X$, such that $\Sigma(Y) = \max\{\Sigma(Z) \mid Z \subseteq X, \Sigma(Z) \leq t\}$.*

Definition 4 (Approximate Subset Sum). *Given a set X, target t and error margin ε, compute a subset $Y \subseteq X$ such that $(1 - \varepsilon) \cdot OPT \leq \Sigma(Y) \leq OPT$, where $OPT = \max\{\Sigma(Z) \mid Z \subseteq X, \Sigma(Z) \leq t\}$.*

By solving SUBSET SUM or its approximate version, one can compute an ε-close set for a given subset $S_i \subseteq A$ as follows.

1. **Closest set $(S_{i,\mathrm{opt}})$ computation**
   Compute the subset sums of set $A \backslash S_i$ with target $\Sigma(S_i)$ and keep the largest non exceeding. This can be achieved by a standard meet in the middle [16] algorithm.
2. **ε-close set $(S_{i,\varepsilon})$ computation**
   Run an approximate SUBSET SUM algorithm [9,19] with error margin ε on set $A \backslash S_i$ with target $\Sigma(S_i)$.

3   Approximation Scheme for a Restricted Version

In this section, we present an FPTAS for the constrained version of the SUBSET SUM RATIO problem where we are only interested in approximating solutions that involve large subset sums. By this, we mean that for at least one of the subsets of the optimal solution, the sum of its large elements must be no less than $\max(A) = a_n$ (assuming that $A = \{a_1, \ldots, a_n\}$ is the *sorted* input set); let r_{opt} denote the subset sum ratio of such an optimal solution. Our FPTAS will return a solution of ratio r, such that $1 \leq r \leq (1 + \varepsilon) \cdot r_{\mathrm{opt}}$, for a given error margin $\varepsilon \in (0,1)$; however, we allow that the sets of the returned solution do not necessarily satisfy the aforementioned constraint (i.e. the sum of their large elements may be less than a_n).

3.1   Outline of the Algorithm

We now present a rough outline of the algorithm, along with its respective pseudocode:

- At first, we search for approximate solutions involving exclusively large elements from $L(A, \varepsilon)$.
- To this end, we produce the subset sums formed by these large elements. If their number exceeds n/ε^2, then we can easily find an approximate solution.
- Otherwise, for each of the produced subsets, we find its corresponding ε'-close set, for some appropriate ε' defined later.
- Then, it suffices to consider only these pairs of subsets when searching for an approximate solution.
- In the case that the optimal solution involves small elements, we can approximate it by adding elements of $M(A, \varepsilon)$ in a greedy way.

Algorithm 1. ConstrainedSSR(A, ε, T)

Input : Sorted set $A = \{a_1, \ldots, a_n\}$, error margin ε and table of partial sums T.
Output : $(1 + \varepsilon)$-approximation of the optimal solution respecting the constraint.
1: Partition A to $M = \{a_i \in A \mid a_i < \varepsilon \cdot a_n\}$ and $L = \{a_i \in A \mid a_i \geq \varepsilon \cdot a_n\}$.
2: Split interval $[0, n \cdot a_n]$ to n/ε^2 bins of size $\varepsilon^2 \cdot a_n$.
3: **while** filling the bins with the subset sums of L **do**
4:     **if** two subset sums correspond to the same bin **then**
5:         **return** an approximation solution based on these. $\triangleright \mathcal{O}(n/\varepsilon^2)$ complexity.
6:     **end if**
7: **end while**
8: $2^{|L|} \leq n/\varepsilon^2 \iff |L| \leq \log(n/\varepsilon^2)$.
9: **for** each subset in a bin **do** $\triangleright \mathcal{O}(n/\varepsilon^2)$ subsets.
10:     Find its ε'-close set. $\triangleright$ Complexity analysis in Section 5.
11:     Add small elements accordingly. $\triangleright \mathcal{O}(\log n)$ complexity, see Subsection 3.3.
12: **end for**

3.2   Regarding only Large Elements

We firstly search for an $(1 + \varepsilon)$-approximate solution with $\varepsilon \in (0, 1)$, without involving any of the elements that are smaller than $\varepsilon \cdot a_n$. Let $M = \{a_i \in A \mid a_i < \varepsilon \cdot a_n\}$ be the set of small elements and $L = A \backslash M = \{a_i \in A \mid a_i \geq \varepsilon \cdot a_n\}$ be the set of large elements.

After partitioning the input set, we split the interval $[0, n \cdot a_n]$ into smaller intervals, called bins, of size $l = \varepsilon^2 \cdot a_n$ each, as depicted in Fig. 1.

Thus, there are a total of $B = n/\varepsilon^2$ bins. Notice that each possible subset of the input set will belong to a respective bin constructed this way, depending on its sum. Additionally, if two sets correspond to the same bin, then the difference of their subset sums will be at most l.

Fig. 1. Split of the interval $[0, n \cdot a_n]$ to bins of size l.

The next step of our algorithm is to generate all the possible subset sums, occurring from the set of large elements L. The complexity of this procedure is $\mathcal{O}(2^{|L|})$, where $|L|$ is the cardinality of set L. Notice however, that it is possible to bound the number of the produced subset sums by the number of bins B, since if two sums belong to the same bin they constitute a solution, as shown in Lemma 1, in which case the algorithm terminates in time $\mathcal{O}(n/\varepsilon^2)$.

Lemma 1. *If two subsets correspond to the same bin, we can find an $(1 + \varepsilon)$-approximation solution.*

Proof. Suppose there exist two sets $L_1, L_2 \subseteq L$ whose sums correspond to the same bin, with $\Sigma(L_1) \leq \Sigma(L_2)$. Notice that there is no guarantee regarding the disjointness of said subsets, thus consider $L_1' = L_1 \backslash L_2$ and $L_2' = L_2 \backslash L_1$, for which it is obvious that $\Sigma(L_1') \leq \Sigma(L_2')$.

Additionally, assume that $L_1' \neq \emptyset$. Then it holds that

$$\Sigma(L_2') - \Sigma(L_1') = \Sigma(L_2) - \Sigma(L_1) \leq l$$

Therefore, the sets L_1' and L_2' constitute an $(1 + \varepsilon)$-approximation solution, since

$$\frac{\Sigma(L_2')}{\Sigma(L_1')} \leq \frac{\Sigma(L_1') + l}{\Sigma(L_1')} = 1 + \frac{l}{\Sigma(L_1')}$$

$$\leq 1 + \frac{\varepsilon^2 \cdot a_n}{\varepsilon \cdot a_n} = 1 + \varepsilon$$

where the last inequality is due to the fact that $L_1' \subseteq L$ is composed of elements $\geq \varepsilon \cdot a_n$, thus $\Sigma(L_1') \geq \varepsilon \cdot a_n$.

It remains to show that $L_1' \neq \emptyset$. Assume that $L_1' = \emptyset$. This implies that $L_1 \subseteq L_2$ and since we consider each subset of L only once and the input is a set and not a multiset, it holds that $L_1 \subset L_2 \implies L_2' \neq \emptyset$. Since L_1 and L_2 correspond to the same bin, it holds that

$$\Sigma(L_2) - \Sigma(L_1) \leq l \implies \Sigma(L_2') - \Sigma(L_1') \leq l \implies \Sigma(L_2') \leq l$$

which is a contradiction, since L_2' is a non empty subset of L, which is comprised of elements greater than or equal to $\varepsilon \cdot a_n$, hence $\Sigma(L_2') \geq \varepsilon \cdot a_n > \varepsilon^2 \cdot a_n = l$, since $\varepsilon < 1$.

Consider an ε' such that $1/(1 - \varepsilon') \leq 1 + \varepsilon$ for all $\varepsilon \in (0, 1)$, for instance $\varepsilon' = \varepsilon/2$ (the exact value of ε' will be computed in Sect. 5). If every produced subset sum of the previous step belongs to a distinct bin, then, we compute their respective ε'-close sets, as described in Sect. 2. We can approximate an optimal solution that involves exclusively large elements using these pairs.

Before we prove the previous statement, observe that, if the optimal solution involves sets $L_1, L_2 \subseteq L$ composed only of large elements, where $\Sigma(L_1) \leq \Sigma(L_2)$, then $\Sigma(L_1) = \Sigma(L_{2,\text{opt}})$, where $L_{2,\text{opt}}$ is a closest set of L_2, with respect to the set $L \backslash L_2$.

Lemma 2. *If the optimal ratio r_{opt} involves sets consisting of only large elements, then there exists an ε'-close pair with ratio $r \leq (1 + \varepsilon) \cdot r_{\text{opt}}$.*

Proof. Assume that the sets $S_1^*, S_2^* \subseteq L$ form the optimal solution (S_2^*, S_1^*) and $\frac{\Sigma(S_2^*)}{\Sigma(S_1^*)} = r_{\text{opt}} \geq 1$ is the optimal ratio. Then, as mentioned, it holds that $\Sigma(S_1^*) = \Sigma(S_{2,\text{opt}}^*)$. For each set of large elements, there exists an ε'-close set and a corresponding ε'-close pair; let $(S_2^*, S_{2,\varepsilon'}^*)$ be this pair for set S_2^*. Then,

$$\Sigma(S_2^*) \geq \Sigma(S_1^*) = \Sigma(S_{2,\text{opt}}^*) \geq \Sigma(S_{2,\varepsilon'}^*) \geq (1 - \varepsilon') \cdot \Sigma(S_1^*)$$

Thus, it holds that

$$1 \leq \frac{\Sigma(S_2^*)}{\Sigma(S_{2,\varepsilon'}^*)} \leq \frac{1}{(1 - \varepsilon')} \cdot \frac{\Sigma(S_2^*)}{\Sigma(S_1^*)} \leq (1 + \varepsilon) \cdot r_{\text{opt}}$$

Therefore, we have proved that in the case where the optimal solution consists of sets comprised of only large elements, it is possible to find an $(1 + \varepsilon)$-approximation solution. This is achieved by computing an ε'-close set for each subset $L_i \subseteq L$ belonging in some bin, using the algorithms described in the preliminaries, with respect to set $L \backslash L_i$ and target $\Sigma(L_i)$. The total cost of these algorithms will be thoroughly analyzed in Sect. 5 and depends on the algorithm used.

It is important to note that by utilizing an (exact or approximation) algorithm for SUBSET SUM, we establish a connection between the complexities of SUBSET SUM and approximating SUBSET SUM RATIO in a way that any future improvement in the first carries over to the second.

3.3 General $(1 + \varepsilon)$-Approximation Solutions

Whereas we previously considered optimal solutions involving exclusively large elements, here we will search for approximations for those optimal solutions that use all the elements of the input set, hence include small elements, and satisfy our constraint. We will prove that in order to approximate those optimal solutions, it suffices to consider only the ε'-close pairs corresponding to each distinct bin and add small elements to them. In other words, instead of considering any two random disjoint subsets consisting of large elements[2] and subsequently adding

---

[2] Note that the number of these random pairs is $3^{|L|}$.

to these the small elements, we can instead consider only the pairs computed in the previous step, the number of which is bounded by the number of bins $B = n/\varepsilon^2$. Moreover, we will prove that it suffices to add the small elements to our solution in a greedy way.

Since the algorithm has not detected a solution so far, due to Lemma 1 every computed subset sum of set L belongs to a different bin. Thus, their total number is bounded by the number of bins B, i.e.

$$2^{|L|} \le \left(\frac{n}{\varepsilon^2}\right) \iff |L| \le \log\left(\frac{n}{\varepsilon^2}\right).$$

We proceed by involving small elements in order to reduce the difference between the sums of ε'-close pairs, thus reducing their ratio.

Lemma 3. *Given the ε'-close pairs, one can find an $(1+\varepsilon)$-approximation solution for the constrained version of* SUBSET SUM RATIO, *in the case that the optimal solution involves small elements.*

Proof (sketch). Due to page limitations, we only give a short sketch of the proof here; the complete proof is included in the full version of the paper.

Let $S_1^* = L_1^* \cup M_1^*$ and $S_2^* = L_2^* \cup M_2^*$ be disjoint subsets that form an optimal solution, where $\Sigma(S_1^*) \le \Sigma(S_2^*)$, $L_1^*, L_2^* \subseteq L$ and $M_1^*, M_2^* \subseteq M$.

For $\Sigma(L_1^*) < \Sigma(L_2^*)$ (respectively $\Sigma(L_2^*) < \Sigma(L_1^*)$), we show that is suffices to add an appropriate subset $M_k \subseteq M$ to $L_{2,\varepsilon'}^*$ (respectively $L_{1,\varepsilon'}^*$) in order to approximate the optimal solution $r_{\text{opt}} = \frac{\Sigma(S_2^*)}{\Sigma(S_1^*)}$, where $M_k = \{a_i \in M \mid i \in [k]\}$ and $k \le |M|$.

Therefore, by adding in a greedy way small elements to an ε'-close set of the set with the largest sum among L_1^* and L_2^*, we can successfully approximate the optimal solution. □

Adding Small Elements Efficiently. Here, we will describe a method to efficiently add small elements to our sets. As a reminder, up to this point the algorithm has detected an ε'-close pair (L_2, L_1), such that $L_1, L_2 \subseteq L$ with $\Sigma(L_1) < \Sigma(L_2)$. Thus, we search for some k such that $\Sigma(L_1 \cup M_k) \le \Sigma(L_2) + \varepsilon \cdot a_n$, where $M_k = \{a_i \in M \mid i \in [k]\}$. Notice that if $\Sigma(M) \ge \Sigma(L_2) - \Sigma(L_1)$, there always exists such a set M_k, since by definition, each element of set M is smaller than $\varepsilon \cdot a_n$. In order to determine M_k, we make use of an array of partial sums $T[k] = \Sigma(M_k)$, where $k \le |M|$. Notice that T is sorted; therefore, since T is already available (see Algorithm 2), each time we need to compute a subset with the desired property, this can be done in $\mathcal{O}(\log k) = \mathcal{O}(\log n)$ time.

4   Final Algorithm

The algorithm presented in the previous section constitutes an approximation scheme for SUBSET SUM RATIO, in the case where at least one of the solution subsets has sum of its large elements greater than, or equal to the max element of the input set. Thus, in order to solve the SUBSET SUM RATIO problem, it suffices to run the previous algorithm n times, where n depicts the cardinality of the input set A, while each time removing the max element of A.

In particular, suppose that the optimal solution involves disjoint sets S_1^* and S_2^*, where $a_k = \max\{S_1^* \cup S_2^*\}$. There exists an iteration for which the algorithm considers as input the set $A_k = \{a_1, \ldots, a_k\}$. In this iteration, the element a_k is the largest element and the algorithm searches for a solution where the sum of the large elements of one of the two subsets is at least a_k. The optimal solution has this property so the ratio of the approximate solution that the algorithm of the previous section returns is at most $(1 + \varepsilon)$ times the optimal.

Consequently, n repetitions of the algorithm suffice to construct an FPTAS for SUBSET SUM RATIO.

Notice that if at some repetition, the sets returned due to the algorithm of Sect. 3 have ratio at most $1 + \varepsilon$, then this ratio successfully approximates the optimal ratio $r_{\mathrm{opt}} \geq 1$, since $1 + \varepsilon \leq (1 + \varepsilon) \cdot r_{\mathrm{opt}}$, therefore they constitute an approximation solution.

Algorithm 2. SSR(A, ε)

Input : Sorted set $A = \{a_1, \ldots, a_n\}$ and error margin ε.
Output : $(1 + \varepsilon)$-approximation of the optimal solution for SUBSET SUM RATIO.
 1: Create array T such that $T[k] = \sum_{i=1}^{k} a_i$. $\triangleright \Theta(n)$ time.
 2: **for** $i = n, \ldots, 1$ **do**
 3:     ConstrainedSSR$(\{a_1, \ldots, a_i\}, \varepsilon, T)$
 4: **end for**

---

5 Complexity

The total complexity of the final algorithm is determined by three distinct operations, over the n iterations of the algorithm:

1. The cost to compute all the possible subset sums occurring from large elements. It suffices to consider the case where this is bounded by the number of bins $B = n/\varepsilon^2$, due to Lemma 1.
2. The cost to find the ε'-close pair for each subset in a distinct bin. The cost of this operation will be analyzed in the following subsection.
3. The cost to include small elements to the ε'-close pairs. There are B ε'-close pairs, and each requires $\mathcal{O}(\log n)$ time, thus the total time required is $\mathcal{O}\left(\frac{n}{\varepsilon^2} \cdot \log n\right)$.

5.1 Complexity to Find the ε'-Close Pairs

Using Exact Subset Sum Computations. The first algorithm we mentioned is a standard meet in the middle algorithm. Here we will analyze its complexity.

Let subset $L' \subseteq L$ such that $|L'| = k$. The meet in the middle algorithm on the set $L\backslash L'$ costs time

$$\mathcal{O}\left(\frac{|L\backslash L'|}{2} \cdot 2^{\frac{|L\backslash L'|}{2}}\right).$$

Notice that the number of subsets of L of cardinality k is $\binom{|L|}{k}$ and that $|L| \le \log(n/\varepsilon^2)$. Additionally,

$$\sum_{k=0}^{|L|} \binom{|L|}{k} \cdot 2^{\frac{|L|-k}{2}} \cdot \frac{|L|-k}{2} = 2^{|L|/2} \cdot \sum_{k=0}^{|L|} \binom{|L|}{k} \cdot 2^{-k/2} \cdot \frac{|L|-k}{2}$$

$$\le 2^{|L|/2} \cdot \frac{|L|}{2} \cdot \sum_{k=0}^{|L|} \binom{|L|}{k} \cdot 2^{-k/2}$$

Furthermore, let $c = (1 + 2^{-1/2})$, where $\log c = 0.7715... < 0.8$. Due to Binomial Theorem, it holds that

$$\sum_{k=0}^{|L|} \binom{|L|}{k} \cdot 2^{-k/2} = (1 + 2^{-1/2})^{|L|} = c^{|L|} \le c^{\log(n/\varepsilon^2)} = (n/\varepsilon^2)^{\log c}$$

Consequently, the complexity to find a closest set for every subset in a bin is

$$\mathcal{O}\left(2^{|L|/2} \cdot \frac{|L|}{2} \cdot (n/\varepsilon^2)^{\log c}\right) = \mathcal{O}\left((n/\varepsilon^2)^{1/2} \cdot \log(n/\varepsilon^2) \cdot (n/\varepsilon^2)^{\log c}\right)$$

$$= \mathcal{O}\left(\frac{n^{1.3}}{\varepsilon^{2.6}} \cdot \log(n/\varepsilon^2)\right)$$

Using approximate Subset Sum computations. Here we will analyze the complexity in the case we run an approximate Subset Sum algorithm in order to compute the ε'-close pairs.

For subset $L_i \subseteq L$ of sum $\Sigma(L_i)$, we run an approximate Subset Sum algorithm ([9,19]), with error margin ε' such that

$$\frac{1}{1-\varepsilon'} \le 1 + \varepsilon \iff \varepsilon' \le \frac{\varepsilon}{1+\varepsilon}$$

By choosing the maximum such ε', we have that

$$\varepsilon' = \frac{\varepsilon}{1+\varepsilon} \implies \frac{1}{\varepsilon'} = \frac{1+\varepsilon}{\varepsilon} = \frac{1}{\varepsilon} + 1 \implies \frac{1}{\varepsilon'} = \mathcal{O}\left(\frac{1}{\varepsilon}\right)$$

Thus, if we use for instance the approximation algorithm[3] presented at [19], the complexity of finding all the ε'-close sets (one for every subset in a bin, for a total of a maximum of $B = n/\varepsilon^2$ subsets) is

---

[3] Of complexity $\mathcal{O}\left(\min\{\frac{n}{\varepsilon}, n + \frac{1}{\varepsilon^2} \cdot \log(1/\varepsilon)\}\right)$ for n elements and error margin ε.

$$\mathcal{O}\left(\frac{n}{\varepsilon^2} \cdot \min\left\{\frac{|L|}{\varepsilon'}, |L| + \frac{1}{(\varepsilon')^2} \cdot \log(1/\varepsilon')\right\}\right) =$$

$$\mathcal{O}\left(\frac{n}{\varepsilon^2} \cdot \min\left\{\frac{|L|}{\varepsilon}, |L| + \frac{1}{\varepsilon^2} \cdot \log(1/\varepsilon)\right\}\right) =$$

$$\mathcal{O}\left(\frac{n}{\varepsilon^2} \cdot \min\left\{\frac{\log(n/\varepsilon^2)}{\varepsilon}, \log(n/\varepsilon^2) + \frac{1}{\varepsilon^2} \cdot \log(1/\varepsilon)\right\}\right)$$

5.2   Total Complexity

The total complexity of the algorithm occurs from the n distinct iterations required and depends on the algorithm chosen to find the ε'-close pairs, since both of the presented algorithms dominate the time of the rest of the operations. Thus, by choosing the fastest one (depending on the relationship between n and ε), the final complexity is

$$\mathcal{O}\left(\min\left\{\frac{n^{2.3}}{\varepsilon^{2.6}} \cdot \log(n/\varepsilon^2), \frac{n^2}{\varepsilon^3} \cdot \log(n/\varepsilon^2), \frac{n^2}{\varepsilon^2}\left(\log(n/\varepsilon^2) + \frac{1}{\varepsilon^2} \cdot \log(1/\varepsilon)\right)\right\}\right)$$

6   Conclusion and Future Work

The main contribution of this paper, apart from the introduction of a new FPTAS for the SUBSET SUM RATIO problem, is the establishment of a connection between SUBSET SUM and approximating SUBSET SUM RATIO. In particular, we showed that any improvement over the classic meet in the middle algorithm [16] for SUBSET SUM, or over the approximation scheme for SUBSET SUM will result in an improved FPTAS for SUBSET SUM RATIO.

Additionally, we establish that the complexity of approximating SUBSET SUM RATIO, expressed in the form $\mathcal{O}((n + 1/\varepsilon)^c)$ has an exponent $c < 5$, which is an improvement over all the previously presented FPTASs for the problem.

It is important to note however, that there is a distinct limit to the complexity that one may achieve for the SUBSET SUM RATIO problem using the techniques discussed in this paper.

As a direction for future research, we consider the notion of the *weak* approximation of SUBSET SUM, as discussed in [9,26], which was used in order to approximate the slightly easier PARTITION problem, and may be able to replace the approximate SUBSET SUM algorithm in the computation of the ε'-close sets.

Another possible direction could be the use of exact SUBSET SUM algorithms parameterized by a *concentration* parameter β, as described in [3,4], where they solve the decision version of SUBSET SUM. See also [15] for a use of this parameter under a pseudopolynomial setting. It would be interesting to investigate whether analogous arguments could be used to solve the optimization version.

References

1. Abboud, A., Bringmann, K., Hermelin, D., Shabtay, D.: Seth-based lower bounds for subset sum and bicriteria path. In: Proceedings of the Thirtieth Annual ACM-SIAM Symposium on Discrete Algorithms, SODA 2019, San Diego, California, 6–9 January 2019, pp. 41–57. SIAM (2019). https://doi.org/10.1137/1.9781611975482. 3
2. Antonopoulos, A., Pagourtzis, A., Petsalakis, S., Vasilakis, M.: Faster algorithms for k-subset sum and variations. In: J., Chen, Li, M., Zhang, G. (eds.): Frontiers of Algorithmics - International Joint Conference, IJTCS-FAW 2021, Beijing, 16–19 August 2021, Proceedings. LNCS, vol. 12874, pp. 37–52. Springer (2021). https://doi.org/10.1007/978-3-030-97099-4_3
3. Austrin, P., Kaski, P., Koivisto, M., Nederlof, J.: Subset sum in the absence of concentration. In: 32nd International Symposium on Theoretical Aspects of Computer Science, STACS 2015, 4–7 March 2015, Garching. LIPIcs, vol. 30, pp. 48–61. Schloss Dagstuhl - Leibniz-Zentrum für Informatik (2015). https://doi.org/10.4230/LIPIcs.STACS.2015.48
4. Austrin, P., Kaski, P., Koivisto, M., Nederlof, J.: Dense subset sum may be the hardest. In: 33rd Symposium on Theoretical Aspects of Computer Science, STACS 2016, 17–20 February 2016. https://doi.org/10.4230/LIPIcs.STACS.2016.13
5. Bazgan, C., Santha, M., Tuza, Z.: Efficient approximation algorithms for the SUBSET-SUMS EQUALITY problem. J. Comput. Syst. Sci. **64**(2), 160–170 (2002). https://doi.org/10.1006/jcss.2001.1784
6. Bellman, R.E.: Dynamic Programming. Princeton University Press, Princeton (1957)
7. Bringmann, K.: A near-linear pseudopolynomial time algorithm for subset sum. In: Proceedings of the Twenty-Eighth Annual ACM-SIAM Symposium on Discrete Algorithms, SODA 2017. pp. 1073–1084. SIAM, Philadelphia (2017). https://doi.org/10.1137/1.9781611974782.69
8. Bringmann, K., Nakos, V.: Top-k-convolution and the quest for near-linear output-sensitive subset sum. In: Proccedings of the 52nd Annual ACM SIGACT Symposium on Theory of Computing, STOC 2020, pp. 982–995. ACM, New York (2020). https://doi.org/10.1145/3357713.3384308
9. Bringmann, K., Nakos, V.: A fine-grained perspective on approximating subset sum and partition. In: Proceedings of the 2021 ACM-SIAM Symposium on Discrete Algorithms, SODA 2021, Virtual Conference, 10–13 January 2021, pp. 1797–1815. SIAM (2021). https://doi.org/10.1137/1.9781611976465.108
10. Cieliebak, M., Eidenbenz, S.J.: Measurement errors make the partial digest problem np-hard. In: LATIN 2004 Theoretical Informatics. 6th Latin American Symposium Lecture Notes in Computer Science, vol. 2976, pp. 379–390. Springer, Berlin, Heidelberg (2004). https://doi.org/10.1007/978-3-540-24698-5_42
11. Cieliebak, M., Eidenbenz, S.J., Pagourtzis, A.: Composing equipotent teams. In: Fundamentals of Computation Theory. 14th International Symposium, FCT 2003 Lecture Notes in Computer Science, vol. 2751, pp. 98–108. Springer, Berlin, Heidelberg (2003). https://doi.org/10.1007/978-3-540-45077-1_10
12. Cieliebak, M., Eidenbenz, S.J., Pagourtzis, A., Schlude, K.: On the complexity of variations of equal sum subsets. Nord. J. Comput. **14**(3), 151–172 (2008)
13. Cieliebak, M., Eidenbenz, S.J., Penna, P.: Noisy data make the partial digest problem NP-hard. In: Algorithms in Bioinformatics, Third International Workshop, WABI 2003. Lecture Notes in Computer Science, vol. 2812, pp. 111–123. Springer, Berlin, Heidelberg (2003). https://doi.org/10.1007/978-3-540-39763-2_9

14. Cygan, M., Mucha, M., Wegrzycki, K., Wlodarczyk, M.: On problems equivalent to (min, +)-convolution. ACM Trans. Algorithms **15**(1), 14:1–14:25 (2019). https://doi.org/10.1145/3293465

15. Dutta, P., Rajasree, M.S.: Efficient reductions and algorithms for variants of subset sum. CoRR abs/2112.11020 (2021). https://arxiv.org/abs/2112.11020

16. Horowitz, E., Sahni, S.: Computing partitions with applications to the knapsack problem. J. ACM **21**(2), 277–292 (1974). https://doi.org/10.1145/321812.321823

17. Jin, C., Wu, H.: A simple near-linear pseudopolynomial time randomized algorithm for subset sum. In: 2nd Symposium on Simplicity in Algorithms (SOSA 2019), vol. 69, pp. 17:1–17:6 (2018). https://doi.org/10.4230/OASIcs.SOSA.2019.17

18. Karp, R.M.: Reducibility among combinatorial problems. In: Complexity of Computer Computations. The IBM Research Symposia Series, pp. 85–103. Springer, Boston (1972). https://doi.org/10.1007/978-1-4684-2001-2_9

19. Kellerer, H., Mansini, R., Pferschy, U., Speranza, M.G.: An efficient fully polynomial approximation scheme for the subset-sum problem. J. Comput. Syst. Sci. **66**(2), 349–370 (2003). https://doi.org/10.1016/S0022-0000(03)00006-0

20. Khan, M.A.: Some problems on graphs and arrangements of convex bodies (2017). https://doi.org/10.11575/PRISM/10182

21. Koiliaris, K., Xu, C.: Faster pseudopolynomial time algorithms for subset sum. ACM Trans. Algorithms **15**(3), 401–4020 (2019). https://doi.org/10.1145/3329863

22. Lipton, R.J., Markakis, E., Mossel, E., Saberi, A.: On approximately fair allocations of indivisible goods. In: Proceedings of the 5th ACM Conference on Electronic Commerce (EC-2004). pp. 125–131. ACM, New York (2004). https://doi.org/10.1145/988772.988792

23. Melissinos, N., Pagourtzis, A.: A faster FPTAS for the subset-sums ratio problem. In: Computing and Combinatorics - 24th International Conference, COCOON 2018. Lecture Notes in Computer Science, vol. 10976, pp. 602–614. Springer, Cham (2018). https://doi.org/10.1007/978-3-319-94776-1_50

24. Melissinos, N., Pagourtzis, A., Triommatis, T.: Approximation schemes for subset sum ratio problems. In: Li, M. (ed.) FAW 2020. LNCS, vol. 12340, pp. 96–107. Springer, Cham (2020). https://doi.org/10.1007/978-3-030-59901-0_9

25. Mucha, M., Nederlof, J., Pawlewicz, J., Wegrzycki, K.: Equal-subset-sum faster than the meet-in-the-middle. In: 27th Annual European Symposium on Algorithms, ESA 2019. LIPIcs, vol. 144, pp. 73:1–73:16 (2019). https://doi.org/10.4230/LIPIcs.ESA.2019.73

26. Mucha, M., Wegrzycki, K., Wlodarczyk, M.: A subquadratic approximation scheme for partition. In: Proceedings of the Thirtieth Annual ACM-SIAM Symposium on Discrete Algorithms, SODA 2019, San Diego, California, 6–9 January 2019, pp. 70–88. SIAM (2019). https://doi.org/10.1137/1.9781611975482.5

27. Nanongkai, D.: Simple FPTAS for the subset-sums ratio problem. Inf. Process. Lett. **113**(19–21), 750–753 (2013). https://doi.org/10.1016/j.ipl.2013.07.009

28. Papadimitriou, C.H.: On the complexity of the parity argument and other inefficient proofs of existence. J. Comput. Syst. Sci. **48**(3), 498–532 (1994). https://doi.org/10.1016/S0022-0000(05)80063-7

29. Pisinger, D.: Linear time algorithms for knapsack problems with bounded weights. J. Algorithms **33**(1), 1–14 (1999). https://doi.org/10.1006/jagm.1999.1034

30. Voloch, N.: Mssp for 2-d sets with unknown parameters and a cryptographic application. Contemp. Eng. Sci. **10**, 921–931 (2017). https://doi.org/10.12988/ces.2017.79101

31. Woeginger, G.J., Yu, Z.: On the equal-subset-sum problem. Inf. Process. Lett. **42**(6), 299–302 (1992). https://doi.org/10.1016/0020-0190(92)90226-L

Faster Algorithm for Finding Maximum 1-Restricted Simple 2-Matchings

Stepan Artamonov[1] and Maxim A. Babenko[2]([⊠])

[1] Moscow State University, Moscow, Russia
[2] National Research University Higher School of Economics (HSE), Yandex LLC, Moscow, Russia
maxim.babenko@gmail.com

Abstract. We revisit the problem of finding a 1-restricted simple 2-matching of maximum cardinality. Recall that, given an undirected graph $G = (V, E)$, a *simple 2-matching* is a subset $M \subseteq E$ of edges such that each node in V is incident to at most two edges in M. Clearly, each such M decomposes into a node-disjoint collection of paths and circuits. M is called *1-restricted* if it contains no isolated edges (i.e. paths of length one).

A combinatorial polynomial algorithm for finding such M of maximum cardinality and also a min-max relation were devised by Hartvigsen. It was shown that finding such M amounts to computing a (not necessarily 1-restricted) simple 2-matching M_0 of maximum cardinality and subsequently altering it into M (with $|M| = |M_0|$) so as to minimize the number of isolated edges. While the first phase (which computes M_0) runs in $O(E\sqrt{V})$ time, the second one (turning M_0 into M) requires $O(VE)$ time.

In this paper we apply the general blocking augmentation approach (initially introduced, e.g., for bipartite matchings by Hopcroft and Karp, and also by Dinic) and present a novel algorithm that reduces the time needed for the second phase to $O(E\sqrt{V})$ thus completely closing the gap between 1-restricted and unrestricted cases.

1 Introduction

Let $G = (V, E)$ be an undirected graph. Recall that a simple 2-matching M is a subset $M \subseteq E$ such that at most two edges of M are incident to any node in V. By an M-*component* we mean a connected component induced by the edge set M; clearly, each such component is either a simple path or a circuit.

Consider the k-*restricted* problem as follows: find a maximum cardinality simple 2-matching M, where each M-component contains more than k edges. The notion of restricted simple 2-matchings is motivated, e.g., by the *Hamiltonian cycle problem*: such cycles are exactly k-restricted simple 2-matchings for $k > |V|/2$. Since deciding if G admits a Hamiltonian cycle is NP-hard, looking for a polynomial algorithm for arbitrary k is essentially hopeless. But the problem remains non-trivial even for small positive values of k.

© Springer Nature Switzerland AG 2022
C. Bazgan and H. Fernau (Eds.): IWOCA 2022, LNCS 13270, pp. 86–100, 2022.
https://doi.org/10.1007/978-3-031-06678-8_7

In the present paper we focus on maximum 1-restricted simple 2-matchings, i.e. matchings without *isolated* edges (hereinafter we shall be using the term *maximum* to denote objects of maximum cardinality). In [2] it was shown that a maximum 1-restricted simple 2-matching can be found in polynomial time. Moreover, the problem is closely related to its unrestricted counterpart as follows:

Theorem 1 ([2], **Th. 6**). *Every graph admits a maximum 1-restricted simple 2-matching that is a subset of some maximum unrestricted simple 2-matching.*

For a subset $X \subseteq E$, let iso(X) be the number of connected components of (V, X) that are isolated edges. Theorem 1 implies that solving the 1-restricted problem amounts to finding (among maximum unrestricted simple 2-matchings) some M minimizing iso(M); then the needed maximum 1-restricted simple 2-matching is just M without its isolated edges.

Let $\nu_0(G)$ (resp. $\nu_1(G)$) be the cardinality of a maximum unrestricted (resp. 1-restricted) simple 2-matching in G. A min-max formula for $\nu_1(G)$ (relating it to $\nu_0(G)$) is also known:

Theorem 2 ([2], **Th. 4**).

$$\nu_0(G) - \nu_1(G) = \max_{W \subseteq V} \left(\text{iso}(G - W) - 2|W| \right).$$

All of the above indicates a deep and intricate connection between 1-restricted and unrestricted problems, at least in the structural sense. As for the algorithmic aspects, Hartvigsen [2] provides a polynomial algorithm for solving the 1-restricted problem as follows. First, solve the unrestricted problem (which can be done in $O(m\sqrt{n})$ time, see, e.g. [1]; hereinafter we define $n := |V|$, $m := |E|$ and assume $m \geq n$) and let M be the resulting simple 2-matching. Second, apply a number of alternating transformations so as to minimize iso(M). The second stage can be done in polynomial time; a careful analysis (omitted in [2]) leads to an $O(mn)$ bound.

The first stage involves blossom manipulations while the second one resembles an $O(mn)$-algorithm for computing maximum bipartite matchings (in particular, it never has to deal with blossoms or similar nested contracted structures). Altogether the algorithm runs in $O(mn)$ time, which is unsatisfactory since the first, seemingly more involved stage, takes just $O(m\sqrt{n})$ time.

In this paper we give a novel approach for implementing the second stage by showing how the general idea of blocking augmentations [3] enables minimizing iso(M) in $O(m\sqrt{n})$ time (Theorem 3). Due to the lack of space, some (mostly technical) proofs are omitted and will appear in the full version of the paper.

2 Algorithm

2.1 Outline

Let $G_0 = (V_0, E_0)$ be an undirected graph; our ultimate goal is to compute a maximum 1-restricted simple 2-matching in G_0. As mentioned earlier, we start by computing a maximum simple 2-matching M_0 in G_0. Then M_0 is updated by applying certain *augmenting* transformations. These transformations preserve $|M_0|$ (hence, in the sense of the unrestricted problem, they are merely alternations) and gradually decrease iso(M_0). At the end, no augmentations remain and a maximum 1-restricted simple 2-matching can be extracted from M_0 by dropping all of its isolated edges. A suitably chosen subset $W \subseteq V_0$ (Lemma 7) certifies that the latter matching is optimal by Theorem 2.

The sequence of augmentations is divided into *phases*. Each phase deals with augmenting paths of certain length, and the latter length strictly increases over phases (Lemma 12). Each phase can be implemented to run in linear time and the number of phases is $O(\sqrt{n})$. Altogether this yields an $O(m\sqrt{n})$-time algorithm (Theorem 3).

2.2   Alternations and Augmentations

We assume that the reader is familiar with the basic ideas and notions of matching theory (see [4] for a survey). Let us discuss transformations that can be applied to a general simple 2-matching to decrease the number of isolated edges while preserving its cardinality. Our transformations are similar in spirit to those described in [2] but for the sake of completeness we present a self-contained description below.

Instead of working with M_0 in G_0 directly it is more enlightening to work with a certain simple 2-matching M in graph G obtained from M_0 and G_0 by contractions as described below. Initially let $M := M_0$ and $G := G_0$. These contractions take a certain node-disjoint collection of edges in G_0 and replace each such edge e with a single node v_e.

First contract each isolated edge e_v of M_0 into a single node v. We call these contracted nodes *sources*.

Second, consider an M_0-component that is a path of length exactly 4 formed by edges e_1, e_2, e_3, e_4, numbered along the path (Remark 1 below explains why these paths are important). We simultaneously contract edges e_1 and e_4 into nodes a' and c' in G and replace the initial path of length 4 with path F of length 2 formed by edges e_2 and e_3. (Hereinafter we identify edges in G with their pre-images in G_0.) Path F is called a *fork*; nodes a' and c' are called the *endpoints* of F. The (unique) common node b of e_2 and e_3 (which is also a node of the original graph G_0) is called the *center* of F. See Fig. 1.

(a) M_0-component of length 4 in G_0

(b) Fork F with center b in G

Fig. 1. Fork contraction

Once all isolated edges of M_0 are contracted into sources and all paths of length 4 are contracted into forks, simple 2-matching M and graph G are ready. The subsequent algorithm updates M in G and returns to G_0 at the very end, when the number of sources in M is minimized. The following invariant is maintained:

(1) For each contracted node v in G, at most one edge in M is incident to v.

Property (1) implies that at any moment any contracted node v in G can be replaced with its pre-image (edge e_v in G_0) and M can be extended by adding e_v. Also if all contracted nodes of G are expanded (turning M into M_0 by adding edges e_v) the isolated edges in the resulting simple 2-matching M_0 exactly correspond to sources in M. This justifies the idea to alter M so as to minimize the number of sources.

Also it turns out crucial to strengthen (1) as follows:

(2) Each contracted node v in G is either not incident to any edge of M (i.e. is a source) or is incident to exactly one such edge e; moreover, in the latter case e belongs to a fork and v is one of its endpoints.

Consider source $s = v_0$ and a sequence of (distinct) forks $F_1, \ldots F_k$ such that s is connected to the center v_1 of F_1, some endpoint v_2 of F_1 is connected to the center v_3 of F_2, and, in general, some endpoint v_{2i} of fork F_i is connected to the center v_{2i+1} of fork F_{i+1} (for $i = 1, \ldots, k-1$). The nodes $v_0, \ldots, v_{2k}$ form a path P of even length called *alternating*. See Fig. 2.

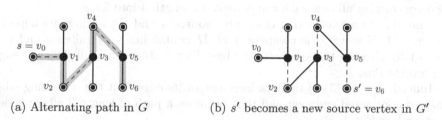

(a) Alternating path in G (b) s' becomes a new source vertex in G'

Fig. 2. Alternation. hereinafter, solid edges belong to the current 2-matching, dashed edges are free from the current 2-matching.

Inspired by the standard matching techniques, one can replace M with a new simple 2-matching M' constructed by adding each first, third, and etc. edge of P to M and removing each second, forth and etc. edge. See Fig. 2.

This transformation clearly preserves the cardinality of M. Forks $F_1, \ldots, F_k$ are, essentially, replaced with new forks $F_1', \ldots, F_k'$ and an M-source s moves to $s' = v_{2k}$, which becomes an M'-source. In a sense, alternating paths enable moving sources of simple 2-matchings across the graph.

Alternations preserve property (2), and also obey the following:

Lemma 1. *Let U and U' be the sets of endpoints of forks before and after alternation along $(v_0, \ldots, v_{2k})$. Then $U' = U \cup \{v_0\} - \{v_{2k}\}$.*

In addition to alternation, the problem admits the notion of *augmentation*. Like for other matching problems, augmentation is just an alternation plus some final local step that improves the solution. In our case, the goal of a single augmentation is to reduce the number of sources by one.

Let t be a (possibly contracted) node in G. We call t a *sink* if one of the following cases applies:

(3) (a) t is an endpoint of a fork F (thus t is contracted); or
    (b) t belongs to a circuit component C of M (and, thus, t is not contracted by (2)); or
    (c) t is an inner node of a non-fork path component P of M (and, thus, t is not contracted by (2) again).

We now describe the structure of source-to-sink augmenting paths and elaborate on how augmentations are performed. Let us start with the following basic case: assume that s is a source and t is its neighbor that happens to be a sink. Then one can add the edge st to M and remove some edge tr from M ((3) implies there is at least one) forming a new simple 2-matching M'.

Unfortunately (2) may cease to hold for M' but its weaker counterpart (1) still holds. Indeed, s becomes incident to exactly one edge of M'. If t happens to be contracted then t must be an endpoint of a fork; the above augmentation replaces edge tr with edge st.

We immediately fix this by first expanding all suffering contracted nodes and then contracting all new path components of length 4 into forks.

The above transformation eliminates source s and thus brings us hope to improve M. However, the component of M containing t gets altered and can potentially give rise to more isolated edges. Fortunately, a careful choice of edge tr prevents this.

Indeed, in case (3)(a) the fork loses one of its edges but the remaining edge enters a contracted endpoint and thus becomes a path of length 2 after uncontraction (see Fig. 3).

In case (3)(b) circuit C turns into a path of length at least 4 after uncontraction (see Fig. 4). If the resulting length is exactly 4 (which happens if the length of C was 3) then we also contract this path into a fork.

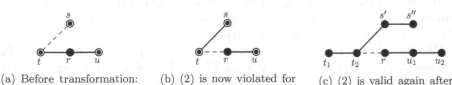

(a) Before transformation: reaching fork endpoint

(b) (2) is now violated for vertices s, t and u

(c) (2) is valid again after uncontractions

Fig. 3. Augmentation, case (3)(a)

(a) Before transformation: reaching circuit C of length 3

(b) Path of length 4 after transformation and uncontraction

(c) Contracting newly created path of length 4 into a fork

Fig. 4. Augmentation, case (3)(b)

Finally, in case (3)(c) let P be the (non-fork) path component of M containing t and let l_1, l_2 be the distances along P from t to the endpoints v_1, v_2 of P (measured in edges). Clearly $l_1 = l_2 = 2$ is impossible (since the length of P is not equal to 4). W.l.o.g. $l_2 \neq 2$. Now pick r as the next node along P from t in the direction to v_2. This way, the augmentation alters P into paths of lengths $l_2 - 1 \neq 1$ and $l_1 + 2 \geq 3$ (after uncontraction). (If $l_2 = 1$ then the former path vanishes. Also one or even both of the resulting paths may be of length 4 and thus must be contracted into a fork.) See Fig. 5.

Remark 1. One can easily see that the above analysis fails to hold for $l_1 = l_2 = 2$ as no choice of r can prevent forming a new isolated edge. This is exactly the reason why paths of length 4 are so special.

Let us introduce another notion that is similar to sinks: call node v *unsaturated* if v is not contracted and is incident to at most one edge of M.

Remark 2. A careful reader may wonder why in (3) we do not regard unsaturated nodes as sinks. In particular, in (3)(c) we restrict t to be an *inner* node of P. The reason is that if t is unsaturated then just adding st to M (and not deleting any edge at all) would produce an unrestricted simple 2-matching of a larger cardinality, which is impossible by the initial choice of M_0 and the fact that its cardinality is preserved during the execution of the algorithm. Hence such edge st cannot appear. Note that for similar reasons the G_0-preimages of

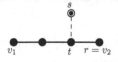

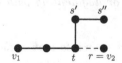

(a) Before transformation: reaching path P of length 3, $l_1 = 2$, $l_2 = 1$

(b) Paths of length 4 and 0 after transformation and uncontraction

(c) Contracting newly created path of length 4 into a fork

Fig. 5. Augmentation, case (3)(c)

contracted edges st and tr in Fig. 3 must share a common endpoint, for otherwise M_0 would not be maximum.

Consider a simple alternating path $(v_0, \ldots, v_{2k})$ as above (given by a sequence of nodes), where v_0 is a source and v_{2k} is either v_0 (for $k = 0$) or an endpoint of a fork (for $k > 0$). Next, let v_{2k} be connected to some sink $t = v_{2k+1}$ (distinct from nodes $v_0, \ldots, v_{2k}$). Then path $P = (v_0, \ldots, v_{2k}, v_{2k+1})$ is called *augmenting*. One can apply P to M by first alternating along $(v_0, \ldots, v_{2k})$ and thus moving the source from s to v_{2k} and then finally applying the transformations explained above (note that t remains a sink even after alternation by Lemma 1). Altogether the augmentation maintains (2), preserves the cardinality of M but decreases the number sources by one.

2.3   Edge Directions

Since our approach incorporates certain blocking augmentations, it should be of no surprise that rather than considering arbitrary augmenting paths one must focus on shortest ones. To this aim, we first introduce edge directions: some (but not all) undirected edges xy are turned into directed ones (x, y). It is possible that some undirected edges xy become directed both as (x, y) and (y, x), i.e. e may give rise to a pair of oppositely directed edges.

The exact rules of assigning directions are as follows:

(4) (a) for a source s and an edge sv such that v is a sink or the center of a fork, sv is directed as (s, v);

   (b) for a fork with edges uv, vw (where u, w are endpoints and v is the center), these edges are directed as (v, u) and (v, w);

   (c) for an endpoint u of a fork F and an edge uv such that v is either a sink or the center of another fork $F' \neq F$, edge uv is directed as (u, v).

In particular, the only case when an edge uv is directed both as (u, v) and (v, u) is when u and v are endpoints of some (possibly coinciding) forks.

Let $\vec{G}$ be the digraph obtained by assigning directions to the edges of G as above (clearly $\vec{G}$ may contain fewer edges than G since not all edges become directed). Note that this graph will be changing as we update M in G. Clearly (4) implies

Lemma 2.   *Each alternating or augmenting path consists of directed edges in $\vec{G}$.*

Lemma 3.   *Each odd directed source-to-sink path $P = (v_0, \ldots, v_{2k}, v_{2k+1})$ in $\vec{G}$, such that $v_1, v_3, \ldots, v_{2k-1}$ are not sinks, is augmenting.*

2.4   Blocking Augmentations

Lemma 3 implies a combinatorial algorithm for computing augmenting paths that extracts a single augmenting path per BFS or DFS run, which is inefficient.

Instead we describe a method for extracting a *blocking* (i.e. inclusion-wise maximal) collection of such paths. Like in most blocking augmentation algorithms, ours executes a series of *phases* each dealing with augmenting paths of some fixed length.

We next focus on a single phase. It first runs a BFS-like algorithm BUILD-LEVELS assigning certain *levels* to nodes (denoted by level(v) for $v \in V$) as follows. Nodes v with assigned level(v) are called *labeled*, others are called *unlabeled*.

Initially, let $D := 0$, define level(s) := 0 for all sources s and run a sequence of iterations. Each iteration considers a node subset $Q \subseteq V$ such that level(v) = 2D for each $v \in Q$. The set Q will consist of all sources for $D = 0$ and of all endpoints of some forks for $D > 0$. Moreover, there will be an alternating path of length $2D$ to each node $u \in Q$.

For each $u \in Q$, the algorithm scans all edges $(u, v) \in E(\overrightarrow{G})$. First note that v cannot be unsaturated since otherwise one could have increased the cardinality of M by applying an alternating path to u and then adding uv.

If v is a sink then BUILD-LEVELS stops as it has discovered an augmenting path. It also labels all currently unlabeled nodes x as level(x) := 2D.

Otherwise (none of v is a sink) all these v must be centers of some forks (either still unlabeled or already labelled). Let R be all such centers v that are not labelled yet. Assign level(v) := 2D+1 for each $v \in R$ and level(w) := 2D+2 for all endpoints w of forks in R, update $D := D + 1$, and repeat.

BUILD-LEVELS also stops if $Q = \emptyset$; in this case we fail to discover an augmenting path. Update $D := D + 1$ and assign level(x) := 2D for all unlabeled nodes x thus leaving level $2D - 1$ empty. Expand all contracted nodes (inserting appropriate edges into M) and terminate claiming that the resulting M is optimal, see Lemma 7 below.

Edges (x, y) in $\overrightarrow{G}$ obeying level(y) = level(x) + 1 are called *geodesic*. Node levels establish the following structure of $\overrightarrow{G}$ described by the invariants below that will be preserved during the phase and that are easily proven to hold right after the invocation of BUILD-LEVELS by induction:

(5) (a) all sources are at level 0, moreover, if $D > 0$ then all nodes at level 0 are sources or endpoints of forks; all sinks that are not endpoints of forks are at level 2D;

    (b) nodes at levels $1, 3, \ldots, 2D - 1$ are exactly centers of forks; nodes at levels $2, 4, \ldots, 2D - 2$ are exactly endpoints of forks;

    (c) for each edge (x, y) in $\overrightarrow{G}$, level(y) $\leq$ level(x) + 1;

    (d) there is no edge (x, y) in $\overrightarrow{G}$ with level(x) = level(y) = $2i$ for some $i = 0, \ldots, D - 1$.

Remark 3. For $D > 0$, level 0 contains just sources after BUILD-LEVELS terminates. However, subsequent augmentations turn some of these sources into endpoints of forks. This explains the somewhat complicated structure of (5)(a).

Remark 4. Among all of $2D + 1$ levels the last one $2D$ is the most "chaotic": if $D = 0$ then it contains all nodes. If $D > 0$ then it may contain virtually anything except for sources: sinks (in particular, endpoints of forks), centers of forks, and also unsaturated nodes.

We call an augmenting path $P = (v_0, \ldots, v_{2k}, v_{2k+1})$ *proper* if $\text{level}(v_i) = i$ for $i = 0, \ldots, 2k$ and $\text{level}(v_{2k}) = \text{level}(v_{2k+1}) = 2D$ (i.e. all of its edges except for the last one are geodesic and follow "downward" and the last one follows "horizontally" within level $2D$).

Lemma 4. *Suppose that* BUILD-LEVELS *discovers an edge* $(u, v) \in E(\overrightarrow{G})$ *from the current node* u *with* $\text{level}(u) = 2D$ *to some sink* v. *Then* $\text{level}(v) = 2D$, *i.e.* BUILD-LEVELS *discovers a proper augmenting path.*

The following lemmas indicate that although levels are constructed w.r.t. $\overrightarrow{G}$, edges of the whole G obey certain properties:

Lemma 5. *There is no edge* $uv \in E(G)$ *such that* $\text{level}(u)$ *and* $\text{level}(v)$ *are both even and less than* $2D$.

Proof. First, let $\text{level}(u) = \text{level}(v) = 0$. Then both u and v are either sources or endpoints of forks; moreover, both u and v were sources at the beginning of the phase by (5)(a). Then uv could have been added to M increasing its cardinality, which is impossible.

Second, suppose $\text{level}(u) = 0$ while $\text{level}(v) \geq 2$ (or vise versa); then by (5)(a, b) u is a source or an endpoint of a fork and v is an endpoint of a fork and, in particular, is a sink; then by (4)(a) $(u, v) \in E(\overrightarrow{G})$ and $\text{level}(v) \leq \text{level}(u) + 1 = 1$ by (5)(a, c), which contradicts $\text{level}(v)$ being positive and even.

Third, suppose $\text{level}(u) \geq 2$ and $\text{level}(v) \geq 2$; then by (5)(a, b) both u and v are endpoints of some (possibly coinciding) forks, and, in particular, both are sinks; then by (4)(c) $(u, v), (v, u) \in E(\overrightarrow{G})$. Hence $|\text{level}(u) - \text{level}(v)| \leq 1$ by (5)(c). Since both levels are even, they must be the same; this contradicts (5)(d). □

Lemma 6. *Consider the beginning of a phase with* $D > 0$. *For each node* u *with* $\text{level}(u) = 2i$, $i = 0, \ldots, D - 1$ *and edge* $uv \in E(G)$, $\text{level}(v) = 2j + 1$ *for some* $j \leq i$.

Proof. First and foremost, v cannot be unsaturated since otherwise, as usual, one could have increased the cardinality of M by first alternating along some even source-to-u path consisting of geodesic edges and then adding uv. (Note that here we rely on the fact that the phase has just started and hence an alternating path ending at u exists.)

Node v cannot be a source by Lemma 5 (all sources are at level 0 by (5)(a)). Therefore, v is either a sink of a center of a fork.

If $i = 0$ then by (5)(a) u is a source or an endpoint of a fork. By (4)(a, c) edge uv is directed as (u, v) and thus $\text{level}(v) \leq 1$ by (5)(c), as required.

Now let $i > 0$. By (5)(b) node u is an endpoint of some fork F. If v is the center of F then $\text{level}(v) = 2i - 1$, and we are done. Otherwise by (4)(c) we have a directed edge (u, v).

Note that $\text{level}(v) < 2D$, otherwise the difference of levels between v and u would be at least 2 in contradiction to (5)(c). Also $\text{level}(v)$ cannot be even by Lemma 5. Now we know that $\text{level}(v) = 2j + 1$ for some $j = 0, \ldots, D - 1$. Finally (5)(c) implies $j \leq i$ due to existence of edge (u, v). $\square$

Lemma 7. *Suppose* BUILD-LEVELS *fails to discover an augmenting path. Then the current M corresponds to a maximum 1-restricted simple 2-matching.*

Proof. Let $V_i := \{v \mid \text{level}(v) = i\}$. Consider $W := V_1 \cup V_3 \cup \ldots V_{2D-1}$. Note that W contains no contracted nodes by (5)(b) and hence may be regarded a node set in G_0. Then by Theorem 2 one has $\nu_0(G_0) - \nu_1(G_0) \geq \text{iso}(G_0 - W) - 2|W|$.

We claim that $\text{iso}(G_0 - W) \geq |V_0| + |V_2| + \ldots + |V_{2D-2}|$. Indeed, after removing W in G nodes in $V_0 \cup V_2 \cup \ldots \cup V_{2D-2}$ become isolated by Lemma 6. Expanding these contracted nodes one gets this many isolated edges.

Also note that $|V_{2i+2}| = 2|V_{2i+1}|$ for all $i = 0, \ldots, D - 2$ (since each fork has one center in V_{2i+1} and two endpoints in V_{2i+2}).

Then plugging these estimates in, one gets $\nu_0(G_0) - \nu_1(G_0) \geq |V_0| + |V_2| + \ldots + |V_{2D-2}| - 2(|V_1| + |V_3| + \ldots + |V_{2D-1}|) = |V_0| - 2|V_{2D-1}|$, hence $|V_0| \leq \nu_0(G_0) - \nu_1(G_0) + 2|V_{2D-1}|$. Recall that in case BUILD-LEVELS discovers no augmenting path, it produces $V_{2D-1} = \emptyset$. Therefore $|V_0| \leq \nu_0(G_0) - \nu_1(G_0)$. Also V_0 is just the set of sources, so after expanding M and G one gets a 1-restricted simple 2-matching M_0 in G_0 with exactly $\nu_0(G_0) - \nu_1(G_0)$ isolated edges, as needed. $\square$

The algorithm extracts augmenting paths in $\vec{G}$ one by one, applies these paths to M, and finally expands and re-contracts the relevant portions of G touched by augmentations as described in Subsect. 2.2. These contractions and expansions are consistent with levels in the following sense. Whenever a contracted node x is expanded into x' and x'', then x', x'' receive the level of x. When we contract certain y' and y'' into y, it is guaranteed that $\text{level}(y') = \text{level}(y'')$ and this number will be regarded as $\text{level}(y)$. Also, the algorithm will be adjusting levels of some nodes by pushing them down to level $2D$.

To describe this in details, we perform a case splitting that is parallel to (3). Consider a proper augmenting path $(s = v_0, \ldots, v_{2k}, v_{2k+1} = t)$ and let $F_1, \ldots, F_k$ be the sequence of forks traversed by P (in the order from s to t).

(6) (a) t is an endpoint of some fork F. Note that it is possible that $F = F_k$ but clearly $F \neq F_1, \ldots, F_{k-1}$ (since endpoints of the latter forks are at levels $< 2D$). Two subcases are possible.

    (a') If $F = F_k$ then forks $F_1, \ldots, F_{k-1}$ are altered into $F'_1, \ldots, F'_{k-1}$ and F_k dissolves. We move nodes $v_{2k-2}, v_{2k-1}, v_{2k}$, and v_{2k+1} to level $2D$ and handle expansions and re-contractions there; see Fig. 6.

    (a") If $F \neq F_1, \ldots, F_k$ then forks $F_1, \ldots, F_k$ are altered into $F'_1, \ldots, F'_k$ and none of them dissolves; all nodes of F are moved to level $2D$ and are similarly expanded and re-contracted if needed; see Fig. 7.

(b) t belongs to a circuit component C of M (and is, thus, not contracted). Forks $F_1, \ldots, F_k$ are altered into $F'_1, \ldots, F'_k$ and none of them dissolves. All nodes of C are already at level $2D$, the needed expansions and re-contractions happen there; see Fig. 8.

(c) t is an inner node of a non-fork path component P of M (and is, thus, also not contracted). Like in (6)(b), forks $F_1, \ldots, F_k$ are altered into $F'_1, \ldots, F'_k$ and none of them dissolves. All nodes of P are already at level $2D$, the needed expansions and re-contractions happen there; see Fig. 9.

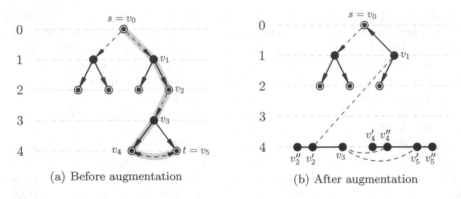

(a) Before augmentation   (b) After augmentation

Fig. 6. Case (6)(a'). We mostly focus on how edge directions and node levels are updated; the transformation is described in detail above.

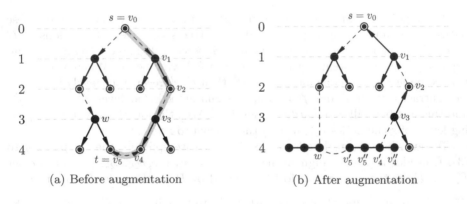

(a) Before augmentation   (b) After augmentation

Fig. 7. Case (6)(a")

Lemma 8. *The above augmentation preserves (5).*

Now let us discuss how paths P are extracted. We start with FIND-ALTERNATING(v) routine that, given a node v with level(v) $= 2D$, aims to find an alternating path of length $2D$ from some source to v. By Lemma 2 and (5)(c) such a path must solely consist of geodesic edges in $\overrightarrow{G}$.

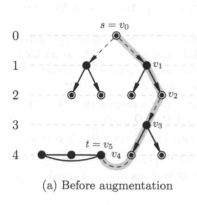

(a) Before augmentation

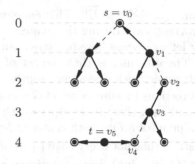

(b) After augmentation. Note that in case of a circuit of length 3 we get a path of length 4 after unconcration, which immediately turns into a fork and some edges get directions.

Fig. 8. Case (6)(b)

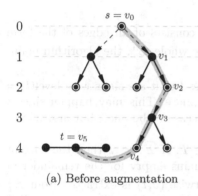

(a) Before augmentation

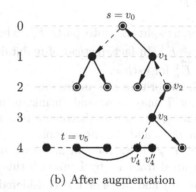

(b) After augmentation

Fig. 9. Case (6)(c)

FIND-ALTERNATING(v) applies backward graph traversal: if v is already a source then we are done. Otherwise, fetch an incoming geodesic edge (u, v). If one is found then reset $v := u$ thus climbing one level up. Otherwise (no incoming geodesic edge exists for v) backtrack.

The algorithm explicitly maintains the set of directed edges $\overrightarrow{E}^0$ in $\overrightarrow{G}$ that are geodesic and uses this set in FIND-ALTERNATING. Note that after augmentation edges belonging to path P found by the above procedure and all newly-added edges are not-geodesic: edges (u, v), where u is a source or is an endpoint of a fork and v is the center of another fork, become parts of altered forks and switch direction to (v, u); edges (u, v), where u is the center and v is an endpoint of some fork, are removed from M; since u is still a fork center these can only be directed as (v, u). Hence, we prune all edges of P from $\overrightarrow{E}^0$ and, moreover, if FIND-ALTERNATING ever backtracks from node u down to node v along geodesic edge (u, v) (after discovering that u is not reachable by alternating paths) we also prune (u, v) from $\overrightarrow{E}^0$. This enables amortizing the time spent by FIND-ALTERNATING calls during the phase.

Now let us discuss how to extract augmenting paths rather than just alternating. The difficulty is that the set of potential final edges of these paths is highly dynamic: such edges may appear and vanish as the phase progresses. We assign directions to some edges of G and introduce the set $\overrightarrow{E}^1$ of "final" directed edges (u, v) such that: (i) level$(u) = 2D$; (ii) u is either a source (for $D = 0$) or is an endpoint of a fork with center at level $2D - 1$ (for $D > 0$).

We stress that not every edge (u, v) of $\overrightarrow{E}^1$ belongs to $\overrightarrow{G}$ (but this is certainly the case if v is a sink). For convenience, we identify edges of $\overrightarrow{E}^1$ in G with their pre-images in G_0; this enables us to compare $\overrightarrow{E}^1$ at various moments during the phase.

Lemma 9. *Augmentations can only decrease $\overrightarrow{E}^1$. Moreover, if $e \in \overrightarrow{E}^1$ is the final edge of an augmenting path P then e vanishes from $\overrightarrow{E}^1$ upon augmenting along P.*

Let us split $\overrightarrow{E}^1$ into parts E_v^1, where $\overrightarrow{E}_v^1$ consists of all edges of the form $(u, v) \in \overrightarrow{E}^1$. In fact, instead of maintaining the whole $\overrightarrow{E}^1$, the algorithm maintains $\overrightarrow{E}_v^1$ for each v.

The algorithm also maintains set T of nodes t that are sinks with level$(t) = 2D$. Set T may grow and shrink during the phase. This may happen since t may change its status (become a new sink or cease to be one), but also due to contractions and uncontractions.

Let $T \neq \emptyset$. Extract an arbitrary node $t \in T$. If $\overrightarrow{E}_t^1 = \emptyset$ then the current t is pruned from T (by Lemma 9 this set remains empty for the remainder of the phase) and another t is considered. Otherwise (v, t) is extracted from $\overrightarrow{E}_t^1$ and FIND-ALTERNATING(v) is invoked. If the latter finds an alternating path P then appending (v, t) we have obtained an augmenting path P by Lemma 3. We update M along it. Note that (v, t) no longer belongs to $\overrightarrow{E}^1$ by Lemma 9.

Otherwise (FIND-ALTERNATING fails to find P) we prune (v, t) from $\overrightarrow{E}_t^1$ and try fetching another edge from there, and so on. The correctness of the algorithm pruning edge (v, t) from $\overrightarrow{E}^1$ is justified by

Lemma 10. *If FIND-ALTERNATING(v) fails then $(v, t) \in \overrightarrow{E}_v^1$ can be ignored for the rest of the phase.*

When the current phase completes, all sets $\overrightarrow{E}_t^1$ are fully pruned for all sinks t with level$(t) = 2D$. This implies

Lemma 11. *When a phase terminates, no proper augmenting path remains w.r.t. the final levels of this phase.*

Lemma 12. *Suppose the next phase runs BUILD-LEVELS and computes D'. Then $D' > D$.*

Proof. Consider the moment the former phase terminates. Suppose $D = 0$ and assume for the sake of contradiction that $D' = 0$. Note that augmenting paths in both of these phases are just source-to-sinks edges and the former phase must have eliminated all of these by Lemma 11.

Now let $D > 0$ and assume for the sake of contradiction that $D' \leq D$ and let $P = (v_0, \ldots, v_{2D'}, v_{2D'+1})$ be an augmenting path of length $2D' + 1$. We follow along P and prove inductively that $\text{level}(v_i) \equiv i \pmod 2$ and $\text{level}(v_i) \leq i$ for all $i = 0, \ldots, 2D'$ (recall that levels are regarded w.r.t. the former phase).

This is obviously true for v_0 since all sources are at level 0 by (5)(a).

Let $\text{level}(v_{2i})$ be even and not exceed $2i$ for some $i = 0, \ldots, D' - 1$. Node v_{2i} is a source or an endpoint of fork while node v_{2i+1} is the center of (another) fork F; hence $(v_{2i}, v_{2i+1}) \in E(\overrightarrow{G})$ by (4)(a, c). This implies $\text{level}(v_{2i+1}) \leq \text{level}(v_{2i}) + 1 \leq 2i + 1$ by (5)(c). Also $\text{level}(v_{2i+1})$ cannot be even since no level $0, 2, \ldots, 2D - 2$ may contain fork centers by (5)(a, b). Hence the induction follows for v_{2i+1}.

Moving from v_{2i+1} to v_{2i+2}, note that v_{2i+2} is some endpoint of fork F (defined above). By (4)(b) $(v_{2i+1}, v_{2i+2}) \in E(\overrightarrow{G})$ and by (5)(c) $\text{level}(v_{2i+2}) \leq \text{level}(v_{2i+1}) + 1 \leq 2i + 2$. It remains to prove that $\text{level}(v_{2i+2})$ is even; this follows from (5)(b) since no odd level may contain endpoints of forks.

Recall that we are assuming (towards contradiction) that $D' \leq D$. If $D' < D$ then the above induction implies that $\text{level}(v_{2D'}) \leq 2D - 2$. By (4)(a, c) $(v_{2D'}, v_{2D'+1}) \in E(\overrightarrow{G})$, hence by (5)(c) $\text{level}(v_{2D'+1}) \leq 2D - 1$. Since $v_{2D'+1}$ is a sink, it must be a fork endpoint by (5)(a, b) and thus $\text{level}(v_{2D'})$ and $\text{level}(v_{2D'+1})$ must both be even and be less than $2D$. Property (5)(d) forbids $\text{level}(v_{2D'+1}) = \text{level}(v_{2D'})$, therefore $\text{level}(v_{2D'+1}) \leq \text{level}(v_{2D'}) - 2$. Now both $v_{2D'}$ and $v_{2D'+1}$ are endpoints of forks (possibly coinciding), hence by (4)(c) one must have $(v_{2D'+1}, v_{2D'}) \in E(\overrightarrow{G})$, contradicting (5)(c) and the above level inequality.

Now suppose $D' = D$. If $\text{level}(v_{2D}) \leq 2D - 2$ then we get a contradiction the same way as above. Hence $\text{level}(v_{2D}) = 2D$ and, moreover, the alternating prefix of P must solely consist of geodesic edges. It remains to prove that $\text{level}(v_{2D+1}) = 2D$ as this would indicate that P is a proper augmenting path, which must have been found by the former phase by Lemma 11.

Suppose $\text{level}(v_{2D+1}) < 2D$. Since levels $1, 3, \ldots, 2D - 1$ only contain fork centers by (5)(b), $\text{level}(v_{2D+1}) = 2i$ for some $i = 0, \ldots, D - 1$; in particular, $\text{level}(v_{2D+1}) \leq 2D - 2 = \text{level}(v_{2D}) - 2$. Now v_{2D+1} and v_{2D} are both endpoints of some (possibly coinciding) forks, hence by (4)(a, c) $(v_{2D+1}, v_{2D}) \in E(\overrightarrow{G})$, contradicting (5)(c) and the above level inequality. □

2.5 Bounding the Number of Phases

Let us establish an $O(\sqrt{n})$ bound for the number of phases. Like in other matching problems, it amounts to showing that once D becomes large enough, the current solution is close to the optimum.

Consider the moment a phase started, computed $D > 0$ by running BFS, and established the levels of nodes. Also recall that G is obtained from the initial graph G_0 by contraction of certain edges.

Lemma 13. *At the beginning of a phase with $D > 0$ the number of sources in M is at most $\nu_0(G_0) - \nu_1(G_0) + 2n/D$.*

Proof. This is, in fact, a refined version of Lemma 7. Let $V_i :=$ $\{v \mid \text{level}(v) = i\}$. Fix some $k = 0, \ldots, D - 1$ and consider $W := V_1 \cup V_3 \cup \ldots V_{2k+1}$. Note that W contains no contracted nodes by (5)(b) and hence may be regarded a node set in G_0. Then by Theorem 2 one has $\nu_0(G_0) - \nu_1(G_0) \geq$ $\text{iso}(G_0 - W) - 2|W|$.

We claim that $\text{iso}(G_0 - W) \geq |V_0| + |V_2| + \ldots + |V_{2k}|$. Indeed, after removing W in G nodes in $V_0 \cup V_2 \cup \ldots \cup V_{2k}$ become isolated by Lemma 6. Expanding these contracted nodes one gets this many isolated edges.

Also note that $|V_{2i+2}| = 2|V_{2i+1}|$ for all $i = 0, \ldots, D - 1$ (since each fork has one center in V_{2i+1} and two endpoints in V_{2i+2}).

Then plugging these estimates in, one gets $\nu_0(G_0) - \nu_1(G_0) \geq |V_0| + |V_2| + \ldots + |V_{2k}| - 2(|V_1| + |V_3| + \ldots + |V_{2k+1}|) = |V_0| - 2|V_{2k+1}|$, hence $|V_0| \leq \nu_0(G_0) - \nu_1(G_0) + 2|V_{2k+1}|$. Since $|V_1| + \ldots + |V_{2D-1}| \leq n$, if one chooses k among $0, \ldots, D - 1$ so as to minimize $|V_{2k+1}|$ then $|V_{2k+1}| \leq n/D$. Also V_0 is just the set of sources, so the needed bound follows. $\square$

Theorem 3. *Converting a maximum unrestricted simple 2-matching into a maximum 1-restricted simple 2-matching can be done in $O(m\sqrt{n})$ time.*

Proof. Run the initial $\lceil \sqrt{n} \rceil$ phases and start a new one (assuming the algorithm did not finish earlier). Then $D \geq \lceil \sqrt{n} \rceil$ by Lemma 12. By Lemma 13 and Theorem 2 the current solution contains at most $O(n/D) = O(\sqrt{n})$ more sources than the optimum one. Therefore additional $O(\sqrt{n})$ phases will suffice (each phase eliminates at least one source). Assuming appropriate implementation, each phase takes $O(m)$ time. Hence the proof follows. $\square$

References

1. Goldberg, A.V., Karzanov, A.V.: Maximum skew-symmetric flows and matchings. Math. Program. **100**(3), 537–568 (2004)
2. Hartvigsen, D.: Maximum cardinality 1-restricted simple 2-matchings. Electron. J. Comb. **14**(1), R73–R73 (2007)
3. Hopcroft, J.E., Karp, R.M.: An $n^{5/2}$ algorithm for maximum matchings in bipartite graphs. SIAM J. Comput. **2**(4), 225–231 (1973)
4. Lovász, L., Plummer, M.D.: Matching Theory. North-Holland, NY (1986)

Lower Bounds on the Performance of Online Algorithms for Relaxed Packing Problems

János Balogh[1], György Dósa[2], Leah Epstein[3(✉)], and Łukasz Jeż[4]

[1] Institute of Informatics, University of Szeged, Szeged, Hungary
baloghj@inf.u-szeged.hu
[2] Department of Mathematics, University of Pannonia, Veszprém, Hungary
dosagy@almos.vein.hu
[3] Department of Mathematics, University of Haifa, Haifa, Israel
lea@math.haifa.ac.il
[4] Institute of Computer Science, University of Wrocław, Wrocław, Poland
lje@cs.uni.wroc.pl

Abstract. We prove new lower bounds for suitable competitive ratio measures of two relaxed online packing problems: online removable multiple knapsack, and a recently introduced online minimum peak appointment scheduling problem. The high level objective in both problems is to pack arriving items of sizes at most 1 into bins of capacity 1 as efficiently as possible, but the exact formalizations differ. In the appointment scheduling problem, every item has to be assigned to a position, which can be seen as a time interval during a workday of length 1. That is, items are not assigned to bins, but only once all the items are processed, the optimal number of bins subject to chosen positions is determined, and this is the cost of the online algorithm. On the other hand, in the removable knapsack problem there is a fixed number of bins, and the goal of packing items, which consists in choosing a particular bin for every packed item (and nothing else), is to pack as valuable a subset as possible. In this last problem it is possible to reject items, that is, deliberately not pack them, as well as to remove packed items at any later point in time, which adds flexibility to the problem.

Keywords: Bin packing · Online algorithms · Competitive ratio

1 Introduction

We study two online problems, for which the offline version is a classic problem, with well-known efficient near-optimal solutions [12,15,18]. Our online problems are not the most natural variants that one can define, but they are more relaxed. This models reality in the sense that often there is some flexibility even in online

Ł. Jeż was supported by the Polish National Science Center grant 2020/39/B/ST6/01679.

C. Bazgan and H. Fernau (Eds.): IWOCA 2022, LNCS 13270, pp. 101–113, 2022.
https://doi.org/10.1007/978-3-031-06678-8_8

environments. Flexibility obviously allows the design of better online algorithms, though such algorithms typically cannot find optimal solutions. Here, we focus on the limitations of online algorithms for relaxed models.

We use the competitive ratio and asymptotic competitive ratio measures for analysis of online algorithms. The competitive ratio for minimization problems is the worst-case ratio between the cost of an online algorithm and the cost of an optimal offline algorithm for the same input. For maximization problems, the roles of the algorithm and an optimal offline solution are reversed. The asymptotic competitive ratio is the supreme limit of the competitive ratio for inputs with optimal costs or profits growing to infinity.

In offline bin packing [12, 17, 18], there are items of indices $1, 2, \ldots, n$, where item j has a rational size $s_j \in (0, 1]$. The goal is to partition the items into the minimum number of sets called bins, where the total size for every bin does not exceed 1. One can see this as a scheduling problem where items are assigned to machines that are available during the time interval $[0, 1)$, but it is not necessary to assign the specific time slots in advance, since this can always be done. The length of the time slot for item j is required to be of length s_j, where the interval has the form $[x, x + s_j)$ for $x \geq 0$ and $x \leq 1 - s_j$. Alternatively, one can assign the time slots, and not the bins, where the assignment to bins can be done by a simple process of coloring an interval graph. In the standard online bin packing problem [2, 5, 6] items are to be assigned to bins sequentially, and it is assumed that items just receive consecutive time slots in the bin, starting from time zero. An alternative online model was defined recently by Escribe, Hu, and Levi [11], where the online feature is the assignment to time slots rather than the bins. The problem is called minimum peak appointment scheduling (MPAS). In both models, items are presented one by one, such that each item is assigned irrevocably before the next item arrives.

In the work by Escribe, Hu, and Levi [11], a randomized algorithm with asymptotic competitive ratio at most 1.5 was designed, which was recently improved to $\frac{16}{11} \approx 1.455$ by Smedira and Shmoys [20]. A lower bound of 1.5 on the competitive ratio of deterministic algorithms was proved [11], while Smedira and Shmoys [20] proved a lower bound of 1.2 for the asymptotic competitive ratio of all randomized (and deterministic) algorithms. These results contrast with those known for the standard online bin packing. While the best known lower bound of 1.54278 [5] holds only for deterministic algorithms, earlier results hold for randomized algorithms, where the best such result is 1.5403 [6]. The current best upper bound for standard online bin packing [2] is 1.57829. A simple way to see the difference between the problems is the following. Consider a large even number of items of size 0.4, possibly followed by the same number of items of size 0.6. A bin packing algorithm has to decide how many pairs of items of size 0.4 to create in order to have a good performance in both cases. However, in the case of MPAS, one can assign half of the items of size 0.4 to the time interval $[0, 0.4)$ and the other half to $[0.6, 1)$. This is optimal if there are no further items, but if there are such items, one can assigned half of them to $[0, 0.6)$ and the other half to $[0.4, 1)$, also obtaining an optimal solution. The problem is not meaningful as a separate offline problem, though generalizations were recently studied as offline problems [10,14,19]. This variant is one of those studied here.

A related problem is the so called *dual bin packing* [1,8], which may also be seen as a variant of the *knapsack problem* with multiple knapsacks available [9,16]: The processing of arriving items is similar, except here the number of available bins (or knapsacks) is fixed in advance, and the goal is rather to maximize the profit associated with those items that are successfully packed. In the multiple knapsack as studied originally, the profit associated with a set of packed items was the *maximum* over all the bins of the total value of all items packed in the bin [16], but later studies [7,9] extended this to the *sum* of values of all the items packed, which is in line with dual bin packing. We are interested in this objective, so we will not specify the results concerned solely with the single bin of maximum value. The packing consists in assigning an item to a particular bin, which is in contrast to MPAS, where instead a "position" or "interval" (on a horizontal axis) within a bin (which is yet to be determined) is specified. An item may also be rejected by an algorithm, i.e., not packed at all. Moreover, in the so called removable online variant of the knapsack problem it is allowed to remove a previously packed item (e.g., to accommodate the one arriving), which from then on counts as rejected. As is the case in vast amount of literature, we will consider two restricted settings, in which there is a particular natural relation between the profit associated with an item and its size (or processing time); note that as we focus on lower bounds, considering these makes our results stronger. The two cases, which we call as in [9], are *proportional*, in which the value of an item equals its size, and *unit*, in which every item is worth 1 regardless of its size.

The dual bin packing problem corresponds to the unit case with no removals, for which no algorithm can attain constant competitive ratio [8]. Thus the studies of this problem focused on "accommodating" instances, in which all items can be packed by the offline solution, for which constant-competitive algorithms were designed. Moreover, it is known that whether an algorithm is allowed to reject an item that it could pack in some bin (thus being "unfair") affects what ratio can be attained [1]. In the later studies of the multiple knapsack problem, it was noted that proportional instances, even non-accommodating, allow constant-competitive ratio, which was eventually determined to be exactly $1 + \ln 2 \approx 1.6903$ [7,9]. Moreover, with removals allowed, a deterministic algorithm of asymptotic competitive ratio at most 3 is known even for general instances, and the proportional and unit instances admit algorithms with much better competitive ratios of 1.6 and 1.5 respectively [9]. The corresponding lower bounds for these two settings, applicable even to randomized algorithms are only $\frac{8}{7} \approx 1.14$ [9] and $\frac{7}{6} \approx 1.17$ [1] respectively. No better lower bounds are known for general instances. For special cases with a small number of bins and the proportional case, lower bounds of $\frac{4}{3}$ and $\frac{6}{5}$ are known for the competitive ratio of deterministic algorithms with two and three bins, respectively [9], and in some special cases the lower bound on the competitive ratio as a function of the number of bins is slightly inferior to the bounds of $\frac{8}{7}$ and $\frac{7}{6}$ [1,9]. This problem is also not of interest as a separate offline problem, though the knapsack problem and its variants are being studied continuously, and a near-optimal solution is known for almost fifty years [15]. Removable knapsack is the second variant studied here.

1.1   Our Results

In this work, we prove the following lower bounds on the performance of online algorithms:

- A lower bound of 1.2287 on the competitive ratio for deterministic algorithms for either the proportional or the unit case of the removable knapsack problem, improving upon the previous bounds of $\frac{8}{7} \approx 1.14$ [9] and $\frac{7}{6} \approx 1.17$ [1] respectively,
- A lower bound of 1.2 on the competitive ratio for randomized algorithms for the proportional case of the removable knapsack problem, also improving upon the previous bound of $\frac{8}{7} \approx 1.14$ [9],
- A lower bound of 1.2691 on the asymptotic competitive ratio for randomized algorithms for the minimum peak appointment scheduling problem (MPAS), improving upon the previous bound of 1.2 [20].

We also consider some special cases for the online removable knapsack problem. For example, our results improve the known lower bound for three bins and deterministic algorithms.

1.2   Adaptive Item Sizing in Designing Hard Instances for Online Algorithms

Those of our lower bounds that are designed specifically for deterministic algorithms employ the "adaptive item sizing" technique [5], which we now describe. This approach and more advanced approaches (in particular, one where there is a multiplicative gap between sizes) were used for other online bin packing problems [3,4,13].

This is a procedure in which a sequence S of items arrives, all with sizes in a predetermined interval $[\alpha, \beta]$, where α and β are parameters. This allows the items from S to be partitioned into a set of *smallish* items with sizes in the interval $[\alpha, \theta)$ and *largish* items with sizes in the interval $(\theta, \beta]$. The threshold θ satisfies $\alpha < \theta < \beta$, and the classification of each item as either smallish or largish occurs immediately after its packing by the deterministic algorithm, but the value of θ is determined later.

Such classification and partitioning can be ensured by a procedure resembling the binary search. Let a and b be variables such that $\alpha \leq a < b \leq \beta$. All items thus far classified as smallish have sizes in $[\alpha, a]$, and all items thus far classified as largish have sizes in $[b, \beta]$. Initially, we let $a = \alpha$ and $b = \beta$. Then, the next item to arrive can have any size in (a, b), e.g., $\frac{a+b}{2}$. Once it is dealt with (packed or rejected) by the algorithm, and hence classified as either smallish or largish, the value of a or b respectively is set to this item's size. Once all items in S are processed, θ can take any value that could be the size of a next item in the sequence and thus separates the sizes of smallish and largish items, e.g., $\theta = \frac{a+b}{2}$ for the final values of the variables a and b.

2   Online Removable Knapsack

In our lower bound proofs, we assume without loss of generality that the online algorithm is lazy in that it defers removing (or rejecting) items as long as the packing is valid. Namely, we assume that upon the arrival of an item e, the algorithm decides to either reject it outright without packing it or chooses any bin B to which e is packed. If the bin B then overflows, the algorithm chooses a subset of items from B, excluding e, for removal; this subset has to be minimal such that its removal makes the total size of the items remaining in bin B at most 1. Additionally, the algorithm is only allowed to reject the item e outright if there is no bin where it fits without removals; note that this does not mean no removals take place when such bin exists, as the algorithm can choose to pack e in another bin.

2.1   Deterministic Online Algorithms for the Proportional and Unit Cases

We start with a deterministic lower bound. The next theorem is valid for both variants (the proportional case and the unit case).

Theorem 1. *Any deterministic online algorithm that works for any number of bins $k \geq 2$, has competitive ratio of at least 1.228713.*

Proof. We focus on the proportional case, and remark that the proof also applies to the unit case because all items in the strategy have sizes that deviate from $\frac{1}{2}$ only by negligible amounts.

The input consists of two phases. The first phase employs the adaptive sizing framework for k items, with $\frac{1}{2} - \varepsilon < \alpha < \beta < \frac{1}{2}$ for arbitrarily small $\varepsilon > 0$. As there are k bins, the algorithm is lazy, and item sizes are slightly below $\frac{1}{2}$, every item will be packed upon arrival (though possibly causing removal of another item), every non-empty bin at all times will contain either one or two items, and the number of items in any bin cannot decrease over time. The items are classified as follows upon packing: The first item to be placed in a bin is largish, the second one to be placed in a bin is smallish, and an item that replaces another inherits the replaced item's class.

As a result, at the end of the first phase, each bin has at most two items, if is it non-empty, then it contains one largish item, and if it contains another one, then that item is smallish. Let Γ denote the number of bins with two items, where $\Gamma \leq \lfloor \frac{k}{2} \rfloor$, since k items were presented. If $\Gamma = 0$, there are no smallish items, but the threshold θ is still well defined.

In the second phase, there are two possible continuations of the input, and the adversary's choice depends on Γ. The first one, applied when Γ is suitably large, is to issue k items of size $1 - \beta > \frac{1}{2}$ each. Since all previous items have sizes at most β, the optimal offline solution packs these items in pairs, and its profit is at least $k \cdot (\frac{1}{2} - \varepsilon) + k \cdot (1 - \beta) > k \cdot (1 - \varepsilon)$. Now consider the algorithm's packing. As each item in the second phase has size strictly larger than $\frac{1}{2}$, any bin that was empty at the beginning of this phase, may contain at most one item at its end. The number of bins for the algorithm with exactly one item before the arrival of the new items is at most $k - 2 \cdot \Gamma$, and therefore the profit of the algorithm is below

$$(1 - \beta) \cdot (k + \Gamma + (k - 2 \cdot \Gamma)) < (\frac{1}{2} + \varepsilon) \cdot (2k - \Gamma) .$$

For $\varepsilon \to 0$, the competitive ratio in this case tends to $\frac{2k}{2k-\Gamma}$.

The other strategy for the second phase, applied when Γ is suitably small, is to issue $\Gamma + \lfloor \frac{k-\Gamma}{2} \rfloor$ items of size $1 - \theta$. Each such item has size slightly above $\frac{1}{2}$, and can be packed together with a smallish item but not with a largish item. Clearly, no bin can have more than one such item. Note that the number of smallish items in the input is at least Γ, and thus the number of largish items is at most $k - \Gamma$, because exactly k items were issued in the first phase. It is possible that the number of smallish items is larger than Γ if the algorithm removed a smallish item to pack another, which is then smallish as well. Offline, it is possible to pack Γ bins by placing one smallish item together with one of size $1 - \theta$ in each bin, $\lfloor \frac{k-\Gamma}{2} \rfloor$ bins with the remaining items of size $1 - \theta$, one per bin, and packing all the now remaining items, i.e., largish and yet unpacked smallish ones in pairs into $\lceil \frac{k-\Gamma}{2} \rceil$ bins; note that if $k - \Gamma$ is odd, one of those last bins will contain only a single item. Again assuming that $\varepsilon \to 0$, in the limit the total size of items packed this way is

$$\Gamma + \frac{1}{2} \left\lfloor \frac{k - \Gamma}{2} \right\rfloor + \left\lfloor \frac{k - \Gamma}{2} \right\rfloor + \frac{1}{2}(k - \Gamma \mod 2) .$$

For the online algorithm, an item of size $1 - \theta$ cannot be added to a bin with only a single item, since such item is largish by the adaptive item sizing used by the adversary. So the algorithm has only Γ bins with total size approximately 1, and its profit is approximately $\frac{k+\Gamma}{2}$ (up to negligible terms). Routine inspection of the cases of odd and even $k - \Gamma$ yields that the competitive ratio is at least

$$\frac{3k + \Gamma - (k - \Gamma \mod 2)}{2k + 2\Gamma} .$$

If $\Gamma \geq k \cdot \frac{\sqrt{33}-5}{2}$, we have $\frac{2k}{2k-\Gamma} \geq 0.75 + \sqrt{33}/12 \approx 1.228713$, and otherwise, $\frac{3k+\Gamma}{2k+2\Gamma} \geq 0.75 + \sqrt{33}/12 \approx 1.228713$. Since for large k, the fraction $\frac{1}{2k+2\Gamma} \leq \frac{1}{2k}$ tends to zero, the lower bound follows.

In the cases $k = 2, 3, 4, 5, 6, 7, 8, 9, 10$ we get:

$$\frac{4}{3}, \frac{5}{4}, \frac{6}{5}, \frac{5}{4}, \frac{5}{4}, \frac{11}{9}, \frac{16}{13}, \frac{5}{4}, \text{ and } \frac{16}{13} ,$$

by testing all values of $0 \leq \Gamma \leq \lfloor \frac{k}{2} \rfloor$, the bounds of the first case and the suitable bound for the second case.

Again, we note that as all items have sizes almost equal to $\frac{1}{2}$, the proofs works also for unit profits (where these profits are approximately twice as large as the proportional profits). $\qquad \Box$

2.2 Randomized Online Algorithms for the Proportional Case

In this section we provide an alternative easier construction, which in general provides weaker bounds, though with the exception of two values of k, for which it improves the result of Theorem 1 (which was for deterministic online algorithms). Unlike that theorem, this one applies to randomized algorithms as well. Another difference is that here the input sequence is not "accommodating", i.e., the optimal offline solution does not always pack all the items. The advantages of this construction are that it is fairly simple, and that it provides a lower bound for the case of randomized online algorithms.

The next theorem is valid for the proportional case.

Theorem 2. *Any randomized online algorithm for $k \geq 2$ bins has competitive ratio of at least* 1.2.

Proof. Let $\varepsilon > 0$ be a very small constant. The input starts with $2k$ items of each of the two sizes: $\frac{2}{3} - \varepsilon$, $\frac{1}{3} + 3\varepsilon$. Since items are removable, and since there are sufficiently many items, we can assume that every bin will either have one item of size $\frac{2}{3} - \varepsilon$ or two items of sizes $\frac{1}{3} + 3\varepsilon$.

Let X be the expected number of bins with one item of size $\frac{2}{3} - \varepsilon$. The input continues in one of two ways. The first one is k items of sizes $\frac{1}{3} + \varepsilon$, and the second one is k items of sizes $\frac{2}{3} - 3\varepsilon$. The optimal offline solution has k full bins in either case. Consider the online algorithm. In the first case, its best approach is to add one item to each bin with an item of size $\frac{2}{3} - \varepsilon$. These bins are full, so there is no better packing for them. For each bin of the other kind, even if some replacements are made, the total size of the items it holds is at most $\frac{2}{3} + 6\varepsilon$. By linearity of expectation, the expected profit of the algorithm is $k - \frac{k-X}{3}$ (letting ε tend to zero and neglecting those terms).

In the other case, there is no reason for the algorithm to replace items of size approximately $\frac{2}{3}$. (Besides, such replacements change only the negligible ε terms.) Every bin without such an item can become full when the algorithm replaces one item of size $\frac{1}{3} + 3\varepsilon$ with the item of size $\frac{2}{3} - 3\varepsilon$. The expected profit of the algorithm is $k - \frac{X}{3}$.

The threshold for X is $\frac{k}{2}$. The first input is used if $X \leq \frac{k}{2}$, and the second one is used otherwise, if $X > \frac{k}{2}$. The competitive ratio is at least 1.2.

For odd values of k, in the deterministic case we can use the integrality of X. The first input is used if $X \leq \frac{k-1}{2}$, and the second one is used otherwise, i.e., for $X \geq \frac{k+1}{2}$. The ratio is at least

$$\frac{k}{k - \frac{k+1}{6}} = \frac{6k}{5k - 1} > 1.2 \ .$$

For $k = 3, 5, 7, 9$ we get lower bounds of $\frac{9}{7}$, 1.25, $\frac{21}{17}$, and $\frac{27}{22}$, which improves slightly upon the lower bound from Theorem 1 for $k = 3$ and $k = 7$.   □

3   The Online Minimum Peak Appointment Scheduling Problem

In this section we study the problem MPAS. On high level, the aim in this problem is the same as in the multiple knapsack problem, i.e., to pack the items efficiently, which possibly explains why our results are similar in spirit and techniques. However, the setup is rather different, and it is a minimization problem rather than maximization. Here, every item has to be packed, and the goal is to minimize the number of the bins. The algorithm is not required to specify the bin for packing, but it has to specify the item's position in any bin it will eventually be packed in, i.e., for an arriving item of size γ (where $0 < \gamma \leq 1$), the algorithm has to specify an interval of the form $[x, x + \gamma)$ such that $0 \leq x \leq 1 - \gamma$.

3.1   Warm-Up: Deterministic Online Algorithms for MPAS

We start with a simple construction for deterministic algorithms that uses the adaptive item sizing technique. Afterwards, we improve it into a lower bound for randomized algorithms.

Theorem 3. *The asymptotic competitive ratio for deterministic online algorithms for MPAS is at least 1.25.*

Proof. The input consists of one or two phases, depending on the algorithm. In the first phase, for sufficiently large $N > 0$, $12N$ items arrive with sizes chosen adaptively using $\alpha = \frac{1}{3} - 2\varepsilon$ and $\beta = \frac{1}{3} - \varepsilon$ for an arbitrarily small value $\varepsilon > 0$. We define the partition into smallish and largish items now. Those items whose intervals after packing contain the point $\frac{1}{2}$ are classified as smallish, and the remaining items (whose intervals do not contain the point $\frac{1}{2}$) are classified as largish. Recall that the adaptive sizing guarantees that there exists a threshold $\theta \in (\alpha, \beta)$ such that smallish items have sizes in $[\alpha, \theta)$ and largish items have sizes in $(\theta, \beta]$. Let Q denote the number of smallish items.

We further classify the largish items as either "low" or "high", depending on whether their intervals lie completely to the left or to the right of the point $\frac{1}{2}$, if the positions are defined on the horizontal axis. Note that all high items contain the point $\frac{3}{4}$ in their intervals and similarly all low items contain the point $\frac{1}{4}$. If $Q \geq 5N$ or $Q \leq 2N$, then a solution that distributes the items evenly, i.e., places $4N$ items in each of the intervals

$$\left[0, \frac{1}{3}\right), \qquad \left[\frac{1}{3}, \frac{2}{3}\right), \qquad \left[\frac{2}{3}, 1\right),$$

and thus has cost of $4N$, proves that the algorithm's competitive ratio is at least $\frac{5}{4}$: For $Q \geq 5N$, there are $Q \geq 5N$ smallish items, all containing the point $\frac{1}{2}$, whereas for $Q \leq 2N$, there are must be at least $5N$ low or at least $5N$ high items (by the pigeonhole principle), which then all contain the point $\frac{1}{4}$ or $\frac{3}{4}$ respectively. Otherwise, when $Q \in (2N, 5N)$, there is a second phase, which depends on Q's relation to $3N$.

If $Q \geq 3N$, $12N$ items of size $\frac{2}{3}$ each are issued. An optimal offline solution places the first phase items in the interval $[0, \frac{1}{3})$, and it places the second phase $12N$ items in the interval $[\frac{1}{3}, 1)$, yielding optimal cost of $12N$. For the algorithm, all $12N$ second phase items must have the point $\frac{1}{2}$ as an internal point of their intervals, as do the Q smallish items, so the cost of the algorithm is at least $15N$. Thus the asymptotic competitive ratio is at least 1.25 in this case.

Finally, if $Q \leq 3N$, Q' items of size $1 - \theta$ are issued, where Q' is divisible by 3 and $Q - 2 \leq Q' \leq Q$. Note that all low items have the point θ as an internal point of their intervals, and similarly, all high items have the point $1 - \theta$ as an internal point. We find that the second phase items have a common point with every interval of the low and high items from the first phase. By the pigeonhole principle, there are at least $\frac{12N-Q}{2}$ low items or at least this many high items. Thus, for at least one of the points θ and $1-\theta$, there are at least $\frac{12N-Q}{2} + Q'$ items whose intervals contain it, for a cost of at least $\frac{12N+Q'}{2} - 1$ for the algorithm. An optimal offline solution has Q' intervals of $[0, \theta)$ for smallish items, Q' intervals of $[\theta, 1)$ for items of size $1 - \theta$, and for the remaining $12N - Q'$ items (where this number is divisible by 3) there are $4N - \frac{Q'}{3}$ intervals of every form out of $[0, \frac{1}{3})$, $[\frac{1}{3}, \frac{2}{3})$, $[\frac{2}{3}, 1)$. Note that all smallish items have sizes not exceeding θ, and the intervals assigned to such items by an optimal offline solution are sometimes slightly too long (because they have lengths of θ).

The cost of an optimal offline solution is therefore at most $\frac{12N+2Q'}{3}$. For large value of N, we can neglect the additive term and find a lower bound on the ratio

$$\frac{(12N + Q')/2}{(12N + 2Q')/3}$$

for $Q' \leq Q < 3N$. This ratio is indeed at least $\frac{5}{4}$ for $Q' < 3N$ since

$$\frac{3(12 + \frac{Q'}{N})}{2(12 + 2\frac{Q'}{N})}$$

is a monotonically decreasing function of $\frac{Q'}{N}$, and for $Q' \leq 3N$ it is minimized for $\frac{Q'}{N} = 3$, in which case

$$\frac{(12N + Q')/2}{(12N + 2Q')/3} = 1.25 \ .$$

As in each of the four cases analyzed in the proof, the competitive ratio is at least 1.25, possibly in the limit as N grows to infinity, 1.25 is a lower bound on asymptotic competitive ratio. $\qquad\square$

3.2   A Lower Bound for Randomized Algorithms for MPAS

Now, instead of using an adaptive classification of items, we show how to use very small items instead. This construction, which yields a superior bound, also applies to randomized algorithms.

Theorem 4. *The deterministic and randomized asymptotic competitive ratios for the peak problem are at least* 1.2691534.

Proof. Let $N, M > 0$ be large integers, such that N is even, and M is divisible by $N!$.

To prove the lower bound for randomized algorithms, we use Yao's approach, where one considers the best deterministic online algorithm for a known probability distribution over inputs. We denote the algorithm by ALG, and its cost for a specific input I by ALG(I). An optimal offline algorithm is defined by OPT, and we denote its cost for a specific input I by OPT(I).

Each input starts with a fixed prefix of $N \cdot M$ items of size $\frac{1}{N}$ each. For every point z such that $0 \leq z < 1$, define $f(z)$ to be the number of items whose assigned intervals contain the point z as an interior point or the left endpoint of the interval. Since the length of every interval is $\frac{1}{N}$, its contribution to the definite integral $\int_0^1 f(z)\, dz$ is $\frac{1}{N}$, and therefore

$$\int_0^1 f(z)\, dz = \frac{MN}{N} = M \ .$$

The integration is possible since the number of discontinuity points of f is at most $2MN$, i.e., a constant for every fixed pair (N, M). In fact, f is constant between every two consecutive discontinuity points, including the boundary points 0 and 1 among those. This implies that we can find the total length of intervals where f has the integer value i for $0 \leq i \leq MN$, which we denote by β_i. Clearly, $\sum_{i=0}^{MN} \beta_i = 1$. Imagine sorting the intervals with fixed values of f, so that, going from right to left along the $[0, 1)$ interval, we have, in this order, a sequence of intervals, the $i+1$-th of which, where $0 \leq i \leq MN$, has length β_i and associated value i. This is captured by a non-increasing step function g defined as follows: For every point z where $0 \leq z < 1$, let $g(z)$ the unique i such that

$$\sum_{j=i+1}^{MN} \beta_j \leq z \quad \text{and} \quad \sum_{j=i}^{MN} \beta_j > z \ ;$$

the function g is well-defined, since naturally $\sum_{j=MN+1}^{MN} \beta_j = 0$. Moreover, it holds that

$$\int_0^1 g(z)\, dz = \sum_{i=0}^{MN} i\beta_i = \int_0^1 f(z)\, dz = M \ . \tag{1}$$

We further note that it follows from the definition of g that any (left-closed) interval of length at least ℓ contains a point z such that $g(z) \geq f(1 - \ell)$, i.e., a point z which is contained in at least $f(1-\ell)$ intervals assigned by the algorithm to the items from the input prefix.

Next, the input may continue in one of many ways. Specifically, consider an eventually fixed t such that $1 \leq t \leq \frac{N}{2} - 1$. Then, for every $q = t, t+1, \ldots, \frac{N}{2} - 1$, there is an input I_q which continues after the prefix with $\frac{MN}{q}$ items of length $1 - \frac{q}{N}$. Since $q \leq \frac{N}{2} - 1$, we have

$$1 - \frac{q}{N} \geq 1 - \frac{\frac{N}{2} - 1}{N} = \frac{1}{2} + \frac{1}{N} > \frac{1}{2} \ .$$

An optimal offline solution assigns all these items the interval $[\frac{q}{N}, 1]$, and it partitions the MN items of size $\frac{1}{N}$ from the prefix into q subsets of $\frac{MN}{q}$, to be assigned the intervals $[\frac{j-1}{N}, \frac{j}{N})$ for $j = 1, 2, \ldots, q$, i.e., all items from a j-th subset are assigned the j-th interval. Clearly, the cost of such solution is $\frac{MN}{q}$, i.e.,

$$\text{OPT}(I_q) = \frac{MN}{q} \ . \tag{2}$$

As for the algorithm, no matter what intervals it assigned to the items of size $1 - \frac{q}{N}$, they must all contain the interval $J_q = [\frac{q}{N}, 1 - \frac{q}{N})$, whose length is $1 - \frac{2q}{N}$. Thus, by aforementioned properties of the function g, there is a point $z \in J_q$ such that $f(z) \geq g\left(\frac{2q}{N}\right)$, which implies that

$$\text{ALG}(I_q) \geq g\left(\frac{2q}{N}\right) + \frac{MN}{q} \ . \tag{3}$$

In addition, we consider the prefix of items by itself, i.e., with no further items released, and denote such instance $I_{N/2}$. The optimal offline solution partitions items into N subsets and uses all intervals $[\frac{j-1}{N}, \frac{j}{N})$ for $q \leq j \leq N$, so

$$\text{OPT}(I_{N/2}) = M \ , \tag{4}$$

while by definition and properties of the function g, for the online algorithm we have

$$\text{ALG}(I_{N/2}) \geq g(0) \ . \tag{5}$$

Suppose that the algorithm is asymptotically R-competitive. Then, for some additive constant C, the following inequality holds for any probability distribution $\{p_q\}_{q=t}^{N/2}$ over the instances $\{I_q\}_{q=t}^{N/2}$:

$$\mathbb{E}_q[\text{ALG}(I_q)] \leq R \cdot \mathbb{E}_q[\text{OPT}(I_q)] + C \ . \tag{6}$$

Plugging in the upper bounds on the costs of OPT, i.e., (2) and (4), as well as the lower bounds on the costs of ALG, i.e., (3) and (5), we get

$$p_{\frac{N}{2}} \cdot g(0) + \sum_{q=t}^{\frac{N}{2}-1} p_q \left(g\left(\frac{2q}{N}\right) + \frac{MN}{q}\right) \leq R\left(p_{\frac{N}{2}} \cdot M + \sum_{q=t}^{\frac{N}{2}-1} p_q \frac{MN}{q}\right) + C \ ,$$

which after moving the terms without g to the right hand side becomes

$$p_{\frac{N}{2}} \cdot g(0) + \sum_{q=t}^{\frac{N}{2}-1} p_q \cdot g\left(\frac{2q}{N}\right) \le R \cdot p_{\frac{N}{2}} \cdot M + (R-1) \sum_{q=t}^{\frac{N}{2}-1} p_q \frac{MN}{q} + C \ .$$

Letting $p_{\frac{N}{2}} = \frac{2t}{N}$ and $p_i = \frac{2}{N}$ for $t \le i < N/2$, the left hand side becomes an upper bound on the integral of g over $[0, 1)$, so by (1),

$$M = \int_0^1 g(t) \, dt \le \frac{2t}{N} \cdot g(0) + \frac{2}{N} \sum_{q=t}^{\frac{N}{2}-1} g\left(\frac{2q}{N}\right) \le \frac{2}{N} \cdot (t \cdot R \cdot M + \sum_{q=t}^{\frac{N}{2}-1} (R-1)\frac{N}{q}) + C \ .$$

Dividing by $2M$, the term $\frac{C}{2M}$ tends to 0 for $M \to \infty$, so letting $\tau = \frac{t}{N}$, this inequality becomes

$$\tau \cdot R + (R-1) \sum_{q=\tau \cdot N}^{\frac{N}{2}-1} \frac{1}{q} \ge \frac{1}{2} \ .$$

With N growing to infinity, the sum $\sum_{q=\tau \cdot N}^{\frac{N}{2}-1} \frac{1}{q}$ tends to $-\ln(2\tau)$. Rearranging, we have

$$(R-1)(\tau - \ln(2\tau)) \ge \frac{1}{2} - \tau \ ,$$

where finally letting $\tau \approx 0.212072$ yields the desired lower bound on R.

For the definition of an asymptotic competitive ratio via additive terms that are not necessarily constant (and we write in the introduction), the additive constant C is replaced by a value of the form $o(\text{OPT})$, where OPT is the cost of an optimal offline solution.

We note that even though we did not specify the probability distribution over instances upfront, it is fixed, and in particular it does not depend on the deterministic online algorithm, which thus may know the distribution a priori, as stipulated by Yao's principle. □

References

1. Azar, Y., Boyar, J., Epstein, L., Favrholdt, L.M., Larsen, K.S., Nielsen, M.N.: Fair versus unrestricted bin packing. Algorithmica **34**(2), 181–196 (2002)
2. Balogh, J., Békési, J., Dósa, G., Epstein, L., Levin, A.: A new and improved algorithm for online bin packing. In: Proceedings of the 26th European Symposium on Algorithms (ESA2018), pp. 5:1–5:14 (2018)
3. Balogh, J., Békési, J., Dósa, G., Epstein, L., Levin, A.: Lower bounds for several online variants of bin packing. Theory Comput. Syst. **63**(8), 1757–1780 (2019)
4. Balogh, J., Békési, J., Dósa, G., Epstein, L., Levin, A.: Online bin packing with cardinality constraints resolved. J. Comput. Syst. Sci. **112**, 34–49 (2020)
5. Balogh, J., Békési, J., Dósa, G., Epstein, L., Levin, A.: A new lower bound for classic online bin packing. Algorithmica **83**(7), 2047–2062 (2021)

6. Balogh, J., Békési, J., Galambos, G.: New lower bounds for certain classes of bin packing algorithms. Theor. Comput. Sci. **440**, 1–13 (2012)
7. Bienkowski, M., Pacut, M., Piecuch, K.: An optimal algorithm for online multiple knapsack. In: Proceedings of the 47th International Colloquium on Automata, Languages, and Programming (ICALP2020), LIPIcs, pp. 13:1–13:17 (2020)
8. Boyar, J., Favrholdt, L.M., Larsen, K.S., Nielsen, M.N.: The competitive ratio for on-line dual bin packing with restricted input sequences. Nord. J. Comput. **8**(4), 463–472 (2001)
9. Cygan, M., Jeż, Ł, Sgall, J.: Online knapsack revisited. Theory Comput. Syst. **58**(1), 153–190 (2014)
10. Deppert, M.A., Jansen, K., Khan, A., Rau, M., Tutas, M.: Peak demand minimization via sliced strip packing. In: Proceedings of the 24th International Conference on Approximation Algorithms for Combinatorial Optimization Problems (APPROX2021), pp. 21:1–21:24 (2021)
11. Escribe, C., Hu, M., Levi, R.: Competitive algorithms for the online minimum peak appointment scheduling. Available at SSRN 3787306, 31 p. (2021)
12. Fernandez de la Vega, W., Lueker, G.S.: Bin packing can be solved within $1 + \varepsilon$ in linear time. Combinatorica **1**(4), 349–355 (1981)
13. Fujiwara, H., Kobayashi, K.: Improved lower bounds for the online bin packing problem with cardinality constraints. J. Comb. Optim. **29**(1), 67–87 (2013)
14. Gálvez, W., Grandoni, F., Ameli, A.J., Khodamoradi, K.: Approximation algorithms for demand strip packing. In: Proceedings of the 24th International Conference on Approximation Algorithms for Combinatorial Optimization Problems (APPROX2021), pp. 20:1–20:24 (2021)
15. Ibarra, O.H., Kim, C.E.: Fast approximation algorithms for the knapsack and sum of subset problems. J. ACM **22**(4), 463–468 (1975)
16. Iwama, K., Taketomi, S.: Removable online knapsack problems. In: Proceedings of the 29th International Colloquium on Automata, Languages, and Programming (ICALP2002), pp. 293–305 (2002)
17. Johnson, D.S., Demers, A., Ullman, J.D., Garey, M.R., Graham, R.L.: Worst-case performance bounds for simple one-dimensional packing algorithms. SIAM J. Comput. **3**, 256–278 (1974)
18. Karmarkar, N., Karp, R.M.: An efficient approximation scheme for the one-dimensional bin-packing problem. In: Proceedings of the 23rd Annual Symposium on Foundations of Computer Science (FOCS1982), pp. 312–320 (1982)
19. Ranjan, A., Khargonekar, P., Sahni, S.: Offline first-fit decreasing height scheduling of power loads. J. Sched. **20**, 527–542 (2017)
20. Smedira, D., Shmoys, D.B.: Scheduling appointments online: the power of deferred decision-making. CoRR, abs/2111.13986 (2021)

An Adjacency Labeling Scheme Based on a Decomposition of Trees into Caterpillars

Avah Banerjee[✉]

Missouri S&T, Rolla, MO 65401, USA
banerjeeav@mst.edu

Abstract. In this paper we look at the problem of adjacency labeling of graphs. Given a family of undirected graphs the problem is to determine an encoding-decoding scheme for each member of the family such that we can decode the adjacency information of any pair of vertices only from their encoded labels. Further, we want the length of each label to be short (logarithmic in n, the number of vertices) and the encoding-decoding scheme to be computationally efficient. We propose a simple tree-decomposition based encoding scheme and use it give an adjacency labeling of size $O(k \log k \log n)$-bits. Here k is the clique-width of the graph family. We also extend the result to a certain family of k-probe graphs.

Keywords: Clique-widths · Hereditary classes · Implicit representation

1 Introduction

Adjacency labeling is a method to store adjacency information implicitly within vertex labels such that we can determine the adjacency between two vertices just from their labels. To be useful in practice we want these labels to be compact and easy to encode-decode. This is a powerful technique for lossless compression of graphs. Since the decoding is fully local, it makes these schemes particularly useful for storing graphs on distributed systems. It is an active area of research to determine adjacency labeling schemes for various graph families of practical importance.

1.1 Preliminaries

Let $G = (V, E)$ be an undirected graph with vertex set V ($|V| = n$) and edge set E ($|E| = m$). We assume G has no self-loops or parallel edges. Let $\mathcal{F}_n$ be a family of graphs on the vertex set V of size n. For any $u, v \in V$, we define $\mathsf{adj}(u, v) = 1$ if $\{u, v\} \in E$ and 0 otherwise.

A. Banerjee–This research was supported in part by the DTIC contract FA8075-14-D-0002/0007.

C. Bazgan and H. Fernau (Eds.): IWOCA 2022, LNCS 13270, pp. 114–127, 2022.
https://doi.org/10.1007/978-3-031-06678-8_9

Definition 1 (modified from [5]). *An L-bit adjacency labeling scheme of a graph family $\mathcal{F}_n$ is a pair of functions* enc $: \mathcal{F}_n \rightarrow (V \rightarrow \{0,1\}^L)$ *and* dec $: \{0,1\}^L \times \{0,1\}^L \rightarrow \{0,1\}$ *such that for all $G = (V, E) \in \mathcal{F}_n$ and for all $u, v \in V$,*

$$\mathsf{adj}(u, v) = \mathsf{dec}(\mathsf{enc}(G)(u), \mathsf{enc}(G)(v)).$$

We say there is an L-bit adjacency labeling for $\mathcal{F}_n$.

We write $\mathsf{enc}(G)(u) = \mathsf{enc}(u)$ when the graph G is clear from the context. According to the above definition a labeling scheme is local; as it determines the adjacency only based on the vertex labels. For a labeling scheme $(\mathsf{enc}, \mathsf{dec})$ to be useful in practice we want both functions, enc and dec, to be efficiently computable. Here we use the qualifier "adjacency" labeling to distinguish it from other types of labeling schemes (see below). However, in their seminal paper, authors in [20] referred to such a scheme simply as an L-labeling of G. In general the $(\mathsf{enc}, \mathsf{dec})_P$ pair may be used as an efficient storage-retrieval scheme for $\mathcal{F}_n$ with respect to some predicate P. For example P could be the predicate that a triple of three vertices forms a triangle in G. Another example is the distance labeling problem [4] where given a pair of vertex labels the decoder outputs the shortest path distance between them.

In this paper we are only concerned with adjacency labeling. There is a simple yet beautiful connection between adjacency labeling and *induced universal graphs* of a hereditary graph family.

Definition 2. *A graph property $\mathcal{P}$ is said to be* hereditary *if it is closed under taking induced subgraphs.*

Definition 3 [1–3]. *A graph $G_\mathcal{P}$ of size $f(n)$ (for some time-constructible[1] function $f : \mathbb{N} \rightarrow \mathbb{N}$) is called* universal *for $\mathcal{P}$ if every graph $G \in \mathcal{P}$ with at most n vertices is an induced subgraph of $G_\mathcal{P}$.*

An adjacency labeling for a hereditary family is said to be *efficient* if $k = O(\log n)$. It is an easy exercise to note that having an efficient adjacency labeling for a hereditary family implies that there is an induced universal graph $G_\mathcal{P}$ with $O(n^{O(1)})$ vertices. In this paper we give an adjacency labeling for graphs parameterized over its *clique-width*. Up to a constant factor, this scheme is efficient for graphs of bounded clique-width.

Definition 4 [14]. *The* clique-width *(denoted by $cw(G)$) of a graph G is the minimum number of distinct labels (of vertices) to construct G using the following four operations:*

1. *Create a vertex in v with label i (denoted by (v, i))*
2. *Disjoint union $G_1 \oplus G_2$[2] of two labeled graphs G_1 and G_2*

---

[1] $f(n)$ can be computed in time $O(f(n))$.
[2] The vertex set of $V(G_1 \oplus G_2)$ of $G_1 \oplus G_2$ is $V(G_1) \cup V(G_2)$ and the edge set $E(G_1 \oplus G_2) = E(G_1) \cup E(G_2)$.

3. *Join operation $\eta_{i,j}$: adds edges between every pair of vertices one with label i and another with label j $(i \neq j)$*
4. *Relabel operation $\rho_{i \to j}$ relabels vertices having label i with label j*

A construction of G using the above operations is known as a k-expression where $cw(G) = k$. A k-expression can be equivalently represented as a rooted binary tree[3] T (called a union tree [19]) as follows. Leaves of T corresponds to the labeled (with their initial labels) vertices (v, i)'s of G. Each internal node corresponds to a union operation. Lastly, each internal node is decorated with a (possibly empty) sequence of join and relabel operations. We use the notation d_z to denote the decorator for the node z.

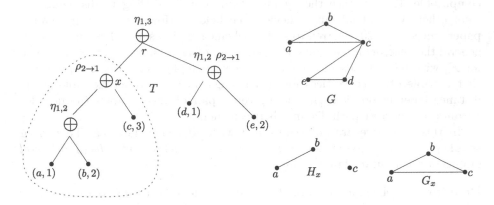

Fig. 1. T is a union tree of the graph G. However, T is not a proper union tree. $G_x = G[\{a, b, c\}]$ has edges ac and bc but H_x, the graph corresponding to the subtree T_x rooted at x, has no such edges.

We say k is the *width* of T. For some internal vertex x of T let T_x be the subtree rooted at x. Let G_x be a induced subgraph of G determined by the leaves of T_x. Then T_x (including any join or relabel operations in d_x) is a union tree for some spanning subgraph[4] H_x of G_x. Borrowing the terminology from [19] we say T is a *proper* union tree of G if for every internal vertex $x \in T$, $H_x = G_x$ (see example in Fig. 1). It is an easy exercise (see lemma 1 in [19]) to show that we can transform any union tree in linear time to a proper one of the same width representing the same graph. Henceforth we shall assume without loss of generality that we are working with proper union trees.

In this paper, we also look at a generalization of k-expressions and study adjacency labeling of the corresponding graph family. Recently, a new width parameter was proposed [10,18].

---

[3] It is a tree and not a DAG as the same graph does not take part in two separate union operations.
[4] H is a spanning subgraph of G if $V(H) = V(G)$ and $E(H) \subseteq E(G)$.

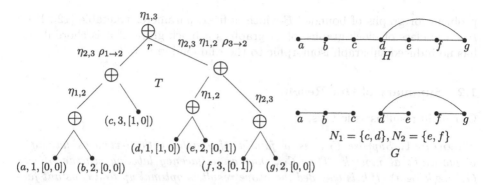

Fig. 2. Top right graph H corresponds to the k-expression $t = \eta_{1,3}(t_1 \oplus t_2)$ where $t_1 = \rho_{1\to2}\eta_{2,3}(\eta_{1,2}((a,1) \oplus (b,2)) \oplus (c,3))$ and $t_2 = \rho_{3\to2}\eta_{1,2}\eta_{2,3}((\eta_{1,2}((d,1) \oplus (e,2))) \oplus (\eta_{2,3}((f,3) \oplus (g,2))))$. G can be embedded into H using two independent sets N_1, N_2 as illustrated by the union tree T of G.

Definition 5 *(from* [12]*).* *Let $\mathcal{F}$ be a family of graphs. The $\mathcal{F}$-width of a graph G is the minimum number k of independent sets $N_1, \ldots, N_k$ in $G = (V, E)$ such that there exists $H = (V, E') \in \mathcal{F}$ where the following holds: 1) G is a spanning subgraph of H and 2) for every edge $(u, v) \in E' \setminus E$ there exists an $i \in [k]$[5] with $u, v \in N_i$.*

A graph which has an $\mathcal{F}$-width of k is known as a *k-probe $\mathcal{F}$-graph*[6]. In this paper we consider the adjacency labeling of w_k-*probe $\mathcal{C}_k$*- graphs. Here $\mathcal{C}_k$ is the family of graphs with clique-width $\leq k$. We can represent a tree decomposition of w_k-*probe $\mathcal{C}_k$*- graphs via a minor modification to the proper union tree structure (Fig. 2). The label of each leaf now has an additional w_k-length binary vector (M_u). Specifically, each leaf corresponds to a tuple (u, i, M_u) where $M_u[j] = 1 \iff u \in N_j$ and i is u's initial label in the k-expression (as before). Adjacency is determined as follows (using Definition 4 and 5). Let $z = lca(u, v)$. Then u, v are adjacent if and only if: 1) there is a join operation in d_z between the current labels of u and v and 2) M_u and M_v do not have a common 1. It should be noted that a w_k-*probe $\mathcal{C}_k$*- graph has a clique-width $\leq k2^{w_k}$.

The motivation for studying adjacency labelling of k-probe $\mathcal{F}$-graphs are threefold. Firstly, they are a generalization of probe-graphs [12], which can model some natural problems. For example a type of DNA mapping problem can be formulated as a recognition problem for probe-graphs of intervals [11]. Secondly, this family of graphs do not have a bounded genus. This may make finding a compact adjacency labeling a challenge; especially if graphs in $\mathcal{F}$ are not "decomposable" (like a union tree). Existing approaches such as those developed in [16] does not extended to graph families whose genus is not bounded. We leave this as an open problem. Finally, depending on $\mathcal{F}$, many computationally hard problems exhibit efficient algorithms when the $\mathcal{F}$-width is bounded. For example, the recognition

---

[5] Here $[n] = \{1, \ldots, n\}$.
[6] Some authors call them probe-k $\mathcal{F}$-graph [12].

problem for graphs of bounded $\mathcal{B}$-width is fixed parameter tractable [12]. Here $\mathcal{B}$ is the class of block graphs [6]. A graph is a block graph if it is chordal and has no induced subgraph isomorphic to the diamond graph $(K_4 - e)$.

1.2   Summary of Our Results

Our main result is as follows.

Theorem 1. *Suppose $\mathcal{C}_{k,n}$ is a family of graphs with n-vertices having a clique-width at most k. Then $\mathcal{C}_{k,n}$ has an adjacency labeling scheme of size $O(k \log k \log n)$. If k is bounded the above result is optimal up to a constant factor. Further, given a union tree the labels can be computed in $O(kn \log k \log n)$ total time and decoding takes time linear in the size of the labels.*

Briefly, we apply a recursive transformation on the union tree to obtain a tree of $O(\log n)$ depth. This transformation preserves the lowest common ancestor relations between the leaves and allows us to encode the adjacency information contained within the internal nodes and the leaves with $O(k \log k)$ bits. We also get the following generalization as a corollary.

Corollary 1. *There is an $O(k \log k \log n + w_k)$-bits labeling scheme for w_k-probe $\mathcal{C}_k$- graphs of size n.*

Proof. This immediately follows from Theorem 1 and the fact that we need an additional w_k-bits to encode the vectors M_u.   □

1.3   Previous and Related Work

Adjacency labeling schemes studied in this paper closely follow the paradigm introduced in [20,23]. However, the study of adjacency labeling schemes goes back more than half a century [7,8]. Since then many results have been discovered for a wide variety of graph classes. A comprehensive overview and some interesting open problems can be found in [24,26] and the references therein. So we restrict our discussion to results which are closely related to ours. A folklore result[7] is that cographs have $O(\log n)$-bit adjacency labeling. This follows from the fact that a cograph is a permutation graph and for which an adjacency labeling follows trivially (for each vertex store $(i, \pi(i))$) [26]. In [17] authors gave a $(\log n + O(k \log \log \frac{n}{k}))$-bits adjacency labeling scheme for graphs of tree-width k.

Only a handful of results are known with respect to the clique-width parameter. There is a parallel line of research based on *ordered binary decision diagrams* (OBDD). OBDD's are a generalization of union trees in the setting of boolean functions. In [21] authors gave an $O(n \frac{k^2}{\log k})$-sized, $O(\log n)$-depth OBDD with an encoding size of $O(\log k \log n)$-bits. This scheme is based on a bottom up tree

---

[7] We thank an anonymous reviewer for pointing this out.

decomposition approach originally introduced in [22]. In contrast our decomposition scheme is top-down. An improvement was proposed based on a tree-decomposition approach similar to ours [19]. Here the author gave an $O(kn)$-sized data structure that supported $O(1)$-time adjacency quires. A more recent result on OBDD-type storage schemes for small clique-width graph can be found in [9]. However, these representations are not local and the adjacency queries are performed with the help of a global data structure (the OBDD or something similar). The result closest to ours can be found in [15]. The paper uses the language of *monodic second-order* logic. There, authors gave an adjacency labeling scheme, which in the language of this paper, translates to a label of size $O(f(k) \log n)$ bits. In their paper authors did not give an explicit expression for $f(k)$. In [26] (chapter 11) the author hinted at an $O(\log n)$ adjacency labeling for graphs with bounded clique-width. The proposal uses a recursive decomposition by successively finding balanced k-modules for any graph with clique-width k. Although explicit bounds were not provided with respect to k, we expect that working out the details can give a bound similar to ours.

2   A Caterpillar-Type Balanced Decomposition

In this section we give a simple balanced decomposition (discussed shortly) of a rooted tree (not necessarily binary). Here we work with a generic rooted tree T having n leaves (hence $\leq n - 1$ internal nodes). We identify the root of T with r. Later in Sect. 3 we will use this result to prove Theorem 1.

It is clear that every tree can be constructed starting from K_1 by repeatedly adding pendent edges. However, it may require $O(n)$ iterations to construct a tree with n vertices. We show that each tree on n leaves can be constructed within $O(\log n)$ iterations using a slightly more relaxed operation. We assume each non-root vertex v has either no children or at least two children (that is, v does not have exactly one child). Let $\mathcal{L}_n$ denote the set of all trees with at most n leaves. See Fig. 3 for all trees in $\mathcal{L}_3$, where the root is colored red.

Fig. 3. Trees in $\mathcal{L}_3$. (Color figure online)

Definition 6. *A caterpillar (see Fig. 4) is a tree for which there exists a root-leaf path P such that all vertices outside P are leaves.*

Let $T_0, T_1, ..., T_k$ be disjoint trees. The operation of *adding* $T_1, ..., T_k$ to T_0 creates a tree obtained by identifying the roots of $T_1, ..., T_k$ with k distinct leaves of T_0, respectively (see Fig. 5). Let $\mathcal{C}_0$ denote the class of all caterpillars. For each

Fig. 4. A rooted caterpillar tree. (Color figure online)

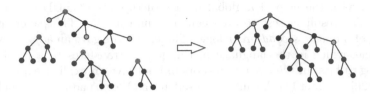

Fig. 5. Adding the three bottom trees with red roots by attaching (identifying) them with the corresponding green leaves of the top tree. (Color figure online)

positive integer p, let $\mathcal{C}_p$ consist of trees obtained by adding trees from $\mathcal{C}_{p-1}$ to trees from $\mathcal{C}_0$. Note that $\mathcal{C}_{p-1} \subseteq \mathcal{C}_p$ since $K_1 \in \mathcal{C}_0$. A path P of a tree T is called an *r-path* if r is an end of P. Let P be an r-path of a tree T. Suppose $T \neq P$. Let $T \backslash P$ denote the graph obtained by deleting $V(P)$ from T. Then each component of $T \backslash P$ must be one of the following two types: those that have no edges, which we call *trivial*, and those that have at least one edge, which we call *nontrivial*. Let T_0 consist of all edges that are incident with at least one vertex of P. Then T_0 is a caterpillar (with root r). Suppose $T \neq T_0$. Then at least one component of $T \backslash P$ is nontrivial. Let $T_1, ..., T_k$ be all such components. Then,

(i) the root of T_i is the vertex of T_i that is closest to r (in T).
(ii) $E(T_0), E(T_1), \ldots, E(T_k)$ form a partition of $E(T)$
(iii) T can be obtained by adding $T_1, \ldots, T_k$ to T_0.

The following theorem gives a structural relationship between $\mathcal{L}_n$ and $\mathcal{C}_p$.

Theorem 2. *For every integer $n \geq 1$, we have $\mathcal{L}_n \subseteq \mathcal{C}_p$, where $p = \lfloor \log_2 n \rfloor$*

To prove the theorem we use the following lemma.

Lemma 1. *Let T be a tree with $n \geq 1$ leaves. Then T has an r-path P such that each component of $G \backslash P$ has at most $n/2$ leaves.*

Proof. If $n = 1$ then the path with only one vertex r satisfies the requirement. So we assume $n \geq 2$. Under this assumption, for any r-path P, $T \backslash P$ must have at least one component. This allows us to define for any r-path P:

$$h(P) = \max\{t \mid T \backslash P \text{ has a component with } t \text{ leaves}\}$$

Let P be an r-path that minimizes $h(P)$. We prove that P satisfies the lemma.

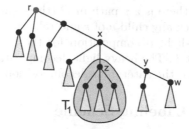

Fig. 6. The tree used in the proof of Lemma 1. (Color figure online)

Suppose on the contrary that P does not satisfy the lemma. That is, $T\backslash P$ has a component T_1 with $n_1 > n/2$ leaves. Let the ends of P be r and w and let the root of T_1 be z, as illustrated in the Fig. 6 above. Let Q be the unique path of T between r and z. We prove that $h(Q) < n_1 \leq h(P)$, which will be a desired contradiction.

To estimate $h(Q)$ we observe that $T\backslash Q$ has two types of components: those that are disjoint from T_1 and those that are contained in T_1. For the ones that are disjoint from T_1, the number of leaves each of them may have is bounded by $n - n_1$, which is smaller than n_1. Next, we consider a component T' of $T\backslash Q$ with $T' \subseteq T_1$. Since T_1 has $n_1 > n/2 \geq 1$ leaves, z is not a leaf of T. By the assumption we made in the beginning of Sect. 2, z has at least two children. It follows that T' does not contain all leaves of T_1, which implies that T' has fewer than n_1 leaves. Therefore, we have shown that every component of $T\backslash Q$ has fewer than n_1 leaves. Consequently, $h(Q) < n_1$, contradicting the choice of P. This contradiction proves the lemma.   □

Proof of Theorem 2.

Proof. As we observed earlier, every tree in $\mathcal{L}_3$ is a caterpillar, so we have $\mathcal{L}_n \subseteq \mathcal{C}_0$, for $n = 1, 2, 3$, and thus the theorem holds for $n = 1, 2, 3$. Suppose the theorem holds for $n - 1$, where $n \geq 4$. We prove that the theorem holds for n, and this would prove the theorem.

Let T be a tree with $n \geq 4$ leaves. We need to show $T \in \mathcal{C}_{\lfloor \log_2 n \rfloor}$. We may assume that T is not a caterpillar because otherwise $T \in \mathcal{C}_0 \subseteq \mathcal{C}_{\lfloor \log_2 n \rfloor}$. By Lemma 1, T has an r-path P such that each component of $T\backslash P$ has at most $n/2$ leaves. Since T is not a caterpillar, $T\backslash P$ has at least one nontrivial component. Let $T_1, ..., T_k$ be all such components. By our induction hypothesis, each T_i belongs to $\mathcal{C}_{\lfloor \log_2(n/2) \rfloor} = \mathcal{C}_{\lfloor \log_2 n \rfloor - 1}$. It follows that $T \in \mathcal{C}_{\lfloor \log_2 n \rfloor}$ since T is obtained by adding $T_1, ..., T_k$ to T_0. This completes our induction and it proves the theorem.   □

Remark 1. The decomposition in the above theorem can be computed in linear time as follows. First we apply depth first search to compute for each node u in T the number of leaves in the subtree T_u. Then we apply a slightly modified heavy-light decomposition (see for example [25]) to obtain a decomposition of T

into disjoint paths $\mathcal{P}$. If there is a r-path in $\mathcal{P}$ (there can be at most one) then use it as P. Otherwise pick any child u of r and take (r, u) as P. We do not need to recompute the heavy-light decomposition for the recursive case and rather use the one computed for T. The heavy-light decomposition can be computed in linear time and hence also the caterpillar decomposition.

3   An Adjacency Labeling Scheme

Recall that a proper union tree T is a rooted binary tree with root r and n leaves. The initial label of a leaf u will be denoted by $c_0(u)$. For an internal node $x \in T$ let L_x be the set of leaves in the subtree T_x rooted at x. We begin with a lemma.

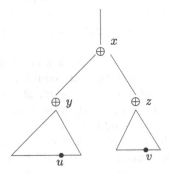

Fig. 7. We want to determine the information needed to compute the adjacency between u and v given we already know their lowest common ancestor x.

Lemma 2. *Suppose G is a graph of clique-width k and T be a proper union tree of G. We consider a node x of T as shown in Fig. 7. Let $x = \mathsf{lca}(u, v)$, where u and v are two leaf nodes. Given $x, c_0(u)$ and $c_0(v)$ we can determine $\mathsf{adj}(u, v)$ with an additional $O(k \log k)$ bits of information stored locally at vertices of G.*

This $O(k \log k)$-bits of information will serve to perform adjacency queries between u and the set L_z. In Theorem 1 we show that we can partition V to $O(\log n)$ such sets for each vertex in V.

Proof. We consider the situation shown in Fig. 7. Let B_x be the set of unique labels assigned to the leaves of the subtree rooted at x after applying d_x. In order to determine $\mathsf{adj}(u, v)$ it is sufficient to know; 1) the labels of u and v after application of the decorators d_y and d_z respectively and 2) the decorator d_x. However, we do not need to know the entirety of d_x but only whether $c_x(v) \in C_x(u)$, which is defined next. Suppose $c_x(u)$ (resp. $c_x(v)$) are the labels of u (resp. v) before applying d_x. From d_x we can easily determine the set of labels $C_x(u) \subseteq B_y \cup B_z$ such that,

$$\forall i \in C_x(u) \; \exists \eta_{i, c_x(u)} \text{ or } \eta_{c_x(u), i} \in d_x$$

It is important to note that when defining the set $C_x(u)$ we consider the labels from the set $B_y \cup B_z$ before any re-labeling due to d_x[8]. As an example, suppose $B_y = \{1,2\}, B_z = \{3,5\}, c_x(u) = 1$ and

$$d_x = \rho_{3 \to 2}\eta_{1,2}\rho_{2 \to 5}\eta_{5,1}$$

then $C_x = \{2,3,5\}$ and not simply $\{2,5\}$. Clearly $|C_x(u)| \le k - 1$ and it takes $O(k \log k)$-bits to store $|C_x(u)|$. Next, we need to retrieve $c_x(v)$ for any $v \in L_z$. This can be done by storing an additional $O(k \log k)$-bits at u. This follows from the fact that, given an initial labeling of L_z, the sub-k-expression induced by T_z (including applying the decorator d_z) is just a re-labeling (more precisely a function in $[k]^{[k]}$[9]). This re-labeling can be stored as a list ($F_x(u)$) of size k where each value is between 1 and k. The $c_0(v)^{th}$ entry of this list gives $c_x(v)$. Finally, we use $O(\log k)$ bits to store $c_x(u)$ at u. □

Going forward, we will describe an encoding of each leaf as an alternating sequence of labels of two types. One containing path information and the other containing adjacency information. For the latter, we will use $C_x(u), F_x(u)$ and $c_x(u)$. We let $A_x(u) = (C_x(u), F_x(u), c_x(u))$.

Theorem 1. *Suppose $\mathcal{C}_{k,n}$ is a family of graphs with n-vertices having a clique-width at most k. Then $\mathcal{C}_{k,n}$ has an adjacency labeling scheme of size $O(k \log k \log n)$. If k is bounded the above result is optimal up to a constant factor. Further, given a union tree the labels can be computed in $O(kn \log k \log n)$ total time and decoding takes time linear in the size of the labels.*

Proof. First we start from the r-path decomposition of T as described in the previous section. Let P be a r-path. From Lemma 1 we know that the subtrees attached to P have $\le n/2$ leaves. Let $\mathcal{T}_{large}$ be a possibly empty collection of subtrees which have between $n/4$ and $n/2$ leaves. These subtrees are identified with light blue color in Fig 8-a. Note that $0 \le |\mathcal{T}_{large}| \le 4$. Consider the sequence(s) of smaller subtrees (we will call them bushes) between the trees in $\mathcal{T}_{large}$. These bushes are highlighted with orange regions in the figure. There may be no such bushes between two large trees. Now we collate the bushes between two successive large subtrees (while descending along P) to create larger bushes until the total number of leaves among them is $\ge n/4$ but $\le n/2$. At this point we call it a super-bush and restart the gathering process on the remaining bushes until we get another super-bush or we reach the end of the bushes. In the latter case we create a super-bush with whatever we have gathered up to that point. That is, we group the bushes (if any) between two successive large subtrees into $\ge n/4$ sized super-bushes (but no larger than $n/2$) except for at most a constant number of groups which can have $< n/4$ leaves. We identify each super-bush with a tree, the root of which is the node closest to r. Further, we attach the tree to P using the node of the super-bush which was closest to r. For example, in Fig. 8-a for the

---

[8] Alternatively, we may assume that all relabeling operations in d_x proceed all join operations [13].

[9] $[k] = \{1, \dots, k\}$.

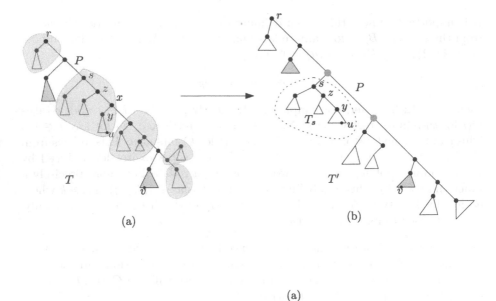

(a)

Fig. 8. The figure shows the tree T' obtained from T after collating the smaller bushes into subtrees. For example, we create the tree T_s, rooted at s from the set of small trees attached to the r-path P via s, z, x. To make T_s a proper union tree, the node x is removed and y is made a child of z. (Color figure online)

super-bush starting from the node s we create a tree T_s with s as the root. We attach T_s to P where the node s was previously located. From our construction, the number of such attachments will also be a constant. The decorators remain with the original vertices and the new (orange vertices in Fig. 8-b) vertices on P do not contain any decorators. The resulting tree, denoted by T', is not necessarily a valid union tree. However, we ensure that each subtree attached to P is a proper union tree (Fig. 8-b).

First we informally describe the decoding scheme; this will give us an idea of what information to encode within the labels. Let u, v be a pair of leaves in T (Fig. 8). Let $x = \mathsf{lca}(u, v)$. From lemma 2 we see that labels of size $O(k \log k)$-bits are sufficient to determine $\mathsf{adj}(u, v)$ given $x, c_0(u)$ and $c_0(v)$. It remains to determine the number of such labels we need to determine adjacency between u and any other vertex in G. Trivially, we can maintain one such label for each node on the root-leaf path (in T) terminating in u. Since a path (in the caterpillar-decomposition) can have arbitrary length we will need $\Omega(\log n)$-bits to locate a node in each level of the caterpillar decomposition. Since there are $O(\log n)$ levels, we may end up needing $O(\log^2 n)$-bits to encode the path information in the final label. To reduce the encoding size and get our claimed bound we make the following crucial observation. It is not necessary to determine the $\mathsf{lca}(u, v)$ explicitly. It suffices to know $A_x(u), c_0(u)$ and $c_0(v)$ to determine $\mathsf{adj}(u, v)$. By taking a recursive approach, we show that we only need to remember $O(\log n)$ many adjacency-type labels per vertex u. Since, each adjacency information requires $O(k \log k)$ bits labels, we get the bound claimed in the theorem. This recursive encoding scheme is determined based on T'. Let $l_1(u)$ be the position of the node (w.r.t. the root r) attaching the subtree, which u is a leaf of, to the path P in T'. For example, in Fig. 8 we have $l_1(u) = 3$ and $l_1(v) = 5$. There are two cases:

(i)($l_1(u) = l_1(v)$) Then $u, v \in L_s$ for some node $s \in P$. To determine adj(u, v), we recurse on the subtree T_s. According to our construction T_s is a proper union tree corresponding to the induced subgraph $G[L_s]$. Then we determine an adjacency labeling scheme for $G[L_s]$ using T_s. This is used to determine adj(u, v). This recursive construction is possible, since any induced subgraph of G has a clique-width $\leq k$ and T is a proper union tree.

(ii)($l_1(u) \neq l_1(v)$) (Fig. 8-b) We assume without loss of generality that $l_1(u) < l_1(v)$ (the case $l_1(u) > l_1(v)$ is symmetric). In this case we use $A_x(u), c_0(u)$ and $c_0(v)$ to determine adj(u, v).

This completes the informal description of the decoding. From this, an encoding scheme emerges naturally.

Encoder: Generate T' from T and for each $u \in V$ we compute $(l_1(u), A_{P(u)}(u))$. Here, $P(u)$ is the lowest ancestor of u on the path P (in Fig. 8-a $P(u) = x$). For notational simplicity we denote $A_{P(u)}(u) = A_1(u)$. Then, perform the encoding recursively on each induced subgraph of G corresponding to the subtrees attached to P. For the vertex u this process gives a sequence of labels $((l_i(u), A_i(u))$'s. Appending to this sequence its initial label $c_0(u)$ gives the final encoding:

$$\mathsf{enc}(u) = (c_0(u), (l_1(u), A_1(u)), \ldots, (l_p(u), A_p(u))),$$

where $p = O(\log n)$ (from Theorem 2). It is clear from the construction that enc uses $O(k \log k \log n)$-bits.

Decoder: Given two strings $\mathsf{enc}(u)$ and $\mathsf{enc}(v)$ first we check the labels $(l_1(u), A_1(u))$ and $(l_1(v), A_1(v))$. If $l_1(u) < l_1(v)$ then we use $A_1(u), c_0(u)$ and $c_0(v)$ to determine adj(u, v). The case $l_1(u) > l_1(v)$ is symmetric. Otherwise, $l_1(u) = l_1(v)$. In this case we proceed to check the next pair of labels $(l_2(u), A_2(u))$ and $(l_2(v), A_2(v))$ and so on. In general, let i be the smallest number such that $l_i(u) \neq l_i(v)$. Then, using either $A_i(u)$ or $A_i(v)$ and $c_0(u), c_0(v)$ we can determine adj(u, v). By our construction there is always such an $i \leq p$ such that $l_i(u) \neq l_i(v)$.

Correctness: We use induction on the depth of the recursive construction. For the base case, we take $p = 0$ and the correctness follows trivially. Assume the encoder-decoder works correctly whenever the decomposition has depth $\leq p-1$. This takes care of the case $l_1(u) = l_1(v)$. For the remaining case assume $l_1(u) < l_1(v)$. Then correctness follows from Lemma 2.

Running Time: Recall from Remark 1, given a union tree T we can determine the recursive decomposition in $O(|T|)$ time. Additionally, $O(k \log k \log n)$ time is spent processing each leaf of T. Thus enc can be computed in $O(kn \log k \log n)$

time. Decoding can be done in linear time in the size of the labels (i.e., in $O(k \log k \log n)$ time). ☐

Acknowledgement. The author would like to thank Guoli Ding for many discussions and considerable advice. In particular for the proof of caterpillar decomposition. We also thank anonymous reviewers for their helpful comments.

References

1. Abrahamsen, M., Alstrup, S., Holm, J., Knudsen, M.B.T., Stöckel, M.: Near-optimal induced universal graphs for bounded degree graphs. arXiv preprint arXiv:1607.04911 (2016)
2. Alon, N.: Asymptotically optimal induced universal graphs. Geom. Funct. Anal. **27**(1), 1–32 (2017). https://doi.org/10.1007/s00039-017-0396-9
3. Alstrup, S., Dahlgaard, S., Knudsen, M.B.T.: Optimal induced universal graphs and adjacency labeling for trees. J. ACM (JACM) **64**(4), 1–22 (2017)
4. Alstrup, S., Dahlgaard, S., Knudsen, M.B.T., Porat, E.: Sublinear distance labeling. arXiv preprint arXiv:1507.02618 (2015)
5. Alstrup, S., Kaplan, H., Thorup, M., Zwick, U.: Adjacency labeling schemes and induced-universal graphs. In: Proceedings of the Forty-Seventh Annual ACM Symposium on Theory of Computing, pp. 625–634 (2015)
6. Brandstädt, A., Le, V.B., Spinrad, J.P.: Graph Classes: A Survey. SIAM (1999)
7. Breuer, M.A., Folkman, J.: An unexpected result in coding the vertices of a graph. J. Math. Anal. Appl. **20**(3), 583–600 (1967)
8. Breuer, M.: Coding the vertexes of a graph. IEEE Trans. Inf. Theory **12**(2), 148–153 (1966)
9. Chakraborty, S., Jo, S., Sadakane, K., Satti, S.R.: Succinct data structures for small clique-width graphs. In: 2021 Data Compression Conference (DCC), pp. 133–142. IEEE (2021)
10. Chandler, D.B., Chang, M.S., Kloks, T., Liu, J., Peng, S.L.: Partitioned probe comparability graphs. Theoret. Comput. Sci. **396**(1–3), 212–222 (2008)
11. Chandler, D.B., Chang, M.S., Kloks, T., Liu, J., Peng, S.L.: On probe permutation graphs. Discret. Appl. Math. **157**(12), 2611–2619 (2009)
12. Chang, M.S., Hung, L.J., Kloks, T., Peng, S.L.: Block-graph width. Theoret. Comput. Sci. **412**(23), 2496–2502 (2011)
13. Courcelle, B., Engelfriet, J.: Graph Structure and Monadic Second-Order Logic: A Language-Theoretic Approach, vol. 138. Cambridge University Press, Cambridge (2012)
14. Courcelle, B., Engelfriet, J., Rozenberg, G.: Handle-rewriting hypergraph grammars. J. Comput. Syst. Sci. **46**(2), 218–270 (1993)
15. Courcelle, B., Vanicat, R.: Query efficient implementation of graphs of bounded clique-width. Discret. Appl. Math. **131**(1), 129–150 (2003)
16. Dujmović, V., Esperet, L., Gavoille, C., Joret, G., Micek, P., Morin, P.: Adjacency labelling for planar graphs (and beyond). In: 2020 IEEE 61st Annual Symposium on Foundations of Computer Science (FOCS), pp. 577–588. IEEE (2020)
17. Gavoille, C., Labourel, A.: Shorter implicit representation for planar graphs and bounded treewidth graphs. In: Arge, L., Hoffmann, M., Welzl, E. (eds.) ESA 2007. LNCS, vol. 4698, pp. 582–593. Springer, Heidelberg (2007). https://doi.org/10.1007/978-3-540-75520-3_52

18. Hung, L.-J., Kloks, T.: On some simple widths. In: Rahman, M.S., Fujita, S. (eds.) WALCOM 2010. LNCS, vol. 5942, pp. 204–215. Springer, Heidelberg (2010). https://doi.org/10.1007/978-3-642-11440-3_19

19. Kamali, S.: Compact representation of graphs of small clique-width. Algorithmica **80**(7), 2106–2131 (2018)

20. Kannan, S., Naor, M., Rudich, S.: Implicat representation of graphs. SIAM J. Discret. Math. **5**(4), 596–603 (1992)

21. Meer, K., Rautenbach, D.: On the OBDD size for graphs of bounded tree-and clique-width. Discret. Math. **309**(4), 843–851 (2009)

22. Miller, G.L., Reif, J.H.: Parallel tree contraction and its application. Technical report, Harvard Univ. Cambridge, MA, Aiken Computation LAB (1985)

23. Muller, J.H.: Local structure in graph classes (1989)

24. Scheinerman, E.: Efficient local representations of graphs. In: Gera, R., Hedetniemi, S., Larson, C. (eds.) Graph Theory. PBM, pp. 83–94. Springer, Cham (2016). https://doi.org/10.1007/978-3-319-31940-7_6

25. Sleator, D.D., Tarjan, R.E.: A data structure for dynamic trees. J. Comput. Syst. Sci. **26**(3), 362–391 (1983)

26. Spinrad, J.: Efficient Graph Representations. Fields Institute Monographs, vol. 19. American Mathematical Society (2003)

Computing Longest (Common) Lyndon Subsequences

Hideo Bannai[1] , Tomohiro I[2] , Tomasz Kociumaka[3] ,
Dominik Köppl[1(✉)] , and Simon J. Puglisi[4]

[1] M&D Data Science Center, Tokyo Medical and Dental University,
Bunkyo City, Japan
{hdbn.dsc,koeppl.dsc}@tmd.ac.jp
[2] Department of Artificial Intelligence, Kyushu Institute of Technology,
Iizuka, Japan
tomohiro@ai.kyutech.ac.jp
[3] University of California, Berkeley, Berkeley, USA
kociumaka@berkeley.edu
[4] Department of Computer Science, Helsinki University, Helsinki, Finland
simon.puglisi@helsinki.fi

Abstract. Given a string T with length n whose characters are drawn
from an ordered alphabet of size σ, its longest Lyndon subsequence is a
longest subsequence of T that is a Lyndon word. We propose algorithms
for finding such a subsequence in $\mathcal{O}(n^3)$ time with $\mathcal{O}(n)$ space, or *online*
in $\mathcal{O}(n^3\sigma)$ space and time. Our first result can be extended to find the
longest common Lyndon subsequence of two strings of length n in $\mathcal{O}(n^4\sigma)$
time using $\mathcal{O}(n^3)$ space.

Keywords: Lyndon word · Subsequence · Dynamic programming

1 Introduction

A recent theme in the study of combinatorics on words has been the generalization of regularity properties from substrings to subsequences. For example,
given a string T over an ordered alphabet, the longest increasing subsequence
problem is to find the longest subsequence of increasing symbols in T [2,25].
Several variants of this problem have been proposed [10,20]. These problems
generalize to the task of finding such a subsequence that is not only present in
one string, but common in two given strings [15,23,26], which can also be viewed
as a specialization of the longest common subsequence problem [17,19,27].

More recently, the problem of computing the longest square word that is
a subsequence [22], the longest palindrome that is a subsequence [6,18], the
lexicographically smallest absent subsequence [21], and longest rollercoasters [4,
11,12] have been considered.

Here, we focus on subsequences that are Lyndon, i.e., strings that are lexicographically smaller than any of its non-empty proper suffixes [24]. Lyndon words

C. Bazgan and H. Fernau (Eds.): IWOCA 2022, LNCS 13270, pp. 128–142, 2022.
https://doi.org/10.1007/978-3-031-06678-8_10

are objects of longstanding combinatorial interest (see e.g., [13]), and have also proved to be useful algorithmic tools in various contexts (see, e.g., [1]). The longest Lyndon *substring* of a string is the longest factor of the Lyndon factorization of the string [5], and can be computed in linear time [9]. The longest Lyndon *subsequence* of a unary string is just one letter, which is also the only Lyndon subsequence of a unary string. A (naive) solution to find the longest Lyndon subsequence is to enumerate all distinct Lyndon subsequences, and pick the longest one. However, the number of distinct Lyndon subsequences can be as large as 2^n considering a string of increasing numbers $T = 1 \cdots n$. In fact, there are no bounds known (except when $\sigma = 1$) that bring this number in a polynomial relation with the text length n and the alphabet size σ [16], and thus deriving the longest Lyndon subsequence from all distinct Lyndon subsequences can be infeasible. In this paper, we focus on the algorithmic aspects of computing this longest Lyndon subsequence in polynomial time without the need to consider all Lyndon subsequences. In detail, we study the problems of computing

1. the lexicographically smallest (common) subsequence for each length online, cf. Sect. 3, and
2. the longest subsequence that is Lyndon, cf. Sect. 4, with two variations considering the computation as online, or the restriction that this subsequence has to be common among two given strings.

The first problem serves as an appetizer. Although the notions of *Lyndon* and *lexicographically smallest* share common traits, our solutions to the two problems are independent, but we will reuse some tools for the online computation.

2   Preliminaries

Let Σ denote a totally ordered set of symbols called *the alphabet*. An element of Σ^* is called a string. The alphabet Σ induces the *lexicographic order* $\prec$ on the set of strings Σ^*. Given a string $S \in \Sigma^*$, we denote its length with $|S|$, its i-th symbol with $S[i]$ for $i \in [1..|S|]$. Further, we write $S[i..j] = S[i] \cdots S[j]$, and we write $S[i..] = S[i..|S|]$ for the suffix of S starting at position i. A *subsequence* of a string S with length ℓ is a string $S[i_1] \cdots S[i_\ell]$ with $i_1 < \ldots < i_\ell$.

Let $\perp$ be the empty string. We stipulate that $\perp$ is lexicographically larger than every string of Σ^+. For a string S, appending $\perp$ to S yields S.

A string $S \in \Sigma^*$ is a *Lyndon word* [24] if S is lexicographically smaller than all its non-empty proper suffixes. Equivalently, a string S is a Lyndon word if and only if it is smaller than all its proper cyclic rotations.

The algorithms we present in the following may apply techniques limited to integer alphabets. However, since the final space and running times are not better than $\mathcal{O}(n)$ space and $\mathcal{O}(n \lg n)$ time, respectively, we can reduce the alphabet of T to an integer alphabet by sorting the characters in T with a comparison based sorting algorithm taking $\mathcal{O}(n \lg n)$ time and $\mathcal{O}(n)$ space, removing duplicate characters, and finally assigning each distinct character a unique rank within $[1..n]$. Hence, we assume in the following that T has an alphabet of size $\sigma \leq n$.

Algorithm 1: Computing the lexicographically smallest subsequence $D[i, \ell]$ in $T[1..i]$ of length ℓ.

---

1 $D[0, 1] \leftarrow \perp$
2 **for** $i = 1$ *to* n **do**   ▷ Initialize $D[\cdot, 1]$
3 $\quad$ $D[i, 1] \leftarrow \min_{j \in [1..i]} T[j] = \min(D[i-1, 1], T[i])$ $\quad$ ▷ $\mathcal{O}(1)$ time per entry

4 **for** $\ell = 2$ *to* n **do**   ▷ Induce $D[\cdot, \ell]$ from $D[\cdot, \ell-1]$
5 $\quad$ **for** $i = 2$ *to* i **do**   ▷ Induce $D[i, \ell]$
6 $\quad\quad$ **if** $\ell < i$ **then** $D[i, \ell] \leftarrow \perp$
7 $\quad\quad$ **else** $D[i, \ell] \leftarrow \min(D[i-1, \ell], D[i-1, \ell-1]T[i])$

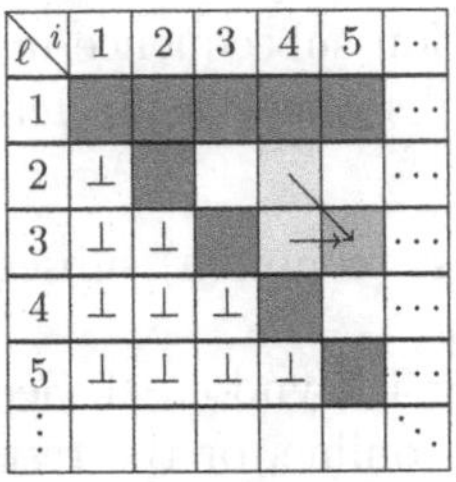

Fig. 1. Sketch of the proof of Lemma 1. We can fill the fields shaded in blue (the first row and the diagonal) in a precomputation step. Further, we know that entries left of the diagonal are all empty. A cell to the right of it (red) is based on its left-preceding and diagonal-preceding cell (green). (Color figure online)

3 Lexicographically Smallest Subsequence

As a starter, we propose a solution for the following related problem: Compute the lexicographically smallest subsequence of T for each length $\ell \in [1..n]$ online.

3.1 Dynamic Programming Approach

The idea is to apply dynamic programming dependent on the length ℓ and the length of the prefix $T[1..i]$ in which we compute the lexicographically smallest subsequence of length ℓ. We show that the lexicographically smallest subsequence of $T[1..i]$ length ℓ, denoted by $D[i, \ell]$ is $D[i-1, \ell]$ or $D[i-1, \ell-1]T[i]$, where $D[0, \cdot] = D[\cdot, 0] = \perp$ is the empty word. See Algorithm 1 for a pseudo code.

Lemma 1. *Algorithm 1 correctly computes $D[i, \ell]$, the lexicographically smallest subsequence of $T[1..i]$ with length ℓ.*

Proof. The proof is done by induction over the length ℓ and the prefix $T[1..i]$. We observe that $D[i, \ell] = \perp$ for $i < \ell$ and $D[i, i] = T[1..i]$ since $T[1..i]$ has only one subsequence of length i. Hence, for (a) $\ell = 1$ as well as for (b) $i \leq \ell$, the claim holds. See Fig. 1 for a sketch.

Now assume that the claim holds for $D[i', \ell']$ with (a) $\ell' < \ell$ and all $i \in [1..n]$, as well as (b) $\ell' = \ell$ and all $i' \in [1..i-1]$. In what follows, we show that the claim also holds for $D[i, \ell]$ with $i > \ell > 1$. For that, let us assume that $T[1..i]$ has a subsequence L of length ℓ with $L \prec D[i, \ell]$.

If $L[\ell] \neq T[i]$, then L is a subsequence of $T[1..i-1]$, and therefore $D[i-1, \ell] \preceq L$ according to the induction hypothesis. But $D[i, \ell] \preceq D[i-1, \ell]$, a contradiction.

If $L[\ell] = T[i]$, then $L[1..\ell-1]$ is a subsequence of $T[1..i-1]$, and therefore $D[i-1, \ell-1] \preceq L[1..\ell-1]$ according to the induction hypothesis. But $D[i, \ell] \preceq D[i-1, \ell-1]T[i] \preceq L[1..\ell-1]T[i] = L$, a contradiction. Hence, $D[i, \ell]$ is the lexicographically smallest subsequence of $T[1..i]$ of length ℓ.

Unfortunately, the lexicographically smallest subsequence of a given length is not a Lyndon word in general, so this dynamic programming approach does not solve our problem finding the longest Lyndon subsequence. In fact, if T has a longest Lyndon subsequence of length ℓ, then there can be a lexicographically smaller subsequence of the same length. For instance, with $T = \text{aba}$, we have the longest Lyndon subsequence ab, while the lexicographically smallest length-2 subsequence is aa.

Analyzing the complexity bounds of Algorithm 1, we need $\mathcal{O}(n^2)$ space for storing the two-dimensional table $D[1..n, 1..n]$. Its initialization costs us $\mathcal{O}(n^2)$ time. Line 7 is executed $\mathcal{O}(n^2)$ time. There, we compute the lexicographical minimum of two subsequences. If we evaluate this computation with naive character comparisons, for which we need to check $\mathcal{O}(n)$ characters, we pay $\mathcal{O}(n^3)$ time in total, which is also the bottleneck of this algorithm.

Lemma 2. *We can compute the lexicographically smallest substring of T for each length ℓ online in $\mathcal{O}(n^3)$ time with $\mathcal{O}(n^2)$ space.*

3.2   Speeding up String Comparisons

Below, we improve the time bound of Lemma 2 by representing each cell of $D[1..n, 1..n]$ with a node in a trie, which supports the following methods:

- insert(v, c): adds a new leaf to a node v with an edge labeled with character c, and returns a handle to the created leaf.
- precedes(u, v): returns true if the string represented by the node u is lexicographically smaller than the string represented by the node v.

Each cell of D stores a handle to its respective trie node. The root node of the trie represents the empty string $\perp$, and we associate $D[0, \ell] = \perp$ with the root node for all ℓ. A node representing $D[i-1, \ell-1]$ has a child representing $D[i, \ell]$ connected with an edge labeled with c if $D[i, \ell] = D[i-1, \ell-1]c$, which is a concept similar to the LZ78 trie. If $D[i, \ell] = D[i-1, \ell]$, then both strings are represented by the same trie node. Since each node stores a constant number of words and an array storing its children, the trie takes $\mathcal{O}(n^2)$ space.

Insert. A particularity of our trie is that it stores the children of a node in the order of their creation, i.e., we always make a new leaf the last among its siblings. This allows us to perform insert in constant time by representing the pointers to the children of a node by a plain dynamic array. When working with the trie, we assure that we do not insert edges into the same node with the same character label (to prevent duplicates).

We add leaves to the trie as follows: Suppose that we compute $D[i, \ell]$. If we can copy $D[i - 1, \ell]$ to $D[i, \ell]$ (Line 7), we just copy the handle of $D[i - 1, \ell]$ pointing to its respective trie node to $D[i, \ell]$. Otherwise, we create a new trie leaf, where we create a new entry of D by selecting a new character ($\ell = 1$), or appending a character to one of the existing strings in D. We do not create duplicate edges since we prioritize copying to the creation of a new trie node: For an entry $D[i, \ell]$, we first default to the previous occurrence $D[i - 1, \ell]$, and only create a new string $D[i - 1, \ell - 1]T[i]$ if $D[i - 1, \ell - 1]T[i] \prec D[i - 1, \ell]$. $D[i - 1, \ell - 1]T[i]$ cannot have an occurrence represented in the trie. To see that, we observe that D obeys the invariants that (a) $D[i, \ell] = \min_{j \in [1..i]} D[j, \ell]$ (where min selects the lexicographically minimal string) and (b) all pairs of rows $D[\cdot, \ell]$ and $D[\cdot, \ell']$ with $\ell \neq \ell'$ have different entries. Since Algorithm 1 fills the entries in $D[\cdot, \ell]$ in a lexicographically non-decreasing order for each length ℓ, we cannot create duplicates (otherwise, an earlier computed entry would be lexicographically smaller than a later computed entry having the same length). The string comparison $D[i-1, \ell-1]T[i] \prec D[i-1, \ell]$ is done by calling precedes, which works as follows:

Precedes. We can implement the function precedes efficiently by augmenting our trie with the dynamic data structure of [7] supporting lowest common ancestor (LCA) queries in constant time and the dynamic data structure of [8] supporting level ancestor queries level-anc(u, d) returning the ancestor of a node u on depth d in amortized constant time. Both data structures conform with our definition of insert that only supports the insertion of *leaves*. With these data structures, we can implement precedes(u, v), by first computing the lowest ancestor w of u and v, selecting the children u' and v' of w on the paths downwards to u and v, respectively, by two level ancestor queries level-anc(u, depth(w) + 1) and level-anc(v, depth(w) + 1), and finally returning true if the label of the edge (w, u') is smaller than of (w, v').

We use precedes as follows for deciding whether $D[i-1, \ell-1]T[i] \prec D[i-1, \ell]$ holds: Since we know that $D[i-1, \ell-1]$ and $D[i-1, \ell]$ are represented by nodes u and v in the trie, respectively, we first check whether u is a child of v. In that case, we only have to compare $T[i]$ with $D[i - 1, \ell][\ell]$. If not, then we know that $D[i - 1, \ell - 1]$ cannot be a prefix of $D[i - 1, \ell]$, and precedes(u, v) determines whether $D[i - 1, \ell - 1]$ or the $\ell - 1$-th prefix of $D[i - 1, \ell]$ is lexicographically smaller.

Theorem 3. *We can compute the table $D[1..n, 1..n]$ in $\mathcal{O}(n^2)$ time using $\mathcal{O}(n^2)$ words of space.*

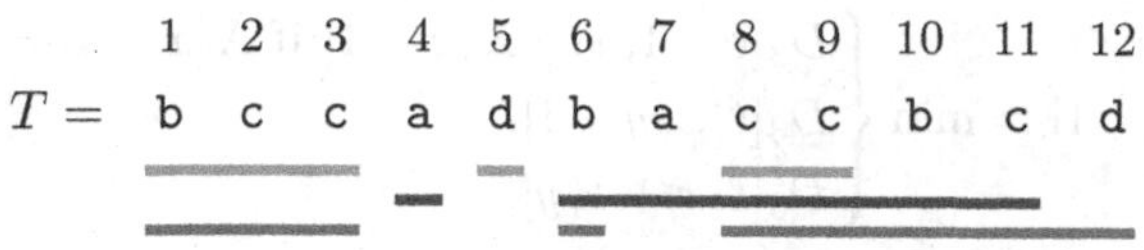

$$
\begin{array}{ccccccccccccc}
 & 1 & 2 & 3 & 4 & 5 & 6 & 7 & 8 & 9 & 10 & 11 & 12 \\
T = & b & c & c & a & d & b & a & c & c & b & c & d
\end{array}
$$

Fig. 2. Longest Lyndon subsequences of prefixes of a text T. The i-th row of bars below T depicts the selection of characters forming a Lyndon sequence. In particular, the i-th row corresponds to the longest subsequence of $T[1..9]$ for $i = 1$ (green), $T[1..11]$ for $i = 2$ (blue), and of $T[1..12]$ for $i = 3$ (red). The first row (green) corresponds also to a longest Lyndon subsequence of $T[1..10]$ and $T[1..11]$ (by extending it with $T[11]$). Extending the second Lyndon subsequence with $T[12]$ gives also a Lyndon subsequence, but is shorter than the third Lyndon subsequence (red). Having only the information of the Lyndon subsequences in $T[1..i]$ at hand seems not to give us a solution for $T[1..i+1]$. (Color figure online)

3.3 Most Competitive Subsequence

If we want to find only the lexicographically smallest subsequence for a fixed length ℓ, this problem is also called to *Find the Most Competitive Subsequence*[1]. For that problem, there are linear-time solutions using a stack S storing the lexicographically smallest subsequence of length ℓ for any prefix $T[1..i]$ with $\ell \le i$. Let top denote the top element of S. The idea is to scan T from left to right linearly. Given we are at a text position i, we recursively pop top as long as (a) S is not empty, (b) $T[\text{top}] > T[i]$, and (c) $n - i \ge (\ell - |S|)$. The last condition ensures that when we are near the end of the text, we still have enough positions in S to fill up S with the remaining positions to obtain a sequence of ℓ text positions. Finally, we put $T[i]$ on top of S if $|S| < \ell$. Since a text position gets inserted into S and removed from S at most once, the algorithm runs in linear time. Consequently, if the whole text T is given (i.e., not online), this solution solves our problem in the same time and space bounds by running the algorithm for each ℓ separately.

Given $T = \text{cba}$ as an example, for $\ell = 3$, we push all three characters of T onto S and output cba. For $\ell = 2$, we first push $T[1] = \text{c}$ onto S, but then pop it and push b onto S. Finally, although $T[3] < T[2]$, we do not discard $T[2] = \text{b}$ stored on S since we need to produce a subsequence of length $\ell = 2$.

3.4 Lexicographically Smallest Common Subsequence

Another variation is to ask for the lexicographically smallest subsequence of each distinct length that is common with two strings X and Y. Luckily, our ideas of Sects. 3.1 and 3.2 can be straightforwardly translated. For that, our matrix D becomes a cube $D_3[1..L, 1..|X|, 1..|Y|]$ with $L := \min(|X|, |Y|)$, and we set

---

[1] https://leetcode.com/problems/find-the-most-competitive-subsequence/.

$$D_3[\ell, x+1, y+1] = \min \begin{cases} D_3[\ell-1, x, y]X[x+1] \text{ if } X[x+1] = Y[y+1], \\ D_3[\ell, x, y+1], \\ D_3[\ell, x+1, y], \end{cases}$$

with $D_3[0, \cdot, \cdot] = D_3[\ell, x, y] = \perp$ for all ℓ, x, y with $\mathrm{LCS}(X[1..x], Y[1..y]) < \ell$, where LCS denotes the length of a longest common subsequence of X and Y. This gives us an induction basis similar to the one used in the proof of Lemma 1, such that we can use its induction step analogously. The table D_3 has $\mathcal{O}(n^3)$ cells, and filling each cell can be done in constant time by representing each cell as a pointer to a node in the trie data structure proposed in Sect. 3.2. For that, we ensure that we never insert a subsequence of D_3 into the trie twice. To see that, let $L \in \Sigma^+$ be a subsequence computed in D_3, and let $D_3[\ell, x, y] = L$ be the entry at which we called insert to create a trie node for L (for the first time). Then $\ell = |L|$, and $X[1..x]$ and $Y[1..y]$ are the shortest prefixes of X and Y, respectively, containing L as a subsequence. Since $D_3[\ell, x, y] = \min_{x' \in [1..x], y' \in [1..y]} D_3[\ell, x', y']$, all other entries $D_3[\ell, x', y'] = L$ satisfy $D_3[\ell, x'-1, y'] = L$ or $D_3[\ell, x', y'-1] = L$, so we copy the trie node handle representing L instead of calling insert when filling out $D_3[\ell, x', y']$.

Theorem 4. *Given two strings X, Y of length n, we can compute the lexicographically smallest common subsequence for each length $\ell \in [1..n]$ in $\mathcal{O}(n^3)$ time using $\mathcal{O}(n^3)$ space.*

4   Computing the Longest Lyndon Subsequence

In the following, we want to compute the longest Lyndon subsequence of T. See Fig. 2 for examples of longest Lyndon subsequences. Compared to the former introduced dynamic programming approach for the lexicographically smallest subsequences, we follow the sketched solution for the most competitive subsequence using a stack, which here simulates a traversal of the trie τ storing all pre-Lyndon subsequences. A *pre-Lyndon subsequence* is a subsequence that is Lyndon or can be extended with characters at its right end to become Lyndon. τ is a subgraph of the trie storing all subsequences, sharing the same root. This subgraph is connected since, by definition, there is no string S such that WS forms a pre-Lyndon word for a non-pre-Lyndon word W (otherwise, we could extend WS to a Lyndon word, and so W, too). We say that the *string label* of a node v is the string read from the edges on the path from root to v. We associate the label c of each edge of the trie with the leftmost possible position such that the string label V of v is associated with the sequence of text positions $i_1 < i_2 < \cdots < i_{|V|}$ and $T[i_1]T[i_2] \cdots T[i_{|V|}] = V$.

4.1   Basic Trie Traversal

Problems already emerge when considering the construction of τ since there are texts like $T = 1 \cdots n$ for which τ has $\Theta(2^n)$ nodes. Instead of building τ, we simulate a preorder traversal on it. With simulation we mean that we enumerate the pre-Lyndon subsequences of T in lexicographic order. For that, we maintain a stack S storing the text positions $(i_1, \ldots, i_\ell)$ with $i_1 < \cdots < i_\ell$ associated with the path from the root to the node v we currently visit i.e., $i_1, \ldots, i_\ell$ are the smallest positions with $T[i_1] \cdots T[i_\ell]$ being the string label of v, which is a pre-Lyndon word. When walking down, we select the next text position $i_{\ell+1}$ such that $T[i_1] \cdots T[i_\ell]T[i_{\ell+1}]$ is a pre-Lyndon word. If such a text position does not exist, we backtrack by popping i_ℓ from S, and push the smallest text position $i'_\ell > i_{\ell-1}$ with $T[i'_\ell] > T[i_\ell]$ onto S and recurse. Finally, we check at each state of S storing the text positions $(i_1, \ldots, i_\ell)$ whether $T[i_1] \cdots T[i_\ell]$ is a Lyndon word. For that, we make use of the following facts:

Facts About Lyndon Words. A Lyndon word cannot have a *border*, that is, a non-empty proper prefix that is also a suffix of the string [9, Prop. 1.1]. A *pre-Lyndon word* is a (not necessarily proper) prefix of a Lyndon word. Given a string S of length n, an integer $p \in [1..n]$ is a *period* of S if $S[i] = S[i + p]$ for all $i \in [1..n - p]$. The length of a string is always one of its periods. We use the following facts:

(Fact 1) Only the length $|S|$ is the period of a Lyndon word S.

(Fact 2) The prefix $S[1..|p|]$ of a pre-Lyndon word S with period p is a Lyndon word. In particular, a pre-Lyndon word S with period $|S|$ is a Lyndon word.

(Fact 3) Given a pre-Lyndon word S with period p and a character $c \in \Sigma$, then
- Sc is a pre-Lyndon word of the same period if and only if $S[|S| - p + 1] = c$ and S is not the largest character in Σ.
- Sc is a Lyndon word if and only if $S[|S| - p + 1] < c$. In particular, if S is a Lyndon word, then Sc is a Lyndon word if and only if $S[1]$ is smaller than c.

Proof. **Fact 1** If S has a period less than $|S|$, then S is bordered.
Fact 2 If $S[1..|p|]$ would not be Lyndon, then there was a suffix X of S with $X \prec S[1..|X|]$, hence $XZ \prec SZ$ for every $Z \in \Sigma^*$, so S cannot be pre-Lyndon.
Fact 3 "$\Rightarrow$": If $T := Sc$ is a pre-Lyndon word with the same period as S, then T has a border $T[p + 1..|T|] = T[1..|T| - p]$. "$\Leftarrow$": Follows from Fact 2 and [9, Corollary 1.4].

Checking Pre-Lyndon Words. Now suppose that our stack S stores the text positions $(i_1, \ldots, i_\ell)$. To check whether $T[i_1] \cdots T[i_\ell]c$ for a character $c \in \Sigma$ is a pre-Lyndon word or whether it is a Lyndon word, we augment each position i_j stored in S with the period of $T[i_1] \cdots T[i_j]$, for $j \in [1..\ell]$, such that we can make use of Fact 3 to compute the period and check whether $T[i_1] \cdots T[i_j]c$ is a pre-Lyndon word, both in constant time, for $c \in \Sigma$.

Trie Navigation. To find the next text position $i_{\ell+1}$, we may need to scan $\mathcal{O}(n)$ characters in the text, and hence need $\mathcal{O}(n)$ time for walking down from a node to one of its children. If we restrict the alphabet to be integer, we can augment each text position i to store the smallest text position i_c with $i < i_c$ for each character $c \in \Sigma$ such that we can visit the trie nodes in constant time per node during our preorder traversal.

This gives already an algorithm that computes the longest Lyndon subsequence with $\mathcal{O}(n\sigma)$ space and time linear to the number of nodes in τ. However, since the number of nodes can be exponential in the text length, we present ways to omit nodes that do not lead to the solution. Our aim is to find a rule to judge whether a trie node contributes to the longest Lyndon subsequence to leave certain subtrees of the trie unexplored. For that, we use the following property:

Lemma 5. *Given a Lyndon word V and two strings U and W such that UW is a Lyndon word, $V \prec U$, and $|V| \geq |U|$, then VW is also a Lyndon word with $VW \prec UW$.*

Proof. Since $V \prec U$ and V is not a prefix of U, $U \succ VW$. In what follows, we show that $S \succ VW$ for every proper suffix S of VW.

- If S is a suffix of W, then $S \succeq UW \succeq U \succ VW$ because S is a suffix of the Lyndon word UW.
- Otherwise, ($|S| > |W|$), S is of the form $V'W$ for a proper suffix V' of V. Since V is a Lyndon word, $V' \succ V$, and V' is not a prefix of V (Lyndon words are border-free). Hence, $V'W \succeq V' \succ VW$.

Note that U in Lemma 5 is a pre-Lyndon word since it is the prefix of the Lyndon word UW.

Our algorithmic idea is as follows: We maintain an array $\mathsf{L}[1..n]$, where $\mathsf{L}[\ell]$ is the smallest text position i such that our traversal has already explored a length-ℓ Lyndon subsequence of $T[1..i]$. We initialize the entries of L with ∞ at the beginning. Now, whenever we visit a node u whose string label is a pre-Lyndon subsequence $U = T[i_1] \cdots T[i_\ell]$ with $\mathsf{L}[\ell] \leq i_\ell$, then we do not explore the children of u. In this case, we call u *irrelevant*. By skipping the subtree rooted at u, we do not omit the solution due to Lemma 5: When $\mathsf{L}[\ell] \leq i_\ell$, then there is a Lyndon subsequence V of $T[1..i_\ell]$ with $V \prec U$ (since we traverse the trie in lexicographically order) and $|V| = |U|$. Given there is a Lyndon subsequence UW of T, then we have already found VW earlier, which is also a Lyndon subsequence of T with $|VW| = |UW|$.

Next, we analyze the complexity of this algorithm, and propose an improved version. For that, we say that a string is *immature* if it is pre-Lyndon but not Lyndon. We also consider a subtree rooted at a node u as pruned if u is irrelevant, i.e., the algorithm does not explore this subtree. Consequently, irrelevant nodes are leaves in the pruned subtree, but not all leaves are irrelevant (consider a Lyndon subsequence using the last text position $T[n]$). Further, we call a node Lyndon or immature if its string label is Lyndon or immature, respectively. (All nodes in the trie are either Lyndon or immature.)

Time Complexity. Suppose that we have the text positions $(i_1, \ldots, i_\ell)$ on S such that $U := T[i_1] \cdots T[i_\ell]$ is a Lyndon word. If $\mathsf{L}[\ell] > i_\ell$, then we lower

$L[\ell] \leftarrow i_\ell$. We can lower an individual entry of L at most n times, or at most n^2 times in total for all entries. If a visited node is Lyndon, we only explore its subtree if we were able to lower an entry of L. Hence, we visit at most n^2 Lyndon nodes that trigger a decrease of the values in L. While each node can have at most σ children, at most one child can be immature due to Fact 3. Since the depth of the trie is at most n, we therefore visit $\mathcal{O}(n\sigma)$ nodes between two updates of L (we pop at most n nodes from the stack, and try to explore at most σ siblings of each node on the stack). These nodes are leaves (of the pruned trie) or immature nodes. Thus, we traverse $\mathcal{O}(n^3\sigma)$ nodes in total.

Theorem 6. *We can compute the longest Lyndon subsequence of a string of length n in $\mathcal{O}(n^3\sigma)$ time using $\mathcal{O}(n\sigma)$ words of space.*

4.2 Improving Time Bounds

We further improve the time bounds by avoiding visiting irrelevant nodes due to the following observation: First, we observe that the number of *relevant* (i.e., non-irrelevant) nodes that are Lyndon is $\mathcal{O}(n^2)$. Since all nodes have a depth of at most n, the total number of relevant nodes in $\mathcal{O}(n^3)$. Suppose we are at a node u, and S stores the positions $(i_1, \ldots, i_\ell)$ such that $T[i_1] \cdots T[i_\ell]$ is the string label of u. Let p denote the smallest period of $T[i_1] \cdots T[i_\ell]$. Then we do not want to consider all σ children of u, but only those whose edges to u have a label $c \geq T[i_{\ell-p+1}]$ such that c occurs in $T[i_\ell + 1..L[\ell + 1] - 1]$ (otherwise, there is already a Lyndon subsequence of length $\ell + 1$ lexicographically smaller than $T[i_1] \cdots T[i_\ell]c$). In the context of our preorder traversal, each such child can be found iteratively using range successor queries: starting from $b = T[i_{\ell-p+1}] - 1$, we want to find the *lexicographically smallest* character $c > b$ such that c occurs in $T[i_\ell+1..L[\ell+1]-1]$. In particular, we want to find the leftmost such occurrence. A data structure for finding c in this interval is the wavelet tree [14] returning the position of the *leftmost* such c (if it exists) in $\mathcal{O}(\lg\sigma)$ time. In particular, we can use the wavelet tree instead of the $\mathcal{O}(n\sigma)$ pointers to the subsequent characters to arrive at $\mathcal{O}(n)$ words of space. Finally, we do not want to query the wavelet tree each time, but only whenever we are sure that it will lead us to a relevant Lyndon node. For that, we build a range maximum query (RMQ) data structure on the characters of the text T in a preprocessing step. The RMQ data structure of [3] can be built in $\mathcal{O}(n)$ time; it answers queries in constant time. Now, in the context of the above traversal where we are at a node u with S storing $(i_1, \ldots, i_\ell)$, we query this RMQ data structure for the largest character c in $T[i_\ell + 1..L[\ell + 1] - 1]$ and check whether the sequence $S := T[i_1] \cdots T[i_\ell]c$ forms a (pre-)Lyndon word.

- If S is not pre-Lyndon, i.e., $T[i_{\ell-p+1}] > c$ for p being the smallest period of $T[i_1] \cdots T[i_\ell]$, we are sure that the children of u cannot lead to Lyndon subsequences [9, Prop. 1.5].
- If S is immature, i.e., $T[i_{\ell-p+1}] = c$, u has exactly one child, and this child's string label is S. Hence, we do not need to query for other Lyndon children.

– Finally, if S is Lyndon, i.e., $T[i_{\ell-p+1}] < c$, we know that there is at least one child of u that will trigger an update in L and thus is a relevant node.

This observation allows us to find all relevant children of u (including the single immature child, if any) by iteratively conducting $\mathcal{O}(k)$ range successor queries, where k is the number of children of u that are relevant Lyndon nodes. Thus, if we condition the execution of the aforementioned wavelet tree query with an RMQ query result on the same range, the total number of wavelet tree queries can be bounded by $\mathcal{O}(n^2)$. This gives us $\mathcal{O}(n^3 + n^2 \lg \sigma) = \mathcal{O}(n^3)$ time for $\sigma = \mathcal{O}(n)$ (which can be achieved by an $\mathcal{O}(n \log n)$ time re-enumeration of the alphabet in a preliminary step).

Theorem 7. *We can compute the longest Lyndon subsequence of a string of length n in $\mathcal{O}(n^3)$ time using $\mathcal{O}(n)$ words of space.*

In particular, the algorithm computes the lexicographically smallest one among all longest Lyndon subsequences: Assume that this subsequence L is not computed, then we did not explore the subtree of the original trie τ (before pruning) containing the node with string label L. Further, assume that this subtree is rooted at an irrelevant node u whose string label is the pre-Lyndon subsequence U. Then U is a prefix of L, and because u is irrelevant (i.e., we have not explored u's children), there is a node v whose string label is a Lyndon word V with $V \prec U$ and $|V| = |U|$. In particular, the edge of v to v's parent is associated with a text position equal to or smaller than the associated text position of the edge between u and u's parent. Hence, we can extend V to the Lyndon subsequence $VL[|U| + 1..]$ being lexicographically smaller than L, a contradiction.

4.3   Online Computation

If we allow increasing the space usage in order to maintain the trie data structure introduced in Sect. 3.2, we can modify our $\mathcal{O}(n^3\sigma)$-time algorithm of Sect. 4.1 to perform the computation online, i.e., with T given as a text stream. To this end, let us recall the trie τ of all pre-Lyndon subsequences introduced at the beginning of Sect. 4. In the online setting, when reading a new character c, for each subsequence S given by a path from τ's root (S may be empty), we add a new node for Sc if Sc is a pre-Lyndon subsequence that is not yet represented by such a path. Again, storing all nodes of τ explicitly would cost us too much space. Instead, we explicitly represent only the visited nodes of the trie τ with an explicit trie data structure τ' such that we can create pointers to the nodes. (In other words, τ' is a lazy representation of τ.) The problem is that we can no longer perform the traversal in lexicographic order, but instead keep multiple fingers in the trie τ' constructed up so far, and use these fingers to advance the trie traversal in text order.

With a different traversal order, we need an updated definition of L$[1..n]$: Now, while the algorithm processes $T[i]$, the entry L$[\ell]$ stores the lexicographically smallest length-ℓ Lyndon subsequence of $T[1..i]$ (represented by a pointer to the corresponding node of τ'). Further, we maintain σ lists storing pointers to nodes of τ'. Initially, τ' consists only of the root node, and each list stores only the root node. Whenever we read a new character $T[i]$ from the text stream, for each node v of the $T[i]$-th list, we add a leaf λ connected to v by an edge with label $T[i]$. Our algorithm adheres to the invariant that λ's string label S is a pre-Lyndon word so that τ' is always a subtree of τ. If S is a Lyndon word satisfying $S \prec L[|S|]$ (which can be tested using the data structure of Sect. 3.2), we further set $L[|S|] := S$. This completes the process of updating $L[1..n]$. Next, we clear the $T[i]$-th list and iterate again over the newly created leaves. For each such leaf λ with label S, we check whether λ is relevant, i.e., whether $S \preceq L[|S|]$. If λ turns out irrelevant, we are done with processing it. Otherwise, we put λ into the c-th list for each character $c \in \Sigma$ such that Sc is a pre-Lyndon word. By doing so, we effectively create new events that trigger a call-back to the point where we stopped the trie traversal.

Overall, we generate exactly the nodes visited by the algorithm of Sect. 4.1. In particular, there are $\mathcal{O}(n^3)$ relevant nodes, and for each such node, we issue $\mathcal{O}(\sigma)$ events. The operations of Sect. 3.2 take constant amortized time, so the overall time and space complexity of the algorithm are $\mathcal{O}(n^3\sigma)$.

Theorem 8. *We can compute the longest Lyndon subsequence online in $\mathcal{O}(n^3\sigma)$ time using $\mathcal{O}(n^3\sigma)$ space.*

5   Longest Common Lyndon Subsequence

Given two strings X and Y, we want to compute the longest common subsequence of X and Y that is Lyndon. For that, we can extend our algorithm finding the longest Lyndon subsequence of a single string as follows. First, we explore in depth-first order the trie of all *common* pre-Lyndon subsequences of X and Y. A node is represented by a pair of positions (x, y) such that, given the path from the root to a node v of depth ℓ visits the nodes $(x_1, y_1), \ldots, (x_\ell, y_\ell)$ with $L = X[x_1] \cdots X[x_\ell] = Y[y_1] \cdots Y[y_\ell]$ being a pre-Lyndon word, L is neither a subsequence of $X[1..x_\ell - 1]$ nor of $Y[1..y_\ell - 1]$, i.e., x_ℓ and y_ℓ are the leftmost such positions. The depth-first search works like an exhaustive search in that it tries to extend L with each possible character in Σ having an occurrence in both remaining suffixes $X[x_\ell + 1..]$ and $Y[y_\ell + 1..]$, and then, after having explored the subtree rooted at v, visits its lexicographically succeeding sibling nodes (and descends into their subtrees) by checking whether $L[1..|L| - 1]$ can be extended with a character $c > L[|L|]$ appearing in both suffixes $X[x_{\ell-1} + 1..]$ and $Y[y_{\ell-1} + 1..]$.

The algorithm uses again the array L to check whether we have already found a lexicographically smaller Lyndon subsequence with equal or smaller ending positions in X and Y than the currently constructed pre-Lyndon subsequence. For that, $L[\ell]$ stores not only one position, but a list of positions (x, y) such that $X[1..x]$ and $Y[1..y]$ have a *common* Lyndon subsequence of length ℓ. Although there can be n^2 such pairs of positions, we only store those that are pairwise non-dominated. A pair of positions (x_1, y_1) is called *dominated* by a pair $(x_2, y_2) \neq$

(x_1, y_1) if $x_2 \leq x_1$ and $y_2 \leq y_1$. A set storing pairs in $[1..n] \times [1..n]$ can have at most n elements that are pairwise non-dominated, and hence $|L[\ell]| \leq n$.

At the beginning, all lists of L are empty. Suppose that we visit a node v with pair (x_ℓ, y_ℓ) representing a common Lyndon subsequence of length ℓ. Then we query whether $L[\ell]$ has a pair dominating (x_ℓ, y_ℓ). In that case, we can skip v and its subtree. Otherwise, we insert (x_ℓ, y_ℓ) and remove pairs in $L[\ell]$ that are dominated by (x_ℓ, y_ℓ). Such an insertion can happen at most n^2 times. Since $L[1..n]$ maintains n lists, we can update L at most n^3 times in total. Checking for domination and insertion into L takes $\mathcal{O}(n)$ time. The former can be accelerated to constant time by representing $L[\ell]$ as an array R_ℓ storing in $R_\ell[i]$ the value y of the tuple $(x, y) \in L[\ell]$ with $x \leq i$ and the lowest possible y, for each $i \in [1..n]$. Then a pair $(x, y) \notin L[\ell]$ is dominated if and only if $R_\ell[x] \leq y$.

Example 9. For $n = 10$, let $L_\ell = [(3,9), (5,4), (8,2)]$. Then all elements in L_ℓ are pairwise non-dominated, and $R_\ell = [\infty, \infty, 9, 9, 4, 4, 4, 2, 2, 2]$. Inserting $(3,2)$ would remove all elements of L_ℓ, and update all entries of R_ℓ. Alternatively, inserting $(7,3)$ would only involve updating $R_\ell[7] \leftarrow 3$; since the subsequent entry $R_\ell[8] = 2$ is less than $R_\ell[7]$, no subsequent entries need to be updated.

An update in $L[\ell]$ involves changing $\mathcal{O}(n)$ entries of R_ℓ, but that cost is dwarfed by the cost for finding the next common Lyndon subsequence that updates L. Such a subsequence can be found while visiting $\mathcal{O}(n\sigma)$ irrelevant nodes during a naive depth-first search (cf. the solution of Sect. 3.1 computing the longest Lyndon sequence of a single string). Hence, the total time is $\mathcal{O}(n^4\sigma)$.

Theorem 10. *We can compute the longest common Lyndon subsequence of a string of length n in $\mathcal{O}(n^4\sigma)$ time using $\mathcal{O}(n^3)$ words of space.*

Open Problems. Since we shed light on the computation of the longest (common) Lyndon subsequence for the very first time, we are unaware of the optimality of our solutions. It would be interesting to find non-trivial lower bounds that would justify our rather large time and space complexities.

Acknowledgments. This work was supported by JSPS KAKENHI Grant Numbers JP20H04141 (HB), JP19K20213 (TI), JP21K17701 and JP21H05847 (DK). TK was supported by NSF 1652303, 1909046, and HDR TRIPODS 1934846 grants, and an Alfred P. Sloan Fellowship.

References

1. Bannai, H., I, T., Inenaga, S., Nakashima, Y., Takeda, M., Tsuruta, K.: The "runs" theorem. SIAM J. Comput. **46**(5), 1501–1514 (2017)
2. de Beauregard Robinson, G.: On the representations of the symmetric group. Am. J. Math. **60**(3), 745–760 (1938)

3. Bender, M.A., Farach-Colton, M., Pemmasani, G., Skiena, S., Sumazin, P.: Lowest common ancestors in trees and directed acyclic graphs. J. Algorithms **57**(2), 75–94 (2005)
4. Biedl, T.C., et al.: Rollercoasters: long sequences without short runs. SIAM J. Discret. Math. **33**(2), 845–861 (2019)
5. Chen, K.T., Fox, R.H., Lyndon, R.C.: Free differential calculus, IV. The quotient groups of the lower central series. Ann. Math. Second Ser. **68**(1), 81–95 (1958). https://www.jstor.org/stable/1970044. Mathematics Department, Princeton University
6. Chowdhury, S.R., Hasan, M.M., Iqbal, S., Rahman, M.S.: Computing a longest common palindromic subsequence. Fundam. Inform. **129**(4), 329–340 (2014)
7. Cole, R., Hariharan, R.: Dynamic LCA queries on trees. SIAM J. Comput. **34**(4), 894–923 (2005)
8. Dietz, P.F.: Finding level-ancestors in dynamic trees. In: Dehne, F., Sack, J.R., Santoro, N. (eds.) Algorithms and Data Structures. WADS 1991. LNCS, vol. 519, pp. 32–40. Springer, Heidelberg (1991). https://doi.org/10.1007/BFb0028247. ISBN 978-3-540-47566-8
9. Duval, J.: Factorizing words over an ordered alphabet. J. Algorithms **4**(4), 363–381 (1983)
10. Elmasry, A.: The longest almost-increasing subsequence. Inf. Process. Lett. **110**(16), 655–658 (2010)
11. Fujita, K., Nakashima, Y., Inenaga, S., Bannai, H., Takeda, M.: Longest common rollercoasters. In: Lecroq, T., Touzet, H. (eds.) SPIRE 2021. LNCS, vol. 12944, pp. 21–32. Springer, Cham (2021). https://doi.org/10.1007/978-3-030-86692-1_3
12. Gawrychowski, P., Manea, F., Serafin, R.: Fast and longest rollercoasters. In: Proceedings of STACS. LIPIcs, vol. 126, pp. 30:1–30:17 (2019)
13. Glen, A., Simpson, J., Smyth, W.F.: Counting Lyndon factors. Electron. J. Comb. **24**(3), P3.28 (2017)
14. Grossi, R., Gupta, A., Vitter, J.S.: High-order entropy-compressed text indexes. In: Proceedings of SODA, pp. 841–850 (2003)
15. He, X., Xu, Y.: The longest commonly positioned increasing subsequences problem. J. Comb. Optim. **35**(2), 331–340 (2017). https://doi.org/10.1007/s10878-017-0170-9
16. Hirakawa, R., Nakashima, Y., Inenaga, S., Takeda, M.: Counting Lyndon subsequences. In: Proceedings of PSC, pp. 53–60 (2021)
17. Hirschberg, D.S.: Algorithms for the longest common subsequence problem. J. ACM **24**(4), 664–675 (1977)
18. Inenaga, S., Hyyrö, H.: A hardness result and new algorithm for the longest common palindromic subsequence problem. Inf. Process. Lett. **129**, 11–15 (2018)
19. Kiyomi, M., Horiyama, T., Otachi, Y.: Longest common subsequence in sublinear space. Inf. Process. Lett. **168**, 106084 (2021)
20. Knuth, D.: Permutations, matrices, and generalized Young tableaux. Pac. J. Math. **34**, 709–727 (1970)
21. Kosche, M., Koß, T., Manea, F., Siemer, S.: Absent subsequences in words. In: Bell, P.C., Totzke, P., Potapov, I. (eds.) RP 2021. LNCS, vol. 13035, pp. 115–131. Springer, Cham (2021). https://doi.org/10.1007/978-3-030-89716-1_8
22. Kosowski, A.: An efficient algorithm for the longest tandem scattered subsequence problem. In: Apostolico, A., Melucci, M. (eds.) SPIRE 2004. LNCS, vol. 3246, pp. 93–100. Springer, Heidelberg (2004). https://doi.org/10.1007/978-3-540-30213-1_13

23. Kutz, M., Brodal, G.S., Kaligosi, K., Katriel, I.: Faster algorithms for computing longest common increasing subsequences. J. Discrete Algorithms **9**(4), 314–325 (2011)

24. Lyndon, R.C.: On Burnside's problem. Trans. Am. Math. Soc. **77**(2), 202–215 (1954)

25. Schensted, C.: Longest increasing and decreasing subsequences. Can. J. Math. **13**, 179–191 (1961)

26. Ta, T.T., Shieh, Y., Lu, C.L.: Computing a longest common almost-increasing subsequence of two sequences. Theor. Comput. Sci. **854**, 44–51 (2021)

27. Wagner, R.A., Fischer, M.J.: The string-to-string correction problem. J. ACM **21**(1), 168–173 (1974)

Structure-Aware Combinatorial Group Testing: A New Method for Pandemic Screening

Thaís Bardini Idalino[1] and Lucia Moura[2(✉)]

[1] Universidade Federal de Santa Catarina, Florianópolis, Brazil
thais.bardini@ufsc.br
[2] University of Ottawa, Ottawa, Canada
lmoura@uottawa.ca

Abstract. Combinatorial group testing (CGT) is used to identify defective items from a set of items by grouping them together and performing a small number of tests on the groups. Recently, group testing has been used to design efficient COVID-19 testing, so that resources are saved while still identifying all infected individuals. Due to test waiting times, a focus is given to non-adaptive CGT, where groups are designed a priori and all tests can be done in parallel. The design of the groups can be done using Cover-Free Families (CFFs). The main assumption behind CFFs is that a small number d of positives are randomly spread across a population of n individuals. However, for infectious diseases, it is reasonable to assume that infections show up in clusters of individuals with high contact (children in the same classroom within a school, households within a neighbourhood, students taking the same courses within a university, people seating close to each other in a stadium). The general structure of these communities can be modeled using hypergraphs, where vertices are items to be tested and edges represent clusters containing high contacts. We consider hypergraphs with non-overlapping edges and overlapping edges (first two examples and last two examples, respectively). We give constructions of what we call *structure-aware* CFF, which uses the structure of the underlying hypergraph. We revisit old CFF constructions, boosting the number of defectives they can identify by taking the hypergraph structure into account. We also provide new constructions based on hypergraph parameters.

1 Introduction

Group testing literature dates back to the Second World War as an efficient way of testing blood samples for syphilis screening [3,4]. The idea consists of grouping blood samples together before testing, so that negative results could save hundreds of individual tests. This idea was then applied to many other areas: screening vaccines for contamination, building clone libraries for DNA sequences, data forensics for altered documents, modification tolerant digital signatures [6,8,14–17,19,20]. Currently, it is considered a promising scheme for

C. Bazgan and H. Fernau (Eds.): IWOCA 2022, LNCS 13270, pp. 143–156, 2022.
https://doi.org/10.1007/978-3-031-06678-8_11

saving time and resources in COVID-19 testing [5,25–27,31]. In fact, several countries, such as China, India, Germany and the United States, have adopted group testing as a way of saving time and resources [25].

In combinatorial group testing (CGT), we are given n items of which at most d are defective (or contaminated). We assume we can test any subset of items, and if the result of the test is positive the subset contains at least one defective (contaminated) item, and if it is negative all items in the subset are non-defective (uncontaminated). The main goal is to minimize the number t of tests for given n and d, while determining all defective items. For a comprehensive treatment, see the text by Du and Hwang [4].

Group testing may be adaptive or non-adaptive [4]. *Adaptive* CGT allows us to decide the next tests according to the results of previous tests. This is the case of the binary spliting algorithm, which meets the information theoretical lower bound of $d \log(n/d)$ tests. In this paper, we focus on *non-adaptive* CGT. Due to test waiting times, non-adaptive CGT is a useful approach, since we decide all groups at once and can run tests in parallel. In addition, in non-adaptive CGT, we can have more balanced sizes of the groups (items in each test), which is limited in some real applications. For COVID-19 screening, researchers are testing how many samples can be grouped together without compromising the detection of positive results [25,31].

Items and tests in CGT can be represented by a binary matrix where items correspond to columns and tests correspond to rows, where a 1 means a test uses an item. A d-cover free family (or d-CFF(t, n)) is a $t \times n$ matrix with special properties that guarantee the identification of d defective items among n items using t tests and a simple decoding algorithm that takes time $O(tn)$ (see Sect. 2).

In this paper, we are interested in applications where the defective items are more likely to appear together in predictable subsets of items, which are given as edges of a hypergraph. For example, if we want to monitor a highly transmissible disease among students in a school, classrooms can be the edges (or regions) where it is more likely that if there is one infected individual we may find many. In this way, outbreaks may be detected early while only a few classrooms have infected students. In this model, we are given a hypergraph where items are vertices and regions are edges such that there are at most r edges that together contain all defective vertices. The objective is still to minimize the number of tests while identifying all defective items. A weaker version of the problem consists of simply identify all infected edges. In this paper we initiate a more systematic study of how to build CFFs for combinatorial group testing under the hypergraph model, which we call *structure-aware* cover-free families.

Recent Related Work. A similar hypergraph model has been recently proposed as group testing in connected and overlapping communities in the context of COVID-19 testing [26,27], as group testing on general set systems [12], and as variable cover-free families motivated by problems in cryptography [14]. The work in [12,26,27] span both adaptive and nonadaptive CGT algorithms, but there is not much emphasis on CGT matrix constructions. Our work is on efficient cover-free family constructions for the hypergraph model. The idea of structure-aware CFF was introduced in the first author's PhD thesis [14] under the name of *variable* CFFs (VCFFs) with an equivalent definition. This was inspired by applications in cryptography, where they would allow for location of clustered modifications in a signed document when using modification-tolerant digital signatures. The hypergraphs considered here and the ones in [26,27] are equivalent to the ones in [12,14], but differ in that the edges directly model the communities and a separate parameter r bounds the number of defective edges.

Our Results and Paper Structure. Basic concepts for cover-free families are given in Sect. 2. The new definitions of structure-aware cover free families and edge-identifying CFFs are given in Sect. 3 along with related decoding algorithms. CFF constructions for hypegraphs with non-overlaping edges are given in Sect. 4. In particular, we revisit known d-CFF constructions (Sperner, product, array group testing, polynomials in finite fields) and show how they can be viewed as a *structure-aware* CFF, allowing a much larger defect identification when items are clustered into conveniently chosen hypergraphs. We exemplify how these hypergraphs relate to realistic community-like structures. In a generalization of the Sperner construction ($r = 1$) we also give results under the more realistic assumption of limited number of samples per tests (Sect. 4.1). CFF constructions for the more general case of hypergraphs with overlapping edges are given in Sect. 5. We give constructions for both $r = 1$ and $r > 1$ using edge-colouring and strong edge-colouring of hypergraphs, to partition the hypergraph into non-overlapping subgraphs that can be constructed using results from the previous section. Some proofs and pictures omitted here are in [18].

2   Cover-Free Families

CFFs were first introduced by Kautz and Singleton [21] in the context of *superimposed codes*. They are equivalent to d-disjunct matrices and strongly selective families [4,29]. We can define d-CFF via a matrix or a set system.

Definition 1 (CFF via matrix). *Let d be a positive integer. A d-cover-free family, denoted d-CFF(t, n), is a $t \times n$ 0–1 matrix where the submatrix given by any set of $d + 1$ columns contains a permutation matrix (each row of an identity of order $d + 1$) among its rows.*

A set system $\mathcal{F} = (X, \mathcal{B})$ consists of a set X and a collection $\mathcal{B}$ of subsets of X. The *set system associate to matrix* $\mathcal{M}$ is the set system $\mathcal{F}_\mathcal{M} = (X, \mathcal{B})$ with X corresponding to rows and $\mathcal{B}$ corresponding to columns of $\mathcal{M}$, where $B_i \subseteq \mathcal{B}$ has column i as its characteristic vector, $1 \leq i \leq n$. A d-CFF can be equivalently defined in terms of its set system $\mathcal{F}_\mathcal{M}$, by specifying that no set of d columns "covers" any other column.

Definition 2 (CFF via set system). *Let d be a positive integer. A d-cover-free family, denoted d-CFF(t, n), is a set system $\mathcal{F} = (X, \mathcal{B})$ with $|X| = t$ and $|\mathcal{B}| = n$ such that for any $d + 1$ subsets $B_{i_0}, B_{i_1}, \ldots, B_{i_d} \in \mathcal{B}$, we have*

$$\left| B_{i_0} \setminus \bigcup_{j=1}^{d} B_{i_j} \right| \geq 1. \tag{1}$$

For a given n and d, we are interested in constructing d-CFFs with the smallest possible t, so we define $t(d, n) = \min\{t : \exists \, d\text{-CFF}(n, t)\}$.

Next we show an example of a 2-CFF(9, 12), which can be used to test $n = 12$ items with $t = 9$ tests and identify up to $d = 2$ defective items.

$$\mathcal{M} = \begin{pmatrix} 1 & 0 & 0 & 1 & 0 & 0 & 1 & 0 & 0 & 1 & 0 & 0 \\ 1 & 0 & 0 & 0 & 1 & 0 & 0 & 1 & 0 & 0 & 1 & 0 \\ 1 & 0 & 0 & 0 & 0 & 1 & 0 & 0 & 1 & 0 & 0 & 1 \\ 0 & 1 & 0 & 1 & 0 & 0 & 0 & 0 & 1 & 0 & 1 & 0 \\ 0 & 1 & 0 & 0 & 1 & 0 & 1 & 0 & 0 & 0 & 0 & 1 \\ 0 & 1 & 0 & 0 & 0 & 1 & 0 & 1 & 0 & 1 & 0 & 0 \\ 0 & 0 & 1 & 1 & 0 & 0 & 0 & 1 & 0 & 0 & 0 & 1 \\ 0 & 0 & 1 & 0 & 1 & 0 & 0 & 0 & 1 & 1 & 0 & 0 \\ 0 & 0 & 1 & 0 & 0 & 1 & 1 & 0 & 0 & 0 & 1 & 0 \end{pmatrix}$$
$X = \{1, 2, \ldots, 9\}$
$B_1 = \{1, 2, 3\}, B_2 = \{4, 5, 6\}, \ldots, B_{12} = \{3, 5, 7\}$
$\mathcal{B} = \{B_1, B_2, \ldots, B_{12}\}$

After running the tests on groups of items according to the rows of a d-CFF matrix $\mathcal{M}$, we can run a simple algorithm to identify the invalid items. When we apply Algorithm 1 with a d-CFF matrix $\mathcal{M}$ and the number of defectives is indeed bounded by d, then after the first loop x has at most d nonzero components. So for d-CFF, the second loop can be removed and substituted by a simple check that the number of 1's in x does not exceed d; in this case, the output will be Boolean, i.e. every component is in $\{0, 1\}$, and correct. We give this more general algorithm, used in Sect. 3. In the case of other types of matrices or when the number of defective items exceeds d, the algorithm classifies the items into three types of defective status (yes, no, maybe) according to the information provided by test results. Assuming correct outcome for group testing, the items with $x_i \in \{0, 1\}$ do not give false positive/negative results.

Algorithm 1. Non-adaptive CGT algorithm to identify invalid items

---

Input: Group testing matrix $\mathcal{M}$ and test result $y = (y_1, \ldots, y_t)$, with $y_i = 1$ iff the i-th test was positive.

Output: $x = (x_1, \ldots, x_n)$, $x_j = 1, 0, 0.5$ if the j-th item is defective, nondefective, unknown, respectively.

$x \leftarrow (1, \ldots, 1)$

for i = 1, $\ldots$, t **do**

    for j = 1, $\ldots$, n **do**

        if $\mathcal{M}_{i,j} = 1$ and $y_i = 0$ **then** $x_j \leftarrow 0$

for j such that $x_j = 1$ **do**

    if $\exists i$ such that $(\mathcal{M}_{i,j} = 1$ and $(x_\ell = 0, \forall \ell \neq j$ with $\mathcal{M}_{i,\ell} = 1))$ **then**

        $x_j \leftarrow 1$ ▷ Item j is on a failing test together with only non-defective items

    else $x_j \leftarrow 0.5$ ▷ Can't guarantee j is the cause of failures but maybe defective

return x

For $d = 1$, Sperner's theorem gives an optimal construction for 1-CFFs. The value t grows as $\log_2 n$ as $n \to \infty$, which meets the information theoretical lower bound. For $d \geq 2$, the best known lower bound on t for d-CFF(t, n) is given by $t(d, n) \geq c \frac{d^2}{\log d} \log n$ for some constant c [9, 30, 32], with c proven to be $\approx 1/4$ in [9] and $\approx 1/8$ in [30]. For $d \geq 2$, there are several approaches to construct d-CFFs using codes and combinatorial designs [24]. Probabilistic methods usually provide the best existence results known, and derandomization techniques yield polynomial time algorithms to construct a d-CFF(n, t) with $t = \Theta(d^2 \log n)$ [2, 10, 11, 29].

3   Structure-Aware Cover-Free Families

In this section, we define structure-aware cover-free families (SCFFs) by adding a hypergraph structure to a CFF. Vertices correspond to columns and edges specify sets of columns where defective items may appear more likely together. We use the assumption that defective items are contained in a small number r of edges inside of which any number of defective items may be found. For example, the outbreak of a disease in a school/university could be detected by associating vertices with students, edges with classrooms/courses; even if the number of infected students is high, the CFF would detect them as long as they are concentrated in a small number of classrooms/courses. A *hypergraph* is a pair (V, S) where V is a finite set called *vertices* and S is a set of nonempty subsets of V called *edges*. We use $[1, n]$ to denote the set $\{1, \ldots, n\}$.

Definition 3 (Structure-aware CFFs). *Let $n, t > 0$ and $r \geq 0$ be integers. Let $\mathcal{H} = ([1, n], S)$ be a hypergraph with n vertices and m edges, and let $\mathcal{M}$ be a $t \times n$ binary matrix with associated set system $\mathcal{F}_{\mathcal{M}} = ([1, t], \mathcal{B})$, $\mathcal{B} = \{B_1, \ldots, B_n\}$. Matrix $\mathcal{M}$ is a* structure-aware cover-free family, *denoted (S, r)-CFF(t, n), if for any r-set of edges $\{S_1, \ldots, S_r\} \subseteq S$, and for any $I \subseteq \cup_{j=1}^r S_j$ and any $i_0 \in [1, n] \setminus I$, we have*

$$\left| B_{i_0} \setminus \left(\bigcup_{i \in I} B_i \right) \right| \geq 1. \tag{2}$$

We observe that a d-CFF(t, n) is equivalent to an (S, d)-CFF(t, n) where edges are singleton vertices $S = \{\{1\}, \{2\}, \ldots \{n\}\}$.

We now consider how the status of edges influence the detectability of defective items. An edge is *defective* if it contains a defective vertex and *non-defective*, otherwise. A set of edges is a *defect cover* if the set of defective vertices is contained in the union of these edges; such a defect cover is *minimal* if no proper subset is a defect cover. A minimal defect cover is always contained in the set of defective edges, but the number of defective edges may be much larger than the size of a defect cover for hypergraphs with overlapping edges. The next proposition shows that a structure-aware CFF ability to detect defectives only depends on the cardinality of a minimum defect cover being bounded by r.

Proposition 1. *Let $\mathcal{H} = ([1, n], S)$ be a hypergraph, $\mathcal{M}$ be an (S, r)-CFF(t, n) and $y \in \{0, 1\}^t$ be the result of tests given by $\mathcal{M}$ on items $1, \ldots, n$. If $\mathcal{H}$ has a defect cover with at most r edges then Algorithm 1 on inputs $(\mathcal{M}, y)$ returns a Boolean output x such that $x_i = 1$ if and only if item i is defective.*

We are also interested in identifying infected edges when the output of Algorithm 1 is not Boolean, which can happen if defective items are spread over too many edges (defective covers have size $> r$). For example, in schools the tests may not provide full information on infected students, but we still may extract information on which classrooms are infected. The following algorithm provides edge information based on ternary vertex information for a hypergraph $\mathcal{H}$.

Algorithm 2. Edge information from vertices

Input: Hypergraph $\mathcal{H} = (V, E)$ with n vertices and m edges; Group testing matrix $\mathcal{M}$, boolean results $y = (y_1, y_2, \ldots, y_t)$; Vector $x = (x_1, x_2, \ldots, x_n)$, $x_j = 1, 0, 0.5$ if the j-th item is defective, non-defective, unknown, respectively.

Output: Vector $z = (z_1, z_2, \ldots, z_m)$, $z_e = 1, 0, 0.5$ if the e-th edge is defective, nondefective, unknown, respectively.

for $s = 1, \ldots, m$ do   ▷ this loop gets edge status from vertices
    $z_s \leftarrow 0$;
    for each vertex v_i in edge e_s do
        if $x_i = 1$ then $z_s \leftarrow 1$
        else
            if $x_i = 0.5$ and $z_s = 0$ then $z_s \leftarrow 0.5$
for $i = 1, \ldots, t$ do   ▷ this loop gets edge status from test results
    if $y_i = 1$ then
        $E = \{j : M_{i,j} = 1 \text{ and } x_j \neq 0\}$
        for $s = 1, \ldots, m$ do
            if $(z_s = 0.5)$ and $(E \subseteq e_s)$ then $z_s \leftarrow 1$
return z

Some CFFs may have a value of r for vertex status identification but have a larger value r for edge status identification (see for example Proposition 7 and Theorem 3). This can be useful for applications, in that infected communities are identifiable even though we do not have perfect individual identification. To capture this property, we define edge-identifying CFFS (ECFFs), which has a weaker coverage requirement than SCFFs.

Definition 4 (Edge-identifying CFFs). *Let r, t, n, $\mathcal{M}$, $\mathcal{H}$ and $\mathcal{F}_\mathcal{M}$ be as in Definition 3. We say $\mathcal{M}$ is an $(\mathcal{S}, r)$-ECFF(t, n) if for any ℓ-subset of edges $\{S_1, \ldots, S_\ell\} \subseteq \mathcal{S}$, $\ell \leq r$, and any $i_0 \notin S = \cup_{j=1}^{\ell} S_j$, we have*

$$\left| B_{i_0} \setminus \left(\bigcup_{i \in S} B_i \right) \right| \geq 1. \tag{3}$$

Proposition 2. *Let $\mathcal{H} = ([1, n], \mathcal{S})$ be a hypergraph, $\mathcal{M}$ be a $(\mathcal{S}, r)$-ECFF(t, n), and y be the test results for $\mathcal{M}$. Let x be the output of Algorithm 1 for inputs $(\mathcal{H}, \mathcal{M}, y)$. Then, if $\mathcal{H}$ has a defect cover with at most r edges then Algorithm 2 applied to $(\mathcal{H}, \mathcal{M}, x, y)$ returns an output z such that $\{S_j \in \mathcal{S} : z_j = 1\}$ forms a defect cover that only contains defective edges.*

For any CFF, structure-aware CFF, or ECFF matrix $\mathcal{M}$ we denote by $L_\mathcal{M}$ the number of ones in each row of $\mathcal{M}$. We keep track of this quantity in some constructions, since we may have limit L_{max} on the number of ones per row, in cases where combining too many samples can result on a false negative.

4 Structure-Aware CFFs: Non-overlapping Edges

We revisit old CFF constructions and show we can boost the number of defectives it can identify by taking a suitable hypergraph structure into account. We also propose some new constructions. Here we consider the case of non-overlapping edges, meaning that items do not participate in more than one edge.

4.1 Sperner-Type Constructions for $r = 1$

A Sperner set system is a set system where no set is contained in any other set in the set system. Sperner's theorem states that the largest Sperner set system on an t-set is formed by taking all subsets of cardinality $\lfloor t/2 \rfloor$. Given n, a 1-CFF(t, n) with minimum t is obtained from Sperner theorem by taking $t = \min\{s : \binom{s}{\lfloor s/2 \rfloor} \leq n\}$ and the corresponding matrix having the characteristic vectors of $\lfloor t/2 \rfloor$-subsets as columns. We note that $t \sim \log n$ and this is the best possible, since being 1-CFF is equivalent to being Sperner.

Some applications have a maximum allowed number of items per test, $L_{\mathcal{M}} \leq L_{max}$, to prevent loss of test precision. A Sperner set system with sets of cardinality $a < t/2$ can be used as a 1-CFF whenever $\binom{t-1}{\lfloor t/2 \rfloor - 1}$ exceeds L_{max}. For nonoverlapping hypergraphs and $r = 1$, we give constructions for SCFF for both unlimited and limited $L_{\mathcal{M}}$.

Proposition 3 ($r = 1$, **unlimited** $L_{\mathcal{M}}$). *Let* $\mathcal{H} = ([1, n], \mathcal{S})$ *be a hypergraph with m disjoint edges of cardinality at most d that span $[1, n]$. Let $\mathcal{M}$ be the vertical concatenation of matrices M_1 and M_2. Let M_1 be obtained from a 1-CFF(t_1, m) matrix A with $t_1 = \min\{s : \binom{s}{\lfloor s/2 \rfloor} \leq m\}$ in such a way that if vertex v_i is incident to edge b_j column i of M_1 repeats column j of A. Let M_2 be a $d \times n$ matrix with an identity matrix of dimension $|S|$ pasted under the items of each edge $S \in \mathcal{S}$. Then, M_1 is an $(\mathcal{S}, 1)$-ECFF(t_1, n) and $\mathcal{M}$ is an $(\mathcal{S}, 1)$-CFF$(t_1 + d, n)$.*

For uniform hypergraphs the construction above gives $t \sim \log m + d = \log n/d + d$, but does not limit $L_{\mathcal{M}}$. The next proposition is useful for limited $L_{\mathcal{M}}$, as shown in the example that follows it.

Proposition 4 ($r = 1$, $L_{\mathcal{M}} \leq L_{max}$). *Let L_{max} be a positive integer that limits the number of 1s in each row of the CFF. Let $\mathcal{H} = ([1, n], \mathcal{S})$ be a hypergraph with m disjoint edges of cardinality at most d that span $[1, n]$, where $d \leq L_{max}$. Let $t_1 = \min\{s : \binom{s}{\lfloor s/2 \rfloor} \leq m\}$. Then,*

1 *If* $d \times \binom{t_1 - 1}{\lfloor t_1/2 \rfloor - 1} \leq L_{max}$ *and* $m \leq L_{max}$ *then* $\mathcal{M}$ *given in Proposition 3 is an* $(\mathcal{S}, 1)$-CFF$(t_1 + d, n)$ *with* $L_{\mathcal{M}} \leq L_{max}$.

2 *Otherwise, let* $q = \lceil m/L_{max} \rceil$. *Take* t, a *such that* $\binom{t}{a} \geq m$ *and* $d \times \binom{t-1}{\lfloor a \rfloor - 1} \leq L_{max}$. *Then, there exists a* $(\mathcal{S}, 1)$-CFF$(t + qd, n)$ *matrix* $\mathcal{M}$ *with* $L_{\mathcal{M}} \leq L_{max}$.

Example 1 (Proposition 4 used for m classrooms with d students each). Suppose n students are divided into m classrooms of size up to d. Then Proposition 4 can be used to identify all infected students, provided they are all in a single classroom ($r = 1$). The table below reports on number of tests for each scenario depending on value of L_{max} for the construction on Proposition 4. The line with $L = \infty$ shows the number of tests for the construction for unlimited L (Proposition 3). The last line shows the lower bound given in [9] for the number of rows t on a d-CFF(t, n) required for location of any set of d infected students, not necessarily concentrated on a single classroom.

$n/100$	.5	1	2	3	1	2	4	6	1.5	3	6	9	2	4	8	12	2.5	5	10	15	3	6	12	18
m	10	10	10	10	20	20	20	20	30	30	30	30	40	40	40	40	50	50	50	50	60	60	60	60
d	5	10	20	30	5	10	20	30	5	10	20	30	5	10	20	30	5	10	20	30	5	10	20	30
$L =$																								
10	11	16	26	36	17	27	47	67	23	38	68	98	28	48	88	128	33	58	108	158	39	69	129	189
15	11	16	26	36	17	27	47	67	18	28	48	68	23	38	68	98	28	48	88	128	29	49	89	129
20	11	16	26	36	12	17	27	37	18	28	48	68	18	28	48	68	23	38	68	98	24	39	69	99
25	10	16	26	36	12	17	27	37	18	28	48	68	18	28	48	68	18	28	48	68	24	39	69	99
30	10	16	26	36	12	17	27	37	14	18	28	38	18	28	48	68	18	28	48	68	19	29	49	69
$L = \infty$	10	15	25	35	11	16	26	36	12	17	27	37	13	18	28	38	13	18	28	38	13	18	28	38
$t(d, n) >$	21	66	180	270	21	66	231	496	21	66	231	496	21	66	231	496	23	66	231	496	25	66	231	496

4.2 Kronecker Product Constructions (General r)

Let A_k be an $m_k \times n_k$ binary matrix, for $k = 1, 2$, and $\mathbf{0}$ be the matrix of all zeroes with same dimension as A_2. The Kronecker product $P = A_1 \otimes A_2$ is a binary matrix formed of blocks $P_{i,j}$ such that $P_{i,j} = A_2$ if $A_{1_{i,j}} = 1$ and $P_{i,j} = \mathbf{0}$, otherwise. We denote by R_k the row matrix with k ones and by I_k the identity matrix of dimension k. The propositions given after each theorem specializes the theorem construction and generalizes to SCFF, boosting the defective detection.

Theorem 1 (Li et al. [24] for $d = 2$, Idalino and Moura [17]). *Let A_1 be a d-CFF(t_1, n_1) and A_2 be a d-CFF(t_2, n_2), then $C = A_1 \otimes A_2$ is a d-CFF$(t_1 t_2, n_1 n_2)$.*

Proposition 5. *Let $\mathcal{H} = ([1, n], \mathcal{S})$ be a hypergraph formed by m disjoint edges of cardinality k, $n = k \times m$. Let r be a positive integer, and let A be an r-CFF(t, m). Then $A \otimes R_k$ is an $(\mathcal{S}, r)$-ECFF(t, km) and $A \otimes I_k$ is an $(\mathcal{S}, r)$-CFF(kt, km) (Fig. 1).*

The *vertical concatenation* of matrices A and B (with the same number of columns) is the matrix consisting of rows of A followed by the rows of B.

Theorem 2 (Li et al. [24] for $d = 2$, Idalino and Moura [17]). *Let $d \geq 2$, A_1 be a d-CFF(t_1, n_1), A_2 be a d-CFF(t_2, n_2), B be a $(d - 1)$-CFF(s, n_2). Let C be the vertical concatenation of $B \otimes A_1$ and $A_2 \otimes R_{n_1}$. Then C is a $d-$CFF$(st_1 + t_2, n_1 n_2)$.*

Proposition 6. *Let $\mathcal{H} = ([1, n], \mathcal{S})$ be a hypergraph formed by m disjoint edges of cardinality k, $n = k \times m$. Let r be a positive integer, A be an r-CFF(t_A, m), and B be an $(r-1)$-CFF(t_B, m). Then the vertical concatenation of $A \otimes R_k$ and $B \otimes I_k$ is an $(\mathcal{S}, r)$-CFF$(t_A + kt_B, km)$. Moreover, if edges have different cardinalities bounded by k, a similar construction yields an $(\mathcal{S}, r)$-CFF$(t_A + kt_B, n)$ (Fig. 1).*

Construction in Proposition 5, using 2-CFF(9,12) A:

$$A = \begin{pmatrix} 1&0&0&1&0&0&1&0&0&1&0&0 \\ 1&0&0&0&1&0&0&1&0&0&1&0 \\ 1&0&0&0&0&1&0&0&1&0&0&1 \\ 0&1&0&1&0&0&0&0&1&0&1&0 \\ 0&1&0&0&1&0&1&0&0&0&0&1 \\ 0&1&0&0&0&1&0&1&0&1&0&0 \\ 0&0&1&1&0&0&0&1&0&0&0&1 \\ 0&0&1&0&1&0&0&0&1&1&0&0 \\ 0&0&1&0&0&1&1&0&0&0&1&0 \end{pmatrix}, \quad I_3 = \begin{pmatrix} 1&0&0 \\ 0&1&0 \\ 0&0&1 \end{pmatrix}, \quad A \otimes I_3 = \begin{pmatrix} I_3&0&0&I_3&0&0&I_3&0&0&I_3&0&0 \\ I_3&0&0&0&I_3&0&0&I_3&0&0&I_3&0 \\ I_3&0&0&0&0&I_3&0&0&I_3&0&0&I_3 \\ 0&I_3&0&I_3&0&0&0&0&I_3&0&I_3&0 \\ 0&I_3&0&0&I_3&0&I_3&0&0&0&0&I_3 \\ 0&I_3&0&0&0&I_3&0&I_3&0&I_3&0&0 \\ 0&0&I_3&I_3&0&0&0&I_3&0&0&0&I_3 \\ 0&0&I_3&0&I_3&0&0&0&I_3&I_3&0&0 \\ 0&0&I_3&0&0&I_3&I_3&0&0&0&I_3&0 \end{pmatrix}.$$

Construction in Proposition 6, using 2-CFF(9,12) A and 1-CFF(6,12) B:

$$B = \begin{pmatrix} 1&1&1&1&1&1&0&0&0&0&0&0 \\ 1&1&1&1&0&0&1&1&0&0&0&0 \\ 1&0&0&0&1&1&0&0&1&1&1&0 \\ 0&1&0&0&1&0&1&0&1&1&0&1 \\ 0&0&1&0&0&1&1&0&1&0&1&1 \end{pmatrix}, \quad R_3 = (1\ 1\ 1), \quad \left(\frac{A \otimes R_3}{B \otimes I_3}\right) = \begin{pmatrix} R_3&0&0&R_3&0&0&R_3&0&0&R_3&0&0 \\ R_3&0&0&0&R_3&0&0&R_3&0&0&R_3&0 \\ R_3&0&0&0&0&R_3&0&0&R_3&0&0&R_3 \\ 0&R_3&0&R_3&0&0&0&0&R_3&0&R_3&0 \\ 0&R_3&0&0&R_3&0&R_3&0&0&0&0&R_3 \\ 0&R_3&0&0&0&R_3&0&R_3&0&R_3&0&0 \\ 0&0&R_3&R_3&0&0&0&R_3&0&0&0&R_3 \\ 0&0&R_3&0&R_3&0&0&0&R_3&R_3&0&0 \\ 0&0&R_3&0&0&R_3&R_3&0&0&0&R_3&0 \\ I_3&I_3&I_3&I_3&I_3&I_3&0&0&0&0&0&0 \\ I_3&I_3&I_3&I_3&0&0&I_3&I_3&0&0&0&0 \\ I_3&0&0&0&I_3&I_3&0&0&I_3&I_3&I_3&0 \\ 0&I_3&0&0&I_3&0&I_3&0&I_3&I_3&0&I_3 \\ 0&0&I_3&0&0&I_3&I_3&0&I_3&0&I_3&I_3 \end{pmatrix}.$$

Fig. 1. Two $(\mathcal{S}, 2)$-CFF$(27, 36)$, $\mathcal{S}$ consists of 12 disjoint edges of size 3. Up to six defective items concentrated within 2 edges can be identified.

4.3   Array and Hypercube Constructions

An array-based scheme for group testing uses an $n_1 \times n_2$ array, where each entry of the array corresponds to an item to be tested and the tests are performed on rows and columns, for a total of $n_1 + n_2$ tests. This can be used on a 2-stage algorithm, where all items at the intersection of a positive row and column should be individually tested in a second stage to solve ambiguities [13, 22, 28]. For $d = 1$ defective item, one stage is enough. Fig. 2(a) shows a 5×5 array with defective items in red. This idea can be generalized to higher dimensions, constructing an $n_1 \times \ldots \times n_k$ hypercube [1, 23], which is a 1-CFF$(n_1 + \ldots + n_k, n_1 \times \ldots \times n_k)$. Figure 2(b) shows a 3-dimensional hypercube, where each point represents an item and tests are given by fixing the value of one dimension. If all defective items are clustered in either a row or a column in a 2-dimensional array, we can precisely identify all of them in one round, thus this is a structure-aware $(\mathcal{S}, 1)$-CFF$(2n, n^2)$ for $\mathcal{S}$ corresponding to rows and columns. We generalize this for higher dimensions in the next proposition. To simplify the notation, we take $n_1 = \ldots = n_k = n$, but the next results are valid for the general case. An $[n]^k$-hypercube group testing matrix is an 1-CFF(kn, n^k) matrix defined as follows. Items are in $\mathbb{Z}_n^k$ and rows/tests are given by $T_{v,a} = \{x \in \mathbb{Z}_n^k : x_v = a\}$, $1 \le v \le k$, $a \in \mathbb{Z}_n$. Denote $x(v) = (x_1, \ldots, x_{v-1}, x_{v+1}, \ldots, x_k)$ for $x \in \mathbb{Z}_n^k$, $1 \le v \le k$.

Proposition 7. *Let A be an $[n]^k$-hypercube group testing matrix. Let $\mathcal{H}_v = ([1,n], \mathcal{S}_v)$ where $\mathcal{S}_v = \{\{x \in \mathbb{Z}_n^k : x(v) = (a_1, \ldots, a_{k-1})\} : (a_1, \cdots, a_{k-1}) \in \mathbb{Z}_n^{k-1}\}$, $1 \le v \le k$, and let $\mathcal{H} = ([1,n], \mathcal{S})$ where $\mathcal{S} = \mathcal{S}_1 \cup \cdots \cup \mathcal{S}_k$. Then, for any $1 \le v \le k$, A is an $(\mathcal{S}_v, 1)$-CFF(kn, n^k) and if $k = 2$, A is also an $(\mathcal{S}_v, |\mathcal{S}_v| = n)$-ECFF$(2n, n^2)$. Moreover, A is an $(\mathcal{S}, 1)$-CFF(kn, n^k).*

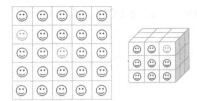

Fig. 2. (a) A 5×5 array GT with 25 items and 10 tests. (b) A $3 \times 3 \times 3$ hypercube GT with 27 items and 9 tests.

Proof (sketch). For ECFF, note that when $k = 2$ each edge is tested in a different test containing only the vertices of the edge. For SCFF, if defectives are contained in an edge of S_v, vertices in all other edges of S_v will be in passing tests, and tests $T_{v,a}$, $a \in \mathbb{Z}_n$, reveal the defective status of each item in S_v.   □

4.4 Construction from Polynomials

Now we look at a construction of d-CFFs from polynomials over finite fields, given by Erdös et al. [7]. Let q be a prime power, k a positive integer, and $\mathbb{F}_q = \{e_1, \ldots, e_q\}$ be a finite field. We define $\mathcal{F} = (X, \mathcal{B})$ as follows, for each polynomial $f \in \mathbb{F}_q[x]_{\leq k}$ of degree at most k: $X = \mathbb{F}_q \times \mathbb{F}_q$, $B_f = \{(e_1, f(e_1)), \ldots, (e_q, f(e_q))\}$, $\mathcal{B} = \{B_f : f \in \mathbb{F}_q[x]_{\leq k}\}$. Then, $\mathcal{F}$ is a d-CFF$(t = q^2, n = q^{k+1})$ for $d \leq \frac{q-1}{k}$.

This d-CFF has an interesting structure, which allows us to discard some rows when smaller values of d are enough [16]. We restrict the CFF matrix to i *blocks* of rows by considering $X = \{e_1, \ldots, e_i\} \times \mathbb{F}_q$, $B_f(i) = \{(e_1, f(e_1)), \ldots, (e_i, f(e_i))\}$ and $\mathcal{B}(i) = \{B_f(i) : f \in \mathbb{F}_q[x]_{\leq k}\}$, which yields the following result.

Proposition 8 (Idalino and Moura [16], Theorem 3.2). *Let q be a prime power, $k \geq 1$ and $q \geq dk + 1$, and let $\mathcal{M}$ be the d-CFF(q^2, q^{k+1}) obtained from the polynomial construction. If we restrict $\mathcal{M}$ to the first $(d'k+1)$ blocks of rows, we obtain a d'-CFF$((d'k+1)q, q^{k+1})$, for any $d' \leq d$.*

For instance, for $q = 5$ and $k = 1$, if we restrict a 4-CFF$(5^2, 5^2)$ to its first 2 blocks of rows, we get a 1-CFF$(2 \times 5, q^2)$, with 3 blocks of rows we get a 2-CFF$(3 \times 5, q^2)$, etc. Next we show that this construction is an structure-aware CFF that can tolerate as many as q defective items with as few as $(k+1)q$ tests.

Theorem 3. *Let $k \geq 1$ and q be a prime power such that $q \geq k + 1$. Let $\mathcal{S} = \{S_1, \ldots, S_{q^k}\}$ be a set-partition of $[1, n]$ such that $|S_i| = q$ for all $1 \leq i \leq q^k$. Then, there exists an $(\mathcal{S}, 1)$-CFF$((k+1)q, q^{(k+1)})$. If $k = 1$, it is also an $(\mathcal{S}, q)$-ECFF$(2q, q^2)$.*

As an example, for $q = 5$ and $k = 1$ we have edges $S_1 = \{0, x, 2x, 3x, 4x\}$, $S_2 = \{1, x+1, 2x+1, 3x+1, 4x+1\}$, $S_3 = \{2, x+2, 2x+2, 3x+2, 4x+2\}$, $S_4 = \{3, x+3, 2x+3, 3x+3, 4x+3\}$, and $S_5 = \{4, x+4, 2x+4, 3x+4, 4x+4\}$. This gives us an $(\mathcal{S}, 1)$-CFF$(2q = 10, q^2 = 25)$ with $\mathcal{S} = \{S_1, S_2, S_3, S_4, S_5\}$, which allows us to find as many as $q = 5$ defective items, as long as they are all in one of the edges S_i. In addition, since $k = 1$, each edge will form a different test, so we can determine which edges are defective.

Note that the construction in Theorem 3 is equivalent to a $[q]^{k+1}$-hypercube, but it is more flexible since we can add more tests (Proposition 8) for a total of $(dk + 1)q$ tests, where $q \geq dk + 1$, to obtain both a $(S, 1)$-CFF$((dk + 1)q, q^{k+1})$ and a d-CFF$((dk + 1)q, q^{k+1})$, so any d defects anywhere or $q > d$ defects inside an edge can be found.

5 Structure-Aware CFFs: Overlapping Edges

Here, edge colouring of hypergraphs is used to partition the edges of the graph into sets of non-overlapping edges (colour classes) allowing the use of previous constructions to deal with each colour class. An ℓ-edge-colouring of a hypergraph $\mathcal{H} = (V, \mathcal{S})$ is a mapping from $\mathcal{S}$ to $\{1, \ldots, \ell\}$ such that no vertex is incident to more than one edge mapping to the same colour. Let $\chi'(\mathcal{H})$ be the edge chromatic number of hypergraph $\mathcal{H}$, which is the minimum ℓ among all ℓ-edge-colourings.

Theorem 4. *Let $\mathcal{H} = ([1, n], \mathcal{S})$ be a hypergraph and let $\mathcal{C}_1, \mathcal{C}_2, \ldots, \mathcal{C}_\ell$ be the sets of edges in each colour class of an ℓ-edge-colouring of $\mathcal{H}$. For each i, $1 \leq i \leq \ell$, let $k_i = \max\{|A| : A \in \mathcal{C}_i\}$ and let $f_i = |\mathcal{C}_i| + \delta_i$, where $\delta_i = 0$ if the $\mathcal{C}_i$ spans $[1, n]$ and $\delta_i = 1$, otherwise. Then, given 1-CFF(t_i, f_i) for $1 \leq i \leq \ell$, we can construct a $(\mathcal{S}, 1)$-CFF(t, n) where $t = \sum_{i=1}^{\ell}(t_i + k_i)$; moreover we can construct an $(\mathcal{S}, 1)$-ECFF(t_i, n).*

Corollary 1. *Let $\mathcal{H} = ([1, n], \mathcal{S})$ be a k-uniform hypergraph. Denote by $t(1, x)$ the number of tests y in the Sperner construction of a 1-CFF(y, x) (See Sect. 4.1). Then, there exists an $(\mathcal{S}, 1)$-CFF(t, n) with $t = \chi'(\mathcal{H}) \times (t(1, \lceil(n/k)\rceil) + k) \sim \chi'(\mathcal{H})((\log n/k) + k)$, and a $(\mathcal{S}, 1)$-ECFF(t', n) with $t' = \chi'(\mathcal{H}) \times t(1, \lceil(n/k)\rceil) \sim \chi'(\mathcal{H}) \log n/k$.*

Example 2. Consider a high school where each student takes P courses per term, in P weekly time periods where in each time period each student attends one courses of their choice. Consider a hypergraph with n vertices corresponding to students and each edge corresponding to students in a course. In this example $\ell = P$, since each time period forms a colour class. We give a tiny example, with $n = 18$ students spread of over $P = 2$ time periods morning/afternoon each with 6 optional courses with 3 students each. This hypergraph has $m = 12$ edges (partitioned into 2 colour classes) and $n = 18$ vertices shown in the table below.

students:	01	02	03	04	05	06	07	08	09	10	11	12	13	14	15	16	17	18
course 1	X	X	X															
course 2				X	X	X												
course 3							X	X	X									
course 4										X	X	X						
course 5													X	X	X			
course 6																X	X	X
course 7	X	X		X														
course 8			X		X		X											
course 9					X	X		X										
course 10									X		X							X
course 11												X		X	X	X		
course 12													X		X		X	X

test 1:	111	111	111	000	000	000
test 2:	111	000	000	111	111	000
test 3:	000	111	000	111	000	111
test 4:	000	000	111	000	111	111
test 5:	100	100	100	100	100	100
test 6:	010	010	010	010	010	010
test 7:	001	001	001	001	001	001
test 8:						
test 9:		permute columns				
test 10:		of above array				
test 11:		so that blocks				
test 12:		of 3 columns are				
test 13:		placed under edges				
test 14:		of second period				

Example 3. Consider the setup of Example 2. Let us consider a more realistic scenario of a high school with students taking 4 courses each term like the ones in Ontario, Canada. Suppose $n = 900$ students take $P = 4$ courses each, each course having 30 students for a total of $m = 120$ courses. We vertically concatenate $2P$ matrices. For time period i we use M_i (built by repeating columns of a 1-CFF$(7, 30 = 120/4)$) and N_i (formed by identities of order 30 side-by-side, pasted under the edges in i). Assume there is an outbreak in a single course, involving any number of students (≤ 30) in that course. We only need $7 \times 4 = 28$ tests to determine the course where the outbreak took place (M_i build from 1-CFF$(7, 30)$, $1 \leq i \leq 4$). A total of $28 + 30 \times 4 = 148$ tests can be used to identify all infected individuals (up to 30) in this set of 900 students. Note that our assumption is that there is $r=1$ course that contains all infected individuals, even thought there may be many infected courses (say up to 90 other courses that the infected students also take in other time periods). In other words the hypergraph is assumed to have a defective cover of size $r = 1$, but it is possible that up to 91 edges are defective.

For $r > 1$, we need to use strong edge-colourings to be able to split the problem according to colour classes without too many infected edges appearing in the same colour class. A strong edge-coloring of a hypergraph $\mathcal{H}$ is an edge-coloring such that any two vertices belonging to distinct edges with the same colour are not adjacent. The strong chromatic index $s'(\mathcal{H})$ is the minimum number of colors in a strong edge-coloring of $\mathcal{H}$.

Theorem 5. *Let $\mathcal{H} = ([1, n], \mathcal{S})$ be a hypergraph and let $r \geq 2$ be an upper bound on the number of edges of a minimal defective cover. Let $\mathcal{C}_1, \mathcal{C}_2, \ldots, \mathcal{C}_{s'}$ be the sets of edges in each colour class of an s'-strong-edge-colouring of $\mathcal{H}$. Let $k_i = \max\{|S| : S \in \mathcal{C}_i\}$. Then there exists an $(\mathcal{S}, r)$-CFF(t, n) with $t \leq \sum_{i=1}^{s'}(t(r, |\mathcal{C}_i|) + k_i t(r - 1, |\mathcal{C}_i|))$.*

Corollary 2. *Let $\mathcal{H} = ([1, n], \mathcal{S})$ be a k-uniform hypergraph and let Δ be the maximum degree of a vertex. Then, we can build a $(\mathcal{S}, r)$-CFF(t, n) with $t \leq s'(H) \times (t(r, \lceil n/k \rceil) + kt(r - 1, \lfloor n/k \rfloor)) \leq (k\Delta + 1) \times (t(r, \lceil n/k \rceil) + kt(r - 1, \lfloor n/k \rfloor))$.*

Example 4. Consider the scenario of Example 2 and $r = 2$. We can find a strong colouring for the hypergraph of that example with $s' = 6$ colours with colour classes: $\{course1, course4\}$, $\{course2, course5\}$, $\{course3, course6\}$, $\{course7, course10\}$, $\{course8, course11\}$, $\{course9, course12\}$. For each colour class we can use identity matrices I_3 as the 2-CFF$(3, 3)$ and 1-CFF$(3, 3)$ required. If there are outbreaks in 2 courses, any set of up to 6 students in these 2 courses can be detected with 36 tests. This is a toy example, and of course 36 tests is more tests than testing the 18 students individually. If we keep the same edge configuration, but multiply the number of students in each box by 6, then each edge has 18 students. The 1-CFF$(3, 3)$ can be substituted by an 1-CFF$(6, 15)$. In this case 54 tests are sufficient to detect up to 36 infected students from an outbreak in 2 classes among these 108 students. To just find which 2 classes cover the outbreak of cases, we only need 18 tests.

Example 5. Consider a venue with 4356 people sitting in a square auditorium of 66 rows with 66 seats per row. Edges are sets of individuals sitting nearby. We consider edges of size 9 consisting of all possible contiguous 3×3 squares. There is a strong colouring with $\ell = 36$ colour classes of $11 \times 11 = 121$ edges each: we need 4 colours to "tile" the room with edges and 9 such tilings to cover all edges. Theorem 5 construction vertically concatenates matrices $M_1, \ldots, M_{36}, N_1, \ldots, N_{36}$. For each colour class i, we use a 2-CFF$(25, 125)$ using the polynomial construction from Proposition 8 for $q = 5$ and $k = 2$ to build each M_i, totalling $36 \times 25 = 900$ tests. For $N_1, \ldots, N_{36}$, each of which is supposed to be a 1-CFF$(t, 121)$ multiplied by I_9, we use instead a single matrix N built as follows. Take A as a 1-CFF$(12, 484)$ obtained from the Sperner construction and do $N = A \otimes I_9$ with 108 rows. Carefully assign vertices in the grid to the columns of matrix N so that each 3×3 square corresponds to a block of identity matrix I_9 in a tiling fashion. This is enough to identify each non-defective vertex that lies inside one of the two defective edges, which is the purpose of N. Therefore with a total of 1008 tests we can screen 4356 people for any 18 infected people that appear within any 2 regions of size 3×3. See pictures for this example in [18].

References

1. Berger, T., Mandell, J.W., Subrahmanya, P.: Maximally efficient two-stage screening. Biometrics **56**(3), 833–840 (2000)
2. Bshouty, N.H.: Linear time constructions of some d-restriction problems. In: Paschos, V.T., Widmayer, P. (eds.) CIAC 2015. LNCS, vol. 9079, pp. 74–88. Springer, Cham (2015). https://doi.org/10.1007/978-3-319-18173-8_5
3. Dorfman, R.: The detection of defective members of large population. Ann. Math. Stat. **14**, 436–440 (1943)
4. Du, D.-Z., Hwang, F.K.: Combinatorial Group Testing and Its Applications, 2nd edn. World Scientific, Singapore (2000)
5. Ellenberg, J.: Five people. One test. This is how you get there. The New York Times. https://www.nytimes.com/2020/05/07/opinion/coronavirus-group-testing.html. Accessed 11 Jan 2022
6. Eppstein, D., Goodrich, M.T., Hirschberg, D.S.: Improved combinatorial group testing algorithms for real-world problem sizes. SIAM J. Comput. **36**(5), 1360–1375 (2007)
7. Erdös, P., Frankl, P., Füredi, Z.: Families of finite sets in which no set is covered by the union of r others. Israel J. Math. **51**, 79–89 (1985)
8. Farach-Colton, M., Kannan, S., Knill, E., Muthukrishnan, S.: Group testing problems with sequences in experimental molecular biology. In: Proceedings of Compression and Complexity of sequences, pp. 357–367. IEEE Press, Washington, DC (1997)
9. Füredi, Z.: On r-cover-free families. J. Combin. Theory Ser. A **73**, 172–173 (1996)
10. Gargano, L., Rescigno, A.A., Vaccaro, U.: Low-weight superimposed codes and their applications. In: Chen, J., Lu, P. (eds.) FAW 2018. LNCS, vol. 10823, pp. 197–211. Springer, Cham (2018). https://doi.org/10.1007/978-3-319-78455-7_15
11. Gargano, L., Rescigno, A.A., Vaccaro, U.: Low-weight superimposed codes and related combinatorial structures: bounds and applications. Theoret. Comput. Sci. **806**, 655–672 (2020)

12. Gonen, M., Langberg, M., Sprintson, A.: Group testing on general set-systems (2022). https://doi.org/10.48550/arXiv.2202.04988
13. Hudgens, M.G., Kim, H.Y.: Optimal configuration of a square array group testing algorithm. Commun. Stat.: Theory Methods **40**(3), 436–448 (2011)
14. Idalino, T.B.: Fault tolerance in cryptographic applications using cover-free families. Ph.D. thesis, University of Ottawa, Ottawa, Canada (2019)
15. Bardini Idalino, T., Moura, L.: Efficient unbounded fault-tolerant aggregate signatures using nested cover-free families. In: Iliopoulos, C., Leong, H.W., Sung, W.-K. (eds.) IWOCA 2018. LNCS, vol. 10979, pp. 52–64. Springer, Cham (2018). https://doi.org/10.1007/978-3-319-94667-2_5
16. Idalino, T.B., Moura, L.: Embedding cover-free families and cryptographical applications. Adv. Math. Commun. **13**(4), 629–643 (2019)
17. Idalino, T.B., Moura, L.: Nested cover-free families for unbounded fault-tolerant aggregate signatures. Theoret. Comput. Sci. **854**, 116–130 (2021)
18. Idalino, T.B., Moura, L.: Structure-aware combinatorial group testing: a new method for pandemic screening (2022). https://doi.org/10.48550/arxiv.2202.09264
19. Idalino, T.B., Moura, L., Adams, C.: Modification tolerant signature schemes: location and correction. In: Hao, F., Ruj, S., Sen Gupta, S. (eds.) INDOCRYPT 2019. LNCS, vol. 11898, pp. 23–44. Springer, Cham (2019). https://doi.org/10.1007/978-3-030-35423-7_2
20. Idalino, T.B., Moura, L., Custódio, R.F., Panario, D.: Locating modifications in signed data for partial data integrity. Inf. Process. Lett. **115**, 731–737 (2015)
21. Kautz, W., Singleton, R.: Nonrandom binary superimposed codes. IEEE Trans. Inf. Theory **10**, 363–377 (1964)
22. Kim, H.-Y., Hudgens, M.G., Dreyfuss, J.M., Westreich, D.J., Pilcher, C.D.: Comparison of group testing algorithms for case identification in the presence of test error. Biometrics **63**, 1152–1163 (2007)
23. Kim, H.Y., Hudgens, M.G.: Three-dimensional array-based group testing algorithms. Biometrics **65**(3), 903–910 (2009)
24. Li, P.C., van Rees, G.H.J., Wei, R.: Constructions of 2-cover-free families and related separating hash families. J. Combin. Designs **14**, 423–440 (2006)
25. Mallapaty, S.: The mathematical strategy that could transform coronavirus testing. Nature **583**(7817), 504–505 (2020)
26. Nikolopoulos, P., Rajan Srinivasavaradhan, S., Guo, T., Fragouli, C., Diggavi, S.: Group testing for connected communities. In: 24th International Conference on Artificial Intelligence and Statistics on Proceedings of Machine Learning Research, pp. 2341–2349. PMLR (2021)
27. Nikolopoulos, P., Rajan Srinivasavaradhan, S., Guo, T., Fragouli, C., Diggavi, S.: Group testing for overlapping communities. In: ICC 2021 - IEEE International Conference on Communications, pp. 1–7. IEEE (2021)
28. Phatarfod, R.M., Sudbury, A.: The use of a square array scheme in blood testing. Stat. Med. **13**(22), 2337–2343 (1994)
29. Porat, E., Rothschild, A.: Explicit nonadaptive combinatorial group testing schemes. IEEE Trans. Inf. Theory **57**, 7982–7989 (2011)
30. Ruszinkó, M.: On the upper bound of the size of the r-cover-free families. J. Combin. Theory Ser. A **66**, 302–310 (1994)
31. Verdun, C.M., et al.: Group testing for SARS-CoV-2 allows for up to 10-fold efficiency increase across realistic scenarios and testing strategies. Front. Public Health **9** (2021)
32. Wei, R.: On cover-free families. Technical report, Lakehead University (2006)

Convex Grid Drawings of Planar Graphs with Constant Edge-Vertex Resolution

Michael A. Bekos[1] ⓘ, Martin Gronemann[2] ⓘ, Fabrizio Montecchiani[3(✉)] ⓘ,
and Antonios Symvonis[4] ⓘ

[1] Department of Mathematics, University of Ioannina, Ioannina, Greece
bekos@uoi.gr
[2] Algorithms and Complexity Group, TU Wien, Vienna, Austria
mgronemann@ac.tuwien.ac.at
[3] Department of Engineering, University of Perugia, Perugia, Italy
fabrizio.montecchiani@unipg.it
[4] School of Applied Mathematical and Physical Sciences, NTUA, Athens, Greece
symvonis@math.ntua.gr

Abstract. We continue the study of the area requirement of convex straight-line grid drawings of 3-connected plane graphs, which has been intensively investigated in the last decades. Motivated by applications, such as graph editors, we additionally require the obtained drawings to have bounded *edge-vertex resolution*, that is, the closest distance between a vertex and any non-incident edge is lower bounded by a constant that does not depend on the size of the graph. We present a drawing algorithm that takes as input a 3-connected plane graph with n vertices and f internal faces and computes a convex straight-line drawing with edge-vertex resolution at least $\frac{1}{2}$ on an integer grid of size $(n-2+a) \times (n-2+a)$, where $a = \min\{n-3, f\}$. Our result improves the previously best-known area bound of $(3n-7) \times (3n-7)/2$ by Chrobak, Goodrich and Tamassia.

Keywords: Graph drawing · Convex grid drawings · Area requirement · Edge-vertex resolution

1 Introduction

Fáry's theorem [20] is a fundamental result in planar graph drawing, as it guarantees the existence of a planar straight-line drawing for every planar graph. In such a drawing, the vertices of the graph are mapped to distinct points of the Euclidean plane in such a way that the edges are straight, non-intersecting line-segments. This central result has been independently proved by several researchers in early works [30,31,36], some of which also suggested corresponding constructive algorithms requiring high-precision arithmetics; see, e.g., [9,33]. In this regard, a breakthrough has been introduced by de Fraysseix, Pach and Pollack [13] in the late 80's, who proposed a method that additionally guarantees the obtained drawings to be on an integer grid (thus making the high-precision

© Springer Nature Switzerland AG 2022
C. Bazgan and H. Fernau (Eds.): IWOCA 2022, LNCS 13270, pp. 157–171, 2022.
https://doi.org/10.1007/978-3-031-06678-8_12

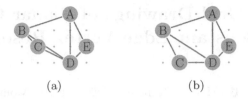

(a)   (b)

Fig. 1. Two planar straight-line grid drawings of the same graph; the drawing in (a) contains an edge-vertex intersection (on vertex C), while the one in (b) does not as it has edge-vertex resolution at least $\frac{1}{2}$.

operations unnecessary). A linear-time implementation of this method was proposed by Chrobak and Payne [12]. Over the years, several works have studied the area requirement of planar graphs under different settings, by providing bounds on the required size of the underlying grid; see, e.g., [16,18,21,24,29]. In the original work by de Fraysseix et al. the size of the underlying grid is $(2n-4) \times (n-2)$ with n being the number of vertices of the graph; such a bound is asymptotically worst-case optimal, as it is known that there exist n-vertex planar graphs that need $\Omega(n) \times \Omega(n)$ area in any of their planar drawings [13,22].

The corresponding best-known[1] upper bound is due to Chrobak and Kant [11], who presented a linear-time algorithm to embed any n-vertex planar graph into a grid of size $(n-2) \times (n-2)$; see also [29]. In contrast to the work by de Fraysseix, Pach and Pollack [13], which requires an augmentation of the input planar graph to maximal planar, the algorithm by Chrobak and Kant [12] requires just 3-connectivity. Furthermore, it guarantees an additional property, which is desired when drawing 3-connected planar graphs (see, e.g., [32]): the obtained drawings are *convex*, i.e., the boundary of each face is a convex polygon.

Back in 1996, Chrobak, Goodrich and Tamassia [10] studied the area requirement of 3-connected planar graphs under an additional requirement, which is essential in practical applications. In particular, they introduced the notion of *edge-vertex resolution*, which measures how close a vertex is to any non-incident edge, and required that the obtained drawings have bounded edge-vertex resolution. This requirement becomes essential in several practical situations, for instance, consider graph editors which usually represent each vertex by an object of a certain size (rather than a point) containing a distinguishing label. Having high edge-vertex resolution allows to avoid potential overlaps between vertices and edges, in particular, having edge-vertex resolution at least $\frac{1}{2}$ allows each vertex to be represented as an open disk of unit diameter, such that overlaps between vertices and non-incident edges are completely avoided, and simultaneously vertices centered at neighboring grid-points do not overlap (although may touch); see Fig. 1. In their work [10], Chrobak, Goodrich and Tamassia claimed that every 3-connected planar graph admits a convex planar straight-line grid

---

[1] Note that improvements on this bound are known but they are obtained by exploiting either the structure of the input graph [8,16,18,37] or higher connectivity [21,24].

drawing on a grid of size $(3n - 7) \times (3n - 7)/2$ with edge-vertex resolution at least $\frac{1}{2}$. However, the details of the algorithm (and of its proof) supporting this claim never appeared in the literature. In this regard, very recently, Bekos et al. [5] referred to the drawings with edge-vertex resolution at least $\frac{1}{2}$ as *disk-link* and proved (among other results) that every planar graph admits a planar straight-line disk-link drawing on a grid of size $(3n - 7) \times (3n - 7)/2$. However, the obtained drawing is not necessarily convex.

We improve both results mentioned above by providing a linear-time algorithm to compute planar straight-line disk-link drawings that are convex and that fit on a grid of size $(n - 2 + a) \times (n - 2 + a)$, where $a = \min\{f, n - 3\}$ and f denotes the number of internal faces of the input graph. In particular, if the input graph is maximal planar (that is, $f = 2n - 5$), our technique yields drawings of area $(2n - 5) \times (2n - 5)$. On the other hand, if the input graph is 3-connected cubic (that is, $f = \frac{n}{2} + 1$), then our technique yields drawings of area $(\frac{3n}{2} - 1) \times (\frac{3n}{2} - 1)$. Our result is summarized in the next theorem.

Theorem 1. *Every 3-connected plane graph with n vertices and f internal faces admits a convex planar straight-line grid drawing with edge-vertex resolution at least $\frac{1}{2}$ on a grid of size $(n - 2 + a) \times (n - 2 + a)$, where $a = \min\{f, n - 3\}$. Also, the drawing can be computed in $O(n)$ time.*

Related Work. Bárány and Rote [2] prove that every 3-connected planar graph has a *strictly* convex drawing on a quartic grid, improving a previous result by Rote [27]. We recall that a planar drawing is strictly convex if each face is bounded by a strictly convex polygon. We point the interested reader to the surveys by Di Battista and Frati [14,15] for additional references and results concerning convex and strictly-convex drawings of planar graphs in small area.

Concerning the edge-vertex resolution requirement there exist multiple related streams of research. A *closed rectangle-of-influence* (closed RI for short) drawing is a planar straight-line drawing such that no vertex lies in the axis-parallel rectangle (including the boundary) defined by the two ends of every edge [1,4,6,7,23,28]. Any closed RI drawing whose vertices are at integer coordinates can be seen as a disk-link drawing. This implies that disk-link drawings (not necessarily convex) in quadratic area exist for several classes of plane graphs [4,6,28]. However, any plane graph with a filled 3-cycle does not admit a closed RI drawing [6]. Another related direction considers drawings where vertices are objects with integer coordinates and the edges are fat segments [3]. In such drawings the edges do not connect the centers of the incident vertex-disks but rather simply enter these vertex-objects through varying angles. Duncan et al. [17] also use fat edges but, in contrast to [3], they do not compute a drawing from scratch but rather try to extend an existing one without modifying the area of the layout. Van Kreveld [34] studies *bold drawings*, in which vertices are drawn as disks of radius r and edges as rectangles of width w, where $r > w/2$. A bold drawing is *good* if all of its vertices and edges are at least partially visible (neither a vertex disk nor an edge-rectangle is completely hidden by overlapping edges). Although disk-link drawings form a special case of bold drawings in which $r = \frac{1}{2} - \varepsilon$ and $w = 2\varepsilon$ (for some sufficiently small $\varepsilon > 0$), the research on bold drawings has mainly focused on finding feasible values of r and w, rather than on area bounds for fixed values of r and w.

2   Preliminaries

Basic Definitions. A *drawing* of a graph maps each vertex to a distinct point of the plane, and each edge to a Jordan arc connecting its endpoints. A drawing is *planar* if no two edges intersect, except possibly at a common endpoint. A planar drawing partitions the plane into topologically connected regions, called *faces*. The unbounded region is called *outer face*; any other face is an *internal face*. A graph is *planar* if it admits a planar drawing. A *planar embedding* of a planar graph is an equivalence class of topologically-equivalent (i.e., isotopic) planar drawings. A planar graph with a given planar embedding is a *plane graph*.

A drawing is *straight-line* if the Jordan arcs representing the edges are straight-line segments. The *slope* of a line ℓ is the tangent of the minimum-angle that a horizontal line needs to be rotated in order to make it overlap with ℓ; a positive slope corresponds to a counter-clockwise rotation, while a negative one corresponds to a clockwise rotation. The *slope* of a segment is the slope of the supporting line containing it. A *grid drawing* of a graph is a straight-line drawing whose vertices are at integer coordinates. We say that the grid size of a grid drawing Γ is $W \times H$ (or, equivalently, the area of Γ is $W \times H$), if the minimum axis-aligned box containing Γ has side lengths $W - 1$ and $H - 1$. Moreover, for a vertex v of a graph G, we denote by $x_\Gamma(v)$ and by $y_\Gamma(v)$ the x- and y-coordinate of v in drawing Γ of G, respectively. When the reference to Γ is clear from the context, we simply write $x(v)$ and $y(v)$.

Disk-Link Drawings. The *edge-vertex resolution* of a grid drawing of a graph is the minimum Euclidean distance between a point representing a vertex and any edge that is not incident to that vertex. A *disk-link drawing* of a graph is a grid drawing of edge-vertex resolution at least $\frac{1}{2}$. Observe that, in a disk-link drawing Γ, for each vertex v one can draw an open disk with radius $\overline{\rho} \leq \frac{1}{2}$ centered at the point of Γ representing v, and this results in a diagram in which no two disks intersect, and no disk is intersected by a non-incident edge. For simplicity, we assume that $\overline{\rho} = \frac{1}{2}$, i.e., the disks have unit diameter. This assumption is not restrictive, since our results carry over for any constant radius up to some multiplicative constant factor for the area.

Canonical Order. Even though we assume familiarity with basic concepts of planar graph drawing [26,35], we recall in this section a key concept that is central in several algorithms for producing planar grid drawings of plane graphs, e.g., [10,13,22]. Namely, the *canonical order* [22] for 3-connected plane graphs, which is defined as follows: Let G be a 3-connected plane graph with n vertices and let $\pi = (P_0, \ldots, P_m)$ be a partition of the vertex-set of G into paths, such that $P_0 = \{v_1, v_2\}$, $P_m = \{v_n\}$, and edges (v_1, v_2) and (v_1, v_n) exist and belong to the outer face of G. For $k = 0, \ldots, m$, let G_k be the subgraph induced by

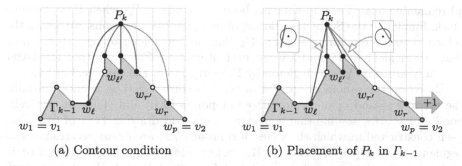

(a) Contour condition   (b) Placement of P_k in Γ_{k-1}

Fig. 2. Introducing a singleton P_k in Γ_{k-1} according to the algorithm by Chrobak and Kant [11]; the white-filled vertices are the critical vertices $w_{\ell'}$ and $w_{r'}$. (Color figure online)

$\cup_{i=0}^{k} P_i$ and denote by C_k the *contour* of G_k defined as follows: If $k = 0$, then C_0 is the edge (v_1, v_2), while if $k > 0$, then C_k is the path from v_1 to v_2 obtained by removing (v_1, v_2) from the cycle delimiting the outer face of G_k. We say that π is a *canonical order* of G if for each $k = 1, \ldots, m - 1$ the following properties hold: **(P.1)** G_k is biconnected and internally 3-connected, **(P.2)** all neighbors of P_k in G_{k-1} are on C_{k-1}, **(P.3)** either P_k is a *singleton* (that is, $|P_k| = 1$), or P_k is a *chain* (that is, $|P_k| > 1$) and the degree of each vertex of P_k is 2 in G_k, and **(P.4)** all vertices of P_k with $0 \leq k < m$ have at least one neighbor in P_j for some $j > k$. A canonical order of G can be computed in linear time [22]. A vertex on contour C_k is called *saturated* in G_k if and only if it is not adjacent to a vertex belonging to a path $P_{k'}$ with $k' > k$.

3   Convex Planar Grid Disk-Link Drawings

In this section, we present our algorithm to compute convex planar grid disk-link drawings of 3-connected plane graphs. As our algorithm builds upon an algorithm by Chrobak and Kant [11] yielding convex planar grid drawings (that are not necessarily disk-link) of 3-connected plane graphs with n vertices on grids of size $(n - 2) \times (n - 2)$, for completeness, we first recall its basic ingredients.

The Algorithm by Chrobak and Kant [11]. This algorithm is incrementally computing a convex planar drawing Γ of a 3-connected plane graph G using a canonical order $\pi = (P_0, \ldots, P_m)$ of G. The drawing Γ has integer grid coordinates and fits in a grid of size $(n - 2) \times (n - 2)$. In order to ease the presentation, we define a Schnyder-like [19,29] 4-coloring of the edges of G based on the canonical order π. G_0 consists of a single edge (v_1, v_2), which is assigned the black color. Assuming that a 4-coloring has been constructed for G_{k-1} with $k = 1, \ldots, m$, we extend it for G_k as follows (see Fig. 2a): We first color the edges of G_k that do not belong to G_{k-1} and are on contour C_k. We color the first such edge encountered in a traversal of C_k from v_1 to v_2 blue, the last one green and all remaining ones (i.e., those having both endpoints in P_k, when P_k is a chain) black. Similar to the Schnyder coloring of maximal planar graphs, we assign the color red to the remaining edges of G_k that do not belong to G_{k-1} (i.e., those that are incident to P_k and are not part of contour C_k). Note that the latter case only arises if P_k is a singleton by Property P.3 of the canonical order.

Based on the canonical order π of G, Γ is constructed as follows: Initially, the vertices v_1 and v_2 of P_0 are placed at points $(0,0)$ and $(1,0)$, respectively. For $k = 1, \ldots, m$, assume that a planar convex grid drawing Γ_{k-1} of G_{k-1} has been constructed in which the edges of contour C_{k-1} are drawn as straight-line segments with slopes 0, -1 or in $[1, +\infty]$ (contour condition; see Fig. 2a); in particular, the slope of each blue edge of C_{k-1} is at least 1, the slope of each black edge of C_{k-1} is 0, while the slope of each green edge of C_{k-1} is -1 (note that C_{k-1} does not contain any red edge by definition). Also, each vertex v in G_{k-1} has been associated with a so-called *shift-set* $S(v)$; the shift-sets of v_1 and v_2 of path P_0 are singletons such that $S(v_1) = \{v_1\}$ and $S(v_2) = \{v_2\}$.

Let $(w_1, \ldots, w_p)$ be the vertices of C_{k-1} from left to right in Γ_{k-1}, where $w_1 = v_1$ and $w_p = v_2$. For the next path $P_k = \{z_1, \ldots, z_q\}$ in π, let w_ℓ and w_r be the leftmost and rightmost neighbors of P_k on C_{k-1} in Γ_{k-1}, where $1 \leq \ell < r \leq p$. For the definition of the shift-set $S(v)$ of each vertex v in P_k, the algorithm identifies two *critical* vertices on the contour C_{k-1}, which we denote by $w_{\ell'}$ and $w_{r'}$, such that $\ell < \ell' \leq r$ and $\ell \leq r' < r$ (refer to the white-filled vertices of Fig. 2); note that it is possible to have $w_{\ell'} = w_{r'}$. Vertex $w_{\ell'}$ is the first vertex encountered in the traversal of C_{k-1} starting from $w_{\ell+1}$ towards w_r that either has a neighbor in P_k or the edge $(w_{\ell'}, w_{\ell'+1})$ is blue or black; note that it is possible to have $w_{\ell'} = w_r$. Symmetrically, vertex $w_{r'}$ is the first vertex encountered in the traversal of C_{k-1} starting from w_{r-1} towards w_ℓ that either has a neighbor in P_k or the edge $(w_{r'-1}, w_{r'})$ is green or black; note that it is possible to have $w_{r'} = w_\ell$. We refer to $w_{\ell'}$ and $w_{r'}$ as the *left-critical* and *right-critical vertices* of P_k. More importantly, since each internal face of Γ_k is convex, in the case where P_k is either a chain or a singleton of degree 2 in G_k, vertices $w_{\ell'}$ and $w_{r'}$ are either consecutive along C_{k-1} or $w_{\ell'} = w_{r'}$ holds. Once $w_{\ell'}$ and $w_{r'}$ have been identified, the algorithm sets the shift-sets of the vertices $z_1, \ldots, z_q$ of P_k as follows:

$$S(z_1) = \{z_1\} \cup \bigcup_{i=\ell'}^{r'} S(w_i), \quad \text{and} \quad S(z_i) = \{z_i\}, \text{ for } i = 2, \ldots, q \quad (1)$$

Furthermore, to guarantee that the resulting drawing is convex, the algorithm updates the shift-sets of w_ℓ and w_r of G_{k-1} as follows:

$$S(w_\ell) = \bigcup_{i=\ell}^{\ell'-1} S(w_i), \quad \text{and} \quad S(w_r) = \bigcup_{i=r'+1}^{r} S(w_i). \quad (2)$$

To compute the drawing Γ_k, the algorithm distinguishes two cases. If w_ℓ is saturated in G_k (i.e., z_1 is the last neighbor of w_ℓ that has not been drawn),

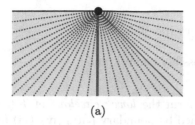

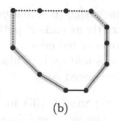

(a)   (b)

Fig. 3. Illustration of (a) Property 1 and (b) Property 2. (Color figure online)

then the x-coordinate of z_1 is the same as the one of w_ℓ, that is, $x(z_1) = x(w_\ell)$. Otherwise, $x(z_1) = x(w_\ell) + 1$. To accommodate the vertices of P_k and to avoid edge-overlaps, the algorithm shifts each vertex in

$$\bigcup_{i=r}^{p} S(w_i). \tag{3}$$

by q units to the right (see Fig. 2b). Then, the algorithm places vertex z_q at $(x(z_1) + q - 1, y(w_r) + x(w_r) - (x(z_1) + q - 1))$, i.e., at the intersection of the line of slope -1 through w_r with the vertical line through point $x(z_1) + q - 1$. Note that this is a grid point above w_ℓ and w_r due to the contour condition and the shifting of $S(w_r)$. For $i = 1, \ldots, q-1$, vertex z_i of P_k is placed $q - i$ units to the left of z_q. Since (w_ℓ, z_1) is blue, (z_q, w_r) is green, and the internal edges of P_k (if any) are black, the contour condition of the algorithm is, by construction, maintained after the placement of the vertices of P_k in Γ_k.

The contour condition together with the shifting procedure described above guarantee Property 1 for the slopes of the edges in Γ_k.

Property 1 (Chrobak and Kant [11]). A shift can only decrease the slope of a blue edge, increase the slope of a green edge, while the black and the red edges are *rigid*, i.e., they maintain their slope. As a result, in Γ_k (see Fig. 3a):

- the slope of each blue edge ranges in $(0, +\infty]$,
- the slope of each black edge is 0,
- the slope of each green edge ranges in $[-1, 0)$, and
- the slope of each red edge ranges in $[-\infty, -1)$.

Since each face of Γ_k is formed when a path $P_{k'}$ with $k' \le k$ of canonical order π is introduced, Property 1 combined with the contour condition and Property P.4 of the canonical order imply the following property for the shape of each face in Γ_k.

Property 2. Let f be a face in Γ_k. Then, a counter-clockwise traversal of f starting from its leftmost vertex that is the bottommost when it is not uniquely defined consists of the following boundary parts (see Fig. 3b):

i. a strictly descendant path of green edges (possibly empty),

ii. a black edge (possible non existent),
iii. a strictly ascendant path of blue edges (possible empty),
iv. a green or red edge,
v. a horizontal path of black edges (possibly empty), and
vi. a blue or red edge.

Boundary parts (i)–(iii) in Property 2 form the *lower envelope* of f (solid in Fig. 3b). The *upper envelope* of f is formed by boundary parts (iv)–(vi) (dotted in Fig. 3b). The latter is introduced in Γ_k when a path of the canonical order is placed. Thus, the upper envelope cannot contain black and red edges simultaneously (by Property P.3 of canonical order). Finally, boundary parts (iii) and (iv) form the *right envelope* of f, while (vi) and (i) form the *left envelope* of f (gray-highlighted in Fig. 3b).

We next state some lemmata regarding the "behavior" of the algorithm by Chrobak and Kant [11] that are employed in the proof of correctness of our modification. Due to space limitations, their proofs are omitted.

Lemma 1. *Let u and v be two distinct vertices of G_k belonging to the same face f of Γ_k. If u and v have the same y-coordinate in Γ_k with $x(u) < x(v)$, then either u and v are connected by a path of black edges of f or the x-coordinate of the bottommost vertex/vertices of f is/are in the interval $(x(u), x(v)]$.*

Lemma 2. *Let $\mathcal{S}_k$ be the vertices of G_{k-1} that were shifted during the introduction of P_k in Γ_k, and let c be a positive integer. Let Γ'_k be the drawing obtained from Γ_{k-1} by first shifting the vertices of $\mathcal{S}_k$ by c units to the right and then attaching P_k as in the algorithm by Chrobak and Kant [11]. Then, Γ'_k is a convex planar grid drawing of G_k.*

Lemma 3. *Let f be a face of Γ_{k-1} that contains a black edge (u, v) at its lower envelope, such that u is to the left of v in Γ_{k-1}, and assume that some vertices of f are not shifted during the introduction of P_k in Γ_k. Then, neither u nor v are shifted, unless (u, v) is the rightmost edge of the lower envelope of f, in which case u is not shifted, while v is shifted.*

Our Modification. We start by placing v_1 and v_2 of path P_0 as in the algorithm by Chrobak and Kant [11], that is, at points $(0, 0)$ and $(1, 0)$, respectively. Assume now that Γ_{k-1} is a convex planar disk-link drawing of G_{k-1}. For placing path P_k in drawing Γ_{k-1}, $k = 1, \ldots, m$, we distinguish two cases. In the first case, P_k is a chain and we proceed as in the algorithm by Chrobak and Kant [11]. Hence, we focus on the more elaborated case, in which P_k is a singleton, i.e., $P_k = \{z_1\}$. In this case, our algorithm first shifts the vertices of Γ_{k-1} appropriately to guarantee that the obtained drawing Γ_k is a disk-link drawing (to be shown in Lemma 5). Let $w_{x_0}, \ldots, w_{x_{\rho+1}}$ be the neighbors of P_k along C_{k-1}, such that $\ell = x_0 < x_1 < \ldots < x_\rho < x_{\rho+1} = r$. Note that, based on this notation, ρ denotes the number of neighbors of P_k between w_ℓ and w_r on C_{k-1}. Besides critical vertices $w_{\ell'}$ and $w_{r'}$, our modification introduces the following $\rho+1$ *pivot vertices* $w_{x'_1}, \ldots, w_{x'_{\rho+1}}$, where $w_{x'_1} = w_{\ell'}$ and $w_{x'_{\rho+1}} = w_r$. For $j = 2, \ldots, \rho$, the pivot vertex $w_{x'_j}$ (with

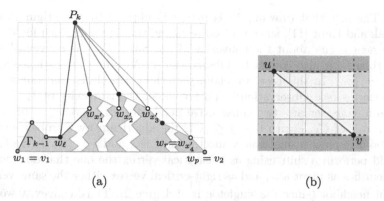

Fig. 4. (a) Introducing a singleton P_k in Γ_{k-1} according to our modification of the algorithm by Chrobak and Kant [11]; the white-filled vertices are the identified pivot vertices, and (b) edge-vertex configurations used in the proof of Lemma 5. (Color figure online)

$x_{j-1} < x'_j \leq x_j$) is defined as the first vertex encountered in the traversal of C_{k-1} starting from $w_{x_{j-1}+1}$ towards w_{x_j} that either is neighboring z_1 or is followed by an edge of C_{k-1} that is blue or black. In other words, pivot vertex $w_{x'_j}$ would be the vertex that the algorithm by Chrobak and Kant [11] identifies as left-critical, when attaching a singleton with exactly two neighbors $w_{x_{j-1}}$ and w_{x_j} on C_{k-1}. The algorithm modifies Γ_{k-1} by performing $\rho+1$ consecutive refinements of the vertex positions. In the j-th refinement, $j = 1, \ldots, \rho+1$, the algorithm shifts each vertex in $\bigcup_{i=x'_j}^{p} S(w_i)$ by one unit to the right; see Fig. 4a. This implies that vertices $w_r, \ldots, w_p$ of C_{k-1} have been shifted in total by $\rho+1$ units to the right. Note that in the algorithm by Chrobak and Kant [11] these vertices would be shifted by only one unit. The next observations follow from our shifting strategy.

Observation 1. *If P_k is a chain, then our shifting strategy and the one by Chrobak and Kant [11] are identical.*

Observation 2. *If P_k is a singleton, then the horizontal distance between any two consecutive neighbors of P_k in C_{k-1} gets increased by one unit in Γ_k, while in the algorithm by Chrobak and Kant [11] this would only be the case for w_{x_ρ} and $w_{x_\rho+1} = w_r$.*

The construction of Γ_k is completed by placing the vertices of P_k as in the algorithm by Chrobak and Kant [11], i.e., we set either $x(z_1) = x(w_\ell)$ or $x(z_1) = x(w_\ell) + 1$ (depending on whether z_1 is saturated or not, respectively), we place z_q at the intersection of the line of slope -1 through w_r with the vertical line through point $x(z_1) + q - 1$ and for $i = 1, \ldots, q-1$, vertex v_i of P_k is placed $q - i$ units to the left of z_q. Thus, the contour condition is maintained in Γ_k.

Lemma 4. *The drawing Γ_k produced by our modification of the algorithm by Chrobak and Kant [11] is planar and convex.*

Proof. The fact that drawing Γ_k is planar is implied by the original proof of Chrobak and Kant [11], since the contour condition is maintained for Γ_k.

We next argue about the convexity of Γ_k. Since Γ_{k-1} is convex, if P_k is a chain, then Γ_k is also convex by Observation 1. Assume that P_k is a singleton, i.e., $P_k = \{z_1\}$. In this case, we claim that the extra shifts that our modification performs (see Observation 2) do not affect the convexity of Γ_k. To prove the claim, consider any two consecutive neighbors w_{x_j} and $w_{x_{j+1}}$ of z_1 along $(w_1, \ldots, w_p)$ of C_{k-1} with $0 \leq j \leq \rho$. If the algorithm by Chrobak and Kant were about to place a singleton connecting only w_{x_j} and $w_{x_{j+1}}$ to derive Γ_k, then it would perform a shift using as left-critical vertex the one that our modification identifies as pivot $w_{x'_j}$, and as right-critical vertex either the same vertex or its right neighbor (since the singleton is of degree 2). Thus, convexity would be maintained. Applying the same reasoning to any pair of consecutive neighbors of P_k, proves that the subdrawing of Γ_k induced by G_{k-1} is indeed convex. In addition, the same reasoning implies that Properties 1 and 2 of the algorithm by Chrobak and Kant [11] also hold. To complete the proof of our claim, we note that the fact that the faces incident to P_k in Γ_k are convex follows using the same approach as in the algorithm by Chrobak and Kant, as the contour condition is maintained. □

Note that since the contour condition is maintained and we do not modify the shift-sets, the fact that Properties 1 and 2 hold in our modification implies that Lemmas 1, 2 and 3 also hold. To complete the proof of correctness of our algorithm, we prove in the following lemma that Γ_k is a disk-link drawing of G_k. To ease the proof, we denote by Γ'_{k-1} the drawing of G_{k-1} obtained after the preparatory shifting in drawing Γ_{k-1} for the introduction of P_k.

Lemma 5. *The following statements hold: (i) the edge-vertex resolution of Γ'_{k-1} is no less than that of Γ_{k-1}, and (ii) introducing the new edges of Γ_k, which are either part of P_k or incident to the endpoints of P_k, preserves the edge-vertex resolution to at least $\frac{1}{2}$.*

Proof. Since the drawing of G_{k-1} is planar in Γ_k, it is sufficient to only consider its faces in order to prove statement (i). To this end, consider any arbitrary face f in Γ'_{k-1}. If either none or all of the vertices of f are shifted by the same amount, then statement (i) obviously holds. Consider now the case where f contains at least one vertex that is shifted and one vertex that remains stationary in Γ'_{k-1}. Suppose, for a contradiction, that f contains an edge (u, v) and a vertex w that is not incident to (u, v) such that (u, v) intersects the disk of w in Γ'_{k-1}. Since Γ'_{k-1} is a grid drawing of G_{k-1}, it follows by Property 1 that (u, v) cannot be a black edge. Assume that the slope of (u, v) is negative (i.e., (u, v) is green or red), as the case in which it is positive (i.e., (u, v) is blue) is similar. W.l.o.g., further assume that u is above v in Γ'_{k-1} (see Fig. 4b).

Since Γ'_{k-1} is a grid drawing of G_{k-1}, it follows that, regardless of whether w was shifted or not, w is neither above nor below the *horizontal* strip delimited by the two horizontal lines through u and v in Γ'_{k-1} (green in Fig. 4b). Similarly, one observes that w is neither to the left nor to the right of the *vertical* strip delimited by the two vertical lines through u and v (blue in Fig. 4b). It follows that w is either in the interior (yellow in Fig. 4b) or on the boundary of the axis-aligned bounding box B_{uv} of the edge (u, v) in Γ'_{k-1}.

We next argue that w can be neither in the interior of B_{uv} nor along its two vertical sides, which implies that w is necessarily on one of the two horizontal sides of B_{uv} (purple in Fig. 4b). To see this, assume for a contradiction that w is in the interior of B_{uv} or along one of its two vertical sides but not at its corners. Since Γ_{k-1} is a disk-link drawing of G_{k-1} while Γ'_{k-1} is not, it follows that the distance between w and (u, v) decreased after the shifting (by one unit) to obtain Γ'_{k-1} from Γ_{k-1}. If the shifting were sufficiently large (and greater than one unit), then w would be on different sides of (u, v) in Γ_{k-1} and in Γ'_{k-1}, violating the planarity of the drawing (which is implied by Lemma 2); a contradiction.

It follows that w is on any of the two horizontal sides of B_{uv}, as we initially claimed. We proceed by considering two subcases depending on whether (u, v) is on the upper or lower envelope of f. Consider first the case where (u, v) is on the upper envelope of f. By Property 2, it follows that w is on the lower envelope of f and, thus, on the lower edge of B_{uv}. If w and v are not adjacent, the fact that w and v have the same y-coordinate implies that the x-coordinate of the bottommost vertex/vertices of f is delimited by w and v (by Lemma 1). This further implies that w and v are on the left and right envelopes of f, respectively. Since not all vertices of f are shifted in Γ'_{k-1}, it follows that, among the vertices of the lower envelope of f, the ones that are shifted are those in the shift-set of the rightmost vertex of the lower envelope of f, which implies that w has not been shifted. On the other hand, if w and v are adjacent, then the edge connecting them is black (by Property 1), and thus by Lemma 3, we conclude again that w is not shifted. In both cases, however, the edge-vertex resolution of Γ'_{k-1} cannot be smaller than the one of Γ_{k-1}; a contradiction.

Consider now the case where (u, v) is on the lower envelope of f. In this case, w can be either on the lower or upper envelope of f. The former case can be ruled out by adopting an argument similar to the one of the previous paragraph. In the latter case, w is on the top side of B_{uv}. By Property 2, u and w are connected by a path of black edges contradicting the fact that (u, v) is green, since, by Property 2.vi, the left edge connecting to a black path on the upper envelope is blue.

We now prove statement (ii). Assume for a contradiction that an edge (u, v) added in Γ_k during the introduction of P_k intersects the disk of a vertex w. Clearly, w belongs to G_{k-1}, since by construction we have no edge-disk intersections between elements of P_k. Thus, one endpoint of (u, v), say u, belongs to P_k and the other, say v, to G_{k-1}, i.e., (u, v) is not a black edge. If (u, v) is green, then its slope in Γ_k is -1, and hence it cannot intersect any non-adjacent disk. Assume first that (u, v) is blue, and observe that the construction is such that the horizontal distance between u and v is either 0 or 1. In the former case, (u, v) is vertical and cannot intersect any non-adjacent vertex-disk. In the latter case, the shifting performed by the algorithm guarantees that the part of the grid column along which vertex u is placed that is contained in the horizontal strip bounded by the horizontal lines through u and v in Γ_k contains no vertex of C_{k-1}, and therefore edge (u, v) again cannot intersect any non-adjacent vertex disk. The argument for the case in which (u, v) is red is analogous.   □

We conclude the proof of our main result by analyzing the area of the produced drawings and the time complexity of the algorithm.

Proof of Theorem 1. Let G be an n-vertex 3-connected plane graph with f internal faces. Let Γ be a planar drawing of G computed by our algorithm. By Lemma 4, drawing Γ is convex, and, by Lemma 5, its edge-vertex resolution is at least $\frac{1}{2}$. By the contour condition, Γ is inside a right isosceles triangle, such that it has a horizontal side (which corresponds to edge (v_1, v_2)) and a vertical side (which contains edge (v_1, v_n)) that have the same length and meet at point $(0, 0)$. In the algorithm by Chrobak and Kant, the value of the width and the height of this triangle is $n - 2$ [11]. The additional unit-shifts due to the introduction of singletons performed by our modification increase the value of the width and the height by the same amount a (see Observations 1 and 2). We focus on the width of Γ, and we distinguish two cases: either $\min\{f, n-3\} = n-3$ or $\min\{f, n - 3\} = f$.

Assume first that $\min\{f, n - 3\} = n - 3$. We develop a charging argument that charges each additional one-unit shift to the red edges of G. In particular, consider a singleton P_k. The additional shifts due to this singleton are two less than its degree in G_k, which equals the number of red edges incident to P_k in G_k. It is immediate to see that each red edge is charged to exactly one additional shift. Hence, the total number of additional shifts is at most the number of red edges in G, which is at most $n - 3$ (recall that the red subgraph of G is a forest with at most $n - 3$ edges). Consequently, in this case $a \leq n - 3$.

Assume now that $\min\{f, n-3\} = f$. In this case we develop a similar charging argument, in which we charge each additional one-unit shift to the internal faces of G, rather than to its red edges. Again, consider a singleton P_k, and observe that the additional shifts due to this singleton are two less than its degree in G_k. This value equals the number of internal faces incident to P_k in G_k minus one, in particular, we can avoid charging the shift to the rightmost internal face incident to P_k. It is not difficult to see that each internal face is charged to at most one additional shift. Hence, the total number of additional shifts is at most the number of internal faces f in G. Consequently, in this case $a \leq f$.

Finally, we discuss the time complexity. The algorithm by Chrobak and Kant can be implemented to run in linear time [11]. In particular, the key ingredient to achieve linear time complexity, is the use of relative coordinates for the vertices, which avoids shifting entire subgraphs. Since our algorithm only requires a linear number of additional one-unit shifts and it does not modify the shift-sets of the vertices, this translates into different relative coordinates and requires neither additional operations nor different data structures. Therefore it can be implemented to also run in linear time.   □

4   Open Problems

In this work, we present improvements upon results in [5,10]. The following research directions naturally stem from our work. (i) Can the bounded edge-vertex resolution requirement be incorporated into an area lower bound so to improve the one given in [13,22]? (ii) Can the area bound of Theorem 1 be improved in the case where the input graph is 4-connected? Note that such graphs admit $W \times H$ drawings with $W + H \leq n - 1$ [25] but their edge-vertex resolution may be arbitrarily small. (iii) Finally, it is of interest to study the edge-vertex resolution requirement for strictly convex drawings.

References

1. Alamdari, S., Biedl, T.: Open rectangle-of-influence drawings of non-triangulated planar graphs. In: Didimo, W., Patrignani, M. (eds.) GD 2012. LNCS, vol. 7704, pp. 102–113. Springer, Heidelberg (2013). https://doi.org/10.1007/978-3-642-36763-2_10

2. Bárány, I., Rote, G.: Strictly convex drawings of planar graphs. Doc. Math. **11**, 369–391 (2006). http://eudml.org/doc/53043

3. Barequet, G., Goodrich, M.T., Riley, C.: Drawing planar graphs with large vertices and thick edges. J. Graph Algorithms Appl. **8**, 3–20 (2004). https://doi.org/10.7155/jgaa.00078

4. Barrière, L., Huemer, C.: 4-labelings and grid embeddings of plane quadrangulations. Discret. Math. **312**(10), 1722–1731 (2012). https://doi.org/10.1016/j.disc.2012.01.027

5. Bekos, M.A., Gronemann, M., Montecchiani, F., Pálvölgyi, D., Symvonis, A., Theocharous, L.: Grid drawings of graphs with constant edge-vertex resolution. Comput. Geom. **98**, 101789 (2021). https://doi.org/10.1016/j.comgeo.2021.101789

6. Biedl, T., Bretscher, A., Meijer, H.: Rectangle of influence drawings of graphs without filled 3-cycles. In: Kratochvíyl, J. (ed.) GD 1999. LNCS, vol. 1731, pp. 359–368. Springer, Heidelberg (1999). https://doi.org/10.1007/3-540-46648-7_37

7. Biedl, T.C., Lubiw, A., Mehrabi, S., Verdonschot, S.: Rectangle-of-influence triangulations. In: Shermer, T.C. (ed.) CCCG, pp. 237–243 (2016)

8. Bonichon, N., Felsner, S., Mosbah, M.: Convex drawings of 3-connected plane graphs. Algorithmica **47**(4), 399–420 (2007). https://doi.org/10.1007/s00453-006-0177-6

9. Chiba, N., Onoguchi, K., Nishizeki, T.: Drawing planar graphs nicely. Acta Inform. **22**, 187–201 (1985). https://doi.org/10.1007/BF00264230

10. Chrobak, M., Goodrich, M.T., Tamassia, R.: Convex drawings of graphs in two and three dimensions (preliminary version). In: Whitesides, S. (ed.) SoCG, pp. 319–328. ACM (1996). https://doi.org/10.1145/237218.237401

11. Chrobak, M., Kant, G.: Convex grid drawings of 3-connected planar graphs. Int. J. Comput. Geom. Appl. **7**(3), 211–223 (1997). https://doi.org/10.1142/S0218195997000144

12. Chrobak, M., Payne, T.H.: A linear-time algorithm for drawing a planar graph on a grid. Inf. Process. Lett. **54**(4), 241–246 (1995). https://doi.org/10.1016/0020-0190(95)00020-D

13. de Fraysseix, H., Pach, J., Pollack, R.: Small sets supporting Fáry embeddings of planar graphs. In: Simon, J. (ed.) STOC, pp. 426–433. ACM (1988). https://doi.org/10.1145/62212.62254

14. Di Battista, G., Frati, F.: Drawing trees, outerplanar graphs, series-parallel graphs, and planar graphs in a small area. In: Pach, J. (ed.) Thirty Essays on Geometric Graph Theory, pp. 121–165. Springer, New York (2013). https://doi.org/10.1007/978-1-4614-0110-0_9

15. Di Battista, G., Frati, F.: A survey on small-area planar graph drawing. CoRR, abs/1410.1006 (2014)

16. Di Battista, G., Tamassia, R., Vismara, L.: Output-sensitive reporting of disjoint paths. Algorithmica 23(4), 302–340 (1999). https://doi.org/10.1007/PL00009264

17. Duncan, C.A., Efrat, A., Kobourov, S.G., Wenk, C.: Drawing with fat edges. Int. J. Found. Comput. Sci. 17(5), 1143–1164 (2006). https://doi.org/10.1142/S0129054106004315

18. Felsner, S.: Convex drawings of planar graphs and the order dimension of 3-polytopes. Order 18(1), 19–37 (2001). https://doi.org/10.1023/A:1010604726900

19. Felsner, S.: Geometric Graphs and Arrangements. Advanced Lectures in Mathematics. Vieweg (2004). https://doi.org/10.1007/978-3-322-80303-0

20. Fáry, I.: On straight lines representation of planar graphs. Acta Sci. Math. (Szeged) 11, 229–233 (1948)

21. He, X.: Grid embedding of 4-connected plane graphs. Discrete Comput. Geom. 17(3), 339–358 (1997). https://doi.org/10.1007/PL00009290

22. Kant, G.: Drawing planar graphs using the canonical ordering. Algorithmica 16(1), 4–32 (1996). https://doi.org/10.1007/BF02086606

23. Miura, K., Matsuno, T., Nishizeki, T.: Open rectangle-of-influence drawings of inner triangulated plane graphs. Discrete Comput. Geom. 41(4), 643–670 (2008). https://doi.org/10.1007/s00454-008-9098-2

24. Miura, K., Nakano, S., Nishizeki, T.: Grid drawings of 4-connected plane graphs. Discrete Comput. Geom. 26(1), 73–87 (2001). https://doi.org/10.1007/s00454-001-0004-4

25. Miura, K., Nakano, S., Nishizeki, T.: Convex grid drawings of four-connected plane graphs. Int. J. Found. Comput. Sci. 17(5), 1031–1060 (2006). https://doi.org/10.1142/S012905410600425X

26. Nishizeki, T., Rahman, M.S.: Planar Graph Drawing. Lecture Notes Series on Computing, vol. 12. World Scientific (2004). https://doi.org/10.1142/5648

27. Rote, G.: Strictly convex drawings of planar graphs. In: SODA, pp. 728–734. SIAM (2005). http://dl.acm.org/citation.cfm?id=1070432.1070535

28. Sadasivam, S., Zhang, H.: Closed rectangle-of-influence drawings for irreducible triangulations. Comput. Geom. 44(1), 9–19 (2011). https://doi.org/10.1016/j.comgeo.2010.07.001

29. Schnyder, W.: Embedding planar graphs on the grid. In: SODA, pp. 138–148. SIAM (1990). http://dl.acm.org/citation.cfm?id=320176.320191

30. Stein, S.K.: Convex maps. Proc. American Math. Soc. 2(3), 464–466 (1951)

31. Steinitz, E., Rademacher, H.: Vorlesungen über die Theorie der Polyeder. Julius Springer, Berlin (1934)

32. Thomassen, C.: A refinement of Kuratowski's theorem. J. Comb. Theory Ser. B 37(3), 245–253 (1984). https://doi.org/10.1016/0095-8956(84)90057-1

33. Tutte, W.T.: How to draw a graph. Proc. London Math. Soc. 13, 743–768 (1963)

34. van Kreveld, M.J.: Bold graph drawings. Comput. Geom. 44(9), 499–506 (2011). https://doi.org/10.1016/j.comgeo.2011.06.002

35. Vismara, L.: Planar straight-line drawing algorithms. In: Tamassia, R. (ed.) Handbook on Graph Drawing and Visualization, pp. 193–222. Chapman and Hall/CRC (2013)
36. Wagner, K.: Bemerkungen zum Vierfarbenproblem. Jahresber. Deutsch. Math.-Verein. **46**, 26–32 (1936)
37. Zhang, H., He, X.: Compact visibility representation and straight-line grid embedding of plane graphs. In: Dehne, F., Sack, J.-R., Smid, M. (eds.) WADS 2003. LNCS, vol. 2748, pp. 493–504. Springer, Heidelberg (2003). https://doi.org/10.1007/978-3-540-45078-8_43

1-Extendability of Independent Sets

Pierre Bergé[✉][iD], Anthony Busson[iD], Carl Feghali[iD], and Rémi Watrigant[iD]

Univ Lyon, CNRS, ENS de Lyon, Université Claude Bernard Lyon 1,
LIP UMR5668, Lyon, France
{pierre.berge,anthony.busson,carl.feghali,remi.watrigant}@ens-lyon.fr

Abstract. In the 70s, Berge introduced 1-extendable graphs (also called
B-graphs), which are graphs where every vertex belongs to a maximum
independent set. Motivated by an application in the design of wireless
networks, we study the computational complexity of 1-EXTENDABILITY,
the problem of deciding whether a graph is 1-extendable. We show that,
in general, 1-EXTENDABILITY cannot be solved in $2^{o(n)}$ time assuming
the Exponential Time Hypothesis, where n is the number of vertices
of the input graph, and that it remains NP-hard in subcubic planar
graphs and in unit disk graphs (which is a natural model for wireless
networks). Although 1-EXTENDABILITY seems to be very close to the
problem of finding an independent set of maximum size (*a.k.a.* MAXIMUM
INDEPENDENT SET), we show that, interestingly, there exist 1-extendable
graphs for which MAXIMUM INDEPENDENT SET is NP-hard. Finally, we
investigate a parameterized version of 1-EXTENDABILITY.

Keywords: 1-extendable graphs · B-graphs · Independent set

1 Introduction and Motivation

1.1 Definitions and Related Work

Understanding the structure of independent sets is among the most studied
subjects in algorithmics and graph theory, and finding graph classes where a
maximum independent set (MIS for short) can be found efficiently is an important theoretical and practical problem. In 1970, Plummer [20] defined the class
of well-covered graphs, which are graphs where every independent set which is
maximal for inclusion is also an MIS. In other words, they are exactly the graphs
for which the greedy algorithm always returns an optimal solution. Well-covered
graphs were studied mostly from an algorithmic perspective: their recognition
was proven coNP-hard [8, 26] in general graphs, but polynomial-time solvable for
claw-free graphs [27], and perfect graphs of bounded clique number [11].

Partially supported by the LABEX MILYON (ANR-10-LABX-0070) of Université
de Lyon, within the program "Investissements d'Avenir" (ANR-11-IDEX-0007), and
the research grant ANR DIGRAPHS ANR-19-CE48-0013-01, operated by the French
National Research Agency (ANR).

© Springer Nature Switzerland AG 2022
C. Bazgan and H. Fernau (Eds.): IWOCA 2022, LNCS 13270, pp. 172–185, 2022.
https://doi.org/10.1007/978-3-031-06678-8_13

A related notion, introduced by Berge [5], is the definition of B-graphs, which are those graphs where every vertex belongs to an MIS. B-graphs were mostly introduced in order to study well-covered graphs [24,25]. Later, the notion of B-graphs was generalized to that of k-extendable graphs [11]: a graph is k-extendable, for a positive integer k, if every independent set of size (exactly) k is contained in an MIS. Thus, B-graphs are exactly the 1-extendable graphs and a graph is well-covered if and only if (iff) it is k-extendable for every $k \in \{1, 2, \ldots, \alpha(G)\}$, where $\alpha(G)$ denotes the size of an MIS of G. Dean and Zito [11] obtained a number of results on 1-extendable graphs; for instance, they proved that a bipartite graph is 1-extendable iff it admits a perfect matching and, hence, bipartite 1-extendable graphs can be recognized in polynomial time. Recently, certain structural properties of k-extendable graphs were stated [2,3].

We should note that the notion of k-extendability was also studied in the context of maximum matchings [21,22]. Recently, it was shown that the recognition of (matching) k-extendable graphs is co-NP-complete [15].

In the remainder, B-graphs will be called 1-extendable graphs, as it is the terminology used by the most recent papers on the topic. In this article, our objective is to determine the computational complexity of the recognition of 1-extendable graphs.

This question is motivated not only by the state of the art described above but also by an application on Wi-Fi networks. Indeed, 1-extendable graphs play an important role in the performance of CSMA/CA (Carrier Sense Multiple Access/Collision Avoidance) networks [17]. Graphs stand as a natural model for CSMA/CA wireless networks. Two vertices, *i.e.* nodes of the CSMA/CA network, are adjacent if the two corresponding nodes are able to detect the transmissions from each others. Transmissions from two vertices can occur in parallel iff they are not adjacent. A set of instantaneous transmitters is thus an independent set of the graph. The *throughput* of a node (the number of bits per second it is able to send) is strongly correlated to the number of MISs it belongs to divided by the total number of MISs of the graph [17]. Hence, the fact, for each vertex, of belonging or not to an MIS is of prior importance in such networks if we aim at ensuring a minimal fairness between the vertices and avoiding *starvation*, *i.e.* very low throughput. Figure 1 shows the throughput obtained in a CSMA/CA network for paths on 4 and 5 vertices using the network simulator ns-3 [19]. The 5-vertex path is not 1-extendable: observe that the two vertices that do not belong to any MIS are in starvation. In the 4-vertex path, which is 1-extendable, there is no starvation.

1.2 Contribution

Most of the outcomes of this paper concern the complexity of 1-EXTENDABILITY, the problem of deciding whether an input graph is 1-extendable. First, we focus on the relationship between MAXIMUM INDEPENDENT SET and 1-EXTENDABILITY. We observe that any polynomial-time algorithm for MAXIMUM INDEPENDENT SET on some hereditary family of graphs $\mathcal{C}$ provides us with a polynomial-time algorithm for 1-EXTENDABILITY on $\mathcal{C}$. Based on this

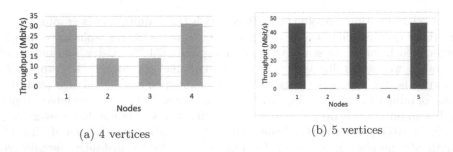

(a) 4 vertices   (b) 5 vertices

Fig. 1. Wi-Fi network simulated with ns-3 for paths on 4 and 5 vertices.

result, we could imagine that, perhaps, MAXIMUM INDEPENDENT SET and 1-EXTENDABILITY are equivalent problems in terms of complexity. However, we show that MAXIMUM INDEPENDENT SET is NP-hard on a certain subfamily of 1-extendable graphs (Theorem 1). This result highlights a gap for the complexity of these two problems.

Second, we establish the hardness of solving 1-EXTENDABILITY. We show that 1-EXTENDABILITY is ETH-hard, *i.e.* cannot be solved in time $2^{o(n)}$ (Corollary 1). Then, we prove that the problem is NP-hard even in planar subcubic graphs (Theorem 3) and in unit disk graphs (Theorem 4), a natural model for CSMA/CA networks.

Eventually, we focus on a parameterized version of 1-EXTENDABILITY, where we ask whether every vertex belongs to an independent set of size at least some parameter k. We show that this problem, PARAM-1-EXTENDABILITY, is W[1]-hard (Theorem 5). Nevertheless, it admits a polynomial kernel if restricted to planar graphs or K_r-free graphs for fixed $r > 0$ (Corollary 2).

1.3 Organization of the Paper

Section 2 is dedicated to the notation, definitions and some basic results; in particular, we explore the relationship between the MAXIMUM INDEPENDENT SET and 1-EXTENDABILITY problems. In Sect. 3, we study three graph transformations and their impact on the 1-extendability property. In Sect. 4, we show that 1-EXTENDABILITY cannot be solved in time $2^{o(n)}$ on n-vertex graphs unless the ETH is false. We also prove that 1-EXTENDABILITY remains NP-hard in planar graphs of maximum degree 3 and in unit disk graphs. Then, Sect. 5 presents the parameterized version PARAM-1-EXTENDABILITY and the results associated with it.

2 Notation and Basic Results

2.1 Notation and Definitions

For a positive integer k, we note $[k] = \{1, \ldots, k\}$. In this paper, all graphs are simple, unweighted and undirected. We denote by $V(G)$ the vertex set of a graph

G and by $E(G)$ its edge set. Edges $(u, v) \in E(G)$ can sometimes be denoted by uv to improve readability. When the identity of the graph considered is clear, we set $n = |V(G)|$ and $m = |E(G)|$. For a vertex $v \in V(G)$, we denote by $N_G(v)$ its set of neighbors (we will sometimes omit the subscript if G is clear from the context). Let $d_G(v) = |N_G(v)|$ be the degree of v. For a subset $R \subseteq V(G)$, let $G[R]$ be the subgraph of G induced by R. A family of graphs $\mathcal{C}$ is called *hereditary* if, for every graph $G \in \mathcal{C}$, every induced subgraph of G also belongs to $\mathcal{C}$. An ℓ-vertex path is denoted by P_ℓ. A *clique cover* of a graph G is a partition of $V(G)$ into sets $C_1, \ldots, C_q$ such that $G[C_i]$ is a clique for every $i \in [q]$. A set of pairwise non-adjacent vertices in a graph is called an *independent set*. A *maximum* independent set (MIS) is an independent set of maximum size. We denote by $\alpha(G)$ the size of an MIS of G. The decision problem of finding an independent set of size at least $k \geq 1$ in a graph G is called MAXIMUM INDEPENDENT SET. A graph G is 1-*extendable* [5] if, for every $u \in V(G)$, there is an MIS S of G such that $u \in S$.

The subject of this paper is to investigate the computational complexity of the following decision problem.

1-EXTENDABILITY
Input: Graph G
Question: Does every vertex of G belong to an MIS of G?

An *embedding* of a graph G is a representation of G in the plane, where vertices are points in the plane and edges are curves which connect their two endpoints. A *plane embedding* of G is an embedding of G where no two edges cross. A graph G is *planar* if it admits a plane embedding.

2.2 Links Between 1-Extendability and Maximum Independent Set

In this section, we investigate to what extent the problems 1-EXTENDABILITY and MAXIMUM INDEPENDENT SET are close to each other. We show a "Turing equivalence" of the two problems in general graphs. More precisely, we prove that solving 1-EXTENDABILITY on an input graph G can be done by solving MAXIMUM INDEPENDENT SET on several induced subgraphs of G, while solving MAXIMUM INDEPENDENT SET on an input graph G can be done by solving 1-EXTENDABILITY on several induced supergraphs of G.

Solving 1-EXTENDABILITY *using* MAXIMUM INDEPENDENT SET. The idea relies on the following lemma, whose straightforward proof is left to the reader.

Lemma 1. *Let G be a graph, and k be a non-negative integer. Then a vertex v of G is contained in an independent set of G of size k iff $G[V(G) \backslash N(v)]$ contains an independent set of size k.*

This lemma allows 1-EXTENDABILITY to inherit many positive results from MAXIMUM INDEPENDENT SET. In particular, it implies that 1-EXTENDABILITY is polynomial-time solvable in any hereditary class where MAXIMUM INDEPENDENT SET is polynomial-time solvable. This is for instance the case for perfect graphs, P_6-free graphs [14], chordal graphs and claw-free graphs. Moreover, it is Fixed-Parameter Tractable (FPT) when parameterized by the tree-width or even the clique-width of the input graph.

Solving MAXIMUM INDEPENDENT SET *using* 1-EXTENDABILITY. The converse of Lemma 1 does not appear to be as straightforward, and we leave as open whether solving 1-EXTENDABILITY in a hereditary graph class C in polynomial-time allows one to solve MAXIMUM INDEPENDENT SET in C in polynomial-time. We can show, however, that this is true if the class satisfies much more conditions than just being hereditary.

Let $G = (V, E)$ be a graph and $r \leq |V|$ be a non-negative integer. Let G_r^+ be the graph obtained from G by adding

- an independent set S of size $|V| - r$ to G,
- for each vertex v of G, a new vertex π_v adjacent to v only, and
- all possible edges between S and the set $T = \{\pi_v : v \in V\}$.

Proposition 1. G_r^+ *is 1-extendable iff* $\alpha(G) = r$.

The ETH [16] states that no algorithm can decide whether a 3SAT formula on n variables is satisfiable in time $2^{o(n)}$. As 3-SAT, MAXIMUM INDEPENDENT SET is ETH-hard. Hence, based on Proposition 1, the same statement holds for 1-EXTENDABILITY.

Corollary 1. *Testing whether an n-vertex graph is 1-extendable cannot be done in time* $2^{o(n)}$ *unless the ETH is false.*

This lower bound is matched by the trivial brute-force algorithm which consists in enumerating all subsets of vertices, and testing whether all MISs cover the entire vertex set.

Another question related to the previous one is whether being 1-extendable helps finding a MIS. The next result suggests that this is unlikely, by showing that 1-EXTENDABILITY and MAXIMUM INDEPENDENT SET can sometimes behave very differently from a computational point of view.

Theorem 1. MAXIMUM INDEPENDENT SET *remains NP-hard and $W[1]$-hard (parameterized by k) in 1-extendable graphs.*

3   Generic Transformations

In this subsection, we present three graph transformations. They are related in some sense to the 1-extendability property: the first one produces a 1-extendable graph, the second one preserves the 1-extendability of the input graph and the third one decreases the maximum degree of the input graph and keeps it 1-extendable. These transformations (or similar ideas) will be used later in some reductions.

Transformation (T_1). The graph $G_{(1)}$ is obtained from G by adding a pendant vertex π_u for any $u \in V(G)$. The vertex π_u has degree one and is adjacent to u. The graph $G_{(1)}$ has thus $2n$ vertices and $m + n$ edges. This provides us with a trivial linear-size vertex-addition scheme to obtain 1-extendable graphs.

Lemma 2. *For any graph G, $G_{(1)}$ is 1-extendable.*

Transformation (T_2). The graph $G_{(2)}$ is obtained from G by subdividing each of its edges an even number of times, *i.e.* each edge becomes an induced $P_{2\ell}$. In fact, (T_2) is a well-known graph transformation which provides, for instance, the proof that MAXIMUM INDEPENDENT SET remains NP-hard on graphs forbidding a fixed graph H as an induced subgraph, for any H different from a path or a subdivided claw [1,23]. This transformation preserves in some sense all independent sets of the input graph G.

Observation 2 ([1,23]). *Consider any MIS X' of $G_{(2)}$ and pick all its vertices $X \subsetneq X'$ which were already present in G, i.e. which do not belong to the subdivided edges. Then X is an MIS of the input graph G. Additionally, the set $X' \backslash X$ contains exactly half of the vertices formed by the subdivisions.*

One can see that $G_{(2)}$ is planar iff G is planar (subdivisions do not influence planar embeddings). Moreover, (T_2) also preserves the 1-extendability.

Lemma 3. *G is 1-extendable iff $G_{(2)}$ is 1-extendable.*

Transformation (T_3). The graph $G_{(3)}$ is obtained from G by replacing each of its vertices by a path in order to decrease the maximum degree of the graph. It is a folklore transformation which also works for other classical problems.

First, we replace each vertex $u \in V(G)$ by an induced odd path P_u of length $\ell = 2\Delta - 1$, where Δ is the maximum degree of G. We denote by $u_1, \ldots, u_\ell$ the vertices of P_u. The vertex set of $G_{(3)}$ is $V(G_{(3)}) = \{u_1, \ldots, u_\ell : u \in V(G)\}$. Second, let $Q_u \subseteq P_u$ be the set of vertices in P_u with odd index: $Q_u = \{u_{2i+1} : 0 \le i \le \Delta - 1\}$. For any $1 \le i \le d(u)$, we assign arbitrarily to each vertex u_{2i+1} of Q_u a neighbor $\rho(u_{2i+1}) \in V(G)$ of u, so that ρ is bijective. There are two types of edges in $G_{(3)}$:

- edges of induced paths P_u, $u \in V(G)$,
- edges $u_{2i+1}v_{2j+1}$ when $\rho(u_{2i+1}) = v$ and $\rho(v_{2j+1}) = u$.

In this way, the maximum degree $G_{(3)}$ is at most 3.

One may observe that Q_u is an independent set of P_u of maximum size Δ. This is the key property which allows us to show that this structure maintains the 1-extendability of the input graph.

Lemma 4. *Let $n = |V(G)|$. We have $\alpha(G_{(3)}) = n(\Delta - 1) + \alpha(G)$. Moreover, if G is 1-extendable, then $G_{(3)}$ is 1-extendable.*

Assume graph G is planar. One can, by defining function ρ in a good way, produce a graph $G_{(3)}$ which is still planar, according to [18]. Unfortunately, one can find examples of graphs G such that $G_{(3)}$ is 1-extendable while G is not.

4   Hardness of 1-Extendability on Subcubic Planar Graphs

The main goal of this section is to study the computational hardness of 1-Extendability. We show that the problem remains NP-hard in subcubic planar graphs and unit disk graphs.

4.1   Gadget

We now focus on restricted graph classes. There exists a well-known gadget [13] which allows, for any graph G, to produce a planar graph G' with $O(n)$ vertices which is equivalent to G for the Maximum Independent Set problem. Concretely, G' is obtained by replacing each crossing appearing in an embedding of G in the plane by this gadget. In this article, we call it the *GJS-gadget* (for Garey-Johnson-Stockmeyer) and denote it by H_{GJS} (see Fig. 2a). Unfortunately, this trick does not work directly for 1-Extendability. In order to make it work, our idea is to define a first reduction producing a non-planar graph, but where the crossings satisfy some interesting properties. Secondly, we add GJS-gadgets on this intermediate graph. Lastly, we use well-known tricks from the literature in order to reduce the maximum degree of the reduced graph, and to obtain both subcubic planar and unit disk graphs.

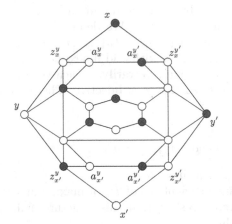

(a) A planar embedding of H_{GJS} together with an MIS of it (in blue)

(b) Largest MISs S of H_{GJS} containing a certain subset

	$|S \cap X| = 0$	$= 1$	$= 2$
$|S \cap Y| = 0$	7	8	8
$|S \cap Y| = 1$	8	9	9
$|S \cap Y| = 2$	7	8	9

Fig. 2. The GJS-gadget [13]

Description of the Gadget. Figure 2a represents H_{GJS}. Let $X = \{x, x'\}$, $Y = \{y, y'\}$, $Z = \left\{ z_x^y, z_x^{y'}, z_{x'}^y, z_{x'}^{y'} \right\}$, $A = \left\{ a_x^y, a_x^{y'}, a_{x'}^y, a_{x'}^{y'} \right\}$. We denote by b_x^y the common neighbor of z_x^y and a_x^y. Vertices $b_x^{y'}, b_{x'}^y, b_{x'}^{y'}$ are defined similarly. We fix $B = \left\{ b_x^y, b_x^{y'}, b_{x'}^y, b_{x'}^{y'} \right\}$. The "$C_6$" of H_{GJS} refers to the vertices which are not in sets X, Y, Z, A, and B. The size of the MIS of H_{GJS} is 9, according to [13]. Blue vertices give an example of such MIS. Vertices x, x', y and y' are called the *endpoints* of H_{GJS}.

Figure 2b indicates the sizes of a largest independent set S we obtain if we fix the intersection size with X and Y. For example, a largest independent set S which contains vertices x, y, y' is of size 8: one of them is such that it also contains $a_x^y, a_x^{y'}$ and 3 vertices from the C_6. Another example: a largest independent set S containing exactly one vertex of X and one vertex of Y has size 9. The blue vertices of Fig. 2a form this kind of independent sets, with $S \cap X = \{x\}$ and $S \cap Y = \{y'\}$.

Consider an embedding of some graph G in the plane, and a crossing consisting of two edges uu' and vv'. By *replacing the crossing by a gadget*, we mean removing the edges uu' and vv', adding a subgraph isomorphic to H_{GJS}, and adding the edges vx, uy', $v'x'$, and $u'y$. By replacing each crossing of G by a gadget, we obtain a graph G_+ which is not only planar, but also equivalent to G for the MAXIMUM INDEPENDENT SET problem, in the sense that G contains an independent set of size k iff G_+ contains an independent set of size $k + 9\lambda$, where λ is the number of crossings in G [13].

The idea behind this statement is the following: if S is an independent set of G, then there exists an independent set S_+ of G_+ of size $|S| + 9\lambda$ which is made up of the vertices of S and 9 vertices per crossing. As S is independent, one can select, for each gadget, one vertex of $\{x, x'\}$ (and one vertex of $\{y, y'\}$) which is not adjacent to an element of S. We know that a largest independent set of H_{GJS} intersecting both $\{x, x'\}$ and $\{y, y'\}$ in exactly one element has size 9, which corresponds to the MIS size of H_{GJS}.

Lemma 5 (Equivalence of G and G_+ for Maximum Independent Set [13]). *Any MIS S of G can be completed into an MIS $S_+ \supseteq S$ of G_+ which contains exactly 9 vertices per crossing gadget. Conversely, given any MIS S^* of G_+, the vertices of S^* which do not belong to a crossing gadget form an MIS of G.*

Preservation of 1-Extendability. Our initial idea was to use the same gadget to transform every graph into a planar one which preserves the 1-extendability of G. Unfortunately, the property described above for MAXIMUM INDEPENDENT SET does not hold for 1-EXTENDABILITY. Indeed, one can find examples of graphs G such that G is 1-extendable and G_+ is not. For this reason, we state a weaker characterization involving the GJS-gadget. We will see further that this result is enough to prove that 1-EXTENDABILITY is NP-hard on planar graphs.

Proposition 2. *Let G be a graph embedded in the plane and $uu' \in E(G)$. Let $v_1v_1', v_2v_2', \ldots, v_\ell v_\ell'$ be the edges of G which cross uu'. Assume, for any $1 \leq i \leq \ell$, the following statements:*

- *there is an MIS $S_u^{(i)}$ of G such that $S_u^{(i)} \cap \{u, u', v_i, v_i'\} = \{u\}$,*
- *there is an MIS $S_{u'}^{(i)}$ of G such that $S_{u'}^{(i)} \cap \{u, u', v_i, v_i'\} = \{u'\}$,*

Let G_+ be the graph obtained from G by replacing each crossing $\{uu', v_iv_i'\}$ with a GJS-gadget. Then, G_+ is 1-extendable iff G is 1-extendable.

Observe that the assumptions concerning the MISs of G_+ are essential if we want pairs $\{x_i, y_i\}$, $\{x_i, y_i'\}$, $\{x_i', y_i\}$, and $\{x_i', y_i'\}$ of each gadget H_i to be covered by MISs. This property is not achieved by all 1-extendable graphs G. Take for instance an embedding of some complete bipartite graph $K_{n,n}$, $n \geq 3$: every MIS intersects each crossing on exactly two vertices.

4.2   Planar Embedding

The GJS-gadget is a key tool in our proof that 1-EXTENDABILITY is NP-hard on subcubic planar graphs. We reduce from an NP-hard variant of 3SAT called PLANAR MONOTONE RECTILINEAR 3SAT, abbreviated PMR 3SAT. Given an input φ of PMR 3SAT, we design a graph G_φ such that φ is satisfiable iff G_φ is 1-extendable. Furthermore, G_φ is planar and its maximum degree is 3. We begin with the construction of G_φ step by step. Then, we show that the 1-extendability of G_φ depends on the satisfiability of the formula φ.

Starting Point of the Reduction. We reduce from PMR 3SAT, which is NP-hard [4]. In this variant of 3SAT, clauses and variables can be represented in the plane in a certain way. The input is a set of variables $X = \{x_1, \ldots, x_n\}$ and a CNF-SAT formula φ over X with exactly three variables per clause. The clauses $C_1, \ldots, C_m$ are *monotone*: they contain either three positive literals or three negative literals. Moreover, φ admits a *rectilinear* representation, that we now explain. Each variable is a point on the x-axis. The positive (resp. negative) clauses are represented by horizontal segments above (resp. below) the x-axis. When a variable x_i appears in a given clause, a vertical edge must connect the point x_i on the x-axis with the segment of this clause (at any point of the segment). Such a representation is rectilinear if no edge crosses a clause segment. Figure 3 provides an example of a formula φ, with $m = 5$, which admits a rectilinear representation.

Let φ be an input of PMR 3SAT provided with its rectilinear representation. The construction of G_φ depends on the rectilinear representation of φ. We proceed with two intermediate steps: first graph G_φ'', second graph G_φ'.

Construction of G_φ''. The first step is inspired from Mohar's reduction [18] for MAXIMUM INDEPENDENT SET. We replace each variable x_i on the x-axis by a cycle. Let r be the number of appearances of x_i (as a literal x_i or $\neg x_i$) in the clauses $C_1, \ldots, C_m$ of φ. The point representing variable x_i becomes a cycle $x_i^1, \bar{x}_i^1, x_i^2, \bar{x}_i^2, \ldots, x_i^r, \bar{x}_i^r$ of length $2r$, drawn as an axis-parallel rectangle (see Fig. 4a). We denote by c^* the total number of vertices in the variable cycles. Each clause $C_j = \ell_j^1 \vee \ell_j^2 \vee \ell_j^3$ is replaced by a triangle T_j of three vertices v_j^1, v_j^2, v_j^3. The edges of these triangles are called *T-edges*. Each vertex of the clause is placed at

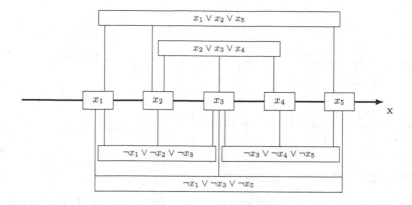

Fig. 3. A rectilinear representation of a PMR 3SAT instance $C_1, \ldots, C_5$

the intersection between the clause segment and vertical edges of the rectilinear representation. In this way, vertices v_j^1, v_j^2 and v_j^3 are aligned horizontally and, w.l.o.g, we assume v_j^1 (resp. v_j^3) is the leftmost (resp. rightmost) vertex of T_j on the clause segment. Edges $v_j^1 v_j^2$ and $v_j^2 v_j^3$ are drawn as straight lines. The third one, $v_j^1 v_j^3$, can be represented as an almost flat curve, passing above (resp. below) vertex v_2^j for positive (resp. negative) clauses. If $\ell_j^q = x_i$ for some $1 \leq j \leq m$ and $q \in \{1, 2, 3\}$, then vertex v_j^q is connected to some cycle vertex $\bar{x}_i^s$ of the top of the rectangle. Otherwise, if $\ell_j^q = \neg x_i$, then vertex v_j^q is connected to some cycle vertex x_i^s of the bottom of the rectangle. For now, the described embedding is planar. Figure 4a shows the embedding of the instance of Fig. 3. Vertices $\bar{x}_i^s$ are drawn in grey to distinguish them from vertices x_i^s (in white).

Less formally, each parity of a variable cycle represents a certain assignation of this variable. Picking up $x_i^1, x_i^2, \ldots$ (resp. $\bar{x}_i^1, \bar{x}_i^2, \ldots$) into an independent set will correspond to assigning x_i to False (resp. True).

We add a "pendant" vertex π_j for any triangle T_j, $1 \leq j \leq m$, that is, π_j is adjacent to all vertices of T_j. The edges created by this operation, *i.e.* all $v_j^q \pi_j$, are called *pendant edges*. Consider the following embedding. We fix two horizontal axes x$^+$ and x$^-$: the first one above the x-axis and all segments of the positive clauses, the second one below the x-axis and all segments of the negative clauses. The pendants issued from the positive clauses are placed on the x$^+$-axis such that every edge (v_j^2, π_j) is vertical. We represent edges (v_j^1, π_j) and (v_j^3, π_j) as straight lines (they cannot be vertical). We proceed similarly with pendants of the negative clauses on the x$^-$-axis. We denote by G_φ'' the obtained graph. Its embedding is not planar. Figure 4b shows graph G_φ'' corresponding to the instance φ of Fig. 3. Pendant edges are drawn in red. We claim that each vertex of G_φ'' belongs to an MIS. This might seem counter-intuitive, but the equivalence between the 1-extendability of the output instance and the satisfaction of φ will appear later (when we will eventually define G_φ).

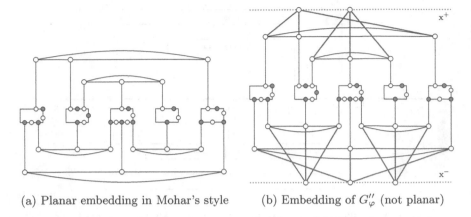

(a) Planar embedding in Mohar's style   (b) Embedding of G''_φ (not planar)

Fig. 4. Graph G''_φ with and without pendant vertices.

Lemma 6. *The graph G''_φ is 1-extendable.*

Construction of G'_φ. The second step consists in transforming G''_φ into some equivalent graph G'_φ which is planar and has maximum degree 3. Two types of crossings appear in the embedding of G''_φ. Each of them necessarily involve pendant edges.

- **Type A:** a pendant edge $v_j^q \pi_j$ crosses a T-edge $v_{j'}^{p'} v_{j'}^{q'}$ (we may have $j = j'$),
- **Type B:** a pendant edge $v_j^q \pi_j$ crosses another pendant edge $v_{j'}^{q'} \pi_{j'}$, $j \neq j'$.

We observe that for any of these types of crossings in the embedding of G''_φ, the assumptions of Proposition 2 are fulfilled.

Lemma 7. *Let $\{uu', vv'\}$ be a crossing of the embedding of G''_φ. There exist two MISs $S_u, S_{u'}$ of G''_φ which intersect $\{u, u', v, v'\}$ respectively in $\{u\}$ and $\{u'\}$.*

As a consequence of Lemma 7 together with Proposition 2, one can replace each crossing of the embedding of G''_φ by a gadget H_{GJS} without altering its 1-extendability. The graph obtained is thus planar and has maximum degree 6 (which is the maximum degree of graph H_{GJS}). Then, we apply transformation (T_3) with $\Delta = 6$ to decrease its maximum degree. Finally, we obtain the graph G'_φ, which is planar and has maximum degree 3.

Lemma 8. *The graph G'_φ is 1-extendable.*

Construction of G_φ. We are now ready to describe the final graph G_φ which consists in a small extension of G'_φ. We add a cycle $z_1, \bar{z}_1, \ldots, z_m, \bar{z}_m$ of size $2m$ to the graph G'_φ. Let $Z = \{z_1, z_2, \ldots, z_m\}$ and $\bar{Z} = \{\bar{z}_1, \bar{z}_2, \ldots, \bar{z}_m\}$. We connect z_j to π_j for every $1 \leq j \leq m$ - concretely, as π_j became an induced path via transformation (T_3), we add an edge between z_j and a vertex of P_{π_j}. The graph obtained is G_φ and its size is polynomial in $|\varphi|$. The graph G_φ is planar: consider the embedding of G'_φ, draw the cycle $Z \cup \bar{Z}$ as a rectangle surrounding it and such that all edges $\pi_j z_j$ are vertical. Its maximum degree is 3.

As shown in details in the proof of Theorem 3 (see Appendix), the graph G_φ is 1-extendable iff φ is satisfiable. Indeed, as G'_φ is 1-extendable, all vertices, except the ones in Z, already belong to a MIS of G_φ, where $\alpha(G_\varphi) = \alpha(G'_\varphi) + m$. Therefore, G_φ is 1-extendable iff it admits a MIS of the form $Z \cup S'$, where S' is a MIS of G'_φ. Such property is fulfilled iff there is a MIS of G''_φ which does not contain pendant vertices and, equivalently, iff there is an assignation of the variables $x_1, x_2, \ldots$ satisfying φ.

Theorem 3. 1-EXTENDABILITY *remains NP-hard on planar graphs of maximum degree* 3.

Unit disk graphs [9] stand as a natural model for wireless networks. Subdivided planar graphs with degree at most 4 can be represented as unit disk graphs [28], hence, combining Theorem 3 with Transformation (T_2), we obtain the following.

Theorem 4. 1-EXTENDABILITY *is NP-hard, even on unit disk graphs.*

5 Parameterized Algorithms

In this section we study a parameterized version of the 1-extendability problem:

PARAM-1-EXTENDABILITY	**Parameter:** k
Input: A graph G, an integer k	
Question: Does every vertex of G belong to an independent set of size k?	

Theorem 5. PARAM-1-EXTENDABILITY *is* $W[1]$-*hard.*

Observe that Lemma 1 does not preserve the existence of polynomial kernels. Hence, it is natural to ask whether 1-EXTENDABILITY admits a polynomial kernel in graph classes where MAXIMUM INDEPENDENT SET does. We answer positively for some of them. We say that a hereditary graph class $\mathcal{C}$ is MIS-(c, t)-*friendly*, for two non-zero constants c and t, if every graph of the class on n vertices contains an independent set of size at least $t \cdot n^c$, and such an independent set can be found in polynomial-time.

Theorem 6. *Let* $\mathcal{C}$ *be an* MIS-(c, t)-*friendly class.* PARAM-1-EXTENDABILITY *on* $\mathcal{C}$ *admits a kernel with* $O(k^{\frac{1}{c} + \frac{1}{c^2}})$ *vertices.*

Since planar graph are MIS-$(1, 1/4)$-friendly, d-degenerate graphs are MIS-$(1, \frac{1}{d+1})$-friendly and K_r-free graphs (for $r \geqslant 3$) are MIS-$\left(\frac{1}{r-1}, 1\right)$-friendly, we have the following:

Corollary 2. PARAM-1-EXTENDABILITY *admits a kernel with* $O(k^2)$ *vertices on planar graphs and more generally d-degenerate graphs for bounded d, and a kernel with* $O(k^{r^2-r})$ *vertices on* K_r-*free graphs for every fixed* $r \geqslant 3$.

6   Conclusion and Further Research

We investigated the computational complexity of 1-EXTENDABILITY. We showed that in general graphs it cannot be solved in subexponential-time unless the ETH fails, and that it remains NP-hard in subcubic planar graphs and in unit disk graphs. Although this behavior seems to be the same as MAXIMUM INDEPENDENT SET, we proved that MAXIMUM INDEPENDENT SET remains NP-hard (and even W[1]-hard) in 1-extendable graphs. It seems challenging to find a larger class of graphs where 1-EXTENDABILITY is polynomial-time solvable (but not trivial) while MAXIMUM INDEPENDENT SET remains NP-hard.

Another interesting subject would be to characterize 1-extendable graphs of graph classes where MAXIMUM INDEPENDENT SET is polynomial-time solvable: *e.g.* chordal graphs, cographs, claw-free graphs. Such outcomes would extend the result of Dean and Zito [11] which state that bipartite graphs are 1-extendable iff they admit a perfect matching.

We also studied PARAM-1-EXTENDABILITY, a parameterized version of 1-EXTENDABILITY and showed that some results for MAXIMUM INDEPENDENT SET could also be obtained for PARAM-1-EXTENDABILITY (although not being as direct). It would be interesting to determine whether this is also the case for other results about MAXIMUM INDEPENDENT SET [6,7,10], for instance: is PARAM-1-EXTENDABILITY W[1]-hard in C_4-free graphs and in $K_{1,4}$-free graphs? Does it admit a polynomial kernel in diamond-free graphs?

Finally, because of its applications in network design, finding an efficient algorithm which works well in practice is of high importance. Toward this, a first step would be to determine in which cases a vertex addition (or deletion) preserves the property of being 1-extendable. We note that such results have already been obtained for the related property of being well-covered [12].

References

1. Alekseev, V.: The effect of local constraints on the complexity of determination of the graph independence number. Combin.-Algebraic Methods Appl. Math. 3–13 (1982)
2. Angaleeswari, K., Sumathi, P., Swaminathan, V.: k-extendability in graphs. Int. J. Pure Appl. Math. **101**(5), 801–809 (2015)
3. Angaleeswari, K., Sumathi, P., Swaminathan, V.: Weakly k-extendable graphs. Int. J. Pure Appl. Math. **109**(6), 35–40 (2016)
4. de Berg, M., Khosravi, A.: Optimal binary space partitions in the plane. In: Thai, M.T., Sahni, S. (eds.) COCOON 2010. LNCS, vol. 6196, pp. 216–225. Springer, Heidelberg (2010). https://doi.org/10.1007/978-3-642-14031-0_25
5. Berge, C.: Some common properties for regularizable graphs, edge-critical graphs and B-graphs. Graph Theory Algorithms Lect. Notes Comput. Sci. **108**, 108–123 (1981)
6. Bonnet, É., Bousquet, N., Charbit, P., Thomassé, S., Watrigant, R.: Parameterized complexity of independent set in H-free graphs. Algorithmica **82**(8), 2360–2394 (2020)

7. Bonnet, É., Bousquet, N., Thomassé, S., Watrigant, R.: When maximum stable set can be solved in FPT time. In: Proceedings of ISAAC, vol. 149, pp. 49:1–49:22 (2019)

8. Chvátal, V., Slater, P.J.: A note on well-covered graphs. In: Quo Vadis, Graph Theory?, Annals of Discrete Mathematics, vol. 55, pp. 179–181. Elsevier (1993)

9. Clark, B.N., Colbourn, C.J., Johnson, D.S.: Unit disk graphs. Discret. Math. **86**(1–3), 165–177 (1990)

10. Dabrowski, K.K., Lozin, V.V., Müller, H., Rautenbach, D.: Parameterized complexity of the weighted independent set problem beyond graphs of bounded clique number. J. Discrete Algorithms **14**, 207–213 (2012)

11. Dean, N., Zito, J.S.: Well-covered graphs and extendability. Discret. Math. **126**(1–3), 67–80 (1994)

12. Finbow, A.S., Whitehead, C.A.: Constructions for well-covered graphs. Aust. J. Combin. **72**(2), 273–289 (2018)

13. Garey, M.R., Johnson, D.S., Stockmeyer, L.J.: Some simplified NP-complete graph problems. Theor. Comput. Sci. **1**(3), 237–267 (1976)

14. Grzesik, A., Klimosová, T., Pilipczuk, M., Pilipczuk, M.: Polynomial-time algorithm for maximum weight independent set on P_6-free graphs. In: Proceedings of SODA, pp. 1257–1271. SIAM (2019)

15. Hackfeld, J., Koster, A.M.C.A.: The matching extension problem in general graphs is co-NP-complete. J. Comb. Optim. **35**(3), 853–859 (2017). https://doi.org/10.1007/s10878-017-0226-x

16. Impagliazzo, R., Paturi, R.: On the complexity of k-SAT. J. Comput. Syst. Sci. **62**(2), 367–375 (2001)

17. Liew, S.C., Kai, C.H., Leung, H.C., Wong, P.: Back-of-the-envelope computation of throughput distributions in CSMA wireless networks. IEEE Trans. Mob. Comput. **9**(9), 1319–1331 (2010)

18. Mohar, B.: Face covers and the genus problem for apex graphs. J. Comb. Theory Ser. B **82**(1), 102–117 (2001)

19. The Network Simulator ns-3. https://www.nsnam.org/. Accessed 30 Sept 2021

20. Plummer, M.D.: Some covering concepts in graphs. J. Combin. Theory **8**(1), 91–98 (1970)

21. Plummer, M.D.: On n-extendable graphs. Discrete Math. **31**, 201–210 (1980)

22. Plummer, M.D.: Extending matchings in graphs: a survey. Discret. Math. **127**(1–3), 277–292 (1994)

23. Poljak, S.: A note on stable sets and colorings in graphs. Commentationes Math. Univ. Carolinae **15**, 307–309 (1974)

24. Ravindra, G.: B-graphs. In: Proceedings of Symposium on Graph Theory, ISI Lecture Notes Calcutta, vol. 4, pp. 268–280 (1976)

25. Ravindra, G.: Well covered graphs. J. Combin. Inform. System Sci. **2**, 20–21 (1977)

26. Sankaranarayana, R.S., Stewart, L.K.: Complexity results for well-covered graphs. Networks **22**(3), 247–262 (1992)

27. Tankus, D., Tarsi, M.: Well-covered claw-free graphs. J. Combin. Theory Ser. B **66**(2), 293–302 (1996)

28. Valiant, L.G.: Universality considerations in VLSI circuits. IEEE Trans. Comput. **30**(2), 135–140 (1981)

Tukey Depth Histograms

Daniel Bertschinger[1]([✉]), Jonas Passweg[1], and Patrick Schnider[2]

[1] Department of Computer Science, ETH Zürich, Zürich, Switzerland
`daniel.bertschinger@inf.ethz.ch`, `jpassweg@student.ethz.ch`
[2] Department of Mathematical Sciences, University of Copenhagen,
Copenhagen, Denmark
`ps@math.ku.dk`

Abstract. Combinatorial representations of point sets play an important role in discrete and computational geometry. In this work, we investigate a new combinatorial quantity of a point set, called *Tukey depth histogram*. The Tukey depth histogram of k-flats in $\mathbb{R}^d$ with respect to a point set P, is a vector $D^{k,d}(P)$, whose i'th entry $D_i^{k,d}(P)$ denotes the number of k-flats spanned by $k+1$ points of P that have Tukey depth i with respect to P. It turns out that several problems in discrete and computational geometry can be phrased in terms of such depth histograms. As our main result, we give a complete characterization of the depth histograms of points, that is, for any dimension d we give a description of all possible histograms $D^{0,d}(P)$. This then allows us to compute the exact number of different histograms of points.

Keywords: Computational geometry · Depth statistics · Tukey depth · Point sets

1 Introduction

Many fundamental problems on point sets, such as the number of extreme points, the number of halving lines, or the crossing number do not depend on the actual location and distances of the points, but rather on some underlying combinatorial structure of the point set. There is a vast body of work of combinatorial representations of point sets, at the beginning of which are the seminal series of papers by Goodman and Pollack [6–8], where many important objects such as *allowable sequences* and *order types* are introduced. In particular, order types have proven to be a very powerful representation of point sets. For many problems however, less information than what is encoded in order types is sufficient. One example for such a problem is the determination of the *depth* of a query point with respect to a planar point set.

The third author has received funding from the European Research Council under the European Unions Seventh Framework Programme ERC Grant agreement ERC StG 716424 - CASe.

C. Bazgan and H. Fernau (Eds.): IWOCA 2022, LNCS 13270, pp. 186–198, 2022.
https://doi.org/10.1007/978-3-031-06678-8_14

Depth measures are a tool to capture how deep a query point lies within a given point set. There is a number of depth measures that have been introduced, most famously *Tukey depth* [18] (also called *halfspace depth*), *Simplicial depth* [10] or *Convex hull peeling depth* (see [1,9] or Chap. 58 in [17] for an overview of depth measures). In this paper, we are mainly concerned with Tukey depth. The Tukey depth of a query point q with respect to a point set P is the minimum number of points of P that lie in a closed halfspace containing q. For Tukey depth (and Simplicial depth), the depth of a query point in the plane can be computed knowing the order type of the point set, but also knowing the *line rotational order* of the points around the query point. The line rotational order of the points of a point set P around a query point q is the order in which a directed line rotating around q passes over the points of P, where we distinguish whether a point of P is passed in front of, or behind q.

In fact, the Tukey depth of a query point q in the plane can be computed using even less information than the line rotational order around q: it suffices to know for each k, how many directed lines through q and a point of P have exactly k points to their left. This defines the *ℓ-vector* of q. The Tukey depth of q is now just the smallest k for which the corresponding entry in the ℓ-vector is non-zero. It turns out, that many other depth measures can also be computed knowing only the ℓ-vector of q [3]. Another quantity that can be computed from this information only is the number of crossing-free perfect matchings on $P \cup \{q\}$, if P is in convex position and q is in the convex hull of P [15]. In [15], a characterization of all possible ℓ-vectors is given, phrased in terms of *frequency vectors*, which is an equivalent object. Knowing the ℓ-vector of every point in a point set P thus still gives us a lot of information about this point set. For example, as this allows us to compute the simplicial depth $\mathrm{sd}(P,q)$ of each point q in P, that is, the number of triangles spanned by $P \backslash \{q\}$ that have q in their interior, this also allows us to compute the crossing number of the complete straight-line graph induced by P, which is just

$$\mathrm{cr}(P) = \binom{n}{4} - \sum_{q \in P} \mathrm{sd}(P,q).$$

In this paper, we study objects that emerge after forgetting yet another piece of information: instead of knowing the ℓ-vector of each point, we only know the sum of all ℓ-vectors. This corresponds to knowing for each j the number of j-*edges* that is, knowing the histogram of j-edges. Note that an edge uv is called a j-*edge* if there are exactly j points to the right of the line uv. The number of j-edges that a point set admits is a fundamental question in discrete geometry and has a rich history, see e.g. [19], Chap. 4 in [5] or Chap. 11 in [11] and the references therein. We generalize the histogram of j-edges to a histogram of j-flats (i.e., affine subspaces of dimension j) in any dimension. For this, we first define the Tukey depth of a flat:

Definition 1. *Let Q be a set of $k+1$ points in $\mathbb{R}^d$, $k < d$, which span a unique k-flat F. The affine Tukey depth of Q with respect to a point set P, denoted by*

$\text{atd}_P(Q)$, is the minimum number of points of P in any closed halfspace containing F. The convex Tukey depth of Q with respect to P, denoted by $\text{ctd}_P(Q)$, is the minimum number of points of P in any closed halfspace containing $\text{conv}(Q)$.

Note that for $k = 0$ both definitions coincide with the standard definition of Tukey depth, and we just write $td_P(q)$ in this case. Further note that if $P \cup Q$ is in convex position, then $\text{atd}_P(Q) = \text{ctd}_P(Q)$. Several results in discrete geometry can be phrased in terms of this generalized Tukey depth. For example, the center transversal theorem [2,20] states that for any $j + 1$ point sets $P_0, \ldots, P_j$ in $\mathbb{R}^d$, there exists a j-flat (not necessarily spanned by points of the point sets) that has affine Tukey depth $\frac{|P_i|}{d+1-j}$ with respect to P_i, for each $i \in \{0, \ldots, j\}$. For $j = 0$ and $j = d - 1$, we retrieve the centerpoint theorem [13] and Ham-Sandwich theorem [16] as boundary cases.

Definition 2. *Let P be a set of points in $\mathbb{R}^d$. The affine Tukey depth histogram of j-flats, denoted by $D^{j,d}(P)$, is a vector whose entries $D_i^{j,d}(P)$ are the number of subsets $Q \subset P$ of size $j+1$ whose affine Tukey depth is i. Similarly, replacing affine Tukey depth with convex Tukey depth, we define the convex Tukey depth histogram of j-flats, denoted by $cD^{j,d}(P)$.*

In the following, we will also call affine Tukey depth histograms just *depth histograms*, that is, unless we specify the *convex*, we always mean an affine Tukey depth histogram. Note however that for $j = 0$ or if P is in convex position, the two histograms coincide.

Many problems in discrete geometry can also be phrased in terms of depth histograms. For example, the number of extreme points of a point set P just corresponds to the entry $D_1^{0,d}(P)$ (note that each (extreme) point of P has Tukey depth at least 1). Further, the number of j-edges or, more generally, j-facets corresponds to the entry $D_j^{d-1,d}(P)$. For a further example consider the following problem, studied in [4,12,14]: let P be a set of n points in general position in the plane. Are there always two points in P such that any circle through both of them contains at least $\frac{n}{4}$ points of P both inside and outside of the circle? Using parabolic lifting, proving that for any point set P of size n in convex position in $\mathbb{R}^d$, we have $D_{n/4}^{1,3}(P) > 0$ would imply a positive answer to the above question [14].

In this paper, we will mainly focus on Tukey depth histograms of points, that is, histograms of the form $D^{0,d}$. We give a complete characterization of possible such histograms for point sets in general position. In particular, we will show the following:

Theorem 1. *A vector $D^{0,d}$ is a depth histogram of a point set of size at least $d + 1$ in general position in $\mathbb{R}^d$ if and only if for all nonzero entries $D_i^{0,d}$ with $i \geq 2$ we have*

$$\sum_{j=1}^{i-1} D_j^{0,d} \geq 2i + d - 3.$$

Note that if a point set has fewer than $d + 1$ points, then they must all be extreme points, as we assume general position. Thus, their depth histogram only has a single non-zero entry $D_1^{0,d}$. In particular, the above theorem covers all interesting cases. In the following, we will always silently assume that P has at least $d + 1$ points.

2   The Condition is Necessary

The goal of this section is to show that the condition $\sum_{j=1}^{i-1} D_j^{0,d} \geq 2i + d - 3$ for all $i \geq 2$ with $D_i^{0,d} > 0$ is necessary for any depth histogram. To prove this, we first give an upper bound on the depth of any single point.

Lemma 1. *For any point set $P \subseteq \mathbb{R}^d$ on at least $d + 1$ points, and any point $p \in P$ we have $\mathrm{td}_P(p) \leq \frac{n-d+2}{2}$.*

Proof. Let $P \subseteq \mathbb{R}^d$ and let $p \in P$ be any point with $td_P(p) = k$. We will show that P consists of at least $2k - 2 + d$ points.

Consider a witnessing halfspace h_p of p and its bounding hyperplane $h \subseteq h_p$. We may assume that h contains p and no other point from P; otherwise we just rotate h a little. As $td_P(p) = k$, we know that h_p contains k points. Now, rotate h in any direction (s.t. p stays on h) until one point, say q, changes halfspaces. If q was in h_p before, then we found a halfspace containing p and $k - 1$ points, contradicting the depth of p. Therefore q was in $\mathbb{R}^d \backslash h_p$. We can repeat this rotating step until there are d points on h (one of them being p). Both halfspaces still need to contain at least $k - 1$ points; thus, we need at least $2(k - 1) + d$ points.   □

A similar line of reasoning can be applied to k-faces and convex Tukey depth:

Proposition 1. *Let P be a point set in $\mathbb{R}^d$ and assume that P spans a k-face F with $\mathrm{ctd}_P(F) = i$. Then P spans at least $2\binom{i-k-1}{m+1}$ m-faces of smaller depth. In other words, for any depth histogram $cD^{k,d}$ and all nonzero entries $cD_i^{k,d}$ with $i \geq 2$ we have*

$$\sum_{j=1}^{i-1} cD_j^{m,d} \geq 2\binom{i-k-1}{m+1}.$$

Proof. Consider a witnessing halfspace h_F of F and its bounding hyperplane h. As $ctd_P(F) = i$ and F is spanned by $k + 1$ points, the halfspace h_F contains $i - k - 1$ other points. Looking at the complement of h_F and translating h, we can find another halfspace h_2 containing $i - k - 1$ points of P, with $h_F \cap h_2 = \emptyset$. We have thus found two disjoint subsets of P, each of size $i - k - 1$. Further, each m-face spanned by $m + 1$ points in a subset has depth at most $i - 1$, as witnessed by a translation of h_F or h_2, respectively.   □

To show the necessity of the condition in Theorem 1 we need to be able to delete points of high depth without changing the depth of points of lower depth.

Lemma 2. *For any point set $P \subseteq \mathbb{R}^d$ and any two points $p, q \in P$ with $\mathrm{td}_P(p) \leq \mathrm{td}_P(q)$ we have $\mathrm{td}_P(p) = \mathrm{td}_{P \setminus q}(p)$.*

Proof. Let h_p be a witnessing halfspace of p with only p on the boundary (following the same reasoning as in the proof of Lemma 1). If $q' \in h_p$ then we translate h_p parallel until q' lies on the boundary. The new halfspace contains at most $|h_p \cap P| - 1$ points of P and so $td_P(q') < td_P(p)$. Hence, for points p, q with $td_P(p) \leq td_P(q)$ we have $q \notin h_p$ and deleting q can never change the depth of p.   □

This lemma has direct applications for histograms of points, that is, by repeatedly applying the lemma one can easily show the following.

Proposition 2. *Let $[a_1, a_2, \ldots, a_{m-1}, a_m]$ be a depth histogram for points, then for $i \leq m$ both $[a_1, a_2, \ldots, a_{i-1}, a_i]$ and if $a_i \geq 1$ then also $[a_1, a_2, \ldots, a_{i-1}, 1]$ are depth histograms.*

We are now able to prove that the condition in Theorem 1 is necessary.

Proof (that the condition in Theorem 1 is necessary). For the sake of contradiction, let us assume that $D^{0,d}$ is a depth histogram with a nonzero entry $D_i^{0,d}$ and $\sum_{j=1}^{i-1} D_j^{0,d} < 2i + d - 3$.

Using Proposition 2, we can "cut off" $D^{0,d}$ at any point, therefore let $D' := [D_1^{0,d}, \ldots, D_{i-1}^{0,d}, 1]$ and denote the corresponding point set as P'. From the assumption we know that there are fewer than $2i+d-3$ points of depth less than i and by definition there is one point of depth i. Thus, we have $|P'| < 2i + d - 2$. But, by Lemma 1, points in P' have depth less than $\frac{(2i+d-2)-d+2}{2} = i$.   □

Similarly, we can say something about k-faces and affine Tukey depth. For a proof, we refer the interested reader to the full version of this paper.

Corollary 1. *For any depth histogram $D^{k,d}$ and all nonzero entries $D_i^{k,d}$ with $i \geq 2$ we have*

$$\sum_{j=1}^{i-1} D_j^{0,d} \geq 2i + d + k - 3.$$

In other words, if a point set P in $\mathbb{R}^d$ spans a k-face F with $atd_P(F) = i$, then P contains at least $2i + d + k - 3$ points of smaller depth.

2.1 Two Special Configurations

Let us revisit Lemma 1 about the maximum possible depth. It is worth noting that the bound given in the lemma is tight. We will give an intuition using point sets in so-called *symmetric configuration* [15]. These point sets will be useful in proving that the condition of Theorem 1 is sufficient.

Definition 3. *A point set $P \subseteq \mathbb{R}^d$ in general position is in*

1. symmetric configuration *if and only if there exists a* central point $c \in P$ *such that every hyperplane through c and $d - 1$ other points of P separates the remaining points into two halves of equal size.*
2. eccentric configuration *if and only if there exists a* central point $c \in P$ *such that every hyperplane through c and $d-1$ other points of P divides the remaining points in two sets with difference in cardinality of at most 1.*

We will denote point sets in symmetric (or eccentric, resp.) configuration as *symmetric* (*eccentric*, resp.). Depending on the dimension and the size of P, only one of the definitions can be applied, see Fig. 1 for an example.

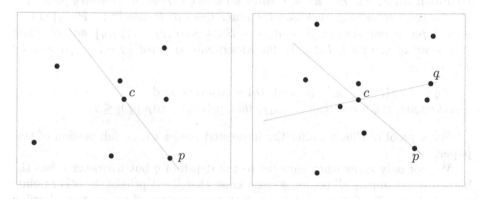

Fig. 1. Two point sets in symmetric and eccentric configuration, respectively. The lines through c and p or c and q, respectively, (almost) divide the remaining point set.

Lemma 3. *The symmetric central point c in a symmetric (or eccentric) point set P has depth* $\mathrm{td}_P(c) = \lfloor \frac{n-d+2}{2} \rfloor$.

For a proof of this, we refer the interested reader to the full version of this paper.

At first glance, it is not clear that symmetric and eccentric point sets of any size exist in any dimension. We will show that they do in the next section.

3 The Condition is Sufficient

To prove that the condition we gave in Theorem 1 is sufficient, we build up point sets according to their histograms by adding points one-by-one. In other words, given a histogram, we start from the points in convex position (as many as there are of depth 1). We then add new points at places, where they have the maximal possible depth, that is, we will add them in the "center" of the point set. We then push them outwards until they have the right depth, without changing the depth of any other point. In this way we successively add all points of depth 2, then the ones of depth 3 and so on. In this section we show what happens to the histogram when pushing points outwards (Sect. 3.1) and where to add new points and in which direction we push them (Sect. 3.2).

3.1 Moving Points

First, we make an easy observation that is key to see how moving points affects the Tukey depth histogram of a point set.

Observation 2. *The depth of a point $q \in P$ can only change if the order type of the point set changes.*

Note that the Tukey depth of q can only change if q is involved in the change in the order type. In other words, q was moved over a hyperplane formed by d other points of the point set.

Proposition 3. *Let $P \in \mathbb{R}^d$ be a point set and $q \in P$ be an arbitrary point. Let q' be a point close to q, such that the order types of P and $P' := P \backslash \{q\} \cup \{q'\}$ only differ in one simplex S, that is in $S := conv\{p_1, \ldots, p_d, q\}$ and S' (with q' instead of q, resp.). Let h be the hyperplane spanned by $p_1, \ldots, p_d$ and let $\hat{q} := h \cap \overline{qq'}$.*

- *If $\hat{q} \notin conv\{p_1, \ldots, p_d\}$, then $td_P(q) = td_{P'}(q')$, and*
- *otherwise, if $\hat{q} \in conv\{p_1, \ldots, p_d\}$, then $|td_P(q) - td_{P'}(q')| \leq 1$.*

For a proof of this, we refer the interested reader to the full version of this paper.

We not only know what happens to the depth of q but whenever q has the highest depth among all points, we also know that the depths of the other points do not change. By first removing q and then reinserting q' we get the following observation as a direct consequence of Lemma 2. Thus, we know how the Tukey depth histogram behaves when moving points of large depths.

Observation 3. *Whenever we have $td_P(q) > td_P(p)$ for all points p in the point set except q, then $td_P(p) = td_{P'}(p)$ (for q, q' and P' as in Proposition 3).*

3.2 Inserting a New Point

We have already seen symmetric (and eccentric) point sets, containing a point of maximum possible depth. These sets will help us placing new points, s.t. they have large depth. Let P be a symmetric (eccentric) point set in general position missing the symmetric central point. We place a new point q at the location of the (previously inexistent) central point, and by Lemmas 1 and 3, we know that q has the maximal possible depth. Now we are able to push q outwards until it has the desired depth and the resulting point set is eccentric (symmetric, resp.) again missing the central point. An example of what happens in dimension two can be found in Fig. 2. It is pretty easy to see that in $\mathbb{R}^2$, this always works.

Lemma 4. *For any symmetric (eccentric) point set $P \subseteq \mathbb{R}^2$ in general position there exists a direction in which we can move the central point, s.t. after adding a new center we have an eccentric (symmetric) point set in general position.*

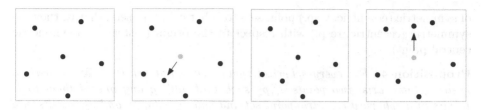

Fig. 2. A symmetric point set (left). After pushing the central point out (second), we arrive at an eccentric point set in missing the symmetric central point. We add a new point at maximum possible depth (third) and by pushing it out again we get a symmetric point set missing the central point (right).

Proof. First, note if P is symmetric, almost any direction does the job. The only crucial bit is ensuring that the resulting point set (after adding a new center) is in general position again; however, this is always possible.

If P is eccentric, then there exist two neighbors in the rotational order of points around q (the central point to be pushed) without a symmetric central line (e.g. a line going through the central point) dividing them. Let us denote these points as p_1 and p_2 and move q outwards on an "opposite" halfline, see Fig. 3, right. This ensures that the point set is in general position and symmetric again (i.e. the line rotational order around the new center is alternating between points passing in front of, and behind q). $\square$

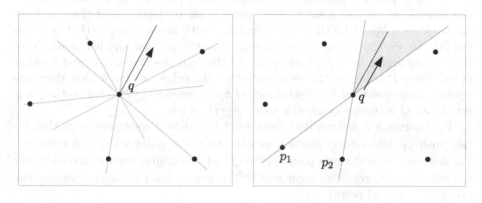

Fig. 3. The central point q to be pushed and the directions in which we push.

In higher dimensions it is not easy to see how to get the directions and why they always exist. We will do an induction argument on the dimension of the point set; to understand the necessary ideas we start in three-dimensional space. Let us further denote a point set as *spherical*, if every point (except maybe one central point) lies on a sphere around the origin. We further extend the definition of symmetric (eccentric, resp.) point sets to spherical point sets, that is, they are symmetric (eccentric, resp.) with respect to the origin (instead of a symmetric central point).

Proposition 4. *For every spherical symmetric point set $P \subseteq \mathbb{R}^3$ in general position there exist two points p_1, p_2 such that adding any one of them to P results in a spherical eccentric point set and adding both, p_1 and p_2, results in a spherical symmetric point set in general position.*

Here general position means that no three points of P lie on a common plane through the origin. In particular, no two points are exactly opposite on the sphere. The reason we add two points at a time is that they heavily depend on one another.

Proof. Let us assume without loss of generality that there is a point p_N at the north pole of the sphere. Note that following assumption of general position this means that there is no point of P at the south pole.

Let us define $P^S := P \cup \{p_S\}$, where p_S is a new point added at the south pole of the sphere. Note that this defines one position to add a point to P; however, the resulting point set is not in general position anymore and therefore not eccentric. To get an eccentric point set, we will therefore slightly move p_N and call the moved point $p_{\tilde{N}}$. We will then be able to add another point p'_N close to the north pole such that the resulting point set in the end is symmetric again.

For this movement, consider the stereographical projection of P at the north pole (that is undefined for the south pole). Let us denote the projection of $p_i \in P$ as q_i and the resulting (two-dimensional) point set as Q. Note that for the symmetric central point $p_c \in P$ there is no projection and thus we have $|Q| = |P| - 1$. We claim that Q is symmetric with respect to q_N (that is at the origin). To see that, consider any line l_Q through q_N and another point. Note that l_Q corresponds to a plane $h_P \subset \mathbb{R}^3$ through the origin, p_N and another point. Since P is symmetric with respect to the origin we know that there are equally many points of P on either side of h_P. Therefore l_Q also halves the point set Q and Q is indeed symmetric with respect to q_N.

By Lemma 4 it follows that there exist two directions w_1, w_2 such that we can push q_N into either direction and the resulting point set $Q^{\tilde{N}}$ is eccentric. Further, we can add a new point $q_{N'}$ to $Q^{\tilde{N}}$ at the origin; push it into the other direction and the resulting point set $Q^{\tilde{N}\tilde{N}'}$ is symmetric once again (missing the symmetric central point).

By reversing the stereographic projection we can map the moved points $q_{\tilde{N}}, q_{\tilde{N}'}$ to the sphere and get points $p_{\tilde{N}}, p_{\tilde{N}'}$ close to the north pole. Note that when we moved q_N we could have moved it in such a way, that $p_{\tilde{N}}$ is very close to p_N. In particular by replacing p_N in P we did not destroy the symmetric configuration of P. Let us denote the resulting point set as P'. Note that we add $p_{\tilde{N}'}$ afterwards very close to the north pole so that it lies above every other point (including $p_{\tilde{N}}$).

It remains to show that $P'' := P' \cup \{p_S, p_{\tilde{N}'}\}$ is a symmetric spherical point set (with respect to the origin). For that we consider every possible plane h

through the origin and another point of P'' and show that there are equally many points on both sides of h.

First, consider a plane h through p_S and another point. Note that h also goes through the north pole and thus in the stereographic projection corresponds to a line through the origin. From the fact that $Q^{\tilde{N}\tilde{N}'}$ is symmetric it follows that the plane h is halving the point set P.

Second, if h does not go through p_S and also not through $p_{\tilde{N}'}$, then since P' was in symmetric position we know that there are equally many points of P' on either side of h. Additionally, we know that p_S is below h and $p_{\tilde{N}'}$ is above h for sure, thus we are fine.

Finally, let h be a plane that does not go through p_S but through $p_{\tilde{N}'}$. This is the hardest case to argue and we will do a proof by contradiction. Let us assume that h does not halve the point set P'' but has more points on one than the other side. Note that by parity the difference is even and at least two, let us assume that on one side there are k points whereas on the other there are $k + 2$ points of P''. We rotate h away from $p_{\tilde{N}'}$ in such a way that $p_{\tilde{N}'}$ falls onto the side with more points. Thus, we get a plane h' going through some point of P'' having $k + 3$ points on one side and k on the other. We rotate further in the same direction until we hit a point again. The resulting plane h'' cannot be halving the point set as it has at least $k + 2$ points on one of its sides. However, h'' is a plane through 2 points and the origin and does not go through $p_{\tilde{N}'}$. Thus, it is covered by one of the previous cases and must be halving P''. We found a contradiction and therefore h has indeed the same number of points on both sides.   □

Note that this proof heavily relied on the facts that we have a spherical point set and that we can find the needed directions for point sets in $\mathbb{R}^2$. While the former condition can easily be avoided (see Lemma 5 below), the latter can be ensured with doing an induction over the dimension.

Formally, for a point set P that is not spherical but in symmetric or eccentric configuration, let S be a surrounding sphere of P with center q (the symmetric central point of P). For all points $p \in P$ such that $p \neq q$, we push p out onto S on the line pq and denote the resulting point set as the *induced spherical point set* P'. Note that P' is clearly spherical and we additionally know the following.

Observation 4. *The induced spherical point set P' of P is symmetric (eccentric, respectively) if and only if P is symmetric (eccentric, respectively).*

This follows from the construction of P' as all necessary hyperplanes remain unchanged and in particular they split the point sets P and P' in the same way.

Lemma 5. *For every symmetric point set $P \subseteq \mathbb{R}^3$, there exist two directions v_1 and v_2 such that we can push the central point into either direction; add a new central point and arrive at an eccentric point set P'. We can then push the newly added point into the other direction, and arrive at a symmetric point set P'' missing the symmetric central point.*

Proof. Let SP be the induced spherical point set of P. By Proposition 4 there exist two positions s_1 and s_2 where we can add new points to SP such that the resulting point sets are eccentric (symmetric, resp.). Define v_1 and v_2 to be the directed lines from the origin to s_1 and s_2, respectively. Pushing the central point of P out yields the same induced spherical point set, independently of how far we push. Hence, using Observation 4, it follows that the point sets P' and P'' are eccentric (symmetric, resp.).   □

The arguments used not only work in $\mathbb{R}^3$ but in any dimension. For a proof, we again refer to the full version of this paper.

Theorem 5. *For every symmetric point set $P \subseteq \mathbb{R}^d$, there exist two directions v_1 and v_2 such that we can push the central point into either direction; add a new central point and arrive at an eccentric point set P'. We can then push the newly added point into the other direction, and arrive at a symmetric point set P'' missing the symmetric central point.*

3.3 Putting Everything Together

We are now able to prove that the condition given in Theorem 1 is sufficient.

Proof (that the condition in Theorem 1 is sufficient). Let $D^{0,d}$ be a vector satisfying the condition. If all entries of $D_i^{0,d}$ with $i \geq 2$ are zero, then we can just put points in general, convex position. Let us therefore assume that there is at least one nonzero entry. Let P be the vertices of a simplex in $\mathbb{R}^d$ around the origin and note that P is (spherical) symmetric. We will now add points to P (in pairs) and maintain the condition that P is symmetric. We first add all points of depth 1, then all points of depth 2 and so on.

Assume that P consists of n points and assume further that there are points missing in P (i.e. P does not have histogram $D_i^{0,d}$). Let us denote the smallest missing depth by j and note that this means that all points in P have depth at most j. We now add a point p in the origin to P. Note that p has depth $\lfloor \frac{(n+1)-d+2}{2} \rfloor$, see Lemma 3. By the condition of the Theorem, we know that $n \geq 2j + d - 3$ and thus $j \leq \frac{n-d+3}{2}$. Therefore, we can push p outwards into a direction given by Theorem 5. We continue pushing p until it has depth j. Proposition 3 and Observation 3 guarantee that the only depth that changed while moving p is the one of point p, as all other points of the point set have lower depth. Theorem 5 gives us not only the needed direction but also shows that we can maintain the property of having symmetric (and eccentric) point sets throughout the entire process.   □

4 Number of Depth Histograms

The characterization of Tukey depth histograms $D^{0,d}(P)$ allows to compute the exact number of different histograms for point sets consisting of n points in $\mathbb{R}^d$. For this, let $D(n,d)$ denote the number of different Tukey depth histograms $D^{0,d}(P)$, for point sets $P \subseteq \mathbb{R}^d$ consisting of n points.

Theorem 6. *For any dimension $d \geq 2$ and any $n \geq d + 1$, we have*

$$D(n,d) = \begin{cases} \frac{2}{n-d+2}\binom{3\frac{n-d}{2}+1}{\frac{n-d}{2}}, & \text{if } n-d \text{ is even and} \\ \frac{3}{n-d+2}\binom{3\frac{n-d}{2}+\frac{1}{2}}{\frac{n-d-1}{2}}, & \text{if } n-d \text{ is odd.} \end{cases} \tag{1}$$

The proof is a single (long) calculation, and can be found in the full version of this paper.

5 Conclusion

We have introduced and studied Tukey depth histograms of j-flats. For histograms of points, we were able to give a full characterization. This characterization allowed us to give an exact number of possible histograms. This is a contrast to other representations of point sets, such as order types, where the exact numbers are not known.

It is an interesting open problem to find better necessary and also sufficient conditions, perhaps even characterizations, of histograms of j-flats for $j > 0$. We hope that the ideas in this paper might be useful in this endeavor. Another interesting open problem is to relate depth histograms to other representations of point sets. For example, in the plane, the order type determines the ℓ-vectors for each point, but not vice versa, that is, there are point sets that have the same sets of ℓ-vectors but different order types. Similarly, the set of ℓ-vectors determines the histograms $D^{0,2}$ and $D^{1,2}$. Is the reverse also true or are there point sets for which both $D^{0,2}$ and $D^{1,2}$ are the same but whose sets of ℓ-vectors are different?

Due to their relation to many problems in discrete geometry, we are convinced that the study of depth histograms has the potential to lead to new insights for many problems.

References

1. Aloupis, G.: Geometric measures of data depth. In: Data Depth: Robust Multivariate Analysis, Computational Geometry and Applications (2003)
2. Dol'nikov, V.: Transversals of families of sets in and a connection between the Helly and Borsuk theorems. Russ. Acad. Sci. Sbornik Math. **79**(1), 93 (1994)
3. Durocher, S., Fraser, R., Leblanc, A., Morrison, J., Skala, M.: On combinatorial depth measures. Int. J. Comput. Geomet. Appl. **28**(04), 381–398 (2018)
4. Edelsbrunner, H., Hasan, N., Seidel, R., Shen, X.J.: Circles through two points that always enclose many points. Geom. Dedicata **32**(1), 1–12 (1989)
5. Felsner, S.: Geometric Graphs and Arrangements: Some Chapters from Combinatorial Geometry. Springer, Heidelberg (2012)
6. Goodman, J., Pollack, R.: A theorem of ordered duality. Geomet. Dedicata **12**, 63–74 (1982)
7. Goodman, J., Pollack, R.: Multidimensional sorting. SIAM J. Comput. **12**, 484–507 (1983)

8. Goodman, J., Pollack, R.: Semispaces of configurations, cell complexes of arrangements. J. Comb. Theory Ser. A **37**, 257–293 (1984)
9. Hugg, J., Rafalin, E., Seyboth, K., Souvaine, D.: An experimental study of old and new depth measures. In: 2006 Proceedings of the Workshop on Algorithm Engineering and Experiments (ALENEX), pp. 51–64 (2006)
10. Liu, R.Y.: On a notion of data depth based on random simplices. Ann. Stat. **18**(1), 405–414 (1990)
11. Matousek, J.: Lectures on Discrete Geometry, vol. 212. Springer, Heidelberg (2013)
12. Neumann-Lara, V., Urrutia, J.: A combinatorial result on points and circles on the plane. Discret. Math. **69**(2), 173–178 (1988)
13. Rado, R.: A theorem on general measure. J. Lond. Math. Soc. **21**, 291–300 (1947)
14. Ramos, P.A., Viaña, R.: Depth of segments and circles through points enclosing many points: a note. Comput. Geom. **42**(4), 338–341 (2009)
15. Ruiz-Vargas, A.J., Welzl, E.: Crossing-free perfect matchings in wheel point sets. In: Loebl, M., Nešetřil, J., Thomas, R. (eds.) A Journey Through Discrete Mathematics, pp. 735–764. Springer, Cham (2017). https://doi.org/10.1007/978-3-319-44479-6_30
16. Stone, A.H., Tukey, J.W.: Generalized "sandwich" theorems. Duke Math. J. **9**(2), 356–359 (1942)
17. Toth, C.D., O'Rourke, J., Goodman, J.E.: Handbook of Discrete and Computational Geometry. Chapman and Hall/CRC (2017)
18. Tukey, J.W.: Mathematics and the picturing of data. In: Proceedings of the International Congress of Mathematicians (Vancouver), pp. 523–531. Canadian Mathematical Congress (1975)
19. Wagner, U.: k-sets and k-facets. Contemp. Math. **453**, 443 (2008)
20. Zivaljević, R.T., Vrećica, S.T.: An extension of the ham sandwich theorem. Bull. Lond. Math. Soc. **22**(2), 183–186 (1990)

An Efficient Algorithm for the Proximity Connected Two Center Problem

Binay Bhattacharya, Amirhossein Mozafari[⊠], and Thomas C. Shermer

School of Computing Science, Simon Fraser University, Burnaby, Canada
{binay,amozafar,shermer}@sfu.ca

Abstract. Given a set P of n points in the plane, the k-center problem is to find k congruent disks of minimum possible radius such that their union covers all the points in P. The 2-center problem is a special case of the k-center problem that has been extensively studied in the recent past [7,20,22]. In this paper, we consider a generalized version of the 2-center problem called *proximity connected* 2-center (PCTC) problem. In this problem, we are also given a parameter $\delta \geq 0$ and we have the additional constraint that the distance between the centers of the disks should be at most δ. Note that when $\delta = 0$, the PCTC problem is reduced to the 1-center(minimum enclosing disk) problem and when δ tends to infinity, it is reduced to the 2-center problem. The PCTC problem first appeared in the context of wireless networks in 1992 [12], but obtaining a nontrivial deterministic algorithm for the problem remained open. In this paper, we resolve this open problem by providing a deterministic $O(n^2 \log n)$ time algorithm for the problem.

1 Introduction

The k-center problem in the plane is a fundamental facility-location problem in which we are given a set of n demand points P and we are going to find a set S of k center points such that $cost(S) := \max_{p \in P} \min_{s \in S} dist(p, s)$ is minimized ($dist(p, s)$ is the Euclidean distance between p and s). The k-center problem is known to be NP-hard [3]. However, there is a simple greedy 2-approximation algorithm for the problem which can not be improved unless $P = NP$ [3]. So, the studies on the problem went in the direction of obtaining polynomial-time algorithms where k is not considered as a part of the problem input. As an example, in 2002, Agarwal and Procopiuc [1] gave a $n^{O(\sqrt{k})}$ time algorithm to solve the k-center problem. Solving the problem for specific values of k like $k = 1$ and $k = 2$ received attention due to the geometric properties that can be applied to solve these problems efficiently. The 1-center problem is indeed equivalent to the problem of covering P with a disk with minimum area. This problem is also called the *minimum enclosing disk (MED)* problem. In 1983, Megiddo [17] used the prune and search technique to give an optimal linear time algorithm to solve the MED problem.

For $k = 2$, Drenzer [9] gave the first nontrivial algorithm for the problem with $O(n^3)$ time complexity. Later in 1994, Agarwal and Sharir [2] improved the time

© Springer Nature Switzerland AG 2022
C. Bazgan and H. Fernau (Eds.): IWOCA 2022, LNCS 13270, pp. 199–213, 2022.
https://doi.org/10.1007/978-3-031-06678-8_15

complexity for the problem to $O(n^2 \log^3 n)$. In 1996, Eppstein [10] gave a randomized algorithm for the problem with $O(n \log^2 n)$ expected running time. In 1997, Katz and Sharir [15] proposed the novel expander-based parametric search technique and showed that applying it to the 2-center problem using the $O(n^2)$ time feasibility test of Hershberger [11], gives an $O(n^2 \log^3 n)$ time algorithm for the problem. Later in the year, Sharir [20] designed an $O(n \log^3 n)$ time algorithm for the decision version of the 2-center problem using the breakthrough observation of breaking the problem into three separate cases(far distant, distant and nearby cases). Next, he parallelized the decision algorithm and put it into the Megiddo's parametric search schema [18] to obtain an $O(n \log^9 n)$ time algorithm. Soon, it turned out that solving the problem in the nearby case is the bottleneck to reduce the time complexity. Later, Sharir's running time was improved by Chan [5] and Wang [22] to $O(n \log^2 n \log^2 \log n)$ and $O(n \log^2 n)$ respectively. Very recently, Choi and Ahn [7] (independently Cho and Oh [6]) obtained an $O(n \log n)$ time algorithm for the nearby case which led to an optimal $O(n \log n)$ time algorithm for the 2-center problem.

We say that a set S of k center points in the plane satisfies the *proximity connectedness condition (PCC)* with respect to a parameter δ if the δ-distance graph of S (the graph with vertex set S such that there is an edge between two vertices if and only if the distance between them is at most δ) is connected. The *proximity connected k-center problem* is defined as a generalized version of the k-center problem for which, in addition to P, a parameter $\delta \geq 0$ is also given. The objective is to find k center points S such that S satisfies the PCC and $cost(S) \leq cost(S')$ for any k center points S' that satisfies PCC ($cost(S)$ is the same cost as in the k-center problem). Note that when δ tends to zero (resp. infinity), the problem reduces to the 1-center (resp. k-center) problem. Also, when δ tends to zero and k tends to infinity the problem becomes the Euclidean Steiner tree problem (connecting the points of P by lines of minimum total length in such a way that any two points can be connected by the lines). This is because in this configuration, the centers should be placed along the lines of the minimum Steiner tree in order to minimize the cost. The Euclidean Steiner tree problem is also NP-hard but it has a PTAS approximation algorithm [4].

In practice, the parameter δ usually specifies the range for which one center can communicate with other centers. So, when S satisfies the PCC, any pair of centers can communicate with each other via the other centers. The proximity connected 2-center (PCTC) problem first emerged in the works of Huang [12] in 1992 while he was studying packet radio networks. In the network terminology, the PCTC problem is the problem of locating two wireless devices as close as possible to the demand points P such that they can send/receive messages between each other. He originally gave an $O(n^5)$ time algorithm for the 2-center problem having proximity constraints between their centers. Later in 2003, Huang *et al.* [14] studied a very close problem to the PCTC problem called *α-connected 2-center problem*. In this problem, instead of δ, a parameter $0 \leq \alpha \leq 1$ is given and the distance between the center of the disks should be at most $2(1 - \alpha)r$ where r is the radius of the disks. They gave an $O(n^2 \log^2 n)$ time algorithm for the decision version (given an r whether it is possible to cover the points with two disks of radius r satisfying the desired conditions) of the problem. Note that this problem is a special case of the PCTC decision problem where $\delta = 2(1 - \alpha)r$. Later in 2006, they gave a randomized algorithm with the same $O(n^2 \log^2 n)$ expected running time to solve the α-*connected* 2-center problem [13]. In this paper, we consider the PCTC problem and propose a deterministic $O(n^2 \log n)$ time algorithm for it.

Here, we need to mention that although we use Sharir's observation [20] of breaking the problem into three different cases(far distant, distant and nearby), the reason we can't get a sub-quadratic algorithm like [5,7,20,22] is that the PCTC problem is structurally different from the 2-center problem. In the 2-center problem, the optimal cost is determined by at most three points of P [20] while in the PCTC problem the cost may be determined by more than three points because of the PCC. This means that our search space has a higher dimension than the search space of the 2-center problem. Also, all the sub-quadratic algorithms for the 2-center problem use Megiddo's [18] or Cole's [8] parametric search schema to reduce the time complexity which makes the resulting algorithm impractical [2] while our algorithm exploits the geometric properties of the problem which make it straightforward to be implemented using standard data structures in computational geometry.

A *solution* for a given PCTC problem instance is defined as a pair of disks whose centers satisfy the PCC and their union covers P. We call a disk with the larger (or equal) radius the *determining disk* of the solution and its radius the *cost* of the solution. An *optimal solution* is a solution with minimum cost among the set of all solutions for the problem. Note that there might be an infinite number of optimal solutions with different pairs of radii because we have freedom on the smaller disk. So, we try to find an optimal solution such that the radius of its smaller disk is minimum among all optimal solutions. We call such a solution a *best optimal solution (BOS)* for the problem. Therefore, if the problem has more than one BOS, they would have the same pair of radii. We can also compare two solutions S_1 and S_2 as follows: we say that S_1 is a better solution than S_2 if $cost(S_1) < cost(S_2)$ and if $cost(S_1) = cost(S_2)$, the radius of the non-determining disk of S_1 is smaller than the radius of the non-determining disk of S_2. In this paper, our algorithm not only gives us an optimal solution but it computes a BOS for the problem.

2   Preliminaries and Definitions

Let (P, δ) be the given PCTC problem instance where P is a set of n demand points in the plane and δ is a given non-negative number. We assume that the points are in general position, by which we mean no four points of P lie on a circle. Let (P_1, P_2) be a partition of P obtained by dividing the plane by a line or two half-lines from a point (henceforth, when we use the term *partition of the plane*, we mean a partition that satisfies this condition). We say that a pair of disks (D_1, D_2) with centers (c_1, c_2) respectively is a *solution for the partition* if D_1 covers P_1, D_2 covers P_2 and $dist(c_1, c_2) \leq \delta$. *Optimal* and *best optimal solutions (BOSs)* for the partition are defined similarly. Let (D_1^*, D_2^*) be a BOS for the partition with centers (c_1^*, c_2^*) respectively. We say that a point $p \in P_1$ is a *dominating point* of D_1^* if (D_1^*, D_2^*) is not a BOS for the partition $(P_1 \backslash p, P_2)$. The dominating points of D_2^* are defined similarly. Note that the dominating points of D_1^* and D_2^* are on their boundaries. By assuming that the points are in general position, if D_1^*(resp. D_2^*) is the MED of P_1(resp. P_2), its dominating points are either three points on the boundary such that their induced triangle contains c_1^*(resp. c_2^*) or two points on the boundary such that their connecting segment passes through c_1^*(resp. c_2^*). In order to simplify the presentation of our algorithm, in the latter case, we consider one of the dominating points as two infinitely close points and so, if D_1^* or D_2^* is the MED of their corresponding points, we assume that it has exactly three dominating points. Similarly, if D_1^*(resp. D_2^*) is not the MED of P_1(resp. P_2), in the case that it only has one dominating point, we can consider it as two infinitely close points. But, if it has three points on its boundary such that their induced triangle does not contain c_1^*, we might have no dominating point for D_1^*. We can assume that such a situation never happens by slightly perturbing the points. So, henceforth, if D_1^*(resp. D_2^*) is not a MED, we assume that it has exactly two dominating points.

We call the problem of computing a BOS for a given partition (P_1, P_2) the *restricted PCTC problem*. We can solve the restricted PCTC using the *intersection hulls* and the *farthest-point Voronoi diagrams* of P_1 and P_2 (the intersection hull of a set of points with respect to some radius r is defined as the intersection of all disks of radius r around the points of the set). Here, we briefly explain the main ideas of our algorithm to solve the restricted PCTC problem. Details can be found in the full version of the paper [19]. We first compute the MED of each part and if the resulting centers satisfy the PCC we are done. Otherwise, we can see that the distance between the centers should be exactly δ. Now, we first compute the optimal cost of the problem and then use this value to build a BOS. In order to obtain the optimal cost, we first build the farthest-point Voronoi diagrams of P_1 and P_2. Next, we design a feasibility test (a procedure that for a given value, determines whether it is smaller, equal or greater than the optimal cost for the partition) and apply it to perform a binary search on the weights of the vertices of the diagrams (the weight of a vertex is its distance to its farthest Voronoi site). Let I be the final interval which contains no vertex weight. The structure of the intersection hulls of P_1 and P_2 at radius r does not change when r varies in I. This property enables us to compute the radius for which the distance between the two intersection hulls becomes δ. We see that this radius is indeed the optimal cost. The time complexity of the algorithm is dominated by the cost of computing the farthest-point Voronoi diagrams which is $O(n \log n)$ [21].

We denote the optimal cost for the PCTC problem by r^* and a BOS for the problem by (D_1^*, D_2^*) with centers (c_1^*, c_2^*) respectively. We can assume that c_1^* and c_2^* lie on the x-axis and c_1^* is on the left side of c_2^*. In [20], Sharir broke the 2-center decision problem (given a parameter r determine whether it is possible to cover the points with two disks of radius r) into three cases -far distant, distant and nearby- with respect to the given parameter r. He showed that providing separate algorithms for these cases will reduce the overall time complexity to solve the decision problem. Although our problem is an optimization problem and the parameter r^* is unknown, we will show that breaking the PCTC problem into the same cases will simplify our algorithm and reduce the overall time complexity. So, our algorithm separately considers each of the following three assumptions about (D_1^*, D_2^*).

1. Nearby: $dist(c_1^*, c_2^*) \leq r^*$.
2. Distant: $r^* < dist(c_1^*, c_2^*) \leq 3r^*$.
3. Far distant: $dist(c_1^*, c_2^*) > 3r^*$.

Denote the smallest cost we can get having the nearby, distant and far distant assumptions by r^{NA}, r^{DA} and r^{FA} respectively. We also use the same notation for a BOS and their corresponding centers having each assumption. So, we can obtain (D_1^*, D_2^*) by comparing (D_1^{NA}, D_2^{NA}), (D_1^{DA}, D_2^{DA}) and (D_1^{FA}, D_2^{FA}) (note that these solutions may not exist or satisfy their corresponding case conditions. For example, $dist(c_1^{NA}, c_2^{NA})$ might be greater than r^{NA} but if (D_1^*, D_2^*) satisfies the nearby case, then $r^{NA} = r^*$ and (D_1^{NA}, D_2^{NA}) would be a BOS for the problem and we have $dist(c_1^{NA}, c_2^{NA}) \leq r^{NA} = r^*$). Henceforth, while studing each of the cases, when we say a BOS, we mean a best solution we can get having the corresponding case assumption. Given two points x and y in the plane, we denote the line passing from x and y by $line(x, y)$. The direction of this line is considered from x to y. Also, we denote the half-line from x passing y by $half\text{-}line(x, y)$ and the line segment with end points x and y by $seg(x, y)$.

3   Computing a BOS in the Nearby Case

First, we can see that if $(D_1^*, D_2^*) \leq r^*$, then there is an optimal partition R^* (may not be unique) such that (D_1^*, D_2^*) is a BOS of R^*. In fact, such a partition can be obtained by considering a point in $D_1^* \cap D_2^*$ and two half-lines from it passing the intersection points of ∂D_1^* (boundary of D_1^*) and ∂D_2^*. In this section, when we say the dominating points of (D_1^*, D_2^*), we mean its dominating points with respect to R^*. Without loss of generality, we can assume that D_2^* is the determining disk. We first compute the $convex\text{-}hull(P)$ and scale the problem such that it fits in a unit square (multiple both x and y coordinates of the points by the greatest constant such that the convex hull remains inside the square). This step can be done in $O(n \log n)$ time. Note that the scaling will not change the solutions.

Proposition 1. *If $(D_1^*, D_2^*) \leq r^*$, then the area of $D_1^* \cap D_2^*$ must be greater than a constant factor of the area of D_2^* (the determining disk).*

Proof. We proceed by contradiction. Suppose that such a factor does not exist. This means that we can build a problem instance such that it has a BOS (D_1^*, D_2^*) in which the radius of the non-determinig disk (D_1^*) becomes infinitely small (because of the nearby assumption and scaling). So, D_1^* should have at least one dominating point that is not covered by D_2^*. Because the radius of D_1^* is infinitely small, δ should tend to $radius(D_2^*)$ (which tends to the radius of the MED of

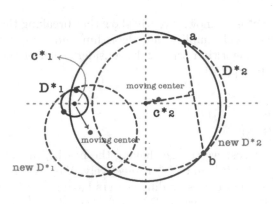

Fig. 1. Enlarging the non-determining disk D_1^* to cover one of the dominating points of D_2^* and get a better solution.

P). Now, D_2^* should have at least one dominating point (point c in Fig. 1) with the x-coordinate less than or equal to c_2^* (otherwise, we can move both c_1^* and c_2^* to the right and reduce the radius of D_2^* which determines the cost). In this configuration, we can enlarge D_1^* by moving c_1^* toward this dominating point of D_2^* while satisfying the PCC in order to cover it and release it from D_2^* (D_1^* does not lose any of its own points and its radius still remains less than the radius of D_2^*). Now, we can reduce the radius of D_2^* which contradicts the optimallity of (D_1^*, D_2^*) (see Fig. 1). $\qquad\square$

Proposition 2. D_1^* *(similarly D_2^*) should have a pair of dominating points such that:*

1. *They lie on different sides of $line(c_1^*, c_2^*)$.*
2. *Their connecting segment does not intersect $seg(c_1^*, c_2^*)$.*

The proof is straightforward using elementary geometry (see the paper's full version for details). Considering the four dominating points in the above proposition, we can say that $D_1^* \cap D_2^*$ should cover at least a constant factor of the area of *convex-hull*(P). Furthermore, $D_1^* \cap D_2^* \cap convex\text{-}hull(P)$ is convex because it is the intersection of convex objects. So, we can build a constant size set of points $\mathcal{M}$ uniformly distributed on *convex-hull*(P) such that (assuming $dist(c_1^*, c_2^*) \leq r^*$) for at least one point $\hat{m} \in \mathcal{M}$, $\hat{m} \in D_1^* \cap D_2^* \cap convex\text{-}hull(P)$. Because $\hat{m}$ is unknown, for each $m \in \mathcal{M}$, we build a BOS (D_1^m, D_2^m) assuming $m \in D_1^* \cap D_2^*$ and finally pick a best solution in $\{(D_1^m, D_2^m) : m \in \mathcal{M}\}$ and set it as (D_1^{NA}, D_2^{NA}). Based on this idea, we present our algorithm to find (D_1^m, D_2^m) for a given point $m \in convex\text{-}hull(P)$.

Let $\mathcal{X}$ be a set of 360 directed lines (each line has a positive direction) passing through m such that the angle between each directed line and its neighbour lines is $1°$. Now, there should be a directed line in $\mathcal{X}$ such that its angel with

$line(c_1^*, c_2^*)$ is at most $1°$ and c_1^* lies on the negative side of c_2^* on the line (note that D_2^* is the determining disk according to our assumption). We call this directed line the *correct directed line* which is unknown. So, we assume each line $l \in \mathcal{X}$ as the correct directed line and compute a BOS $(D_1^{m,l}, D_2^{m,l})$ having this assumption and finally pick the best one as (D_1^m, D_2^m).

So, assume that a directed line $l \in \mathcal{X}$ called the m-line is given. Here we explain how to compute $(D_1^{m,l}, D_2^{m,l})$. The m-line divides the points of P into two disjoint sets one on the right side and the other on the left side of the m-line. We sort these sets according to the polar angles of their points (from m) with respect to the positive direction of the m-line. These angles should lie between $-180°$ and $180°$ and we sort them by increasing magnitude (see Fig. 2 for an illustration). Based on these orders, we denote the two sequences of points on the left and right side of the m-line by $(p_1, \ldots, p_{n'})$ and $(q_1, \ldots, q_{n''})$ respectively. We call a point p-*type* (resp. q-*type*) if it is in the first(resp. second) sequence. We also call a half-line from m that separates $\{p_1, \ldots, p_i\}$ from $\{p_{i+1}, \ldots, p_{n'}\}$ an ith-*separator* of the p-type points. A jth-separator of q-type points is defined similarly (we assume that the 0th and n'^{th}(resp. n''^{th}) separators have the entire p-type(resp. q-type) points in one side). The ith and j^{th} separators of the p-type and q-type points partition the plane into two parts. We call this partition the (i, j)-*partition* of the plane. One part of this partition contains the positive direction of the m-line which we call it the *positive side* of the partition and we call the other part the *negative side* of the partition.

Observation 1. *If $dist(c_1^*, c_2^*) \leq r^*$, $m = \hat{m}$ and the m-line is correct, then an (i, j)-partition can be considered as R^* and (D_1^*, D_2^*) is its BOS.*

Note that in the above observation, the two separators from m passing the intersection points of D_1^* and D_2^* give us the desired (i, j)-partition. We denote the set of points in the positive and negative sides of the partition by $P_+^{i,j}$ and $P_-^{i,j}$ respectively. Based on our algorithm for restricted PCTC problem, a BOS for an (i, j)-partition can be computed in $O(n \log n)$ time. Let $(D_-^{i,j}, D_+^{i,j})$ (with centers $(c_-^{i,j}, c_+^{i,j})$ respectively) be the output of this algorithm for the (i, j)-partition (see Fig. 2 for an example). We refer to the first(resp. second) disk the *negative disk*(resp. *positive disk*) of the partition. A naive approach to obtain (D_1^*, D_2^*) is to apply our restricted PCTC problem algorithm to each of the (i, j)-partitions and pick the best one. This will give us an $O(n^3 \log n)$ time complexity as there are quadratic partitions. In the following we show how we can get $(D_1^{m,l}, D_2^{m,l})$ by evaluating a sub-quadratic number of partitions. The idea is first computing $r^{m,l}$ which is the best cost we can get assuming m and l are correct. Then, we use it to compute $(D_1^{m,l}, D_2^{m,l})$.

3.1   Computing $r^{m,l}$

Let's define M^+ as a $(n' + 1) \times (n'' + 1)$ matrix whose $[i, j]$-element ($0 \leq i \leq n'$ and $0 \leq j \leq n''$) is $radius(D_+^{i,j})$. We call $M^+[i, j]$ *non-critical* if $D_+^{i,j}$ is the MED of $P_+^{i,j}$. Otherwise, we call it *critical*. We call $M^+[i, j]$ a *valid* element if

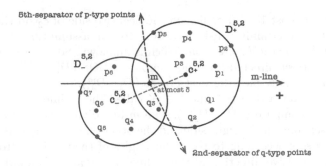

Fig. 2. An (i,j)-partition of a set of points and its corresponding BOS.

$M^+[i,j] \geq radius(D^{i,j}_-)$ and we call it *non-valid* otherwise. Because we assumed that l is correct, we can assume that positive disks determine $r^{m,l}$. This means that $r^{m,l}$ is indeed the minimum valid element of M^+.

Proposition 3. *For any $0 \leq i \leq n'$ and $0 \leq j \leq n''$, we have:*

1. *If $M^+[i,j]$ is non-critical, then $M^+[i',j'] \geq M^+[i,j]$ for all $i' \geq i$ and $j' \geq j$.*
2. *If $M^+[i,j] > radius(D^{i,j}_-)$, $M^+[i,j]$ is non-critical.*
3. *If $M^+[i,j]$ is valid and critical, then $M^+[i,j] = radius(D^{i,j}_-)$ and $dist(c^{i,j}_-, c^{i,j}_+) = \delta$.*

Briefly, case 1 is clear and if each of the cases 2 or 3 is not true, by moving the centers we can get a better solution. We search M^+ to find $r^{m,l}$ as follows: we maintain a set of *candidate values*. During the search, when we evaluate an element $M^+[i,j]$ (computing $(D^{i,j}_-, D^{i,j}_+)$ and its dominating points), if $M^+[i,j]$ is valid, we add it to the candidate values and finally we set $r^{m,l}$ as the minimum candidate value.

In order to search M^+, we maintain two variables I and J where I stores the index of the current row that we are searching and J stores the column index for which we can discard any column with index greater than that. Initially, we set $I = 0$, $J = n''$ (n'' is the number of columns of M^+). We search the Ith-row by evaluating its elements backward starting from its Jth-element (if $J = -1$, the matrix search is done) toward its first element. Because we are looking for a minimum valid element of the matrix, we can use Proposition 3 to improve our search as follows: during the traversal of the row, if $M^+[I,j]$ is valid and non-critical, we set $J = j - 1$ (because $D^{I,j}_+$ is the MED of $P^{I,j}_+$, when we add more points to the positive side we can't get a smaller positive disk). We finish traversing the row and increase I by one if either the row is exhausted or we reach an index j such that $M^+[I,j]$ becomes non-valid. Note that in this case, $D^{I,j}_-$ is the MED of $P^{I,j}_-$ (similar to Proposition 3 part 3). Here we might have a valid element on some index $j' < j$ but, the cost of this solution can not be less than $radius(D^{I,j}_-)$ (we add points to the negative side as we move left wise on a row). In order to make sure that we will count such costs in our algorithm, we can add $radius(D_-^{I,j})$ to the candidate set of the directed line in $\mathcal{X}$ with the opposite direction of the current m-line.

We continue this procedure until no element is left. Note that when none of $D_-^{i,j}$ and $D_+^{i,j}$ are a MED, we can't discard any element from the matrix because it is possible that when we move a point from one side to the other, the radii of both disks become greater or smaller while they remain equal (this situation can happen because of the PCC). So the number of evaluations in the above schema might be still quadratic. Next, we explain how to fix this problem.

Proposition 4. *If $M^+[i,j]$ is valid-critical and q_j is not a dominating point of $D_+^{i,j}$, then $M^+[i,j-1] \geq M^+[i,j]$.*

Proof. Because $D_+^{i,j}$ is not a MED, its center can't get closer to its farthest points in $P_+^{i,j}$ (dominating points of $D_+^{i,j}$) namely d_1 and d_2 because of the PCC. Now, by adding q_j to $P_-^{i,j}$, $D_-^{i,j-1}$ needs to cover more points. If its radius gets bigger, the proposition follows. Otherwise, according to the fact that q_j is not a dominating point of $D_+^{i,j}$, it is not possible to put $c_-^{i,j-1}$ on a place such that allow $c_+^{i,j-1}$ to get closer to d_1 and d_2 due to best optimality of $(D_-^{i,j}, D_+^{i,j})$.   □

Note that in this proposition, if q_j does not become a dominating point of $D_-^{i,j-1}$, then $M^+[i,j-1] = M^+[i,j]$. A similar statement is also correct for two consecutive valid-critical elements in a column. Based on the above proposition, we can improve our matrix search as follows: while traversing a row (left wise), when we hit a valid-critical element $M^+[i,j]$, if both dominating points of $D_+^{i,j}$ are p-type, we discard the rest of the row (because by traversing a row, only q-type points will move to the other part of the partition) and continue the search on the next row. Similarly, if both dominating points of $D_-^{i,j}$ are q-type, we can discard the rest of its column. Otherwise, we jump to the first (largest index) element of the row for which a q-type dominating point of $D_+^{i,j}$ moves to the negative side and discard all the elements in between (because of Proposition 4). Similarly, we discard the portion of the rest of the column of $M^+[i,j]$ with row index smaller than the index of the p-type dominating point(s) of $D_-^{i,j}$ (applying the column version of Proposition 4).

When we evaluate a valid-critical element $M^+[i,j]$, if we didn't discard the entire rest of its row or column, we *mark* the portion of its column that is not discarded after the evaluation of $M^+[i,j]$. Now, when we traverse the rows, we ignore and jump discarded and marked elements. Specially, if after evaluating an element $M^+[i,j]$, the largest index of the q-type dominating point of $D_+^{i,j}$ is j' and $M^+[i,j']$ is marked, we continue searching from the first(biggest index) unmarked or undiscarded element of the ith-row after $M^+[i,j']$. Applying this marking schema in the matrix search will guarantee that the number of evaluations is linear. The problem of our matrix search with marking schema is that we may mark the minimum valid element of M^+ and so get an incorrect $r^{m,l}$. In the rest, we will show how to overcome this problem.

We call the above matrix search *initial search* of M^+ from *top-right*. Another way of searching M^+ is starting the search from $M^+[n', 0]$ (the first element of the last row). But this time, instead of traversing the rows from right to left, we traverse the columns from bottom to top. The way we search the matrix is exactly symmetrical to the top-right search but here we mark sub-rows instead of sub-columns. We call this matrix search the initial search of M^+ from bottom-left. After performing two initial searches on M^+ one from the top-right and one from the bottom-left, still there might be some elements that are marked in both initial searches. We call these elements as *doubly-marked* elements. The next theorem enables us to search the doubly-marked elements in an efficient way which leads us to find $r^{m,l}$. Lets denote the doubly-marked elements of M^+ by $Doubly\text{-}Marked(M^+)$.

Theorem 1. *By evaluating a doubly-marked element $M^+[i,j]$, we can discard one of the following sub-rows or sub-columns of M^+:*

1. *Elements above $[i,j]$ ($M^+[i',j]$ with $i' \leq i$).*
2. *Elements below $[i,j]$ ($M^+[i',j]$ with $i' \geq i$).*
3. *Elements in front of $[i,j]$ ($M^+[i,j']$ where $j' \geq j$).*

Suppose that $M^+[\hat{i},\bar{j}]$ is a given doubly-marked element which is marked when we evaluate $M^+[\bar{i},\bar{j}]$ and $M^+[\hat{i},\hat{j}]$ in the initial top-right and bottom-left search respectively. When we evaluate $M^+[\hat{i},\bar{j}]$, we get $D_+^{\hat{i},\bar{j}}$ and $D_-^{\hat{i},\bar{j}}$ and their dominating points. For the sake of simplicity, let's denote the first disk by D'_+ and the second disk by D'_-. If D'_- is MED, then either $radius(D'_-) \geq radius(D'_+)$ or $radius(D'_-) < radius(D'_+)$. In the former, case 1 in Theorem 1 happens and in the latter, D'_+ should be MED (otherwise we can reduce its cost and the solution can't be optimal) and so cases 2 and 3 of the theorem happen. We have a similar argument when D'_+ is a MED. So, the only left case is when none of the disks is a MED. Note that in this case each of D'_+ and D'_- has exactly two dominating points. Let h_1, h_2 be the dominating points of D'_+ and h'_1, h'_2 be the dominating points of D'_-. If both h'_1 and h'_2 are p-type, case 3 happens (when we traverse the $\hat{i}^{th}$-row from left to right, we only add q-type points to the positive side). Also, if they are both q-type, case 2 happens. The bottleneck of proving Theorem 1 is when h'_1 and h'_2 have different types. In order to prove Theorem 1 in this special case, we use two key properties. First $M^+[\hat{i},\bar{j}]$ should be doubly-marked and second, m should be inside the convex hull of the points. We leave this proof for the full version of the paper [19] and in the rest, we focus on how to use Theorem 1 to search $Doubly\text{-}Marked(M^+)$ efficiently.

3.2   Searching the Doubly-Marked Elements

For simplicity, we assume that $n' = 2^g - 1$ for some integer value $g > 1$ (so the number of rows is a power of 2). We define the kth-division of M^+ as the sub-matrix consisting of the rows from $n'/2^k$ to $n'/2^{k-1} - 1$ (we search the first row independently by evaluating all of its doubly-marked elements). We search

---

Algorithm 1. SEARCH-DM(M)

1: Let M be a $n \times m$ matrix.
2: Split M into $\log n$ divisions $\{DIV_1, \ldots, DIV_{\log n}\}$.
3: **for** $k = 1, \ldots, \log n$ **do** // We search the divisions in order.
4:     Set $I = J = 1$.
5:     **repeat**
6:         Evaluate $DIV_k[I, J]$ and discard the portion of M according to Theorem 1.
7:         **if** (case 1 or 2 happens) **and** $J < m$ **then** $J = J + 1$.
8:         **else**
9:             $I = I + 1$.
10:        **end if**
11:    **until** $I > n/2^k$ // number of rows in DIV_k.
12:    SEARCH-DM(DIV_k) if DIV_k has an unevaluated/undiscarded element.
13: **end for**
14: Evaluate all non-discarded elements of the first row of M. // Until the case 3 happens.

---

the divisions of M^+ in order from its first division. Let's denote the kth-division sub-matrix by DIV_k. Here, we explain how to search DIV_k. Let I and J be the row and column indices (with respect to DIV_k) of the element that we are processing at each time. Initially, we have $I = J = 1$ (the first row and column of DIV_k). We evaluate the non-discarded elements of the Ith-row from left to right starting from the column index J. If the result of an evaluation is case 1 or 2 in Theorem 1, we discard the corresponding portion of M^+ (in all divisions) and increase J by one. But if case 3 happens, we go to the next row and increase I (we always move rightwise). After we proceed with all divisions, some elements might left unevaluated and undiscarded in each division due to the occurrence of case 1. We recursively perform the entire above process on these unevaluated elements in each division until all elements are either discarded or evaluated. So, if only doubly-marked elements remained in M^+ (we have discarded all other elements in the two initial searches), then the procedure SEARCH-DM(M^+) in Algorithm 1 will give us a minimum valid element of M^+.

Theorem 2. *SEARCH-DM(M^+) evaluates $O(n \log n)$ elements of M^+.*

Proof. First, if only cases 2 and 3 happen in the algorithm, then we don't need the recursion part and so the total number of evaluations becomes $O(n \log n)$ (in each iteration of searching DIV_k either I or J would be increased). Now, suppose that any of the cases 1, 2, or 3 can happen. Note that the number of case 3s in all divisions of a same recursion level (the original $\log n$ divisions has recursion level zero and the level of the divisions in the recursion part of the algorithm is defined based on their depth in the recursion tree) is at most n because two divisions of a same level have disjoint rows. So, because we have $O(\log n)$ levels, the total number of case 3 evaluations is $O(n \log n)$. Now, if after the evaluation of some $DIV_k[i, j]$, case 1 happens, we can't discard any new element from DIV_k but all the elements above $DIV_k[i, j]$ in M should be discarded. This means that while searching each of the divisions $DIV_{k+1}, \ldots, DIV_{\log n}$ and the first row, we don't need to evaluate the jth-column. On the other hand, DIV_k has $\log(n/2^k) = \log n - k$ divisions and a row. Each of these divisions can have at most one cases 1 or 2 in the jth-column. So, we can have a correspondence between the extra cases 1 and 2 evaluations in searching the divisions and the first row of DIV_k (not its recursion part) and the matrix elements that we didn't evaluate in $DIV_{k+1}, \ldots, DIV_{\log n}$. So, the total number of evaluations would remain $O(n \log n)$.   □

Note that in a constant time, we can check whether an element is discarded or not. Because in each recursion level, the divisions are disjoint, at each level we check each element of M at most once and because we have $O(\log n)$ levels, the total cost of matrix element checking would be $O(n^2 \log n)$. On the other hand, our algorithm to solve the restricted PCTC problem costs $O(n \log n)$, if we directly use it to evaluate matrix elements, the total time complexity of SEARCH-DM(M^+) becomes $O(n^2 \log^2 n)$. As we mentioned in Sect. 2, the bottleneck of solving the restricted PCTC problem is computing the farthest-point Voronoi diagram of each part of the partition which costs $O(n \log n)$. So, if we can reduce this cost by performing a preprocessing step, we can reduce the overall time complexity of SEARCH-DM(M^+). In order to speed up matrix element evaluation, we use the following lemma from [16]:

Lemma 1 [16]. *If X and Y are arbitrary sets of points in the plane, then $\mathcal{F}(X \cup Y)$ can be constructed from $\mathcal{F}(X)$ and $\mathcal{F}(Y)$ in $O(|X| + |Y|)$ time ($\mathcal{F}(X)$ represents the farthest-point Voronoi diagram of X).*

The Preprocessing Step: Let (X_i^+, X_i^-) (resp. (Y_j^+, Y_j^-)) be the partition of the p-type (resp. q-type) points induced by the ith-separator (resp. j^{th}-separator). In the preprocessing phase, we compute the farthest-point Voronoi diagram of all X_i^+, X_i^-, Y_j^+ and Y_j^- for $0 \leq i \leq n'$ and $0 \leq j \leq n''$. This step can be done in $O(n^2)$ using Lemma 1 because as i or j increases or decreases by one, a point from one side would be added to the other side.

Now, we can reduce the cost of matrix element evaluation as follows: In order to evaluate $M^+[i, j]$, we construct $\mathcal{F}(P_+^{i,j})$(resp. $\mathcal{F}(P_-^{i,j})$) in $O(n)$ time by applying Lemma 1 to $\mathcal{F}(X_i^+)$ and $\mathcal{F}(Y_j^+)$ (resp. $\mathcal{F}(X_i^-)$ and $\mathcal{F}(Y_j^-)$). So, the total complexity of matrix evaluation would be $O(n)$. This reduces the time complexity of SEARCH-DM(M^+) and so the cost of finding $r^{m,l}$ to $O(n^2 \log n)$.

3.3   Obtaining $(D_-^{m,l}, D_+^{m,l})$ Having $r^{m,l}$

Note that we already have an initial solution that is optimal and its cost is $r^{m,l}$ (from our search for $r^{m,l}$). But, there might be another optimal solution with the same cost and a smaller non-determining disk that we discarded during the search. If this initial solution is not best optimal, then the non-determining disk of a BOS must be strictly smaller than its determining disk. So, we can assume that the positive disk of the BOS should be the MED of the points in the positive side. Consider a matrix $\bar{M}$ for which its $(i,j)^{th}$-element is the radius of the MED of the points on the positive side of the (i,j)-partition. We search $\bar{M}$ from the last element of its first row and traverse the rows backwards (similar to the initial top-right search). After evaluating an element (i,j) of the matrix (which can be done in linear time according to [17]), if it is bigger than $r^{m,l}$, we discard all elements (i,j') of the matrix with $j' \geq j$ because they are all greater than $r^{m,l}$ and if it is less than $r^{m,l}$, we discard the elements with $i' \leq i$ because they are all less than $r^{m,l}$. But, when it is exactly $r^{m,l}$, we compute its non-determining disk using the restricted PCTC problem algorithm (costs $O(n \log n)$) and store its radius. Here, we can also discard all elements (i',j) of the matrix with $i' \leq i$. This is because as we advance more left into the row, we would have more points on the negative side and so if there is any optimal solution on the left of (i,j) in the row, its non-determining disk should cover more points and thus can't give us a better solution. So, by each evaluation, we discard a row or a column of the matrix which means that the total number of evaluations is linear. Therefore, the total complexity of finding a BOS given $r^{m,l}$ would be $O(n^2 \log n)$. Combining it with the complexity of computing $r^{m,l}$ gives us the total time complexity $O(n^2 \log n)$ to obtain (D_1^{NA}, D_2^{NA}).

4   Computing a BOS in the Far Distant and Distant Cases

For the far distant case, we assume that $dist(c_1^*, c_2^*) > 3r^*$. In this situation, the approach of Sharir's far distant case [20] for the decision 2-center problem still works as follows: set an arbitrary point in the plane as the origin and build 360 directed lines $\mathcal{X}$ passing from the origin such that the degree between each line and its neighbours is $1°$. Then for one unknown correct line $\vec{x}_c \in \mathcal{X}$, the angle between $line(c_1^*, c_2^*)$ and $\vec{x}_c$ is at most $1°$. Suppose that we set $\vec{x}_c$ is the x-axis and sort the x-coordinates of the points in P as a sequence $(x_1, \ldots, x_n)$. Now, if we consider the set of lines $\mathcal{L}_F^{\vec{x}_c} = \{x_i \perp x_{i+1} : 1 \leq i < n\}$ ($x_i \perp x_{i+1}$ is the vertical line on $\vec{x}_c$ at the mid-point of $[x_i, x_{i+1}]$), at least one $l \in \mathcal{L}_F^{\vec{x}_c}$ will separate D_1^* from D_2^*. Because $\vec{x}_c$ is unknown, we build $\mathcal{L}_F^{\vec{x}}$ for all $\vec{x} \in \mathcal{X}$ and set $\mathcal{L}_F = \bigcup_{\vec{x} \in \mathcal{X}} \mathcal{L}_F^{\vec{x}}$. Note that the number of lines in $\mathcal{L}_F$ is linear. Here, each line $l \in \mathcal{L}_F$ induces a partition on P. We apply our algorithm for the restricted PCTC problem to each of such partitions and set the best one as (D_1^{FA}, D_2^{FA}). So, the time complexity of the far distant case would be $O(n^2 \log n)$.

For the distant case, here we provide the main ideas of our $O(n^2 \log n)$ time algorithm, leaving the complete presentation and analysis for the full version of the paper [19]. We first compute r^{DA} and then use it to compute (D_1^{DA}, D_2^{DA}). To do this, we first impose that the optimal disks should be congruent and build a constant size set of vertical lines $\mathcal{L}_D$ such that for at least one $l \in \mathcal{L}_D$, all points of P on the left side of l namely P_l^- should be covered by D_1^*. We see that c_1^* should be on the boundary of an intersection hull of P_l^-. Let's denote the intersection hull of P_l^- at radius r by $H^-(r)$. Now, the optimal cost r^{DA} is the smallest radius r for which there is a point $x \in \partial H^-(r)$ such that the distance between x and $H_x^+(r)$ is at most δ where $H_x^+(r)$ is the intersection hull of the points in P not covered by $disk(x, r)$ at radius r ($disk(x, r)$ is the disk with center x and radius r). The idea to find r^{DA} is first designing a feasibility test for it and use it to do a binary search on the weights of the farthest-point Voronoi diagram of P_l^- to get some interval I containing r^{DA}. So, the structure of $H^-(r)$ (its consisting arcs) does not change when r varies in I. Next, we define partitioning $\partial H^-(r)$ into a set of sub-regions such that for each sub-region R, all disks $\{disk(x, r) : x \in R\}$ cover a same set of points. Note that the endpoints of these sub-regions change when we change r. We compute a sequence of radii $(r_1, \ldots, r_t) \subset I$ such that at each r_i ($1 \leq i \leq t$), a sub-region emerges or vanishes on the boundary of the intersection hull of P_l^-. Finally, we do another round of binary search on this sequence to find another interval $I' \subseteq I$ such that when r varies in I', no such sub-regions appear or disappear on $\partial H^-(r)$. Now for each sub-region, it is straightforward to compute the smallest radius such that its distance from intersection hull of the points not covered by it at most δ. Comparing all such radii and pick the smallest one will give us r^{DA} which enables us to compute (D_1^{DA}, D_2^{DA}).

References

1. Agarwal, P.K., Procopiuc, C.M.: Exact and approximation algorithms for clustering. Algorithmica **33**(2), 201–226 (2002)
2. Agarwal, P.K., Sharir, M.: Planar geometric location problems. Algorithmica **11**(2), 185–195 (1994)
3. Agarwal, P.K., Sharir, M.: Efficient algorithms for geometric optimization. ACM Comput. Surv. (CSUR) **30**(4), 412–458 (1998)
4. Arora, S.: Nearly linear time approximation schemes for Euclidean TSP and other geometric problems. In: Proceedings 38th Annual Symposium on Foundations of Computer Science, pp. 554–563. IEEE, 20 October 1997
5. Chan, T.M.: More planar two-center algorithms. Comput. Geom. **13**(3), 189–198 (1999)
6. Cho, K., Oh, E.: Optimal algorithm for the planar two-center problem. arXiv preprint arXiv:2007.08784, 17 July 2020
7. Choi, J., Ahn, H.K.: Efficient planar two-center algorithms. Comput. Geom. **2**, 101768 (2021)
8. Cole, R.: Slowing down sorting networks to obtain faster sorting algorithms. J. ACM (JACM) **34**(1), 200–208 (1987)
9. Drezner, Z.: The planar two-center and two-median problems. Transp. Sci. **18**(4), 351–361 (1984)
10. Eppstein, D.: Faster construction of planar two-centers. In: Proceedings of the 8th Annual ACM-SIAM Symposium on Discrete Algorithms (SODA), pp. 131–138 (1997)
11. Hershberger, J.: A faster algorithm for the two-center decision problem. Inf. Process. Lett. **47**(1), 23–29 (1993)
12. Huang, C.H.: Some problems on radius-weighted model of packet radio network. Doctoral dissertation, Ph.D. Dissertation, Department of Computer Science, Tsing Hua University, Hsinchu, Taiwan (1992)
13. Huang, P.H., Tsai, Y.T., Tang, C.Y.: A near-quadratic algorithm for the alpha-connected two-center problem. J. Inf. Sci. Eng. **22**(6), 1317 (2006)

14. Huang, P.H., Te Tsai, Y., Tang, C.Y.: A fast algorithm for the alpha-connected two-center decision problem. Inf. Process. Lett. **85**(4), 205–10 (2003)
15. Katz, M.J., Sharir, M.: An expander-based approach to geometric optimization. SIAM J. Comput. **26**(5), 1384–1408 (1997)
16. Gowda, I., Kirkpatrick, D., Lee, D., Naamad, A.: Dynamic voronoi diagrams. IEEE Trans. Inf. Theory **29**(5), 724–731 (1983)
17. Megiddo, N.: Linear-time algorithms for linear programming in $\mathbb{R}^3$ and related problems. SIAM J. Comput. **12**(4), 759–776 (1983)
18. Megiddo, N.: Applying parallel computation algorithms in the design of serial algorithms. J. ACM (JACM) **30**(4), 852–865 (1983)
19. Bhattacharya, B., Mozafari, A., Shermer, T.C.: An efficient algorithm for the proximity connected two center problem. arXiv preprint arXiv:2204.08754, 19 Apr 2022
20. Sharir, M.: A near-linear algorithm for the planar 2-center problem. Discret. Comput. Geom. **18**(2), 125–34 (1997)
21. Toth, C.D., O'Rourke, J., Goodman, J.E. (eds.): Handbook of discrete and computational geometry. CRC Press, Boca Raton (2017)
22. Wang, H.: On the planar two-center problem and intersection hulls. arXiv preprint arXiv:2002.07945, 9 February 2020

A New Temporal Interpretation
of Cluster Editing

Cristiano Bocci[1] , Chiara Capresi[1] , Kitty Meeks[2] ,
and John Sylvester[2]([⊠])

[1] Dipartimento di Ingegneria dell'Informazione e Scienze Matematiche,
Università degli Studi di Siena, Siena, Italy
cristiano.bocci@unisi.it, capresi3@student.unisi.it
[2] School of Computing Science, University of Glasgow, Glasgow, UK
{kitty.meeks,john.sylvester}@glasgow.ac.uk

Abstract. The NP-complete graph problem CLUSTER EDITING seeks
to transform a static graph into disjoint union of cliques by making the
fewest possible edits to the edge set. We introduce a natural interpretation of this problem in the setting of temporal graphs, whose edge-sets
are subject to discrete changes over time, which we call EDITING TO
TEMPORAL CLIQUES. This problem is NP-complete even when restricted
to temporal graphs whose underlying graph is a path, but we obtain two
polynomial-time algorithms for special cases with further restrictions.
In the static setting, it is well-known that a graph is a disjoint union
of cliques if and only if it contains no induced copy of P_3; we demonstrate that there can be no universal characterisation of cluster temporal
graphs in terms of subsets of at most four vertices. However, subject to a
minor additional restriction, we obtain a characterisation involving forbidden configurations on five vertices. This characterisation gives rise to
an FPT algorithm parameterised simultaneously by the permitted number of modifications and the lifetime of the temporal graph, which uses
a simple search-tree strategy.

Keywords: Temporal graphs · Cluster editing · Graph clustering ·
Parameterised complexity

1 Introduction

The CLUSTER EDITING problem encapsulates one of the simplest and best-
studied notions of graph clustering: given a graph G, the goal is to decide whether
it is possible to transform G into a disjoint union of cliques – a *cluster graph* – by
adding or deleting a total of at most k edges. While this problem is known to be
NP-complete in general [2,11,15,20], it has been investigated extensively through
the framework of parameterised complexity, and admits efficient parameterised
algorithms with respect to several natural parameters [1,3,6–8,12,16] (for more
details see Sect. 1.1).

© Springer Nature Switzerland AG 2022
C. Bazgan and H. Fernau (Eds.): IWOCA 2022, LNCS 13270, pp. 214–227, 2022.
https://doi.org/10.1007/978-3-031-06678-8_16

Motivated by the fact that many real-world networks of interest are subject to discrete changes over time, there has been much research in recent years into the complexity of graph problems on *temporal graphs*, which provide a natural model for networks exhibiting these kinds of changes in their edge-sets. A first attempt to generalise CLUSTER EDITING to the temporal setting was made by Chen, Molter, Sorge and Suchý [9], who recently introduced the problem TEMPORAL CLUSTER EDITING: here the goal is to ensure that graph appearing at each timestep is a cluster graph, subject to restrictions on both the number of modifications that can be made at each timestep and the differences between the cluster graphs created at consecutive timesteps. A dynamic version of the problem, DYNAMIC CLUSTER EDITING, has also recently been studied by Luo, Molter, Nichterlein and Niedermeier [17]: here we are given a solution to a particular instance, together with a second instance (that which will be encountered at the next timestep) and are asked to find a solution for the second instance that does not differ too much from the first. One drawback of previous approaches is that they require each snapshot to be a cluster graph. In the static case, the notion of cluster graph is far too rigid for any meaningful application to community detection [22], as it is unreasonable that all pairs in a community are linked by an edge. For temporal graphs this assumption is even more restrictive.

We take a different approach, using a notion of temporal clique that already exists in the literature [13, 21]. Under this notion, a temporal clique is specified by a vertex-set and a time-interval, and we require that each pair of vertices is connected by an edge frequently, but not necessarily continuously, during the time-interval. An example could be emails within a company, where vertices are employees and there is an edge at time t between two employees if they are senders/recipients of an email at time t. Pairs of employees may correspond more or less frequently, however each pair is included in regular company-wide circular emails. We say that a temporal graph is a *cluster temporal graph* if it is a union of temporal cliques that are pairwise *independent*: here we say that two temporal cliques are independent if either their vertex sets are disjoint, or their time intervals are sufficiently far apart (similar to the notion of independence used to define temporal matchings [19]). Equipped with these definitions, we introduce a new temporal interpretation of cluster editing, which we call EDITING TO TEMPORAL CLIQUES (ETC): the goal is to add/delete a collection of at most k edge appearances so that the resulting graph is a cluster temporal graph.

We prove that ETC is NP-hard, even when the underlying graph is a path; this reduction, however, relies on edges appearing at many distinct timesteps, and we show that, when restricted to paths, ETC is solvable in polynomial time when the maximum number of timesteps at which any one edge appears in the graph is bounded. It follows immediately from our hardness reduction that the variant of the problem in which we are only allowed to delete, but not add, edge appearances, is also NP-hard in the same setting. On the other hand, the corresponding variant in which we can only add edges, which we call COMPLETION TO TEMPORAL CLIQUES (CTC); admits a polynomial-time algorithm on arbitrary inputs.

In the static setting, a key observation – which gives rise to a simple FPT search-tree algorithm for CLUSTER EDITING parameterised by the number of modifications – is the fact that a graph is a cluster graph if and only if it contains no induced copy of the three-vertex path P_3 (sometimes referred to as a *conflict triple* [5]). We demonstrate that cluster temporal graphs cannot be fully characterised by any local condition that involves only sets of at most four vertices; however, in the most significant technical contribution of this paper, we prove that (subject to a minor additional restriction on the relationship between the spacing parameters that define temporal cliques and independence) a temporal graph is a cluster temporal graph if and only if every subset of at most five vertices induces a cluster temporal graph. Using this characterisation, we obtain an FPT algorithm for ETC parameterised simultaneously by the number of modifications and the lifetime (# of timesteps) of the input temporal graph.

1.1   Related Work

CLUSTER EDITING is known to be NP-complete [2,11,15,20], even for graphs with maximum degree six and when at most four edge modifications incident to each vertex are allowed [14]. On the positive side, the problem can be solved in polynomial time if the input graph has maximum degree two [6] (recently improved to degree three [3]) or is a unit interval graph [18]. Further complexity results and heuristic approaches are discussed in a survey article [5].

Variations of the problem in which only deletions or additions of edges respectively are allowed have also been studied. The version in which edges can only be added is trivially solvable in polynomial time, since an edge must be added between vertices u and v if and only if u and v belong to the same connected component of the input graph but are not already adjacent. The deletion version, on the other hand, remains NP-complete even on 4-regular graphs, but is solvable in polynomial time on graphs with maximum degree three [14].

CLUSTER EDITING has received substantial attention from the parameterised complexity community, with many results focusing on the natural parameterisation by the number k of permitted modifications. Fixed-parameter tractability with respect to this parameter can easily be deduced from the fact that a graph is a cluster graph if and only if it contains no induced copy of the three-vertex path P_3, via a search tree argument; this approach has been refined repeatedly in non-trivial ways, culminating in an algorithm running in time $\mathcal{O}(1.76^k + m + n)$ for graphs with n vertices and m edges [6]. More recent work has considered as a parameter the number of modifications permitted above the lower bound implied by the number of modification-disjoint copies of P_3 (copies of P_3 such that no two share either an edge or a non-edge) [16]. Other parameters that have been considered include the number of clusters [12] and a lower bound on the permitted size of each cluster [1].

1.2 Organisation of the Paper

We begin in Sect. 2 by introducing some notation and definitions, and formally defining the ETC problem. In Sect. 3 we collect several results and fundamental lemmas which are either used in several other sections or may be of independent interest. In Sect. 4 we restrict to temporal graphs which have a path as the underlying graph: in Sect. 4.1 we show that ETC is NP-hard even in this setting, however in Sect. 4.2 we then show that if we further restrict temporal graphs on paths to only have a bounded number of appearances of each edge then ETC is solvable in polynomial time. In Sect. 5 we consider a variant of the ETC problem where edges can only be added, and show that this can be solved in polynomial time on any temporal graph. Finally in Sect. 6 we present our main result which gives a characterisation of cluster temporal graphs by induced temporal subgraphs on five vertices. We prove this result in Sect. 6.1 before applying it in Sect. 6.2 to show that (subject to minor additional restrictions) ETC is in FPT when parameterised by the lifetime of the temporal graph and number of permitted modifications. Due to space constraints, many proofs are omitted but can be found in the full version of the paper [4].

2 Preliminaries

In this section we first give basic definitions and introduce some new notions that are key to the paper, before formally specifying the ETC problem.

2.1 Notation and Definitions

Elementary Definitions. Let $\mathbb{N}$ denote the natural numbers (with zero) and $\mathbb{Z}^+$ denote the positive integers. We refer to a set of consecutive natural numbers $[i, j] = \{i, i+1, \ldots, j\}$ for some $i, j \in \mathbb{N}$ with $i \leq j$ as an *interval*, and to the number $j - i + 1$ as the *length* of the interval. Given an undirected (static) graph $G = (V, E)$ we denote its vertex-set by $V = V(G)$ and edge-set by $E = E(G) \subseteq \binom{V}{2}$. We work in the word RAM model of computation, so that arithmetic operations on integers represented using a number of bits logarithmic in the total input size can be carried out in time $\mathcal{O}(1)$. We use standard notions from parameterised complexity, following the notation of [10].

Temporal Graphs. A *temporal graph* $\mathcal{G} = (G, \mathcal{T})$ is a pair consisting of a static (undirected) underlying graph $G = (V, E)$ and a labeling function $\mathcal{T} : E \to 2^{\mathbb{Z}^+} \setminus \{\emptyset\}$. For a static edge $e \in E$, we think of $\mathcal{T}(e)$ as the set of time appearances of e in $\mathcal{G}$ and let $\mathcal{E}(\mathcal{G}) := \{(e, t) \mid e \in E \text{ and } t \in \mathcal{T}(e)\}$ denote the set of edge appearances, or *time-edges*, in a temporal graph $\mathcal{G}$. We consider temporal graphs $\mathcal{G}$ with *finite lifetime* given by $T(\mathcal{G}) := \max\{t \in \mathcal{T}(e) \mid e \in E\}$, that is, there is a maximum label assigned by $\mathcal{T}$ to an edge of G. We assume w.l.o.g. that $\min\{t \in \mathcal{T}(e) \mid e \in E\} = 1$. We denote the lifetime of $\mathcal{G}$ by T when $\mathcal{G}$ is clear from the context. The *snapshot* of $\mathcal{G}$ *at time* t is the static graph on V with edge set $\{e \in E \mid t \in \mathcal{T}(e)\}$. Given temporal graphs $\mathcal{G}_1$ and $\mathcal{G}_2$, let $\mathcal{G}_1 \triangle \mathcal{G}_2$ be the set of time-edges appearing in exactly one of $\mathcal{G}_1$ or $\mathcal{G}_2$. For the purposes of computation, we assume that $\mathcal{G}$ is given as a list of (static) edges together with the list of times $\mathcal{T}(e)$ at which each static edge appears, so the size of $\mathcal{G}$ is $\mathcal{O}(\max\{|\mathcal{E}|, |V|\}) = \mathcal{O}(|V|^2 T)$.

Templates and Cliques. For an edge $e \in E(G)$ in the underlying graph of a temporal graph $\mathcal{G} = (G, \mathcal{T})$, an interval $[a, b]$, and $\Delta_1 \in \mathbb{Z}^+$, we say that e is Δ_1-*dense* in $[a, b]$ if for all $\tau \in [a, \max\{a, b - \Delta_1 + 1\}]$ there exists a $t \in \mathcal{T}(e)$ with $t \in [\tau, \tau + \Delta_1 - 1]$. This formalises the idea of two vertices being connected 'frequently, but not continuously' from the introduction. We define a *template* to be a pair $C = (X, [a, b])$ where X is a set of vertices and $[a, b]$ is an interval. For a set S of time-edges we let $V(S)$ denote the set of all endpoints of time-edges in S, and the *lifetime* $L(S) = [s, t]$, where $s = \min\{s : (e, s) \in S\}$ and $t = \max\{t : (e, t) \in S\}$. We say that S *generates* the template $(V(S), L(S))$. A set $S \subset \mathcal{E}(\mathcal{G})$ forms a Δ_1-*temporal clique* if for every pair $x, y \in V(S)$ of vertices in the template $(V(S), L(S))$ generated by S, the edge xy is Δ_1-dense in $L(S)$. We can assume that the lifetime of any template generated by a set S is minimal, that is, a time-edge from S occurs at each end-point of $L(S)$.

Independence and Cluster Temporal Graphs. For $\Delta_2 \in \mathbb{Z}^+$ we say that two templates $(X, [a, b])$ and $(Y, [c, d])$ are Δ_2-*independent* if

$$X \cap Y = \emptyset \quad \text{or} \quad \min_{s \in [a,b], t \in [c,d]} |s - t| \geq \Delta_2.$$

Thus, two templates are independent if they share no vertices, or their time intervals are at least Δ_2 time steps apart. Let $\mathfrak{T}(\mathcal{G}, \Delta_2)$ be the class of all collections of pairwise Δ_2-independent templates where each $(X, [a, b]) \in \mathfrak{T}(\mathcal{G}, \Delta_2)$ satisfies $X \subseteq V(G)$ and $1 \leq a \leq b \leq T(\mathcal{G})$. Two sets S_1, S_2 of time-edges are Δ_2-*independent* if the templates they generate are Δ_2-independent. As a special case of this, two time-edges (e, t), (e', t') are Δ_2-*independent* if $e \cap e' = \emptyset$ or $|t - t'| \geq \Delta_2$. A temporal graph $\mathcal{G}$ *realises a collection* $\{(X_i, [a_i, b_i])\}_{i \in [k]} \in \mathfrak{T}(\mathcal{G}, \Delta_2)$ of pairwise Δ_2-independent templates if

- for each $(xy, t) \in \mathcal{E}$ there exists $i \in [k]$ such that $x, y \in X_i$ and $t \in [a_i, b_i]$,
- for each $i \in [k]$ and $x, y \in X_i$, the edge xy is Δ_1-dense in $[a_i, b_i]$.

The first condition specifies that every time-edge of $\mathcal{G}$ is contained in a single template. The second states that for any template, and any pair of vertices in vertex set of the template, there is a time edge between the vertices contained in any time window of length Δ_1 contained in the lifetime of the template.

If there exists some $C \in \mathfrak{T}(\mathcal{G}, \Delta_2)$ such that $\mathcal{G}$ realises C then we call $\mathcal{G}$ a (Δ_1, Δ_2)-*cluster temporal graph*. Throughout we assume that $\Delta_2 > \Delta_1$, since if $\Delta_2 \leq \Delta_1$ then one Δ_1-temporal clique can realise many different sets of Δ_2-independent templates. For example, if $\Delta_2 = \Delta_1$ then the two time-edges (e, t) and $(e, t + \Delta_1)$ are Δ_2-independent but also e is Δ_1-dense in the interval $[t, t + \Delta_1]$.

Induced, Indivisible, and Saturated Sets. Let S be a set of time edges and A be a set of vertices, then we let $S[A] = \{(xy, t) \in S : x, y \in A\}$ be the set of all the time-edges in S *induced* by A. Similarly, given a temporal graph $\mathcal{G}$ and $A \subset V$, we let $\mathcal{G}[A]$ be the temporal graph with vertex set A and temporal edges $\mathcal{E}[A]$. For an interval $[a, b]$ we let $\mathcal{G}|_{[a,b]}$ be the temporal graph on V with the set of time-edges $\{(e, t) \in \mathcal{E}(\mathcal{G}) : t \in [a, b]\}$. We will say that a set S of time-edges is Δ_2-*indivisible* if there does not exist a pairwise Δ_2-independent collection $\{S_1, \ldots, S_k\}$ of time-edge sets satisfying $\cup_{i \in [k]} S_i = S$. A Δ_2-indivisible set S is said to be Δ_2-*saturated in* $\mathcal{G}$ if after including any other time-edge of $\mathcal{E}(\mathcal{G})$ it would cease to be Δ_2-indivisible.

2.2 Problem Specification

Editing to Temporal Cliques. We can now introduce the ETC problem which, given as input a temporal graph $\mathcal{G}$ and natural numbers $k, \Delta_1, \Delta_2 \in \mathbb{Z}^+$, asks whether it is possible to transform $\mathcal{G}$ into a (Δ_1, Δ_2)-cluster temporal graph by applying at most k modifications (addition or deletion) of time-edges. Given any temporal graph $\mathcal{G}$, the set Π of time-edges which are added to or deleted from $\mathcal{G}$ is called the *modification set*. We note that the modification set Π can be defined as the symmetric difference between the time-edge set $\mathcal{E}(\mathcal{G})$ of the input graph and that of the same graph after the modifications have been applied. More formally, our problem can be formulated as follows.

EDITING TO TEMPORAL CLIQUES (ETC):
Input: A temporal graph $\mathcal{G} = (G, \mathcal{T})$ and positive integers $k, \Delta_1, \Delta_2 \in \mathbb{Z}^+$.
Question: Does there exist a set Π of time-edge modifications, of cardinality at most k, such that the modified temporal graph is a (Δ_1, Δ_2)-cluster temporal graph?

We begin with a simple observation about the hardness of ETC which shows we can only hope to gain tractability in settings where the static version is tractable. However, we shall see in Sect. 4.1 that ETC is hard on temporal graphs with paths as their underlying graphs, and thus the converse is false.

Proposition 1. *Let $\mathcal{C}$ be a class of graphs on which* CLUSTER EDITING *is* NP-*complete. Then* ETC *is* NP-*complete on the class of temporal graphs* $\{(G, \mathcal{T}) : G \in \mathcal{C}\}$.

3   Basic Observations on ETC

In this section we collect many fundamental results on the structure of temporal graphs and the cluster editing problem. We will use many of these results frequently throughout the proofs of results in this paper; we include all lemma statements here as they provide some insight into the behaviour of (Δ_1, Δ_2)-cluster temporal graphs and may be of use in the further study of cluster editing in the temporal setting.

Lemma 1 shows that there is a way to uniquely partition any temporal graph.

Lemma 1. *For any $\Delta_2 \in \mathbb{Z}^+$, any temporal graph $\mathcal{G}$ has a unique decomposition of its time-edges into Δ_2-saturated subsets.*

The next three elementary lemmas are useful for relating indivisible sets to Δ_1-temporal cliques to clusters in the proof of the characterisation, Theorem 4.

Lemma 2. *If two Δ_2-indivisible sets S_1 and S_2 of time-edges satisfy $S_1 \cap S_2 \neq \emptyset$, then $S_1 \cup S_2$ is Δ_2-indivisible.*

Lemma 3. *Let $\mathcal{G}$ be a temporal graph, $S \subseteq \mathcal{E}(\mathcal{G})$ be a Δ_2-saturated set of time-edges, and $\mathcal{K}$ a Δ_1-temporal clique such that $\mathcal{K} \subseteq \mathcal{E}(\mathcal{G})$ and $\mathcal{K} \cap S \neq \emptyset$. Then $\mathcal{K} \subseteq S$.*

Lemma 4. *Let $\mathcal{G}$ be any (Δ_1, Δ_2)-cluster temporal graph. Then, any Δ_2-indivisible set $S \subseteq \mathcal{E}(\mathcal{G})$ must be contained within a single Δ_1-temporal clique.*

However, the first application of these lemmas is the following result, which shows that the partition from Lemma 2 can be found in polynomial time.

Lemma 5. *Let $\mathcal{G} = (G = (V, E), \mathcal{T})$ be a temporal graph, and let $\mathcal{E} = \{(e, t) : e \in E, t \in \mathcal{T}(e)\}$ be the set of time-edges of $\mathcal{G}$. Then, there is an algorithm which finds the unique partition of $\mathcal{E}$ into Δ_2-saturated subsets in time $\mathcal{O}(|\mathcal{E}|^3 |V|)$.*

Since any temporal graph has a unique decomposition into Δ_2-saturated sets by Lemma 1, and using the fact that any pair of Δ_2-saturated sets is Δ_2-independent by definition, we obtain the following corollary to Lemma 4.

Lemma 6. *A temporal graph $\mathcal{G}$ is a (Δ_1, Δ_2)-cluster temporal graph if and only if every Δ_2-saturated set of time-edges forms a Δ_1-temporal clique.*

Lemmas 5 and 6 allow us to deduce the following result.

Lemma 7. *Let $\mathcal{G} = (G = (V, E), \tau)$ be a temporal graph, and let $\mathcal{E} = \{(e, t) : e \in E, t \in \mathcal{T}(e)\}$ be the set of time-edges of $\mathcal{G}$. Then, we can determine in time $\mathcal{O}(|\mathcal{E}|^3 |V|)$ whether $\mathcal{G}$ is a (Δ_1, Δ_2)-cluster temporal graph.*

The next three Lemmas concern induced structures in cluster temporal graphs.

Lemma 8. *Let $\mathcal{G}$ be a (Δ_1, Δ_2)-cluster temporal graph and $S \subseteq V(\mathcal{G})$. Then, $\mathcal{G}[S]$ is also a (Δ_1, Δ_2)-cluster temporal graph.*

Lemma 9. *Let $\mathcal{G} = (G, \mathcal{T})$ be a temporal graph, and $\mathcal{C} \in \mathfrak{T}(\mathcal{G}, \Delta_2)$ be a collection minimising $\min_{\mathcal{G}_C \text{ realises } \mathcal{C}} |\mathcal{G} \bigtriangleup \mathcal{G}_C|$. Then, for any template $C = (X, [a, b]) \in \mathcal{C}$, the static underlying graph of $\mathcal{G}[X]|_{[a,b]}$ is connected.*

Lemma 10. *Let $\mathcal{G}$ be a temporal graph. Then, there exists a (Δ_1, Δ_2)-cluster temporal graph $\mathcal{G}'$, minimising the edit distance $|\mathcal{G} \bigtriangleup \mathcal{G}'|$ between $\mathcal{G}$ and $\mathcal{G}'$, such that the lifetime of $\mathcal{G}'$ is a subset of the lifetime of $\mathcal{G}$.*

4   ETC on Paths

Throughout P_n will denote the path on $V(P_n) = \{v_1, \ldots, v_n\}$ with $E(P_n) = \{v_i v_{i+1} : 1 \leq i < n\}$. Define $\mathfrak{F}_n$ be the set of all temporal graphs $\mathcal{P}_n = (P_n, \mathcal{T})$ on n vertices which have the path P_n as the underlying static graph.

4.1   NP-Completeness

Clearly, ETC is in NP because, for any input instance $(\mathcal{G}, \Delta_1, \Delta_2, k)$, a non-deterministic algorithm can guess (if one exists) the modification set Π and, using Lemma 7, verify that the modified temporal graph is a (Δ_1, Δ_2)-cluster temporal graph in time polynomial in k and the size of $\mathcal{G}$. We show that ETC is NP-hard even for temporal graphs with a path as underlying graph.

Theorem 1. ETC *is* NP-*complete, even if the underlying graph G of the input temporal graph $\mathcal{G}$ is a path.*

To prove this result we construct a reduction to ETC from the NP-complete problem TEMPORAL MATCHING. For a fixed $\Delta \in \mathbb{Z}^+$, a Δ-*temporal matching* $\mathcal{M}$ of a temporal graph $\mathcal{G}$ is a set of time-edges of $\mathcal{G}$ which are pairwise Δ-independent. It is easy to note that if $\mathcal{G} = \mathcal{M}$, then $\mathcal{G}$ is a (Δ_1, Δ)-cluster temporal graph for any value of $\Delta_1 \geq 1$, because then each time-edge in $\mathcal{G}$ can be considered as a Δ_1-temporal clique with unit time interval, and these cliques are by definition Δ-independent. We can now state this problem formally.

TEMPORAL MATCHING (TM):
Input: A temporal graph $\mathcal{G} = (G, \mathcal{T})$ and two positive integers $k, \Delta \in \mathbb{Z}^+$.
Question: Does $\mathcal{G}$ admit a Δ-temporal matching $\mathcal{M}$ of size k?

It was shown in [19] that TEMPORAL MATCHING is NP-complete even if $\Delta = 2$ and the underlying graph G is a path. The reduction fixes $\Delta_1 = 1$ and $\Delta_2 = 5$. It then takes our input temporal graph $\mathcal{P}_n$ and transforms it into an new instance $\mathcal{P}'_n$ by adding empty "filler" snapshots between each snapshot $\mathcal{P}_n$, see Fig. 1. It is shown that a matching in the original instance corresponds to a $(1,5)$-cluster temporal graph, which gives one direction of the reduction. We then show that, if enough filler snapshots are added, then there exists an optimal solution to ETC where time-edges are only deleted from $\mathcal{P}'_n$. We can then deduce from this that, since the underlying graph is a path, a solution to ETC using only deletions must be a matching of the required size.

4.2   Bounding the Number of Edge Appearances

We now show that, if additionally the number of appearances of each edge in $\mathcal{P}_n$ is bounded by a fixed constant, then ETC can be solved in time polynomial in the size of the input temporal graph.

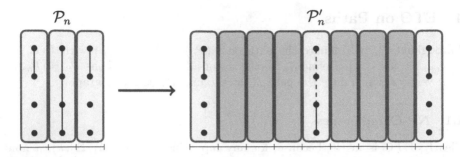

Fig. 1. An instance $\mathcal{P}$ of TM is shown on the left and the stretched graph $\mathcal{P}'_n$ on which we solve EDITING TO TEMPORAL CLIQUES is on the right. Non-filler snapshots are shown in white and filler snapshots are grey. Dotted edges show edges that were removed to leave a $(1,5)$-temporal cluster graph (which is also a 5-temporal matching, and corresponds to a 2-temporal matching in $\mathcal{P}$).

Theorem 2. *Let $(\mathcal{P}_n, \Delta_1, \Delta_2, k)$ be any instance of* ETC *where $\mathcal{P}_n \in \mathfrak{F}_n$ and there exists a natural number σ such that $|\mathcal{T}(e)| \leq \sigma$ for any $e \in E(P_n)$. Then,* ETC *on $(\mathcal{P}_n, \Delta_1, \Delta_2, k)$ is solvable in time $\mathcal{O}(T^{4\sigma}\sigma^2 \cdot n^{2\sigma+1})$.*

This theorem is proved using a fairly standard dynamic programming approach, where we go along the underlying path P_n uncovering one vertex in each step. In particular, at the ith vertex we try to extend the current set of templates on the first $i - 1$ vertices of the path to an optimal set of templates also including the ith vertex.

5   Completion to Temporal Cluster Graphs

In this section we consider the following variant of ETC, in which we are only allowed to add time-edges.

COMPLETION TO TEMPORAL CLIQUES (CTC):
Input: A temporal graph $\mathcal{G} = (G, \mathcal{T})$ and positive integers $k, \Delta_1, \Delta_2 \in \mathbb{Z}^+$.
Question: Does there exist a set Π of time-edge additions, of cardinality at most k, such that the modified temporal graph is a (Δ_1, Δ_2)-cluster temporal graph?

The main result of this section is to show that the above problem can be solved in time polynomial in the size of the input temporal graph.

Theorem 3. *There is an algorithm solving* COMPLETION TO TEMPORAL CLIQUES *on any temporal graph $\mathcal{G}$ in time $\mathcal{O}(|\mathcal{E}(\mathcal{G})|^3|V|)$.*

As observed in [20], the cluster completion problem is also solvable in polynomial time on static graphs. In this case the optimum solution is obtained by transforming each connected component of the input graph into a complete graph. The situation is not quite so simple in temporal graphs, however a similar phenomenon holds with Δ_2-saturated sets taking the place of connected components; our algorithm relies heavily on the fact (Lemma 5) that we can find these Δ_2-saturated sets efficiently.

6   A Local Characterisation of Cluster Temporal Graphs

In Sect. 6.1 we give a characterisation of cluster temporal graphs. We then use this characterisation in Sect. 6.2 to give an FPT algorithm for ETC.

6.1   The Five-Vertex Characterisation

In this section we show the following characterization of (Δ_1, Δ_2)-cluster temporal graphs in terms of their induced five-vertex subgraphs. The characterisation relies on a fairly natural additional condition which says clusters cannot appear too close to each other in time. We discuss the potential to improve this characterisation in more detail in Sect. 7.

Theorem 4. *Let $\Delta_2 > 2\Delta_1$. Then any temporal graph $\mathcal{G}$ is a (Δ_1, Δ_2)-cluster temporal graph if and only if $\mathcal{G}[S]$ is a (Δ_1, Δ_2)-cluster temporal graph for every set $S \subseteq V(\mathcal{G})$ of at most five vertices.*

One direction of Theorem 4 follows easily from Lemma 8. The other direction is far more challenging. The following lemma illustrates a key idea in the proof of this more challenging direction of Theorem 4: the five vertex condition allows us to 'grow' certain sets of time-edges.

Lemma 11. *Let $\mathcal{G}$ be any temporal graph satisfying the property that $\mathcal{G}[S]$ is a (Δ_1, Δ_2)-cluster temporal graph for every set $S \subseteq V(\mathcal{G})$ of at most five vertices. Let $\mathcal{H}$ be a Δ_1-temporal clique realising the template $(H, [c, d])$, and $x, y \in H$. Suppose that xy is Δ_1-dense in the set $[a, b] \supseteq [c, d]$ and let $r_1 = \min(\mathcal{T}(xy) \cap [a, b])$ and $r_2 = \max(\mathcal{T}(xy) \cap [a, b])$. Then there exists a Δ_1-temporal clique $\mathcal{H}'$ which realises the template $(H, [r_1, r_2])$ where $[r_1, r_2] \supseteq [a + \Delta_1 - 1, b - \Delta_1 + 1]$.*

We are now ready to prove the final direction of Theorem 4; full details of this proof, including proofs of the claims, can be found in the full version [4].

Lemma 12. *Let $\mathcal{G}$ be any temporal graph such that $\mathcal{G}[S]$ is a (Δ_1, Δ_2)-cluster temporal graph for every set $S \subseteq V(\mathcal{G})$ of at most five vertices. Then $\mathcal{G}$ is a (Δ_1, Δ_2)-cluster temporal graph.*

Proof. Let $\mathcal{P}_{\mathcal{G}}$ be the partition of $\mathcal{E}(\mathcal{G})$ into Δ_2-saturated subsets; we know that this partition exists and is unique by Lemma 1. Fix any subset $S \in \mathcal{P}_{\mathcal{G}}$ and denote $L(S) = [s, t]$. We want to show that S forms a Δ_1-temporal clique in $\mathcal{G}$.

To prove this, we introduce a collection $\varkappa_S = \{S_1, \ldots, S_m\}$ of subsets of S, such that each S_i is a Δ_1-temporal clique for any $i \in \{1, \ldots, m\}$, $S = \bigcup_{i=1}^m S_i$ and for any $S_i \in S$ there does not exist any other Δ_1-temporal clique $K \subseteq S$ such that $S_i \subset K$; we will say that each S_i is a *maximal Δ_1-temporal clique within S*. First of all, we note that this collection exists: in fact, because even the singleton set containing any time-edge is a Δ_1-temporal clique, every time-edge in S belongs to at least one Δ_1-temporal clique. Note that the subsets S_i with $i \in \{1, \ldots, m\}$ are not required to be pairwise disjoint.

We will assume for a contradiction that $m \geq 2$. Because S is Δ_2-saturated, it is not possible that all the Δ_1-temporal cliques contained in $\varkappa_S$ are pairwise Δ_2-independent, since this would imply that S is not Δ_2-indivisible. Thus, as we assume $m \geq 2$, let us consider any distinct $S_i, S_j \in \varkappa_S$ that are not Δ_2-independent. We shall then show that they must both be contained within a larger Δ_1-temporal clique, which itself is contained in S, contradicting maximality. It will then follow that $m = 1$ and thus S is itself a Δ_1-temporal clique, which establishes the theorem.

The next claim shows that if two maximal Δ_1-temporal cliques in S are not Δ_2-independent, then there is a small sub-graph witnessing this non-independence.

Claim 1. *Let $S_i, S_j \in \varkappa_S$ be any pair of Δ_1-temporal cliques which are not Δ_2-independent. Then, there exists a set $W \subseteq V(S_i) \cup V(S_j)$ with $|W| \leq 5$ such that $(S_i \cup S_j)[W]$ is Δ_2-indivisible and contains at least one time-edge from each of S_i and S_j.*

Recall that S_i and S_j are both Δ_2-indivisible as they are Δ_1-temporal cliques. It therefore follows from Claim 1 and Lemma 2 that both $S_i[W] \cup S_j$ and $S_j[W] \cup S_i$ are Δ_2-indivisible. As their intersection is $(S_i \cup S_j)[W] \neq \emptyset$, invoking Lemma 2 once again gives that $S_i \cup S_j$ is Δ_2-indivisible.

Claim 2. *Let S_i and S_j be as above with $L(S_i) = [s_i, t_i]$ and $L(S_j) = [s_j, t_j]$. Then, there exists some $K \subseteq V$ and $s', t' \in \mathbb{Z}^+$ such that $\mathcal{G}$ contains a Δ_1-temporal clique $\mathcal{K}$ realising the template $(K, [s', t'])$ where:*

- $s' \in [s, \min\{s_i, s_j\} + \Delta_1 - 1]$ *and* $t' \in [\max\{t_i, t_j\} - \Delta_1 + 1, t]$,
- *there exist* $x, y \in K$ *and a time* $r_i \in [s', t']$ *such that* $(xy, r_i) \in S_i$, *and*
- *there exist* $w, z \in K$ *and a time* $r_j \in [s', t']$ *such that* $(wz, r_j) \in S_j$.

Let us now consider $\mathcal{K}$, the Δ_1-temporal clique of Claim 2. From this we want to extend S_i and S_j to a Δ_1-temporal clique with vertex set $V(S_i) \cup V(S_j)$ and lifetime at least $L(\mathcal{K}) \cup L(S_i) \cup L(S_j) = [\min\{s_i, s_j, s'\}, \max\{t_i, t_j, t'\}]$. We do this in two stages, via the following claims.

Claim 3. *There exist $h_1 \leq h_2$ satisfying*

$$[h_1, h_2] \supseteq [\min\{s_i, s_j, s'\} + \Delta_1 - 1, \max\{t_i, t_j, t'\} - \Delta_1 + 1]$$

such that $(V(S_i) \cup V(S_j), [h_1, h_2])$ clma Δ_1-temporal clique.

Claim 4. $(V(S_i) \cup V(S_j), [\min\{s_i, s_j, s'\}, \max\{t_i, t_j, t'\}])$ *is a* Δ_1*-temporal clique in* $\mathcal{G}$.

Observe that Claim 4 contradicts the initial assumption that S_i and S_j were maximal in S. Thus the assumption that $m \geq 2$ must be incorrect and thus S consists of a single Δ_1-temporal clique. Because S was a generic set of the partition $\mathcal{P}_{\mathcal{G}}$ of the given temporal graph $\mathcal{G}$ into Δ_2-saturated subsets, then $\mathcal{G}$ must be a (Δ_1, Δ_2)-cluster temporal graph by Lemma 6, giving the result. □

6.2 A Search-Tree Algorithm

Using the characterisation from the previous section, we are now able to prove the following result using a standard bounded search tree technique.

Theorem 5. *Let* $\Delta_2 > 2\Delta_1$. *Then* ETC *can be solved in time* $(10T)^k \cdot T^3 |V|^5$.

7 Conclusions and Open Problems

In this paper we introduced a new temporal variant of the cluster editing problem, ETC, based on a natural interpretation of what it means for a temporal graph to be divisible into "clusters". We showed hardness of this problem even in the presence of strong restrictions on the input, but identified two special cases in which polynomial-time algorithms exist: firstly, if underlying graph is a path and the number of appearances of each edge is bounded by a constant, and secondly if we are only allowed to add (but not delete) time-edges. One natural open question arising from the first of these positive results is whether bounding the number of appearances of each edge can lead to tractability in a wider range of settings: we conjecture that the techniques used here can be generalised to obtain a polynomial-time algorithm when the underlying graph has bounded pathwidth, and it may be that they can be extended even further.

Our main technical contribution was Theorem 4, which gives a characterisation of (Δ_1, Δ_2)-cluster temporal graphs in terms of five vertex subsets, whenever the condition $\Delta_2 > 2\Delta_1$ holds. The assumption that $\Delta_2 > 2\Delta_1$ is needed in two places in the proof of Theorem 4, but we believe that with care it may be possible to modify the proof so that this condition is not required. If it is indeed possible to remove this condition on Δ_1 and Δ_2, then the resulting characterisation would be best possible, as the graph illustrated in Fig. 2 demonstrates that no such characterisation involving only four-vertex subsets can exist.

In addition to providing substantial insight into the structure of (Δ_1, Δ_2)-cluster temporal graphs, Theorem 4 also gives rise to a simple search tree algorithm, which is an FPT algorithm parameterised simultaneously by the number k of permitted modifications and the lifetime of the input temporal graph. An interesting direction for further research would be to investigate whether this result can be strengthened: does there exist a polynomial kernel with respect to this dual parameterisation, and is ETC in FPT parameterised by the number of permitted modifications alone?

226   C. Bocci et al.

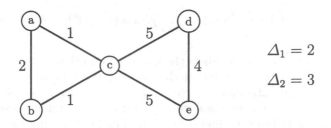

Fig. 2. A temporal graph which is not a $(2,3)$-cluster temporal graph, whose every induced temporal subgraph on at most four vertices is a $(2,3)$-cluster temporal graph.

Acknowledgements. Meeks and Sylvester gratefully acknowledge funding from the Engineering and Physical Sciences Research Council (ESPRC) grant number EP/T004878/1 for this work, while Meeks was also supported by a Royal Society of Edinburgh Personal Research Fellowship (funded by the Scottish Government).

References

1. Abu-Khzam, F.N.: The multi-parameterized cluster editing problem. In: Widmayer, P., Xu, Y., Zhu, B. (eds.) COCOA 2013. LNCS, vol. 8287, pp. 284–294. Springer, Cham (2013). https://doi.org/10.1007/978-3-319-03780-6_25
2. Bansal, N., Blum, A., Chawla, S.: Correlation clustering. Mach. Learn. **56**(1–3), 89–113 (2004)
3. Bartier, V., Bathie, G., Bousquet, N., Heinrich, M., Pierron, T., Prieto, U.: PACE solver description: μsolver - heuristic track. In: Golovach, P.A., Zehavi, M. (eds.) 16th International Symposium on Parameterized and Exact Computation, IPEC 2021, 8–10 September 2021, Lisbon, Portugal, volume 214 of LIPIcs, pp. 33:1–33:3. Schloss Dagstuhl - Leibniz-Zentrum für Informatik (2021)
4. Bocci, C., Capresi, C., Meeks, K., Sylvester, J.: A new temporal interpretation of cluster editing. CoRR, abs/2202.01103 (2022)
5. Böcker, S., Baumbach, J.: Cluster editing. In: Bonizzoni, P., Brattka, V., Löwe, B. (eds.) CiE 2013. LNCS, vol. 7921, pp. 33–44. Springer, Heidelberg (2013). https://doi.org/10.1007/978-3-642-39053-1_5
6. Böcker, S.: A golden ratio parameterized algorithm for cluster editing. J. Discrete Algorithms **16**, 79–89 (2012). Selected papers from the 22nd International Workshop on Combinatorial Algorithms (IWOCA 2011)
7. Cao, Y., Chen, J.: Cluster editing: kernelization based on edge cuts. Algorithmica **64**(1), 152–169 (2012)
8. Chen, J., Meng, J.: A 2k kernel for the cluster editing problem. J. Comput. Syst. Sci. **78**(1), 211–220 (2012)
9. Chen, J., Molter, H., Sorge, M., Suchý, O.: Cluster editing in multi-layer and temporal graphs. In: Hsu, W.-L., Lee, D.-T., Liao, C.-S. (eds.) 29th International Symposium on Algorithms and Computation, ISAAC 2018, 16–19 December 2018, Jiaoxi, Yilan, Taiwan, volume 123 of LIPIcs, pp. 24:1–24:13. Schloss Dagstuhl - Leibniz-Zentrum für Informatik (2018)
10. Cygan, M., et al.: Parameterized Algorithms. Springer, Cham (2015). https://doi.org/10.1007/978-3-319-21275-3

11. Delvaux, S., Horsten, L.: On best transitive approximations to simple graphs. Acta Inform. **40**(9), 637–655 (2004)
12. Fomin, F.V., Kratsch, S., Pilipczuk, M., Pilipczuk, M., Villanger, Y.: Tight bounds for parameterized complexity of cluster editing with a small number of clusters. J. Comput. Syst. Sci. **80**(7), 1430–1447 (2014)
13. Himmel, A.-S., Molter, H., Niedermeier, R., Sorge, M.: Adapting the Bron-Kerbosch algorithm for enumerating maximal cliques in temporal graphs. Soc. Netw. Anal. Min. **7**(1), 35:1–35:16 (2017)
14. Komusiewicz, C., Uhlmann, J.: Cluster editing with locally bounded modifications. Discret. Appl. Math. **160**(15), 2259–2270 (2012)
15. Krivánek, M., Morávek, J.: NP -hard problems in hierarchical-tree clustering. Acta Inform. **23**(3), 311–323 (1986)
16. Li, S., Pilipczuk, M., Sorge, M.: Cluster editing parameterized above modification-disjoint P_3-packings. In: Bläser, M., Monmege, B. (eds.) 38th International Symposium on Theoretical Aspects of Computer Science, STACS 2021, 16–19 March 2021, Saarbrücken, Germany (Virtual Conference), volume 187 of LIPIcs, pp. 49:1–49:16. Schloss Dagstuhl - Leibniz-Zentrum für Informatik (2021)
17. Luo, J., Molter, H., Nichterlein, A., Niedermeier, R.: Parameterized dynamic cluster editing. Algorithmica **83**(1), 1–44 (2021)
18. Mannaa, B.: Cluster editing problem for points on the real line: a polynomial time algorithm. Inf. Process. Lett. **110**(21), 961–965 (2010)
19. Mertzios, G.B., Molter, H., Niedermeier, R., Zamaraev, V., Zschoche, P.: Computing maximum matchings in temporal graphs. In: Paul, C., Bläser, M. (eds.) 37th International Symposium on Theoretical Aspects of Computer Science, STACS 2020, 10–13 March 2020, Montpellier, France, volume 154 of LIPIcs, pp. 27:1–27:14. Schloss Dagstuhl - Leibniz-Zentrum für Informatik (2020)
20. Shamir, R., Sharan, R., Tsur, D.: Cluster graph modification problems. Discret. Appl. Math. **144**(1-2), 173–182 (2004)
21. Viard, T., Latapy, M., Magnien, C.: Computing maximal cliques in link streams. Theor. Comput. Sci. **609**, 245–252 (2016)
22. Yang, Z., Algesheimer, R., Tessone, C.J.: A comparative analysis of community detection algorithms on artificial networks. Sci. Rep. **6**, article no. 30750 (2016)

List Covering of Regular Multigraphs

Jan Bok[1]($\boxtimes$)(iD), Jiří Fiala[2](iD), Nikola Jedličková[2](iD), Jan Kratochvíl[2](iD),
and Paweł Rzążewski[3,4](iD)

[1] Computer Science Institute, Faculty of Mathematics and Physics,
Charles University, Prague, Czech Republic
bok@iuuk.mff.cuni.cz
[2] Department of Applied Mathematics, Faculty of Mathematics and Physics,
Charles University, Prague, Czech Republic
{fiala,jedlickova,honza}@kam.mff.cuni.cz
[3] Warsaw University of Technology, Warsaw, Poland
p.rzazewski@mini.pw.edu.pl
[4] University of Warsaw, Warsaw, Poland

Abstract. A graph covering projection, also known as a locally bijective homomorphism, is a mapping between vertices and edges of two graphs which preserves incidencies and is a local bijection. This notion stems from topological graph theory, but has also found applications in combinatorics and theoretical computer science.

It has been known that for every fixed simple regular graph H of valency greater than 2, deciding if an input graph covers H is NP-complete. In recent years, topological graph theory has developed into heavily relying on multiple edges, loops, and semi-edges, but only partial results on the complexity of covering multigraphs with semi-edges are known so far. In this paper we consider the list version of the problem, called LIST-H-COVER, where the vertices and edges of the input graph come with lists of admissible targets. Our main result reads that the LIST-H-COVER problem is NP-complete for every regular multigraph H of valency greater than 2 which contains at least one semi-simple vertex (i.e., a vertex which is incident with no loops, with no multiple edges and with at most one semi-edge). Using this result we almost show the NP-co/polytime dichotomy for the computational complexity of LIST-H-COVER of cubic multigraphs, leaving just five open cases.

1 Introduction

Graph Covering Projections and Related Notions. For simple graphs G and H, a covering projection from G to H is a mapping $f : V(G) \cup E(G) \to V(H) \cup E(H)$, such that (i) vertices are mapped to vertices and edges are mapped to edges, (ii) incidencies are retained, and (iii) f is bijective in the neighborhood of each vertex. The last condition means that if for some $v \in V(G)$ and $x \in V(H)$ we have $f(v) = x$, then for each edge e of H containing x there must be exactly one edge containing v that is mapped to e. For a fixed graph H, in the H-COVER problem we ask if an instance graph G admits a covering projection to H.

© Springer Nature Switzerland AG 2022
C. Bazgan and H. Fernau (Eds.): IWOCA 2022, LNCS 13270, pp. 228–242, 2022.
https://doi.org/10.1007/978-3-031-06678-8_17

The notion of a graph covering projection, as a natural discretization of the covering projection used in topology, originates (not surprisingly) in topological graph theory. However, since then it has found numerous applications elsewhere. Covering projections were used for constructing highly symmetrical graphs [4, 9,21,24], embedding complete graphs in surfaces of higher genus [37], and for analyzing a model of local computations [2].

Graph covering projections are also known as *locally bijective homomorphisms* and as such they fall into a family of *locally constrained homomorphisms*. Other problems from this family are locally surjective and locally injective graph homomorphisms, where we ask for the existence of a homomorphism that is, respectively, surjective or injective in the neighborhood of each vertex. Locally surjective homomorphisms play an important role in social sciences [19] (there this problem is called the Role Assignment Problem). On the other hand, a prominent special case of the locally injective homomorphism problem is the well-studied $L(2,1)$-labeling problem [23] and, more generally, $H(p,q)$-coloring [14,29].

Computational Complexity. The complexity of finding locally constrained homomorphisms was studied by many authors. For locally surjective homomorphisms we know a complete dichotomy [19]. The problem is polynomial-time solvable if the target graph H either (a) has no edge, or (b) has a component that consists of a single vertex with a loop, or (c) is simple and bipartite, with at least one component isomorphic to K_2. In all other cases the problem is NP-complete.

The dichotomy for locally injective homomorphisms is still unknown, despite some work [11,17]. However, we understand the complexity of the *list* variant of the problem [12]: it is polynomial-time solvable if every component of the target graph has at most one cycle, and NP-complete otherwise.

To the best of our knowledge, Abello et al. [1] were the first to ask about the computational complexity of H-COVER. Note that in order to map a vertex of G to a vertex of H, they must be of the same degree, a natural interesting special case is when H is regular. It is known that for every $k \geq 3$, the H-COVER problem is NP-complete for every simple k-regular graph H [26], [18]. Some other partial results are known, mostly focusing on small graphs H [13,27,28]. Let us point out that in all the above results it was assumed that H has no multiple edges.

Recall further that there is also some more work concerning the complexity of locally surjective and injective homomorphisms if G is assumed to come from some special class [3,5,8,15,36]. We also refer the reader to the survey concerning various aspects of locally constrained homomorphisms [18].

(Multi)graphs with Semi-edges. In the course of development of topological graph theory, it became standard to consider loops and multiedges, but recently also semi-edges are playing a more and more important role. Intuitively, a semi-edge (sometimes also called a *half-edge* or a *fin*) is an edge with just one end (this is in contrast with a loop, which has two ends, both in the same vertex). To name just a few most significant examples, Malnič et al. [32] considered semi-edges during their study of abelian covers to allow for a broader range of

applications. Furthermore, the concept of graphs with semi-edges was introduced independently and naturally in mathematical physics [22]. It is also natural to consider semi-edges in the mentioned framework of local computations (we refer to the introductory section of [6] for more details). Finally, a well-known Leighton's theorem [31] on finite common covers has been recently generalized to the semi-edge setting in [38,39]. To highlight a few other contributions, the reader is invited to consult [33,35], the surveys [30,34], and finally for more recent results, the series of papers [15,16,20] and the introductions therein. From now on, when we talk about graphs, we allow multiple edges, loops and semi-edges without explicitly stating it.

The complexity study of H-COVER for graphs H that allow semi-edges has been initiated only very recently in [6,7]. We continue this line of research. In particular, our far-reaching goal is to prove the following conjecture.

Strong Dichotomy Conjecture. *For every H, the H-COVER problem is either polynomial-time solvable for general graphs, or NP-complete for simple graphs.*

Our Results. The goal of this paper is to push further the understanding of the complexity of H-COVER for regular graphs. Recall that the problem is known to be NP-complete for every fixed k-regular *simple* graph H of valency $k \geq 3$ [26]. Though it was known already from [25] that in order to fully understand the complexity of covering general simple graphs, it is necessary (and sufficient) to prove a complete characterization for colored mixed multigraphs, the result of [26] was formulated and proved only for simple graphs. In this paper we revisit the method developed in [26] and we conclude that though it does not seem to work for multigraphs in general, it is possible to modify it and – under certain assumptions – prove hardness for the *list* variant of the problem, LIST-H-COVER, where the vertices and edges of the instance graph are given lists of admissible targets. Our main result is the following theorem (a vertex is *semi-simple* if it belongs to no loops nor multiple edges, and is incident to at most one semi-edge).

Theorem 1. *Let $k \geq 3$ and let H be a k-regular graph. If H contains a semi-simple vertex, then LIST-H-COVER is NP-complete for simple input graphs.*

We do believe that the Strong Dichotomy Conjecture holds true for LIST-H-COVER.

The second goal of the current paper is to show how Theorem 1 could be used to prove the Strong Dichotomy Conjecture for cubic graphs. Recall that for the closely related locally injective homomorphism problem, introducing lists was helpful in obtaining the full complexity dichotomy [12]. In Theorem 4 we fully characterize the computational complexity of LIST-H-COVER for almost all cubic graphs, and identify just five exceptionally stubborn graphs H for which the complexity of the problem is still open.

2   Preliminaries

In the sequel, a *graph* is allowed to have loops, multiple edges and semi-edges, and all these objects are referred to as *edges*. Edges are thus distinguished to be of three types: *ordinary edges* that are incident with two distinct vertices, *loops* that have two ends, both in the same vertex, and *semi-edges* that have only one end. By saying that we allow *multiedges* we mean that our graph may have more edges with the same set of endpoints (so we may have multiple loops at the same vertex, multiple semi-edges at the same vertex, or multiple ordinary edges incident with the same pair of vertices). Given a graph G and a vertex $u \in V(G)$, the set of edges of G incident with u will be denoted by $E_G(u)$.

The *degree* (or *valency*) of a vertex u is the number of edge endpoints equal to u. In particular, each ordinary edge and each semi-edge contribute 1 to the degree of each of its vertices, and each loop contributes 2. A graph is regular if all of its vertices have the same degree. We further say that:

- a vertex is *semi-simple* if it belongs to no loops, no multiple edges and at most one semi-edge,
- a graph is *semi-simple* if each of its vertices is semi-simple,
- a vertex is *simple* if it is semi-simple and is incident with no semi-edges,
- a graph is *simple* if each of its vertices is simple,
- a graph is *bipartite* if it has no loops, no semi-edges and no odd cycles.

Given graphs G and H, a mapping $f : V(G) \cup E(G) \longrightarrow V(H) \cup E(H)$ is a *graph covering projection* if vertices of G are mapped onto vertices of H, edges of G are mapped onto edges of H so that incidences are retained, and in such a way that the preimage of a loop is a disjoint union of cycles spanning the preimage of the vertex incident with the loop, the preimage of a semi-edge is a disjoint union of semi-edges and ordinary edges spanning the preimage of the vertex incident with this semi-edge, and the preimage of a ordinary edge is a matching spanning the preimage of the two vertices incident with this edge.

The computational problem of deciding whether an input graph G covers a fixed graph H is denoted by H-COVER.

The mapping $f : V(G) \cup E(G) \longrightarrow V(H) \cup E(H)$ is a *partial covering projection* when the preimages are not required to be spanning subgraphs, but all other properties are fulfilled. I.e., the vertex- and edge-mappings are both surjective and the incidences are retained, the preimage of a ordinary edge connecting vertices say u and v is a matching consisting of edges each connecting a vertex from $f^{-1}(u)$ to a vertex from $f^{-1}(v)$, the preimage of a semi-edge incident with vertex say u is a disjoint union of semi-edges and ordinary edges all incident only with vertices from $f^{-1}(u)$, and the preimage of a loop incident with a vertex say u is a disjoint union of cycles (including loops) and paths whose all edges are incident only with vertices from $f^{-1}(u)$.

In the LIST-H-COVER problem the input graph G is given with lists $\mathcal{L} = \{L_u, L_e : u \in V(G), e \in E(G)\}$, such that $L_u \subseteq V(H)$ for every $u \in V(G)$ and $L_e \subseteq E(H)$ for every $e \in E(G)$. A covering projection $f : G \longrightarrow H$ *respects* the lists of $\mathcal{L}$ if $f(u) \in L_u$ for every $u \in V(G)$ and $f(e) \in L_e$ for every $e \in E(G)$.

3   Proof of Theorem 1

In the first two subsections we will prove the theorem for the case when H is bipartite (and hence does not contain loops) and has no semi-edges. By the celebrated König-Hall theorem, such a graph is k-edge-colorable.

3.1   Multicovers

The following construction will be used to build gadgets for the hardness proof.[1]

Proposition 1 (♠). *Let H be a connected k-regular k-edge-colorable graph with no loops or semi-edges. Let x, y be two adjacent vertices of H. Then there exists a connected simple k-regular k-edge-colorable graph G and $u \in V(G)$, such that*

(a) *for any bijection from $E_G(u)$ onto $E_H(x)$, there exists a covering projection from G to H which extends this bijection and maps u to x, and*

(b) *for any bijection from $E_G(u)$ onto $E_H(y)$ there exists a covering projection from G to H which extends this bijection and maps u to y.*

The main building block of our reduction is the graph G_u obtained from G by splitting vertex u into k pendant vertices of degree 1. For each edge e of G incident with u, we formally keep this edge with the same name in G_u, denote its pendant vertex of degree 1 by u_e and denote by w_e the other endpoint of e. (Thus, with this slight abuse of notation, $E_G(u) = \bigcup_{e \in E_G(u)} E_{G_u}(u_e)$.) Then we have the following proposition (Fig. 1).

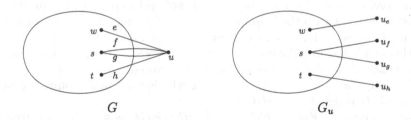

Fig. 1. An illustration to the construction of G_u.

Proposition 2 (♠). *The graph G_u constructed from the multicover G of H as above satisfies the following:*

(a) *for every bijection $\sigma_x \colon E_G(u) \to E_H(x)$, there exists a partial covering projection of G_u onto H that extends σ_x and maps each $u_e, e \in E_G(u)$ to x;*

(b) *for every bijection $\sigma_y \colon E_G(u) \to E_H(y)$, there exists a partial covering projection of G_u onto H that extends σ_y*

---

[1] The proofs of statements marked with (♠) will appear in the journal version of the paper.

(c) in every partial covering projection from G_u onto H, the pendant vertices $u_e, e \in E_G(u)$ are mapped onto the same vertex of H;

(d) in every partial covering projection from G_u onto H, the pendant edges are mapped onto different edges (incident with the image of the pendant vertices).

3.2 Reduction from Hypergraph Coloring

The reduction is exactly the same as in [26], but the proof for the case when multiple edges are allowed needs some extra analysis. Hence we need to describe the reduction in full detail in here. We reduce from k-edge-colorability of $(k-1)$-uniform k-regular hypergraphs. In the wording of the incidence graph of the hypergraph, suppose we are given a simple bi-regular bipartite graph $K = (A \cup B, E)$ such that all vertices in A (which represent the edges of the hypergraph) have degree $k - 1$ and all vertices in B (which represent the vertices of the hypergraph) have degree k. The question is if the vertices of A can be colored by k colors so that the neighborhood of each vertex from B is rainbow colored (i.e., each vertex from B sees all k colors on its neighbors, each color exactly once). This problem is NP-complete for every fixed $k \geq 3$ [26] (Fig. 2).

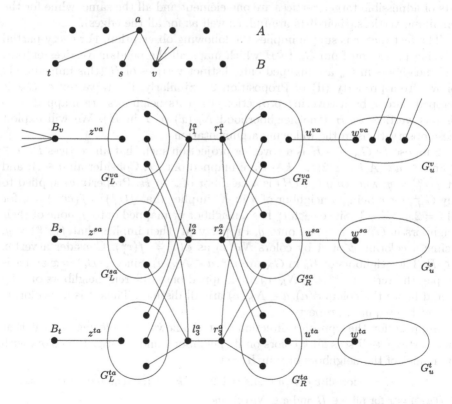

Fig. 2. An illustration to the construction of G_K for $k = 4$.

Given such a graph K, we build an input graph G_K by local replacements. Recall that we are working with a k-edge-colorable k-regular graph H with a simple vertex x, and with respect to this vertex (and one of its neighbors y) we are guaranteed the existence of a graph G_u which satisfies the properties stated in Proposition 2. This G_u will be a key building block in our construction.

First, every vertex $v \in B$ will be replaced by a copy of the so-called vertex gadget, which is a disjoint union of a copy G_u^v of G_u and a single vertex B_v. For every neighbor $a \in A$ of v, one of the pendant vertices of G_u^v will be denoted by u^{va}, and its neighbor within G_u^v will be denoted by w^{va}.

The hyperedge gadgets used to replace the vertices of A are more complicated. This gadget consists of $2(k-1)$ copies of G_u linked together in the following way. Let $a \in A$. We take $2(k-1)$ vertices $\ell_i^a, r_i^a, i = 1, 2, \ldots, k-1$, and for every neighbor v of a, we take two copies G_L^{va} and G_R^{va} of G_u. The pendant vertices of G_L^{va} will be unified with B_v and $\ell_1^a, \ell_2^a, \ldots, \ell_{k-1}^a$, while the pendant vertices of G_R^{va} will be unified with w^{va} and $r_1^a, r_2^a, \ldots, r_{k-1}^a$. The neighbor of B_v in G_L^{va} will be denoted by z^{va}. Lastly, the matching $\ell_i^a r_i^a, i = 1, 2, \ldots, k-1$ is added.

The resulting graph G_K is k-regular. To make it an instance of the LIST-H-COVER problem, we prescribe that the vertices $B_v, v \in B$ and $\ell_i^a, a \in A, i = 1, 2, \ldots, k-1$ are all mapped onto x (this means, that for these vertices, their lists of admissible target vertices are one-element and all the same, while for the remaining vertices, their lists are full, as well as for all the edges).

The fact that x is simple implies the following observation: For every partial covering projection from G_u to H which maps all the pendant vertices onto x, their neighbors in G_u are mapped onto distinct vertices of H (this immediately follows from property (d) of Proposition 2). Similarly, if any vertex of G_K is mapped onto x by a covering projection, then its neighbors are mapped onto distinct vertices of H (the neighborhood $N_H(x)$ of x in H). We will exploit these observations in the following argumentation.

Suppose $f : G_K \longrightarrow H$ is a covering projection such that all vertices $B_v, v \in B$ and $\ell_i^a, a \in A, i = 1, 2, \ldots, k-1$ are mapped onto x. Consider an $a \in A$, and let $f(r_1^a) = y$, whence $y \in V(H)$ is a neighbor of x in H. Property c) applied to any G_R^{va}, for v being a neighbor of a in K, implies that $f(r_i^a) = f(w^{va}) = y$ for all $i = 2, \ldots, k-1$. Since each ℓ_i^a has a neighbor r_i^a mapped onto y, none of their neighbors in G_L^{va} is mapped onto y. Property (d) then implies that $f(z^{va}) = y$. Define a coloring ϕ of A by colors $N_H(x)$ as $\phi(a) = f(r_1^a)$. Consider a vertex $v \in B$. The neighbors of B_v in G_K are $z^{va}, a \in N_K(v)$. Since $f(B_v) = x$ and x is simple, the vertices $z^{va}, a \in N_K(v)$ are mapped onto different neighbors of x by f, and hence the colors $\phi(a), a \in N_K(v)$ are all distinct. Thus ϕ is a k-coloring of A of the required property.

Suppose for the opposite direction that A allows a k-coloring ϕ such that each vertex $v \in B$ sees all k colors on its neighbors, and identify the colors with the names of the neighbors of x in H. Set

$f(B_v) = \ell_i^a = x$ for all $v \in B, a \in A, i = 1, 2, \ldots, k-1$ (as required by the lists),

$f(u^{va}) = x$ for all $v \in B$ and $a \in N_K(v)$, and

$f(r_i^a) = f(w^{va}) = f(z^{va}) = \phi(a)$ for all $a \in A$ and $v \in N_K(a)$.

and define f on the edges incident to ℓ_i^a (r_i^a, respectively) so that for every i, these edges are mapped onto different edges incident to x (to $\phi(a)$, respectively), and on the other hand for every $v \in N_K(a)$, the pendant edges of G_L^{va} (of G_R^{va}, respectively) are mapped onto distinct edges incident to x (to $\phi(a)$, respectively). (This is a simple exercise.) The properties (a) and (b) of Proposition 2 imply that this mapping can be extended to partial covering projections within each copy of G_u used in the construction of G_K. To see that they altogether provide a covering projection from G_K to H, note that for each $v \in B$, the edges incident with the vertex B_v are mapped onto different edges because their other endpoints are $\phi(a), a \in N_K(v)$, and hence all different by the assumption on the coloring ϕ, and also each copy G_u^v has its pendant edges mapped onto different edges incident to x, since the pendant vertices $u^{va}, a \in N_K(v)$ are all mapped onto x and their neighbors w^{va} in G_u^v are mapped onto distinct vertices $\phi(a), a \in N_K(v)$. This concludes the proof of the case of bipartite H.

3.3   The Non-bipartite Case

Suppose the graph H is not k-edge-colorable (this includes the case when H contains loops and/or semi-edges). Consider $H' = H \times K_2$. This H' may still continue multiple edges (the product of a multiple ordinary edge with K_2 is again a multiple edge, but also the product of a loop with K_2 is a double ordinary edge, and the product of a multiple semi-edge with K_2 results in a multiple ordinary edge as well), but it is bipartite (and thus has neither semi-edges nor loops) and therefore is k-edge-colorable. In the product with K_2, every semi-simple vertex of H results in two simple vertices of H'. Hence, by the result of the preceding subsection, LIST-H'-COVER is NP-complete.

It is proved in [18] that for simple graphs, G covers $H \times K_2$ if and only if G is bipartite and covers H. This proof readily extends to graphs that allow loops, semi-edges and multiple edges. The proof for the list version of the problem may get more complicated in general. However, the list version that we have proven NP-complete in the preceding subsection is very special: the lists of all edges are full, and so are the lists of all the vertices except for those which are prescribed to be mapped onto the same simple vertex, say x'. If we take such an instance of LIST-H'-COVER, this x' is a copy of a semi-simple vertex $x \in V(H)$, and all vertices of the input graph G that are prescribed to be mapped onto x' are from the same class of its bipartition. We just prescribe them to be mapped onto x as an instance of LIST-H-COVER. It is easy to see that this mapping can be extended to a covering projection to H if and only if G allows a covering projection to H' in which all these prescribed vertices are mapped onto x'. This concludes the proof of Theorem 1.

4   Sausages and Rings

In this section we consider two special classes of cubic graphs. These graphs play a special role in the classification in Theorem 4. The k-*ring* is the cubic graph obtained from the cycle of length $2k$ by doubling every second edge. We call a *k-sausage* every cubic graph that is obtained from a path on k vertices by doubling every other edge and adding loops or semi-edges to the end-vertices of the path to make the graph 3-regular. Note that while for every k, the k-ring is defined uniquely, there are several types of k-sausages, as depicted in Fig. 3.

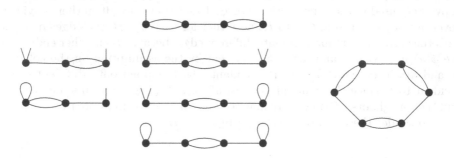

Fig. 3. The two non-isomorphic 3-sausages (left), the four non-isomorphic 4-sausages (middle), and the 3-ring (right).

Proposition 3. *For every $k \geq 2$, let S_k be a k-sausage. Then $S_k \times K_2$ is isomorphic to the k-ring.*

Proof. The product $H \times K_2$ is a bipartite graph with no loops or semi-edges, in which every ordinary edge in H gives rise to a pair of ordinary edges of the same multiplicity, a loop, or a pair of semi-edges incident to the same vertex of H gives rise to a double ordinary edge, and a single semi-edge in H gives rise to a simple ordinary edge in $H \times K_2$. Thus $S_k \times K_2$ has a cyclic structure and the number of double edges is equal to the number of vertices of S_k, see Fig. 4.   □

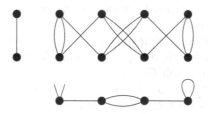

Fig. 4. The product of a 4-sausage with K_2 is isomorphic to the 4-ring.

In the following two theorems we show that for every $k \neq 4$, the LIST-k-RING-COVER problem is NP-complete in simple graphs.

Theorem 2 (♠). *The K-RING-COVER problem is NP-complete for simple input graphs for every $k = 2^\alpha(2\beta + 3)$ such that α and β are non-negative integers.*

Theorem 3 (♠). *The* LIST-k-RING-COVER *problem is NP-complete for simple input graphs for every* $k = 2^\alpha$ *such that* $\alpha \geq 3$ *is an integer.*

The following observation shows that hardness for k-rings implies the hardness for k-sausages.

Proposition 4 (♠). *For every* $k \geq 2$ *and every* k-*sausage* S_k, k-RING-COVER $\propto S_k$-COVER *and* LIST-k-RING-COVER $\propto$ LIST-S_k-COVER.

Proof. A graph G covers $H \times K_2$ if and only if it is bipartite and covers H. Since bipartiteness can be tested in polynomial time, testing if G covers the k-ring polynomially reduces to testing if G covers S_k.

The proof for the list version is a bit more complicated and we defer it to the journal version of the paper. $\square$

5 Towards Strong Dichotomy for Cubic Graphs

In this section we are getting close to proving the Strong Dichotomy Conjecture for cubic graphs.

Theorem 4. *Let* H *be a connected cubic graph which is neither the* 4-*ring nor a* 4-*sausage. Then* LIST-H-COVER *is polynomial-time solvable for general graphs when* H *has only one vertex and at most one semi-edge, and it is NP-complete even for simple input graphs otherwise.*

Proof. The proof is divided into several cases, depending on the structure of H.

Case 1: $|V(H)| = 1$. We distinguish two subcases.

Case 1A - H Has One Semi-edge and One Loop. The preimage of the semi-edge should be a disjoint union of the semi-edges of the input graph G and of a perfect matching on the vertices not incident to a semi-edge. Then the remaining edges of G form a spanning collection of cycles (including loops) which form the preimage of the loop. The existence of a spanning subgraph of G that is a preimage of the semi-edge can be tested in polynomial time.

If lists are present as part of the input, the situation gets a little more tricky. We start with a preprocessing phase. We check the below conditions:

(a) G has a vertex or an edge with an empty list.
(b) G has a vertex incident to two or more semi-edges,
(c) G has a semi-edge whose list does not contain the semi-edge of H,
(d) G has a vertex incident to a semi-edge and an edge, whose list does not contain the loop of H,
(e) G has a vertex incident to two ordinary edges, whose lists do not contain the loop of H,
(f) G has a loop whose list does not contain the loop of H.

It is clear that if any of the above conditions is satisfied, then $(G, \mathcal{L})$ is a no-instance. Thus we reject and quit.

Now we shall construct an auxiliary graph G'. We start our construction with G and perform the following steps.

1. If some vertex v is incident to a semi-edge, then delete v with all its edges.
2. If some edge e does not have the semi-edge of H in its list, remove e from the graph.
3. If some edge e does not have the loop of H in the list, leave e, but remove all edges incident to e.

Let G' be the graph after the exhaustive application of steps 1, 2, and 3. It is straightforward to verify that steps 1 and 2 ensure that the union of a perfect matching in G' and the semi-edges removed in step 1. can be a preimage of the semi-edge of H. Furthermore, by step 3 we ensure that if some edge has to be mapped to the semi-edge, then it will be so.

We can verify in polynomial time if G' has a perfect matching. If not, we reject and quit. So let M be a perfect matching in G', and let M' be the union of M and the set of semi-edges removed in step 1. Observe that the graph $G - M$ is 2-regular, i.e., is a disjoint union of cycles (including loops). Furthermore, every edge of $G - M$ has the loop of H in its list, this is guaranteed by step 3 and the preprocessing phase. Thus in this case we report a yes-instance.

Case 1B - H Has Three Semi-edges. In this case already H-COVER is NP-complete, as it is equivalent to 3-edge-colorability of cubic graphs.

Case 2: $|V(H)| = 2$. If H has neither loops nor semi-edges, then H is a bipartite graph formed by a triple edge between two vertices. Only bipartite graphs can cover a bipartite one. Hence a covering projection corresponds to a 3-edge-coloring of the input graph. Thus H-COVER is polynomial-time solvable (every cubic bipartite graph is 3-edge-colorable), but LIST-H-COVER is NP-complete, because LIST 3-COLORING is NP-complete for line graphs of cubic bipartite graphs [10]. If H has a loop or a semi-edge, then it is one of the four graphs in Fig 5, and for each of these already the H-COVER problem is NP-complete [6].

Fig. 5. The non-bipartite 2-vertex graphs.

Case 3: $|V(H)| \geq 3$. Here we split into several subcases.

Case 3A - H is Acyclic. If we shave all semi-edges off from H, we get a tree with at least three vertices. At least one of them has degree greater than 1, and such vertex is semi-simple in H. Thus LIST-H-COVER is NP-complete by Theorem 1.

Case 3B - H Has a Cycle of Length Greater than 2 Which Does not Span all of its Vertices. Then H has a vertex outside of this cycle, and thus H has a semi-simple vertex and LIST-H-COVER is NP-complete.

Case 3C - H Has a Cycle of Length Greater than 2 with a Diagonal. Then again H has a semi-simple vertex and LIST-H-COVER is NP-complete.

Case 3D - H Has a Cycle of Length Greater than 2, but None of the Previous Cases Apply. Then H is the k-ring for some $k \geq 2$. If $k = 2$, 2-RING-COVER is NP-complete by [6]. If $k \neq 2^\alpha$ for some $\alpha \geq 2$, k-RING-COVER is NP-complete by Theorem 2. In the case of $k = 2^\alpha$ with $\alpha \geq 3$, the LIST-k-RING-COVER problem is NP-complete by Theorem 3. The case of $k = 4$ remains open.

Case 3E - H Has a Cycle, but all Cycles are of Length One or Two. If, in addition, H has no semi-simple vertex, then H is a k-sausage for some $k \geq 2$. If $k \neq 4$, the NP-completeness of LIST-H-COVER follows from Case 3D via Proposition 4. The case of $k = 4$ remains open. □

6 Concluding Remarks

We have studied the complexity of the LIST-H-COVER problem in the setting of graphs with multiple edges, loops, and semi-edges for regular target graphs. We have shown in Theorem 1 a general hardness result under the assumption that the target graph contains at least one semi-simple vertex. It is worthwhile to note that in fact we have proved the NP-hardness for the more specific H-PRECOVERING EXTENSION problem, when all the lists are either one-element, or full. Actually, we proved hardness for the even more specific VERTEX H-PRECOVERING EXTENSION version, when only vertices may come with prescribed covering projections, but all edges have the lists full.

On the contrary, the nature of the NP-hard cases that appear in the almost complete characterization of the complexity of LIST-H-COVER of cubic graphs given by Theorem 4 is more varied. Some of them are NP-hard already for H-COVER, some of them are NP-hard for H-PRECOVERING EXTENSION, but apart the VERTEX H-PRECOVERING EXTENSION version in applications of Theorem 1, this time we also utilize the EDGE H-PRECOVERING EXTENSION version for the case of the bipartite 2-vertex graph formed by a triple edge between two vertices. Finally, for the cases of sausages and rings of length power of two, nontrivial lists are required to make our proof technique work.

Needless to say, we are leaving the following problem open. An affirmative answer would imply the hardness of LIST H-COVER for 4-sausages H, and thus prove the list variant of the Strong Dichotomy Conjecture for cubic graphs.

Conjecture. *The 4-RING-COVER problem is NP-complete for simple input graphs.*

Note Added in Proof

After submitting the paper to IWOCA 2022, we have proved the above-stated Conjecture. The proof will appear in the journal version of the paper. A preliminary full version is available at http://arxiv.org/abs/2204.04280.

Acknowledgments. Jan Bok and Nikola Jedličková were supported by research grant GAČR 20-15576S of the Czech Science Foundation and by SVV–2020–260578 and GAUK 1580119. Jiří Fiala and Jan Kratochvíl were supported by research grant GAČR 20-15576S of the Czech Science Foundation. Paweł Rzążewski was supported by the Polish National Science Centre grant no. 2018/31/D/ST6/00062. The last author is grateful to Karolina Okrasa and Marta Piecyk for fruitful and inspiring discussions.

References

1. Abello, J., Fellows, M.R., Stillwell, J.C.: On the complexity and combinatorics of covering finite complexes. Aust. J. Combin. **4**, 103–112 (1991)
2. Angluin, D.: Local and global properties in networks of processors. In: Proceedings of the 12th ACM Symposium on Theory of Computing, pp. 82–93 (1980)
3. Bard, S., Bellitto, T., Duffy, C., MacGillivray, G., Yang, F.: Complexity of locally-injective homomorphisms to tournaments. Discret. Math. Theor. Comput. Sci. **20**(2) (2018). http://dmtcs.episciences.org/4999
4. Biggs, N.: Algebraic Graph Theory. Cambridge University Press, Cambridge (1974)
5. Bílka, O., Jirásek, J., Klavík, P., Tancer, M., Volec, J.: On the complexity of planar covering of small graphs. In: Kolman, P., Kratochvíl, J. (eds.) WG 2011. LNCS, vol. 6986, pp. 83–94. Springer, Heidelberg (2011). https://doi.org/10.1007/978-3-642-25870-1_9
6. Bok, J., Fiala, J., Hliněný, P., Jedličková, N., Kratochvíl, J.: Computational complexity of covering multigraphs with semi-edges: small cases. In: Bonchi, F., Puglisi, S.J. (eds.) 46th International Symposium on Mathematical Foundations of Computer Science, MFCS 2021, 23–27 August 2021, Tallinn, Estonia. LIPIcs, vol. 202, pp. 21:1–21:15. Schloss Dagstuhl - Leibniz-Zentrum für Informatik (2021). https://doi.org/10.4230/LIPIcs.MFCS.2021.21
7. Bok, J., Fiala, J., Jedličková, N., Kratochvíl, J., Seifrtová, M.: Computational complexity of covering disconnected multigraphs. In: Bampis, E., Pagourtzis, A. (eds.) FCT 2021. LNCS, vol. 12867, pp. 85–99. Springer, Cham (2021). https://doi.org/10.1007/978-3-030-86593-1_6
8. Chaplick, S., Fiala, J., van't Hof, P., Paulusma, D., Tesař, M.: Locally constrained homomorphisms on graphs of bounded treewidth and bounded degree. Theor. Comput. Sci. **590**, 86–95 (2015)
9. Djoković, D.Ž: Automorphisms of graphs and coverings. J. Combin. Theory B **16**, 243–247 (1974)
10. Fiala, J.: NP completeness of the edge precoloring extension problem on bipartite graphs. J. Graph Theory **43**(2), 156–160 (2003)
11. Fiala, J., Kratochvíl, J.: Complexity of partial covers of graphs. In: Eades, P., Takaoka, T. (eds.) ISAAC 2001. LNCS, vol. 2223, pp. 537–549. Springer, Heidelberg (2001). https://doi.org/10.1007/3-540-45678-3_46

12. Fiala, J., Kratochvíl, J.: Locally injective graph homomorphism: lists guarantee dichotomy. In: Fomin, F.V. (ed.) WG 2006. LNCS, vol. 4271, pp. 15–26. Springer, Heidelberg (2006). https://doi.org/10.1007/11917496_2

13. Fiala, J.: Locally injective homomorphisms. Ph.D. thesis, Charles University, Prague (2000)

14. Fiala, J., Heggernes, P., Kristiansen, P., Telle, J.A.: Generalized H-coloring and H-covering of trees. Nordic J. Comput. **10**(3), 206–224 (2003)

15. Fiala, J., Klavík, P., Kratochvíl, J., Nedela, R.: Algorithmic aspects of regular graph covers with applications to planar graphs. In: Esparza, J., Fraigniaud, P., Husfeldt, T., Koutsoupias, E. (eds.) ICALP 2014. LNCS, vol. 8572, pp. 489–501. Springer, Heidelberg (2014). https://doi.org/10.1007/978-3-662-43948-7_41

16. Fiala, J., Klavík, P., Kratochvíl, J., Nedela, R.: 3-connected reduction for regular graph covers. Eur. J. Comb. **73**, 170–210 (2018)

17. Fiala, J., Kratochvíl, J.: Partial covers of graphs. Discuss. Math. Graph Theory **22**, 89–99 (2002)

18. Fiala, J., Kratochvíl, J.: Locally constrained graph homomorphisms – structure, complexity, and applications. Comput. Sci. Rev. **2**(2), 97–111 (2008)

19. Fiala, J., Paulusma, D.: A complete complexity classification of the role assignment problem. Theoret. Comput. Sci. **1**(349), 67–81 (2005)

20. Fiala, J., Klavík, P., Kratochvíl, J., Nedela, R.: Algorithmic aspects of regular graph covers (2016). https://arxiv.org/abs/1609.03013

21. Gardiner, A.: Antipodal covering graphs. J. Combin. Theory B **16**, 255–273 (1974)

22. Getzler, E., Kapranov, M.M.: Modular operads. Compos. Math. **110**(1), 65–125 (1998)

23. Griggs, J.R., Yeh, R.K.: Labelling graphs with a condition at distance 2. SIAM J. Discret. Math. **5**(4), 586–595 (1992). https://doi.org/10.1137/0405048

24. Gross, J.L., Tucker, T.W.: Generating all graph coverings by permutation voltage assignments. Discret. Math. **18**, 273–283 (1977)

25. Kratochvíl, J., Proskurowski, A., Telle, J.A.: Complexity of colored graph covers I. Colored directed multigraphs. In: Möhring, R.H. (ed.) WG 1997. LNCS, vol. 1335, pp. 242–257. Springer, Heidelberg (1997). https://doi.org/10.1007/BFb0024502

26. Kratochvíl, J., Proskurowski, A., Telle, J.A.: Covering regular graphs. J. Combin. Theory Ser. B **71**(1), 1–16 (1997)

27. Kratochvíl, J., Proskurowski, A., Telle, J.A.: Complexity of graph covering problems. Nordic J. Comput. **5**, 173–195 (1998)

28. Kratochvíl, J., Telle, J.A., Tesař, M.: Computational complexity of covering three-vertex multigraphs. Theoret. Comput. Sci. **609**, 104–117 (2016)

29. Kristiansen, P., Telle, J.A.: Generalized H-coloring of graphs. In: Goos, G., Hartmanis, J., van Leeuwen, J., Lee, D.T., Teng, S.-H. (eds.) ISAAC 2000. LNCS, vol. 1969, pp. 456–466. Springer, Heidelberg (2000). https://doi.org/10.1007/3-540-40996-3_39

30. Kwak, J.H., Nedela, R.: Graphs and their coverings. Lect. Notes Ser. **17**, 118 (2007)

31. Leighton, F.T.: Finite common coverings of graphs. J. Combin. Theory B **33**, 231–238 (1982)

32. Malnič, A., Marušič, D., Potočnik, P.: Elementary abelian covers of graphs. J. Algebraic Combin. **20**(1), 71–97 (2004)

33. Malnič, A., Nedela, R., Škoviera, M.: Lifting graph automorphisms by voltage assignments. Eur. J. Comb. **21**(7), 927–947 (2000)

34. Mednykh, A.D., Nedela, R.: Harmonic Morphisms of Graphs: Part I: Graph Coverings. Vydavatelstvo Univerzity Mateja Bela v Banskej Bystrici, 1st edn. (2015)

35. Nedela, R., Škoviera, M.: Regular embeddings of canonical double coverings of graphs. J. Combin. Theory Ser. B **67**(2), 249–277 (1996)
36. Okrasa, K., Rzążewski, P.: Subexponential algorithms for variants of the homomorphism problem in string graphs. J. Comput. Syst. Sci. **109**, 126–144 (2020). https://doi.org/10.1016/j.jcss.2019.12.004
37. Ringel, G.: Map Color Theorem, vol. 209. Springer, Berlin (1974). https://doi.org/10.1007/978-3-642-65759-7
38. Shepherd, S., Gardam, G., Woodhouse, D.J.: Two generalisations of Leighton's theorem (2019). https://arxiv.org/abs/1908.00830
39. Woodhouse, D.J.: Revisiting Leighton's theorem with the Haar measure. Math. Proc. Camb. Philos. Soc. **170**(3), 615–623 (2021). https://doi.org/10.1017/S0305004119000550

The Slotted Online One-Sided Crossing Minimization Problem on 2-Regular Graphs

Elisabet Burjons$^{(\boxtimes)}$ ⓘ, Janosch Fuchs$^{(\boxtimes)}$ ⓘ, and Henri Lotze$^{(\boxtimes)}$ ⓘ

Department of Computer Science, RWTH Aachen University,
Ahornstr. 55, 52074 Aachen, Germany
{burjons,fuchs,lotze}@cs.rwth-aachen.de

Abstract. In the area of graph drawing, the One-Sided Crossing Minimization Problem (OSCM) is defined on a bipartite graph with both vertex sets aligned parallel to each other and all edges being drawn as straight lines. The task is to find a permutation of one of the node sets such that the total number of all edge-edge intersections, called *crossings*, is minimized. Usually, the degree of the nodes of one set is limited by some constant k, with the problem then abbreviated to OSCM-k.

In this work, we study an online variant of this problem, in which one of the node sets is already given. The other node set and the incident edges are revealed iteratively as requests and each node has to be inserted into placeholders, which we call *slots*. The number of slots coincides with the number of requests and their order is fixed. The goal is again to minimize the number of crossings in the final graph. Minimizing crossings in an online way is related to the more empirical field of dynamic graph drawing. Note that the *slotted* OSCM problem is harder to solve for an online algorithm but in the offline case it is equivalent to the version without slots.

We show that the online slotted OSCM-k is not competitive for any $k \geq 2$ and subsequently limit the graph class to that of 2-regular graphs, for which we show a lower bound of 4/3 and an upper bound of 5 on the competitive ratio.

Keywords: Online algorithms · Crossing minimization · Graph drawing

1 Introduction

Online algorithms were introduced by Sleator and Tarjan [17] to solve problems for which the instance is piecewise revealed to an algorithm, which must make some irrevocable decision before the next element of the instance is presented. Online algorithms are classically analyzed using competitive analysis, where the performance of an online algorithm is compared to that of an optimal offline algorithm working on the same instance. The worst-case ratio between any online

© Springer Nature Switzerland AG 2022
C. Bazgan and H. Fernau (Eds.): IWOCA 2022, LNCS 13270, pp. 243–256, 2022.
https://doi.org/10.1007/978-3-031-06678-8_18

algorithm and the optimal offline solution is the competitive ratio of a problem. For a deeper introduction to online algorithms and competitive analysis we refer the reader to the reference books [2,9].

In graph drawing problems, one usually wants to embed a given graph into some space with limited dimensions. The most common and practical examples are on the Euclidean plane. It is also usual to try to embed such graphs in a way that minimizes the number of edges that cross each other, i.e., their depictions overlap in a point that is not occupied by a vertex. If a graph can be embedded in the Euclidean plane without any crossings, we say the graph is planar. A survey on graph drawing and crossing minimization can be found in [1,15].

One common way to depict bipartite graphs is by arranging the vertices in each partition on a straight (horizontal) line, making the lines for the two partition sides parallel. In this scenario, the edges are drawn from one side of the partition to the other as straight segments. The problem of minimizing the crossings in this scenario is reduced, thus, to properly ordering the vertices in each partition. However, in some practical applications it is enough to restrict ourselves to ordering one set of the partition (the free side), while the other set remains fixed (the fixed side). It is also usual to restrict the degree of the vertices in the free side [10,11]. This (one sided) problem is formally defined as follows.

Definition 1. *Given a bipartite Graph $G = (S \dot\cup V, E)$. Let the nodes of S and V be aligned in some ordering on straight lines parallel to each other, where S is on the top line and V on the bottom line. Let the edges E be drawn as straight lines only. Let the degree of the nodes of S be bound by some $k \in \mathbf{N}$. The One-Sided Crossing Minimization Problem (OSCM-k) is defined as the problem of finding a total ordering of the nodes of S such that the number of resulting edge crossings in the graph is minimized.*

We will assume that the ordering of V is part of the instance and fixed, such that we can label and reference the nodes of V with ascending natural numbers, starting from the "left".

1.1 Related Work

The OSCM problem has already been extensively studied in the past under different names, such as *bipartite crossing number* [7,15], *crossing problem* [5], *fixed-layer bipartite crossing minimization* [10] and others. Eades and Wormald [5] showed that the OSCM problem is NP-complete for dense graphs, while Muñoz et al. [11] showed NP-completeness for sparse graphs. Muñoz et al. also introduced the OSCM-k and showed that the OSCM-2 can be solved optimally using the barycenter heuristic.

Li and Stallmann [10] showed that the approximation ratio of the barycenter heuristic is in $\Omega(\sqrt{n})$ on general bipartite graphs and also proved that OSCM-k admits a tight $k-1$ approximation. This latter problem definition coincides with the definition of the OSCM-k. Nagamochi presented a randomized approximation algorithm for general graphs [12] and another approximation algorithm for bipartite graphs of large degree [13].

Further researching the complexity, Dujmović and Whitesides [4] first showed that OSCM is fixed parameter tractable, i.e., it can be solved in $f(k)n^{\mathcal{O}(1)}$, where the parameter k is the number of crossings. The currently best known FPT running time is $\mathcal{O}(3^{\sqrt{2k}+n})$ and was given by Kobayashi and Tamaki [8].

To the best of our knowledge, the field of online analysis on crossing minimization is hardly researched. A closely related problem arises in the field of graph drawing, called *dynamic graph drawing*. Here, the task is to visually arrange a graph that is iteratively expanded over time. The visualization follows certain empirical criteria to make the data comprehensible, where crossing minimization is one of these criteria. For a survey regarding dynamic graph drawing, see [16]. Dynamic graph drawing has many applications, for instance, Frishman and Tal [6] present an algorithm to compute online layouts for a sequence of graphs and its application in discussion thread visualization and social network visualization. In another example, North and Woodhull [14] focus on hierarchical graph drawing, a more restricted graph class that needs to be visualized in a tree-like fashion, which overlaps with our topic regarding applications. While one of the most mentioned applications of the offline OSCM is wire crossing minimization in VLSI this is arguably less applicable when looking at an online version of the problem. However, the results of an online analysis can be helpful for the application fields of graph drawing, e.g., software visualization, decision support systems and interactive graph editors.

While dynamic graph drawing and online graph problems are similar in that parts of the graph are revealed in an iterative fashion and not previously known, a central difference is that in dynamic graph drawing the manipulation of previous decisions is usually allowed. This is not the case in the classical online model. Thus, while theory and practice are looking at similar problems, with the goal of aesthetic graph drawings, the methods to achieve this goal are different.

1.2 Our Contribution

In this paper, we look at the online version of the OSCM-k problem. Observe, that the online version of OSCM-k can be defined in two different ways. The first version is the online *free* OSCM-k, where given a bipartite graph $(S \cup V, E)$, an algorithm initially sees a fixed set of vertices V, and then, in each step a request appears for a subset of vertices $R_i \subseteq V$, which must be adjacent to a vertex in S. Thus, after the arrival of the request R_i, one has to place a vertex $s_i \in S$ on the top line and adjacent to the vertices in R_i. In this version, one chooses the partial ordering of s_i with respect to the other vertices already present in S.

The online free OSCM-k problem is solvable with a competitive ratio of at most $k - 1$, using the same barycenter algorithm as in the offline case [10].

In this paper, we focus on a different version of this problem, which we call the online *slotted* OSCM-k, which is formally defined as follows.

Definition 2. *Given a vertex set V, a request sequence for online slotted OSCM-k is a sequence $R_1, \ldots, R_n$ of subsets of V, each of size at most k. The set of vertices S is initiated as $S = \{s_1, \ldots, s_n\}$. Initially there are no edges between S*

and V. *Once a request $R_j \subseteq V$ arrives, an online algorithm solving online slotted OSCM-k chooses a vertex s_i without any edges, and places an edge between s_i and every vertex in R_j. The goal is to minimize the total number of crossings.*

The slotted OSCM-k is a model that follows the aesthetic paradigms of the area of dynamic graph drawing, where the so-called *mental map* and human readability is sustained. The term *mental map* describes the goal to make current visualization of the graph recognizable in later iterations of the graph. Compared to the free OSCM-k no upper or lower bound on the competitive ratio is known.

We call the vertices $s_i \in S$ *slots* moving forward. If a request $R_j \subseteq V$ is fulfilled by adding edges between every vertex in R_j and slot s_i we say that request R_j is *assigned* to slot s_i. Moreover, we call a slot s_i *unfulfilled* or *free* if no request has been satisfied using this slot, thus the slot has no edges yet. Correspondingly, a *fulfilled* slot s_i is a slot in S with edges to a subset $R_j \subseteq V$.

Online slotted OSCM-k has the advantage of knowing in advance the number of requests. However, one has the distinct constraint that, once two consecutive slots are fulfilled, the algorithm will not be able to assign any request to a vertex between the fulfilled slots, as such a vertex does not exist.

We prove that online slotted OSCM-k is not competitive for any $k \geq 2$ in general graph classes. However, if we focus on 2-regular graphs, we prove that this problem has a constant competitive ratio. In particular, we prove a lower bound of 4/3 in this case, and then present an algorithm with a competitive ratio of at most 5 as an upper bound.

Due to space constraints most proofs and figures can only be found in the full version [3].

2   Lower Bounds on General Graphs

We begin by looking at online slotted OSCM-k on general graphs, and show that for every value of $k \geq 2$, there is no algorithm with a constant competitive ratio.

Theorem 1. *There is no online algorithm with a constant competitive ratio for online slotted OSCM-k, for any $k \geq 2$.*

Proof. Let us consider an algorithm A solving online slotted OSCM-k. Given the initial sets of vertices $V = \{v_1, \ldots, v_n\}$ and slots $S = \{s_1, \ldots, s_n\}$, A is presented the following request sequence: $\{v_1, v_2\}, \{v_2, v_3\}, \ldots, \{v_{n-1}, v_n\}$. Assume without loss of generality that A has assigned these requests to slots in S without producing a single crossing. Since we have n requests to fill n slots with, and A has only one unfulfilled slot s_i for some $i \in \{1, \ldots, n\}$, the last request will be assigned to s_i. We assume, without loss of generality, that $i \leq \lceil \frac{n}{2} \rceil$. The adversary now presents the request $\{v_{n-1}, v_n\}$ as the last request of the input. This results in at least $2 \cdot 2 \cdot (\frac{n}{2} - 1)$ crossings as opposed to the optimal solution, which only results in a single crossing as depicted in Fig. 1. The competitive ratio is thus at least $\frac{2 \cdot 2 \cdot (\frac{n}{2} - 1)}{1} = 2n - 4$ and therefore not bounded by any fixed constant c. $\square$

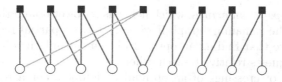

Fig. 1. Theorem 1: an algorithm is presented the requests colored in blue first. Some slot has to be left open for which the request associated with the red edges is given. (Color figure online)

The proof relies on the adversary being able to freely choose the degrees of the vertices in V. If we require the degree of the vertices in V to be defined in advance, the same strategy does not work. Thus, it makes sense to look at graph classes where the degree of the vertices in the graph is fixed, i.e., regular graphs.

In what follows, we focus on online slotted OSCM-2 on 2-regular graphs, as this particular case is already hard to analyze, and we prove that the competitive ratio is within the range between 4/3 and 5.

We conjecture that for higher degrees, online slotted OSCM-k on k-regular graphs also has a constant competitive ratio, with the constant depending on k. A higher vertex degree means that even optimal solutions must have a lot of crossings. Thus, even when an online algorithm makes a sub-optimal choice, the crossings of the optimal solution that it is compared to compensate the mistakes.

3   Lower Bound for 2-Regular Graphs

It is important to note that an offline algorithm can find an optimal solution in a greedy fashion, as we will see in Lemma 1. In the following lower bound, we prove that online algorithms cannot find an optimal solution, greedily or otherwise. The difficulty is that a request cannot be assigned in between two consecutive fulfilled slots. Thus, an online algorithm has to fulfill a request by assigning it to a sub-optimal slot. We can use this fact to construct a lower bound for online slotted OSCM-2 on 2-regular graphs as follows.

Theorem 2. *Every deterministic online algorithm, solving the slotted OSCM-2 on 2-regular graphs, has a competitive ratio of at least* $4/3 - \varepsilon$.

This lower bound proves that no online algorithm for online slotted OSCM-2 on 2-regular graphs can perform optimally on all instances. In the following, we introduce some notions that are used to prove an upper bound for the competitive ratio in the same setting.

4   Preliminaries and Notation

In order to prove upper bounds for online slotted OSCM-2 on 2-regular graphs, we need to extract some structural properties of this problem. First, we introduce the notion of *propagation arrows*, which helps us to lower bound the total number of crossings of the remaining graph if we only have a partial request sequence. Then, we observe that finding an optimal placement only involves the order of every pair of requests relative to each other.

The number of crossings of an optimal assignment for a request sequence is the *number of unavoidable crossings* of the request sequence. The difference between the number of crossings incurred by an algorithm A, and the number of unavoidable crossings is consequently the *number of avoidable crossings* of A on that request sequence.

Consider a 2-regular instance for online slotted OSCM-k with slots $S = \{s_1, \ldots, s_n\}$ and vertices $V = \{v_1, \ldots, v_n\}$, and a request sequence $R_1, \ldots, R_n$. Let us assume that at some point after the k-th request has been fulfilled by algorithm A, there are fulfilled slots, and the vertices in V have degree 2, 1 or 0, depending on how many times these vertices have appeared in requests. Because we know that the final graph will be 2-regular, for those vertices in V with degree less than two we are still expecting a request that contains the vertex, and for any unfulfilled slot, there will be a request which will be fulfilled using this slot.

Intuitively, we use propagation arrows to greedily match unfulfilled vertices to available slots in a way that minimizes the number of crossings. For instance, in an empty graph every vertex v_i in V will have two propagation arrows to the slot s_i, but once some slots are occupied, we take the leftmost vertex with degree less than two and assign a propagation arrow to the left-most unfulfilled slot. We know that the instance is 2-regular, so for every pair of missing edges of vertices in V there must be an empty slot. We can define the propagation arrows formally as follows.

First, we know that after k requests for a 2-regular graph, there are $n - k$ unfulfilled requests, which corresponds to $2(n - k)$ missing edges. We will double count the missing edges with the following two lists.

The *list of unfulfilled vertices* L_V of an instance after the k-th request, is an ordered list that contains every vertex $v_i \in V$ from smallest to largest at most twice. L_V will contain no copies of a vertex $v_i \in V$ if it already has appeared twice in the request sequence $R_1, \ldots, R_k$, i.e., if v_i has degree 2, L_V will contain $v_i \in V$ once if v_i has appeared only once in $R_1, \ldots, R_k$, i.e., if v_i has degree 1 in the partially fulfilled graph, finally, L_V contains a vertex v_i twice if v_i does not appear in $R_1, \ldots, R_k$, and thus has degree 0 at that point.

Observe, that the size of the list of missing edges is twice the number of unfulfilled slots by an easy application of the handshaking lemma. We can, thus, analogously consider the *list of unfulfilled slots* L_S as an ordered list that contains each unfulfilled slot twice, again from smallest to largest. From the previous observation it should be clear that $|L_V| = |L_S|$.

Definition 3. *Consider a 2-regular instance for online slotted OSCM-k with slots $\{s_1, \ldots, s_n\} = S$ and vertices $\{v_1, \ldots, v_n\} = V$, and a request sequence $R_1, \ldots, R_n$. Let A be an algorithm that has fulfilled k requests. Let us consider the corresponding L_V and L_S for this request. There is a propagation arrow* *from vertex v to slot s if both occupy the same place in the ordered lists L_V and L_S, i.e., if v is the i-th element of L_V and s is i-th element of L_S for some $i \in [2(n-k)]$.*

Observe that propagation arrows do not cross one another by construction. If we count the crossings of a partial graph including the crossings between graph edges and propagation arrows, we have a lower bound on the number of crossings that the graph will have after the request sequence is completely fulfilled.

The next lemma observes that an instance is optimally solved if an only if, for every pair of requests, the relative order of their slot assignments is optimal, i.e., if the placement of these two requests is such that there are fewer crossings between them than otherwise. This basically means, that a crossing is unavoidable, if and only if, the relative order of the two requests involved in this crossing is optimal, regardless of any other placement of any other request within the graph. This provides us with a very powerful tool to analyze the performance of online algorithms solving online slotted OSCM-2 on 2-regular graphs.

Lemma 1. *Given two requests $R_x = \{x_1, x_2\}$ and $R_y = \{y_1, y_2\}$ assigned to slots s_x and s_y. Without loss of generality assume that $x_1 \leq y_1$ and $x_2 \leq y_2$. An assignment where $s_x < s_y$ generates fewer or equally many crossings in the final graph than an assignment where $s_y < s_x$ if every other assigned slot remains unchanged.*

Lemma 1 plainly states that for each pair of requests, the optimal ordering gives the left-most request a slot that is to the left of the slot assigned to the right-most request. The notion of left and right requests only means here, that if the requests are not for identical pairs of vertices, the left request contains the left-most distinguished vertex.

In order to find an upper bound on the competitive ratio, we only have to see that any pair of requests is either placed optimally or otherwise bound the number of crossings generated by that pair with the number of unavoidable crossings in the optimal solution.

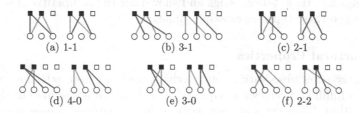

(a) 1-1   (b) 3-1   (c) 2-1

(d) 4-0   (e) 3-0   (f) 2-2

Fig. 2. Case distinction for step one of Lemma 1. Each case is depicted before and after the untangling. The request s_x is drawn in red and s_y in blue. (Color figure online)

Algorithm 1. Chooses in each step the insertion with the lowest number of additional edge-edge and edge-propagation arrow crossings.

1: $free\_slots = \{1, \dots, n\}$;
2: **for** $request$ in $input$ **do**
3:     $least\_crossings := \infty$;
4:     $best\_slot := 0$;
5:     **for** $slot$ in $free\_slots$ **do**
6:         $G.simulate\_node\_insertion(slot, request)$;
7:         $new\_crossings = G.edge\_edge\_crossings() + G.edge\_prop\_crossings()$;
8:         **if** $new\_crossings < least\_crossings$ **then**
9:             $least\_crossings = new\_crossings$;
10:            $best\_slot = slot$;
11:         $G.revert\_simulated\_insertion(slot, request)$;
12:     $G.insert\_node(best\_slot, request)$;
13:     $free\_slots := free\_slots \backslash best\_slot$;

---

5 Upper Bound for 2-Regular Graphs

We now present Algorithm 1 that, given a request, selects the slot that minimizes the total number of crossings – including crossings between edges and propagation arrows – among all available slots.

Note that analyzing an algorithm in this setting is not completely trivial. Our approach is to show that the types of crossings between two requests produced by our algorithm are good-natured. Specifically, we look at pairs of requests for which the crossings can be completely avoided if they are appropriately ordered, i.e., 3-0 or 4-0 crossings as depicted in Fig. 2(e) and (d) respectively. This type of crossing is either not produced by Algorithm 1 or we can show that a number of unavoidable crossings is necessary to produce this configuration. With this, we can upper bound the competitive ratio. Note that this is a relatively rough estimate, but even this estimate already requires a lot of structural analysis.

First, we present some lemmas outlining relevant structural properties of assignments made by Algorithm 1, then we consider each type of critical crossing, 4-0 crossings and then 3-0 crossings and show that the competitive ratio is still bounded when these types of crossings appear.

5.1 Structural Properties

We first make a few observations on the changes of the propagation arrows after a request is fulfilled. Consider a request $\{x_1, x_2\}$, which is assigned to slot s_x by some algorithm. Before this request arrived, there were two propagation arrows from vertices y_1 and y_2 going to slot s_x (note that it is possible that $y_1 = y_2$). After the request is assigned to s_x the propagation arrows pointing to s_x have to be shifted, as slot is not available anymore. Simultaneously, one propagation arrow of each x_1 and x_2 disappears as the request is fulfilled. The rest of the propagation arrows have to reflect this movement out of s_x and into the two empty positions left by x_1 and x_2, and they do so in the following way.

Observation 1. *Let $R = \{x_1, x_2\}$ be a request assigned to slot s_x. And let $y_1 \leq y_2$ be the vertices (or vertex) whose propagation arrows point to s_x before this request arrived. Only propagation arrows connected to nodes between the leftmost vertex of x_1 and y_1 and the rightmost vertex of x_2 and y_2 will be shifted.*

Observe that there are no propagation arrows connected to nodes between y_1 and y_2 as otherwise these would be connected to slots other than s_x and produce crossings between propagation arrows, which is impossible by definition.

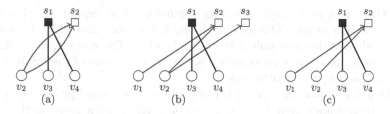

Fig. 3. Types of crossings avoided by Algorithm 1, mentioned in Lemma 2. The propagation arrows are in blue and the edges already in the graph are in black. (Color figure online)

While Observation 1 is not specific to Algorithm 1, we can use it in the proofs to come. We continue with a lemma that allows us to shorten a lot of case distinctions in the following proofs.

Lemma 2. *There is no instance during which two propagation arrows connected to a slot s_2 cross both edges adjacent to a fulfilled slot s_1 when using Algorithm 1.*

Lemma 2 forbids specific configurations, depicted in Fig. 3, of the propagation arrows when applying Algorithm 1. The following lemma uses a counting argument to guarantee that a specific request between two (far apart) vertices must eventually appear in a specific setting. Such requests from vertices that are far apart, always guarantee the appearance of unavoidable crossings as depicted in Fig. 2(f). The appearance of such requests guarantees, in later proofs, the existence of such unavoidable crossings, which can be counted in a way that bounds the competitive ratio.

Lemma 3. *Let there be two request $\{x_1, x_2\}$ and $\{y_1, y_2\}$ that are assigned to slots s_x and s_y, with $x_1 < x_2 < y_1 < y_2$ and no free slot between s_x and s_y. If there are two neighboring vertices u, v, with $x_2 \leq u < v \leq y_1$ and propagation arrows pointing to two different slots s_l, s_r, with $s_l < s_x < s_y < s_r$, and the request $\{u, v\}$ appears, then there must be a future request $\{a, b\}$, with $a \leq x_2$ and $y_1 \leq b$, which unavoidably crosses all edges of u and v.*

Figure 4 depicts the situation described in the statement of Lemma 3.

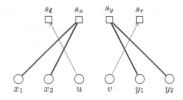

Fig. 4. Sketch of the situation described in the statement of Lemma 3.

Proof. Our proof is a simple counting argument. The request $\{u, v\}$ removes two propagation arrows. One points to the left of the filled block between s_x and s_y and the other one points to the right of it. The request, depending on its placement, pushes one propagation arrow from one side of the fulfilled block between s_x and s_y to the other one.

W.l.o.g. we assume that $\{u, v\}$ is placed on s_l. The second propagation arrow pointing to s_l comes from x_2 (if $u \neq x_2$) or a vertex even more to the left. It is not possible that is comes from a vertex between x_2 and v due to Lemma 2. When the request $\{u, v\}$ is placed, it pushes this second propagation arrow to the slot s_r. This propagation arrow represents a mismatch between open slots and "open/remaining" edges. The number of "open/remaining" edges to the left of u and to the right of v is odd, but the slots always consume two of these "open/remaining" edges. This has to be compensated by some request $\{a, b\}$ that is placed right of s_y, where a is to the left hand side of u and b is to the right hand side of v. This request crosses all edges of u and v. □

Where Lemma 2 and 3 are applicable for specific configurations, the following lemma provides a tool that gives a set of edges or propagation arrows that are necessary to make a local configuration (e.g., a crossing of two requests) feasible in the context of the remaining graph.

Lemma 4. *For every edge or propagation arrow, starting at a vertex v_i of V and pointing to a slot s_j with $i < j$ (analogously $j < i$), there is one edge or propagation arrow pointing from a vertex v_k to a slot s_l with $i < k$ and $l \leq i$ (analogously $k < i$ and $i \leq l$).*

With our structural observations regarding the propagation arrows we can now start to analyze the critical crossings depicted in Fig. 2(d) and (e). These crossings are critical in the sense that they have only avoidable crossings and no unavoidable ones. So, they decrease the performance of our algorithm and do not guarantee a constant competitive ratio like the other crossings depicted in Fig. 2. In the following sections, we overcome this problem by showing that for each of these critical crossings there must exist some other request that unavoidably crosses one of the requests, involved in the critical crossing.

5.2   The 4-0 Crossings

Recall that, by Lemma 1, the optimal solution for a 2-regular instance of the online OSCM-2 consists of minimizing crossings between every pair of requests. Thus, we can look at a pair of requests and exhaustively classify them as depicted in Fig. 2, and analyze the competitive ratio of an algorithm depending on how many of these types of crossings appear. In particular, if no 3-0 crossings (Fig. 2(e)) or 4-0 crossings (Fig. 2(d)) were produced by an algorithm, the algorithm would be 3-competitive at worst, as any sub-optimal placement would be trivially compensated by at least one unavoidable crossing. Thus, we only have to look at 3-0 and 4-0 crossings.

Using Lemma 2 we can now prove that Algorithm 1 will not make too many mistakes when producing 4-0 crossings. First we prove that Algorithm 1 will never produce 4-0 crossings with gaps, i.e., unfulfilled slots between the 2 slots generating the 4-0 crossing.

Lemma 5. *Algorithm 1 never generates 4-0 crossings with gaps in between. More precisely, for each pair s_i, s_j with $i < j$ assigned by Algorithm 1 that generate a 4-0 crossing, every s_k with $i < k < j$ is already full.*

If we have a request for a pair of vertices, such that every available slot generates at least one 4-0 crossing, we call it a forced 4-0 crossing. We prove now that Algorithm 1 only generates 4-0 crossings when they are forced or in a very specific configuration. Observe, that it is possible that more than one 4-0 crossing is forced by the same request.

Lemma 6. *If Algorithm 1 is used, for every forced 4-0 crossing there is at least one uniquely identifiable and unavoidable crossing.*

We just proved that forced 4-0 incur in one additional unavoidable crossing, this means that we can consider 4-0 crossings as if they were, in a sense 5-1 crossings instead, with a competitive ratio of 5 instead of being unbounded. However, this is not enough, there can be 4-0 crossings produced by Algorithm 1 that are not forced. In the following lemma we prove that non-forced 4-0 crossings are only produced in a very specific configuration. Then we will proceed to look at the number of uniquely identifiable unavoidable crossings of that configuration.

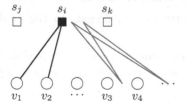 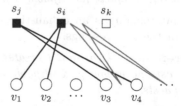

Fig. 5. If there is more than one slot positioned like the red ones (between the slots s_i and s_k with one vertex between v_3 and v_4, and one to the right of v_4 each), Algorithm 1 may choose slot s_j generating a 4-0 crossing. (Color figure online)

Lemma 7. *Given a request for a pair of vertices in a graph, whose 4-0 crossings have either been forced or were served because any alternative placement would cause two 3-1 crossings as sketched in Fig. 5. If a slot is available that will not generate any 4-0 crossings this slot will be selected by Algorithm 1 over any slot that will generate a 4-0 crossing, unless there are two additional requests resulting in two 3-1 crossings for the alternative placement as depicted in Fig. 5.*

The 4-0 crossings described in Lemma 7, also have uniquely identifiable unavoidable crossings, just as the forced 4-0 crossings from Lemma 6.

Lemma 8. *Any 4-0 crossing incurred by Algorithm 1, where any alternative placement would result in two 3-1 crossings as sketched in Fig. 5, has two uniquely identifiable unavoidable crossings.*

We can finally conclude, using Lemmas 6 and 8, that any 4-0 crossings incurred by Algorithm 1 have at least one unavoidable crossing.

Theorem 3. *Forced and non-forced 4-0 crossings incurred by Algorithm 1 have at least one unavoidable crossing.*

5.3   The 3-0 Crossings

It remains to prove that Algorithm 1 only generates a 3-0 crossing – depicted in Fig. 2(e) – if there is at least one unavoidable crossing for one of the two requests responsible for the 3-0 crossing. The following proofs use case distinctions like in the proofs from the previous section, handling the 4-0 crossings.

Similar to the 4-0 case, we start by proving that Algorithm 1 never produces a 3-0 crossing with a gap.

Lemma 9. *Algorithm 1 never generates 3-0 crossings with gaps in between. More precisely, for each pair s_j, s_i assigned by Algorithm 1 with $j < i$ that generate a 3-0 crossing, every slot s_k with $j < k < i$ is already full.*

Next, we explore the situation that 3-0 crossings happen at the edge of the graph, i.e., a placement on any remaining slot causes a 3-0 crossing.

Lemma 10. *Given two requests $\{v_1, v_2\}$ and $\{v_2, v_3\}$ with $v_1 < v_2 < v_3$. Assume w.l.o.g. that Algorithm 1 creates a 3-0 crossing between these requests, with the first request for vertices $\{v_1, v_2\}$ being placed in slot s_i and during the placement of the second request there is no available slot $s_k > s_i$. Then there is at least one uniquely identifiable unavoidable crossing with the request $\{v_2, v_3\}$.*

What remains is an exhaustive case distinction analogous to the analysis done for the 4-0 crossings.

Theorem 4. *If Algorithm 1 creates a 3-0 crossing between two requests, there is at least one uniquely identifiable unavoidable crossing for at least one of the two requests.*

This theorem shows that 3-0 crossings incurred by Algorithm 1 only happen in conjunction with two extra unavoidable crossings with the request generating the 3-0 crossing, this means, that any 3-0 crossing is in effect a 5-2 crossing, which would be better than 5-competitive.

5.4   The Upper Bound

We can now put all results together to give an upper bound for the competitive ratio of Algorithm 1 to solve online slotted OSCM-2 on 2-regular graphs.

Theorem 5. *Algorithm 1 solves the online slotted OSCM-2 on 2-regular graphs with a competitive ratio of at most 5.*

Proof. In order to calculate the competitive ratio of Algorithm 1 we simply compare for every pair of requests, what the optimal placement compared to the placement chosen by Algorithm 1 would be.

We exhaustively look at possible placements of pairs of requests, as depicted in Fig. 2. Observe, that except for the 3-0 crossings and 4-0 crossings, the rest of possible request pairs are no worse than 3-competitive regardless of the algorithm used. Moreover, Theorem 3 ensures that for every 4-0 crossing incurred by Algorithm 1, there is at least one uniquely identifiable unavoidable crossing, meaning that the number of crossings incurred by Algorithm 1 is 5, but optimally there must be at least 1 unavoidable crossing. Finally, Theorem 4 guarantees that there are also two uniquely identifiable unavoidable crossings for every occurrence of a 3-0 crossing. Thus, Algorithm 1 is at most 5-competitive.   □

6   Conclusion

In this work we have shown that the general slotted OSCM-k is not competitive for any $k \geq 2$, which led us to analyze the case of the slotted OSCM-k on 2-regular graphs. On this graph class, we have given a construction which proved a lower bound on the competitive ratio of 4/3. Algorithm 1, which utilizes the information of the remaining space and unavoidable crossings in the graph in the form of our so-called *propagation arrows*, was proven to be at most 5-competitive. This was done by limiting the number of total crossings generated by pairs of requests that do not cross one another in an optimal solution.

There are several open questions which we were not able to answer in the scope of this work. First, there is still a considerable gap between the lower and upper bound of the competitive ratio that we have given. We assume that Algorithm 1 performs better than analyzed and that the upper bound can be made tighter.

While Theorem 1 proves non-competitiveness on general graphs for any $k \geq 2$, the case of regular graphs with degree 3 or higher is still open. We suggest to analyze this graph class further.

References

1. Battista, G.D., Eades, P., Tamassia, R., Tollis, I.G.: Algorithms for drawing graphs: an annotated bibliography. Comput. Geom. **4**, 235–282 (1994)
2. Borodin, A., El-Yaniv, R.: Online Computation and Competitive Analysis. Cambridge University Press, Cambridge (1998)

3. Burjons, E., Fuchs, J., Lotze, H.: The slotted online one-sided crossing minimization problem on 2-regular graphs. CoRR abs/2201.04061 (2022)
4. Dujmovic, V., Whitesides, S.: An efficient fixed parameter tractable algorithm for 1-sided crossing minimization. Algorithmica **40**(1), 15–31 (2004)
5. Eades, P., Wormald, N.C.: Edge crossings in drawings of bipartite graphs. Algorithmica **11**(4), 379–403 (1994)
6. Frishman, Y., Tal, A.: Online dynamic graph drawing. IEEE Trans. Vis. Comput. Graph. **14**(4), 727–740 (2008)
7. Garey, M.R., Johnson, D.S.: Crossing number is NP-complete. SIAM J. Algebraic Discrete Methods **4**(3), 312–316 (1983)
8. Kobayashi, Y., Tamaki, H.: A fast and simple subexponential fixed parameter algorithm for one-sided crossing minimization. Algorithmica **72**(3), 778–790 (2015)
9. Komm, D.: An Introduction to Online Computation - Determinism, Randomization, Advice. Texts in Theoretical Computer Science. An EATCS Series, Springer, Cham (2016). https://doi.org/10.1007/978-3-319-42749-2
10. Li, X.Y., Stallmann, M.F.M.: New bounds on the barycenter heuristic for bipartite graph drawing. Inf. Process. Lett. **82**(6), 293–298 (2002)
11. Muñoz, X., Unger, W., Vrt'o, I.: One sided crossing minimization is NP-hard for sparse graphs. In: Mutzel, P., Jünger, M., Leipert, S. (eds.) GD 2001. LNCS, vol. 2265, pp. 115–123. Springer, Heidelberg (2002). https://doi.org/10.1007/3-540-45848-4_10
12. Nagamochi, H.: An improved bound on the one-sided minimum crossing number in two-layered drawings. Discret. Comput. Geom. **33**(4), 569–591 (2005)
13. Nagamochi, H.: On the one-sided crossing minimization in a bipartite graph with large degrees. Theor. Comput. Sci. **332**(1–3), 417–446 (2005)
14. North, S.C., Woodhull, G.: Online hierarchical graph drawing. In: Mutzel, P., Jünger, M., Leipert, S. (eds.) GD 2001. LNCS, vol. 2265, pp. 232–246. Springer, Heidelberg (2002). https://doi.org/10.1007/3-540-45848-4_19
15. Schaefer, M.: The graph crossing number and its variants: a survey. Electron. J. Combin. (2012)
16. Shannon, R., Quigley, A.J.: Considerations in dynamic graph drawing: a survey. Comput. Sci. Inform. (2007)
17. Sleator, D.D., Tarjan, R.E.: Amortized efficiency of list update and paging rules. Commun. ACM **28**(2), 202–208 (1985)

Perfect Matching Cuts Partitioning a Graph into Complementary Subgraphs

Diane Castonguay[1], Erika M. M. Coelho[1], Hebert Coelho[1],
Julliano R. Nascimento[1(✉)], and Uéverton S. Souza[2,3]

[1] Instituto de Informática, Universidade Federal de Goiás, Goiânia, Brazil
{diane,erikamorais,hebert,jullianonascimento}@inf.ufg.br
[2] Instituto de Computação, Universidade Federal Fluminense, Niterói, Brazil
[3] Institute of Informatics, University of Warsaw, Warsaw, Poland
ueverton@ic.uff.br

Abstract. In PARTITION INTO COMPLEMENTARY SUBGRAPHS (COMP-SUB) we are given a graph $G = (V, E)$, and an edge set property Π, and asked whether G can be decomposed into two graphs, H and its complement $\overline{H}$, for some graph H, in such a way that the edge cut $[V(H), V(\overline{H})]$ satisfies the property Π. Motivated by previous work, we consider COMP-SUB(Π) when the property $\Pi = \mathscr{PM}$ specifies that the edge cut of the decomposition is a perfect matching. We prove that COMP-SUB($\mathscr{PM}$) is GI-hard when the graph G is $\{C_{k \geq 7}, \overline{C}_{k \geq 7}\}$-free. On the other hand, we show that COMP-SUB($\mathscr{PM}$) is polynomial time solvable on *hole*-free graphs and on P_5-free graphs. Furthermore, we present characterizations of COMP-SUB($\mathscr{PM}$) on chordal, distance-hereditary, and extended P_4-laden graphs.

Keywords: Graph partitioning · Complementary subgraphs · Perfect matching · Matching cut · Graph isomorphism

1 Introduction

Finding graph partitions with some special properties has been a topic of extensive research. Several combinatorial problems can be viewed as partition problems, such as VERTEX COLORING and CLIQUE COVER. In addition, many graph classes, e.g. bipartite and split graphs, can also be defined through a partition of its vertex set. In particular, the class of *complementary prisms* [11] are defined over complementary parts. The *complementary prism* $G\overline{G}$ of a graph G arises from the disjoint union of the graph G and its complement $\overline{G}$ by adding the edges

This research has received funding from Rio de Janeiro Research Support Foundation (FAPERJ) under grant agreement E-26/201.344/2021, National Council for Scientific and Technological Development (CNPq) under grant agreement 309832/2020-9, and the European Research Council (ERC) under the European Union's Horizon 2020 research and innovation programme under grant agreement CUTACOMBS (No. 714704).

C. Bazgan and H. Fernau (Eds.): IWOCA 2022, LNCS 13270, pp. 257–269, 2022.
https://doi.org/10.1007/978-3-031-06678-8_19

of a perfect matching between vertices with same label in G and $\overline{G}$. Studies concerning the computational complexity of classical graph problems restricted to the class of complementary prisms graphs can be found in [4,9].

We say that a graph $G = (V, E)$ is decomposed into two graphs G_1 and G_2 if $V(G)$ can be partitioned into V_1 and V_2, where $G[V_1] = G_1$ and $G[V_2] = G_2$. The edge cut $[V_1, V_2]$ is called the edge cut of this decomposition.

As a generalization of complementary prisms, Nascimento, Souza and Szwarcfiter [16] introduced the problem defined as follows.

PARTITION INTO COMPLEMENTARY SUBGRAPHS (COMP-SUB)

Instance: A graph $G = (V, E)$, and an edge set property Π.

Question: Can G be decomposed into two graphs, H and its complement $\overline{H}$, for some graph H, in such a way that the edge cut M of the decomposition satisfies the property Π?

For short, we abbreviate PARTITION INTO COMPLEMENTARY SUBGRAPHS with the edge set property Π as COMP-SUB(Π). We write $G \in$ COMP-SUB(Π) to denote that G is a *yes*-instance of COMP-SUB(Π) and we call $(H, \overline{H})$ as a *complementary decomposition* of G.

The COMP-SUB(Π) problem also finds motivation in parameterized complexity. Recognizing whether a graph has a complementary decomposition can be useful for solving problems in FPT-time, as pointed out in [16]. Nascimento, Souza and Szwarcfiter [16] considered the cases where the edge cut M is empty or induces a complete bipartite graph. They also presented some remarks when Π is a general edge set property. In particular, when M is empty, they make some links between COMP-SUB(Π) and the GRAPH ISOMORPHISM problem, from which they show that COMP-SUB(Π) is GI-hard.

It is known that the recognition of complementary prisms can be done in polynomial time [5]. This implies that, when the property Π is a perfect matching M between corresponding vertices in H and $\overline{H}$, the COMP-SUB(Π) problem is polynomial-time solvable. So, a natural question is the study of COMP-SUB(Π) when Π specifies that M is any perfect matching. In this context, two related problems arise: MATCHING CUT [13,17] and PERFECT MATCHING CUT [12]. A *(perfect) matching cut* is a partition of vertices of a graph into two parts such that the set of edges crossing between the parts forms a (perfect) matching. Considering $\Pi = \mathscr{PM}$ as the property of being a perfect matching, COMP-SUB($\mathscr{PM}$) can be seen as a variant of PERFECT MATCHING CUT with the additional restriction that the two parts must induce complementary subgraphs. Note that studies regarding matchings satisfying particular constraints have received wide attention in the literature (c.f. [10,14,15,19,20]).

Motivated by Nascimento, Souza and Szwarcfiter [16], in this paper we deal with COMP-SUB(Π), when $\Pi = \mathscr{PM}$ considers M as a perfect matching. We show that COMP-SUB($\mathscr{PM}$) is GI-hard when the graph G is $\{C_{k \geq 7}, \overline{C}_{k \geq 7}\}$-free. On the other hand, we present polynomial time algorithms able to solve COMP-SUB($\mathscr{PM}$) when the input graph G is *hole*-free or P_5-free. In addition, we characterize graphs $G \in$ COMP-SUB($\mathscr{PM}$) when G is chordal, distance-hereditary, or extended P_4-laden. Although extended P_4-laden graphs generalize cographs, we also show a simpler characterization for cographs.

The paper is organized as follows. Section 2 contains some fundamental concepts and an auxiliary result. Sections 3 and 4 contains our results on some *cycle*-free graphs and graphs with few P_4's, respectively. Further discussions are presented in Sect. 5.

2 Preliminaries

We consider only finite, simple, and undirected graphs, and we use standard terminology and notation. See [1] for graph-theoretic terms not defined here.

Let G be a graph. For a vertex $v \in V(G)$, we denote its *open neighborhood* by $N_G(v)$, and its *closed neighborhood*, denoted by $N_G[v] := N_G(v) \cup \{v\}$. For a set $U \subseteq V(G)$, let $N_G(U) = \bigcup_{v \in U} N_G(v) \setminus U$, and $N_G[U] = N_G(U) \cup U$. The subgraph of G *induced* by U, denoted by $G[U]$, is the graph whose vertex set is U and whose edge set consists of all the edges in $E(G)$ that have both endvertices in U.

Let G be a graph. A set $U \subseteq V(G)$ is called a *clique* (resp. *independent set*) if the vertices in U are pairwise adjacent (resp. nonadjacent). We denote by K_n a *complete graph*, I_n an independent set, P_n a *path graph*, and C_n a *cycle graph* on n vertices. Let r be a positive integer. An *r-partite graph* is one whose vertex set can be partitioned into r subsets, in such a way that no edge has both ends in the same subset. An r-partite graph is *complete* if any two vertices in different subsets are adjacent. When r is not specified, we simply say *(complete) multipartite*. A *split graph* G is one whose vertex set admits a partition $V(G) = C \cup I$ into a clique C and an independent set I. The *complement* $\overline{G}$ of a graph G is the graph defined by $V(\overline{G}) = V(G)$ and $uv \in E(\overline{G})$ if and only if $uv \notin E(G)$.

Let $P = v_1 v_2 \ldots v_n$ be a path. We call $v_2, \ldots, v_{n-1}$ as *inner vertices* of P. Two or more paths in a graph are *independent* if none of them contains an inner vertex of another. A graph G is *ℓ-connected* if any two of its vertices can be joined by ℓ independent paths. A 2-connected graph is called *biconnected*.

A vertex v in a graph G is a *cutvertex* or *cutpoint*, if $G \setminus \{v\}$ is disconnected. A maximal connected subgraph without a cutpoint is a *block*. The *block-cutpoint tree* of a graph G is a bipartite graph whose vertex set consists of the set of cutpoints of G and the set of blocks of G. A cutpoint is adjacent to a block whenever the cutpoint belongs to the block in G.

Two graphs $G = (V, E)$ and $G' = (V', E')$ are *isomorphic*, denoted as $G \simeq H$, if and only if there is a bijection, called *isomorphism function*, $\varphi : V \to V'$ such that $uv \in E$ if and only if $\varphi(u)\varphi(v) \in E'$, for every $u, v \in V$. A graph G is *self-complementary* if $G \simeq \overline{G}$. The GRAPH ISOMORPHISM problem receives as input two graphs G and G' and asks whether $G \simeq G'$. We denote by GI the class of problems that admit a polynomial-time reduction to GRAPH ISOMORPHISM.

A problem Q is GI-*complete* if the two conditions are satisfied: (i) Q is a member of GI; and (ii) Q is GI-*hard*, that is, for every problem $Q' \in$ GI, Q' is polynomially reducible to Q.

We denote the set of positive integers $\{1,\ldots,k\}$ by $[k]$. Let G and $G_1,\ldots,G_k$ be graphs. We say that G is $\{G_1,\ldots,G_k\}$-*free* if G does not contain G_i as an induced subgraph, for every $i \in [k]$.

Let G and H be two graphs such that $V(G) \cap V(H) = \emptyset$. The *disjoint union* of G and H, denoted by $G \cup H$, is the graph with $V(G \cup H) = V(G) \cup V(H)$ and $E(G \cup H) = E(G) \cup E(H)$. The *join* of G and H, denoted by $G + H$, is the graph with $V(G + H) = V(G) \cup V(H)$ and $E(G + H) = E(G) \cup E(H) \cup \{uv : u \in V(G) \text{ and } v \in V(H)\}$.

Let G be a graph and $\mathscr{C}$ a class of graphs. A set $S \subseteq V(G)$ is a $\mathscr{C}$-*modulator* if $G \setminus S$ belongs to $\mathscr{C}$. We define the *distance* of G to class $\mathscr{C}$ as the size of a minimum S which is a $\mathscr{C}$-modulator.

Let G be a graph that has a complementary decomposition (G_1, G_2) with perfect matching cut $M = \{u_1v_1, \ldots, u_nv_n\}$, where $u_i \in V(G_1)$ and $v_i \in V(G_2)$, $i \in [n]$. We say that u_i (resp. v_i) is the *corresponding vertex* of v_i (resp. u_i), for every $i \in [n]$. For $X \subseteq V(G_1)$, we call $X^{G_2} = \{v_i \in V(G_2) : u_i \in X\}$ as the *corresponding set* of X over G_2. Similarly, for $X \subseteq V(G_2)$, we call $X^{G_1} = \{u_i \in V(G_1) : v_i \in X\}$ as the *corresponding set* of X over G_1.

Next, we present an auxiliary result, defined for COMP-SUB($\mathscr{PM}$) with a restriction on the graphs of the decomposition. A *cograph* is a P_4-free graph.

Lemma 1. *Let G be a graph. The problem of determining whether G can be decomposed into two graphs, G_1, and its complement G_2, such that G_1 is a cograph and the edge cut of the decomposition is a perfect matching, can be solved in polynomial time.*

Proof. Let $\mathscr{C}$ be the class of cographs and G a $2n$-vertex graph. Suppose that G is decomposable into complementary subgraphs G_1 and G_2, such that $G_1 \in \mathscr{C}$ and the edge cut M of the decomposition is a perfect matching.

Since $\mathscr{C}$ is closed under complement, we have that $G_2 \in \mathscr{C}$. Given that GRAPH ISOMORPHISM is linear-time solvable on cographs [7], we perform a brute force algorithm to check every *relevant* partition $V(G_1), V(G_2)$ of $V(G)$. For that, we propose Algorithm 1, explained in sequel.

Algorithm 1: PARTITION-INTO-COMPLEMENTARY-COGRAPHS(G)

Input: A graph G.
Output: Whether G admits a complementary decomposition such that the edge cut of the decomposition is a perfect matching.

1 **forall** $x_1, x_2, y_1, y_2 \in V(G)$ **do**
2 | $V(G_1) := N_G[\{x_1, x_2\}] \setminus \{y_1, y_2\}$
3 | $V(G_2) := V(G) \setminus V(G_1)$
4 | $M := \{xy \in E(G) : x \in V(G_1), y \in V(G_2)\}$
5 | **if** M *is a perfect matching* **and** G_1 *is a cograph* **and** G_2 *is a cograph* **and** $G_1 \simeq \overline{G}_2$ **then**
6 | | **return** yes

7 **return** no

We know that a cograph is connected if and only if its complement is disconnected [8]. Consequently, if a complementary decomposition (G_1, G_2) exists, then either G_1 or G_2 is disconnected, say G_2. Then G_1 can be obtained by a join between the corresponding connected components of G_2. Thus, there exist two adjacent vertices $x_1, x_2 \in V(G_1)$, such that $N_{G_1}[\{x_1, x_2\}] = V(G_1)$. Furthermore, the edge set M of the decomposition implies that there exist $y_1, y_2 \in V(G_2)$ such that $x_1 y_1, x_2 y_2 \in M$.

By the above arguments, it is possible to find $V(G_1)$ by means of $N_G[\{x_1, x_2\}]$ except for two vertices $y_1, y_2 \in N_G[\{x_1, x_2\}]$ that must belong to $V(G_2)$. This way, $V(G_2)$ is obtained by $\{y_1, y_2\} \cup \{v \in V(G) : v \notin N_G[\{x_1, x_2\}]\}$. Once found $V(G_1), V(G_2)$, and M, we test whether M is a perfect matching and whether G_1 and G_2 are cographs. If so, we compute $\overline{G}_2$ and then we check isomorphism between G_1 and $\overline{G}_2$.

The correctness of the algorithm follows from the fact that all the possible relevant partitions (for the emergence of the cographs, if any) are considered.

Now, we show that Algorithm 1 runs in polynomial time.

For enumerating every 4-tuple of vertices $x_1, x_2, y_1, y_2 \in V(G)$ it is required $O(n^4)$ time. After, in $O(n + m)$ time we can check whether M is a perfect matching, as well as checking whether G_1 and G_2 are cographs. Finally, for computing $\overline{G}_2$ and checking isomorphism between G_1 and $\overline{G}_2$ is also required $O(n + m)$ time. Therefore, the running time of Algorithm 1 takes $O(n^5 + n^4 m)$ time. $\qquad\square$

3 Results on Some C_k-Free Graphs

We begin by showing a hardness result, in Theorem 1.

Theorem 1. COMP-SUB($\mathscr{PM}$) *is* GI*-hard on* $\{C_{k \geq 7}, \overline{C}_{k \geq 7}\}$*-free graphs.*

Proof. Given that GRAPH ISOMORPHISM is GI-hard on split graphs [6], we show a polynomial-time reduction from such a problem to COMP-SUB($\mathscr{PM}$).

Note that a split graph is connected if and only if it does not contain isolated vertices. Therefore, we may assume that the instances of GRAPH ISOMORPHISM on split graphs are pairs of connected split graphs.

Let A and B be connected split graphs such that $|V(A)| = |V(B)| = n$, for some $n \geq 3$. From an instance (A, B) of GRAPH ISOMORPHISM, we construct an instance G of COMP-SUB($\mathscr{PM}$).

Let G arise from the disjoint union between A, $\overline{B}$, K_n, and I_n. Denote K_n by K and I_n by I. We make every vertex in $V(A)$ adjacent to every vertex in $V(K)$. Furthermore, we add an arbitrary perfect matching between $V(A)$ and $V(I)$ and between $V(K)$ and $V(\overline{B})$. An example of graph G follows in Fig. 1. Additionally, let $H_1 = G[V(A) \cup V(K)]$ and $H_2 = G[V(\overline{B}) \cup V(I)]$. Clearly, the construction can be done in polynomial time.

We first show that G is $\{C_{k \geq 7}, \overline{C}_{k \geq 7}\}$-free.

Claim 1. *Let G be the graph obtained from the construction. It holds that G is a $\{C_{k \geq 7}, \overline{C}_{k \geq 7}\}$-free graph.*

Proof (of Claim 1). We prove that (I) G is $C_{k \geq 7}$-free, and (II) G is $\overline{C}_{k \geq 7}$-free.

(I) Suppose by contradiction that G contains a $C_{k \geq 7}$, denoted as C, as induced subgraph. We may assume that k is minimum.

By construction, H_1 and H_2 are split graphs and it is clear that H_1 and H_2 are $C_{\ell+4}$-free, for every $\ell \geq 0$. Then $V(C) \not\subseteq V(H_1)$ and $V(C) \not\subseteq V(H_2)$. So, we assume that $V(C) \cap V(H_1) \neq \emptyset$ and $V(C) \cap V(H_2) \neq \emptyset$. Since I is a set of vertices with degree one in G, we have that $V(C) \cap I = \emptyset$. So, we may suppose that $V(C) \cap V(\overline{B}) \neq \emptyset$ and since C is a cycle, $|V(C) \cap V(\overline{B})| \geq 2$. Since $\overline{B}$ is split, we have that $|V(C) \cap V(\overline{B})| \leq 4$.

Since C is a cycle and K is a complete graph, C must contain exactly two vertices from K and no vertex of A. Then, $|V(C)| \geq 7$ implies that $|V(C) \cap V(\overline{B})| \geq 5$, a contradiction.

(II) Suppose by contradiction that G contains a $\overline{C}_{k \geq 7}$, denoted as D, as induced subgraph. Let $V(D) = \{d_1, \ldots, d_\ell\}$, for some $\ell \geq 7$, and $E(D) = \{d_i d_j : 1 \leq i < j \leq \ell\} \setminus (\{d_i d_{i+1} : 1 \leq i \leq \ell - 1\} \cup \{d_\ell d_1\})$.

By definition of D, $\{d_1, d_2, d_4, d_5\}$ induces a C_4. Then, since H_1 and H_2 are split graphs, $V(D) \not\subseteq V(H_1)$ and $V(D) \not\subseteq V(H_2)$. So, we assume that $V(D) \cap V(H_1) \neq \emptyset$ and $V(D) \cap V(H_2) \neq \emptyset$. Then, there exists $i, j \in [\ell]$ such that $d_i \in V(H_1)$, $d_j \in V(H_2)$ and $d_i d_j \in E(D)$.

Without loss of generality, suppose that $i = 1$. Since $\{d_1, d_3, d_5\}$ induces a K_3, we may assume that $\{d_1, d_3, d_5\} \subseteq V(H_1)$. Thus, $d_1 d_j \in E(D)$, for some $j \in \{4, 6, \ldots, \ell - 1\}$. Notice that, if $j = 4$ (resp. $j \geq 6$), then $\{d_1, d_4, d_6\}$ (resp. $\{d_1, d_3, d_j\}$) induces a K_3 which intersects both $V(H_1)$ and $V(H_2)$, a contradiction. Therefore G is $\overline{C}_{k \geq 7}$-free.   □

In what follows, we prove that (A, B) is a *yes*-instance of GRAPH ISOMORPHISM if and only if G is a *yes*-instance of COMP-SUB($\mathscr{PM}$).

Suppose that $A \simeq B$. Since $I_n = \overline{K}_n$, $\overline{B} \simeq A$, and there is no edge between a vertex in I and a vertex in $V(\overline{B})$, it is easy to see that H_1 and $\overline{H}_2$ are isomorphic. Therefore, G is a *yes*-instance of COMP-SUB($\mathscr{PM}$).

For the converse, we suppose that G is a *yes*-instance of COMP-SUB($\mathscr{PM}$). Let (V', V'') be a partition of $V(G)$ into complementary parts such that $[V', V'']$ is a perfect matching. Since I is a set of vertices with degree one in G and A is connected, it holds that either $(I \subset V'$ and $V(A) \subset V'')$ or $(V(A) \subset V'$ and $I \subset V'')$. Suppose that $V(A) \subset V'$. This implies that $V' = V(A) \cup K$ and $V'' = V(\overline{B}) \cup I$. Since $G[V']$ and $G[V'']$ are complementary, we have that $G[V'] \simeq \overline{G[V'']}$. Hence, due to the automorphism of universal vertices, it holds that $A \simeq B$.   □

See in Fig. 1 an example of the construction presented in Theorem 1.

Despite the hardness results presented in Theorem 1, next we show that COMP-SUB($\mathscr{PM}$) can be solved in polynomial time on *hole*-free graphs. Recall that a *hole* is a cycle on 5 or more vertices.

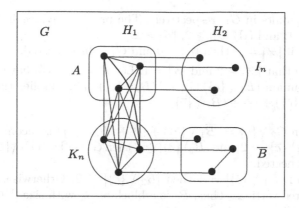

Fig. 1. Graph G constructed for Theorem 1.

Theorem 2. COMP-SUB($\mathscr{PM}$) *is polynomial-time solvable on hole-free graphs.*

Proof. Let G be a *hole*-free graph having $2n$-vertices. We assume that n is at least 5; otherwise, the problem can be solved in $O(1)$ time.

Suppose that $G \in$ COMP-SUB($\mathscr{PM}$), then G is decomposable into complementary subgraphs G_1 and G_2, such that the edge cut M of the decomposition is a perfect matching.

Recall that G_1 or G_2 is a connected graph. Thus, we assume that G_1 is connected.

We go through the proof by analysing the structure of the graphs of the decomposition by means of their connectivity (Claims 1 and 2), and we conclude by showing how to find that decomposition when it exists.

Claim 1. *Let G_1 be a connected graph with at least five vertices and $F \subseteq V(G_1)$. If $G[F]$ is biconnected, then $G[F^{G_2}]$ is a cluster graph.*

Proof (of Claim 1). Suppose, by contradiction, that $G[F^{G_2}]$ is not a cluster graph and let $v_1v_2v_3$ be a P_3 in $G[F^{G_2}]$. Since $G[F]$ is 2-connected, there exist two independent paths between any two vertices in F. Consider $u_1, u_2, u_3 \in F$ as the corresponding vertices of v_1, v_2, v_3, respectively. Let P and P' be two independent paths between u_1 and u_3 in F. Since P and P' are independent, u_2 does not belong to P or P', say P. Then $P \cup \{v_1, v_2, v_3\}$ induces a *hole* in G, a contradiction. □

Next, we see more on the structure of G_1 and G_2.

Claim 2. *Let G_1 be a connected graph having at least five vertices. If G_1 is non-biconnected, then either there is $S \subset V(G_1)$ with $|S| \leq 2$ such that $G_1 \setminus S$ is biconnected; or there is $S' \subset V(G_2)$ with $|S'| \leq 2$ such that $G_2 \setminus S'$ is biconnected.*

Proof (of Claim 2). Suppose that G_1 is non-biconnected and let T be a block-cut-point tree of G_1. Let $\mathscr{B} = \{B_1, \ldots, B_s\}$ and $\mathscr{C} = \{c_1, \ldots, c_t\}$ be the sets of blocks and cutpoints in G_1, respectively. The proof is divided in two cases: (I) $|\mathscr{B}| \geq 2$, $|\mathscr{C}| = 1$; and (II) $|\mathscr{B}| \geq 2$, $|\mathscr{C}| \geq 2$.

Recall that if $|\mathscr{B}| = 1$, then $|\mathscr{C}| = 0$ and G_1 is biconnected.

(I) Suppose that $|\mathscr{B}| \geq 2$ and $|\mathscr{C}| = 1$. Let $\mathscr{C} = \{c\}$. We have that $G_1 \setminus \{c\}$ is the disjoint union $(B_1 \setminus \{c\}) \cup \cdots \cup (B_s \setminus \{c\})$. This implies that $\overline{G}_1 \setminus \{\overline{c}\}$ is the join $(\overline{B}_1 \setminus \{\overline{c}\}) + \cdots + (\overline{B}_s \setminus \{\overline{c}\})$.

- If $s \geq 3$ then $G_2 \setminus \{\overline{c}\} = (\overline{B}_1 \setminus \{\overline{c}\}) + \cdots + (\overline{B}_s \setminus \{\overline{c}\})$ is biconnected.
- If $s = 2$, $|\overline{B}_1 \setminus \{\overline{c}\}| \geq 2$, and $|\overline{B}_2 \setminus \{\overline{c}\}| \geq 2$ then $G_2 \setminus \{\overline{c}\} = (\overline{B}_1 \setminus \{\overline{c}\}) + (\overline{B}_2 \setminus \{\overline{c}\})$ is also biconnected.
- If $s = 2$ and $|\overline{B}_1 \setminus \{\overline{c}\}| = 1$ then $|\overline{B}_2 \setminus \{\overline{c}\}| \geq 2$. Otherwise, G_1 (and G_2) has only three vertices. Thus, B_2 is a block of G_1 with size $|V(G_1)| - 1$, and $S = B_1 \setminus \{c\}$ is as required.

(II) Now, consider that $|\mathscr{B}| \geq 2$ and $|\mathscr{C}| \geq 2$. Let $B, B' \in \mathscr{B}$ two distinct leaves in T and $c, c' \in \mathscr{C}$ be two distinct cutpoints such that $Bc, B'c' \in E(T)$.

Let $D = V(G_1) \setminus (B \cup B')$. Since B (resp. B') is a leaf in T, we have that $V(B) \setminus \{c\}$ (resp. $V(B') \setminus \{c'\}$) is not adjacent to $B' \cup D$ (resp. $B \cup D$). This implies that $\overline{G}_1 \setminus \{\overline{c}, \overline{c}'\}$ is the join $(\overline{B} \setminus \{\overline{c}\}) + (\overline{B}' \setminus \{\overline{c}'\}) + \overline{D}$.

- If $D \neq \emptyset$, we have that $(\overline{B} \setminus \{\overline{c}\}) + (\overline{B}' \setminus \{\overline{c}'\}) + \overline{D}$ is biconnected. Thus, $G_2 \setminus \{\overline{c}, \overline{c}'\}$ is biconnected as required.
- If $D = \emptyset$, $|B \setminus \{c\}| \geq 2$, and $|B' \setminus \{c'\}| \geq 2$, then $G_2 \setminus \{\overline{c}, \overline{c}'\} = (\overline{B} \setminus \{\overline{c}\}) + (\overline{B}' \setminus \{\overline{c}'\})$ is also biconnected.
- If $D = \emptyset$ and $|B \setminus \{c\}| = 1$, then $|B' \setminus \{c'\}| \geq 2$. Otherwise, G_1 (and G_2) has only four vertices. Thus, $G_1 \setminus B$ is biconnected (notice that $|B| = 2$).

This completes the proof of Claim 2.   □

By Claim 1, if G_1 is biconnected, then G_2 is a cluster graph. Since $G_1 \simeq \overline{G}_2$, we have that G_1 is a complete multipartite graph. Hence, G_1 and G_2 are cographs and, by Lemma 1, we can find the complementary partition of G in polynomial time.

Now, if G_1 is non-biconnected, recall that by Claim 2, either there is $S \subset V(G_1)$ with $|S| \leq 2$ such that $G_1 \setminus S$ is biconnected; or there is $S' \subset V(G_2)$ with $|S'| \leq 2$ such that $G_2 \setminus S'$ is biconnected.

Thus, there is a fixed number of vertices (at most 2) such that removing from G_1 or G_2 leaves a biconnected graph. We deal with the case that there exist $c, c' \in V(G_2)$ such that $G_2 \setminus \{c, c'\}$ is biconnected. The approach for the other case is similar.

If there exist $c, c' \in V(G_2)$ such that $G_2 \setminus \{c, c'\}$ is 2-connected, by Claim 1 (dual), we have that the graph induced by $(V(G_2) \setminus \{c, c'\})^{G_1}$ is a cluster graph. Then G_1 and $\overline{G}_2$ have distance to cluster equals 2. We proceed by Algorithm 2.

Algorithm 2: PARTITION-INTO-COMPLEMENTARY-SUBGRAPHS(G)

Input: A graph G.
Output: Whether G is partitionable into two complementary graphs G_1
    and G_2 such that G_1 and $\overline{G}_2$ have distance to cluster equals 2
    and the edge cut of the decomposition is a perfect matching.

1 **forall** $x_1, \ldots, x_4, y_1, \ldots, y_4 \in V(G)$ **do**
2 $\quad$ $V(G_2) := N_G[\{y_1, \ldots, y_4\}] \setminus \{x_1, \ldots, x_4\}$
3 $\quad$ $V(G_1) := V(G) \setminus V(G_2)$
4 $\quad$ $M := \{xy \in E(G) : x \in V(G_1), y \in V(G_2)\}$
5 $\quad$ **if** M *is a perfect matching* **then**
6 $\quad\quad$ **forall** *cluster-modulator* S_1 *of* G_1, *such that* $|S_1| \leq 2$ **do**
7 $\quad\quad\quad$ **forall** *cluster-modulator* S_2 *of* $\overline{G}_2$, *such that* $|S_2| = |S_1|$ **do**
8 $\quad\quad\quad\quad$ **forall** *mapping* $f : S_1 \mapsto S_2$ **do**
9 $\quad\quad\quad\quad\quad$ **if** f *can be extended to an isomorphism from* G_1 *to* $\overline{G}_2$
        then
10 $\quad\quad\quad\quad\quad\quad$ **return** yes

---

11 **return** no

Since G_2 has distance to complete multipartite equals 2, there exist four
vertices $y_1, \ldots, y_4 \in V(G_2)$ such that $N_{G_2}[\{y_1, \ldots, y_4\}] = V(G_2)$. Then, if a
complementary decomposition (G_1, G_2) exists, we have that $|N_G[\{y_1, \ldots, y_4\}]| =$
$n + 4$. Thence, it is possible to find $V(G_2)$ which is $N_G[\{y_1, \ldots, y_4\}]$ except for
four vertices $x_1, \ldots, x_4 \in N_G[\{y_1, \ldots, y_4\}]$. We put $x_1, \ldots, x_4$ in $V(G_1)$ as well as
the remaining vertices $\{v \in V(G) : v \notin N_G[\{y_1, \ldots, y_4\}]\}$. Given $V(G_1), V(G_2)$,
and M, we check whether M is a perfect matching. If so, we compute $\overline{G}_2$ and we
proceed to the step of finding cluster-modulators S_1 for G_1 and S_2 for $\overline{G}_2$, that
are done by Lines 6–7. In a naive manner, all the possible pair of modulators
can be found in $O(n^4)$, but we show how to find them in a more efficient way.

We first find a $P_3 = w_1 w_2 w_3$ in G_1. We know that at least one vertex in
$\{w_1, w_2, w_3\}$ must be included in a cluster-modulator for G_1. Then, for every
$w \in \{w_1, w_2, w_3\}$ we put $w \in S_1$ and we branch by searching (if any) for a
$P_3 = w'_1 w'_2 w'_3$ in $G_1 \setminus \{w\}$. Again, given that at least one vertex in $\{w'_1, w'_2, w'_3\}$
must be included in a cluster-modulator for G_1, for every $w' \in \{w_1, w_2, w_3\}$ we
put $w' \in S_1$. If $G_1 \setminus S_1$ is a cluster graph, we proceed to finding, in the same
manner, a cluster-modulator S_2 for $\overline{G}_2$. Note that this is basically a bounded
search tree algorithm for finding cluster vertex deletion sets.

Given a pair of modulators S_1 and S_2 such that $|S_1| = |S_2|$, and a mapping
from S_1 to S_2, the final task is checking if such a mapping can be extended to an
isomorphism between G_1 and $\overline{G}_2$. Note that, by the bounded search tree technique, the number of pairs of modulators and mappings that must be considered
is bounded by a constant.

Recall that G_1 (resp. $\overline{G}_2$) is a disjoint union of complete graphs $H_1 \cup \cdots \cup H_p$
(resp. $H'_1 \cup \cdots \cup H'_p$), for some $p \geq 2$, with the addition of two vertices w, w'

(resp. z, z') arbitrarily adjacent to $H_1 \cup \cdots \cup H_p$ (resp. $H_1' \cup \cdots \cup H_p'$). With this structure, an isomorphism from G_1 to $\overline{G}_2$ can be determined as follows.

For a mapping $w \mapsto z$, $w' \mapsto z'$, we can map H_i to H_j', $i, j \in [p]$, if and only if

$$|V(H_i)| = |V(H_j')| \text{ and}$$
$$|N_{H_i}(w) \setminus N_{H_i}(w')| = |N_{H_j'}(z) \setminus N_{H_j'}(z')| \text{ and}$$
$$|N_{H_i}(w') \setminus N_{H_i}(w)| = |N_{H_j'}(z') \setminus N_{H_j'}(z)| \text{ and}$$
$$|N_{H_i}(w) \cap N_{H_j}(w')| = |N_{H_j'}(z) \cap N_{H_j'}(z')|.$$

Therefore, each mapping $w \mapsto z$, $w' \mapsto z'$ defines "types" of cliques, from which the mapping can be extended to an isomorphism from G_1 to $\overline{G}_2$ if and only if G_1 and $\overline{G}_2$ have the same number of cliques per type.

Next, we analyse the running time of Algorithm 2.

First, in Line 1, we check every 8-tuple of vertices in $V(G)$ to separate those $x_1, \ldots, x_4 \in V(G_1)$ and $y_1, \ldots, y_4 \in V(G_2)$, which requires $O(n^8)$ time. Lines 2–4 define $V(G_1), V(G_2)$, and M, which run in $O(n + m)$ time. Checking whether M is a perfect matching (Line 5) can be done in $O(n + m)$ time.

Recall that a P_3 in G can be found in $O(n + m)$ time. By the method previously described, Line 6 can be done by finding a $P_3 = w_1 w_2 w_3$ in G_1; for every $w \in \{w_1, w_2, w_3\}$ finding a $P_3 = w_1' w_2' w_3'$ in $G_1 \setminus \{w\}$ in G_1; and finally, for every $w' \in \{w_1, w_2, w_3\}$, checking whether $G_1 \setminus S_1 = \{w, w'\}$ is a cluster graph. This produces a ternary search tree with height equals 2. Hence with 9 leaf nodes, that are at most 9 possible cluster-modulators $\{w, w'\}$ for G_1. This requires a running time of $O(m + n)$. For every of those possible cluster-modulators for G_1 we proceed to finding every cluster-modulator for $\overline{G}_2$ (Line 7) by the same method. This gives an amount of at most 81 possible 4-tuples w, w', z, z' that must be checked, hence Lines 6–8 run in $O(n + m)$ time.

Finally, for Line 9, checking whether an isomorphism from G_1 to $\overline{G}_2$ can be extended from f can be done by checking sizes of cliques and neighborhoods, which can be done in $O(n + m)$ time.

Therefore, the overall running time of Algorithm 2 is of order $O(n^8(n+m)) = O(n^9 + n^8 m)$. □

Next, it follows a characterization for COMP-SUB($\mathcal{PM}$) in the class of distance hereditary graphs, which is a subclass of *hole*-free graphs. A *distance-hereditary* graph is a $\{domino, house, gem, hole\}$-free graph. See a *domino*, a *house* and a *gem* in Fig. 2. For the next result, let ϱ be the graph in Fig. 2.

Proposition 1. *Let G be a distance-hereditary graph of order $2n$. It holds that $G \in$ COMP-SUB($\mathcal{PM}$) if and only if $G \in \{K_n \overline{K}_n, \varrho\}$.*

We close this section with a characterization of COMP-SUB($\mathcal{PM}$) on chordal graphs. Recall that a chordal graph is a $C_{k \geq 4}$-free graph.

Proposition 2. *Let G be a chordal graph of order $2n$. Then, $G \in$ COMP-SUB($\mathcal{PM}$) if and only if $G = K_n \overline{K}_n$.*

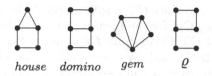

house   domino   gem ϱ

Fig. 2. Some small subgraphs.

4   Results on Some P_k-Free Graphs

In this section, we still consider $\varPi = \mathscr{PM}$ as the property that considers M as a perfect matching. We begin by showing how to solve COMP-SUB($\mathscr{PM}$) in polynomial time when the input graph G is P_5-free.

Theorem 3. COMP-SUB($\mathscr{PM}$) *is polynomial-time solvable on P_5-free graphs.*

Proof. Let G be a $2n$-vertex P_5-free graph. Recall that if $G \in$ COMP-SUB($\mathscr{PM}$), then G is decomposable into complementary subgraphs G_1 and G_2, such the edge cut M of the decomposition is a perfect matching. Since G is P_5-free, the existence of M implies that G_1 and G_2 are P_4-free, that is, G_1 and G_2 are cographs. Then, the conclusion follows by applying Lemma 1.   □

A graph is *extended P_4-laden* if every induced subgraph with at most six vertices that contains more than two induced P_4's is $\{2K_2, C_4\}$-free. Extended P_4-laden graphs generalize cographs, P_4-sparse, P_4-lite, P_4-laden and P_4-tidy graphs, and they were considered under the perspective of partitioning. For instance, Bravo et al. [2] show that partitioning an extended P_4-laden graph into at most k independent sets and at most ℓ cliques is linear-time solvable, for $k, \ell \geq 1$ and Bravo et al. [3] show a linear time algorithm for recognizing graphs that can be partitionable into a clique and a forest. In addition, Pedrotti and De Mello [18] describe a linear-time algorithm that lists the minimal separators of extended P_4-laden graphs.

Another related result to partitioning is implied by considering that extended P_4-laden graphs are P_6-free. The result on 3-colorability by Randerath and Schiermeyer [21] implies that the problem of partitioning a graph into 3 independent sets is polynomial-time solvable on extended P_4-laden graphs.

We present in Proposition 3 a characterization concerned to COMP-SUB($\mathscr{PM}$) on extended P_4-laden graphs.

Proposition 3. *Let G be an extended P_4-laden graph of order $2n$. It holds that $G \in$ COMP-SUB($\mathscr{PM}$) if and only if $G = K_n \overline{K}_n$.*

Proof. Let $G = K_n \overline{K}_n$. We analyse the subgraphs of G with at most 6 vertices to show that G is an extended P_4-laden graph. Let G' be a subgraph of G such that $|V(G')| \leq 6$. If G' is a subgraph of K_n or $\overline{K}_n$, it is clear that G' does not have induced P_4's. Then, we suppose that $V(G')$ intersects both $V(K_n)$ and $V(\overline{K}_n)$. Notice that two induced P_4's arise in G' only if $|V(G') \cap V(\overline{K}_n)| \geq 3$

and $|V(G') \cap V(K_n)| \geq 3$. Since G is a split graph, G' is also a split graph. This implies that G' is $\{2K_2, C_4\}$-free and hence, G is extended P_4-laden.

Now, we show that $G \in \text{COMP-SUB}(\mathscr{PM})$ implies that $G = K_n \overline{K_n}$. Suppose that $G \in \text{COMP-SUB}(\mathscr{PM})$, and, by contradiction, that $G \neq K_n \overline{K_n}$. Since $G \in \text{COMP-SUB}(\mathscr{PM})$ there exist a complementary decomposition (G_1, G_2) of G, such that the edge cut M of the decomposition is a perfect matching. Let $M = \{u_1 v_1, \ldots, u_n v_n\}$ where $u_i \in V(G_1)$ and $v_i \in V(G_2)$, for every $i \in [n]$.

Given that $G \neq K_n \overline{K_n}$, let $u_1 u_2 u_3$ be a P_3 in G_1 and $G' = G[\{u_i, v_i : i \in [3]\}]$.

Since $u_i v_i \in E(G')$, for every $i \in [3]$, we have that $\{u_1, v_1, u_3, v_3\}$ induces a $2K_2$ in G'. Then, we may suppose that $v_1 v_3 \in E(G')$. Notice that $\{u_2, v_2, u_1, v_3\}$ induces a $2K_2$ in G', then we consider that $v_1 v_2 \in E(G')$ or $v_2 v_3 \in E(G')$. In both possibilities we have an induced C_4 in G', by $\{u_1, v_1, u_2, v_2\}$ in the first, and by $\{u_2, v_2, u_3, v_3\}$ in the latter, a contradiction.   □

Our last result characterizes cographs *yes*-instances of $\text{COMP-SUB}(\mathscr{PM})$. Recall that a *cograph* is a P_4-free graph.

Proposition 4. *Let G be a cograph of order $2n$. Then, $G \in \text{COMP-SUB}(\mathscr{PM})$ if and only if $G = K_2$.*

5   Concluding Remarks

We have considered $\text{COMP-SUB}(\mathscr{PM})$ problem when $\mathscr{PM}$ states the edge cut of the decomposition as a perfect matching. We have presented polynomial-time algorithms for solving $\text{COMP-SUB}(\mathscr{PM})$ when the input graph G is *hole*-free or P_5-free and we have shown characterizations on chordal, distance-hereditary, and extended P_4-laden graphs.

With respect to complexity results, despite its resemblance with the NP-complete problem PERFECT MATCHING CUT, we show that $\text{COMP-SUB}(\mathscr{PM})$ is GI-hard when the given input graph G is $\{C_{k \geq 7}, \overline{C}_{k \geq 7}\}$-free.

We remark that our results by Theorem 1 and Theorem 2 address the cases when G is a $C_{\ell \geq k}$-free graph, for every $k \geq 3$, except for $k = 6$. Then, we leave the following conjecture.

Conjecture 1. $\text{COMP-SUB}(\mathscr{PM})$ is GI-hard on $C_{k \geq 6}$-free graphs.

We also leave the complexity of $\text{COMP-SUB}(\mathscr{PM})$ on C_5-free graphs and P_6-free graphs open. Furthermore, we still do not know whether $\text{COMP-SUB}(\mathscr{PM})$ is GI-complete.

References

1. Bondy, J.A., Murty, U.S.R., et al.: Graph Theory with Applications, vol. 290. Macmillan, London (1976)
2. Bravo, R.S.F., Klein, S., Nogueira, L.T., Protti, F., Sampaio, R.M.: Partitioning extended P_4-laden graphs into cliques and stable sets. Inf. Process. Lett. **112**(21), 829–834 (2012). https://doi.org/10.1016/j.ipl.2012.07.011

3. Bravo, R.S.F., Klein, S., Nogueira, L.T., Protti, F.: Clique cycle transversals in graphs with few P_4's. Discret. Math. Theor. Comput. Sci. **15**, 13–20 (2013). https://doi.org/10.46298/dmtcs.616
4. Camargo, P.P., Souza, U.S., Nascimento, J.R.: Remarks on k-clique, k-independent set and 2-contamination in complementary prisms. Int. J. Found. Comput. Sci. **32**(01), 37–52 (2021). https://doi.org/10.1142/S0129054121500027
5. Cappelle, M.R., Penso, L., Rautenbach, D.: Recognizing some complementary products. Theor. Comput. Sci. **521**, 1–7 (2014). https://doi.org/10.1016/j.tcs.2013.11.006
6. Chung, F.R.K.: On the cutwidth and the topological bandwidth of a tree. SIAM J. Algebraic Discret. Methods **6**(2), 268–277 (1985). https://doi.org/10.1137/0606026
7. Colbourn, C.J.: On testing isomorphism of permutation graphs. Networks **11**, 13–21 (1981). https://doi.org/10.1002/net.3230110103
8. Corneil, D.G., Lerchs, H., Stewart Burlingham, L.: Complement reducible graphs. Discret. Appl. Math. **3**, 163–174 (1981). https://doi.org/10.1016/0166-218X(81)90013-5
9. Duarte, M.A., Penso, L., Rautenbach, D., dos Santos Souza, U.: Complexity properties of complementary prisms. J. Comb. Optim. **33**(2), 365–372 (2015). https://doi.org/10.1007/s10878-015-9968-5
10. Duarte, M.A., Joos, F., Penso, L.D., Rautenbach, D., Souza, U.S.: Maximum induced matchings close to maximum matchings. Theor. Comput. Sci. **588**, 131–137 (2015). https://doi.org/10.1016/j.tcs.2015.04.001
11. Haynes, T.W., Henning, M.A., Slater, P.J., van der Merwe, L.C.: The complementary product of two graphs. Bull. Inst. Combin. Appl. **51**, 21–30 (2007)
12. Heggernes, P., Telle, J.A.: Partitioning graphs into generalized dominating sets. Nordic J. Comput. **5**(2), 128–142 (1998)
13. Kratsch, D., Le, V.B.: Algorithms solving the matching cut problem. Theor. Comput. Sci. **609**, 328–335 (2016). https://doi.org/10.1016/j.tcs.2015.10.016
14. Lima, C.V.G.C., Rautenbach, D., Souza, U.S., Szwarcfiter, J.L.: On the computational complexity of the bipartizing matching problem. Ann. Oper. Res. 1–22 (2021). https://doi.org/10.1007/s10479-021-03966-9
15. Lima, C.V., Rautenbach, D., Souza, U.S., Szwarcfiter, J.L.: Decycling with a matching. Inf. Process. Lett. **124**, 26–29 (2017). https://doi.org/10.1016/j.ipl.2017.04.003
16. Nascimento, J.R., Souza, U.S., Szwarcfiter, J.L.: Partitioning a graph into complementary subgraphs. Graphs Comb. **37**(4), 1311–1331 (2021). https://doi.org/10.1007/s00373-021-02319-4
17. Patrignani, M., Pizzonia, M.: The complexity of the matching-cut problem. In: Brandstädt, A., Le, V.B. (eds.) WG 2001. LNCS, vol. 2204, pp. 284–295. Springer, Heidelberg (2001). https://doi.org/10.1007/3-540-45477-2_26
18. Pedrotti, V., De Mello, C.P.: Minimal separators in extended p4-laden graphs. Discret. Appl. Math. **160**(18), 2769–2777 (2012). https://doi.org/10.1016/j.dam.2012.01.025
19. Penso, L.D., Rautenbach, D., Souza, U.S.: Graphs in which some and every maximum matching is uniquely restricted. J. Graph Theory **89**(1), 55–63 (2018). https://doi.org/10.1002/jgt.22239
20. Protti, F., Souza, U.S.: Decycling a graph by the removal of a matching: new algorithmic and structural aspects in some classes of graphs. Discret. Math. Theor. Comput. Sci. **20**(2) (2018). https://doi.org/10.23638/DMTCS-20-2-15
21. Randerath, B., Schiermeyer, I.: 3-Colorability $\in$ P for P_6-free graphs. Discret. Appl. Math. **136**(2–3), 299–313 (2004). https://doi.org/10.1016/S0166-218X(03)00446-3

On the Intractability Landscape
of Digraph Intersection Representations

Andrea Caucchiolo and Ferdinando Cicalese[(✉)] [iD]

Department of Computer Science, University of Verona, Verona, Italy
{andrea.caucchiolo,ferdinando.cicalese}@univr.it

Abstract. We study the classical graph intersection number problem
[Erdős et al., CJM1966] for directed acyclic graphs as recently proposed
in [Kostochka et al., ISIT'2019]. We prove a strong inapproximability
result for arbitrary DAGs. We show that the problem is NP-hard when
restricted to arborescences, which strongly contrasts with the existence of
a trivial linear time solution for the corresponding problem on undirected
trees. For the restriction of the problem to the case of arborescences,
we complement the hardness result with an asymptotic FPTAS, which
significantly improves on a previously known 2-approximation algorithm.

Keywords: Intersection number · NP-hardness · Arborescences ·
Inapproximability · Asymptotic fully polynomial time approximation
schemes

1 Introduction

Motivated by the study of networks of webpages generated by their information
content, and of cardinality dependent generative models of networks [6,19], Kostochka et al. [11] (see also [13]) introduced the novel notion of *directed intersection representation* of a directed acyclic graph (DAG). In this model, a directed
acyclic graph $G = (V(G), E(G))$ is represented by a family of subsets of a ground
set $\mathcal{C}$ via an assignment $\varphi : V(G) \mapsto 2^{\mathcal{C}}$ such that $(u, v) \in E(G)$ if and only if
$\varphi(u) \cap \varphi(v) \neq \emptyset$ and $|\varphi(u)| < |\varphi(v)|$. The authors of [11] studied the corresponding notion of *directed intersection number* of a DAG G, denoted by $DIN(G)$,
defined as the minimum cardinality of a ground set $\mathcal{C}$ such that there exists a
directed intersection representation $\varphi : V(G) \mapsto 2^{\mathcal{C}}$.

The main results of [11] are about extremal values of $DIN(G)$, more pre-
cisely: (i) $DIN(G) \leq \frac{5n^2}{8} - \frac{3n}{4} + 1$ for every DAG G with n vertices; and (ii)
for each n there exist DAGs G with n vertices such that $DIN(G) \geq \frac{n^2}{2}$. These
results, however, only bound the extremal approximation one can get in the
worst possible case for a given size $n = |V(G)|$. The results of [11] leave open
the question of the tractability of the problem and the authors limited themselves to observing that $DIN(G)$ is lower bounded by the length of the longest
path in G, which is, however, an easy problem on DAGs. In [2], Caucchiolo and
Cicalese started to study the computational complexity of determining $DIN(G)$

© Springer Nature Switzerland AG 2022
C. Bazgan and H. Fernau (Eds.): IWOCA 2022, LNCS 13270, pp. 270–284, 2022.
https://doi.org/10.1007/978-3-031-06678-8_20

and showed that the problem is in fact NP-hard for general DAGs. They also considered the restricted variant of computing $DIN(G)$ when the input graph G is an arborescence, providing a 2-approximation algorithm. This result, however, left open the question whether for this restricted version of the problem an exact polynomial time algorithm exists.

Our Results. In this paper, we contribute to the study of the computational complexity landscape of the DIN problem. We significantly strengthen both results of [2]. We first show a strong inapproximability result on the computation of the directed intersection number of an arbitrary DAG. Then, in the quest for islands of tractability, following [2], we focus on arborescence graphs. In strong contrast to the case of undirected graphs, where the intersection number of a tree is trivially given by the cardinality of its edge set, we show that computing the directed intersection number is NP-hard even when the input graph is restricted to be an arborescence. Moreover, on the positive/algorithmic side, we are able to complement this negative result with a strong approximation guarantee. In fact, we give an asymptotic FPTAS for computing the DIN of an arborescence.

Related Work. Every finite undirected graph is representable by a family of finite sets such that each vertex is associated to one of the sets of the family and two vertices are adjacent if and only if their associated sets intersect. The *intersection number* $(IN(G))$ of an undirected graph G, is the minimum cardinality of a set U such that G is representable by a family of subsets of U. Erdős, Goodman and Pósa [7] showed that $IN(G)$ equals the minimum number of cliques needed to cover the edges of G, i.e., the size of a minimum edge clique cover of G. Determining the size of a minimum edge clique cover—and equivalently the intersection number—was proved to be NP-hard in [16] (see also [12]). By [8,14], both problems are not approximable within a factor of $|V|^\epsilon$ for some $\epsilon > 0$ unless $P = NP$. On the other hand, by the result of [9], it follows that computing the intersection number of a graph is fixed parameterized tractable (with respect to the intersection number as parameter). Several applications of clique covers were discovered over the years in areas as diverse as computational geometry, matrix factorization, compiler optimization applied statistics, etc.: see, e.g., the survey papers [17,18], and the comprehensive introduction of [4].

Several analogues of the above concepts have been proposed for the case of directed graphs [1,5,15], based on the representation of a digraph by identifying each vertex v with a pair of subsets S_v, T_v of a ground set U, with $(u, v) \in E$ if and only if $S_u \cap T_v \neq \emptyset$. In [5], a characterization is provided on the intersection number of a digraph, analogous to the one for undirected graphs given in [7].

2   Notation and Basic Definitions

Given a DAG G, we use coloring as metaphor of intersection representation according to the definition of [11]. We say that $\varphi : V(G) \mapsto 2^{\mathcal{C}_\varphi}$ is a *proper coloring* of a DAG G if for each $u, v \in V(G)$, it holds that $(u, v) \in E(G)$ if and only if $\varphi(v) \cap \varphi(u) \neq \emptyset$ and $|\varphi(u)| < |\varphi(v)|$. We denote by $\Phi(G)$ the set of proper colorings of G. For a proper coloring φ, we denote by $|\varphi|$ the cardinality of the color ground set $\mathcal{C}_\varphi$ and refer to $|\varphi| = |\mathcal{C}_\varphi|$ as the size of the coloring φ. We are interested in the problem of determining $DIN(G) = \min\limits_{\varphi \in \Phi(G)} |\varphi|$.

We say that φ is an *optimal coloring* of G if φ is a proper coloring of G and $|\varphi| = DIN(G)$.

For a set of vertices $V' \subseteq V$, we define $\varphi(V') = \cup_{v \in V'} \varphi(v)$. Moreover, given a proper coloring φ of graph $G = (V, E)$, and a subgraph $G' = (V', E')$ of G, we denote by $\varphi(G')$ the set of colors used in G', i.e., $\varphi(G') = \varphi(V')$.

Given a graph $G = (V, E)$ and a subset of vertices $V' \subseteq V$, we denote by $G[V']$ the subgraph induced by V', i.e., $G[V'] = (V', E')$ with $E' = \{(u, v) \in E \mid u, v \in V'\}$.

For an integer $n \geq 1$ we use $[n]$ to denote the set $\{1, \ldots, n\}$.

Fact 1. *Let G be a directed graph and G' an induced subgraph of G. Let φ_G be a proper coloring of G and $\varphi_{G'}$ be an optimal coloring of G'. Then, $|\varphi_G| \geq |\varphi_{G'}|$.*

Proof. The claim follows by the observation that every proper coloring for G is also a proper coloring for G'.

Note that the fact does not hold when the subgraph G' is not induced.

Arborescences. Recall that a directed graph T is an *arborescence* if there is a vertex r called the *root* such that for all other vertices v there is exactly one path from r to v. Equivalently, T is a directed rooted tree where all edges are directed away from the root. A *leaf* is a vertex with out-degree 0; vertices with out-degree different from 0 are referred to as *internal vertices*. For a vertex v, we use $d(v)$ to denote the distance of v from the root. For an edge (u, v), we refer to u as the parent of v and v as a child of u.

The following lemma implies a simple lower bound on the cardinality of any coloring restricted to a subgraph of an arborescence, based on the number of internal nodes. It will be used later in order to obtain more precise lower bounds by considering separately different parts of the arborescence.

Lemma 1. *Let T be an arborescence and φ a proper coloring of T. Then, for each internal vertex v of T, there is a color $c_v \subseteq \varphi(v)$ such that, for each vertex w which is not a child of v, it holds that $c_v \notin \varphi(w)$.*

3 Hardness of Approximation for General DAGs

The first result we present is a strengthening of the NP-hardness of computing the DIN of an arbitrary DAG, which was proved in [2]. We show that under the standard complexity assumption $P \neq NP$, the problem of computing the directed intersection number of a DAG $G = (V, E)$ does not admit an $|V|^{1-\epsilon}$ approximation algorithm. The proof employs a gap-reduction from the problem:

BICLIQUE COVER: **Given** a bipartite undirected graph $G = (V, E)$, **find** a minimum cardinality collection $\{B_i = (V_i, E_i)\}_{i \in [k]}$ of k bicliques (bipartite complete subgraph) of G that cover the edges of G, i.e., $\cup_i E_i \supseteq E$.

Given a bipartite graph G we use $bc(G)$ to denote the size of the minimum biclique cover of G. We will use the following result from [3].

Theorem 1. *Let $\eta > 0$. It is NP-hard to approximate* BICLIQUE COVER *within a factor of $n^{1-\eta}$. In particular, there exists a polynomial algorithm that takes a SAT instance ϕ and produces a bipartite graph $G_\phi = (V, E)$ with $|V| = |\phi|^{O(1)}$ such that the following properties hold: (i) If ϕ is satisfiable, then $bc(G) \le |V|^\eta$; (ii) if ϕ is not satisfiable, then $bc(G) \ge |V|^{1-\eta}$*

The following theorem gives the desired hardness of approximability of the DIN computation.

Theorem 2. *Let $\epsilon > 0$. It is NP-hard to approximate* DIN *within a factor of $n^{1-\epsilon}$. In particular, there exists a polynomial algorithm that takes a SAT instance ϕ and produces a bipartite graph $D_\phi = (V, E)$ with $|V| = |\phi|^{O(1)}$ such that the following properties hold: (i) If ϕ is satisfiable, then $DIN(D) \le |V|^{\frac{\epsilon}{2}}$; (ii) if ϕ is not satisfiable, then $DIN(D) \ge |V|^{1-\frac{\epsilon}{2}}$*

Proof. Let $\eta = \eta(\epsilon) > 0$ such that $\eta \le \epsilon/2$ and for all sufficiently large instances ϕ of SAT, the bipartite graph $G_\phi = (V_G, E_G)$ (instance of BICLIQUE COVER) guaranteed by Theorem 1 satisfies $3n^\eta + 1 \le n^{\frac{\epsilon}{2}}$, where $n = |V_G|$. From the bipartite graph G_ϕ on parts A, B, create a directed graph D_ϕ by orienting the edges from A to B. Then the number of vertices of D is also n.

We have that $DIN(D) \ge bc(G)$: For each edge uv there is a colour in common for u and v. Moreover, each color shared by a vertex in A and a vertex in B induces a biclique. The bicliques associated to the colours shared by at least a vertex of A and a vertex of B (i.e. taking the set of vertices that have such a colour c in their list) must cover the edges of G. In particular, if $bc(G) \ge n^{1-\eta}$ then $DIN(D) \ge n^{1-\eta} \ge n^{1-\frac{\epsilon}{2}}$.

For the other direction, it suffices to show that if $bc(G) \le n^\eta$, then $DIN(D) \le 3n^\eta + 1 \le n^{\frac{\epsilon}{2}}$. For this, we start by giving each vertex a distinct colour for each biclique it is in. Then we use at most n^η additional colours for increasing the cardinality of the color set of each color in A to n^η and at most $n^\eta + 1$ additional colours for having the cardinality of the color set of each vertex in B equal to $n^\eta + 1$. This ensures the cardinality of the color sets of the vertices in A (resp B) are all of the same size, with that of the vertices in A being strictly smaller than that of the vertices in B. Therefore, the obtained coloring ensures there are no edges within A or B, but all edges are correctly from A to B as $|\phi(u)| < |\phi(v)|$ for all $uv \in E$.

4 Hardness for Arborescences

In this section we show that the computation of the directed intersection number is NP-hard also when the input graph is an arborescence. The main tools at the basis of the proof are a special class of arborenscences given in Definition 1 and depicted in Fig. 1, and the characterization of optimal colorings of such arborescences obtainable via Algorithm 1.

Algorithm 1: Canonical Coloring Algorithm

Input: A non-empty arborescence graph T with root $r = root(T)$; a level
function $level : V \mapsto \mathbb{N}$ such that: (i)
$level(v) \geq \max\{level(parent(v)), d(v)\} + 1$, where $d(v)$ is the distance of node
v from the root; (ii) $level(u) = level(w)$ if $parent(u) = parent(w)$.
Output: a proper coloring φ of T such that for each v it holds $|\varphi(v)| = level(v)$.
Set $L \leftarrow \max_{v \in V} level(v)$; $C^{(1)} = \emptyset$; $\Xi = \{level(v) \mid v \in V\}$;
for $\ell \in \Xi \setminus \{1\}$ **do**

> **Define a new set of colors for level** ℓ: $C^{(\ell)} = \{a_1^{(\ell)}, \ldots, a_{\ell-2}^{(\ell)}\}$;
> **if** there is at least one leaf f with $level(f) = \ell$
> **then Define a new color** $\lambda(\ell)$;

For each internal node v **Define a new color** $c(v)$;
Set $\varphi(root(T)) \leftarrow c(root(T)) \cup C^{level(root(T))}$;
for each vertex $v \neq root(T)$ **do**

> **if** v is not a leaf **then Set** $\varphi(v) \leftarrow c(v) \cup C^{(level(v))} \cup c(parent(v))$;
> **else Set** $\varphi(v) \leftarrow \lambda(level(v)) \cup C^{(level(v))} \cup c(parent(v))$

return φ

For the sake of the analysis, it is useful to think of the cardinality of the
color set that a coloring φ assigns to a node u as the *level used by* φ *for* u. The
following algorithm takes in input for each v the value $level(v)$ and outputs a
coloring φ that uses $level(v)$ as the level for v, i.e., such that $|\varphi(v)| = level(v)$.

Lemma 2. *Fix an arborescence T and a level function $level : V \mapsto \mathbb{N}$ satisfying
the condition in Input of Algorithm 1. Let ι be the number of internal vertices
of T. Let $\Lambda = |\{level(f) \mid f$ is a leaf of $T\}|$, be the number of levels assigned
to at least one leaf. Let $\Xi = \{level(v) \mid v \in V\}$ be set of levels assigned to
vertices of T. Then, Algorithm 1, on input $T, level$ produces a coloring φ with
$|\varphi| = \sum_{\ell \in \Xi \setminus \{1\}} (\ell - 2) + \iota + \Lambda$, in linear time (plus the time to write down the
sets of colors). In particular, if there exists h such that $\Xi = \{1, 2, \ldots, h\}$ then
$|\varphi| = \frac{(h-2)(h-1)}{2} + \iota + \Lambda$.*

Let $T = (V, E)$ be an arborescence. We say that a proper coloring φ of T is
canonical if there exists a choice of the level assignment, $level : V \mapsto \mathbb{N}$, such
that up to renaming of the colors, φ coincides with the coloring produced by
Algorithm 1. In particular, for any pair of sibling nodes, φ uses the same level.

For any $h \geq 1$, we denote by $P_{3,h}$ the directed graph consisting of three node-
disjoint paths with h nodes each, denoted by $A = a_2, \ldots, a_{h+1}$; $B = b_2, \ldots, b_{h+1}$;
$C = c_2, \ldots, c_{h+1}$, and an additional node r which is a in-neighbour of each
one of the stating vertices of the three paths. Therefore $P_{3,h} = (V, E)$, where
$V = \{r, a_2, \ldots, a_{h+1}, b_2, \ldots, b_{h+1}, c_2, \ldots, c_{h+1}\}$ and $E = \{(r, a_2), (r, b_2), (r, c_2)\} \cup$
$\bigcup_{i=1}^{h-1} \{(a_i, a_{i+1}), (b_i, b_{i+1}), (c_i, c_{i+1})\}$. We also refer to the common origin r as a_1
(resp. b_1, c_1). Figure 1 contains an example of a $P_{3,h}$, for $h = 12$. For a $P_{3,h}$
graph, we can precisely characterize optimal colorings (details in the full version
of the paper) which will be important in the construction of the arborescence
realizing our reduction.

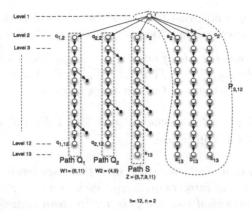

Fig. 1. The structure $T(W_1,\ldots,W_n,Z,P_{3,h})$, in the particular case $h = 13, n = 2$, and a choice of the sets Q_1,Q_2,W, as indicated in the figure. The picture shows the substructure denoted $P_{3,12}$ (on the left); and the paths Q_1,Q_2,S. On the right, it is also shown the way levels are counted assuming a coloring for which $|\varphi(v)| = d(v) + 1$.

Definition 1. *Fix positive integers h,n, as well as $n+1$ sets $W_1,\ldots,W_n,Z$, all subsets of $\{4,\ldots,h-1\}$. Let $T(W_1,\ldots,W_n,Z,P_{3,h})$ be the arborescence graph obtained by the following procedure:* • *start with graph $P_{3,h}$;* • *create n new paths with $h - 1$ nodes, each. For $i = 1,\ldots,n$, denote by $Q_i = q_{i2},\ldots,q_{ih}$ the ith one of such paths;* • *for each $i = 1,\ldots,n$, add an edge (r,q_{i2}) from the root of $P_{3,h}$ to the first node of Q_i;* • *create a new path with h nodes, denoted by $S = s_2, s_3, \ldots, s_{h+1}$ and add an edge (r,s_2) from the root of $P_{3,h}$ to the starting vertex of S;* • *for each $i = 1,\ldots,n$ and $j \in W_i$, create a new leaf node ℓ_{ij} and an edge $(q_{ij-1},\ell_{i,j})$, i.e., the leaf $\ell_{i,j}$ is a sibling of the jth node of path Q_i;* • *for each $j \in Z$, create a new node ℓ_{Sj} and an edge $(s_{ij-1},\ell_{S,j})$, i.e., the leaf ℓ_{Sj} is a sibling of the jth node of path S; For each $i = 1,\ldots,n$, we define $\mathbf{F}_i = \{\ell_{i,j} \mid j \in W_i\}$, i.e., the set of leaves with siblings in the path Q_i. We define $\mathbf{F}_S = \{\ell_{S,j} \mid j \in Z\}$, i.e., the set of leaves with siblings in the path S. We let $\mathbf{F} = \mathbf{F}_S \cup \bigcup_{i=1}^n \mathbf{F}_i$. Then $\mathbf{F}$ is the set of all leaves of $T(W_1,\ldots,W_n,Z,P_{3,h})$ with a sibling which is an internal node.*

The following theorem says that some optimal colorings of arborescence $T = T(W_1,\ldots,W_n,Z,P_{3,h})$ can be attained also by a canonical coloring.

Theorem 3. *Fix integers $h,n \geq 1$ and sets $W_1,W_2,\ldots,W_n,Z$. Let T be equal to $T(W_1,\ldots,W_n,Z,P_{3,h})$, the arborescence as given by Definition 1. For every proper coloring φ of T such that $|\varphi| \leq \frac{h^2+5h-2}{2} + \frac{h}{2} + (n+1)(h-2)$ there is a canonical coloring φ' of T such that $|\varphi'| \leq |\varphi|$ and uses the levels $\{1,2,\ldots,h+1\}$. In particular, for any pair of sibling nodes, φ' uses the same level.*

We can show (details in the full version) that (up to renaming of the colors) a canonical coloring of $T = T(W_1,\ldots,W_n,Z,P_{3,h})$ that uses levels $\{1,2,\ldots,h+1\}$ is determined (via Algorithm 1) by the values $g_1,\ldots,g_n$, where g_i is the level from

$\{1, \ldots, h+1\}$ which is not used for any of the nodes of path Q_i. This justifies the following parameterized definition of a canonical coloring of T.

Definition 2. *For any n-tuple of integers $g_1, \ldots, g_n$, we denote by $\varphi^{can}(g_1, \ldots, g_n)$ a canonical coloring of $T = \mathcal{T}(W_1, \ldots, W_n, Z, P_{3,h})$ that is output by Algorithm 1 when the level assignment of the nodes is given by:*

$$level(v) = \begin{cases} d(v) + 2 & \text{if for some } i \in [n] \ v \in Q_i \cup \mathbf{F}_i \text{ and } d(v) + 1 \geq g_i \\ d(v) + 1 & \text{otherwise.} \end{cases} \quad (1)$$

Dually, for any canonical coloring φ of T that uses levels $\{1, \ldots, h+1\}$, we let $\mathbf{g}_\varphi^T$ be the sequence of integers $g_1, \ldots, g_n$ such that $\varphi = \varphi^{can}(g_1, \ldots, g_n)$.

We say that a canonical coloring φ is a $\mathbf{g}$-Boolean canonical coloring if for each $i = 1, \ldots, n$, it holds that $g_i \in \{3, h\}$.

If φ is a $\mathbf{g}$-Boolean canonical coloring we let $\mathbf{a}_\varphi = a_1, \ldots, a_n \in \{true, false\}^n$ to be the Boolean assignments defined by $a_i = true$ (resp. $a_i = false$) if $g_i = h$ (resp. $g_i = 3$).

The Structure of the Reduction. The reduction is from MIN-2-SAT: given a 2-CNF formula ϕ, find an assignment for ϕ that minimizes the number of satisfied clauses.

We are going to show how we choose the sets $W_1, \ldots, W_n, Z$ in order to encode a 2-CNF ϕ with n variables and m clauses (an instance of MIN-2-SAT) as an instance of our problem given by an arborescence $T_\phi = \mathcal{T}(W_1, \ldots, W_n, Z, P_{3,h})$, with $h \geq 16mn + 3m$. In particular we will guarantee that by Theorem 3 the arborescence T_ϕ has an optimal coloring φ which is canonical.

Let $\phi = C_1 \wedge C_2 \wedge \cdots \wedge C_m$ be a 2-CNF over the variables $x_1, \ldots, x_n$. We assume that all clauses have two distinct literals of two distinct variables, i.e., for each $i = 1, \ldots, m$, it holds that $C_i = \ell_1^{(i)} \vee \ell_2^{(i)}$ where, for $t = 1, 2$, we have $\ell_t^{(i)} \in \{x_1, \ldots, x_n, \overline{x_1}, \ldots, \overline{x_n}\}$, and $\ell_1^{(i)} \notin \{\ell_2^{(i)}, \overline{\ell_2^{(i)}}\}$. We write $x_j \in C_i$ (resp. $\overline{x_j} \in C_i$) if there is $t \in \{1, 2\}$ such that $x_j = \ell_t^{(i)}$ (resp. $\overline{x_j} = \ell_t^{(i)}$).

The idea of the reduction is to have that for each $i = 1, \ldots, n$, the path Q_i represents the variable x_i. The relationship between each variable x_i and the clauses C_j is encoded by the alignment of the leaves in $\mathbf{F}_i$ and the leaves in $\mathbf{F}_S$, i.e., respectively, the leaves with siblings in Q_i and the leaves with siblings in path S.

The idea is that the paths $Q_1, \ldots, Q_n$ behave like sliding locks that can be shifted by just one level down. A canonical optimal coloring chooses which one of such locks to shift down in order to minimize the number of leaves (stemming out of the corresponding path) that are not aligned with the leaves of the path S. Having path Q_i shifted down corresponds to set x_i to false because of the way leaves in the middle of Q_i get aligned with the leaves in the middle of S (see also Fig. 2 for a pictorial explanation).

For $i = 1, \ldots, n$, let us define sets H_i, T_i, M_i, whose union will be used as W_i, the set that represent the depths of leaves in $\mathbf{F}_i$:

$$H_i = \{4m(i-1) + 2(j+1) \mid j \in [2m]\}, \quad T_i = \{4m(i-1) + 4mn + 3m + 2(j+1) \mid j \in [2m]\},$$

$$M_i = \bigcup_{j : \overline{x_i} \in C_j} \{4mn + 3j + 1\} \cup \bigcup_{j : x_i \in C_j} \{4mn + 3j + 2\}.$$

We set $W_i = H_i \cup T_i \cup M_i$.

Moreover, for each $g \in \{3, 4, \ldots, h\}$ we define the following variants of H_i, T_i, M_i:

$$H_{i|g} = \{j \mid j \in H_i, j < g\} \cup \{j+1 \mid j \in H_i, j \geq g\}, \qquad H_i^{\perp} = H_{i|3}$$

$$T_{i|g} = \{j \mid j \in T_i, j < g\} \cup \{j+1 \mid j \in T_i, j \geq g\}, \qquad T_i^{\perp} = T_{i|3}$$

$$M_{i|g} = \{j \mid j \in M_i, j < g\} \cup \{j+1 \mid j \in M_i, j \geq g\}, \qquad M_i^{\perp} = M_{i|3}.$$

We have that $H_{i|g} \cup T_{i|g} \cup M_{i|g}$ corresponds to the sets of levels used for the leaves in $\mathbf{F}_i$, by a canonical coloring φ such that $\varphi = \varphi^{can}(g_1, \ldots, g_n)$ and $g_i = g$. We let

$$M_S = \bigcup_{j=1}^{m} \{4mn + 3j + 1, 4mn + 3j + 3\}, \text{ and we set } Z = M_S \cup \bigcup_{i=1}^{n}(H_i^{\perp} \cup T_i).$$

Finally, we let $h = 16mn + 3m$ and $T_\phi = \mathcal{T}(W_1, \ldots, W_n, Z, P_{3,h})$.

The key relationship between a **g**-*Boolean* canonical coloring φ of T_ϕ and the corresponding assignment $\mathbf{a} = \mathbf{a}_\varphi$ can be summarized as follows: (i) the number of clauses of ϕ satisfied by $\mathbf{a}$ equals the number of leaves of $\cup_i \mathbf{F}_i$ of depth in $\cup_i M_i$ for which φ does not use a level used for some leaf of $\mathbf{F}_S$. Intuitively, for each one of such leaves, a new color has to be accounted for by φ besides the number of colors already necessary for the leaves in $\mathbf{F}_S$.; (ii) there is an exact correspondence between the total number of colors used by the **g**-*Boolean* canonical coloring φ and the number of clauses of ϕ satisfied by the corresponding assignment $\mathbf{a}_\varphi$.

We then show that there exist optimal colorings which are **g**-Boolean canonical, hence we restrict the analysis to such colorings.

Finally, since MIN-2-SAT is NP-hard [10], from the following theorem and the polynomiality of constructing of T_ϕ we have that DIRINTNUM on arborescences is NP-hard.

Theorem 4. *There exists an assignment for ϕ that satisfies at most κ clauses if and only if there is a proper coloring φ of $T = T_\phi = \mathcal{T}(W_1, \ldots, W_n, Z, P_{3,h})$ with $|\varphi| \leq \dfrac{h^2 + 5h}{2} + (h-2)(n+1) + 6mn + 2m + \kappa$.*

5 An Asymptotic FPTAS for Arborescence Graphs

Our algorithmic result, presented in this section, is an asymptotic FPTAS for computing the DIN of an arborescence. To the best of our knowledge, the best approximation result known to date is the 2-approximation algorithm of

[2]. In the context of our paper, the algorithm of 2 coincides with running our Algorithm 1, with $level(v) = d(v) + 1$ for each v.

In order to discuss our strategy, let us keep on using the level terminology introduced in the previous section, i.e., given a coloring φ of the input arborescence, we refer to the size of the color set $\varphi(v)$ as the level assigned to v by φ. In this perspective, we can think of a coloring algorithm as a two step procedure that first fixes the level of each vertex and then chooses the color sets, trying to optimize the overall number of colors among all colorings that agree on the levels choices made in the first step.

In the quest of an good algorithms following this scheme, we try to employ some ideas from the analyses leading to the hardness results of the previous sections. In particular, from the characterization of optimal colorings of the arborescence $P_{3,h}$ used in the previous section, we can derive a sort of amortized lower bound on the level-wise cost of a coloring for an arborescence: Assuming level assignments have been fixed, an optimal coloring needs to use $\sim i$ new colors for each level i that is used for more than one node, and $\sim i/2$ new colors for each level used for exactly one node (Lemma 3 gives a new lower bound based on this intuition). Let us call a level *private* if it is used for only one node, and otherwise, we call the level *public*.

With this new terminology, we can say that Algorithm 1 (like the 2-approximation of [2]) treats all levels as public, by reserving $i - 2$ new colors for each level (independently of whether there is a single node or more nodes assigned to it).

The idea at the basis of our algorithm is to refine the construction of Algorithm 1 in two ways: employing the two phase approach in order to optimize in the selection of public and private levels and, hence, reducing the colors used for nodes that are on a private level in order to more closely follow the amortized lower bound estimate above.

We start with identifying a longest root-to-leaf path (let h be its length) and assigning ith vertex on such path to level i. Then, the algorithm defines which of the h levels will be public. The goal is to minimize the total number of public levels and to have them among the ones that are closest to the root, since each level i costs $\sim i$ if public and $\sim i/2$ if private. Optimizing over the exponentially many possible choices of the subset of public levels is generally hard, then we look for an optimal set of public levels that consists of a fixed number z (a parameter depending on the approximation guarantee we aim at) of disjoint intervals, and whose cost (according to the above lower bound) is minimum. It is not hard to see that such a set of public levels can be computed in polytime. Moreover, we can show that its cost can be guaranteed to be close to the size of an optimal coloring (Lemma 4).

Notation and the Choice of Public Levels. Let $\mathcal{G}$ be an arborescence rooted in r, and let $\hat{P} = \{\hat{p_0} = r, \hat{p_1}, \ldots, \hat{p}_{h-1}\}$ be a longest root-to-leaf path in $\mathcal{G}$. Fix a constant $\frac{2}{h} < \alpha \leq 1$, and let $z = \frac{1}{\alpha}$. For the sake of simplifying notation we assume this is an integer value. All the arguments are easily seen to hold if we apply $\lceil\ \rceil$ operators whenever necessary to guarantee integrality. We use

$[z]_0 = \{0, 1, \ldots, z\}$. For each $i = 1, \ldots, z$, we say that the interval of vertices p_j with $j \in \{h\alpha(i-1)+1, h\alpha(i-1)+2, \ldots, h\alpha i\}$ is the ith segment of $\hat{P}$. For each $i = 1, \ldots, z$, let $P^i = \{p_0^i, p_1^i, \ldots, p_{l_i}^i\}$ be a longest path starting from a node p_0^i within the ith segment of $\hat{P}$, i.e., $p_0^i \in \{\hat{p}_{(i-1)\alpha h+1}, \ldots, \hat{p}_{i\alpha h}\}$. Then, p_0^i is a node of the longest path $\hat{P}$, p_1^i is the first node on $P^i \setminus \hat{P}$, and l_i is the number of nodes stemming out of $\hat{P}$ on the path P^i. We say that p_0^i is the stem of P^i. Recall that any coloring φ of $\mathcal{G}$ can be seen as assigning a level to the vertices, namely the cardinality of the color set used for that vertex. We define h_i to be the minimum possible level that can be used for p_0^i, hence $h_0 = 1$ and $h_i = h_0 + d(p_0^i)$ where $d(p_0^i)$ denotes the distance of p_0^i from the root $\hat{p}_0$. We also set $h_{z+1} = h$ for the sake of definiteness in later computations.

Let $\mathcal{G}'$ be the subgraph of $\mathcal{G}$ induced by the vertices $V(\mathcal{G}') = \hat{P} \cup \sum_{i=1}^{z} P^i$.

Lemma 3. *For all proper colorings φ of $\mathcal{G}'$, it holds that*

$$|\varphi(\mathcal{G}')| \geq \sum_{i=1}^{h} \frac{i}{2} + \left(\min_{\substack{I_1, \ldots, I_z \\ I_k \subseteq \{h_k+1, \ldots, h_{k+1}\} \\ l_k \leq \sum_{j=k}^{z} |I_j|}} \sum_{j=1}^{z} \sum_{i \in I_j} \frac{i}{2} \right) - \sum_{k=1}^{z} \frac{|\varphi(p_0^k)|}{2}, \qquad (2)$$

where for each $k = 1, \ldots, z$, the set I_k is an interval of $\{1, \ldots, h\}$.

Let $\hat{I} = \{\hat{I}_1, \ldots, \hat{I}_z\}$ be a set of intervals, one for each segment of $\hat{P}$ of length $h\alpha$, that achieves the minimum of the term in bracket in the right hand side of (2), i.e., $\hat{I} = \{\hat{I}_1, \ldots, \hat{I}_z\} = \text{argmin}_{\substack{I_1, \ldots, I_z \\ I_k \subseteq \{h_k+1, \ldots, h_{k+1}\} \\ l_k \leq \sum_{j=k}^{z} |I_j|}} \sum_{j=1}^{z} \sum_{i \in I_j} \frac{i}{2}$.

Remark 1. The set $\hat{I}$ can be constructed with a greedy approach; for each k such that $I_k \neq \emptyset$, the minimum in I_k is $h_k + 1$; the condition $l_k \leq \sum_{j=k}^{z} |I_j|$ guarantees that the levels in $\cup_{j \geq k} I_k$ suffice to accommodate the nodes of P^k.

Let us now define a new set of intervals $\tilde{I} = \{\tilde{I}_1, \ldots, \tilde{I}_z\}$. We will construct a coloring that uses levels $\{1, \ldots, h\}$ for the nodes in $\hat{P}$ and levels in $\cup_{j=k}^{z} \tilde{I}_k$ for the nodes in the paths stemming from the k-th segment of $\hat{P}$. As by Remark 1, the levels defined by intervals in $\hat{I}$ are by definition sufficient for accommodating the nodes of the stemmings $P^1, \ldots, P^z$. However, if in $\mathcal{G}$ there is some other path Q^k stemming from $\hat{P}$ in segment k at some level below the stem of P^k, the conditions on $\hat{I}$ do not guarantee that the levels in $\cup_{j=h}^{z} \hat{I}_j$ are enough to accommodate all the vertices of Q^k.

For solving this issue, the idea of our algorithm is to shift down the intervals in $\hat{I}$, just as much as it is needed to guarantee that all vertices of $\mathcal{G} \setminus \{\hat{P}\}$ can be accommodated on them.

Let $\Delta = \max_{1 \leq k \leq z} h\alpha k - (h_k + 1)$. Recall that each non-empty $\hat{I}_k$ starts at $h_k + 1 \in \{h\alpha(k-1) + 1, \ldots, h\alpha k\}$. Hence, we have $\Delta \leq \alpha h$.

For each $k = z, z - 1, \ldots, 1$, in decreasing order and setting $shift_{z+1} = 0$, we define $shift_k = \min\{shift_{k+1} + h_{k+1} - \max(I_k), \Delta\}$, and let $\tilde{I}_k = \{j + shift_k \mid j \in \hat{I}_k\}$. Each interval $\tilde{I}_k \in \tilde{I}$ has the same cardinality as $\hat{I}_k$ and intervals in $\tilde{I}$ are pairwise disjoint, as it was the case for intervals in $\hat{I}$. Moreover, we can show that they suffice to accommodate all the vertices in $\mathcal{G} \setminus \{\hat{P}\}$: Fix a segment k. Let $Q^k = q_0^k, q_1^k, \ldots, q_{l'}^k$ be a path stemming out of $\hat{P}$ at some vertex q_0^k within segment k. Let $h' = d(q_0^k)$. By definition $|Q^k| \leq |P^k|$. Hence, $\bigcup_{j=k}^z \tilde{I}_j$ contains enough levels to accommodate the vertices of Q^k. In particular, if $shift_k = \Delta$ or $\min(\tilde{I}_k) \geq h' + 1$, then the first level in $\bigcup_{j=k}^z \tilde{I}_j$ is not before the first possible level to accommodate the vertices of $Q^k \setminus \hat{P} = \{q_1^k, \ldots, q_{l'}^k\}$.

On the other hand, if $\min(\tilde{I}_k) < h' + 1$ and $shift_k < \Delta$ then, because of the latter, it must also hold that $shift_j < \Delta$ for all $j > k$. Therefore, $\bigcup_{j=k}^z \tilde{I}_j = \{t, t + 1, \ldots, h\}$, for some $h_k + 1 \leq t < h' + 1$. Since $\hat{P}$ is a longest path, then $h - h' \geq l'$, and again we can conclude that $\bigcup_{j=k}^z \tilde{I}_j$ can accommodate all the nodes in Q^k. We then compare the cost of levels in $\tilde{I}$ to an optimal coloring.

Lemma 4. *Let* $\tilde{LB} = \sum_{j=1}^h \frac{j}{2} + \sum_{j=1}^z \sum_{k \in \tilde{I}_j} \frac{k}{2}$. *Then, letting* ι *be the number of internal vertices of* $\mathcal{G}$ *not in* $\mathcal{G}'$. *we have* $\tilde{LB} \leq \left(OPT(\mathcal{G}) - \iota + \dfrac{h}{2\alpha} \right)(1 + 2\alpha)$.

The Algorithm and Its Analysis. We only sketch the steps of the algorithm in order to argue its correctness, complexity and approximation guarantee. A complete analysis and pseudocode will be given in the full version of the paper.

The set $\hat{I}$ can be computed by a straightforward greedy implementation of the definition given above. Then, the intervals in $\hat{I}$ are shifted in order to define $\tilde{I}$, as described in the text above. Let Pub be the union of the intervals in $\tilde{I}$. We refer to the elements of Pub as public levels, in the sense that the levels whose index is in Pub will be shared by more than one node. The remaining levels $\{1, \ldots, h\} \setminus Pub$ are referred to as private levels, meaning that each one of these levels will be used by a single node of $\hat{P}$.

The algorithm uses the level partition (public/private) defined above to assign, top-down, each node not in $\hat{P}$ to the closest public level below the level assigned to its parent. A node $v \notin \hat{P}$ assigned to a public level i is then colored using: (i) a distinct color $c(v)$ specifically created for v, the color $c(parent(v))$ distinctly created for v's parent and $U(i)$, a set of $i - 2$ colors created to be shared by all nodes on level i. Nodes of $\hat{P}$ on public levels are colored following an analogous method, with the only difference that in order to create the intersection between a node's and its parent's color set, a different distinct color $\chi(v)$ is defined and used for each $v \in \hat{P}$.

For a vertex $v \notin \hat{P}$ all the children of v are assigned to the same level, hence they can share the color $c(v)$ even if they have no edge between them. Conversely, a node v in $\hat{P}$ on level i has a child $v' \in \hat{P}$ which is on level $i + 1$ and any other children w of v is on the first public level $> i$ which is possibly $> i + 1$. Therefore, we use $c(v)$ for the intersection between $\varphi(v)$ and $\varphi(w)$ and $\chi(v)$ for the intersection between $\varphi(v)$ and $\varphi(v')$. For otherwise, having $\varphi(v') \cap \varphi(w) \neq \emptyset$ could clash with the absence of edges between these nodes.

For leaves we define a distinct color $f(i)$ shared by all leaves assigned to level i. Then, a leaf ℓ on level i is assigned the colors $c(parent(\ell))$ created for its parent, the leaf-level color $\hat{f}(i)$ and the level colors $U(i)$.

For nodes v of $\hat{P}$ assigned to a private level i, a separate set $R(i)$ of colors is used. The peculiarity of a private level is that a node v on a private level can share the set R of its parent node if this is also on a private level. Hence, since we want $|\varphi(v)| = i$, it is enough to have $|R(i)| = i - |R(i-1)| - 3$, and set $\varphi(v) = \{\chi(v), \chi(parent(v)), c(v)\} \cup R(i) \cup R(i-1)$. For a level i that is not private we set $R(i) = \emptyset$.

We can now bound the number of colors used by the coloring φ produced by the algorithm as the sum of the following quantities: (i) $\tilde{\imath} = |\{c(\nu) \mid \nu \text{ is an internal node of } \mathcal{G}\}|$ to color each internal node ν with $c(\nu)$, where $\tilde{\imath}$ denotes the number of internal vertices of $\mathcal{G}$; (ii) at most $\sum_{i=1}^{1/\alpha} \sum_{k \in \tilde{I}_i}(k-2)$ for the shared colors U of the public levels; (iii) at most h colors, one for each public level where there is at least one leaf; (iv) h colors, one for each vertex $\nu \in \hat{P}$, to create the color $\chi(\nu)$; (v) at most $\sum_{i \notin Pub} \frac{i}{2} + \frac{h}{\alpha}$ colors for the sets $R(i)$ for each private level i, that are used for the vertices of $\hat{P}$.

Finally, it is not difficult to see that all the steps can be implemented to run in polynomial time. In fact, we can show a more precise time bound $O(\frac{|V|}{\alpha} + \frac{1}{\alpha^2})$.

The Approximation Guarantee. Let us write OPT for $OPT(\mathcal{G})$. Let φ^A be the coloring given by the algorithm. Then, we have

$$|\varphi^A| \leq \sum_{i \notin Pub} \frac{i}{2} + \frac{h}{\alpha} + 2h + \tilde{\imath} + \sum_{i=1}^{1/\alpha} \sum_{k \in \tilde{I}_i}(k-2) \leq \sum_{i=1}^{h} \frac{i}{2} + \sum_{i=1}^{1/\alpha} \sum_{k \in \tilde{I}_i} \frac{k}{2} + \tilde{\imath} + (2 + \frac{1}{\alpha})h$$

$$\leq \tilde{LB} + \iota + h\left(3 + \frac{3}{2\alpha}\right) \leq \left(OPT - \iota + \frac{h}{2\alpha}\right)(1 + 2\alpha) + \iota + 3(1 + 2\alpha)\frac{h}{2\alpha} \qquad (3)$$

$$\leq OPT(1 + 2\alpha) + 2(1 + 2\alpha)\frac{h}{\alpha} \leq OPT(1 + 2\alpha) + \frac{1 + 2\alpha}{\alpha}\sqrt{OPT}, \qquad (4)$$

where: the first inequality of (3) follows from $\tilde{\imath} \leq \sum_{i=1}^{1/\alpha} hi\alpha + h + \iota \leq \iota + h\left(1 + \frac{1}{2\alpha}\right)$, where ι denotes the number of internal vertices of $\mathcal{G}$ that are not vertices of $\mathcal{G}'$; the second inequality in (3) follows from Lemma 4; the last inequality follows from the lower bound $OPT \geq \frac{h^2}{4}$ due to the fact that we have a path of length h and the bound on the number of colors needed for a path from [2,11].

Having shown that $\varphi^{ALGO}(\mathcal{G}) \leq OPT(1 + 2\alpha) + \sqrt{OPT}\frac{1 + 2\alpha}{\alpha}$ we can now provide an upper bound on the approximation guarantee. Arguing for h large, so that OPT is also large, for any constant $\epsilon > 0$, with $\alpha = \frac{1}{2}\epsilon$,

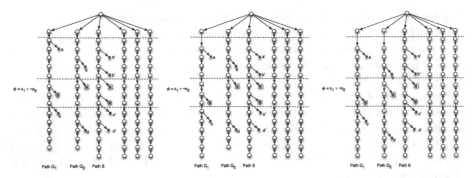

Fig. 2. • The left figure shows the arborescence T_ϕ for $\phi = x_1 \vee \neg x_2$. It also shows the levels assigned to the nodes by a canonical coloring $\varphi = \varphi^{can}(g_1 = h, g_2 = h)$, i.e. where the paths Q_1, Q_2 are not shifted down. Note that (i) the leaves attached to the path Q_1 and Q_2 in the upper part are not aligned to the leaves of the upper part of the S path; (ii) the leaves attached to the path Q_1 and Q_2 in the lower part are aligned to the leaves of the lower part of the S path; (iii) since $x_1 = true$ satisfies ϕ, then the leaf of the central part of Q_1 is not aligned to any of the central leaves of S; (iv) since $x_2 = true$ does not satisfies ϕ, then the leaf of the central part of Q_2 is aligned to the first leaf of the central part of S. • The central figure shows the arborescence T_ϕ for a canonical coloring $\varphi = \varphi^{can}(g_1 = 3, g_2 = h)$, i.e., a coloring that shifts down Q_1 starting from its 3rd node. Now (i) the leaves attached to the path Q_1 in the upper part are aligned to the leaves of the upper part of the S path, while the leaves attached to the path Q_1 in the lower part are not aligned to the leaves of the lower part of the S path; (iii) the leaf of the central part of Q_1 is now aligned to the second central leaf of S; this is consistent with setting $x_1 = false$ and the fact that such assignment does not satisfies ϕ. • The rightmost figure shows the arborescence T_ϕ for a canonical coloring $\varphi = \varphi^{can}(g_1 = 3, g_2 = 3)$, i.e., a coloring that shifts down both Q_1 and Q_2 starting from their 3rd node. Now (i) the leaves attached to both path Q_1 and Q_2 in the upper part are aligned to the leaves of the upper part of the S path, while the leaves attached to the path Q_1, Q_2 in the lower part are not aligned to the leaves of the lower part of the S path; (iii) the leaf of the central part of Q_1 is now aligned to the second central leaf of S; this is consistent with setting $x_1 = false$ and the fact that such assignment does not satisfies ϕ, but the leaf of the central part of Q_2 is now not aligned anymore to a leaf of the central part of S, which is consistent with setting $x_2 = false$. Sliding down a path Q_i does not change the total number of leaves in the upper and lower part that are aligned to leaves of S. It can be useful to optimize the number of leaves of the central part that are aligned with leaves in S. Putting the gap on levels in the central part of the path Q_i is always less efficient than a complete shifting ($g_i = 3$) or a no shifting ($g_i = h$) of the path.

$$\frac{\varphi^{ALGO}}{OPT} \leq \frac{OPT(1+\epsilon) + \sqrt{OPT}\frac{2(1+\epsilon)}{\epsilon}}{OPT} = (1+\epsilon) + \frac{2(1+\epsilon)}{\epsilon\sqrt{OPT}}. \qquad (5)$$

Hence, we have an approximation guarantee $1 + \epsilon$ for all sufficiently large values of OPT. This together with the polynomiality in the size of the instance and $\frac{1}{\epsilon}$, implies that our algorithm is an asymptotic FPTAS.

6 Open Problems

A natural question left open by our investigation is whether we can further narrow the gap between the two results for arborescences: Is it possible to strengthen the hardness result also for arborescences and show APX-hardness? Or, conversely, can we improve the approximation guarantee possibly achieving a (non-asymptotic) PTAS for the problem? Another direction for further investigation is in the realm of fixed parameterized complexity.

References

1. Beineke, L.W., Zamfirescu, C.: Connection digraphs and second-order line digraphs. Discret. Math. **39**(3), 237–254 (1982)
2. Caucchiolo, A., Cicalese, F.: On the complexity of directed intersection representation of DAGs. In: Kim, D., Uma, R.N., Cai, Z., Lee, D.H. (eds.) COCOON 2020. LNCS, vol. 12273, pp. 554–565. Springer, Cham (2020). https://doi.org/10.1007/978-3-030-58150-3_45
3. Chalermsook, P., Heydrich, S., Holm, E., Karrenbauer, A.: Nearly tight approximability results for minimum biclique cover and partition. In: Schulz, A.S., Wagner, D. (eds.) ESA 2014. LNCS, vol. 8737, pp. 235–246. Springer, Heidelberg (2014). https://doi.org/10.1007/978-3-662-44777-2_20
4. Cygan, M., Pilipczuk, M., Pilipczuk, M.: Known algorithms for edge clique cover are probably optimal. SIAM J. Comput. **45**(1), 67–83 (2016)
5. Das, S., Sen, M.K., Roy, A.B., West, D.B.: Interval digraphs: an analogue of interval graphs. J. Graph Theory **13**(2), 189–202 (1989)
6. Dau, H., Milenkovic, O.: Latent network features and overlapping community discovery via Boolean intersection representations. IEEE/ACM Trans. Network. **25**(5), 3219–3234 (2017)
7. Erdős, P., Goodman, A.W., Pósa, L.: The representation of a graph by set intersections. Can. J. Math. **18**, 106–112 (1966)
8. Feige, U., Kilian, J.: Zero knowledge and the chromatic number. J. Comput. Syst. Sci. **57**(2), 187–199 (1998)
9. Gramm, J., Guo, J., Hüffner, F., Niedermeier, R.: Data reduction and exact algorithms for clique cover. ACM J. Exp. Algorithmics **13**, 2 (2008)
10. Kohli, R., Krishnamurti, R., Mirchandani, P.: The minimum satisfiability problem. SIAM J. Discret. Math. **7**(2), 275–283 (1994)
11. Kostochka, A.V., Liu, X., Machado, R., Milenkovic, O.: Directed intersection representations and the information content of digraphs. In: IEEE International Symposium on Information Theory, ISIT 2019, Paris, France, 7–12 July 2019, pp. 1477–1481. IEEE (2019)
12. Kou, L.T., Stockmeyer, L.J., Wong, C.K.: Covering edges by cliques with regard to keyword conflicts and intersection graphs. Commun. ACM **21**(2), 135–139 (1978)
13. Liu, X., Machado, R.A., Milenkovic, O.: Directed intersection representations and the information content of digraphs. IEEE Trans. Inf. Theory **67**(1), 347–357 (2021)
14. Lund, C., Yannakakis, M.: On the hardness of approximating minimization problems. J. ACM **41**(5), 960–981 (1994)
15. Maehara, H.: A digraph represented by a family of boxes or spheres. J. Graph Theory **8**(3), 431–439 (1984)

16. Orlin, J.: Contentment in graph theory: covering graphs with cliques. Indagationes Mathematicae (Proc.) **80**(5), 406–424 (1977)
17. Pullman, N.J.: Clique coverings of graphs—a survey. In: Casse, L.R.A. (ed.) Combinatorial Mathematics X. LNM, vol. 1036, pp. 72–85. Springer, Heidelberg (1983). https://doi.org/10.1007/BFb0071509
18. Roberts, F.S.: Applications of edge coverings by cliques. Discret. Appl. Math. **10**(1), 93–109 (1985)
19. Tsourakakis, C.: Provably fast inference of latent features from networks: with applications to learning social circles and multilabel classification. In: Proceedings of the 24th International Conference on World Wide Web, pp. 1111–1121 (2015)

The RED-BLUE SEPARATION Problem on Graphs

Subhadeep Ranjan Dev[1], Sanjana Dey[1], Florent Foucaud[2,3]([⊠]), Ralf Klasing[4], and Tuomo Lehtilä[5,6]

[1] ACM Unit, Indian Statistical Institute, Kolkata, India
[2] Université Clermont-Auvergne, CNRS, Mines de Saint-Étienne, Clermont-Auvergne-INP, LIMOS, 63000 Clermont-Ferrand, France
florent.foucaud@uca.fr
[3] Univ. Orléans, INSA Centre Val de Loire, LIFO EA 4022, 45067 Orléans Cedex 2, France
[4] Université de Bordeaux, Bordeaux INP, CNRS, LaBRI, UMR 5800, Talence, France
[5] Univ Lyon, Université Claude Bernard, CNRS, LIRIS - UMR 5205, 69622 Lyon, France
[6] Department of Mathematics and Statistics, University of Turku, Turku, Finland

Abstract. We introduce the RED-BLUE SEPARATION problem on graphs, where we are given a graph $G = (V, E)$ whose vertices are colored either red or blue, and we want to select a (small) subset $\mathcal{S} \subseteq V$, called *red-blue separating set*, such that for every red-blue pair of vertices, there is a vertex $s \in \mathcal{S}$ whose closed neighborhood contains exactly one of the two vertices of the pair. We study the computational complexity of RED-BLUE SEPARATION, in which one asks whether a given red-blue colored graph has a red-blue separating set of size at most a given integer. We prove that the problem is NP-complete even for restricted graph classes. We also show that it is always approximable in polynomial time within a factor of $2 \ln n$, where n is the input graph's order. In contrast, for triangle-free graphs and for graphs of bounded maximum degree, we show that RED-BLUE SEPARATION is solvable in polynomial time when the size of the smaller color class is bounded by a constant. However, on general graphs, we show that the problem is $W[2]$-hard even when parameterized by the solution size plus the size of the smaller color class. We also consider the problem MAX RED-BLUE SEPARATION where the coloring is not part of the input. Here, given an input graph G, we want to determine the smallest integer k such that, *for every possible red-blue-coloring* of G, there is a red-blue separating set of size at most k. We derive tight bounds on the cardinality of an optimal solution of

This study has been carried out in the frame of the "Investments for the future" Programme IdEx Bordeaux - SysNum (ANR-10-IDEX-03-02).

F. Foucaud—Research financed by the French government IDEX-ISITE initiative 16-IDEX-0001 (CAP 20-25) and by the ANR project GRALMECO (ANR-21-CE48-0004-01).

T. Lehtilä—Research supported by the Finnish Cultural Foundation and by the Academy of Finland grant 338797.

C. Bazgan and H. Fernau (Eds.): IWOCA 2022, LNCS 13270, pp. 285–298, 2022.
https://doi.org/10.1007/978-3-031-06678-8_21

MAX RED-BLUE SEPARATION, showing that it can range from logarithmic in the graph order, up to the order minus one. We also give bounds with respect to related parameters. For trees however we prove an upper bound of two-thirds the order. We then show that MAX RED-BLUE SEPARATION is NP-hard, even for graphs of bounded maximum degree, but can be approximated in polynomial time within a factor of $O(\ln^2 n)$.

1   Introduction

We introduce and study the RED-BLUE SEPARATION problem for graphs. Separation problems for discrete structures have been studied extensively from various perspectives. In the 1960s, Rényi [24] introduced the SEPARATION problem for set systems (a set system is a collection of sets over a set of vertices), which has been rediscovered by various authors in different contexts, see e.g. [2, 6, 17, 23]. In this problem, one aims at selecting a solution subset S of sets from the input set system to separate every pair of vertices, in the sense that the subset of S corresponding to those sets to which each vertex belongs to, is unique. The graph version of this problem (where the sets of the input set system are the closed neighborhoods of a graph), called IDENTIFYING CODE [18], is also extensively studied. These problems have numerous applications in areas such as monitoring and fault-detection in networks [26], biological testing [23], and machine learning [20]. The RED-BLUE SEPARATION problem which we study here is a red-blue colored version of SEPARATION, where instead of all pairs we only need to separate red vertices from blue vertices.

In the general version of the RED-BLUE SEPARATION problem, one is given a set system $(V, \mathcal{S})$ consisting of a set $\mathcal{S}$ of subsets of a set V of vertices which are either blue or red; one wishes to separate every blue from every red vertex using a solution subset $\mathcal{C}$ of $\mathcal{S}$ (here a set of $\mathcal{C}$ separates two vertices if it contains exactly one of them). Motivated by machine learning applications, a geometric-based special case of RED-BLUE SEPARATION has been studied in the literature, where the vertices of V are points in the plane and the sets of $\mathcal{S}$ are half-planes [7]. The classic problem SET COVER over set systems generalizes both GEOMETRIC SET COVER problems and graph problem DOMINATING SET (similarly, the set system problem SEPARATION generalizes both GEOMETRIC DISCRIMINATING CODE and the graph problem IDENTIFYING CODE). It thus seems natural to study the graph version of RED-BLUE SEPARATION.

Problem Definition. In the graph setting, we are given a graph G and a red-blue coloring $c : V(G) \rightarrow \{\text{red}, \text{blue}\}$ of its vertices, and we want to select a (small) subset S of vertices, called *red-blue separating set*, such that for every red-blue pair r, b of vertices, there is a vertex from S whose closed neighborhood contains exactly one of r and b. Equivalently, $N[r] \cap S \neq N[b] \cap S$, where $N[x]$ denotes the closed neighborhood of vertex x; the set $N[x] \cap S$ is called the *code* of x (with respect to S), and thus all codes of blue vertices are different from all codes of red vertices. The smallest size of a red-blue separating set of (G, c) is denoted by $\text{sep}_{\text{RB}}(G, c)$. Note that if a red and a blue vertex have the same closed neighborhood, they cannot be separated. Thus, for simplicity, we will consider only *twin-free* graphs, that is, graphs where no two vertices have the same closed neighborhood. Also, for a twin-free graph, the vertex set $V(G)$ is always a red-blue separating set as all the vertices have a unique subset of neighbors. We have the following associated computational problem.

RED-BLUE SEPARATION
Input: A red-blue colored twin-free graph (G, c) and an integer k.
Question: Do we have $\text{sep}_{\text{RB}}(G, c) \le k$?

It is also interesting to study the problem when the red-blue coloring is not part of the input. For a given graph G, we thus define the parameter max-$\text{sep}_{\text{RB}}(G)$ which denotes the largest size, over each possible red-blue coloring c of G, of a smallest red-blue separating set of (G, c). The associated decision problem is stated as follows.

MAX RED-BLUE SEPARATION
Input: A twin-free graph G and an integer k.
Question: Do we have max-$\text{sep}_{\text{RB}}(G) \le k$?

In Fig. 1, to note the difference between sep_{RB} and max-sep_{RB}, a path of 6 vertices P_6 is shown, where the vertices are colored red or blue.

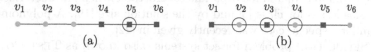

Fig. 1. A path of 6 vertices where (a) $\text{sep}_{\text{RB}}(P_6, c) = 1$ and (b) max-$\text{sep}_{\text{RB}}(P_6) = 3$; the members of the red-blue separating set are circled. Square vertices are blue, round vertices are red. (Color figure online)

Our Results. We show that RED-BLUE SEPARATION is NP-complete even for restricted graph classes such as planar bipartite sub-cubic graphs, in the setting where the two color classes[1] have equal size. We also show that the problem is NP-hard to approximate within a factor of $(1 - \epsilon) \ln n$ for every $\epsilon > 0$, even for split graphs[2] of order n, and when one color class has size 1. On the other hand, we show that RED-BLUE SEPARATION is always approximable in polynomial time within a factor of $2 \ln n$. In contrast, for triangle-free graphs and for graphs of bounded maximum degree, we prove that RED-BLUE SEPARATION is solvable in polynomial time when the smaller color class is bounded by a constant (using algorithms that are in the parameterized class XP, with the size of the smaller color class as parameter). However, on general graphs, the problem is shown to be $W[2]$-hard even when parameterized by the solution size plus the size of the smaller color class. (This is in contrast with the geometric version of separating points by half-planes, for which both parameterizations are known to be fixed-parameter tractable [3, 19]).

---

[1] One class consists of vertices colored *red* and the other class consists of vertices colored *blue*.

[2] A graph $G = (V, E)$ is called a *split graph* when the vertices in V can be partitioned into an independent set and a clique.

As the coloring is not specified, max-sep$_{RB}(G)$ is a parameter that is worth studying from a structural viewpoint. In particular, we study the possible values for max-sep$_{RB}(G)$. We show the existence of tight bounds on max-sep$_{RB}(G)$ in terms of the order n of the graph G, proving that it can range from $\lfloor \log_2 n \rfloor$ up to $n - 1$ (both bounds are tight). For trees however we prove bounds involving the number of support vertices (i.e. which have a leaf neighbor), which imply that max-sep$_{RB}(G) \leq \frac{2n}{3}$. We also give bounds in terms of the (non-colored) separation number. We then show that the associated decision problem MAX RED-BLUE SEPARATION is NP-hard, even for graphs of bounded maximum degree, but can be approximated in polynomial time within a factor of $O(\ln^2 n)$.

Related Work. RED-BLUE SEPARATION has been studied in the geometric setting of red and blue points in the Euclidean plane [3,5,22]. In this problem, one wishes to select a small set of (axis-parallel) lines such that any two red and blue points lie on the two sides of one of the solution lines. The motivation stems from the DISCRETIZATION problem for two classes and two features in machine learning, where each point represents a data point whose coordinates correspond to the values of the two features, and each color is a data class. The problem is useful in a preprocessing step to transform the continuous features into discrete ones, with the aim of classifying the data points [7,19,20]. This problem was shown to be NP-hard [7] but 2-approximable [5] and fixed-parameter tractable when parameterized by the size of a smallest color class [3] and by the solution size [19]. A polynomial time algorithm for a special case was recently given in [22].

The SEPARATION problem for set systems (also known as TEST COVER and DISCRIMINATING CODE) was introduced in the 1960s [24] and widely studied from a combinatorial point of view [1,2,6,17] as well as from the algorithmic perspective for the settings of classical, approximation and parameterized algorithms [8,10,23]. The associated graph problem is called IDENTIFYING CODE [18] and is also extensively studied (see [21] for an online bibliography with almost 500 references as of January 2022); geometric versions of SEPARATION have been studied as well [9,15,16]. The SEPARATION problem is also closely related to the VC DIMENSION problem [27] which is very important in the context of machine learning. In VC DIMENSION, for a given set system $(V, \mathcal{S})$, one is looking for a (large) set X of vertices that is *shattered*, that is, for every possible subset of X, there is a set of $\mathcal{S}$ whose trace on X is the subset. This can be seen as "perfectly separating" a subset of $\mathcal{S}$ using X; see [4] for more details on this connection.

Structure of the Paper. We start with the algorithmic results on RED-BLUE SEPARATION in Sect. 2. We then present the bounds on max-sep$_{RB}$ in Sect. 3 and the hardness result for MAX RED-BLUE SEPARATION in Sect. 4. Due to space constraints, we have omitted some proofs or parts of proofs.

2   Complexity and Algorithms for RED-BLUE SEPARATION

We will prove some algorithmic results for RED-BLUE SEPARATION by reducing to or from the following problems.

SET COVER
Input: A set of elements U, a family $\mathcal{S}$ of subsets of U and an integer k.
Question: Does their exist a cover $\mathcal{C} \subseteq \mathcal{S}$, with $|\mathcal{C}| \leq k$ such that $\bigcup_{C \in \mathcal{C}} C = U$?

DOMINATING SET
Input: A graph $G = (V, E)$ and an integer k.
Question: Does there exist a set $D \subseteq V$ of size k with $\forall v \in V, N[v] \cap D \neq \emptyset$?

2.1 Hardness

Theorem 1. RED-BLUE SEPARATION *cannot be approximated within a factor of* $(1 - \epsilon) \cdot \ln n$ *for any* $\epsilon > 0$ *even when the smallest color class has size* 1 *and the input is a split graph of order* n, *unless* $P = NP$. *Moreover,* RED-BLUE SEPARATION *is W[2]-hard when parameterized by the solution size together with the size of the smallest color class, even on split graphs.*

Proof. For an instance $((U, \mathcal{S}), k)$ of SET-COVER, we construct in polynomial time an instance $((G, c), k)$ of RED-BLUE SEPARATION where G is a split graph and one color class has size 1. The statement will follow from the hardness of approximating MIN SET COVER proved in [11], and from the fact that SET COVER is W[2]-hard when parameterised by the solution size [12].

We create the graph (G, c) by first creating vertices corresponding to all the sets and the elements. We connect a vertex u_i corresponding to an element $i \in U$ to a vertex v_j corresponding to a set $S_j \in \mathcal{S}$ if $u_i \in S_j$. We color all these vertices blue. We add two isolated blue vertices b and b'. We connect all the vertices of type $u_i \in U$ to each other. Also, we add a red vertex r and connect all vertices $u_i \in U$ to r. Now, note that the vertices $U \cup \{r\}$ form a clique whereas the vertices v_j along with b and b' form an independent set. Thus, our constructed graph (G, c) with the coloring c is a split graph. See Fig. 2.

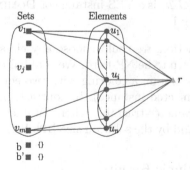

Fig. 2. Reduction from SET COVER to RED-BLUE SEPARATION of Theorem 1. All vertices are blue, except vertex r, which is red. (Color figure online)

Claim 1. $\mathcal{S}$ has a set cover of size k if and only if G has a red-blue separating set of size at most $k + 1$. □

Theorem 2. RED-BLUE SEPARATION *is NP-hard for bipartite planar sub-cubic graphs of girth at least 12 when the color classes have almost the same size.*

Proof. We reduce from DOMINATING SET, which is NP-hard for bipartite planar sub-cubic graphs with girth at least 12 that contain some degree-2 vertices [28]. We reduce any instance (G, k) of DOMINATING SET to an instance $((H, c), k')$ of RED-BLUE SEPARATION, where $k' = k + 1$ and the number of red and blue vertices in c differ by at most 2.

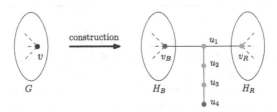

Fig. 3. Reduction from DOMINATING SET to RED-BLUE SEPARATION of Theorem 2. Vertices v_B and u_4 are blue, the others are red. (Color figure online)

Construction. We create two disjoint copies of G namely H_B and H_R and color all vertices of H_B blue and all vertices of H_R red. Select an arbitrary vertex v of degree-2 in G (we may assume such a vertex exists in G by the reduction of [28]) and look at its corresponding vertices $v_R \in V(H_R)$ and $v_B \in V(H_B)$. We connect v_R and v_B with the head of the path u_1, u_2, u_3, u_4 as shown in Fig. 3. The tail of the path, i.e. the vertex u_4, is colored blue and the remaining three vertices u_1, u_2 and u_3 are colored red. Our final graph H is the union of H_R, H_B and the path u_1, u_2, u_3, u_4 and the coloring c as described. Note that if G is a connected bipartite planar sub-cubic graph of girth at least g, then so is H (since v was selected as a vertex of degree-2). We make the following claim.

Claim 2. The instance (G, k) is a YES-instance of DOMINATING SET if and only if $\text{sep}_{\text{RB}}(H, c) \le k' = k + 1$.   □

In the previous reduction, we could choose any class of instances for which DOMINATING SET is known to be NP-hard. We could also simply take two copies of the original graph and obtain a coloring with two equal color class sizes (but then we obtain a disconnected instance). In contrast, in the geometric setting, the problem is fixed-parameter-tractable when parameterised by the size of the smallest color class [3], and by the solution size [19]. It is also 2-approximable [5].

2.2   Positive Algorithmic Results

We start with a reduction to SET COVER implying an approximation algorithm.

Proposition 3. RED-BLUE SEPARATION *has a polynomial time $(2 \ln n)$-factor approximation algorithm.*

Proposition 4. *Let (G, c) be a red-blue colored triangle-free and twin-free graph with R, B the two color classes. Then, $sep_{RB}(G, c) \leq 3 \min\{|R|, |B|\}$.*

Proof. Without loss of generality, we assume $|R| \leq |B|$. We construct a red-blue separating set S of (G, c). First, we add all red vertices to S. It remains to separate every red vertex from its blue neighbors. If a red vertex v has at least two neighbors, we add (any) two such neighbors to S. Since G is triangle-free, no blue neighbor of v is in the closed neighborhood of both these neighbors of v, and thus v is separated from all its neighbors. If v had only one neighbor w, and it was blue, then we separate w from v by adding one arbitrary neighbor of w (other than v) to S. Since G is triangle-free, v and w are separated. Thus, we have built a red-blue separating set S of size at most $3|R|$.   □

Proposition 5. *Let (G, c) be a red-blue colored twin-free graph with maximum degree $\Delta \geq 3$. Then, $sep_{RB}(G, c) \leq \Delta \min\{|R|, |B|\}$.*

Proof. Without loss of generality, let us assume $|R| \leq |B|$. We construct a red-blue separating set S of (G, c). Let v be any red vertex. If there is a blue vertex w whose closed neighborhood contains all neighbors of v (w could be a neighbor of v), we add both v and w to S. If v is adjacent to w, since they cannot be twins, there must be a vertex z that can separate v and w; we add z to S. Now, v is separated from every blue vertex in G.

If such a vertex w does not exist, then we add all neighbors of v to S. Now again, v is separated from every vertex of G. Thus, we have built a red-blue separating set S of size at most $\Delta|R|$.   □

The previous propositions imply that RED-BLUE SEPARATION can be solved in XP time for the parameter "size of a smallest color class" on triangle-free graphs and on graphs of bounded degree (by a brute-force search algorithm). This is in contrast with the fact that in general graphs, it remains hard even when the smallest color class has size 1 by Theorem 1.

Theorem 6. RED-BLUE SEPARATION *on graphs whose vertices belong to the color classes R and B can be solved in time $O(n^{3 \min\{|R|,|B|\}})$ on triangle-free graphs and in time $O(n^{\Delta \min\{|R|,|B|\}})$ on graphs of maximum degree Δ.*

3   Extremal Values and Bounds for max-sep$_{RB}$

We denote by $sep(G)$ the smallest size of a (non-colored) separating set of G, that is, a set that separates *all* pairs of vertices. We will use the relation max-sep$_{RB}(G) \leq sep(G)$, which clearly holds for every twin-free graph G.

3.1   Lower Bounds for General Graphs

We can have a large twin-free colored graph with solution size 2 (for example, in a large blue path with a single red vertex, two vertices suffice). We show that in every twin-free graph, there is always a coloring that requires a large solution.

Theorem 7. *For any twin-free graph G of order $n \geq 1$ and $n \notin \{8, 9, 16, 17\}$, we have max-sep$_{RB}(G) \geq \lfloor \log_2(n) \rfloor$.*

Proof. Let G be a twin-free graph of order n with max-sep$_{RB}(G) = k$. There are 2^n different red-blue colorings of G. For each such coloring c, we have sep$_{RB}(G, c) \leq k$. Consider the set of vertex subsets of G which are separating sets of size k for some red-blue colorings of G. Notice that each red-blue coloring has a separating set of cardinality k. There are at most $\binom{n}{k} \leq n^k$ such sets.

Consider such a separating set S and consider the set $I(S)$ of subsets S' of S for which there exists a vertex v of G with $N[v] \cap S = S'$. Let i_S be the number of these subsets: we have $i_S \leq 2^{|S|} \leq 2^k$. If S is a separating set for (G, c), then all vertices having the same intersection between their closed neighborhood and S must receive the same color by c. Thus, there are at most $2^{i_S} \leq 2^{2^k}$ red-blue colorings of G for which S is a separating set. Overall, we thus have $2^n \leq \binom{n}{k} 2^{2^k} \leq n^k 2^{2^k}$, and thus $n \leq k \log_2(n) + 2^k$.

We now claim that this implies that $k \geq \log_2(n - \log_2(n) \log_2(n))$. Suppose to the contrary that this is not the case, then we would obtain:

$$n < \log_2(n - \log_2(n) \log_2(n)) \log_2(n) + n - \log_2(n) \log_2(n)$$
$$n < \log_2(n) \log_2(n) + n - \log_2(n) \log_2(n)$$

And thus $n < n$, a contradiction. Since k is an integer, we actually have $k \geq \lceil \log_2(n - \log_2(n) \log_2(n)) \rceil$. To conclude, one can check that whenever $n \geq 70$, we have $\lceil \log_2(n - \log_2(n) \log_2(n)) \rceil \geq \lfloor \log_2(n) \rfloor$. Moreover, if we compute values for $2^n - \binom{n}{k} 2^{2^k}$ when $1 \leq n \leq 69$ and $k = \lfloor \log_2(n) \rfloor - 1$, then we observe that this is negative only when $n \in \{8, 9, 16, 17\}$. Thus, $\lfloor \log_2(n) \rfloor$ is a lower bound for max-sep$_{RB}(G)$ as long as $n \notin \{8, 9, 16, 17\}$. $\square$

The bound of Theorem 7 is tight for infinitely many values of n.

Proposition 8. *For any integers $k \geq 1$ and $n = 2^k$, there exists a graph G of order n with max-sep$_{RB}(G) = k$.*

We next relate parameter max-sep$_{RB}$ to other graph parameters.

Theorem 9. *Let G be a graph on n vertices. Then, $sep(G) \leq \min\{\lceil \log_2(n) \rceil \cdot max\text{-}sep_{RB}(G), \lceil \log_2(\Delta(G) + 1) \rceil \cdot max\text{-}sep_{RB}(G) + \gamma(G)\}$, where $\gamma(G)$ is the domination number of G and $\Delta(G)$ its maximum degree.*

Proof. Let G be a graph on $2^{k-1} + 1 \leq n \leq 2^k$ vertices for some integer k. We denote each vertex by a different k-length binary word $x_1 x_2 \cdots x_k$ where each $x_i \in \{0, 1\}$. Moreover, we give k different red-blue colorings $c_1, \ldots, c_k$ such that vertex $x_1 x_2 \cdots x_k$ is red in coloring c_i if and only if $x_i = 0$ and blue otherwise. For each i, let S_i be an optimal red-blue separating set of (G, c_i). We have $|S_i| \leq$ max-sep$_{RB}(G)$ for each i. Let $S = \bigcup_{i=1}^{k} S_i$. Now, $|S| \leq k \cdot$max-sep$_{RB}(G) = \lceil \log_2(n) \rceil \cdot$ max-sep$_{RB}(G)$. We claim that S is a separating set of G. Assume to the contrary that for two vertices $x = x_1 x_2 \cdots x_k$ and $y = y_1 y_2 \cdots y_k$, $N[x] \cap S =$

$N[y] \cap S$. For some i, we have $y_i \neq x_i$. Thus, in coloring c_i, vertices x and y have different colors and hence, there is a vertex $s \in c_i$ such that $s \in N[y] \triangle N[x]$, a contradiction which proves the first bound.

Let S be an optimal red-blue separating set for such a coloring c and let D be a minimum-size dominating set in G; $S \cup D$ is also a red-blue separating set for coloring c. At most $\Delta(G) + 1$ vertices of G may have the same closed neighborhood in D. Thus, we may again choose $\lceil \log_2(\Delta(G) + 1) \rceil$ colorings and optimal separating sets for these colorings, each coloring (roughly) halving the number of vertices having the same vertices in the intersection of separating set and their closed neighborhoods. Since each of these sets has size at most max-sep$_{RB}(G)$, we get the second bound.   □

We do not know whether the previous bound is reached, but as seen next, there are graphs G such that $sep(G) = 2$max-sep$_{RB}(G)$.

Proposition 10. *Let* $G = K_{k_1,\ldots,k_t}$ *be a complete t-partite graph for $t \geq 2$, $k_i \geq 5$ odd for each i. Then* $sep(G) = n - t$ *and* $max\text{-}sep_{RB}(G) = (n-t)/2$.

3.2 Upper Bound for General Graphs

We will use the following classic theorem in combinatorics to show that we can always spare one vertex in the solution of MAX RED-BLUE SEPARATION.

Theorem 11 (Bondy's Theorem [2]). *Let V be an n-set with a family $\mathcal{A} = \{\mathcal{A}_1, \mathcal{A}_2, \ldots, \mathcal{A}_n\}$ of n distinct subsets of V. There is an $(n-1)$-subset X of V such that the sets $\mathcal{A}_1 \cap X, \mathcal{A}_2 \cap X, \mathcal{A}_3 \cap X, \ldots, \mathcal{A}_n \cap X$ are still distinct.*

Corollary 12. *For any twin-free graph G on n vertices, we have $max\text{-}sep_{RB}(G) \leq sep(G) \leq n - 1$.*

This bound is tight for every even n for complements of *half-graphs* (studied in the context of identifying codes in [14]).

Definition 13 (Half-graph [13]). *For any integer $k \geq 1$, the half-graph H_k is the bipartite graph on vertex sets $\{v_1, \ldots, v_k\}$ and $\{w_1, \ldots, w_k\}$, with an edge between v_i and w_j if and only if $i \leq j$.*
The complement $\overline{H_k}$ of H_k thus consists of two cliques $\{v_1, \ldots, v_k\}$ and $\{w_1, \ldots, w_k\}$ and with an edge between v_i and w_j if and only if $i > j$.

Proposition 14. *For every $k \geq 1$, we have $max\text{-}sep_{RB}(\overline{H_k}) = 2k - 1$.*

3.3 Upper Bound for Trees

We will now show that a much better upper bound holds for trees.

Degree-1 vertices are called *leaves* and the set of leaves of the tree T is $L(T)$. Vertices adjacent to leaves are called *support vertices*, and the set of support vertices of T is denoted $S(T)$. We denote $\ell(T) = |L(T)|$ and $s(T) = |S(T)|$. The set of support vertices with exactly i adjacent leaves is denoted $S_i(T)$ and the set of leaves adjacent to support vertices in $S_i(T)$ is denoted $L_i(T)$. Observe that $|L_1(T)| = |S_1(T)|$. Moreover, let $L_+(T) = L(T) \setminus L_1(T)$ and $S_+(T) = S(T) \setminus S_1(T)$. We denote the sizes of these four types of sets $s_i(T), \ell_i(T), s_+(T)$ and $\ell_+(T)$.

To prove our upper bound for trees, we need Theorems 15 and 16.

Theorem 15. *For any tree T of order $n \geq 5$, we have $max\text{-}sep_{RB}(T) \leq \frac{n+s(T)}{2}$.*

Proof. Observe that the claim holds for stars (select the vertices of the smallest color class among the leaves, and at least two leaves). Thus, we assume that $s(T) \geq 2$. Let c be a coloring of T such that $max\text{-}sep_{RB}(T) = sep_{RB}(T, c)$.

We build two separating sets C_1 and C_2; the idea is that one of them is small. We choose a non-leaf vertex x and add to the first set C_1' every vertex at odd distance from x and every leaf. If there is a support vertex $u \in S_1(T) \cap C_1'$ and an adjacent leaf $v \in L_1(T) \cap N(u)$, we create a separating set C_1 from C_1' by shifting the vertex away from leaf v to some vertex $w \in N(u) \setminus L(T)$. We construct in a similar manner sets C_2' and C_2, except that we add the vertices at even distance from x to C_2' and do the shifting when $u \in S_1(T)$ has even distance to x.

Claim 3. Both C_1 and C_2 are separating sets.

Let us denote by $NS_3(T)$ a smallest set of vertices in T such that for each vertex $v \in S_3(T)$ which has $N(v) \cap S_+(T) = \emptyset$, we have at least one vertex $u \in N(v) \setminus L(T)$ in $NS_3(T)$ (such a set exists since T is not a star).

We assume that out of the two sets C_1' and C_2', C_a' ($a \in \{1, 2\}$) has less vertices among the vertices in $V(T) \setminus (L(T) \cup S_+(T) \cup NS_3(T))$. In particular, it contains at most half of those vertices and we have $|C_a' \setminus (L(T) \cup S_+(T) \cup NS_3(T))| \leq (n - \ell(T) - s_+(T) - |NS_3(T)|)/2$. Now, we will construct set C from C_a'. Let us start by having each vertex in C_a' be in C. Let us then, for each support vertex $u \in S_+(T)$, remove from C every adjacent leaf $w \in L_+(T) \cap N(u)$ such that w is in the more common color class within the vertices in $N(u) \cap L_+(T)$ in coloring c. We then add some vertices to C as follows. For $u \in S_i(T)$, $i \geq 4$, we add u to C and some leaves so that there are at least two vertices in $N(u) \cap C$. We have at most $|L(T) \cap N[u]|/2 + 1$ vertices in $C \cap (N[u] \cap L(T) \cup \{u\})$.

For $i = 3$, we add u and any $v \in NS_3(T) \cap N(u) \setminus C$, depending on which one already belongs to C. Then, if all leaves in $N(u)$ have the same color, we add one of them to C. Hence, we have $|C \cap (L_2(T) \cup NS_3(T))|/s_3(T) \leq 2$.

Finally, for $i = 2$, if the two leaves have same color and $u \notin C_a'$, we add u and one of the two leaves to C. If the two leaves have the same color and $u \in C_a'$, we add a non-leaf neighbor of u to C. If the leaves have different colors, one of them, say v, has the same color as u. We add u to C and shift the vertex in C in the leaves so that v is in C. We added at most two vertices to C in this case. Notice that now we have $S_+(T) \subseteq C$.

Each time, we added to C at most half the considered vertices in $N(u)$, and at most one additional vertex. After these changes, we shift some vertices in C

away from $L_1(T)$ the same way we built C_a from C'_a. As $|C'_a \setminus (L(T) \cup S_+(T) \cup NS_3(T))| \leq (n - \ell(T) - s_+(T) - |NS_3(T)|)/2$, we get:

$$|C| \leq \frac{n - \ell(T) - s_+(T) - |NS_3(T)|}{2} + \ell_1(T) + \frac{\ell_+(T) + |NS_3(T)|}{2} + s_+(T)$$

$$= \frac{n + \ell_1(T) + s_+(T)}{2} = \frac{n + s(T)}{2}.$$

Claim 4. C is a red-blue separating set for coloring c.   □

The upper bound of Theorem 15 is tight. Consider, for example, a path on eight vertices. Also, the trees presented in Proposition 18 are within $1/2$ from this upper bound. In the following theorem, we offer another upper bound for trees which is useful when the number of support vertices is large.

Theorem 16. *For any tree T of order $n \geq 5$, $sep(T) \leq n - s(T)$.*

The following corollary is a direct consequence of Theorems 15 and 16. Indeed, we have max-sep$_{RB}(T) \leq \min\{n - s(T), (n + s(T))/2\}$.

Corollary 17. *For any tree T of order $n \geq 5$, we have max-sep$_{RB}(T) \leq \frac{2n}{3}$.*

We next show that Corollary 17 (and Theorem 15) is not far from tight.

Proposition 18. *For any $k \geq 1$, there is a tree T of order $n = 5k + 1$ with max-sep$_{RB}(T) = \frac{3(n-1)}{5} = \frac{n+s(T)-1}{2}$.*

4   Algorithmic Results for MAX RED-BLUE SEPARATION

The problem MAX RED-BLUE SEPARATION does not seem to be naturally in the class NP (it is in the second level of the polynomial hierarchy). Nevertheless, we show that it is NP-hard by reduction from a special version of 3-SAT [25].

3-SAT-2L
Input: A set of m clauses $C = \{c_1, \ldots, c_m\}$ each with at most three literals, over n Boolean variables $X = \{x_1, \ldots, x_n\}$, and each literal appears at most twice.
Question: Is there an assignment of X where each clause has a true literal?

Theorem 19. MAX RED-BLUE SEPARATION *is NP-hard even for graphs of maximum degree 12.*

Proof. To show hardness we reduce from the 3-SAT-2L problem. Given an instance σ of 3-SAT-2L with m clauses and n variables, we create an instance (G, k) of MAX RED-BLUE SEPARATION as follows.

First let us explain the construction of a domination gadget and its properties. A *domination gadget* on vertices v_1 and v_2 is represented in Fig. 4. The vertices v_1 and v_2 may be connected to each other or to some other vertices which is represented by the dashed edges. Both v_1 and v_2 are also connected to

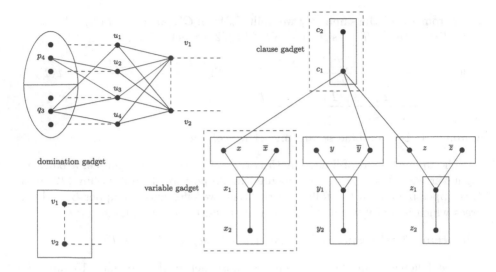

Fig. 4. Reduction from 3-SAT-2L to MAX RED-BLUE SEPARATION.

the vertices u_1, u_2, u_3 and u_4 as shown in the figure. Next we have a clique K_{10} consisting of the vertices $\{p_1, \ldots, p_6, q_1, \ldots, q_4\}$. Every vertex p_i is connected to a unique pair of vertices from $\{u_1, u_2, u_3, u_4\}$ and every vertex q_j is connected to a unique triple of vertices from $\{u_1, u_2, u_3, u_4\}$. For example in the figure we have p_4 connected with the pair of vertices u_2 and u_3 and q_3 connected with the triplet of vertices u_1, u_3 and u_4.

Let $H(v_1, v_2)$ be a subgraph of some graph G such that H is connected to the rest of G only by the vertices v_1 and v_2. We define a worst-coloring of G as any red-blue coloring of G where $\mathrm{sep}_{\mathrm{RB}}(G, c) = \mathrm{max\text{-}sep}_{\mathrm{RB}}(G)$. We make the following claim.

Claim 5. For any worst-coloring c of G the optimal red-blue separating code of (G, c) will always contain the vertices u_1, u_2, u_3 and u_4.

The *variable gadget* for a variable x consists of the graph $H(x_1, x_2)$ and $H(x, \overline{x})$ with additional edges $(x_1, x_2), (x_1, x)$ and $(x_1, \overline{x})$. If x_1 and x_2 are colored differently, then either x or $\overline{x}$ needs to be in the red-blue separating set. Selecting at least one of x or $\overline{x}$ also separates x and $\overline{x}$ themselves. The *clause gadget* for a clause $c = (x \vee \overline{y} \vee z)$ is $H(c_1, c_2)$, where c_1 is connected to the vertices $x, \overline{y}$ and z. If c_1 and c_2 are colored differently, then the red-blue separating set should contain at least one of $x, \overline{y}$ or z in order to separate them. This is used to show the following, and complete the proof.

Claim 6. σ is satisfiable if and only if $\mathrm{max\text{-}sep}_{\mathrm{RB}}(G) \leq k = 4m + 9n$. $\square$

We can use Theorem 9 and a reduction to SET COVER to show the following.

Theorem 20. MAX RED-BLUE SEPARATION *can be approximated within a factor of $O((\ln n)^2)$ on graphs of order n in polynomial time.*

5   Conclusion

We have initiated the study of RED-BLUE SEPARATION and MAX RED-BLUE SEPARATION on graphs, problems which seem natural given the interest that their geometric version has gathered, and the popularity of its "non-colored" variants IDENTIFYING CODE on graphs or TEST COVER on set systems.

When the coloring is part of the input, the solution size of RED-BLUE SEPARATION can be as small as 2, even for large instances; however, we have seen that this is not possible for MAX RED-BLUE SEPARATION since max-sep$_{RB}(G) \geq \lfloor \log_2(n) \rfloor$ for twin-free graphs of order n. max-sep$_{RB}(G)$ can be as large as $n - 1$ in general graphs, yet, on trees, it is at most $2n/3$ (we do not know if this is tight, or if the upper bound of $3n/5$, which would be best possible, holds). It would also be interesting to see if other interesting upper or lower bounds can be shown for other graph classes.

We have shown that sep$(G) \leq \lceil \log_2(n) \rceil \cdot$ max-sep$_{RB}(G)$. Is it true that sep$(G) \leq 2$max-sep$_{RB}(G)$? As we have seen, this would be tight.

We have also shown that MAX RED-BLUE SEPARATION is NP-hard, yet it does not naturally belong to NP. Is the problem actually hard for the second level of the polynomial hierarchy?

References

1. Bollobás, B., Scott, A.D.: On separating systems. Eur. J. Comb. **28**, 1068–1071 (2007)
2. Bondy, J.A.: Induced subsets. J. Comb. Theory Ser. B **12**(2), 201–202 (1972)
3. Bonnet, É., Giannopoulos, P., Lampis, M.: On the parameterized complexity of red-blue points separation. J. Comput. Geom. **10**(1), 181–206 (2019)
4. Bousquet, N., Lagoutte, A., Li, Z., Parreau, A., Thomassé, S.: Identifying codes in hereditary classes of graphs and VC-dimension. SIAM J. Discret. Math. **29**(4), 2047–2064 (2015)
5. Călinescu, G., Dumitrescu, A., Karloff, H.J., Wan, P.: Separating points by axis-parallel lines. Int. J. Comput. Geom. Appl. **15**(6), 575–590 (2005)
6. Charbit, E., Charon, I., Cohen, G., Hudry, O., Lobstein, A.: Discriminating codes in bipartite graphs: bounds, extremal cardinalities, complexity. Adv. Math. Commun. **2**(4), 403–420 (2008)
7. Chlebus, B.S., Nguyen, S.H.: On finding optimal discretizations for two attributes. In: Polkowski, L., Skowron, A. (eds.) RSCTC 1998. LNCS (LNAI), vol. 1424, pp. 537–544. Springer, Heidelberg (1998). https://doi.org/10.1007/3-540-69115-4_74
8. Crowston, R., Gutin, G., Jones, M., Muciaccia, G., Yeo, A.: Parameterizations of test cover with bounded test sizes. Algorithmica **74**(1), 367–384 (2016)
9. Dey, S., Foucaud, F., Nandy, S.C., Sen, A.: Discriminating codes in geometric setups. In: Proceedings of the 31st International Symposium on Algorithms and Computation (ISAAC 2020). Leibniz International Proceedings in Informatics, vol. 181, pp. 24:1–24:16 (2020)
10. De Bontridder, K.M.J., et al.: Approximation algorithms for the test cover problem. Math. Program. Ser. B **98**, 477–491 (2003)
11. Dinur, I., Steurer, D.: Analytical approach to parallel repetition. In: ACM Symposium on Theory of Computing, vol. 46, pp. 624–633 (2014)

12. Downey, R.G., Fellows, M.R.: Parameterized Complexity. Springer, Heidelberg (1999)
13. Erdös, P.: Some combinatorial, geometric and set theoretic problems in measure theory. In: Kölzow, D., Maharam-Stone, D. (eds.) Measure Theory Oberwolfach 1983. LNM, vol. 1089, pp. 321–327. Springer, Heidelberg (1984). https://doi.org/10.1007/BFb0072626
14. Foucaud, F., Guerrini, E., Kovše, M., Naserasr, R., Parreau, A., Valicov, P.: Extremal graphs for the identifying code problem. Eur. J. Comb. **32**(4), 628–638 (2011)
15. Gledel, V., Parreau, A.: Identification of points using disks. Discret. Math. **342**, 256–269 (2019)
16. Har-Peled, S., Jones, M.: On separating points by lines. Discret. Comput. Geom. **63**, 705–730 (2020)
17. Henning, M.A., Yeo, A.: Distinguishing-transversal in hypergraphs and identifying open codes in cubic graphs. Graphs Comb. **30**(4), 909–932 (2014)
18. Karpovsky, M.G., Chakrabarty, K., Levitin, L.B.: On a new class of codes for identifying vertices in graphs. IEEE Trans. Inf. Theory **44**, 599–611 (1998)
19. Kratsch, S., Masařík, T., Muzi, I., Pilipczuk, M., Sorge, M.: Optimal discretization is fixed-parameter tractable. In: Proceedings of the 32nd ACM-SIAM Symposium on Discrete Algorithms (SODA 2021), pp. 1702–1719 (2021)
20. Kujala, J., Elomaa, T.: Improved algorithms for univariate discretization of continuous features. In: Kok, J.N., Koronacki, J., Lopez de Mantaras, R., Matwin, S., Mladenič, D., Skowron, A. (eds.) PKDD 2007. LNCS (LNAI), vol. 4702, pp. 188–199. Springer, Heidelberg (2007). https://doi.org/10.1007/978-3-540-74976-9_20
21. Lobstein, A.: Watching systems, identifying, locating-dominating and discriminating codes in graphs: a bibliography. https://www.lri.fr/~lobstein/debutBIBidetlocdom.pdf
22. Misra, N., Mittal, H., Sethia, A.: Red-blue point separation for points on a circle. In: Proceedings of the 32nd Canadian Conference on Computational Geometry (CCCG 2020), pp. 266–272 (2020)
23. Moret, B.M.E., Shapiro, H.D.: On minimizing a set of tests. SIAM J. Sci. Stat. Comput. **6**(4), 983–1003 (1985)
24. Rényi, A.: On random generating elements of a finite Boolean algebra. Acta Scientiarum Mathematicarum Szeged **22**, 75–81 (1961)
25. Tovey, C.A.: A simplified NP-complete satisfiability problem. Discret. Appl. Math. **8**(1), 85–89 (1984)
26. Ungrangsi, R., Trachtenberg, A., Starobinski, D.: An implementation of indoor location detection systems based on identifying codes. In: Aagesen, F.A., Anutariya, C., Wuwongse, V. (eds.) INTELLCOMM 2004. LNCS, vol. 3283, pp. 175–189. Springer, Heidelberg (2004). https://doi.org/10.1007/978-3-540-30179-0_16
27. Vapnik, V.N., Chervonenkis, A.Y.: On the uniform convergence of relative frequencies of events to their probabilities. Theory Probab. Appl. **16**(2), 264–280 (1971)
28. Zvervich, I.E., Zverovich, V.E.: An induced subgraph characterization of domination perfect graphs. J. Graph Theory **20**(3), 375–395 (1995)

Harmless Sets in Sparse Classes

Pål Grønås Drange[1] , Irene Muzi[2] , and Felix Reidl[2(✉)]

[1] University of Bergen, Bergen, Norway
Pal.Drange@uib.no
[2] Birkbeck College, University of London, London, UK
{i.muzi,f.reidl}@bbk.ac.uk

Abstract. In the classic TARGET SET SELECTION problem, we are asked to minimise the number of nodes to *activate* so that, after the application of a certain propagation process, all nodes of the graph are active. Bazgan and Chopin [*Discrete Optimization*, 14:170–182, 2014] introduced the opposite problem, named HARMLESS SET, in which they ask to maximise the number of nodes to activate such that not a single additional node is activated.

In this paper we investigate how sparsity impacts the tractability of HARMLESS SET. Specifically, we answer two open questions posed by the aforementioned authors, namely a) whether the problem is FPT on planar graphs and b) whether it is FPT parametrised by treewidth. The first question can be answered in the positive using existing meta-theorems on sparse classes, and we further show that HARMLESS SET not only admits a polynomial kernel, but that it can be solved in subexponential time. We then answer the second question in the negative by showing that the problem is W[1]-hard when parametrised by a parameter that upper bounds treewidth.

1 Introduction

How information and cascading events spread through social and complex networks is an important measure of their underlying systems, and is a well-researched area in network science. The dynamic processes governing the diffusion of information and "word-of-mouth" effects have been studied in many fields, including epidemiology, sociology, economics, and computer science [3,14,19,20].

A classic propagation problem is the TARGET SET SELECTION problem, first studied by Domingos and Richardson [10,27], and later formalised in the context of graph theory by Chen [3,6]. Chen defines the problem as how to find k initial *seed* vertices that when activated cascade to a maximum; this model is called *standard independent cascade model of network diffusion*. It has also been studied under the name of INFLUENCE MAXIMIZATION [23,24] in the context of lies spreading through a network [4,5], bio-terrorism [12], and the spread of fires [28]. Information propagation is modelled as an *activation process* where each individual is activated if a sufficient number of its neighbours are active. Sufficient here means that the number of active neighbours of an individual v

C. Bazgan and H. Fernau (Eds.): IWOCA 2022, LNCS 13270, pp. 299–312, 2022.
https://doi.org/10.1007/978-3-031-06678-8_22

exceeds a given threshold $t(v)$ which is assigned to each individual to capture their resilience to being influenced.

Motivated by *cascading of information* we study vertices that are *harmless*, i.e., a set of vertices that can be activated without any cascades whatsoever. However, activating all vertices in a graph is a trivial solution in the standard diffusion model, since we cannot cascade further. We therefore want to differentiate between *initially activated vertices* and vertices that have been *activated by a cascade*. In this setting, we can therefore say that we want a largest possible set of initially activated vertices that do not cascade at all, even to itself. It was first studied by Bazgan and Chopin [1] under the name HARMLESS SET, who showed that it is W[2]-complete in general and W[1]-complete if thresholds are bounded by a constant. They observe (see Observation 1 below) that one can bound the maximum threshold by the solution size and thus obtain a simple FPT algorithm when parametrised by the solution size k *and* the treewidth. Bazgan and Chopin conclude their work with the following open questions: (1) Is HARMLESS SET fixed-parameter tractable on general graphs with respect to the parameter treewidth? (2) Is HARMLESS SET fixed-parameter tractable on planar graphs with respect to the solution size?

Here we answer both these problems and simultaneously discover surprising connections between HARMLESS SET and DOMINATING SET in sparse graphs.

Our Results. Let us distinguish two flavours of this problem: p-BOUNDED HARMLESS SET, where we consider the bound p a constant, and HARMLESS SET where the threshold is unbounded.

HARMLESS SET

Input: A graph G with a threshold function $t\colon V(G) \to \mathbb{N}_{>0}$ and an integer k.

Problem: Is there a vertex set $S \subseteq V(G)$ of size at least k such that every vertex $v \in G$ has fewer than $t(v)$ neighbours in S?

---

p-BOUNDED HARMLESS SET

Input: A graph G with a threshold function $t\colon V(G) \to [p]$ and an integer k.

Problem: Is there a vertex set $S \subseteq V(G)$ of size at least k such that every vertex $v \in G$ has fewer than $t(v)$ neighbours in S?

Note that harmless sets are hereditary in the sense that if S is a harmless set of an instance (G, t), then any subset $S' \subseteq S$ is also harmless for (G, t). Therefore instead of searching for a harmless set of size at least k, we can equivalently search for a harmless set of size exactly k. In this scenario we can replace all thresholds above k with $k + 1$:

Observation 1. HARMLESS SET *parametrised by* k *is equivalent to* $(k + 1)$-BOUNDED HARMLESS SET *parametrised by* k.

It turns out that a simple of application the powerful machinery of first-order model checking[1] in sparse classes [18] answers the first question of Bazgan and Chopin in the positive.

Proposition 1 ($\star$[2])**.** HARMLESS SET *parametrised by k is fixed-parameter tractable in nowhere dense classes.*

These previous results and our observation regarding tractability in sparse classes leave two important questions for us. First, does the problem admit a polynomial kernel in sparse classes? And second, can the problem be solved on e.g. graphs of bounded treewidth without parametrising by the solution size? In the following we answer the kernelization question in the affirmative:

Theorem 1. HARMLESS SET *admits a polynomial sparse kernel in classes of bounded expansion. p-*BOUNDED HARMLESS SET*, for any constant p, admits a linear sparse kernel in these classes.*

Classes with bounded expansion include planar graphs (and generally graphs of bounded genus), graphs of bounded degree, classes excluding a (topological) minor, and more. The term *sparse kernel* is explained below in Sect. 2.1; It alludes to the fact that the constructed kernel does not necessarily belong to the original graph class but is guaranteed to be "almost as sparse".

Bazgan and Chopin give an algorithm for HARMLESS SET parametrised by treewidth and the solution size running in time $O(k^{O(\mathrm{tw})} \cdot n)$, when provided a tree decomposition as part of the input[3]. They conclude by asking whether the problem is "fixed-parameter tractable on general graphs with respect to the parameter treewidth [alone]" [1]. We answer this question in the negative:

Theorem 2. HARMLESS SET *is W[1]-hard when parametrised by a modulator to a 2-spider-forest[4].*

Since a 2-spider-forest has treedepth, pathwidth, and treewidth at most 3, a graph with a modulator M to a 2-spider-forest has treedepth, pathwidth, and treewidth at most $|M| + 3$. This very strong structural parametrisation means that the problem is not only hard on general sparse graphs, but indeed also W[1]-hard for parameters like treewidth, pathwidth, and even treedepth. We complement this result by showing that a slightly stronger parameter, the vertex cover number, does indeed make the problem tractable:

Theorem 3 ($\star$)**.** HARMLESS SET *is fixed-parameter tractable when parametrised by the vertex cover number of the input graph.*

---

[1] There exist some intricacies regarding the type of nowhere dense class and whether the resulting FPT algorithm is uniform or not. This is just a technicality in our context and we refer the reader to Remark 3.2 in [18] for details.

[2] Omitted proofs can be found in the full version available at https://arxiv.org/abs/2111.11834.

[3] This can be relaxed using a constant factor, linear time approximation for computing tree decompositions [2].

[4] A 1-spider-forest is a starforest, and a 2-spider-forest is a subdivided starforest.

Note. We obtained our results simultaneously with and independent from those by Gaikwad and Maity [17]. They provide an explicit and potentially practical FPT algorithm for planar graphs while we show that the problem is not only FPT on planar graphs, but indeed on a much more general class of graphs, namely those of bounded expansion. We also show that on apex-minor-free graphs (which include planar graphs), there exists a subexponential time algorithm for the problem. That is, we show the following results, which improves on Gaikwad and Maity's $2^{O(k \log k)} n^{O(1)}$ algorithm for planar graphs:

Theorem 4. HARMLESS SET *is solvable in time* $2^{o(k)} \cdot n$ *on apex-minor-free graphs.*

2 Preliminaries

2-spider   A 2-spider is a graph obtained from a star by subdividing every edge at most once. A 2-spider-forest is the disjoint union of arbitrarily many 2-spiders.

$f(G)$,
$f(X)$,
$N(X)$,
$N^r(\bullet)$,
$N^r[\bullet]$
For functions $f \colon V(G) \to \mathbb{R}$ we will often use the shorthands $f(X) := \sum_{u \in X} f(x)$ and $f(G) := f(V(G))$. Similarly, we use the shorthand $N(X) := \left(\bigcup_{u \in X} N(u) \right) \setminus X$ for all neighbours of a vertex set X. The rth neighbourhood $N^r(u)$ contains all vertices at distance exactly r from u, the closed r^{th} neighbourhood $N^r[u]$ all vertices at distance at most r from u (also known as the r-ball of u). This corresponds to $N(u) = N^1(u)$ and $N[u] = N^1[u]$. We refer to the textbook by Diestel [9] for more on graph theory notation.

r-scat-tered,
r-dom-inating,
dom$_r$(G),
dom$_r$
(G, X)
A vertex set $X \subseteq V(G)$ is *r-scattered* if for $x_1 \in X$ and $x_2 \in X$, $N^r[x_1] \cap N^r[x_2] = \varnothing$. A vertex set $D \subseteq V(G)$ is *r-dominating* if $N^r[D] = V(G)$ and we write $\mathbf{dom}_r(G)$ to denote the minimum size of such a set. Similarly, we say that D *r-dominates* another vertex set $X \subseteq V(G)$ if $X \subseteq N^r[D]$ and we write $\mathbf{dom}_r(G, X)$ for the minimum size of such a set. In both cases we will omit the subscript r for the case of $r = 1$.

X-avoiding,
r-pro-jection
Given a set $X \subseteq V(G)$ we call a path *X-avoiding* if its internal vertices are not contained in X. A *shortest X-avoiding path* between vertices x, y is shortest among all X-avoiding paths between x and y.

Definition 1 (r-projection). *For a vertex set $X \subseteq V(G)$ and a vertex $u \notin X$ we define the r-projection of u onto X as the set $P^r_X(u) := \{ v \in X \mid \text{there exists an X-avoiding u-v-path of length} \leqslant r \}$.*

Two vertices with the same r-projection onto X do not, however, necessarily have the same (short) distances to X. To distinguish such cases, it is useful to consider the *projection profile* of a vertex to its projection:

Definition 2 (r-projection profile). *For a vertex set $X \subseteq V(G)$ and a vertex $u \notin X$ we define the r-projection profile of u onto X as a function $\pi^r_{G,X}[u] \colon X \to [r] \cup \infty$ where $\pi^r_{G,X}[u](v)$ for $v \in X$ is the length of a shortest X-avoiding path from u to v if such a path of length at most r exists and ∞ otherwise.*

2.1   Bounded Expansion Classes and Kernels

Nešetřil and Ossona de Mendez [22] introduced bounded expansion as a general-isation of many well-known sparse classes like planar graphs, graphs of bounded genus, bounded-degree graphs, classes excluding a (topological) minor, and more. The original definition of bounded expansion classes made use of the concept of *shallow minors* inspired by the work of Plotkin, Rao, and Smith [25].

Definition 3. *A graph H is an r-shallow minor of G, written as $H \preccurlyeq_m^r G$, if H can be obtained from G by contracting disjoint sets of radius at most r.*

Definition 4. *The* greatest-reduced average degree *(grad) ∇_r of a graph G is defined as $\nabla_r(G) = \sup_{H \preccurlyeq_m^r G} \frac{\|H\|}{|H|}$.*   *grad,*
$\nabla_r(\bullet)$

Definition 5. *A graph class $\mathcal{G}$ has* bounded expansion *if there exists a function f such that $\nabla_r(G) \leqslant f(r)$ for all $G \in \mathcal{G}$.*

In the following we will often make use of the property that the grad of a graph does not change much under the addition of a few high-degree vertices: if G is a graph and G' is obtained from G by adding an apex-vertex, then $\nabla_r(G') \leqslant \nabla_r(G) + 1$.

One principal issue with designing kernels for bounded expansion classes is the uncertainty of whether certain gadget constructions preserve the class. When working with concrete classes like planar graphs we can be certain that e.g. adding pendant vertices will result in a planar graph; but when working with an arbitrary bounded expansion class this is not necessarily possible. In such cases, the addition of a pendant vertex takes us outside of the class even though the   *sparse*
grad did not increase. We resolve this issue as proposed in the paper [15]. Let Π *kernel*
be a parametrised problem over graphs. A *sparse kernel* of Π is a kernelization for which there exists a function g that, given an instance with graph G, outputs a graph G' that besides the usual constraints on the size $|G| + \|G'\|$ further satisfies that $\nabla_r(G') \leqslant g(\nabla_r(G))$ for all $r \in \mathbb{N}$. Therefore if the input graphs are taken from a bounded expansion class $\mathcal{G}$, the outputs will also belong to some bounded expansion class $\mathcal{G}'$.

2.2   The Bounded Expansion Toolkit

In our kernelization result we will attempt to construct suitable scattered sets which we can leverage to create a "win-win" argument. To that end, we use Dvořák's algorithm [13] which provides us either with a small r-dominating set or a large r-scattered set. The following variant of the original algorithm is called the *warm-start* variant (see e.g. [15]):

Theorem 5 (Dvořák's algorithm [13]). *For every bounded expansion class $\mathcal{G}$ and $r \in \mathbb{N}$ there exists a polynomial-time algorithm that, given a vertex set $X \subseteq V(G)$, computes an r-dominating set D of X and an r-scattered set $I \subseteq D \cap X$ with $|D| = O(|I|)$.*

Note that since an r-scattered set $I \subseteq X$ provides a lower bound for the r-domination of X we have that $|D| = O(\mathbf{dom}_r(G, X))$. We further need the following fundamental property of bounded expansion classes which refines their characterisation by "neighbourhood complexity" [26]:

Lemma 1 (Adapted from [11,21]). *For every bounded expansion class $\mathcal{G}$ and $r \in \mathbb{N}$ there exists a constant c_r^{proj} such that for every $G \in \mathcal{G}$ and $X \subseteq V(G)$, the number of r-projection profiles realised on X is at most $c_r^{proj}|X|$.*

2.3 Waterlilies

Reidl and Einarson introduced the notion of *waterlilies* as a structure which is very useful in constructing kernels [15]. We simplify the definition here as we do not need it in its full generality.

Definition 6 (Waterlily). *A* waterlily *of radius r and depth $d \leqslant r$ in a graph G is a pair (R, C) of disjoint vertex sets with the following properties: (1) C is r-scattered in $G - R$, (2) $N_{G-R}^r[C]$ is d-dominated by R in G.*

We call R the roots, *C the* centres, *and the sets $\{N_{G-R}^r[x]\}_{x \in C}$ the* pads *of the waterlily. A waterlily is* uniform *if all centres have the same d-projection onto R, e.g. $\pi_R^d[x]$ is the same function for all $x \in C$.*

We will frequently talk about the *ratio* of a waterlily which we define as a guaranteed lower bound of $|C|$ in terms of $|R|$, e.g. a waterlily of *ratio* $2|R| + 1$ satisfies $|C| \geqslant 2|R| + 1$. The authors in [15] used waterlilies with a constant ratio, but a modification of their proof lets us improve this ratio to any polynomial.

Lemma 2 ($\star$). *For every bounded expansion class $\mathcal{G}$ and $r, d \in \mathbb{N}$, $d \leqslant r$, the following holds. There exists a polynomial p_r such that for every $G \in \mathcal{G}$, $t \in \mathbb{N}$ and $A \subseteq V(G)$ with $|A| \geqslant p_r(t)\mathbf{dom}_d(G, A)$ there exists a uniform waterlily $(R, C \subseteq A)$ with depth d, radius r, and with $|R| = O(1)$ and $|C| \geqslant t$, moreover, such a waterlily can be computed in polynomial time.*

3   A Sparse Kernel for p-BOUNDED HARMLESS SET

In order to give a sparse kernel we first show how to construct a bikernel into the following annotated problem.

ANNOTATED p-BOUNDED HARMLESS SET

Input: A graph G with a threshold function $t\colon V(G) \to [p]$, an integer k, and a subset $K \subseteq V(G)$.

Problem: Is there a vertex set $S \subseteq K$ of size at least k such that every vertex $v \in G$ has fewer than $t(v)$ neighbours in S?

---

solution core

We call the set K the *solution core* of the instance (see [15] for a general definition). Next, we present two lemmas whose application will step-wise construct smaller annotated instances. The first lemma lets us reduce the size of the solution core, the second the size of the graph. Afterwards, we demonstrate how these two reduction rules serve to construct a bikernel.

In the following, we often need to treat vertices with a threshold equal to one *fragile* differently. For brevity, we will call these vertices *fragile*; observe that a fragile vertex can be part of a solution but none of its neighbours can.

Lemma 3. *Let (G, t, k, K) be an instance of* ANNOTATED p-BOUNDED HARM-LESS SET *where G is taken from a bounded expansion class and K is a solution core. There exists a polynomial $q(p)$ such that the following holds: If $|K| \geqslant q(p) \cdot k$, then in polynomial time we either find that (G, t, k, K) is a YES-instance or we identify a vertex $x \in K$ such that $K \setminus \{x\}$ is a solution core.*

Sketch of proof. If there is a vertex $x \in K$ with a fragile neighbour $u \in N(x)$. Then x of course cannot be in any solution and $K \setminus \{x\}$ is a solution core.

Assume that no vertex in K has a fragile neighbour. We use Dvořák's algorithm (Theorem 5) to compute a 1-dominating set for K; let D be the resulting dominating set and $I \subseteq D \cap K$ the promised 1-scattered set, i.e., with $|I| = \Omega(|D|)$. Since the neighbourhoods of vertices in I are pairwise disjoint and no vertex in I has a fragile neighbour, it follows that I itself is a harmless set. So if $|I| \geqslant k$ we conclude that (G, t, k, K) is a YES-instance.

Otherwise $|I| < k$ and by Theorem 5 $\mathbf{dom}(G, K) = O(k)$. We apply Lemma 2 to compute a waterlily for the set K at depth 1 and with radius 2. Where we choose $q(k)$ large enough to ensure that the following arguments go through.

Let $(R, C \subseteq K)$ be the resulting uniform waterlily with $|C| \geqslant \kappa$, where κ is an appropriately large value that we choose later. For the centres $v \in C$, define signature $\sigma(v) = \{(t(u), N(u) \cap R) \mid u \in N_{G-R}(v)\}$ and the equivalence relation $\sim_\sigma$ over C via $v \sim_\sigma w$ iff $\sigma(v) = \sigma(w)$. Recall that, by Lemma 1 the number of 1-projections onto R is at most $c_1^{\mathrm{proj}}|R|$. Therefore we can picture $\sigma(v)$ as a string of length at most $c_1^{\mathrm{proj}}|R|$ over the alphabet $\{0, \ldots, p\}$ where 0 indicates that a certain neighbourhood is not contained in $\sigma(v)$ and any non-zero value $a \in [p]$ indicates that this neighbourhood is realised by one of v's neighbours with weight a. Accordingly, we can bound the index of $\sim_\sigma$ by $|C/ \sim_\sigma | \leqslant (p+1)^{c_1^{\mathrm{proj}}|R|}$ and thus by averaging there exists an equivalence class $C' \in C/ \sim_\sigma$ of size at least $|C|/(p+1)^{c_1^{\mathrm{proj}}|R|}$.

We choose $|C|$ big enough so that $|C'| > (p-1)|R|$. Since the vertices in C' are uniform under σ, any vertex of C' can be safely removed from C. □

Lemma 4 ($\star$). *Let (G, t, k, K) be an instance of our* ANNOTATED p-BOUNDED HARMLESS SET *problem where G is taken from a bounded expansion class. Then, if $|K| < |G|/(c_1^{\mathrm{proj}} + 1)$, then there exists a vertex $v \in V(G) \setminus K$ such that $(G - v, t|_{V(G)-v}, k, K)$ is an equivalent instance.*

With these two reduction rules in hand, we can finally prove the main result of this section.

Theorem 6 (⋆). p-BOUNDED HARMLESS SET *over bounded expansion classes admits a bikernel into* ANNOTATED p-BOUNDED HARMLESS SET *of size* $f(p) \cdot k$, *for some polynomial f.*

Corollary 1 (⋆). HARMLESS SET *admits a polynomial sparse kernel.*

Corollary 2 (⋆). p-BOUNDED HARMLESS SET *for any constant p admits a linear sparse kernel.*

4   Sparse Parametrisation

In this section we first prove Theorems 2 and 3, namely that HARMLESS SET is intractable when parametrised by the size of a modulator to a 2-*spider-forest* but is FPT when parametrised by the vertex cover number of the input graph. We then show that a simple application of the bidimensionality framework [8,16] proves Theorem 4, i.e. that HARMLESS SET can be solved in subexponential linear FPT time on graphs excluding an apex-minor.

The first result of this section, whose proof has been omitted, is that HARM-LESS SET is FPT parameterized on the size of a minimum vertex cover:

Theorem 3 (⋆). HARMLESS SET *is fixed-parameter tractable when parametrised by the vertex cover number of the input graph.*

4.1   Modulator to 2-Spider-Forest

An instance of MULTICOLOURED CLIQUE consists of a k-partite graph $G = (V_1, \ldots, V_k, E)$. The task is to find a clique that intersects each colour V_i in exactly one vertex. Since MULTICOLOURED CLIQUE is W[1]-hard [7], our reduction establishes the same for HARMLESS SET.

In the following, we fix an instance $(V_1, \ldots, V_k, E)$ of MULTICOLOURED CLIQUE. By a simple padding argument, we can assume that the sizes of the sets V_i are all the same and we will denote this cardinality by n (thus the graph has a total of nk vertices). For convenience, we let $v_1^i, \ldots, v_n^i$ be the vertices of the set V_i. For indices $1 \leqslant i < j \leqslant k$ we denote by $m_{ij} = |E(V_i, V_j)|$ the number of edges between colours V_i and V_j. We further let m be the total number of edges.

Forbidden Vertices. Let $F \subseteq V(G)$ be a set of vertices that we want to prevent from being in any solution. To that end, we construct a global *forbidden set* gadget which enforces that no vertex from F can be selected. The construction is similar to the *forbidden edge* gadget by Bazgan and Chopin [1]:

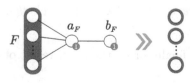 We add two vertices a_F and b_F with threshold one to the graph and connected them. Then we connect a_F to every vertex in F. In the following gadgets we will often mark vertices as "forbidden". We will denote this graphically by drawing a thick red border around these vertices.

Observation 2. *Let F, a_F, b_F be vertices as above in some instance (G, t, k) of* HARMLESS SET. *Then for every harmless set S of (G, t) it holds that $S \cap (F \cup \{a_F, b_F\}) = \varnothing$.*

XOR Gadget. We construct an *XOR gadget* for vertices u and v by adding a new forbidden vertex x with threshold two and adding the edges xu and xv to the graph. To simplify the drawing of the following gadgets, we will simply draw a thick red edge between two vertices to denote that they are connected by an XOR gadget.

Observation 3. *Let u, x, v be as above in some instance (G, t, k) of* HARMLESS SET. *Then for every harmless set S of (G, t) it holds that $|S \cap \{u, v\}| \leqslant 1$.*

Selection Gadget

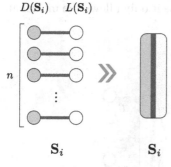

The role of a selection gadget $\mathbf{S}_i$ will be to select a single vertex from one coloured set V_i. The final construction will therefore contain k of these gadgets $\mathbf{S}_1, \dots, \mathbf{S}_k$. The gadget consists of n pairs of vertices d_s, l_s, $s \in [n]$, where each pair is connected by an XOR gadget. We call the set $D(\mathbf{S}_i) = \{d_1, \dots, d_n\}$ the *dark* vertices and $L(\mathbf{S}_i) = \{l_1, \dots, l_n\}$ the *light* vertices. We make two simple observations about the behaviour of this gadget:

Observation 4. *Let $\mathbf{S}_i$ be as above in some instance (G, t, k) of* HARMLESS SET. *Then for every harmless set S of (G, t) it holds that $|S \cap (D(\mathbf{S}_i) \cup L(\mathbf{S}_i))| \leqslant n$.*

Observation 5. *Let $\mathbf{S}_i$ be as above in some instance (G, t, k) of* HARMLESS SET. *Then for every harmless set S of (G, t) with $|S \cap (D(\mathbf{S}_i) \cup L(\mathbf{S}_i))| = n$ it holds that $|S \cap D(\mathbf{S}_i)| + |S \cap L(\mathbf{S}_i)| = n$.*

Port Gadget

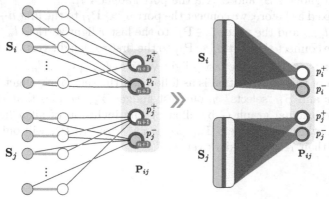

For every pair of selection gadgets $\mathbf{S}_i$, $\mathbf{S}_j$ we need to communicate the choices these gadgets encode to further gadgets (described below) which verify that this choice corresponds to an edge in $E(V_i, V_j)$.

The port gadget $\mathbf{P}_{ij}$ responsible for the pair $\mathbf{S}_i$, $\mathbf{S}_j$ consists of four forbidden *port vertices* p_i^+, p_i^-, p_j^+, and p_j^-, each with a threshold of $n + 1$. For $\ell \in \{i, j\}$, we connect the port vertex p_ℓ^+ to the light vertices $L(\mathbf{S}_\ell)$ and the port vertex p_ℓ^- to the dark vertices $D(\mathbf{S}_\ell)$. Note that every selection gadget will be connected to $k - 1$ port gadgets in this manner and our naming scheme of the variables $p_\bullet^+$, $p_\bullet^-$ does not reflect that. However, we will in the following only ever talk about a single port gadget and therefore it will always be clear to which vertices we refer.

Test Gadget. The final gadget $\mathbf{T}_{xy}$ exists to test whether two selection gadgets $\mathbf{S}_i$, $\mathbf{S}_j$ selected the edge $v_x^i v_y^j \in E(V_i, V_j)$. If that is the case, the gadget allows the inclusion of n vertices into the solution; otherwise it only allows the inclusion of a single vertex.

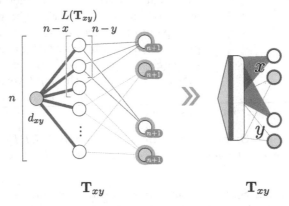

The gadget consists of n ordered light vertices $L(\mathbf{T}_{xy}) = \{l_1, \ldots, l_n\}$ which are all connected to a single dark vertex d_{xy} via XOR gadgets. This already concludes the structure of the gadget itself, but we need to discuss how it will be wired to the selection gadgets $\mathbf{S}_i$ and $\mathbf{S}_j$ via the port gadget $\mathbf{P}_{ij}$.

For i, j fixed as before, we connect the port $p_i^+ \in \mathbf{P}_{ij}$ to the first $n-x$ light vertices $l_1, \ldots, l_{n-x}$ and the port $p_i^- \in \mathbf{P}_{ij}$ to the last x light vertices $l_{n-x+1}, \ldots, l_n$. Similarly, we connect the port $p_j^+ \in \mathbf{P}_{ij}$ to the first $n-y$ light vertices $l_1, \ldots, l_{n-y}$ and the port $p_j^- \in \mathbf{P}_{ij}$ to the last y light vertices $l_{n-y+1}, \ldots, l_n$.

The idea of this construction is as follows: If the selection gadget $\mathbf{S}_i$ "selects" the vertex x and $\mathbf{S}_j$ "selects" y, our test gadget $\mathbf{T}_{xy}$ verifies that the edge xy exists in the original graph G by allowing the inclusion of all n light vertices $L(\mathbf{T}_{xy})$. All other test gadgets $\mathbf{T}_{uv}$, $uv \neq xy$, wired to $\mathbf{P}_{ij}$ will only allow the inclusion of their respective dark vertex d_{uv}.

Full Construction. The full construction for the reduction looks as follows.

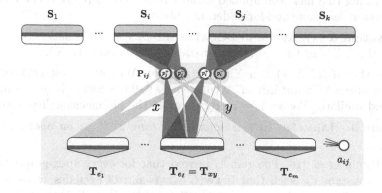

Given the instance $G = (V_1 \uplus \cdots \uplus V_k, E)$ of MULTICOLOURED CLIQUE, we construct an instance (H, t) of HARMLESS SET as follows:

1. We add k selection gadgets $\mathbf{S}_1, \ldots, \mathbf{S}_k$.
2. For every pair of indices $1 \leqslant i < j \leqslant k$:
   - We add the port gadget $\mathbf{P}_{ij}$ and connect it to $\mathbf{S}_i$ and $\mathbf{S}_j$ as described above.
   - We add $m_{ij} := |E(V_i, V_j)|$ test gadgets $\{\mathbf{T}_{xy}\}_{xy \in E(V_i, V_j)}$.
   - We wire each test gadget $\mathbf{T}_{xy}$ to $\mathbf{P}_{ij}$ as described above.
   - We add a forbidden vertex a_{ij} to H with threshold $n + 1$ and connect it to all light vertices $\bigcup_{xy \in E(V_i, V_j)} L(\mathbf{T}_{xy})$.
3. Finally, we add the vertices a_F and b_F to H and connect a_F to all vertices marked as "forbidden" in the gadgets as well as to b_F.

Lemma 5 ($\star$). *We can delete* $5\binom{k}{2} + 1$ *vertices from H to obtain a 2-spider forest.*

Lemma 6 ($\star$). *If G contains a multi-coloured clique on k vertices, then (H, t) has a harmless set of size* $\binom{k}{2}(n - 1) + kn + m$.

Lemma 7 ($\star$). *If (H, t) has a harmless set of size* $\binom{k}{2}(n - 1) + kn + m$, *then G contains a multi-coloured clique on k vertices.*

Lemmas 5, 6, and 7 together prove Theorem 2.

4.2 Subexponential Time Algorithm

In order to apply the bidimensionality framework we introduce an annotated AVOIDING 1-SCATTERED SET which takes an additional "forbidden" vertex subset as input and asks to find a solution without forbidden vertices. We say that a vertex $x \in X$ is *simplicially forbidden* if it is forbidden and all its neighbours are forbidden. Observe that we may safely remove any simplicially forbidden vertices for either of the two problems. We will assume in the following that this preprocessing rule has been applied exhaustively and therefore every forbidden vertex has at least one non-forbidden neighbour.

Observation 6 ($\star$). AVOIDING 1-SCATTERED SET *is closed under contractions, that is, if* $(G, X)/uv$ *has a solution of size* k *then so does* (G, X).

Observe that if (G, X, k) is a YES-instance of AVOIDING 1-SCATTERED SET then it is also a YES-instance of AVOIDING HARMLESS SET, where this problem is defined similarly. We are now ready to apply the bidimensionality framework.

Theorem 4. HARMLESS SET *is solvable in time* $2^{o(k)} \cdot n$ *on apex-minor-free graphs.*

Proof. Fomin *et al.* [16, Theorem 1] proved that for every apex-graph H there exists a constant c_H such that if $\mathbf{tw}(G) \geqslant k$ and G excludes H as a minor, then G has the graph $\Gamma_{c_H \cdot k}$ as a contraction minor. Here Γ_t is the triangulated $t \times t$ grid where additionally one corner vertex is attached to all border vertices of the grid.

So assume that our input instance (G, X, k) has treewidth $\mathbf{tw}(G) \geqslant (5\sqrt{k} + 10)/c_H$, then G contains Γ_t as a contraction minor with $t = 5\sqrt{k} + 10$. Let $X' \subseteq V(\Gamma_t)$ be the contracted forbidden vertices as defined above. As we observed earlier, every vertex in X' has at least one neighbour in Γ_t which is not in X'.

Claim ($\star$). Γ_t contains a 1-scattered set that avoids X' of size at least k.

We conclude that if G has treewidth at least $w := (5\sqrt{k} + 10)/c_H$, then (G, X, k) is a YES-instance. Using the single-exponential 5-approximation for treewidth [2], we can in time $2^{O(w)}n = 2^{O(\sqrt{k})}n$ either find that G has treewidth at least w or we obtain a tree decomposition of width no larger than $5w$. In the latter case, we use the algorithm by Bazgan and Chopin to solve the problem in time $k^{O(w)}n = 2^{O(\sqrt{k}\log k)}n$. Note that the total running time is bounded by $2^{o(k)} \cdot n$, as claimed. □

5   Conclusion

We observed that the problem HARMLESS SET is in FPT for sparse graph classes and we investigated its tractability in the kernelization sense. We found that HARMLESS SET admits a polynomial and p-BOUNDED HARMLESS SET a linear sparse kernel. We expect these results to extend to nowhere dense classes. When the graph class is restricted to apex-minor-free graphs, we are also able to solve HARMLESS SET in subexponential parameterized time $2^{o(k)}n$.

On the negative side, we demonstrated that sparseness alone does not make the problem tractable. While the problem is in FPT when parametrised by e.g. treewidth and solution size, we showed that it is in fact W[1]-hard when only parametrised by treewidth. Our reduction shows even more, namely that most sparse parameters (treedepth, pathwidth, feedback vertex set) can be ruled out as the problem is already hard when parametrised by a modulator to a 2-spider-forest. We conjecture—and leave as an interesting open problem—that HARMLESS SET is already hard when parametrised by a modulator to a starforest.

References

1. Bazgan, C., Chopin, M.: The complexity of finding harmless individuals in social networks. Discret. Optim. **14**, 170–182 (2014)
2. Bodlaender, H.L., Drange, P.G., Dregi, M.S., Fomin, F.V., Lokshtanov, D., Pilipczuk, M.: A $c^k n$ 5-approximation algorithm for treewidth. SIAM J. Comput. **45**(2), 317–378 (2016)
3. Chen, N.: On the approximability of influence in social networks. SIAM J. Discret. Math. **23**(3), 1400–1415 (2009)
4. Chen, W., et al.: Influence maximization in social networks when negative opinions may emerge and propagate. In: Proceedings of the 2011 SIAM International Conference on Data Mining, pp. 379–390. SIAM (2011)
5. Chen, W., Lakshmanan, L.V.S., Castillo, C.: Information and influence propagation in social networks. Synth. Lect. Data Manag. **5**(4), 1–177 (2013)
6. Chen, W., Wang, Y., Yang, S.: Efficient influence maximization in social networks. In: Proceedings of the 15th ACM SIGKDD International Conference on Knowledge Discovery and Data Mining, pp. 199–208 (2009)
7. Cygan, M., et al.: Parameterized Algorithms. Springer, Heidelberg (2015)
8. Demaine, E.D., Hajiaghayi, M.T.: The bidimensionality theory and its algorithmic applications. Comput. J. **51**(3), 292–302 (2008)
9. Diestel, R.: Graph Theory. Graduate Texts in Mathematics, vol. 173, 5th edn. Springer, Heidelberg (2016)
10. Domingos, P., Richardson, M.: Mining the network value of customers. In: Proceedings of the Seventh ACM SIGKDD International Conference on Knowledge Discovery and Data Mining, pp. 57–66 (2001)
11. Drange, P.G., et al.: Kernelization and sparseness: the case of dominating set. In: 33rd Symposium on Theoretical Aspects of Computer Science, STACS 2016. LIPIcs, vol. 47 pp. 31:1–31:14. Schloss Dagstuhl - Leibniz-Zentrum für Informatik (2016)
12. Dreyer, P.A., Jr., Roberts, F.S.: Irreversible k-threshold processes: graph-theoretical threshold models of the spread of disease and of opinion. Discret. Appl. Math. **157**(7), 1615–1627 (2009)
13. Dvořák, Z.: Constant-factor approximation of the domination number in sparse graphs. Eur. J. Comb. **34**(5), 833–840 (2013)
14. Easley, D., Kleinberg, J.: Networks, Crowds, and Markets, vol. 8. Cambridge University Press, Cambridge (2010)
15. Einarson, C., Reidl, F.: A general kernelization technique for domination and independence problems in sparse classes. In: 15th International Symposium on Parameterized and Exact Computation, IPEC 2020, Hong Kong, China, 14–18 December 2020 (Virtual Conference). LIPIcs, vol. 180, pp. 11:1–11:15. Schloss Dagstuhl - Leibniz-Zentrum für Informatik (2020)
16. Fomin, F.V., Golovach, P., Thilikos, D.M.: Contraction bidimensionality: the accurate picture. In: Fiat, A., Sanders, P. (eds.) ESA 2009. LNCS, vol. 5757, pp. 706–717. Springer, Heidelberg (2009). https://doi.org/10.1007/978-3-642-04128-0_63
17. Gaikwad, A., Maity, S.: The harmless set problem. CoRR, abs/2111.06267 (2021)
18. Grohe, M., Kreutzer, S., Siebertz, S.: Deciding first-order properties of nowhere dense graphs. J. ACM **64**(3), 17:1–17:32 (2017)
19. Kempe, D., Kleinberg, J., Tardos, É.: Maximizing the spread of influence through a social network. In: Proceedings of the ninth ACM SIGKDD International Conference on Knowledge Discovery and Data Mining, pp. 137–146 (2003)

20. Kempe, D., Kleinberg, J., Tardos, É.: Influential nodes in a diffusion model for social networks. In: Caires, L., Italiano, G.F., Monteiro, L., Palamidessi, C., Yung, M. (eds.) ICALP 2005. LNCS, vol. 3580, pp. 1127–1138. Springer, Heidelberg (2005). https://doi.org/10.1007/11523468_91

21. Kreutzer, S., Rabinovich, R., Siebertz, S.: Polynomial kernels and wideness properties of nowhere dense graph classes. ACM Trans. Algorithms 15(2), 24:1-24:19 (2019)

22. Nešetřil, J., de Mendez, P.O.: Grad and classes with bounded expansion I. Decompositions. Eur. J. Comb. 29(3), 760–776 (2008)

23. Nguyen, H.T., Thai, M.T., Dinh, T.N.: Stop-and-stare: optimal sampling algorithms for viral marketing in billion-scale networks. In: Proceedings of the 2016 International Conference on Management of Data, pp. 695–710 (2016)

24. Nguyen, N.P., Yan, G., Thai, M.T., Eidenbenz, S.: Containment of misinformation spread in online social networks. In: Proceedings of the 4th Annual ACM Web Science Conference, pp. 213–222 (2012)

25. Plotkin, S.A., Rao, S., Smith, W.D.: Shallow excluded minors and improved graph decompositions. In: Sleator, D.D. (ed.) Proceedings of the Fifth Annual ACM-SIAM Symposium on Discrete Algorithms, Arlington, Virginia, USA, 23–25 January 1994, pp. 462–470. ACM/SIAM (1994)

26. Reidl, F., Villaamil, F.S., Stavropoulos, K.: Characterising bounded expansion by neighbourhood complexity. Eur. J. Comb. 75, 152–168 (2019)

27. Richardson, M., Domingos, P.: Mining knowledge-sharing sites for viral marketing. In: Proceedings of the Eighth ACM SIGKDD International Conference on Knowledge Discovery and Data Mining, pp. 61–70 (2002)

28. Roberts, F.S.: Graph-theoretical problems arising from defending against bioterrorism and controlling the spread of fires. In: Proceedings of the DIMACS/DIMATIA/Renyi Combinatorial Challenges Conference, Piscataway, NJ (2006)

The Parameterized Complexity of s-Club with Triangle and Seed Constraints

Jaroslav Garvardt, Christian Komusiewicz(iD), and Frank Sommer$^{(\boxtimes)}$(iD)

Fachbereich Mathematik und Informatik, Philipps-Universität Marburg,
Marburg, Germany
{garvardt,komusiewicz,fsommer}@informatik.uni-marburg.de

Abstract. The s-CLUB problem asks, for a given undirected graph G, whether G contains a vertex set S of size at least k such that $G[S]$, the subgraph of G induced by S, has diameter at most s. We consider variants of s-CLUB where one additionally demands that each vertex of $G[S]$ is contained in at least ℓ triangles in $G[S]$, that $G[S]$ contains a spanning subgraph G' such that each edge of $E(G')$ is contained in at least ℓ triangles in G', or that S contains a given set W of seed vertices. We show that in general these variants are W[1]-hard when parameterized by the solution size k, making them significantly harder than the unconstrained s-CLUB problem. On the positive side, we obtain some FPT algorithms for the case when $\ell = 1$ and for the case when $G[W]$, the graph induced by the set of seed vertices, is a clique.

1 Introduction

Finding cohesive subgroups in social or biological networks is a fundamental task in network analysis. A classic formulation of cohesiveness is based on the observation that cohesive groups have small diameter. This observation led to the s-club model originally proposed by Mokken [15]. An s-*club* in a graph $G = (V, E)$ is a set of vertices S such that $G[S]$, the subgraph of G induced by S has diameter at most s. The 1-clubs are thus precisely the cliques and the larger the value of s, the more the clique-defining constraint of having diameter one is relaxed. In the s-CLUB problem we aim to decide whether G contains an s-club of size at least k.

A big drawback of s-clubs is that the largest s-clubs are often not very cohesive with respect to other cohesiveness measures such as density or minimum degree. This behavior is particularly pronounced for $s = 2$: the largest 2-club in a graph is often the vertex v of maximum degree together with its neighbors [10]. To avoid these so-called hub-and-spoke structures, it has been proposed to augment the s-club definition with additional constraints [5,14,16,18].

One of these augmented models, proposed by Carvalho and Almeide [5], asks that every vertex is part of a triangle. This property was later generalized to

---

F. Sommer—Supported by the DFG, project EAGR (KO 3669/6-1).

C. Bazgan and H. Fernau (Eds.): IWOCA 2022, LNCS 13270, pp. 313–326, 2022.
https://doi.org/10.1007/978-3-031-06678-8_23

the *vertex-ℓ-triangle* property, which asks that every vertex of S is in at least ℓ triangles in $G[S]$ [1].

VERTEX TRIANGLE s-CLUB

Input: An undirected graph $G = (V, E)$, and two integers $k, \ell \geq 1$.

Question: Does G contain an s-club S of size at least k that fulfills the vertex-ℓ-triangle property?

The vertex-ℓ-triangle constraint entails some desirable properties for cohesive subgraphs. For example, in a vertex-ℓ-triangle s-club, the minimum degree is larger than $\sqrt{2\ell}$. However, some undesirable behavior of hub-and-spoke structures remains. For example, the graph consisting of two cliques of size $d+1$ that are connected via one edge is a vertex-$\binom{d}{2}$-triangle 3-club but it can be made disconnected via one edge deletion. Thus, vertex-ℓ-triangle s-clubs are not robust with respect to edge deletions.

To overcome this problem, we introduce a new model where we put triangle constraints on the edges of the s-club instead of the vertices. More precisely, we say that a vertex set S of a graph G fulfills the *edge-ℓ-triangle* property if $G[S]$ contains a spanning subgraph $G' := (S, E')$ such that every edge in $E(G')$ is in at least ℓ triangles in G' and the diameter of G' is at most s.

EDGE TRIANGLE s-CLUB

Input: An undirected graph $G = (V, E)$, and two integers $k, \ell \geq 1$.

Question: Does G contain a vertex set S of size at least k that fulfills the edge-ℓ-triangle property?

Note that in this definition, the triangle and diameter constraints are imposed on a spanning subgraph of $G[S]$. In contrast, for VERTEX TRIANGLE s-CLUB, they are imposed directly on $G[S]$. The reason for this distinction is that we would like to have properties that are closed under edge insertions. Properties which are closed under edge insertions are also well-motivated from an application point of view since adding a new connection within a group should not destroy this group. If we would impose the triangle constraint on the induced subgraph $G[S]$ instead, then an edge-ℓ-triangle s-club S would not be robust to edge additions. For example, consider a graph G consisting of clique C to which two vertices u and v are attached in such a way that both u and v have exactly 2 neighbors in C which are distinct. The $V(G)$ is an edge-1-triangle 3-club, but other adding the edge uv, the edge uv is contained in no triangle and thus any edge-1-triangle 3-club cannot contain both u and v.

Observe that every set that fulfills the edge-ℓ-triangle property also fulfills the vertex-ℓ-triangle property. Moreover, each vertex $v \in S$ has at least $\ell + 1$ neighbors in S: Consider an arbitrary edge uv. Since uv is in at least ℓ triangles $\{u, v, w_1\}, \ldots, \{u, v, w_\ell\}$ we thus conclude that u and v have degree at least ℓ. We can show an even stronger statement: an edge-ℓ-triangle s-club S is robust against up to ℓ edge deletions, as desired.

Proposition 1. *Let $G = (V, E)$ be a graph and let S be an edge-ℓ-triangle s-club in G. More precisely, let G' be a spanning subgraph of $G[S]$ such that every edge* *in $E(G')$ is in at least ℓ triangles in G' and the diameter of G' is at most s. If ℓ edges are removed from G', then S is still an $(s + \ell)$-club and a $(2s)$-club.*

Proof. We show that if ℓ edges are removed from G', the diameter of the resulting graph $\widetilde{G}$ increases by at most ℓ. Let $P = (v_1, \ldots, v_{s+1})$ be a path of length s in G'. Since G' is an edge-ℓ-triangle s-club, every edge $v_i v_{i+1}$ of P is part of at least ℓ triangles in G'. Thus, for two vertices v_i and v_{i+1} in P there is a path of length at most two from v_i to v_{i+1} in G', either directly through the edge $v_i v_{i+1}$ or via a vertex u that forms one of the ℓ triangles with v_i and v_{i+1} in G'. Thus, $\text{dist}(v_i, v_{i+1})$ increases by at most 1 after one edge deletion and only if $v_i v_{i+1}$ is removed. Since at most ℓ of the edges in P are removed, we have $\text{dist}(v_1, v_{s+1}) \leq \text{dist}(v_1, v_2) + \ldots + \text{dist}(v_s, v_{s+1}) \leq s + \ell$ in $\widetilde{G}$. By the same arguments, we also have $\text{dist}(v_1, v_{s+1}) \leq 2s$.

Thus, after deleting ℓ edges in G', S is an $(s + \ell)$-club and a $(2s)$-club. $\square$

The following further variant of s-CLUB is also practically motivated but not necessarily by concerns about the robustness of the s-club. Here the difference to the standard problem is simply that we are given a set of seed vertices W and aim to find a large s-club that contains all seed vertices.

SEEDED s-CLUB

Input: An undirected graph $G = (V, E)$, a subset $W \subseteq V$, and an integer $k \geq 1$.

Question: Does G contain an s-club S of size at least k such that $W \subseteq S$?

This variant has applications in community detection, where we are often interested in finding communities containing some set of fixed vertices [12,19].

In this work, we study the parameterized complexity of the three above-mentioned problems with respect to the standard parameter solution size k. Our goal is to determine whether FPT results for s-CLUB [6,17] transfer to these practically motivated problem variants.

Known Results. The s-CLUB problem is NP-hard for all $s \geq 1$ [4], even when the input graph has diameter $s + 1$ [2]. For $s = 1$, s-CLUB is equivalent to CLIQUE and thus W[1]-hard with respect to k. In contrast, for every $s > 1$, s-CLUB is fixed-parameter tractable (FPT) with respect to the solution size k [6,17]. This fixed-parameter tractability can be shown via a Turing kernel with $\mathcal{O}(k^2)$ vertices for even s and $\mathcal{O}(k^3)$ vertices for odd s [6,17]. The complexity of s-CLUB has been also studied with respect to different classes of input graphs [9] and with respect to structural parameters of the input graph [11]. The s-CLUB problem can be solved efficiently in practice, in particular for $s = 2$ [4,6,10].

VERTEX TRIANGLE s-CLUB is NP-hard for all $s \geq 1$ and for all $\ell \geq 1$ [1, 5]. We are not aware of any algorithmic studies of EDGE TRIANGLE s-CLUB or SEEDED s-CLUB. NP-hardness of EDGE TRIANGLE s-CLUB for $\ell = 1$ can be shown via the reduction for VERTEX TRIANGLE s-CLUB for $\ell = 1$ [1]. Also, the NP-hardness of SEEDED s-CLUB for $W \neq \emptyset$ follows directly from the fact that an algorithm for the case where $|W| = 1$ can be used as a black box to

Table 1. Overview of our results of the parameterized complexity of the three problems with respect to the parameter solution size k.

	VERTEX TRIANGLE s-CLUB	EDGE TRIANGLE s-CLUB	SEEDED s-CLUB
FPT	$\ell = 1$ and $s \geq 4$	$\ell = 1$ for each s	W is a clique
W[1]-h	$\ell = 1$ and $s \leq 3$	$\ell \geq 2$ for each s	$s = 2$ and $G[W]$ contains at least two non-adjacent vertices
	$\ell \geq 2$ for each s		$s \geq 3$ and $G[W]$ contains at least 2 connected components

solve s-CLUB. Further robust models of s-clubs, which are not considered in this work, include t-hereditary s-clubs [16], t-robust s-clubs [18], and t-connected s-clubs [14,20]. For an overview on clique relaxation models and complexity issues for the corresponding subgraph problems we refer to the relevant surveys [13,16].

Our Results. An overview of our results is given in Table 1. For VERTEX TRIANGLE s-CLUB and EDGE TRIANGLE s-CLUB, we provide a complexity dichotomy for all interesting combinations of s and ℓ, that is, for all $s \geq 2$ and $\ell \geq 1$, into cases that are FPT or W[1]-hard with respect to k, respectively. Our W[1]-hardness reduction for EDGE TRIANGLE s-CLUB for $\ell \geq 2$ also shows the NP-hardness of this case. The FPT-algorithms are obtained via adaptions of the Turing kernelization for s-CLUB. Interestingly, VERTEX TRIANGLE s-CLUB with $\ell = 1$ is FPT only for larger s, whereas EDGE TRIANGLE s-CLUB with $\ell = 1$ is FPT for all s. In our opinion, this means that the edge-ℓ-triangle property is preferable not only from a modelling standpoint but also from an algorithmic standpoint as it allows to employ Turing kernelization as a part of the solving procedure, at least for $\ell = 1$. It is easy to see that standard problem kernels of polynomial size are unlikely to exist for VERTEX TRIANGLE s-CLUB and EDGE TRIANGLE s-CLUB: s-clubs are necessarily connected and thus taking the disjoint union of graphs gives a trivial or-composition and, therefore, a polynomial problem kernel implies coNP $\subseteq$ NP/poly [3].

For SEEDED s-CLUB, we provide a kernel with respect to k for clique seeds W and W[1]-hardness with respect to k for some other cases. For $s = 2$, our results provide a dichotomy into FPT and W[1]-hardness with respect to k in terms of the structure of the seed.

Our W[1]-hardness results, in particular those for SEEDED s-CLUB, show that the FPT results for s-CLUB are quite brittle since the standard argument that we may assume $k \geq \Delta$ fails and that adding even simple further constraints makes finding small-diameter subgraphs much harder. Due to lack of space, the proofs of several results (marked with $(\star)$) are deferred to a full version of this article.

Preliminaries. For integers p, q, we denote $[p, q] := \{p, p+1, \ldots, q\}$ and $[q] := [1, q]$. For a graph G, we let $V(G)$ denote its vertex set and $E(G)$ its edge set. We let n and m denote the order of G and the number of edges in G, respectively. A *path of length p* is a sequence of pairwise distinct vertices $v_1, \ldots, v_{p+1}$ such that $v_i v_{i+1} \in E(G)$ for each $i \in [p]$. The *distance* $\text{dist}_G(u,v)$ is the length of a shortest path between vertices u and v. Furthermore, we define $\text{dist}_G(u, W) := \min_{w \in W} \text{dist}(u,w)$. We denote by $\text{diam}_G(G) := \max_{u,v \in V(G)} \text{dist}_G(u,v)$ the *diameter* of G. Let $S \subseteq V(G)$ be a vertex set. We denote by $N_i(S) := \{u \in V \mid \text{dist}(u,S) = i\}$ the *open i-neighborhood* of S and by $N_i[S] := \bigcup_{j \leq i} N_i(S) \cup S$ the *closed i-neighborhood* of S. For a vertex $v \in V(G)$, we write $N_i(v) := N_i(\{v\})$ and $N_i[v]'' := N_i[\{v\}]$. A graph $G' := (V', E')$ with $V' \subseteq V$, and $E' \subseteq E(G[V'])$ is a *subgraph* of G. By $G[S] := (S, \{uv \in E(G) \mid u,v \in S\})$ we denote the *subgraph induced by S*. Furthermore, by $G - S := G[V \setminus S]$ we denote the induced subgraph obtained after the deletion of the vertices in S. A vertex set such that each pair of vertices is adjacent is called a *clique* and a clique consisting of three vertices is a *triangle*.

For the definitions of parameterized complexity theory, we refer to the standard monographs [7,8]. All of our hardness results are shown by a reduction from CLIQUE which is the special case of S-CLUB with $s = 1$ and which is known to be W[1]-hard with respect to k [7,8].

2   Vertex Triangle s-Club

First, we show that VERTEX TRIANGLE s-CLUB is fixed-parameter tractable when $\ell = 1$ and $s \geq 4$. Afterwards, we show W[1]-hardness for all remaining cases. The first step of the FPT algorithm is to apply the following rule.

Reduction Rule 1. *Let (G, k) be an instance of VERTEX TRIANGLE s-CLUB. Delete all vertices from G which are not part of any triangle.*

Clearly, Reduction Rule 1 is correct and can be exhaustively applied in polynomial time. The application of Reduction Rule 1 has the following effect: if some vertex v is close to many vertices, then (G, k) is a trivial yes-instance.

Lemma 1. *Let (G, k) be an instance of VERTEX TRIANGLE s-CLUB with $\ell = 1$ and $s \geq 4$ to which Reduction Rule 1 is applied. Then, (G, k) is a yes-instance if $|N_{\lfloor s/2 \rfloor - 1}[v]| \geq k$ for some vertex $v \in V(G)$.*

Proof. Let $v \in V(G)$ be a vertex such that $|N_{\lfloor s/2 \rfloor - 1}[v]| \geq k$. We construct a vertex-1-triangle s-club T of size at least $|N_{\lfloor s/2 \rfloor - 1}[v]| \geq k$. Initially, we set $T := N_{\lfloor s/2 \rfloor - 1}[v]$. Now, for each vertex $w \in N_{\lfloor s/2 \rfloor - 1}(v)$ we do the following: Since Reduction Rule 1 is applied, we conclude that there exist two vertices x and y such that $G[\{w, x, y\}]$ is a triangle. We add x and y to the set T. We call the set of vertices added in this step the *T-expansion*.

Next, we show that T is indeed a vertex-1-triangle s-club for $s \geq 4$. Observe that each vertex in T is either in $N_{\lfloor s/2 \rfloor - 1}[v]$ or a neighbor of a vertex in $N_{\lfloor s/2 \rfloor - 1}(v)$. Hence, each vertex in T has distance at most $\lfloor s/2 \rfloor$ to vertex v. Thus, T is an s-club. It remains to show that each vertex of T is in a triangle. Observe that for each vertex $w \in N_{\lfloor s/2 \rfloor - 2}[v]$ we have $N(w) \subseteq N_{\lfloor s/2 \rfloor - 1}[v]$.

Recall that since Reduction Rule 1 is applied, each vertex in G is contained in a triangle. Thus, each vertex of $N_{\lfloor s/2 \rfloor -2}[v]$ is contained in a triangle in T. Furthermore, all vertices in $N_{\lfloor s/2 \rfloor -1}(v) \cup (T \setminus N_{\lfloor s/2 \rfloor -1}[v])$ are in a triangle because of the T-expansion. Since $|T| \geq |N_{\lfloor s/2 \rfloor -1}[v]| \geq k$, the statement follows. $\square$

Next, we show that Lemma 1 implies the existence of a Turing kernel. This in turn implies that the problem is fixed-parameter tractable.

Theorem 2. VERTEX TRIANGLE s-CLUB *for* $\ell = 1$ *admits a* k^4-*vertex Turing kernel if* $s = 4$ *or* $s = 7$, *a* k^5-*vertex Turing kernel if* $s = 5$, *and a* k^3-*vertex Turing kernel if* $s = 6$ *or* $s \geq 8$.

Proof. First, we apply Reduction Rule 1. Because of Lemma 1 we conclude that (G, k) is a trivial yes-instance if $|N_{\lfloor s/2 \rfloor -1}[v]| \geq k$ for any vertex $v \in V(G)$. Thus, in the following we can assume that $|N_{\lfloor s/2 \rfloor -1}[v]| < k$ for each vertex $v \in V(G)$. We use this fact to bound the size of $N_s[v]$ in non-trivial instances.

Note that if $s = 4$ or $s = 5$ we have $\lfloor s/2 \rfloor - 1 = 1$. Hence, in this case from $|N_{\lfloor s/2 \rfloor -1}[v]| < k$ we obtain that the size of the neighborhood of each vertex is bounded. Thus, we obtain a k^4-vertex Turing kernel for $s = 4$ and a k^5-vertex Turing kernel for $s = 5$. Furthermore, if $s = 7$ we have $\lfloor s/2 \rfloor - 1 = 2$. Thus, we obtain a k^4-vertex Turing kernel for $s = 7$ since $N_7[v] \subseteq N_8[v] = N_2[N_2[N_2[N_2[v]]]]$. The cases $s = 6$ and $s \geq 8$ can be shown similarly. $\square$

Note that $s \geq 4$ is necessary to ensure $\lfloor s/2 \rfloor - 1 \geq 1$. In our arguments to obtain a Turing kernel $\ell = 1$ is necessary for the following reason: if $\ell \geq 2$, then the remaining vertices of the other triangles of a vertex in the T-expansion may be contained in $N_{\lfloor s/2 \rfloor +1}$ and, thus, adding them will not necessarily give an s-club. This argument can be extended to show that using $N_t[v]$ for some $t < \lfloor s/2 \rfloor - 1$ does not help.

Next, we provide W[1]-hardness for the remaining cases.

Theorem 3 ($\star$). VERTEX TRIANGLE s-CLUB *is* W[1]-*hard for parameter* k *if* $\ell \geq 2$, *and if* $\ell = 1$ *and* $s \in \{2, 3\}$.

We prove the theorem by considering four subcases where we distinguish different combinations of s and ℓ. The proofs for the four cases all use a reduction from the W[1]-hard CLIQUE problem. In these constructions, each vertex v of the CLIQUE instance is replaced by a vertex gadget T^v such that every vertex-ℓ-triangle s-club S either contains T^v completely or contains no vertex of T^v. This property is obtained since each vertex in T^v will be in exactly ℓ triangles and each of these triangles is within T^v. The idea is that if $uv \notin E(G)$ then there exists a vertex $x \in T^u$ and a vertex $y \in T^v$ such that $\text{dist}(x, y) \geq s + 1$.

Here, we only provide a proof for the case $s = 2$. The proofs for all remaining cases are deferred to the full version of this article.

Construction 4. Let (G, k) be an instance of CLIQUE and let c be the smallest number such that $\binom{c-1}{2} \geq \ell$. We construct an instance $(G', c(k+1))$ of VERTEX TRIANGLE 2-CLUB as follows. For each vertex $v \in V(G)$, we add a clique $T^v :=$

$\{x_1^v, \ldots, x_c^v\}$ of size c to G'. Furthermore, for each edge $vw \in E(G)$, we connect the cliques T^v and T^w by adding the edge $x_{2i-1}^v x_{2i}^w$ and $x_{2i-1}^w x_{2i}^v$ to G' for each $i \in [\lfloor c/2 \rfloor]$. Next, we add a clique $Y := \{y_1, \ldots, y_c\}$ of size c to G'. We also add, for each $i \in [c]$ and each $v \in V(G)$, the edge $x_i^v y_i$ to G'.

Note that the clique size c ensures that each vertex $x \in V(G')$ is contained in at least $\binom{c-1}{2} \geq \ell$ triangles in G'. Furthermore, note that the clique Y is only necessary when c is odd to ensure that the vertices x_c^v and x_c^w have a common neighbor. We add the clique Y in both cases to unify the construction and the correctness proof. Next, we show that for each vertex gadget T^v the intersection with each vertex-ℓ-triangle 2-club is either empty or T^v.

Lemma 2 ($\star$). *Let S be a vertex-ℓ-triangle 2-club in G'. Then,*

a) $S \cap T^v \neq \emptyset \Leftrightarrow T^v \subseteq S$, and
b) $S' := S \cup Y$ is also a vertex-ℓ-triangle 2-club in G'.

Now, we prove the correctness of Construction 4. If $\ell = \binom{c-1}{2}$ for some integer c, then Lemma 3 also holds for the restriction that each vertex is contained in exactly ℓ triangles in the input graph G'.

Lemma 3. *For each $\ell \in \mathbb{N}$, the VERTEX TRIANGLE 2-CLUB problem parameterized by k is W[1]-hard.*

Proof. We prove that G contains a clique of size k if and only if G' contains a vertex-ℓ-triangle 2-club of size at least $c(k+1)$.

Let C be a clique of size at least k in G. We argue that $S := Y \cup \bigcup_{v \in C} T^v$ is a vertex-ℓ-triangle 2-club of size $c(k+1)$ in G'. Note that for each vertex $v \in T^v$ we have $|T^v| = c$. Since T^v is a clique, we conclude that each vertex in T^v is contained in exactly $\binom{c-1}{2} \geq \ell$ triangles. The same is true for each vertex in Y. Hence, each vertex in S is contained in at least ℓ triangles. Thus, it remains to show that S is a 2-club. Consider the vertices x_i^v and x_j^w for $v, w \in C$, $i \in [c-1]$, and $j \in [c]$. If i is odd, then $x_{i+1}^v \in N(x_i^v) \cap N(x_j^w)$. Otherwise, if i is even, $x_{i-1}^v \in N(x_i^v) \cap N(x_j^w)$. In both cases, we obtain $\text{dist}(x_i^v, x_j^w) \leq 2$. Next, consider two vertices x_c^v and x_c^w in S. Observe that $y_c \in N(x_c^v) \cap N(x_c^w)$. Since Y is a clique, it remains to consider vertices x_i^v and y_j in S for $i \in [c]$ and $j \in [c]$. Observe that $x_j^v \in N(y_j) \cap N[x_i^v]$. Thus, G' contains a vertex-ℓ-triangle 2-club of size at least $c(k+1)$.

Conversely, suppose that G' contains a vertex-ℓ-triangle 2-club S of size at least $c(k+1)$. By Lemma 2, we can assume that $Y \subseteq S$ and for each vertex gadget $T^v \in G'$ we either have $T^v \subseteq S$ or $T^v \cap S = \emptyset$. Hence, S contains at least k cliques of the form T^v. Assume towards a contradiction that S contains two cliques T^v and T^w such that $vw \notin E(G)$ and consider the two vertices $x_1^v \in T^v$ and $x_2^w \in T^w$. Note that these vertices always exist since $c \geq 3$. Observe that $N[x_1^v] = T^v \cup \{x_2^u \mid uv \in E(G)\} \cup \{y_1\}$ and $N[x_2^w] = T^w \cup \{x_1^u \mid uw \in E(G)\} \cup \{y_2\}$. Thus, $\text{dist}(x_1^v, x_2^w) \geq 3$, a contradiction. Hence, for each two distinct vertex gadgets T^v and T^w that are contained in S, we observe that $vw \in E(G)$. Consequently, the set $\{v \mid T^v \subseteq S\}$ is a clique of size at least k in G. $\square$

3   Edge Triangle s-Club

Recall that a vertex set S is an edge-ℓ-triangle s-club if $G[S]$ contains a spanning subgraph $G' = (S, E')$ such that each edge in $E(G')$ is contained in at least ℓ triangles within G' and the diameter of G' is at most s. First, we show that EDGE TRIANGLE s-CLUB is FPT with respect to k when $\ell = 1$ irrespective of the value of s by providing a Turing kernel. To show this, it is sufficient to delete edges which are not part of a triangle.

Theorem 5 ($\star$). EDGE TRIANGLE s-CLUB for $\ell = 1$ admits a k^2-vertex Turing kernel if s is even and a k^3-vertex Turing kernel if s is odd and $s \geq 3$.

Now, for the remaining cases we show W[1]-hardness.

Theorem 6. EDGE TRIANGLE s-CLUB is W[1]-hard for parameter k if $\ell \geq 2$.

Next, we describe the reduction to prove Theorem 6 (see Fig. 1).

Construction 7. Let (G, k) be an instance of CLIQUE with $k \geq 3$. We construct an equivalent instance (G', k') of EDGE-TRIANGLE s-CLUB for some fixed $\ell \geq 2$ as follows. Let $\ell^* := \lceil \ell/2 \rceil$ and let $x := 6 \cdot \ell^*(s - 1) + \lfloor \ell/2 \rfloor$. For each vertex $v \in V(G)$, we construct the following vertex gadget T^v. For better readability, all sub-indices of the vertices in T^v are considered modulo x. Our construction distinguishes even and odd values of ℓ. First, we describe the part of the construction which both cases have in common.

1. We add vertex sets $A_v := \{a_i^v \mid i \in [0, x]\}$ and $B_v := \{b_i^v \mid i \in [0, x]\}$ to G'.
2. We add the edges $a_i^v a_{i+j}^v$, and $b_i^v b_{i+j}^v$ for each $i \in [0, x]$ and each $j \in [-3\ell^*, 3\ell^*] \setminus \{0\}$ to G'.
3. We add the edge $a_i^v b_{i+j}^v$ for each $i \in [0, x]$ and each $j \in [-3\ell^*, 3\ell^*]$ to G'.

In other words, an edge $a_i^v b_j^v$ is added if the indices differ by at most $3\ell^*$. For even ℓ, this completes the construction of T^v. For odd ℓ, we extend T^v as follows:

0-1 We add the vertex set $C_v := \{c_i^v \mid i \in [0, x]$ and $i \equiv 0 \mod \ell^*\}$ to G'. Note that C_v consists of exactly $6s - 5$ vertices.
0-2 We add the edges $c_i^v a_{i+j}^v$ and $c_i^v b_{i+j}^v$ for each $i \in [0, x]$ such that $i \equiv 0 \mod \ell^*$ and each $j \in [-3\ell^*, 3\ell^*]$ to G'.
0-3 Also, we add the edge $c_i^v c_{i+j}^v$ to G' for each $i \in [0, x]$ such that $i \equiv 0 \mod \ell^*$ and each $j \in [-3\ell^*, 3\ell^*] \setminus \{0\}$ to G' if the corresponding vertex c_{i+j}^v exists.

In other words, an edge between c_i^v and a_j^v, b_j^v, or c_j^v is added if the indices differ by at most $3\ell^*$. Now, for each edge $uv \in E(G)$, we add the following to G':

0-4 We add the edges $a_i^v b_{i+j}^u$ and $a_i^u b_{i+j}^v$ for each $i \in [0, x]$ and $j \in [0, \lfloor \ell/2 \rfloor]$.
0-5 If ℓ is odd, we also add the edges $c_i^v b_{i+j}^u$ and $c_i^u b_{i+j}^v$ for each $i \in [0, x]$ such that $i \equiv 0 \mod \ell^*$ and each $j \in [0, \lfloor \ell/2 \rfloor]$ to G'.
Observe that each vertex b_{i+j}^u is adjacent to *exactly* one vertex in C_v.

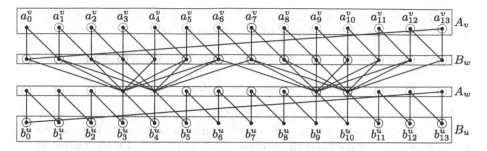

Fig. 1. Construction for Theorem 6 when $s = 3$ and $\ell = 2$ and G is a P_3 on $\{u, v, w\}$ with $uv \notin E(G)$. Only the gadgets A_v, B_w, A_w, and B_u are shown. For simplicity, no edges within A_v, B_w, A_w, and B_u are drawn and edges between B_w and A_w are only drawn if one endpoint is a_3^w, a_4^w, a_9^w, or a_{10}^w. Blue encircled vertices are neighbors of a_0^v, red encircled vertices have distance 2 to a_0^v, and black encircled vertices have distance 3 to a_0^v. Thus, a_0^v and $b_7^u = b_{0+1+3\cdot1\cdot2}^u = b_{0+\lfloor \ell/2 \rfloor + 3\ell^*(s-1)}^u$ have distance at least 4. (Color figure online)

In other words, an edge between a_i^v or c_i^v and b_j^v is added if j exceeds i by at most $\lfloor \ell/2 \rfloor$. Finally, if ℓ is even, we set $k' := 2(x+1)k = (\ell(6s-5)+2)\cdot k$, and if ℓ is odd, we set $k' := (2(x+1)+6s-5)k = (\ell+2)(6s-5)\cdot k$.

Construction 7 has two key mechanisms: First, if $uv \notin E(G)$ then for each vertex $a \in A_v$ there is at least one vertex $b \in B_u$ such that $\text{dist}(a, b) > s$. Second, each edge with one endpoint in A_v and one endpoint in B_u is contained in *exactly* ℓ triangles. Furthermore, if ℓ is odd, then this also holds for each edge with one endpoint in C_v and one in B_u. Consider an edge-ℓ-triangle s-club S and let $\widetilde{G} = (S, \widetilde{E})$ be a spanning subgraph of $G[S]$ with the maximal number of edges, such that each edge of $\widetilde{E}$ is contained in at least ℓ triangles in $\widetilde{G}$ and the diameter of $\widetilde{G}$ is s. As we will show, the two mechanics ensure that an edge with one endpoint in A_v (or C_v) and the other endpoint in B_u is contained in $\widetilde{E}$ if and only if S contains all vertices of A_v (and C_v) and B_u. We call this the *enforcement property*. The proof of this property is deferred to the full version.

Proof (of Theorem 6). We show that G contains a clique of size at least k if and only if G' contains an edge-ℓ-triangle s-club of size at least k'.

Let K be a clique of size k in G. Then $S := \{u \in V(T^v) \mid v \in K\}$ is an edge-ℓ-triangle s-club of size at least k'; the proof is deferred to the full version.

Conversely, let S be an edge-ℓ-triangle s-club of size at least k' in G'. More precisely, let $\widetilde{G}$ be a maximal spanning subgraph of $G[S]$ which has diameter at most s and such that each edge in $E(\widetilde{G})$ is contained in at least ℓ triangles is $\widetilde{G}$. We show that G contains a clique of size at least k. Here, we only consider the case that s is even; the case that s is odd is deferred to the full version. First, we show that for each vertex $x \in A_v \cup B_v$ there exists a vertex $y \in A_u \cup B_u$ such that $\text{dist}(x, y) \geq s + 1$ if $uv \notin E(G)$. For this, recall that by construction each two vertices with sub-indices i' and j' are not adjacent if their difference (modulo x) is larger than $3\ell^*$.

Claim 1. In G' we have $\mathrm{dist}(x_i, y_j) \geq s+1$ for each $i \in [0, x]$, $j := i + \lfloor \ell/2 \rfloor + 3\ell^*(s-1)$, $x_i \in \{a_i^v, b_i^v\}$, and $y_j \in \{a_j^u, b_j^u\}$ if $uv \notin E(G)$.

The next statement follows from Claim 1 and the enforcement property.

Claim 2. If $A_v \subseteq S$ and $B_u \subseteq S$ then $uv \in E(G)$.

We now use Claims 1 and 2 to show that G contains a clique of size at least k. We distinguish the cases whether S contains only parts of one of the gadgets A_v or B_v, or whether S contains all vertices of the gadgets A_v or B_v completely.

First, assume that for some vertex $v \in V(G)$ we have $A_v \cap S \neq \emptyset$ and $A_v \not\subseteq S$. In the following, we show that S only contains vertices of gadget T^v and from gadgets T^u such that $uv \in E(G)$. Since $A_v \not\subseteq S$, we conclude that in $\widetilde{G}$ we have $N_{\widetilde{G}}(A_v \cap S) \subseteq B_v$: Otherwise, vertex a_v has a neighbor $b_u \in B_u$ and by the enforcement property we would obtain $A_v \subseteq S$, a contradiction to the assumption $A_v \not\subseteq S$. If $B_v \not\subseteq S$, then by the enforcement property no vertex in B_v can have a neighbor a_i^w for some $w \neq v$. Hence, $S \cap T^v$ would be a connected component of size at most $2(x+1)$, a contradiction to the size of S since $k \geq 3$. Thus, we may assume that $B_v \subseteq S$. Observe that if $a_i^w \in S$ for some $w \in V(G)$ such that $vw \in E(\widetilde{G})$, that is, also $vw \in E(G')$, then we have $A_w \subseteq S$ by the enforcement property since each vertex a_i^w has a neighbor in B_v. Let $W := \{w_1, \ldots, w_t\}$ denote the set of vertices w_j such that $vw_j \in E(G)$ and $A_{w_j} \subseteq S$. If $w_x w_y \notin E(G)$ for some $x, y \in [t]$ with $x \neq y$, then $a_0^{w_x}$ and $a_{\lfloor \ell/2 \rfloor + 3\ell^*(s-1)}^{w_y}$ have distance at least $s+1$ by Claim 1. Thus $w_x w_y \in E(G)$ for each $x, y \in [t]$ with $x \neq y$.

Assume towards a contradiction that $a_i^p \in S$ for some $p \in V(G \setminus W)$ with $p \neq v$. Note that $pv \notin E(G)$ since otherwise $p \in W$ by the definition of W. Observe that since $B_v \subseteq S$ we also have $b_{i+\lfloor \ell/2 \rfloor + 3\ell^*(s-1)}^v \in S$. But since $pv \notin E(G)$ we obtain from Claim 1 that $\mathrm{dist}(a_i^p, b_{i+\lfloor \ell/2 \rfloor + 3\ell^*(s-1)}^v) \geq s+1$, a contradiction. We conclude that S does not contain any vertex a_i^p with $p \neq v$ or $p \neq w_j$ for $j \in [t]$.

Next, assume towards a contradiction that $b_i^p \in S$ for some $p \in V(G)$ with $p \neq v$ and $p \notin W$. If $pv \notin E(G)$, then b_i^p and $b_{i+\lfloor \ell/2 \rfloor + 3\ell^*(s-1)}^v$ have distance at least $s+1$ again by Claim 1. Thus, we can assume that $pv \in E(G)$. Recall that $b_{i+\lfloor \ell/2 \rfloor + 3\ell^*(s-1)}^v \in S$. As defined by Claim 1, each shortest path from b_i^p to $b_{i+\lfloor \ell/2 \rfloor + 3\ell^*(s-1)}^v$ can swap at most once between different vertex gadgets. In this case, there is exactly one swap from T^p to T^v. From the above we know that $A_p \cap S = \emptyset$. Thus, each shortest path from b_i^p to $b_{i+\lfloor \ell/2 \rfloor + 3\ell^*(s-1)}^v$ uses at least one vertex in A_v. Since at least one edge with endpoint A_v is contained in S, we conclude from the enforcement property that $A_v \subseteq S$, a contradiction to the assumption $A_v \not\subseteq S$.

Hence, there is no vertex $p \neq v$ and $p \notin W$ such that $T^p \cap S \neq \emptyset$. In other words, S contains only vertices from the gadget T^v and from gadgets T^u with $vu \in E(G)$. Thus, $S \subseteq T^v \cup \bigcup_{j=1}^t T^{w_j}$. By definition of k', we have $t \geq k-1$ and we conclude that G contains a clique of size at least k. The case that we have $B_v \cap S \neq \emptyset$ and $B_v \not\subseteq S$ for some vertex $v \in V(G)$ can be handled similarly.

Second, consider the case that for each set A_v with $A_v \cap S \neq \emptyset$ we have $A_v \subseteq S$, and that for each set B_v with $B_v \cap S \neq \emptyset$ we have $B_v \subseteq S$. Let $W_A :=$ $\{w_A^j \mid A_{w_j} \subseteq S\}$ and $W_B := \{w_B^j \mid B_{w_j} \subseteq S\}$. If $W_A = \emptyset$ or $W_B = \emptyset$ then each connected component in $G'[S]$ has size at most $x + 1 < k'$. Thus, we may assume that $W_A \neq \emptyset$ and that $W_B \neq \emptyset$. By Claim 2, we have $w_A^i w_B^j \in E(G)$ for each $w_A^i \in W_A$ and $w_B^j \in W_B$. Furthermore, by Claim 1, we have $w_B^j w_B^{j'} \in E(G)$ for $w_B^j, w_B^{j'} \in W_B$ and $w_A^i w_A^{i'} \in E(G)$ for $w_A^i, w_A^{i'} \in W_A$. Hence, we obtain that $\min(|W_A|, |W_B|) \geq k$ and thus G contains a clique of size k. $\qquad\square$

4   Seeded s-Club

In this section we study the parameterized complexity of SEEDED s-CLUB with respect to the standard parameter solution size k. For clique seeds, we provide the following kernel.

Theorem 8. SEEDED s-CLUB admits a kernel with $\mathcal{O}(k^{2|W|+1})$ vertices if W is a clique.

To prove the kernel, we first remove all vertices with distance at least $s + 1$ to any vertex in W.

Reduction Rule 9. Let (G, W, k) be an instance of SEEDED s-CLUB. If G contains a vertex u such that $\text{dist}(u, w) \geq s + 1$ for some $w \in W$, then remove u.

Clearly, Reduction Rule 9 is correct and can be applied in polynomial time. Next, we show that if the remaining graph is sufficiently large then (G, W, k) is a yes-instance of SEEDED s-CLUB.

Lemma 4 ($\star$). An instance (G, W, k) of SEEDED s-CLUB with $|N_{s-1}[W]| \geq k^2$ is a yes-instance.

Finally, we bound the size of $N_s(W)$. There we assume that $|N_{s-1}[W]| < k^2$ by Lemma 4 and that Reduction Rule 9 is applied. Afterwards, Theorem 8 follows directly from Lemmas 4 and 5.

Lemma 5. An instance (G, W, k) of SEEDED s-CLUB with $|N_s(W)| \geq k^{2|W|+1}$ which is reduced with respect to Reduction Rule 9 is a yes-instance.

Proof. Since Reduction Rule 9 has been applied exhaustively, each vertex $p \in N_s(W)$ has distance exactly s to each vertex in W. In other words, for each vertex $w_\ell \in W$ there exists a vertex $u_{s-1}^\ell \in N_{s-1}(w_\ell)$ such that $pu_{s-1}^\ell \in E(G)$. Note that $N_{s-1}(w_\ell) \subseteq N_{s-1}[W]$. Moreover, by Lemma 4 we may assume that $|N_{s-1}[W]| < k^2$. In particular: $|N_{s-1}(W)| < k^2$. Since $|N_s(W)| \geq k^{2|W|+1}$, by the pigeonhole principle there exists a set $\{u_{s-1}^1, u_{s-1}^2, \ldots, u_{s-1}^{|W|}\}$ with $u_{s-1}^\ell \in N_{s-1}(w_\ell)$ for $\ell \in [|W|]$ such that the set $P := N_s(W) \cap \bigcap_{\ell \in [|W|]} N(u_{s-1}^\ell)$ has size at least k. The size bound follows from the observation that each $N_{s-1}(w_\ell)$ has size at most k^2 and we have exactly $|W|$ many of these sets. By the definition of vertex u_{s-1}^{ℓ}, there exists for each $i \in [s-2]$ a vertex $u_i^{\ell} \in N_i(w_{\ell})$ such that $w_{\ell}, u_1^{\ell}, \ldots, u_{s-1}^{\ell}$ is a path of length $s-1$ in G. We define the set $U := \{u_i^{\ell} \mid \ell \in [\|W\|], i \in [s-1]\}$. Next, we show that $Z := P \cup W \cup U$ induces an s-club.

First, observe that all vertices in P have distance at most 2 to each other since they have the common neighbor u_{s-1}^1. Second, note that the vertices w_{ℓ}, $u_1^{\ell}, \ldots, u_{s-1}^{\ell}, p, u_{s-1}^{j}, \ldots, u_1^{j}, w_j$ form a cycle with $2s+1$ vertices, for each $p \in P$ and each two indices $j, \ell \in [\|W\|]$. Each vertex in this cycle has distance at most s to each other vertex in that cycle. Hence, Z is indeed an s-club.   □

Now, we show hardness for some of the remaining cases.

Theorem 10 ($\star$). *Let H be a fixed graph. SEEDED s-CLUB is $W[1]$-hard parameterized by k even if $G[W]$ is isomorphic to H, when*

- *$s = 2$ and H contains at least two non-adjacent vertices, or if*
- *$s \geq 3$ and H contains at least two connected components.*

In this extended abstract, we only show $W[1]$-hardness for the case $s \geq 3$ when the seed contains at least two connected components. Fix a graph H with at least two connected components. We show $W[1]$-hardness for $s \geq 3$ even if $G[W]$ is isomorphic to H.

Construction 11. Let (G, k) be an instance of CLIQUE. We construct an equivalent instance (G', k') of SEEDED s-CLUB as follows. Initially, we add the set W to G', and add edges such that $G'[W]$ is isomorphic to H. Let D_1 be one connected component of $G'[W]$. By assumption, $D_2 := W \setminus D_1$ is not empty. Next, we add two copies G_1 and G_2 of G to G'. Then, we add edges to G' such that each vertex in D_1 is adjacent to each vertex in $V(G_1)$ and such that each vertex in D_2 is adjacent to each vertex in $V(G_2)$. Furthermore, we add a path $(p_1, \ldots, p_{s-1})$ consisting of exactly $s - 1$ new vertices to G', make p_1 adjacent to each $u \in D_1$, and make p_{s-1} adjacent to each $v \in D_2$. By $P := \{p_i \mid i \in [s-1]\}$ we denote the set of these newly added vertices. Now, for each $x \in V(G)$ we do the following. Consider the copies $x_1 \in V(G_1)$ and $x_2 \in V(G_2)$ of vertex $x \in V(G)$. We add a path $(x_1, q_1^x, \ldots, q_{s-2}^x, x_2)$ consisting of $s - 2$ new vertices to G'. By $Q_x := \{q_i^x \mid i \in [s-2]\}$ we denote the set of the new internal path vertices. Finally, we set $k' := sk + |W| + s - 1$.

Now, we prove the correctness of Construction 11.

Lemma 6. *Let H be a fixed graph with at least two connected components. SEEDED s-CLUB parameterized by k is $W[1]$-hard even if $G[W]$ is isomorphic to H.*

Proof. We show that G contains a clique of size k if and only if G' contains a W-seeded s-club of size at least $k' = sk + |W| + s - 1$. Let K be a clique of size k in G and let K_1 and K_2 be the copies of K in G_1 and G_2. Then $S := W \cup P \cup K_1 \cup K_2 \cup \bigcup_{x \in K} Q_x$ is a W-seeded s-club of size at least k'; the proof of this fact is deferred to the full version of this article.

Conversely, suppose that G' contains a W-seeded s-club S of size at least k'. Let $Q'_v := \{v_1, q_1^v, \ldots, q_{s-2}^v, v_2\}$ for each $v \in V(G)$. We show that $Q'_v \cap S \neq \emptyset$ if and only if $Q'_v \subseteq S$. Assume towards a contradiction, that $Q'_v \cap S \neq \emptyset$ for some $v \in V(G)$ such that $Q'_v \nsubseteq S$. If $v_1 \notin S$, and also $v_2 \notin S$, then no vertex in $S \cap Q'_v$ is connected to any vertex in $S \setminus Q'_v$. Thus, without loss of generality assume that $v_1 \in S$. Note that $N(D_2) = V(G_2) \cup \{p_{s-1}\}$. Observe that $\text{dist}(v_1, p_{s-1}) = s$, that $\text{dist}(v_1, q_{s-2}^v) = s-2$, and that $\text{dist}(v_1, q_{s-2}^u) \geq s-1$ for each $u \in V(G) \setminus \{v\}$. Thus, the unique path of length at most s from v_1 to D_2 contains all vertices in Q'_v. Hence, $Q'_v \cap S \neq \emptyset$ if and only if $Q'_v \subseteq S$. By the definition of k' we may thus conclude that $Q'_v \subseteq S$ for at least k vertices $v \in V(G)$.

Assume towards a contradiction that $Q'_u \subseteq S$ and $Q'_v \subseteq S$ such that $uv \notin E(G)$. We consider the vertices v_1 and u_2. Observe that by construction each path from v_1 to u_2 containing any vertex p_i has length at least $s+1$. Hence, each shortest path from v_1 to u_2 contains the vertex set of Q'_w for some $w \in V(G)$. Since the path induced by each Q'_w has length $s-1$, we conclude that $w = u$ or $w = v$. Assume without loss of generality that $w = v$. Hence, the $(s-1)$th vertex on the path from v_1 is v_2. Since $uv \notin E(G)$ we have by construction that $u_2 v_2 \notin E(G')$. Hence, $\text{dist}(v_1, u_2) \geq s+1$, a contradiction to the fact that S is an s-club. Thus, $\{v \mid Q'_v \subseteq S\}$ is a clique of size at least k in G. □

5   Conclusion

We provided a complexity dichotomy for VERTEX TRIANGLE s-CLUB and EDGE TRIANGLE s-CLUB for the standard parameter solution size k with respect to s and ℓ. We also provided a complexity dichotomy for SEEDED 2-CLUB for k in terms of the structure of $G[W]$. In contrast, it remains an open question whether SEEDED s-CLUB with $s \geq 3$ admits an FPT algorithm when $G[W]$ is connected but W is not a clique. It is particularly interesting to study seeds of constant size since this seems to be the most interesting case for applications.

For future work, it seems interesting to study the complexity of the considered variants of s-CLUB with respect to further parameters, for example the treewidth of G. Additionally, the parameterized complexity of further robust variants of s-CLUB such as t-HEREDITARY s-CLUB [14,16] with respect to k remains open. It is also interesting to study other problems for detecting communities with seed constraints. One prominent example is s-PLEX. This problem is also NP-hard for $W \neq \emptyset$, since an algorithm for the case when $|W| = 1$ can be used as a black box to solve the unseeded variants. From a practical perspective, we plan to implement combinatorial algorithms for all three problem variants for the most important special case $s = 2$. Based on experience with previous implementations for 2-CLUB [10] and some of its robust variants [14] we are optimistic that these problems can be solved efficiently on sparse real-world instances.

References

1. Almeida, M.T., Brás, R.: The maximum l-triangle k-club problem: complexity, properties, and algorithms. Comput. Oper. Res. **111**, 258–270 (2019)

2. Balasundaram, B., Butenko, S., Trukhanov, S.: Novel approaches for analyzing biological networks. J. Comb. Optim. **10**(1), 23–39 (2005)
3. Bodlaender, H.L., Downey, R.G., Fellows, M.R., Hermelin, D.: On problems without polynomial kernels. J. Comput. Syst. Sci. **75**(8), 423–434 (2009)
4. Bourjolly, J., Laporte, G., Pesant, G.: An exact algorithm for the maximum k-club problem in an undirected graph. Eur. J. Oper. Res. **138**(1), 21–28 (2002)
5. Carvalho, F.D., Almeida, M.T.: The triangle k-club problem. J. Comb. Optim. **33**(3), 814–846 (2017)
6. Chang, M., Hung, L., Lin, C., Su, P.: Finding large k-clubs in undirected graphs. Computing **95**(9), 739–758 (2013)
7. Cygan, M., et al.: Parameterized Algorithms. Springer, Cham (2015). https://doi.org/10.1007/978-3-319-21275-3
8. Downey, R.G., Fellows, M.R.: Fundamentals of Parameterized Complexity. TCS, Springer, London (2013). https://doi.org/10.1007/978-1-4471-5559-1
9. Golovach, P.A., Heggernes, P., Kratsch, D., Rafiey, A.: Finding clubs in graph classes. Discret. Appl. Math. **174**, 57–65 (2014)
10. Hartung, S., Komusiewicz, C., Nichterlein, A.: Parameterized algorithmics and computational experiments for finding 2-clubs. J. Graph Algorithms Appl. **19**(1), 155–190 (2015)
11. Hartung, S., Komusiewicz, C., Nichterlein, A., Suchý, O.: On structural parameterizations for the 2-club problem. Discret. Appl. Math. **185**, 79–92 (2015)
12. Kanawati, R.: Seed-centric approaches for community detection in complex networks. In: Meiselwitz, G. (ed.) SCSM 2014. LNCS, vol. 8531, pp. 197–208. Springer, Cham (2014). https://doi.org/10.1007/978-3-319-07632-4_19
13. Komusiewicz, C.: Multivariate algorithmics for finding cohesive subnetworks. Algorithms **9**(1), 21 (2016)
14. Komusiewicz, C., Nichterlein, A., Niedermeier, R., Picker, M.: Exact algorithms for finding well-connected 2-clubs in sparse real-world graphs: theory and experiments. Eur. J. Oper. Res. **275**(3), 846–864 (2019)
15. Mokken, R.J., et al.: Cliques, clubs and clans. Qual. Quant. **13**(2), 161–173 (1979)
16. Pattillo, J., Youssef, N., Butenko, S.: On clique relaxation models in network analysis. Eur. J. Oper. Res. **226**(1), 9–18 (2013)
17. Schäfer, A., Komusiewicz, C., Moser, H., Niedermeier, R.: Parameterized computational complexity of finding small-diameter subgraphs. Optim. Lett. **6**(5), 883–891 (2012)
18. Veremyev, A., Boginski, V.: Identifying large robust network clusters via new compact formulations of maximum k-club problems. Eur. J. Oper. Res. **218**(2), 316–326 (2012)
19. Whang, J.J., Gleich, D.F., Dhillon, I.S.: Overlapping community detection using seed set expansion. In: Proceedings of the 22nd ACM International Conference on Information and Knowledge Management (CIKM 2013), pp. 2099–2108. ACM (2013)
20. Yezerska, O., Pajouh, F.M., Butenko, S.: On biconnected and fragile subgraphs of low diameter. Eur. J. Oper. Res. **263**(2), 390–400 (2017)

Space-Efficient B Trees
via Load-Balancing

Tomohiro I[2] and Dominik Köppl[1(✉)]

[1] M&D Data Science Center, Tokyo Medical and Dental University,
Bunkyo City, Japan
koeppl.dsc@tmd.ac.jp
[2] Department of Artificial Intelligence, Kyushu Institute of Technology,
Iizuka, Japan
tomohiro@ai.kyutech.ac.jp

Abstract. We study succinct variants of B trees in the word RAM model that require $s + o(s)$ bits of space, where s is the number of bits essentially needed for storing keys and possibly other satellite values. Assuming that elements are sorted by keys (not necessarily in the order of their integer representations), our B trees support standard operations such as searching, insertion and deletion of elements. In some applications it is useful to associate a satellite value to each element, and to support aggregate operations such as computing the sum of values, the minimum/maximum value in a given range, or search operations based on those values. We propose a B tree representation storing n elements in $s + \mathcal{O}(s/\lg n)$ bits of space and supporting all mentioned operations in $\mathcal{O}(\lg n)$ time.

Keywords: B tree · Succinct data structure · Predecessor data structure

1 Introduction

A B tree [1] is the most ubiquitous data structure found for relational databases and is, like the balanced binary search tree in the pointer machine model, the most basic search data structure in the external memory model. A lot of research has already been dedicated for solving various problems with B trees, and various variants of the B tree have already been proposed (cf. [12] for a survey). Here, we study a space-efficient variant of the B tree in the word RAM model under the context of a dynamic predecessor data structure, which provides the following methods:

predecessor(K) returns the predecessor of a given key K (or K itself if it is already stored);
insert(K) inserts the key K; and
delete(K) deletes the key K.

© Springer Nature Switzerland AG 2022
C. Bazgan and H. Fernau (Eds.): IWOCA 2022, LNCS 13270, pp. 327–340, 2022.
https://doi.org/10.1007/978-3-031-06678-8_24

We call these three operations *B tree operations* in the following. Nowadays, when speaking about B trees we actually mean B+ trees [4, Sect. 3] (also called *leaf-oriented B-tree* [2]), where the leaves store the actual data (i.e., the keys). We stick to this convention throughout the paper. Another variant we want to focus on in this paper is the B* tree [16, Sect. 6.2.4], where a node split on inserting a key into a full node has chances to be deferred by balancing the loads of this node with one of its siblings.

1.1   Related Work

The standard B tree as well its B+ and B* tree variants support the above methods in $\mathcal{O}(\lg n)$ time, while taking $\mathcal{O}(n)$ words of space for storing n keys. Even if each key uses only $k = o(\lg n)$ bits, the space requirement remains the same since its pointer-based tree topology already needs $\mathcal{O}(n)$ pointers. To improve the space while retaining the operational time complexity in the word RAM model is main topic of this article. However, this is not a novel idea:

The earliest approach we are aware of is due to Blandford and Blelloch [3] who proposed a representation of the leaves as blocks of size $\Theta(\lg n)$. Assuming that keys are integer of k bits, they store the keys not in their plain form, but by their differences encoded with Elias-γ code [7]. Their search tree takes $\mathcal{O}(n \lg((2^k + n)/n))$ bits while conducting B tree operations in $\mathcal{O}(\lg n)$ time.

More recently, Prezza [19] presented a B tree whose leaves store between $b/2$ and b keys for $b = \lg n$. Like [2, Sect. 3] or [6, Thm. 6], the main aim was to provide prefix-sums by augmenting each internal node of the B tree with additional information about the leaves in its subtree such as the sum of the stored values. Given m is the sum of all stored keys plus n, the provided solution uses $2n \left(\lg(m/n) + \lg \lg n + \mathcal{O}(\lg m / \lg n)\right)$ bits of space and supports B tree operations as well as prefix-sum in $\mathcal{O}(\lg n)$ time. This space becomes $2nk + 2n \lg \lg n + o(n)$ bits if we store each key in plain k bits.

Data structures computing prefix-sums are also important for dynamic string representations [13, 17, 18]. For instance, He and Munro [13] use a B tree as underlying prefix-sum data structure for efficient deletions and insertions of characters into a dynamic string. If we omit the auxiliary data structures on top of the B tree to answer prefix-sum queries, their B tree uses $nk + \mathcal{O}(nk/\sqrt{\lg n})$ bits of space while supporting B tree operations in $\mathcal{O}(\lg n / \lg \lg n)$ time, an improvement over the $\mathcal{O}(\lg n)$ time of the data structure of González and Navarro [11, Thm. 1] sharing the same space bound. In the static case, Delpratt et al. [5] studied compression techniques for a static prefix-sum data structure.

Asides from prefix-sums, another problem is to maintain a set of strings, where each node v is augmented with the length of the longest common prefix (LCP) among all strings stored as satellite values in the leaves of the subtree rooted at v [9].

Next, there is a line of research on implicit data structures supporting B tree operations: Under the assumption that all keys are distinct, the data structure of González and Navarro [10] supports $\mathcal{O}(\lg n)$ time for predecessor and $\mathcal{O}(\lg n)$ amortized time for updates (delete and insert) while using only constant number of words of extra space to a dynamic array A of size kn bits storing the keys. However, they assume a more powerful model of computation, where expanding or contracting A at its end can be done in constant time. This model is more powerful in the sense that the standard RAM model only supports the real-location of a new array and copying the contents of the old array to the new array, thus taking time linear in the size of the two arrays. In the standard RAM model, arrays with such operations (extension or contraction at their ends) are called *extendible arrays*, and the best solution in this model (we are aware of) uses $nk + \mathcal{O}(w + \sqrt{knw})$ bits of space for supporting constant-time access and constant-time amortized updates [20, Lemma 1]. Allowing duplicate keys, Kata-jainen and Rao [15] presented a data structure with the same time bounds as [10] but using $\mathcal{O}(n \lg \lg n / \lg n)$ bits of extra space.

With respect to similar techniques but different aim, we can point out the succinct dynamic tree representation of Farzan and Munro [8, Thm 2] who pro-pose similar techniques like rebuilding substructures after a certain amount of updates (cf. Sect. 4.1), or storing satellite data in blocks (cf. Sect. 4). They also have a need for space-efficient prefix-sum data structures.

In what follows, we present a solution for B trees based on different known techniques for succinct data structures such as [20] and the aforementioned B tree representations.

1.2 Our Contribution

Our contribution (cf. Sect. 3) is a combination of a generalization of the rear-rangement strategy of the B* tree with the idea to enlarge the capacity of the leaves similarly to some approaches listed in the related work. With these tech-niques we obtain:

Theorem 1. *There is a B tree representation storing n keys, each of k bits, in $nk + \mathcal{O}(nk/ \lg n)$ bits of space, supporting all B tree operations in $\mathcal{O}(\lg n)$ time.*

We stress that this representation does not compress the keys, which can be advantageous if keys are not simple data types but for instance pointers to complex data structure such that equality checking cannot be done by merely comparing the bit representation of the keys, but still can be performed in con-stant time. In this setting of incompressible keys, the space of a *succinct* data structure supporting predecessor, insert, and delete is $nk + o(nk)$ bits for storing n keys.

We present our space-efficient B tree in Sect. 3. Additionally, we show that we can augment our B tree with auxiliary data such that we can address the prefix-sum problem and LCP queries without worsening the query time (cf. Sect. 4).

2 Preliminaries

Our computational model is the word RAM model with a word size of w bits. We assume that each key uses $k = \mathcal{O}(w)$ bits, and that we can compare two keys in

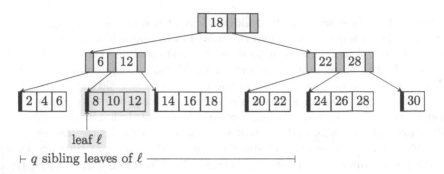

Fig. 1. A B+ tree with degree $t = 3$ and height 3. A leaf can store at most $b = t = 3$ children. A child pointer is a gray field in the internal nodes. An internal node v stores $t - 1$ integers in an array I_v where the value $I_v[i]$ regulates that only those keys of at most $I_v[i]$ go to the children in the range from the first up to the i-th child. In what follows (Fig. 2), we consider inserting the key 9 into the full leaf ℓ (storing the keys 8, 10, and 12), and propose a strategy different from splitting ℓ by considering its $q = 3$ siblings.

$\mathcal{O}(1)$ time. More precisely, we support the comparison to be more complex than just comparing the k-bit representation bitwise as long as it can be evaluated within constant time. Let $n = \mathcal{O}(2^w) \cap \Omega((w \lg^2 n)/k)$ be the number of keys we store at a specific, fixed time.

A B+ tree of degree t for a constant $t \geq 3$ is a rooted tree whose nodes have an out-degree between $\lceil t/2 \rceil$ and t. See Fig. 1 for an example. All leaves are on the same height, which is $\Theta(\lg n)$ when storing n keys. The number of keys each leaf stores is between $\lfloor t/2 \rfloor$ and t (except if the root is a leaf). Each leaf is represented as an array of length t; each entry of this array has k bits. We call such an array a *leaf array*. Each leaf additionally stores a pointer to its preceding and succeeding leaf. Each internal node v stores an array of length t for the pointers to its children, and an integer array I_v of length $t - 1$ to distinguish the children for guiding a top-down navigation. In more detail, $I_v[i]$ is a key-comparable integer such that all keys of at most $I_v[i]$ are stored in the subtrees rooted (a) at the i-th child u of v or (b) at u's left siblings. Since the integers of I_v are stored in ascending order (with respect to the order imposed on the keys), to know in which subtree below v a key is stored, we can perform a binary search on I_v.

A root-to-leaf navigation can be conducted in $\mathcal{O}(\lg n)$ time, since there are $\mathcal{O}(\lg n)$ nodes on the path from the root to any leaf, and selecting a child of a node can be done with a linear scan of its stored keys in $\mathcal{O}(t) = \mathcal{O}(1)$ time.

Regarding space, each leaf stores at least $t/2$ keys. So there are at most $2n/t$ leaves. Since a leaf array uses kt bits, the leaves can use up to $2nk$ bits. This is at most twice the space needed for storing all keys in a plain array. In what follows, we provide a space-efficient variant.

3   Space-Efficient B Trees

To obtain a space-efficient B tree variant, we apply two ideas. We start with the idea to share keys among several leaves (Sect. 3.1) to maintain the space of the leaves more economically. Subsequently, we can adapt this technique for leaves maintaining a *non-constant* number of keys efficiently (Sect. 3.2), leading to the final space complexity of our proposed data structure (Sect. 3.3) and Theorem 1.

3.1   Resource Management by Distributing Keys

Our first idea is to keep the leaf arrays more densely filled. For that, we generalize the idea of B* trees [16, Sect. 6.2.4]: The B* tree is a variant of the B tree (more precisely, we focus on the B+ tree variant) with the aim to defer the split of a full leaf on insertion by rearranging the keys with a dedicated sibling leaf. On inserting a key into a full leaf, we try to move a key of this leaf to its dedicated sibling. If this sibling is also full, we split both leaves up into three leaves, each having $2/3 \cdot b$ keys on average [16, Sect. 6.2.4], where $b = t$ is the maximum number of keys a leaf can store. Consequently, we have the number of leaves is at most $3n/2b$. We can generalize this bound by allowing a leaf to share its keys with $q \in \Theta(\lg n)$ siblings. For that, we introduce the following invariant:

   Among the q siblings of every non-full leaf, there is at most one other non-full leaf.

   We can leave it open to precisely specify which q siblings are assigned to which leaf. For instance, the following is possible: we can assign $q/2$ leaves to the right and to the left side to each leaf. However, if the leaf in question has $o(q)$ left siblings like the leftmost leaf, we take more of its right siblings in considerations (and by symmetry if the leaf has $o(q)$ right siblings), such that each leaf gets q siblings assigned. For this to work, we need at least q leaves, which is granted by $n = \Omega(bq)$ as stated in Sect. 2. We note that it is possible to also accommodate smaller numbers with our techniques; we defer this analysis to Sect. 3.4.

   Let us first see why this invariant helps us to improve the upper bound on the number of leaves; subsequently we show how to sustain the invariant while retaining our operational time complexity of $\mathcal{O}(\lg n)$: By definition, for every q subsequent leaves, there are at most two leaves that are non-full. Consequently, these q subsequent leaves store at least $qb - 2b$ keys. Hence, the number of leaves is at most $\lambda := nq/(qb - 2b)$, and all leaves of the tree use up to

$$\lambda bk = nqbk/(qb - 2b) = nkq/(q - 2) = nk + 2nk/(q - 2)$$
$$= nk + \mathcal{O}(nk/\lg n) \text{ bits for } q \in \Theta(\lg n).$$

$$(1)$$

To obey the aforementioned invariant, we need to take action whenever we delete a key from a full leaf or try inserting a key into a full leaf:

Deletion. When deleting a key from a full leaf ℓ having a non-full leaf ℓ' as one of its q siblings, we shift a key from ℓ' to ℓ such that ℓ is still full after the deletion. If ℓ' becomes empty, then we delete it.

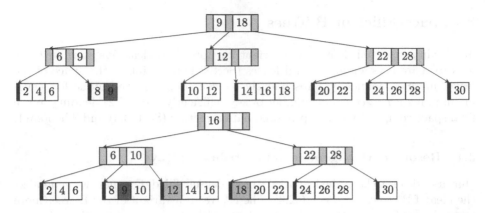

Fig. 2. Figure 1 after inserting the key 9 into the leaf ℓ. Top: The standard B+ and B* variants split ℓ on inserting 9, causing its parent to split, too. Bottom: In our proposed variant (cf. Sect. 3) for $q \geq 3$, we shift the key 12 of ℓ to its succeeding leaf, from which we shift the key 18 to the next succeeding leaf, which was not yet full.

Insertion. Suppose that we want to insert a key into a leaf ℓ that is full. Given that one of the q sibling leaves of ℓ, say ℓ', is not full, then we shift a key from ℓ to ℓ' such that ℓ can store the new key. If there is no such ℓ', then we split ℓ. In that case, we create two new leaves, each inheriting half of the keys of the old leaf. In particular, these two leaves are the only non-full leaves among their q siblings.

It is left to analyze the time for the shifting of a key: Since each leaf stores up to $b = t = \mathcal{O}(1)$ keys, shifting a key to one of the q siblings takes $\mathcal{O}(bq) = \mathcal{O}(\lg n)$ time. That is because, for shifting a key from the i-th leaf to the j-th leaf with $i < j$, we need to move the largest key stored in the g-th leaf to the $(g+1)$-th leaf for $g \in [i..j]$ (the moved key becomes the smallest key stored in the $(g+1)$-th leaf, cf. Fig. 2). Since a shift changes the entries of $\mathcal{O}(q)$ leaves, we have to update the information of those leaves' ancestors. By updating an ancestor node v we mean to update its integer array I_v as described in Sect. 2, which can be done in $\mathcal{O}(t)$ time. There are at most $\sum_{h=1}^{\lg n} \lceil q(t/2)^{-h} \rceil = \mathcal{O}(\lg n + q)$ many such ancestors, and all of them can be updated in time linear to the tree height, which is $\mathcal{O}(\lg n)$ for B trees with constant degree $t = \mathcal{O}(1)$. Thus, we obtain a B* tree variant with the same time complexities, but higher occupation rates of the leaves.

3.2 Shifting Keys Among Large Leaves

Next, we want to reduce the number of internal nodes. For that, we increase the number of elements a leaf can store up to $b := (w \lg n)/k$. Since a leaf now maintains a large number of keys, shifting a key to one of its q neighboring sibling leaves takes $\mathcal{O}(bqk/w) = \mathcal{O}(\lg^2 n)$ time. That is because, for an insertion into a leaf array, we need to shift the stored keys to the right to make space for the key we want to insert. We do not shift the keys individually (that would take

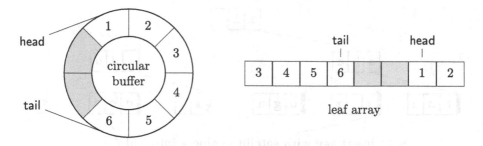

Fig. 3. A circular buffer representation of a leaf array capable of storing 8 keys. The pointers head and tail support prepending a key, removing the first key, appending a key, and removing the last key, all in constant time. The right figure shows that the circular buffer is actually implemented as a plain array with two pointers.

$\mathcal{O}(b)$ total time). Instead, we can shift $\Theta(w/k)$ keys in constant time by using word-packing, yielding $\mathcal{O}(bk/w)$ time for an insertion or deletion of a key in a leaf array. In what follows, we combine the word-packing technique with *circular buffers* representing the leaf arrays to improve the time bounds to $\mathcal{O}(\lg n)$.

A *circular buffer* supports, additionally to removing or adding the last element in constant time like a standard (non-resizable) array, the same operations for the first element in constant time as well. See Fig. 3 for a visualization. For an insertion or deletion elsewhere, we still have to shift the keys to the right or to the left. This can be done in $\mathcal{O}(bk/w) = \mathcal{O}(\lg n)$ time with word-packing as described in the previous paragraph for the plain leaf array (only extra care has to be taken when we are at the borders of the array representing the circular buffer). Finally, on inserting a key into a full leaf ℓ, we pay $\mathcal{O}(bk/w) = \mathcal{O}(\lg n)$ time for the insertion into this full leaf, but subsequently can shift keys among its sibling leaves in constant time per leaf. Similarly, on deleting a key of a full leaf ℓ, we first rearrange the circular buffer of ℓ in $\mathcal{O}(bk/w) = \mathcal{O}(\lg n)$ time, and subsequently shift a key among the $\mathcal{O}(q)$ circular buffers of ℓ's siblings to keep ℓ full, which takes also $\mathcal{O}(q) = \mathcal{O}(\lg n)$ time.

3.3 Final Space Complexity

Finally, we can bound the number of internal nodes by the number of leaves λ defined in Sect. 3.1: Since the minimum out-degree of an internal node is $t/2$, there are at most

$$\lambda \sum_{i=1}^{\infty} (2/t)^i = 2\lambda/(t-2) = \mathcal{O}(n(q+1)/(qtb)) = \mathcal{O}(n/tb) \text{ internal nodes.}$$

Since an internal node stores t pointers to its children, it uses $\mathcal{O}(tw)$ bits. In total we can store the internal nodes in

$$\mathcal{O}(twn/tb) = \mathcal{O}(wn/b) = \mathcal{O}(nk/\lg n) \text{ bits.} \tag{2}$$

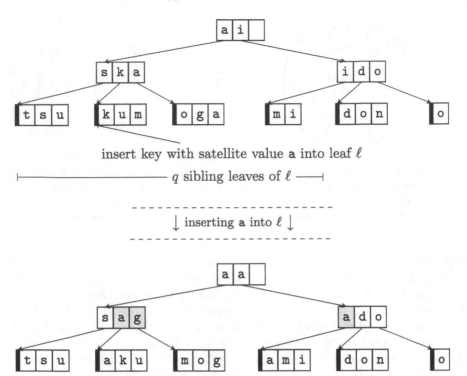

Fig. 4. Change of aggregate values on shifting keys. A shift causes the need to recompute the aggregate values of the satellite values stored in a leaf whose contents changed due to the shift. The example uses the same B tree structure as Fig. 1, but depicts the satellite values (plain characters) instead of the keys. Here, we used the minimum on the canonical Latin alphabet order as aggregate function.

Each circular buffer (introduced in Sect. 3.2) requires $\Theta(\lg b)$ bits (for the pointers in Fig. 3). The additional total space is $\lambda\, \mathcal{O}(\lg b) = \mathcal{O}(nq \lg b/(bq-2b)) = \mathcal{O}(n \lg b/b) = o(n)$ bits. Together with Eq. (1), we finally obtain Theorem 1.

3.4 Low Number of Keys

For our B tree, we require that $n = \Omega((w \lg^2 n)/k)$. When $n = \mathcal{O}((w \lg^2 n)/k)$ but $k = o(n/\lg^2 n)$, we can still provide a succinct solution within the same operational time complexity, which consists of a single internal node (i.e., the root node) governing the leaves. The leaves are defined as before, except that we set the maximum number of keys a leaf can store to $b = (w \lg n)/k^2$. Consequently, the root maintains $\mathcal{O}(k \lg n)$ leaves, and for each leaf ℓ the root stores a key to delegate a search to ℓ. By maintaining this key-leaf delegation in a binary search tree, we can search and update these keys in the root in $\mathcal{O}(\lg(k \lg n)) = \mathcal{O}(\lg n)$ time. We only keep at most one non-full leaf costing us $(w \lg n)/k$ bits. The distribution of the keys among the leaves is performed as before, except that we consider, when shifting keys, all leaves instead of just the q siblings. In total, we have an overhead of $(w \lg n)/k + \mathcal{O}(k \lg^2 n) = o(n)$ bits.

4   Augmenting with Aggregate Values

As highlighted in the related work section (Sect. 1.1), B trees are often augmented with auxiliary data to support prefix sum queries or LCP queries when storing strings. We present a more abstract solution covering these cases with *aggregate values*, i.e., values composed of the satellite values stored along with the keys in the leaves. In detail, we augment each node v with an *aggregate value* that is the return value of a decomposable aggregate function applied on the satellite values stored in the leaves of the subtree rooted at v. A *decomposable aggregate function* [14, Sect. 2A] such as the sum, the maximum, or the minimum, is a function f on a subset of satellite values with a constant-time merge operation $\cdot_f$ such that, given two disjoint subsets X and Y of satellite values, $f(X \cup Y) = f(X) \cdot_f f(Y)$, and the left-hand and the right-hand side of the equation can be computed in the same time complexity. We further assume that each aggregate value produced by f is storable in $\mathcal{O}(w)$ bits to fit into the $\mathcal{O}(tw)$-bits space bound of an internal node.

While sustaining the methods described in the introduction like predecessor for *keys*, we enhance insert to additionally take a value as argument, and provide access to the aggregate values:

insert(K, V) inserts the key K with satellite value V;
access(v) returns the aggregate value of the node v; and
access(K) returns the satellite value of the key K.

To make use of access(v), the B tree also provides access to the root, and a top-down navigation based on the way predecessor(K) works, for a key K as search parameter. To keep things simple, we assume that all keys are distinct[1] (i.e., we allow no duplicates).

For the computational analysis, let us assume that every satellite value uses $\mathcal{O}(k)$ bits, and that we can evaluate the given aggregate function f bit-parallel such that it can be evaluated in $\mathcal{O}(bk/w) = \mathcal{O}(\lg n)$ time for a leaf storing $b = \Theta(w \lg n/k)$ values.

Under this setting, we claim that we can obtain $\mathcal{O}(bk/w) = \mathcal{O}(\lg n)$ time for every B tree operation while maintaining the aggregate values, even if we distribute keys among q leaves on (a) an insertion of a key into a full leaf or (b) the deletion of a key. This is nontrivial: For instance, when maintaining minima as aggregate values, if we shift the key with minimal value of a leaf ℓ to its sibling, we have to recompute the aggregate value of ℓ (cf. Fig. 4), which we need to do from scratch (since we do not store additional information about finding the next minimum value). So a shift of a key to a leaf costs $\mathcal{O}(bk/w) = \mathcal{O}(\lg n)$ time, resulting in $\mathcal{O}(qbk/w) = \mathcal{O}(\lg^2 n)$ overall time for an insertion.

---

[1] Note that if all keys are distinct, then $k \geq \lg n$ by the pigeonhole principle.

Our idea is to decouple the satellite values from the leaf arrays where they are actually stored. To explain this idea, let us *conceptually* think of the leaf arrays as a global array—meaning that these arrays are still represented by their respective circular buffers *individually*. Given our B tree has λ leaves, we partition this global array into λ blocks, where the i-th block with $i \in [1..\lambda]$ starts initially at entry position $1 + (i-1)b$, corresponds to the i-th leaf, and has initially the size equal to the capacity of the circular buffer of its corresponding leaf. The crucial change is that we let the aggregate value of a leaf depend on its corresponding block instead of its leaf array. While leaf arrays (represented by circular buffers) have a fixed capacity, we can move block boundaries freely to extend or shrink the size of a block.

Now suppose that we want to insert an element e into a full leaf ℓ, and that one of its q siblings is not full. Hence, we can redistribute one element of ℓ's leaf array by shifting one element across $\mathcal{O}(q)$ leaf arrays as explained in Sect. 3.2. After the redistribution, without the blocks, we would have to update the aggregate values of $\mathcal{O}(q)$ siblings. Instead of that, we just enlarge the block of ℓ to cover e, and update the aggregate value of ℓ with e. This allows us to process an update operation by $\mathcal{O}(q)$ block boundary updates, and a constant number of updates of the aggregate values stored in the leaves. In summary, we can decouple the aggregate values from the leaf arrays with the aid of the blocks in the global array, and therefore can use the techniques introduced in Sect. 3.2, where we shift keys among $q + 1$ sibling leaves, without the need to recompute the aggregate values of the sibling leaves when shifting keys.

Example 1. Let us assume for simplicity that $b = 3$ and that the keys are the values. Suppose that our B tree consists of exactly three leaves ℓ_i for $i = 1, 2, 3$. Each leaf ℓ_i has a leaf array A_i with the following contents: $A_1 = (1, 2, 4)$, $A_2 = (5, 6, 7)$, and $A_3 = (8, 9, \perp)$, where $\perp$ denotes an empty slot. Further assume that our aggregate function f is min such that $f(A_1) = 1, f(A_2) = 5, f(A_3) = 8$. Now suppose that we want to insert 3 into A_1. Without the block reassignment, we would shift 4 to A_2 and 7 to A_3 such that we need to update the aggregate values of ℓ_2 and ℓ_3 to $f(A_2) = 4, f(A_3) = 7$. Now, with the block reassignment, we do the following: We think of the A_i's as a single array $A[1..9] = (1, 2, 4, 5, 6, 7, 8, 9, \perp)$ and partition it initially into blocks of equal length $B_1 = A[1..3], B_2 = A[4..6], B_3 = A[7..9]$. The blocks are basically just pointers into A such that updates of A automatically update the contents of the B_i's. Now the aggregate values of the leaves are no longer based on the A_i's, but on the B_i's, i.e., $f(B_1) = 1, f(B_2) = 5, f(B_3) = 8$. If we perform the same insertion as above inserting 3 into A_1, we perform the shifting as before such that $A[1..9] = (1, 2, 3, 4, 5, 6, 7, 8, 9)$, but additionally increase the size of B_1 and shift B_2 and B_3 to the right such that $B_1 = A[1..4], B_2 = A[5..7], B_3 = A[8..9]$. Consequently, the aggregate values of ℓ_1's siblings do not have to be updated. The key observation is that while the leaf array A_1 of ℓ_1 governs $b = 3$ elements, the block B_1 of ℓ_1 is allowed to contain more/fewer than b elements.

To track the boundaries of the blocks, we augment each leaf ℓ with an offset value and the current size of its block. The offset value stores the relative offset

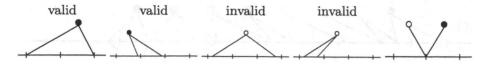

Fig. 5. Valid and invalid blocks according to the definition given in Sect. 4.2. The (conceptual) global array is symbolized by a horizontal line. The leaf arrays are intervals of the global array separated by vertical dashes. A dot symbolizes a leaf ℓ and the intersection of the triangle spawning from ℓ with the global array symbolizes the block of ℓ. A node has an invalid block if its dot is hollow. The rightmost picture shows the border case that a block is invalid if its offset is b, while a block can be valid even if it is empty.

of the block with respect to the initial starting position of the block (equal to the starting position of ℓ's leaf array) within the global array. We decrement the offset by one if we shift a key from ℓ to ℓ's preceding sibling, while we increment its offset by one if we shift a key of ℓ's preceding leaf to ℓ.

If we only care about insertions (and not about deletions and blocks becoming too large) we are done since we can update $f(X)$ to $f(X \cup \{x\})$ in constant time for a new satellite value $x \notin X$ per definition. However, deletions pose a problem for the running time because we usually cannot compute $f(X \setminus \{x\})$ from $f(X)$ with $x \in X$ in constant time. Therefore, we have to recompute the aggregate value of a block by considering all its stored satellite values. However, unlike leaf arrays whose sizes are upper bounded by b, blocks can grow beyond $\omega(b)$. Supporting deletions, we cannot ensure with our solution up so far to recompute the aggregate value of a block in $\mathcal{O}(\lg n)$ time. In what follows, we show that we can retain logarithmic update time, first with a simple solution taking $\mathcal{O}(\lg n)$ time amortized, and subsequently with a solution taking $\mathcal{O}(\lg n)$ time in the worst case.

4.1 Updates in Batch

Our amortized solution takes action after a node split occurs, where it adjusts the blocks of all $q + 2$ nodes that took part in that split (i.e., the full node, its q full siblings and the newly created node). The task is to evenly distribute the block sizes, reset the offsets, and recompute the aggregate values. We can do all that in $\mathcal{O}(q(bk/w + \lg n)) = \mathcal{O}(\lg^2 n)$ time, since

- there are $\mathcal{O}(q)$ leaves involved,
- each leaf stores at most b values, whose aggregate value can be computed in $\mathcal{O}(bk/w) = \mathcal{O}(\lg n)$ time, and
- each leaf has $\mathcal{O}(\lg n)$ ancestors whose aggregate values may need to be recomputed.

Although the obtained $\mathcal{O}(\lg^2 n)$ time complexity seems costly, we have increased the total capacity of the $b + 2$ nodes involved in the update by $\Theta(b)$ keys in total. Consequently, before splitting one of those nodes again, we perform at

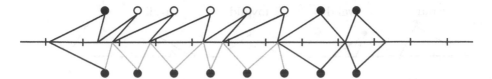

Fig. 6. Revalidation of multiple invalid blocks. The figure uses the same pictography as Fig. 5, but additionally shows on the bottom (vertically mirrored) the outcome of our algorithm fixing the invalid blocks (Sect. 4.2), where we empty the rightmost invalid block and swap the blocks until we find a block that can be merged with the previous block.

least $b = \Omega(\lg n)$ insertions (remember that we split a node only if it and its q siblings are full). Now, whenever a block becomes larger than $2b$, we can afford the above rearrangement costing $\mathcal{O}(\lg n)$ amortized time.

4.2 Updates by Merging

To improve the time bound to $\mathcal{O}(\lg n)$ worst case time, our trick is to merge blocks and reassign the ownership of blocks to sibling leaves. For the former, a merge of two blocks means that we have to combine two aggregate values, but this can be done in constant time by the definition of the decomposable aggregate function. To keep the size of the blocks within $\mathcal{O}(b)$, we watch out for blocks whose shape underwent too much changes, which we call invalid (see Fig. 5 for a visualization). We say a block is *valid* if it covers at most $2b$ keys, it has an offset in $(-b..b)$ (i.e., the block starts within the leaf array of the preceding leaf or of its corresponding leaf), and the sum of offset and size is in $[0..2b)$ (i.e., the block ends within the leaf array of its corresponding leaf or its succeeding leaf). Initially, all blocks are valid because they have size b and offset 0. If one of those conditions for a block becomes violated, we say that the block is *invalid*, and we take action to restore its validity. Blocks can become invalid when changing their sizes by one, or when shifting their boundaries by one. A shift can cause $\mathcal{O}(q)$ blocks to become invalid (i.e., the number of siblings considered when distributing keys). Suppose that a block B_i has become invalid due to a tree update, which already costed $\mathcal{O}(\lg n)$ time (the time for a root-to-leaf traversal). Our goal is to rectify the invalid block B_i within the same time bound. B_i has become invalid because of the events that (a) it covers $2b + 1$ keys, (b) has offset $-b$ or the sum of offset and size are negative, or (c) has offset $+b$ or the sum of offset and size are at least $2b$.

For event (a), we redistribute sizes and offsets of B_i with B_{i-1} and B_{i+1}. This is possible since at least one block B_{i-1} or B_{i+1} has less than $2b$ keys. Otherwise, since they are valid blocks[2], there is no space left for B_i to have $2b + 1$ keys. It is therefore possible for B_i to consign at least one key to its neighbors without

---

[2] More precisely, these blocks were valid at least before the enlargement of B_i, which could have triggered a shifting that invalidated either B_{i-1} or B_{i+1}.

making them overfull. We finish by recomputing the aggregate values of the three nodes and their ancestors, costing $\mathcal{O}(\lg n)$ total time.

The events (b) and (c) can happen when shifting blocks by one to the left (b) or to the right (c). Given B_i is the rightmost (for (b)) or the leftmost (for (c)) invalid block, we swap boundaries with the preceding blocks (for (b)) or succeeding blocks (for (c)) of B_i until finding a block whose boundaries can be extended to cover the shifted part without becoming invalid. The number of blocks we take into consideration is $\mathcal{O}(q)$, since we stop at a block B_j with $|B_j| + |B_{j+1}| \leq 2b$; if there are more blocks that do not satisfy this condition, then more than q consecutive siblings leaves are full, and a leaf split must have had occurred.

To solve (b), we proceed as follows—(c) can be solved symmetrically. First, we put B_i on a stash S (storing B_i's boundaries and its aggregate value), and empty B_i. Next, we check whether B_{i-1} can be extended to cover S without becoming invalid. If this is possible, we let B_{i-1} cover S, update the aggregate value of B_{i-1}, and terminate. Otherwise (B_{i-1} would become invalid), we swap B_{i-1} with S. Now B_{i-1} stores the boundaries of S. By doing so, B_{i-1} does not become invalid since the offset of B_i was $-b$ (and thus the offset of B_{i-1} becomes 0), or the sum of offset and size was in $[-b..0)$ (which becomes $[0..b)$), while the changed offset poses no problem, since the sum of offset and size of B_{i-1} is now at most $2b - 1$. Finally, we iteratively select the next preceding block B_{i-2} to check whether it is mergeable with the stash S without becoming invalid (cf. Fig. 6). Since each visit of a block takes constant time (either swapping or merging contents), and we visit $\mathcal{O}(q)$ blocks, fixing all invalid blocks takes $\mathcal{O}(q) = \mathcal{O}(\lg n)$ time.

5   Conclusion

We provided a space-efficient variation of the B tree that retains the time complexity of the standard B tree. It achieves succinct space when the keys are considered to be incompressible. Our main tools were the following: First, we generalized the B* tree technique to exchange keys not only with a dedicated sibling leaf but with up to q many sibling leaves. Second, we let each leaf store $\Theta(b)$ elements represented by a circular buffer such that moving a largest (resp. smallest) element of a leaf to its succeeding (resp. preceding) sibling can be performed in constant time. Additionally, we could augment each node with an aggregate value and maintain these values, either with a batch update weakening the worst case time complexities to amortized time, or with a blocking of the leaf arrays that can be maintained within the worst case time complexities.

Acknowledgment. This work was supported by JSPS KAKENHI Grant Numbers JP19K20213 (TI), JP21K17701 and JP21H05847 (DK).

References

1. Bayer, R., McCreight, E.M.: Organization and maintenance of large ordered indexes. In: Proceedings of the SIGFIDET, pp. 107–141 (1970)
2. Bille, P., et al.: Dynamic relative compression, dynamic partial sums, and substring concatenation. Algorithmica **80**(11), 3207–3224 (2018)
3. Blandford, D.K., Blelloch, G.E.: Compact representations of ordered sets. In: Proceedings of the SODA, pp. 11–19 (2004)
4. Comer, D.: The ubiquitous B-tree. ACM Comput. Surv. **11**(2), 121–137 (1979)
5. Delpratt, O.N., Rahman, N., Raman, R.: Compressed prefix sums. In: van Leeuwen, J., Italiano, G.F., van der Hoek, W., Meinel, C., Sack, H., Plášil, F. (eds.) SOFSEM 2007. LNCS, vol. 4362, pp. 235–247. Springer, Heidelberg (2007). https://doi.org/10.1007/978-3-540-69507-3_19
6. Dietz, P.F.: Optimal algorithms for list indexing and subset rank. In: Dehne, F., Sack, J.-R., Santoro, N. (eds.) WADS 1989. LNCS, vol. 382, pp. 39–46. Springer, Heidelberg (1989). https://doi.org/10.1007/3-540-51542-9_5
7. Elias, P.: Efficient storage and retrieval by content and address of static files. J. ACM **21**(2), 246–260 (1974)
8. Farzan, A., Munro, J.I.: Succinct representation of dynamic trees. Theor. Comput. Sci. **412**(24), 2668–2678 (2011)
9. Ferragina, P., Grossi, R.: The string B-tree: a new data structure for string search in external memory and its applications. J. ACM **46**(2), 236–280 (1999)
10. Franceschini, G., Grossi, R.: Optimal implicit dictionaries over unbounded universes. Theory Comput. Syst. **39**(2), 321–345 (2006)
11. González, R., Navarro, G.: Rank/select on dynamic compressed sequences and applications. Theor. Comput. Sci. **410**(43), 4414–4422 (2009)
12. Graefe, G.: Modern B-tree techniques. Found. Trends Databases **3**(4), 203–402 (2011)
13. He, M., Munro, J.I.: Succinct representations of dynamic strings. In: Chavez, E., Lonardi, S. (eds.) SPIRE 2010. LNCS, vol. 6393, pp. 334–346. Springer, Heidelberg (2010). https://doi.org/10.1007/978-3-642-16321-0_35
14. Jesus, P., Baquero, C., Almeida, P.S.: A survey of distributed data aggregation algorithms. IEEE Commun. Surv. Tutor. **17**(1), 381–404 (2015)
15. Katajainen, J., Rao, S.S.: A compact data structure for representing a dynamic multiset. Inf. Process. Lett. **110**(23), 1061–1066 (2010)
16. Knuth, D.E.: The Art of Computer Programming, Volume 3: Sorting and Searching. Addison Wesley, Redwood City (1998)
17. Munro, J.I., Nekrich, Y.: Compressed data structures for dynamic sequences. In: Bansal, N., Finocchi, I. (eds.) ESA 2015. LNCS, vol. 9294, pp. 891–902. Springer, Heidelberg (2015). https://doi.org/10.1007/978-3-662-48350-3_74
18. Navarro, G., Nekrich, Y.: Optimal dynamic sequence representations. SIAM J. Comput. **43**(5), 1781–1806 (2014)
19. Prezza, N.: A framework of dynamic data structures for string processing. In: Proceedings of the SEA. LIPIcs, vol. 75, pp. 11:1–11:15 (2017)
20. Raman, R., Rao, S.S.: Succinct dynamic dictionaries and trees. In: Baeten, J.C.M., Lenstra, J.K., Parrow, J., Woeginger, G.J. (eds.) ICALP 2003. LNCS, vol. 2719, pp. 357–368. Springer, Heidelberg (2003). https://doi.org/10.1007/3-540-45061-0_30

An Additive Approximation Scheme for the Nash Social Welfare Maximization with Identical Additive Valuations

Asei Inoue and Yusuke Kobayashi[✉]

Kyoto University, Kyoto, Japan
yusuke@kurims.kyoto-u.ac.jp

Abstract. We study the problem of efficiently and fairly allocating a set of indivisible goods among agents with identical and additive valuations for the goods. The objective is to maximize the Nash social welfare, which is the geometric mean of the agents' valuations. While maximizing the Nash social welfare is NP-hard, a PTAS for this problem is presented by Nguyen and Rothe. The main contribution of this paper is to design a first additive PTAS for this problem, that is, we give a polynomial-time algorithm that maximizes the Nash social welfare within an additive error $\varepsilon v_{\max}$, where ε is an arbitrary positive number and $v_{\max}$ is the maximum utility of goods. The approximation performance of our algorithm is better than that of a PTAS. The idea of our algorithm is simple; we apply a preprocessing and then utilize an additive PTAS for the target load balancing problem given recently by Buchem et al. However, a nontrivial amount of work is required to evaluate the additive error of the output.

Keywords: Additive approximation algorithm · Nash social welfare · Target load balancing problem

1 Introduction

1.1 Nash Social Welfare Maximization

We study the problem of efficiently and fairly allocating a set of indivisible goods among agents with identical and additive valuations for the goods. There are many ways to measure the quality of the allocation in the literature, and in this paper, we aim to maximize the Nash social welfare [15], which is the geometric mean of the agents' valuations in the allocation.

Suppose we are given a set of agents $\mathcal{A} = \{1, 2, \ldots, n\}$ and a set of goods $\mathcal{G} = \{1, 2, \ldots, m\}$ with a utility $v_j > 0$ for each $j \in \mathcal{G}$. An *allocation* is a partition $\pi = (\pi_1, \ldots, \pi_n)$ of $\mathcal{G}$ where $\pi_i \subseteq \mathcal{G}$ is a set of goods assigned to agent i. For an allocation $\pi = (\pi_1, \ldots, \pi_n)$, let $v(\pi_i)$ be the valuation of i that is defined as the

This work is partly supported by JSPS, KAKENHI grant numbers JP18H05291, JP19H05485, and JP20K11692, Japan.

© Springer Nature Switzerland AG 2022
C. Bazgan and H. Fernau (Eds.): IWOCA 2022, LNCS 13270, pp. 341–354, 2022.
https://doi.org/10.1007/978-3-031-06678-8_25

sum of the utility of the goods assigned to i, i.e., $v(\pi_i) = \sum_{j \in \pi_i} v_j$. The goal is to find an allocation π that maximizes the function

$$f(\pi) = \left(\prod_{i \in \mathcal{A}} v(\pi_i) \right)^{1/n},$$

which is called the *Nash social welfare* [4,22]. In this paper, we refer to this problem as IDENTICAL ADDITIVE NSW.

IDENTICAL ADDITIVE NSW
Input: A set of agents $\mathcal{A} = \{1, 2, \ldots, n\}$ and a set of goods $\mathcal{G} = \{1, 2, \ldots, m\}$ with a utility $v_j > 0$ for each $j \in \mathcal{G}$.
Output: An allocation π that maximizes the Nash social welfare $f(\pi)$.

The Nash social welfare can be defined in a more general setting where the valuation of each agent i is determined by a set function $v_i \colon 2^{\mathcal{G}} \to \mathbb{R}_{\geq 0}$. In such a case, the Nash social welfare of an allocation $\pi = (\pi_1, \ldots, \pi_n)$ is defined as $\left(\prod_{i \in \mathcal{A}} v_i(\pi_i) \right)^{1/n}$. In IDENTICAL ADDITIVE NSW, we focus on the case where the valuation function is additive and independent of the agent. Note that, by removing goods with zero utility, we can assume that $v_j > 0$ without loss of generality.

The Nash social welfare was named after John Nash, who introduced and studied the Nash social welfare in the context of bargaining in the 1950s s [21]. Later, the same concept was independently studied in the context of competitive equilibria with equal incomes [23] and proportional fairness in networking [16]. It has traditionally been studied in the economics literature for divisible goods [20]. For divisible goods, an allocation maximizing the Nash social welfare can be computed in polynomial time when the valuation functions are additive [10].

In the context of goods allocation, the Nash social welfare is a measure that captures efficiency and fairness at the same time. To see this, for a parameter $q \in \mathbb{R}$ and for an allocation π, one can define the *generalized mean* of the valuation of each agent as $f_q(\pi) = \left(\frac{1}{n} \sum_{i=1}^{n} v_i(\pi_i)^q \right)^{1/q}$. The generalized mean can be a variety of mean functions depending on the value of q. When $q = 1$, $f_q(\pi)$ is the average valuation of the agents, and hence maximizing $f_q(\pi)$ is equivalent to maximizing the social welfare. In this case, $f_q(\pi)$ is a measure of the efficiency of an allocation. When $q \to -\infty$, $f_q(\pi)$ is the minimum value of $v_i(\pi_i)$, namely the valuation of the least satisfied agent. In this case, an allocation maximizing $f_q(\pi)$ can be considered fair in a sense. It is known that in the limit as $q \to 0$, $f_q(\pi)$ coincides with the geometric mean, which is the Nash social welfare (see [7]). Therefore, maximizing the Nash social welfare (i.e., $q \to 0$) can be viewed as a compromise between Maximum Social Welfare (i.e., $q = 1$) and Max-Min Welfare (i.e., $q \to -\infty$).

The Nash social welfare is closely related to other concepts EF1 and Pareto optimality that describe fairness and efficiency, respectively, which also supports the importance of the Nash social welfare. An allocation is said to be *EF1* (*envy-free up to at most one good*) if each agent prefers its own bundle over the bundle of any other agent up to the removal of one good. An allocation is called *Pareto optimal* if no one else's valuation can be increased without sacrificing someone else's valuation. Caragiannis et al. [8] showed that an allocation that maximizes the Nash social welfare is both EF1 and Pareto optimal when agents have additive valuations for the goods. This motivates studying the problem of finding an allocation that maximizes the Nash social welfare.

1.2 Our Contribution: Approximation Algorithm

The topic of this paper is the approximability of the Nash social welfare maximization. By an easy reduction from the Subset Sum problem, we can see that maximizing the Nash social welfare is NP-hard even in the case of two agents with identical additive valuations. That is, IDENTICAL ADDITIVE NSW is NP-hard even when $n = 2$. Furthermore, maximizing the Nash social welfare is APX-hard for multiple agents with non-identical valuations even when the valuations are additive [17].

On a positive side, several approximation algorithms are proposed for maximizing the Nash social welfare, and the difficulty of the problem depends on the class of valuations v_i. Under the assumption that the valuation set function is monotone and submodular, Li and Vondrák [18] recently proposed a constant factor approximation algorithm based on an algorithm for Rado valuations [12]. Better constant factor approximation algorithms are known for subclasses of submodular functions [2,9,11,19]. When the valuation function is additive, a 1.45-approximation algorithm is known [3], and this is the current best approximation ratio. When the valuation functions are additive and identical, the situation is much more tractable. Indeed, for IDENTICAL ADDITIVE NSW, it is known that a polynomial-time approximation scheme (PTAS) exists [22] and a simple fast greedy algorithm achieves a 1.061-approximation guarantee [4].

For IDENTICAL ADDITIVE NSW, the above results show the limit of the approximability and so no further improvement seems to be possible in terms of the approximation ratio. Nevertheless, a better approximation algorithm may exist if we evaluate the approximation performance in a fine-grained way. The main contribution of this paper is to show that this is indeed the case if we evaluate the approximation performance by using the additive error. Formally, our result is stated as follows.

Theorem 1. *For an instance of* IDENTICAL ADDITIVE NSW, *let* $v_{\max} = \max_{j \in \mathcal{G}} v_j$ *and let* OPT *be the optimal value. For any* $\varepsilon > 0$, *there is an algorithm* A_ε *for* IDENTICAL ADDITIVE NSW *that runs in* $(nm/\varepsilon)^{O(1/\varepsilon)}$ *time and returns an allocation* π *such that* $f(\pi) \geq \text{OPT} - \varepsilon v_{\max}$.

Recall that a PTAS for IDENTICAL ADDITIVE NSW is an algorithm that returns an allocation π with $f(\pi) \geq \frac{\text{OPT}}{1+\varepsilon}$. Since $\frac{\text{OPT}}{1+\varepsilon} \approx (1 - \varepsilon)\text{OPT}$, the additive error of a PTAS is roughly εOPT, which can be much greater than $\varepsilon v_{\max}$. Furthermore, as we will see in Proposition 2, our algorithm given in the proof of Theorem 1 is also a PTAS. In this sense, we can say that our algorithm is better than a PTAS if we evaluate the approximation performance in a fine-grained way.

We also note that there is no polynomial-time algorithm for finding an allocation π with $f(\pi) \geq \mathrm{OPT}-\varepsilon$ unless P = NP. This is because the additive error can be arbitrarily large by scaling the utility unless we obtain an optimal solution. Therefore, parameter $v_{\max}$ is necessary to make the condition scale-invariant.

1.3 Related Work: Additive PTAS

The algorithm in Theorem 1 is called an *additive PTAS* with parameter $v_{\max}$, and so our result has a meaning in a sense that it provides a new example of a problem for which an additive PTAS exists. In this subsection, we describe known results on additive PTASs, some of which are used in our argument later.

An additive PTAS is a framework for approximation guarantees that was recently introduced by Buchem et al. [5,6]. For any $\varepsilon > 0$, an additive PTAS returns a solution whose additive error is at most ε times a certain parameter.

Definition 1. *For an optimization problem, an* additive PTAS *is a family of polynomial-time algorithms* $\{A_\varepsilon \mid \varepsilon > 0\}$ *with the following condition: for any instance I and for every $\varepsilon > 0$, A_ε finds a solution with value $A_\varepsilon(I)$ satisfying* $|A_\varepsilon(I) - \mathrm{OPT}(I)| \leq \varepsilon h$, *where h is a suitably chosen parameter of instance I and $\mathrm{OPT}(I)$ is the optimal value.*

In some cases, an additive PTAS is immediately derived from an already known algorithm. For example, by setting the error factor appropriately, a fully polynomial-time approximation scheme (FPTAS) for the knapsack problem [13] is also an additive PTAS where the parameter is the maximum utility of a good. However, evaluating the additive error is difficult in general, and so additive PTASs are known for only a few problems. In the pioneering paper on additive PTASs by Buchem et al. [5,6], an additive PTAS was proposed for the completion time minimization scheduling problem, the Santa Claus problem, and the envy minimization problem. In order to derive these additive PTASs, they introduced the *target load balancing problem* and showed that it is possible to determine whether a solution exists by only slightly violating the constraints.

In the *target load balancing problem*, we are given a set of jobs $\mathcal{J}$ with a processing time $v_j > 0$ for each $j \in \mathcal{J}$ and a set of machines $\mathcal{M}$ with real values l_i and u_i for each $i \in \mathcal{M}$. The goal is to assign each job $j \in \mathcal{J}$ to a machine $i \in \mathcal{M}$ such that for each machine $i \in \mathcal{M}$ the load of i (i.e., the sum of the processing times of the jobs assigned to i) is in the interval $[l_i, u_i]$. In a similar way to IDENTICAL ADDITIVE NSW, an assignment is represented by a partition $\pi = (\pi_i)_{i \in \mathcal{M}}$ of $\mathcal{J}$. Let $v_{\max} = \max_{j \in \mathcal{J}} v_j$ and let K denote the number of types of machines, that is, $K = |\{(l_i, u_i) \mid i \in \mathcal{M}\}|$. While the target load balancing problem is NP-hard, Buchem et al. [5,6] showed that it can be solved in polynomial time if we allow a small additive error and K is a constant.

Theorem 2 (Buchem et al. [5, Theorem 12]). *For the target load balancing problem and for any $\varepsilon > 0$, there is an algorithm (called LOADBALANCING) that either*

1. *concludes that there is no feasible solution for a given instance, or*
2. *returns an assignment* $\pi = (\pi_i)_{i \in \mathcal{M}}$ *such that the total load* $\sum_{j \in \pi_i} v_j$ *is in* $[l_i - \varepsilon v_{\max}, u_i + \varepsilon v_{\max}]$ *for each* $i \in \mathcal{M}$

in $|\mathcal{M}|^{K+1}(\frac{|\mathcal{J}|}{\varepsilon})^{O(1/\varepsilon)}$ *time.*

Note that the algorithm in this theorem is used as a subroutine in our additive PTAS for IDENTICAL ADDITIVE NSW. Note also that the term "assignment" is used in this theorem by following the convention, but it just means a partition of the jobs. Therefore, we do not distinguish "assignment" and "allocation" in what follows in this paper.

1.4   Technical Highlights

In this subsection, we describe the outline of our algorithm for IDENTICAL ADDITIVE NSW and explain two technical issues that are peculiar to additive errors.

The basic strategy of our algorithm is simple; we guess the valuation $v(\pi_i^*)$ of each agent i in an optimal solution π^*, and then seek for an allocation π such that $|v(\pi_i) - v(\pi_i^*)| \leq \varepsilon v_{\max}$ for each $i \in \mathcal{A}$ by using LOADBALANCING in Theorem 2.

The first technical issue is that even if the additive error of $v(\pi_i)$ is at most $\varepsilon v_{\max}$ for each $i \in \mathcal{A}$, the additive error of $f(\pi)$ is not easily bounded by $\varepsilon v_{\max}$. This is in contrast to the case of the multiplicative error (i.e., if $v(\pi_i) \geq v(\pi_i^*)/(1 + \varepsilon)$ for each $i \in \mathcal{A}$, then $f(\pi) \geq f(\pi^*)/(1 + \varepsilon)$). The first technical ingredient in our proof is to bound the additive error of $f(\pi)$ under the assumption that $v_{\max}$ is at most the average valuation of the agents; see Lemma 5 for a formal statement. In order to apply this argument, we modify a given instance so that $v_{\max}$ is at most the average valuation of the agents by a naive preprocessing. In the preprocessing, we assign a good j with high utility to an arbitrary agent i and remove i and j from the instance, repeatedly (see Sect. 2.1 for details).

The second technical issue is that the preprocessing might affect the additive error of the output, whereas it does not affect the optimal solutions of the instance (see Lemma 1). Suppose that an instance I is converted to an instance I' by the preprocessing, and suppose also that we obtain an allocation π' for I'. Then, by recovering the agents and the goods removed in the preprocessing, we obtain an allocation π for I from π'. The issue is that the additive error of the objective function value might be amplified by this recovering process, which makes the evaluation of the additive error of $f(\pi)$ hard. Nevertheless, we show that the additive error of $f(\pi)$ is bounded by $O(\varepsilon v_{\max})$ with the aid of the mean-value theorem for differentiable functions (see Proposition 3), which is the second technical ingredient in our proof. It is worth noting that the differential calculus plays a crucial role in the proof, whereas the problem setting is purely combinatorial.

The remaining of this paper is organized as follows. In Sect. 2, we describe our algorithm for IDENTICAL ADDITIVE NSW. Then, in Sect. 3, we show its approximation guarantee and prove Theorem 1.

2   Description of the Algorithm

As we mentioned in Sect. 1.4, in our algorithm, we first apply a preprocessing so that $v_{\max}$ is at most the average valuation of the agents. Then, we guess the valuation of each agent in an optimal solution, and then seek for an allocation that is close to the optimal solution by using LOADBALANCING, which is the main procedure. We describe the preprocessing and the main procedure in Sects. 2.1 and 2.2, respectively.

2.1   Preprocessing

Consider an instance $I = (\mathcal{A}, \mathcal{G}, \mathbf{v})$ of IDENTICAL ADDITIVE NSW where $\mathbf{v} = (v_1, \ldots, v_m)$. If $|\mathcal{A}| > |\mathcal{G}|$, then the optimal value is zero, and hence any allocation is optimal. Therefore, we may assume that $|\mathcal{A}| \leq |\mathcal{G}|$, which implies that the optimal value is positive. Let $\mu(I)$ be the average valuation of agents, that is, $\mu(I) = \frac{1}{|\mathcal{A}|} \sum_{j \in \mathcal{G}} v_j$. When I is obvious, we simply write μ for $\mu(I)$. The objective of the preprocessing is to modify a given instance so that $v_j < \mu$ for any $j \in \mathcal{G}$.

Our preprocessing immediately follows from the fact that an agent who receives a valuable good does not receive other goods in an optimal solution. Note that similar observations were shown in previous papers (see e.g. [1, 22]).

Lemma 1 ($\star^1$). *Let $j \in \mathcal{G}$ be an item with $v_j \geq \mu$. In an optimal solution π^*, an agent who receives j cannot receive any goods other than j.*

For $\mathcal{A}_0 \subseteq \mathcal{A}$ and $\mathcal{G}_0 \subseteq \mathcal{G}$, let $I \setminus (\mathcal{A}_0, \mathcal{G}_0)$ denote the instance obtained from I by removing $\mathcal{A}_0$ and $\mathcal{G}_0$, that is, $I \setminus (\mathcal{A}_0, \mathcal{G}_0) = (\mathcal{A} \setminus \mathcal{A}_0, \mathcal{G} \setminus \mathcal{G}_0, \mathbf{v} \setminus \mathcal{G}_0)$, where $\mathbf{v} \setminus \mathcal{G}_0 = (v_j)_{j \in \mathcal{G} \setminus \mathcal{G}_0}$. In the preprocessing, we assign a good $j \in \mathcal{G}$ with $v_j \geq \mu$ to some agent $i \in \mathcal{A}$ and remove i and j from the instance, repeatedly. A formal description is shown in Algorithm 1.

Algorithm 1. PREPROCESSING

Input: instance $I = (\mathcal{A}, \mathcal{G}, \mathbf{v})$ where $\mathbf{v} = (v_1, v_2, \ldots, v_m)$
Output: subsets $\mathcal{A}_0 \subseteq \mathcal{A}$ and $\mathcal{G}_0 \subseteq \mathcal{G}$
1: Initialize $\mathcal{A}_0$ and $\mathcal{G}_0$ as $\mathcal{A}_0 = \mathcal{G}_0 = \emptyset$.
2: **while** there exists a good $j \in \mathcal{G} \setminus \mathcal{G}_0$ with $v_j \geq \mu(I \setminus (\mathcal{A}_0, \mathcal{G}_0))$ **do**
3:      $\mathcal{G}_0 \leftarrow \mathcal{G}_0 \cup \{j\}$
4:      Choose $i \in \mathcal{A} \setminus \mathcal{A}_0$ arbitrarily and add it to $\mathcal{A}_0$.
5: **return** $\mathcal{A}_0, \mathcal{G}_0$

Let $\mathcal{A}_0$ and $\mathcal{G}_0$ be the output of PREPROCESSING. Lemma 1 shows that if we obtain an optimal solution for $I \setminus (\mathcal{A}_0, \mathcal{G}_0)$, then we can immediately obtain an optimal solution for I by assigning goods in $\mathcal{G}_0$ to agents in $\mathcal{A}_0$. Note that the

---

[1] ($\star$) indicates that the proof is given in the full version of this paper [14].

inequality $v_j < \mu(I \setminus (\mathcal{A}_0, \mathcal{G}_0))$ holds for all $j \in \mathcal{G} \setminus \mathcal{G}_0$ after the preprocessing. Thus, the maximum utility of a good is less than the average valuation of the agents in the instance $I \setminus (\mathcal{A}_0, \mathcal{G}_0)$. Note also that, since the number of while loop iterations is at most $|\mathcal{A}|$, PREPROCESSING runs in polynomial time.

2.2 Main Procedure

We describe the main part of the algorithm, in which we guess the valuation of each agent in an optimal solution and then apply LOADBALANCING. In order to obtain a polynomial-time algorithm, we have the following difficulties: the number of guesses has to be bounded by a polynomial and the number of machine types K has to be a constant when we apply LOADBALANCING. To overcome these difficulties, we get good upper and lower bounds on the valuation of each agent in an optimal solution, which is a key observation in our algorithm. We prove the following lemma by tracing the proof of Lemma 1.

Lemma 2 ($\star$). *For any instance I of* IDENTICAL ADDITIVE NSW*, let π^* be an optimal allocation of I. Then, $\mu - v_{\max} < v(\pi_i^*) < \mu + v_{\max}$ holds for any $i \in \mathcal{A}$.*

We are now ready to describe our algorithm. Suppose we are given an instance $I = (\mathcal{A}, \mathcal{G}, \mathbf{v})$ with $v_{\max} < \mu$. To simplify the description, suppose that $1/\varepsilon$ is an integer.

Our idea is to guess $v(\pi_i^*)$ with an additive error $\varepsilon v_{\max}$ for each $i \in \mathcal{A}$, where π^* is an optimal solution. By Lemma 2, we already know that the value of an optimal solution is in the interval of width $2v_{\max}$. Let L be the set of points delimiting this interval with width $\varepsilon v_{\max}$, that is, $L = \{\mu - v_{\max} + t\varepsilon v_{\max} \mid t \in \{0, 1, 2, \ldots, 2/\varepsilon - 1\}\}$. Let $L^{\mathcal{A}}$ be the set of all the maps from $\mathcal{A}$ to L. For $\tau, \tau' \in L^{\mathcal{A}}$, we denote $\tau \sim \tau'$ if τ' is obtained from τ by changing the roles of the agents, or equivalently $|\{i \in \mathcal{A} \mid \tau(i) = x\}| = |\{i \in \mathcal{A} \mid \tau'(i) = x\}|$ for each $x \in L$. In such a case, since each agent is identical, we can identify τ and τ'. This motivates us to define $D := L^{\mathcal{A}}/\sim$, where $\sim$ is the equivalence relation defined as above.

For each $\tau \in D$, we apply LOADBALANCING in Theorem 2 to the following instance of the target load balancing problem: $\mathcal{M} := \mathcal{A}$, $\mathcal{J} := \mathcal{G}$, the processing time of $j \in \mathcal{J}$ is v_j, and the target interval is $[\tau(i), \tau(i) + \varepsilon v_{\max}]$ for each $i \in \mathcal{M}$. Then, LOADBALANCING either concludes that no solution exists or returns an assignment (allocation) π^τ such that $v(\pi_i^\tau) \in [\tau(i) - \varepsilon v_{\max}, \tau(i) + 2\varepsilon v_{\max}]$ for each $i \in \mathcal{M}$.

Among all solutions π^τ returned by LOADBALANCING, our algorithm chooses an allocation with the largest objective function value. A pseudocode of our algorithm is shown in Algorithm 2.

Algorithm 2. MAINPROCEDURE

Input: instance $I = (\mathcal{A}, \mathcal{G}, \mathbf{v})$ such that $v_{\max} < \mu$
Output: allocation π
1: Initialize π as an arbitrary allocation.
2: **for** $\tau \in D$ **do**
3:     Apply LOADBALANCING with the target interval $[\tau(i), \tau(i) + \varepsilon v_{\max}]$ for $i \in \mathcal{A}$.
4:     **if** LOADBALANCING returns an allocation π^τ **then**
5:         **if** $f(\pi) < f(\pi^\tau)$ **then**
6:             $\pi \leftarrow \pi^\tau$
7: **return** π

Proposition 1. *The running time of* MAINPROCEDURE *is* $(nm/\varepsilon)^{O(1/\varepsilon)}$.

Proof. To obtain an upper bound on the number of for loop iterations, we estimate the number of elements in D. Since each $\tau \in D$ is determined by the number of agents $i \in \mathcal{A}$ such that $\tau(i) = x$ for $x \in L$, we obtain $|D| \leq |\{0, 1, \ldots, n\}|^L \leq (n+1)^{2/\varepsilon} = n^{O(1/\varepsilon)}$.

We next estimate the running time of LOADBALANCING. Since $|\mathcal{M}| = n$, $|\mathcal{J}| = m$, and the number of machine types K is at most $|L| = 2/\varepsilon$, the running time of LOADBALANCING is $n^{2/\varepsilon+1}(m/\varepsilon)^{O(1/\varepsilon)}$ by Theorem 2.

Thus, the total running time of MAINPROCEDURE is $(nm/\varepsilon)^{O(1/\varepsilon)}$. □

The entire algorithm for IDENTICAL ADDITIVE NSW consists of the following steps: apply PREPROCESSING, apply MAINPROCEDURE, and recover the removed sets. A pseudocode of the entire algorithm is shown in Algorithm 3.

Algorithm 3. MAXNASHWELFARE

Input: instance $I = (\mathcal{A}, \mathcal{G}, \mathbf{v})$
Output: allocation π'
1: Apply PREPROCESSING to I and obtain $\mathcal{A}_0$ and $\mathcal{G}_0$.
2: Apply MAINPROCEDURE to $I \setminus (\mathcal{A}_0, \mathcal{G}_0)$ and obtain π.
3: Let σ be a bijection from $\mathcal{A}_0$ to $\mathcal{G}_0$.
4: Set $\pi'_i = \{\sigma(i)\}$ for $i \in \mathcal{A}_0$ and set $\pi'_i = \pi_i$ for $i \in \mathcal{A} \setminus \mathcal{A}_0$.
5: **return** π'

Since the most time consuming part is MAINPROCEDURE, the running time of MAXNASHWELFARE is $(nm/\varepsilon)^{O(1/\varepsilon)}$ by Proposition 1.

3   Analysis of Approximation Performance

In this section, we show that MAXNASHWELFARE returns a good approximate solution for IDENTICAL ADDITIVE NSW and give a proof of Theorem 1. We first analyze the performance of MAINPROCEDURE in Sect. 3.1, and then analyze the effect of PREPROCESSING in Sect. 3.2.

3.1   Approximation Performance of MainProcedure

In this subsection, we consider an instance $I = (\mathcal{A}, \mathcal{G}, \mathbf{v})$ of Identical Additive NSW such that $v_{\max} < \mu$. The following lemma is easy, but useful in our analysis of MainProcedure.

Lemma 3. *Assume that $v_{\max} < \mu$. Let π be the allocation returned by MainProcedure. For any optimal solution π^*, there exists an allocation π^τ such that*

1. $|v(\pi_i^\tau) - v(\pi_i^*)| \leq 2\varepsilon v_{\max}$ *for each $i \in \mathcal{A}$, and*
2. $f(\pi^\tau) \leq f(\pi)$.

Proof. Let π^* be a given optimal solution. Take $\tau^* \in L^{\mathcal{A}}$ so that the valuation $v(\pi_i^*)$ is in the interval $[\tau^*(i), \tau^*(i) + \varepsilon v_{\max}]$ for each $i \in \mathcal{A}$. Note that such τ^* always exists by Lemma 2. Since we apply LoadBalancing with $l_i = \tau(i)$ and $u_i = \tau(i) + \varepsilon v_{\max}$ in MainProcedure for some τ with $\tau \sim \tau^*$, we obtain an allocation π^τ that corresponds to τ. Then, the inequality $|v(\pi_i^\tau) - v(\pi_i^*)| \leq 2\varepsilon v_{\max}$ holds by reordering the agents appropriately. By the choice of π in MainProcedure, the inequality $f(\pi^\tau) \leq f(\pi)$ holds. $\square$

In preparation for the analysis, we show another bound on the valuation of an agent in an optimal solution. Note that a similar result is shown by Alon et al. [1] for a different problem, and our proof for the following lemma is based on their argument.

Lemma 4 ($\star$). *Assume that $v_{\max} < \mu$. Let π^* be an optimal allocation of goods. Then, $\frac{\mu}{2} < v(\pi_i^*) < 2\mu$ holds for any $i \in \mathcal{A}$.*

We are now ready to evaluate the performance of MainProcedure.

Lemma 5. *Assume that $v_{\max} < \mu$ and $0 < \varepsilon \leq 1/5$. Let π be the allocation returned by MainProcedure and let OPT be the optimal value. Then, it holds that $f(\pi) \geq \mathrm{OPT} - 48\varepsilon v_{\max}$.*

Proof. By Lemma 3, there exist an allocation π^τ and an optimal solution π^* such that

$$|v(\pi_i^\tau) - v(\pi_i^*)| \leq 2\varepsilon v_{\max}, \tag{1}$$

and $f(\pi^\tau) \leq f(\pi)$. Let $S = f(\pi^\tau)$. Since $S \leq f(\pi)$, in order to obtain $f(\pi) \geq \mathrm{OPT} - 48\varepsilon v_{\max}$, it suffices to show that $\mathrm{OPT} - S \leq 48\varepsilon v_{\max}$.

We first evaluate the ratio between OPT and S as follows:

$$\frac{\mathrm{OPT}}{S} = \left(\prod_{i \in \mathcal{A}} \frac{v(\pi_i^*)}{v(\pi_i^\tau)} \right)^{1/n} \leq \frac{1}{n} \sum_i \frac{v(\pi_i^*)}{v(\pi_i^\tau)}$$

$$\leq \frac{1}{n} \sum_i \left(1 + \frac{2\varepsilon v_{\max}}{v(\pi_i^\tau)} \right) = 1 + \frac{2\varepsilon v_{\max}}{n} \sum_i \frac{1}{v(\pi_i^\tau)}, \tag{2}$$

where we use the inequality of arithmetic and geometric means (AM-GM inequality) in the first inequality and use (1) in the second inequality. By using (2), the difference between OPT and S can be evaluated as follows:

$$\text{OPT} - S = S\left(\frac{\text{OPT}}{S} - 1\right) \leq 2\varepsilon v_{\max}\left(\frac{1}{n}\sum_i \frac{S}{v(\pi_i^\tau)}\right) = 2\varepsilon v_{\max}\left(\frac{S}{H}\right), \quad (3)$$

where we define $1/H = \frac{1}{n}\sum_i 1/v(\pi_i^\tau)$, that is, H is the harmonic mean of $v(\pi_i^\tau)$. Therefore, to obtain an upper bound on $\text{OPT}-S$, it suffices to give upper bounds on S and $1/H$.

By (1), Lemma 4, and $v_{\max} < \mu$, we see that

$$v(\pi_i^\tau) \leq v(\pi_i^*) + 2\varepsilon v_{\max} \leq 2\mu + 2\varepsilon\mu,$$
$$v(\pi_i^\tau) \geq v(\pi_i^*) - 2\varepsilon v_{\max} \geq \mu/2 - 2\varepsilon\mu$$

for each agent $i \in \mathcal{A}$. Since S and H are the arithmetic mean and the harmonic mean of $v(\pi_i^\tau)$, respectively, we obtain $S \leq 2\mu + 2\varepsilon\mu$ and $H \geq \mu/2 - 2\varepsilon\mu$, where we note that $v(\pi_i^\tau) \geq \mu/2 - 2\varepsilon\mu > 0$ if $\varepsilon \leq 1/5$. Therefore, for $\varepsilon \leq 1/5$, we obtain

$$\frac{S}{H} \leq \frac{2\mu + 2\varepsilon\mu}{\mu/2 - 2\varepsilon\mu} = \frac{4(1+\varepsilon)}{1 - 4\varepsilon} \leq 24. \quad (4)$$

Hence, it holds that $\text{OPT} - S \leq 48\varepsilon v_{\max}$ by (3) and (4), which completes the proof. $\qquad\square$

This lemma shows that MainProcedure is an additive PTAS for Identical Additive NSW under the assumption that $v_{\max} < \mu$.

It is worth noting that MainProcedure is not only an additive PTAS, but also a PTAS in the conventional sense.

Lemma 6. *Assume that* $v_{\max} < \mu$ *and* $0 < \varepsilon \leq 1/5$. *Let* π *be the allocation returned by* MainProcedure *and let* OPT *be the optimal value. Then, it holds that* $f(\pi) \geq \frac{\text{OPT}}{1+20\varepsilon}$.

Proof. Let $S = f(\pi^\tau)$ be the value as in the proof of Lemma 5. According to inequality (2), we obtain

$$\frac{\text{OPT}}{S} \leq 1 + \frac{2\varepsilon v_{\max}}{H} \qquad\qquad \text{(by (2))}$$

$$\leq 1 + \frac{4\varepsilon v_{\max}}{\mu(1 - 4\varepsilon)} \qquad\qquad \text{(by } H \geq \mu/2 - 2\varepsilon\mu)$$

$$\leq 1 + \frac{4\varepsilon}{1 - 4\varepsilon} \qquad\qquad \text{(by } v_{\max} < \mu)$$

$$\leq 1 + 20\varepsilon. \qquad\qquad \text{(by } 0 < \varepsilon \leq 1/5)$$

Since $f(\pi) \geq S$, this shows that $f(\pi) \geq \text{OPT}/(1 + 20\varepsilon)$. $\qquad\square$

3.2   Approximation Performance of MaxNashWelfare

We have already seen in the previous subsection that MainProcedure is a PTAS and an additive PTAS for Identical Additive NSW under the assumption that $v_{\max} < \mu$. In this subsection, we analyze the effect of Preprocessing and show that MaxNashWelfare is a PTAS and an additive PTAS. We first show that MaxNashWelfare is a PTAS in the conventional sense.

Proposition 2. *Let* $I = (\mathcal{A}, \mathcal{G}, \mathbf{v})$ *be an instance of* Identical Additive NSW *and suppose that* $0 < \varepsilon \leq 1/5$. *Let* π *be the allocation returned by* MaxNashWelfare *and let* OPT *be the optimal value. Then, it holds that* $f(\pi) \geq \frac{\text{OPT}}{1+20\varepsilon}$.

Proof. Let π^* be an optimal allocation. Let $\mathcal{A}_0$ and $\mathcal{G}_0$ be the set of agents and goods removed in Preprocessing respectively. Set $I' = I \setminus (\mathcal{A}_0, \mathcal{G}_0)$. In an optimal solution π^*, for each good $j \in \mathcal{G}_0$ there exists an agent i that satisfies $\pi_i^* = \{j\}$ by Lemma 1. By rearranging the agents and the goods appropriately, we can assume that $\pi_i^* = \pi_i$ for each $i \in \mathcal{A}_0$. Then the following holds:

$$\frac{\text{OPT}}{f(\pi)} = \left(\prod_{i \in \mathcal{A}} \frac{v(\pi_i^*)}{v(\pi_i)} \right)^{1/n} = \left(\prod_{i \in \mathcal{A} \setminus \mathcal{A}_0} \frac{v(\pi_i^*)}{v(\pi_i)} \right)^{1/n}.$$

Let $A(I')$ be the objective function value of the solution returned by MainProcedure for instance I', and let $\text{OPT}(I')$ be the optimal value of instance I'. Set $k = |\mathcal{A}_0|$. Then, we obtain

$$\left(\prod_{i \in \mathcal{A} \setminus \mathcal{A}_0} \frac{v(\pi_i^*)}{v(\pi_i)} \right)^{1/n} = \left(\frac{\text{OPT}(I')}{A(I')} \right)^{(n-k)/n} \leq (1 + 20\varepsilon)^{(n-k)/n} \leq 1 + 20\varepsilon$$

by Lemma 6, which completes the proof.   □

The proof of Proposition 2 is easy, because the multiplicative error is not amplified when we recover the agents and goods removed in Preprocessing. However, this property does not hold when we consider the additive error, which makes the situation harder. Nevertheless, we show that the additive error of $f(\pi)$ is bounded by $O(\varepsilon v_{\max})$ with the aid of the mean-value theorem for differentiable functions.

Proposition 3. *Let* $I = (\mathcal{A}, \mathcal{G}, \mathbf{v})$ *be an instance of* Identical Additive NSW *and suppose that* $0 < \varepsilon \leq 1/192$. *Let* π *be the allocation returned by* MaxNashWelfare *and let* OPT *be the optimal value. Then, it holds that* $f(\pi) \geq \text{OPT} - 192\varepsilon v_{\max}$.

Proof. Let $A = f(\pi)$ and let π^* be an optimal allocation. Let $\mathcal{A}_0$ and $\mathcal{G}_0$ be the set of agents and goods removed in Preprocessing, respectively. Set $k = |\mathcal{A}_0|$. In an optimal solution π^*, for each good $j \in \mathcal{G}_0$ there exists an agent i that satisfies $\pi_i^* = \{j\}$ by Lemma 1. By rearranging the agents and the goods appropriately, we can assume that $\pi_i^* = \pi_i$ for each $i \in \mathcal{A}_0$. Set $I' = I \setminus (\mathcal{A}_0, \mathcal{G}_0)$. Let $A(I')$ be the objective function value of the solution returned by MAIN-PROCEDURE for instance I', and let $\mathrm{OPT}(I')$ be the optimal value of instance I'.

We define a function $g: \mathbb{R} \to \mathbb{R}$ as

$$g(x) = \left(\prod_{j \in \mathcal{G}_0} v_j \right)^{1/n} x^{(n-k)/n}.$$

By using g, the expression to be evaluated can be written as follows:

$$\mathrm{OPT} - A = g(\mathrm{OPT}(I')) - g(A(I')). \tag{5}$$

Since g is differentiable, by the mean value theorem, there exists a real number c such that

$$A(I') \le c \le \mathrm{OPT}(I'), \tag{6}$$
$$g(\mathrm{OPT}(I')) - g(A(I')) = (\mathrm{OPT}(I') - A(I'))g'(c). \tag{7}$$

By (5), (7), and Lemma 5, we obtain

$$\mathrm{OPT} - A \le 48\varepsilon v_{\max}(I')g'(c), \tag{8}$$

where $v_{\max}(I') = \max_{j \in \mathcal{G} \setminus \mathcal{G}_0} v_j$. Therefore, all we need to do is to evaluate $g'(c)$. For this purpose, we first give a lower bound on c as follows:

$$
\begin{aligned}
c &\ge A(I') & \text{(by (6))} \\
&\ge \mathrm{OPT}(I') - 48\varepsilon v_{\max}(I') & \text{(by Lemma 5)} \\
&= \left(\prod_{i \in \mathcal{A} \setminus \mathcal{A}_0} v(\pi_i^*) \right)^{1/(n-k)} - 48\varepsilon v_{\max}(I') \\
&\ge \left(\prod_{i \in \mathcal{A} \setminus \mathcal{A}_0} \frac{\mu(I')}{2} \right)^{1/(n-k)} - 48\varepsilon v_{\max}(I') & \text{(by Lemma 4)} \\
&\ge \frac{\mu(I')}{2} - 48\varepsilon v_{\max}(I') & \text{(by } |\mathcal{A} \setminus \mathcal{A}_0| = n - k) \\
&\ge \left(\frac{1}{2} - 48\varepsilon \right) v_{\max}(I'). & \text{(by } \mu(I') > v_{\max}(I')) \quad (9)
\end{aligned}
$$

By using this inequality, we obtain the following upper bound on $g'(c)$:

$$g'(c) = \frac{n-k}{n} \left(\prod_{j \in \mathcal{G}_0} v_j \right)^{1/n} \left(\frac{1}{c} \right)^{k/n}$$

$$\leq \left(\frac{v_{\max}}{c} \right)^{k/n} \qquad \text{(by } |\mathcal{G}_0| = k \text{ and } v_j \leq v_{\max})$$

$$\leq \left(\frac{v_{\max}}{(1/2 - 48\varepsilon)v_{\max}(I')} \right)^{k/n} \qquad \text{(by (9))}$$

$$\leq \left(\frac{4v_{\max}}{v_{\max}(I')} \right)^{k/n} \qquad \text{(by } 0 < \varepsilon \leq 1/192)$$

$$\leq \frac{4v_{\max}}{v_{\max}(I')}. \qquad \text{(by } v_{\max} \geq v_{\max}(I')) \qquad (10)$$

Hence, we obtain $\mathrm{OPT} - A \leq 192\varepsilon v_{\max}$ from (8) and (10), which completes the proof. $\qquad\square$

By setting ε appropriately, Theorem 1 follows from Proposition 3.

Proof (Proof of Theorem 1). Suppose that we are given an instance of IDEN-TICAL ADDITIVE NSW and a real value $\varepsilon > 0$. Define ε' as the largest value subject to $1/\varepsilon'$ is an integer and $\varepsilon' \leq \min(1/192, \varepsilon/192)$. That is, $\varepsilon' := 1/\lceil \max(192, 192/\varepsilon) \rceil$. Then, apply MAXNASHWELFARE in which ε is replaced with ε'. Since $0 < \varepsilon' \leq 1/192$, MAXNASHWELFARE returns an allocation π such that $f(\pi) \geq \mathrm{OPT} - 192\varepsilon' v_{\max} \geq \mathrm{OPT} - \varepsilon v_{\max}$ by Proposition 3. As described in Sect. 2, the running time of MAXNASHWELFARE is $(nm/\varepsilon')^{O(1/\varepsilon')}$, which can be rewritten as $(nm/\varepsilon)^{O(1/\varepsilon)}$. This completes the proof of Theorem 1. $\qquad\square$

References

1. Alon, N., Azar, Y., Woeginger, G.J., Yadid, T.: Approximation schemes for scheduling on parallel machines. J. Sched. 1(1), 55–66 (1998)
2. Anari, N., Mai, T., Gharan, S.O., Vazirani, V.V.: Nash social welfare for indivisible items under separable, piecewise-linear concave utilities. In: Proceedings of the Twenty-Ninth Annual ACM-SIAM Symposium on Discrete Algorithms (SODA 2018), pp. 2274–2290 (2018)
3. Barman, S., Krishnamurthy, S.K., Vaish, R.: Finding fair and efficient allocations. In: Proceedings of the 2018 ACM Conference on Economics and Computation, pp. 557–574 (2018)
4. Barman, S., Krishnamurthy, S.K., Vaish, R.: Greedy algorithms for maximizing Nash social welfare. In: Proceedings of the 17th International Conference on Autonomous Agents and MultiAgent Systems, pp. 7–13 (2018)
5. Buchem, M., Rohwedder, L., Vredeveld, T., Wiese, A.: Additive approximation schemes for load balancing problems. arXiv preprint arXiv:2007.09333 (2020)

6. Buchem, M., Rohwedder, L., Vredeveld, T., Wiese, A.: Additive approximation schemes for load balancing problems. In: Proceedings of the 48th International Colloquium on Automata, Languages, and Programming (ICALP 2021), pp. 42:1–42:17 (2021)
7. Bullen, P.S.: Handbook of Means and Their Inequalities, vol. 560. Springer, Heidelberg (2013)
8. Caragiannis, I., Kurokawa, D., Moulin, H., Procaccia, A.D., Shah, N., Wang, J.: The unreasonable fairness of maximum Nash welfare. ACM Trans. Econ. Comput. (TEAC) **7**(3), 1–32 (2019)
9. Chaudhury, B.R., Cheung, Y.K., Garg, J., Garg, N., Hoefer, M., Mehlhorn, K.: On fair division for indivisible items. In: Proceedings of the 38th IARCS Annual Conference on Foundations of Software Technology and Theoretical Computer Science (FSTTCS 2018), pp. 25:1–25:17 (2018)
10. Eisenberg, E., Gale, D.: Consensus of subjective probabilities: the pari-mutuel method. Ann. Math. Stat. **30**(1), 165–168 (1959)
11. Garg, J., Hoefer, M., Mehlhorn, K.: Approximating the Nash social welfare with budget-additive valuations. In: Proceedings of the Twenty-Ninth Annual ACM-SIAM Symposium on Discrete Algorithms (SODA 2018), pp. 2326–2340 (2018)
12. Garg, J., Husić, E., Végh, L.A.: Approximating Nash social welfare under Rado valuations. In: Proceedings of the 53rd Annual ACM SIGACT Symposium on Theory of Computing (STOC 2021), pp. 1412–1425 (2021)
13. Ibarra, O.H., Kim, C.E.: Fast approximation algorithms for the knapsack and sum of subset problems. J. ACM **22**(4), 463–468 (1975)
14. Inoue, A., Kobayashi, Y.: An additive approximation scheme for the Nash social welfare maximization with identical additive valuations. arXiv:2201.01419 (2022)
15. Kaneko, M., Nakamura, K.: The Nash social welfare function. Econometrica **47**(2), 423–435 (1979)
16. Kelly, F.: Charging and rate control for elastic traffic. Eur. Trans. Telecommun. **8**(1), 33–37 (1997)
17. Lee, E.: APX-hardness of maximizing Nash social welfare with indivisible items. Inf. Process. Lett. **122**, 17–20 (2017)
18. Li, W., Vondrák, J.: A constant-factor approximation algorithm for Nash social welfare with submodular valuations. arXiv preprint arXiv:2103.10536 (2021)
19. Li, W., Vondrák, J.: Estimating the Nash social welfare for coverage and other submodular valuations. In: Proceedings of the Thirty-Second Annual ACM-SIAM Symposium on Discrete Algorithms (SODA 2021), pp. 1119–1130 (2021)
20. Moulin, H.: Fair Division and Collective Welfare. MIT Press, Cambridge (2003)
21. Nash, J.F.: The bargaining problem. Econometrica **18**(2), 155–162 (1950)
22. Nguyen, T.T., Rothe, J.: Minimizing envy and maximizing average Nash social welfare in the allocation of indivisible goods. Discret. Appl. Math. **179**, 54–68 (2014)
23. Varian, H.R.: Equity, envy, and efficiency. J. Econ. Theory **9**(1), 63–91 (1974)

Controlling Weighted Voting Games by Deleting or Adding Players with or Without Changing the Quota

Joanna Kaczmarek[✉] and Jörg Rothe

Heinrich-Heine-Universität Düsseldorf, Düsseldorf, Germany
{Joanna.Kaczmarek,rothe}@hhu.de

Abstract. Weighted voting games are a well-studied class of succinct simple games that can be used to model collective decision-making in, e.g., legislative bodies such as parliaments and shareholder voting. Power indices [5,10,23,28] are used to measure the influence of players in weighted voting games. In such games, it has been studied how a distinguished player's power can be changed, e.g., by merging or splitting players (the latter is a.k.a. false-name manipulation) [2,24], by changing the quota [31], or via structural control by adding or deleting players [25]. We continue the work on the structural control initiated by Rey and Rothe [25] by solving some of their open problems. In addition, we also modify their model to a more realistic setting in which the quota is indirectly changed during the addition or deletion of players (in a different sense than that of Zuckerman et al. [31] who manipulate the quota directly without changing players' set), and we study the corresponding problems in terms of their computational complexity.

1 Introduction

Weighted voting games are an important class of compactly representable simple games and have been thoroughly studied in cooperative game theory (see, e.g., the textbooks [9,22,29] and the book chapter [11]). Most crucially, WVGs have been analyzed in terms of power indices that describe how much influence a player has in a game. Well-known power indices are the *normalized Penrose-Banzhaf index* due to Penrose [23] and Banzhaf [5], the *probabilistic Penrose-Banzhaf index* due to Dubey and Shubik [10], and the *Shapley-Shubik index* due to Shapley and Shubik [28]. We will focus on the latter two. There are many applications of WVGs. They can be used for collective decision-making in legislative bodies (e.g., in parliamentary voting), in order to analyze the voting structures of the European Union Council of Ministers and the International Monetary Fund [14,19], they are applied in joint stock companies where each shareholder gets votes in proportion to the ownership of a stock and in automated stock-trading systems [1,15], and widely used in many practical application areas beyond social choice theory and game theory. Just as for voting rules in computational social choice [8,12,27], for judgment aggregation procedures [7], and for algorithms and protocols in fair division, strategic behavior has attracted much attention for WVGs. Bachrach and Elkind [4] were the first to study the complexity of *false-name*

© Springer Nature Switzerland AG 2022
C. Bazgan and H. Fernau (Eds.): IWOCA 2022, LNCS 13270, pp. 355–368, 2022.
https://doi.org/10.1007/978-3-031-06678-8_26

manipulation (i.e., changing the players' power indices by splitting a player into several players and distributing the weight among them) or by merging several players into one. These problems have then been further analyzed by Aziz et al. [2,3], Faliszewski and Hemaspaandra [13], and Rey and Rothe [24]. Zuckerman et al. [31] studied the problem of influencing power indices in WVGs by *manipulating the quota.*

Inspired by electoral control of voting rules [6,18], Rey and Rothe [25] introduced problems of *structural control by adding players to and by deleting players from WVGs* and studied them in terms of their computational complexity. Continuing their analysis, in Sect. 3 we solve some of their open problems regarding control by deleting players from WVGs, also fixing a minor flaw in their paper [25] for bounds of how much the Shapley-Shubik index can change by deleting players.

In Sect. 4, we modify the model presented by Rey and Rothe [25] in a natural way: While they assume that the quota remains the same even though players have been added to or deleted from a weighted voting game, we will assume that the quota will change accordingly in the modified game, i.e., the quota will be a fraction of the players' total weight. This way of modifying the quota, however, differs from the model of Zuckerman et al. [31] who manipulate the quota directly. We define the corresponding problems of control by adding or deleting players *with changing the quota,* with the goal to increase, to decrease, or to maintain a distinguished player's power index. We study these problems for the probabilistic Penrose-Banzhaf index and the Shapley-Shubik index in terms of their computational complexity.

We conclude in Sect. 5 and mention some open problems for future work. All proofs except one are omitted due to space limitations.

2   Preliminaries

In this section, we provide the needed notions from cooperative game theory and computational complexity theory.

Definition 1. *A* coalitional game *is a pair* $G = (N, v)$, *where* $N = \{1, 2, \ldots, n\}$ *is a set of players and* $v : 2^N \to \mathbb{R}$, *with* $v(\emptyset) = 0$, *is a characteristic function that assigns a payoff to every coalition of players (i.e., subset of N). $G = (N, v)$ is called* simple *if* $v(C) \in \{0, 1\}$ *for every coalition* $C \subseteq N$ *and v is* monotonic, *i.e.,* $v(A) \leq v(B)$ *whenever* $A \subseteq B \subseteq N$.

We focus on a special class of simple coalitional games: weighted voting games (WVGs, for short).

Definition 2. *A WVG* $G = (w_1, \ldots, w_n; q)$ *is a simple coalitional game that consists of a quota* $q \in \mathbb{R}_+$ *and weights* $w_i \in \mathbb{R}_+$, *where* w_i *is the i-th player's weight,* $i \in N$. *For each coalition* $S \subseteq N$, *letting* $w_S = \sum_{i \in S} w_i$, *S wins if* $w_S \geq q$, *and loses otherwise:*

$$v(S) = \begin{cases} 1 \ \textit{if } w_S \geq q, \\ 0 \ \textit{otherwise.} \end{cases}$$

In Sect. 4, we will use the quota depending on the players' total weight as $q = r \sum_{i \in N} w_i$ for a parameter $r \in (0, 1]$.

We now define two of the most popular power indices that can be used to measure a player's significance in a simple game, the *probabilistic Penrose-Banzhaf index* (introduced by Dubey and Shapley [10] as an alternative to the normalized Penrose-Banzhaf index that was originally introduced by Penrose [23] and later re-invented by Banzhaf [5]) and the *Shapley-Shubik index* due to Shapley and Shubik [28].

Definition 3. *Let n be the number of players in a simple game $G = (N, v)$ and let $i \in N$ be a player. The* probabilistic Penrose-Banzhaf index of player i in G is defined by

$$\beta(G, i) = \frac{\sum_{S \subseteq N \setminus \{i\}} (v(S \cup \{i\}) - v(S))}{2^{n-1}}.$$

The Shapley-Shubik index of player i in G is defined by

$$\varphi(G, i) = \frac{\sum_{S \subseteq N \setminus \{i\}} |S|!(n-1-|S|)!(v(S \cup \{i\}) - v(S))}{n!}.$$

If $v(S \cup \{i\}) - v(S) = 1$, we say that i is pivotal *for S. If a player is pivotal for all coalitions, we call it a* dictator, *and if it is not pivotal for any set, we call it a* dummy *player.*

We will study structural control by adding and deleting players in WVGs, and we adopt the notation of Rey and Rothe [25] who introduced these concepts. For control by adding players, let $G = (w_1, \ldots, w_n; q)$ be a given WVG and $N = \{1, \ldots, n\}$ and let $M = \{n+1, \ldots, n+m\}$ be a set of m unregistered players with weights $w_{n+1}, \ldots, w_{n+m}$. Adding M to G yields a new WVG that is denoted by $G_{\cup M} = (w_1, \ldots, w_{n+m}; q)$. Similarly, if $M \subseteq N$, deleting M from G yields a new WVG $G_{\setminus M} = (w_{j_1}, \ldots, w_{j_{n-m}}; q)$, where $\{j_1, \ldots, j_{n-m}\} = N \setminus M$. For more background on cooperative game theory, we refer to the books by Chalkiadakis et al. [9], Peleg and Sudhölter [22], and Taylor and Zwicker [29], and to the chapter by Elkind and Rothe [11].

We assume familiarity with the most fundamental notions of computational complexity, in particular with the complexity classes P (deterministic polynomial time), NP (nondeterministic polynomial time), and PP (probabilistic polynomial time). Moreover, we will also use the well-known complexity classes DP (consisting of differences of NP sets, as introduced by Papadimitriou and Yannakakis [21]) and Θ_2^p (a.k.a. $P^{NP[\log]}$, the class of sets accepted by a P algorithm accessing its NP oracle logarithmically often, see [17]). The notion of *hardness* for these classes is based on the *polynomial-time many-one reducibility*: $X \leq_m^p Y$ if there is a polynomial-time computable, total function f such that for each input x, $x \in X$ if and only if $f(x) \in Y$. We refer the reader to the textbooks by Garey and Johnson [16], Papadimitriou [20], and Rothe [26] for more background on complexity theory.

We use the following two well-known NP-complete problems (see, e.g., [16]). In PARTITION, given a set $I = \{1, \ldots, n\}$, a function $a : I \to \mathbb{N} \setminus \{0\}$, $i \mapsto a_i$, such that $\sum_{i=1}^n a_i$ is even, we ask whether there exists a partition of I into two subsets of equal weight, that is, whether there exists a subset $I' \subseteq I$ such that $\sum_{i \in I'} a_i = \sum_{i \in I \setminus I'} a_i$. In SUBSETSUM, we are given a set $I = \{1, \ldots, n\}$, a function $a : I \to \mathbb{N} \setminus \{0\}$, $i \mapsto a_i$, and a positive integer q, and we ask whether there exists a subset $I' \subseteq I$ such that $\sum_{i=I'} a_i = q$.

We also use the following two PP-complete problems that Rey and Rothe [24] used in their work on false-name manipulation in WVGs.

COMPARE-#SUBSETSUM-RR

Given: A set $I = \{1,\ldots,n\}$, a function $a : I \to \mathbb{N} \setminus \{0\}, i \mapsto a_i$, where $\alpha = \sum_{i=1}^{n} a_i$.

Question: Is the number of subsets of I with values summing up to $\frac{\alpha}{2} - 2$ greater than the number of subsets of I with values summing up to $\frac{\alpha}{2} - 1$, i.e., #SUBSETSUM$((a_1,\ldots,a_n), \frac{\alpha}{2} - 2)$ > #SUBSETSUM$((a_1,\ldots,a_n), \frac{\alpha}{2} - 1)$?

COMPARE-#SUBSETSUM-ЯЯ

Given: A set $I = \{1,\ldots,n\}$, a function $a : I \to \mathbb{N} \setminus \{0\}, i \mapsto a_i$, where $\alpha = \sum_{i=1}^{n} a_i$.

Question: Is the number of subsets of I with values summing up to $\frac{\alpha}{2} - 2$ smaller than the number of subsets of I with values summing up to $\frac{\alpha}{2} - 1$, i.e., #SUBSETSUM$((a_1,\ldots,a_n), \frac{\alpha}{2} - 2)$ < #SUBSETSUM$((a_1,\ldots,a_n), \frac{\alpha}{2} - 1)$?

In the NP-complete problem X3C, an input consists of a set of elements $\mathcal{B}$, $|\mathcal{B}| = 3k$ for $k \in \mathbb{N}$, and a family of its three-element subsets $\mathcal{S}$, and the question is whether there exists a subfamily $\mathcal{S}^*$ of $\mathcal{S}$ such that each element from $\mathcal{B}$ is contained in exactly one set in $\mathcal{S}^*$. Faliszewski and Hemaspaandra [13] proved the following useful property about X3C (also using the fact that there exists a reduction from X3C to SUBSETSUM), applied by them and by Rey and Rothe [24] and to be applied here as well later on.

Lemma 1. *Every* X3C *instance* $(\mathcal{B}', \mathcal{S}')$ *can be transformed into an* X3C *instance* $(\mathcal{B}, \mathcal{S})$, *where* $|\mathcal{B}| = 3k$ *and* $|\mathcal{S}| = n$, *such that* $\frac{k}{n} = \frac{2}{3}$ *without changing the number of solutions. Consequently, we can assume that the size of each solution in a* SUBSETSUM *instance is* $\frac{2n}{3}$, *that is, each subsequence summing up to the given quota contains the same number of elements.*

In our proofs, we will also apply the following two lemmas due to Wagner [30].

Lemma 2. *Let A be some* NP-*complete problem and let B be an arbitrary problem. If there exists a polynomial-time computable function f such that, for all input strings x_1 and x_2 for which $x_2 \in A$ implies $x_1 \in A$, we have $(x_1 \in A \land x_2 \notin A) \iff f(x_1, x_2) \in B$, then B is* DP-*hard.*

Lemma 3. *Let A be some* NP-*complete set, and let B be any set. If there exists a polynomial-time computable function g such that, for all $k \geq 1$ and all input strings $x_1, \ldots, x_{2k}$ satisfying $\chi_A(x_1) \geq \cdots \geq \chi_A(x_{2k})$ (where $\chi_A(x_i) = 1$ if $x_i \in A$, and $\chi_A(x_i) = 0$ if $x_i \notin A$), it holds that $|\{i \mid x_i \in A\}|$ is odd $\iff g(x_1, \ldots, x_{2k}) \in B$, then B is* Θ_2^p-*hard.*

We consider the following decision problem introduced by Rey and Rothe [25] for a given power index PI as well as its analogous variants where the goal is to decrease, to nonincrease, to nondecrease, or to maintain a power index by deleting players or by adding players:

CONTROL BY DELETING PLAYERS TO INCREASE PI

Given: A WVG $\mathcal{G}$ with players $N = \{1,\ldots,n\}$, a distinguished player $p \in N$, and a positive integer k.

Question: Can at least one and at most k players $M \subseteq N \setminus \{p\}$ be deleted from $\mathcal{G}$ such that for the new game $\mathcal{G}_{\setminus M}$, it holds that $\mathrm{PI}(\mathcal{G}_{\setminus M}, p) > \mathrm{PI}(\mathcal{G}, p)$?

3 Deleting Players from WVGs Without Changing the Quota

In this section, we consider the model of structural control by (adding or) deleting players where the goal is to increase, to decrease, to nonincrease, to nondecrease, or to maintain a power index, as proposed by Rey and Rothe [25]. First, we will show upper and lower bounds of how much the Penrose-Banzhaf index and the Shapley-Shubik index can change when players are deleted. Then we will study the problems CONTROL BY DELETING PLAYERS TO INCREASE PI and CONTROL BY DELETING PLAYERS TO DECREASE PI in terms of their complexity, solving open problems from their work [25].

3.1 Change of Power Indices by Deleting Players

Rey and Rothe [25] analyzed how deleting players can change the Penrose-Banzhaf and the Shapley-Shubik index, by providing upper and lower bounds for both power indices. Unfortunately, their result on the lower bound of the Shapley-Shubik index is not correct[1] and we fix it in Theorem 1 below (which, for completeness, also contains the correct upper bound for the Shapley-Shubik index and both bounds for the Penrose-Banzhaf index due to Rey and Rothe [25]).

Theorem 1. *After deleting the players of a subset $M \subseteq N \setminus \{i\}$ of size $m \geq 1$ from a WVG G with $n = |N|$ players, the difference between player i's old and new*

1. *Penrose-Banzhaf index is at most $1 - 2^{-m}$ and is at least $-1 + 2^{-m}$ (as shown by Rey and Rothe [25]);*
2. *Shapley-Shubik index is at most $1 - \frac{(n-m+1)!}{2n!}$ and is at least $-1 + \frac{(n-m+1)!}{2n!}$.*

Let us look at the counter-examples for the (wrong) value of $\frac{(n-m-1)!}{2(n-2)!}$ (see Footnote 1) for the difference between a player's old and new Shapley-Shubik power index.

Example 1. Firstly, let us consider the game $G = (2,2,2;2)$ and let 1 be a distinguished player. Then, $\varphi(G, 1) = \frac{1}{3}$ and if we remove two other players from the game, the index will increase to 1, so $\varphi(G, 1) - \varphi(G_{\setminus\{2,3\}}, 1) = -\frac{2}{3}$. The lower bound due to Theorem 1 equals $-1 + \frac{(3-2+1)!}{2\cdot3!} = -\frac{5}{6} < -\frac{2}{3}$, so the difference belongs to the new interval. The old bound would equal $-1 + \frac{(3-2-1)!}{2(3-2)!} = -\frac{1}{2} > -\frac{2}{3}$, so the difference would be outside the interval.

Let us consider now the game $\mathcal{H} = (4,1,1;5)$ and the player 1. Then, $\varphi(\mathcal{H}, 1) = \frac{2}{3}$ and if we remove the other players, the index will decrease to 0, so $\varphi(\mathcal{H}, 1) - \varphi(\mathcal{H}_{\setminus\{2,3\}}, 1) = \frac{2}{3}$. If we consider the upper bound $1 - \frac{(3-2+1)!}{2\cdot3!} = \frac{5}{6}$, the index will belong to the interval. If we assumed that $1 - \frac{(3-2-1)!}{2(3-2)!} = \frac{1}{2}$ is the upper bound, our example would be outside the interval, so we would get a contradiction.

---

[1] Under the assumptions of Theorem 1, their incorrect lower bound of the Shapley-Shubik index [25] is $-1 + \frac{(n-m)!}{2(n-2)!}.$.

The previous theorem describes the bounds of how much the power indices can change depending only on the number of deleted players. In the next theorems, we will see the bounds of changes for a given player which depend not only on the number of deleted players but also on the power indices of the given player and of the deleted players from the initial game. We start with the lower bounds.

Theorem 2. *Let $G = (w_1, \ldots, w_n; q)$ be a WVG with the set of the players N. Let $M \subseteq N \setminus \{i\}$ be a set of players which are going to be deleted and $m = |M|$.*

1. $\beta(G,i) - \beta(G_{\setminus M}, i) \geq \max((1 - 2^m)\beta(G,i), \beta(G,i) - 1)$.
2. $\varphi(G,i) - \varphi(G_{\setminus M}, i) \geq \max((1 - \binom{n}{m})\varphi(G,i), \varphi(G,i) - 1)$.

The following theorem shows how much smaller the power indices can be in new games after deleting players.

Theorem 3. *Let $G = (w_1, \ldots, w_n; q)$ be a WVG with the set of the players N. Let $M \subseteq N \setminus \{i\}$ be a set of players which are going to be deleted and $m = |M|$.*

1. $\beta(G,i) - \beta(G_{\setminus M}, i) \leq \min\left(\beta(G,i), \sum_{j \in M} \beta(G,j) + \frac{(2^m - 1)^2}{2^{n-1}}\right)$.
2. $\varphi(G,i) - \varphi(G_{\setminus M}, i) \leq \min\left(\varphi(G,i), \sum_{j \in M} \varphi(G,j) + \frac{1}{(n-m)!}\right)$.

Proof Sketch. The idea of the proof is to consider the situation where player i shares with the players from M as many coalitions as possible for which i is pivotal but, at the same time, each player from M is not pivotal for the same coalitions as the others. So, while deleting M from the game, we delete as many coalitions counted in i's power index as possible. □

Example 2. Let $G = (4,2,1,1,1; 4)$ be a WVG. We are going to remove the subset $M = \{5\}$ from the set of players. Let us consider player 2, whose old and new Penrose-Banzhaf indices are $\beta(G,2) = \frac{1}{4}$ and $\beta(G_{\setminus M}, 2) = \frac{1}{8}$, so the index decreases by $\frac{1}{8}$. The upper bound from Theorem 1 is $\beta(G,2) - \beta(G_{\setminus M}, 2) \leq 1 - \frac{1}{2} = \frac{1}{2}$ and that from Theorem 3 is $\beta(G,2) - \beta(G_{\setminus M}, 2) \leq \min(\frac{1}{4}, \frac{1}{8} + \frac{1}{16}) = \frac{3}{16}$, so both upper bounds are greater than the actual difference but the second one is more exact. Now, let us consider player 2's old and new Shapley-Shubik index: $\varphi(G,2) = \frac{11}{60}$ and $\varphi(G_{\setminus M}, 2) = \frac{5}{60}$, so it decreases by $\frac{1}{10}$. The upper bound from Theorem 1 is $\varphi(G,2) - \varphi(G_{\setminus M}, 2) \leq 1 - \frac{(5-1+1)!}{2 \cdot 5!} = \frac{1}{2}$ and that from Theorem 3 is $\varphi(G,2) - \varphi(G_{\setminus M}, 2) \leq \min(\frac{11}{60}, \frac{1}{10} + \frac{1}{4!}) = \frac{17}{120}$, which are greater again, but the second one is much closer to the true difference.

3.2 Control by Deleting Players

In this section, we will consider control by deleting players in WVGs, with the goal of increasing or decreasing a distinguished player's power index. Rey and Rothe [25] analyzed these problems in terms of their complexity. While they obtained many results for the case of control by adding players, they left many problems open for control by deleting players. In the next two theorems, we solve two of these open problems: one for increasing the Penrose-Banzhaf index and the other for decreasing the Shapley-Shubik index. In particular, Rey and Rothe [25] showed that the problem of control by deleting a single player to increase a distinguished player's Shapley-Shubik index is NP-hard. We show the following result for the Penrose-Banzhaf index.

Theorem 4. *Control by deleting players to increase a distinguished player's Penrose-Banzhaf index in a WVG is DP-hard.*

Rey and Rothe [25] also showed that the problem of control by deleting a single player to decrease a distinguished player's Penrose-Banzhaf index is coNP-hard and we improve this lower bound to Θ_2^p-hardness with the following theorem.

Theorem 5. *Control by deleting players to decrease a distinguished player's Penrose-Banzhaf index in a WVG is Θ_2^p-hard.*

Proof. Let us define a reduction using the NP-complete problem PARTITION (which we will call A, just as the set from Lemma 3). Let $x_i = (a_{i,1}, \ldots, a_{i,m_i})$ be an instances of PARTITION for $i \in \{1, \ldots, 2n\}$ and $n \in \mathbb{N}$, let $\alpha_i = \sum_{j=1}^{m_i} a_{i,j}$, and let ξ_i be the number of x_i's solutions for PARTITION.

Let $\ell_1, \ldots, \ell_{2n} \in \mathbb{N}$ be chosen such that for all $i \in \{1, \ldots, 2n-1\}$, we have

$$10^{\ell_i} > \sum_{j=1}^{2n-i} \alpha_{2n+1-j} \cdot 10^{i+1},$$

let $y_1 = 1$, $y_2 = 2$, and for all $i \in \{3, \ldots, 2n\}$, let

$$y_i = \begin{cases} \sum_{j=1}^{\frac{i-1}{2}} y_{2j} & \text{if } i \text{ is odd,} \\ y_{i-1} & \text{if } i \text{ is even.} \end{cases}$$

Furthermore, choose $z \in \mathbb{N}$ so that $y_{2n} \cdot z < 10^{\ell_{2n}}$, and define

$$q = \frac{\alpha_1}{2} \cdot 10^{\ell_1} + \frac{\alpha_2}{2} \cdot 10^{\ell_2} + \cdots + \frac{\alpha_{2n}}{2} \cdot 10^{\ell_{2n}} + z + 1$$

and $q' = q - 1$. Consider the weighted voting game

$$G = \Big(1, a_{1,1} \cdot 10^{\ell_1}, \ldots, a_{1,m_1} \cdot 10^{\ell_1}, \ldots, a_{2n,1} \cdot 10^{\ell_{2n}}, \ldots, a_{2n,m_{2n}} \cdot 10^{\ell_{2n}},$$

$$x, r_1, r_2, r_2, \underbrace{r_3, \ldots, r_3}_{y_3}, \ldots, \underbrace{r_{2n-1}, \ldots, r_{2n-1}}_{y_{2n-1}}, \underbrace{r_{2n}, \ldots, r_{2n}}_{y_{2n}}; q\Big)$$

with $\tilde{n} = \sum_{i=1}^{2n}(m_i + y_i) + 2$ players, where $x \in \mathbb{N}$, $x < z$, and for all $i \in \{1, \ldots, 2n\}$,

$$r_i = \begin{cases} q' - (\sum_{j=1}^{i} \frac{\alpha_j}{2} \cdot 10^{\ell_j}) - x & \text{if } i \text{ is odd,} \\ q' - \sum_{j=1}^{i} \frac{\alpha_j}{2} \cdot 10^{\ell_j} & \text{if } i \text{ is even.} \end{cases}$$

Let the first player be the distinguished player and let the deletion limit be $k = 1$. Assume that $\chi_A(x_1) \geq \chi_A(x_2) \geq \cdots \geq \chi_A(x_{2n})$. We will now prove that

$$(\exists i \in \{2, \ldots, \tilde{n}\}) \left[\beta(G, 1) - \beta(G_{\setminus\{i\}}, 1) > 0\right] \iff |\{i \mid \chi_A(x_i) = 1\}| \text{ is odd.}$$

First, suppose that $|\{i \mid \chi_A(x_i) = 1\}|$ is even. If $|\{i \mid \chi_A(x_i) = 1\}| = 0$, then for all $i \in \{2, \ldots, \tilde{n}\}, \beta(\mathcal{G}, 1) = \beta(\mathcal{G}_{\setminus\{i\}}, 1) = 0$. If $|\{i \mid \chi_A(x_i) = 1\}| > 0$, then there exists some i such that $\chi_A(x_{2i}) = 1$ and $\chi_A(x_{2i+1}) = 0$ (or $i = 2n$) and

$$\beta(\mathcal{G}, 1) = \frac{\xi_1 + 2\xi_1\xi_2 + \cdots + y_{2i}\xi_1 \cdots \xi_{2i}}{2^{\tilde{n}-1}}.$$

If we delete any player j with weight $a_k^j \cdot 10^{\ell_j}$ or r_j for $j > 2i$, then the index will increase:

$$\beta(\mathcal{G}_{\setminus\{j\}}, 1) = \frac{\xi_1 + 2\xi_1\xi_2 + \cdots + y_{2i}\xi_1 \cdots \xi_{2i}}{2^{\tilde{n}-2}}.$$

If we delete any player j with weight $a_k^j \cdot 10^{\ell_j}$ for $j \le 2i$, then

$$
\begin{aligned}
\beta(\mathcal{G}_{\setminus\{j\}}, 1) &= \frac{\xi_1 + \cdots + y_j\xi_1 \cdots \frac{\xi_j}{2} + \cdots + y_{2i}\xi_1 \cdots \frac{\xi_j}{2} \cdots \xi_{2i}}{2^{\tilde{n}-2}} \\
&= \frac{\xi_1 + \cdots + y_{j-1}\xi_1 \cdots \xi_{j-1}}{2^{\tilde{n}-2}} + \frac{y_j\xi_1 \cdots \xi_j + \cdots + y_{2i}\xi_1 \cdots \xi_{2i}}{2^{\tilde{n}-1}} \ge \beta(\mathcal{G}, 1).
\end{aligned}
$$

If we remove any player j with weight r_j for $j \le 2i$, then

$$\beta(\mathcal{G}_{\setminus\{j\}}, 1) = \frac{\xi_1 + \cdots + (y_j - 1)\xi_1 \cdots \xi_j + \cdots + y_{2i}\xi_1 \cdots \xi_{2i}}{2^{\tilde{n}-2}},$$

so the index does not decrease because $2y_j - 2 \ge y_j$ for $j \ge 2$, as $y_j \ge 2$ and $2\xi_1\xi_2 > \xi_1$. Finally, if we delete the player with weight x, we have

$$
\begin{aligned}
\beta(\mathcal{G}_{\setminus\{\sum_{j=1}^{2n} m_j + 2\}}, 1) &= \frac{2\xi_1\xi_2 + y_4\xi_1\xi_2\xi_3\xi_4 + \cdots + y_{2i}\xi_1 \cdots \xi_{2i}}{2^{\tilde{n}-2}} \\
&= \frac{2\xi_1\xi_2 + 2\xi_1\xi_2 + \cdots + y_{2i}\xi_1 \cdots \xi_{2i} + y_{2i}\xi_1 \cdots \xi_{2i}}{2^{\tilde{n}-1}} > \beta(\mathcal{G}, 1).
\end{aligned}
$$

Summing up, if $|\{i \mid \chi_A(x_i) = 1\}|$ is even, the Penrose-Banzhaf power index of the first player increases or stays the same after removing a player from the game.

Let us assume now that $|\{i \mid \chi_A(x_i) = 1\}|$ is odd. If $|\{i \mid \chi_A(x_i) = 1\}| = 1$, then $\beta(\mathcal{G}, 1) = \frac{\xi_1}{2^{\tilde{n}-1}}$, and after removing the player with weight x, the index decreases to 0. If $|\{i \mid \chi_A(x_i) = 1\}| > 1$, there exists some i such that $\chi_A(x_{2i-1}) = 1$ and $\chi_A(x_{2i}) = 0$ and

$$\beta(\mathcal{G}, 1) = \frac{\xi_1 + 2\xi_1\xi_2 + \cdots + y_{2i-1}\xi_1 \cdots \xi_{2i-1}}{2^{\tilde{n}-1}}.$$

After removing the player with weight x, we have

$$\beta(\mathcal{G}_{\setminus\{\sum_{j=1}^{2n} m_j + 2\}}, 1) = \frac{2\xi_1\xi_2 + y_4\xi_1\xi_2\xi_3\xi_4 + \cdots + y_{2i-2}\xi_1 \cdots \xi_{2i-2}}{2^{\tilde{n}-2}}$$

and

$$
\begin{aligned}
&\beta(\mathcal{G}, 1) - \beta(\mathcal{G}_{\setminus\{\sum_{j=1}^{2n} m_j + 2\}}, 1) \\
&= \frac{\xi_1 - 2\xi_1\xi_2 + y_3\xi_1\xi_2\xi_3 - \cdots - y_{2i-2}\xi_1 \cdots \xi_{2i-2} + y_{2i-1}\xi_1 \cdots \xi_{2i-1}}{2^{\tilde{n}-1}} \\
&> \frac{\xi_1 + \cdots + y_{2i-1}\xi_1 \cdots \xi_{2i-1} - \sum_{j=1}^{i-1} y_{2j}\xi_1 \cdots \xi_{2i-1}}{2^{\tilde{n}-1}} > 0,
\end{aligned}
$$

since $y_{2i-1} = \sum_{j=1}^{i-1} y_{2j}$. Therefore, if $|\{i \mid \chi_A(x_i) = 1\}|$ is odd, it is possible to decrease the Penrose-Banzhaf index of the first player. ☐

For the Shapley-Shubik, we show NP-hardness for this problem.

Theorem 6. *Control by deleting players to decrease a player's Shapley-Shubik index in a WVG is NP-hard.*

4 A New Model: Control Problems with Changing the Quota

From now on, we define the quota of a WVG depending on the players' total weight. With this assumption, we modify the model of Rey and Rothe [25] in a natural way: While they assume that the quota remains the same after players have been added or deleted, we now assume that the quota will change accordingly in the modified game.

4.1 Change of Power by Adding or Deleting Players with Changing the Quota

As we have already mentioned in the introduction, Zuckerman et al. [31] studied manipulation of the quota in WVGs without any structural changes in the set of players. They presented upper and lower bounds for how much the power index of a single player can change when the quota is manipulated.

Our next two theorems present the bounds in situations when quotas are changed not directly but they change as a consequence of adding or deleting players: Recall that from now on, in a WVG $G = (w_1, \ldots, w_n; q)$, the quota will depend on the players' total weight as $q = r \sum_{i=1}^{n} w_i$ for a parameter $r \in (0, 1]$, thus changing the quota by adding or deleting players. In these cases, the power of a player can change much more extremely than in games where the quota remains the same after our manipulation – for example, a player with no power can become the most powerful one and the other way around.

We start with the case when we add some new players to a WVG. Theorem 7 shows how the power indices can change depending on the number of added players.

Theorem 7. *Let $G = (w_1, \ldots, w_n; q_1)$ be a WVG with set N of players and quota $q_1 = r \sum_{j=1}^{n} w_j$ for some $r \in (0, 1]$. Let M, $m = |M|$, be a set of players that are to be added to the game G. Let $G_{\cup M}$ be the new game with players $N \cup M$ and quota $q_2 = r \sum_{j \in N \cup M} w_j$. Then, for $i \in N$:*

1. $-1 + 2^{-m} \leq \beta(G, i) - \beta(G_{\cup M}, i) \leq 1$,
2. $-1 + \frac{(n+1)!}{2(n+m)!} \leq \varphi(G, i) - \varphi(G_{\cup M}, i) \leq 1$.

Interestingly, it is possible for the strongest player to become a dummy by adding even one new player but it is impossible to turn a dummy into a dictator. The following example shows an extreme change of a player's power in a game.

Example 3. Let $G = (5, 1, 1; 4)$ be a WVG with $r = \frac{4}{7}$. It is easy to see that player 1 is a dictator, i.e., $\beta(G, 1) = \varphi(G, 1) = 1$. Let us add to the game a new player with weight 10. In this way, we get a new game: $G_{\cup\{4\}} = (5, 1, 1, 10; \frac{68}{7})$ and the new quota is equivalent to 10. Therefore, the new player becomes the new dictator in the game $G_{\cup\{4\}}$ and player 1's power indices decrease to 0.

The changes of the power indices by deletion of players were presented by Rey and Rothe [25]. Those changes were derived for the case of the structural manipulation without changing the quota of a game. As we can see in Theorem 8, the Penrose-Banzhaf index and the Shapley-Shubik index can decrease by at most the same value with and without the change of quotas while the indices can increase more when the quota changes.

Theorem 8. *Let* $G = (w_1, \ldots, w_n; q_1)$ *be a WVG with set N of players and quota* $q_1 = r\sum_{j=1}^{n} w_j$ *for some* $r \in (0, 1]$. *Let* $M \subseteq N \setminus \{i\}$, $m = |M|$, *be a set of players that are to be deleted from* G. *Let* $G_{\setminus M}$ *be the new game with players* $N \setminus M$ *and quota* $q_2 = r\sum_{j \in N \setminus M} w_j$. *Then, for* $i \in N$:

1. $-1 \leq \beta(G, i) - \beta(G_{\setminus M}, i) \leq 1 - 2^{-m}$,
2. $-1 \leq \varphi(G, i) - \varphi(G_{\setminus M}, i) \leq 1 - \frac{(n-m+1)!}{2n!}$.

Analogously to control by adding new players to a WVG, it is possible for a dummy player to become a dictator when we delete some other players from a game. We now give an example that illustrates how the power indices can change when we delete some players from a game and the new game has an accordingly changed quota.

Example 4. Let $G = (5, 5, 3, 3, 1, 1; 10)$ be a WVG with $r = \frac{5}{9}$. Let us start with the Penrose-Banzhaf indices of the players: $\beta(G, 1) = \beta(G, 2) = \frac{1}{2}$, $\beta(G, 3) = \beta(G, 4) = \frac{1}{4}$, and $\beta(G, 5) = \beta(G, 6) = \frac{1}{8}$. Now, we are going to create a new game by deleting one player with weight 5 and one player with weight 3, so $G_{\setminus\{1,3\}} = (5, 3, 1, 1; \frac{50}{9})$ with the new quota equivalent to 6. The Penrose-Banzhaf indices in the new game are as follows: $\beta(G_{\setminus\{1,3\}}, 1) = \frac{7}{8}$, $\beta(G_{\setminus\{1,3\}}, 2) = \frac{1}{8}$, and $\beta(G_{\setminus\{1,3\}}, 3) = \beta(G_{\setminus\{1,3\}}, 4) = \frac{1}{8}$. The index of the player with weight 5 has increased by $\frac{3}{8}$ and at the same time the index of the player with weight 3 has decreased by $\frac{1}{8} < 1 - 2^{-2} = \frac{3}{4}$. Finally, although the new quota is smaller than the old one, the Penrose-Banzhaf index of the players with weight 1 is unchanged.

Let us now analyze the Shapley-Shubik indices in these two games. The indices in G are: $\varphi(G, 1) = \varphi(G, 2) = \frac{3}{10}$, $\varphi(G, 3) = \varphi(G, 4) = \frac{2}{15}$, and $\varphi(G, 5) = \varphi(G, 6) = \frac{1}{15}$; and in $G_{\setminus\{1,3\}}$: $\varphi(G_{\setminus\{1,3\}}, 1) = \frac{3}{4}$ and $\varphi(G_{\setminus\{1,3\}}, 2) = \varphi(G_{\setminus\{1,3\}}, 3) = \varphi(G_{\setminus\{1,3\}}, 4) = \frac{1}{12}$. The Shapley-Shubik indices of the player with weight 5 and of the players with weight 1 have increased, whereas the index of the player with weight 3 has decreased.

4.2 Control by Adding or Deleting Players with Changing the Quota

First, we define problems of control by adding and by deleting players *with changing the quota* in WVGs, where the goals are to increase, to decrease, or to maintain a distinguished player's power. Specifically, for a power index PI, we consider the following decision problems that slightly modify the problems introduced and studied by Rey and Rothe [25]:

CONTROL BY ADDING PLAYERS WITH CHANGING QUOTA TO INCREASE PI

Given: A WVG G with players $N = \{1,\dots,n\}$, a quota $r\sum_{i=1}^{n} w_i$ ($r \in (0,1]$), a set M
of unregistered players with weights $w_{n+1},\dots,w_{n+m}$, a distinguished player
$p \in N$, and a positive integer k.

Question: Can at least one and at most k players $M' \subseteq M$ be added to G such that for the
new game $G_{\cup M}$ with the new quota $r\sum_{i\in N\cup M'} w_i$, it holds that $\text{PI}(G_{\cup M},p) >$
$\text{PI}(G,p)$?

CONTROL BY DELETING PLAYERS WITH CHANGING QUOTA TO INCREASE PI

Given: A WVG G with players $N = \{1,\dots,n\}$, a quota $r\sum_{i=1}^{n} w_i$ ($r \in (0,1]$), a distin-
guished player $p \in N$, and a positive integer $k < |N|$.

Question: Can at least one and at most k players $M \subseteq N \setminus \{p\}$ be deleted from G such
that for the new game $G_{\setminus M}$ with the new quota $r\sum_{i\in N\setminus M'} w_i$, it holds that
$\text{PI}(G_{\setminus M},p) > \text{PI}(G,p)$?

Analogously, we consider decreasing and maintaining a distinguished player's
power, in relation to the original game.

As one can assume, an additionally varying parameter will not make the decision problems easier: The problems with changing quotas caused by structural control
remain hard when the original problems were hard. Table 1 presents a summary of our
complexity results. We now state and prove one of our results from Table 1.

Table 1. Overview of complexity results of control problems in WVGs with quota change with
respect to the Shapley-Shubik (SSI) and the probabilistic Penrose-Banzhaf (PBI) index.

Goal	Control by adding a player	Control by deleting a player
Decrease PI	PP-hard (PBI & SSI)	DP-hard (PBI)
		NP-hard (SSI)
Increase PI	PP-hard (PBI & SSI)	DP-hard (PBI)
		NP-hard (SSI)
Maintain PI	coNP-hard (PBI & SSI)	coNP-hard (PBI & SSI)

Theorem 9. *For both the Penrose-Banzhaf and the Shapley-Shubik index, control by
adding players to decrease and to increase a distinguished player's power index in a
WVG with changing the quota is* PP-*hard.*

Proof.   We show PP-hardness of control to decrease the two indices by means of a
reduction from the COMPARE-#SUBSETSUM-RR (PP-hardness of control to increase
these indices can be proven analogously with exactly the same reduction but starting
from COMPARE-#SUBSETSUM-ЯЯ instead).

Let $(a_1,\dots,a_n)$ be a COMPARE-#SUBSETSUM-RR instance with $\alpha = \sum_{i=1}^{n} a_i$. Let
ξ_1 and ξ_2, respectively, be the number of SUBSETSUM solutions for $((a_1,\dots,a_n),\frac{\alpha}{2}-1)$

and $((a_1, \ldots, a_n), \frac{\alpha}{2} - 2)$, respectively. Now, construct the control problem instance consisting of a game $G = (1, a_1, \ldots, a_n; \frac{\alpha}{2} - 1)$ with $n + 1$ players, $r = \frac{\frac{\alpha}{2} - 1}{\alpha + 1}$, and distinguished player $p = 1$. Its power indices are $\beta(G, 1) = \frac{\xi_2}{2^n} = \frac{2\xi_2}{2^{n+1}}$ and, using Lemma 1, we can assume that each coalition for which 1 is pivotal has the same size, so $\varphi(G, 1) = \xi_2 \frac{t!(n-t)!}{(n+1)!}$. Let the addition limit be $k = 1$, and let $n + 2$ be the new player with weight $w_{n+2} = 1$. So, the quota in the new game after adding the player $n + 2$ is equivalent to $\frac{\alpha}{2}$, since all players' weights are integers.

We will show that $\mathrm{PI}(G_{\cup\{n+2\}}, 1) - \mathrm{PI}(G, 1) < 0 \iff \xi_1 < \xi_2$.

Assume that $\xi_1 < \xi_2$. Then, after adding the new player, the indices will change to $\beta(G_{\cup\{n+2\}}, 1) = \frac{\xi_1 + \xi_2}{2^{n+1}} < \frac{2\xi_2}{2^{n+1}} = \beta(G, 1)$ and $\varphi(G_{\cup\{n+2\}}, 1) = \xi_1 \frac{(t+1)!(n-t)!}{(n+2)!} + \xi_2 \frac{t!(n-t+1!)}{(n+2)!} < \xi_2 \frac{t!(n-t)!}{(n+2)!}(t+1+n-t+1) = \varphi(G, 1)$, so they both decrease.

Conversely, assume now that $\xi_1 \geq \xi_2$. Then we have $\beta(G_{\cup\{n+2\}}, 1) \geq \beta(G, 1)$ and $\varphi(G_{\cup\{n+2\}}, 1) \geq \varphi(G, 1)$. So both power indices do not decrease.   ☐

Rey and Rothe [25] presented also the upper bound for the problems which they were studying. Exactly the same argumentation[2] is valid also for weighted voting games with changing the quota. Therefore, we get the following upper bound.

Remark 1. Control by adding and deleting players to decrease, to increase, and to maintain both Penrose-Banzhaf power index and Shapley-Shubik power index in a weighted voting game with changing quota is in $\mathrm{NP^{PP}}$.

5   Conclusions and Future Research

We have continued the work on structural control by adding or deleting players in WVGs initiated by Rey and Rothe [25]. In particular, we have solved two of their open problems and have fixed a minor flaw for a lower bound of how much the Shapley-Shubik index can change by deleting players. We have also modified their model in a natural way by making the quota of a new WVG resulting from adding or deleting players dependent on the total weight of the players and have initiated the complexity analysis in this model. Still, many problems remain open for future work. For example, it would be interesting to study the goals of *nonincreasing* or *nondecreasing* a distinguished player's power in our new model. Also, many of the existing complexity results provide only lower bounds – can we find matching upper bounds or can we raise these lower bounds, for example to PP-hardness as it was done by Rey and Rothe [25] for some of their problems and as we succeeded to do for some of our problems in the proof of Theorem 9?

Acknowledgments. We thank the reviewers for helpful comments. This work was supported in part by Deutsche Forschungsgemeinschaft under grant RO 1202/21-1.

---

[2] The result comes from the fact that computing the numerator of the Penrose-Banzhaf index is #P-parsimonious-complete and computing the numerator of the Shapley-Shubik index is #P-many-one-complete.

References

1. Arcaini, G., Gambarelli, G.: Algorithm for automatic computation of the power variations in share trading. Calcolo **23**(1), 13–19 (1986)
2. Aziz, H., Bachrach, Y., Elkind, E., Paterson, M.: False-name manipulations in weighted voting games. J. Artif. Intell. Res. **40**, 57–93 (2011)
3. Aziz, H., Paterson, M.: False name manipulations in weighted voting games: splitting, merging and annexation. In: Proceedings of the 8th International Conference on Autonomous Agents and Multiagent Systems, pp. 409–416. IFAAMAS, May 2009
4. Bachrach, Y., Elkind, E.: Divide and conquer: false-name manipulations in weighted voting games. In: Proceedings of the 7th International Conference on Autonomous Agents and Multiagent Systems, pp. 975–982. IFAAMAS, May 2008
5. Banzhaf, J., III.: Weighted voting doesn't work: a mathematical analysis. Rutgers Law Rev. **19**, 317–343 (1965)
6. Bartholdi, J., III., Tovey, C., Trick, M.: How hard is it to control an election? Math. Comput. Model. **16**(8/9), 27–40 (1992)
7. Baumeister, D., Rothe, J., Selker, A.: Strategic behavior in judgment aggregation. In: Endriss, U. (ed.) Trends in Computational Social Choice, chap. 8, pp. 145–168. AI Access Foundation (2017)
8. Brandt, F., Conitzer, V., Endriss, U., Lang, J., Procaccia, A. (eds.): Handbook of Computational Social Choice. Cambridge University Press (2016)
9. Chalkiadakis, G., Elkind, E., Wooldridge, M.: Computational Aspects of Cooperative Game Theory. Synthesis Lectures on Artificial Intelligence and Machine Learning. Morgan and Claypool Publishers (2011)
10. Dubey, P., Shapley, L.: Mathematical properties of the Banzhaf power index. Math. Oper. Res. **4**(2), 99–131 (1979)
11. Elkind, E., Rothe, J.: Cooperative game theory. In: Rothe, J. (ed.) Economics and Computation. STBE, pp. 135–193. Springer, Heidelberg (2016). https://doi.org/10.1007/978-3-662-47904-9_3
12. Endriss, U. (ed.): Trends in Computational Social Choice. AI Access Foundation (2017)
13. Faliszewski, P., Hemaspaandra, L.: The complexity of power-index comparison. Theor. Comput. Sci. **410**(1), 101–107 (2009)
14. Felsenthal, D., Machover, M.: The Treaty of Nice and qualified majority voting. Soc. Choice Welfare **18**(2), 431–464 (2001)
15. Gambarelli, G.: Power indices for political and financial decision making: a review. Ann. Oper. Res. **51**, 163–173 (1994)
16. Garey, M., Johnson, D.: Computers and Intractability: A Guide to the Theory of NP-Completeness. W.H. Freeman and Company (1979)
17. Hemachandra, L.: The strong exponential hierarchy collapses. J. Comput. Syst. Sci. **39**(3), 299–322 (1989)
18. Hemaspaandra, E., Hemaspaandra, L., Rothe, J.: Anyone but him: the complexity of precluding an alternative. Artif. Intell. **171**(5–6), 255–285 (2007)
19. Leech, D.: Voting power in the governance of the international monetary fund. Ann. Oper. Res. **109**(1–4), 375–397 (2002)
20. Papadimitriou, C.: Computational Complexity, 2nd edn. Addison-Wesley, Boston (1995)
21. Papadimitriou, C., Yannakakis, M.: The complexity of facets (and some facets of complexity). J. Comput. Syst. Sci. **28**(2), 244–259 (1984)
22. Peleg, B., Sudhölter, P.: Introduction to the Theory of Cooperative Games. Kluwer Academic Publishers (2003)

23. Penrose, L.: The elementary statistics of majority voting. J. Roy. Stat. Soc. **109**(1), 53–57 (1946)
24. Rey, A., Rothe, J.: False-name manipulation in weighted voting games is hard for probabilistic polynomial time. J. Artif. Intell. Res. **50**, 573–601 (2014)
25. Rey, A., Rothe, J.: Structural control in weighted voting games. B.E. J. Theor. Econ. **18**(2), 1–15 (2018)
26. Rothe, J.: Complexity Theory and Cryptology. An Introduction to Cryptocomplexity. EATCS Texts in Theoretical Computer Science, Springer, Heidelberg (2005)
27. Rothe, J. (ed.): Economics and Computation An Introduction to Algorithmic Game Theory, Computational Social Choice, and Fair Division. Springer Texts in Business and Economics, Springer, Heidelberg (2015)
28. Shapley, L., Shubik, M.: A method of evaluating the distribution of power in a committee system. Am. Polit. Sci. Rev. **48**(3), 787–792 (1954)
29. Taylor, A., Zwicker, W.: Simple Games: Desirability Relations, Trading, Pseudoweightings. Princeton University Press, Princeton (1999)
30. Wagner, K.: More complicated questions about maxima and minima, and some closures of NP. Theor. Comput. Sci. **51**(1–2), 53–80 (1987)
31. Zuckerman, M., Faliszewski, P., Bachrach, Y., Elkind, E.: Manipulating the quota in weighted voting games. Artif. Intell. **180–181**, 1–19 (2012)

Practical Space-Efficient Index
for Structural Pattern Matching

Sung-Hwan Kim[⊠] and Hwan-Gue Cho

Pusan National University, 46241 Busan, South Korea
{sunghwan,hgcho}@pusan.ac.kr

Abstract. In structural pattern matching, two n-length strings X and Y over Σ are said to match, if there exists a one-to-one function $f : \Sigma \to \Sigma$ such that (i) for $0 \le i < n$, $f(X[i]) = Y[i]$ and (ii) for any $x, y \in \Sigma$ whose complements are x' and y', respectively, if $f(x) = y$ then $f(x') = y'$. In this paper, we present a $2n \lg \sigma + 2n + o(n)$-bit index for this problem. Although it does not theoretically achieve the succinctness for a general alphabet, the proposed method is more practical and the space requirement can be smaller than the previous succinct solution especially when σ is small. A source code is available at: https://github.com/sunghwank/spmindex.

Keywords: Compact data structure · String matching · Suffix array · FM-index · LF-mapping

1 Motivation

Structural pattern matching was introduced by Shibuya [14,15] to address a string matching problem on RNA sequences regarding their secondary structure. In this problem, matching of two strings is defined differently from the standard string matching problem (see Sect. 2.1). An encoding method was used to transform the suffixes into a certain form so that indexing the encoded suffixes with a suffix tree can resolve the problem. In order to reduce the space requirement, which is excessively large for the suffix tree-based indexing methods, Beal and Adjeroh [1] proposed the use of a suffix array as well as its construction method. However, indexing an n-length string still requires $\Theta(n \lg n)$ space in bits, which is quite far from $n \lg \sigma$ bits, the space required to represent the string, where σ is the alphabet size.

Recently, Ganguly et al. [5] presented the first succinct data structure for this problem, which requires $n \lg \sigma + \mathcal{O}(n)$ bits of space. Although this method dramatically reduces the space requirement, it relies on several data structures that are theoretically optimal but hard to implement in practice, such as a multiary wavelet tree [3,7,8], and a fully-functional succinct tree supporting constant-time queries [12,13].

This paper is devoted to present a data structure, which is practically implementable as well as efficient in time and space. Comparison with the existing works is shown in Table 1. The proposed index uses $2n \lg \sigma + 2n + o(n)$ bits where

C. Bazgan and H. Fernau (Eds.): IWOCA 2022, LNCS 13270, pp. 369–382, 2022.
https://doi.org/10.1007/978-3-031-06678-8_27

Table 1. Comparison with other works.

Method	Space (in bits)	Query time (counting)
Suffix tree [14,15]	$\Theta(n \lg n)$	$\mathcal{O}(m)$
Suffix array [1]	$\Theta(n \lg n)$	$\mathcal{O}(m + \lg n)$
sBWT [5]	$n \lg \sigma + \mathcal{O}(n)$	$\mathcal{O}(m \lg \sigma)$
Proposed	$2n \lg \sigma + 2n + o(n)$	$\mathcal{O}(m \lg \sigma)$

n is the text length and σ is the alphabet size, and it can count the number of occurrences of an m-length pattern in $\mathcal{O}(m \lg \sigma)$ time. It is also practical in the sense that it uses bitvector dictionaries [2,9], wavelet trees [8] and range maximum query index [4], which have practical implementations available in public software libraries such as sdsl-lite [6].

As mentioned in [5], the main challenge in using the suffix-encoding method described in [14] for space-efficient indexing is that prepending a single character can affect more than one positions in its encoded string. In this paper, we address this issue by transforming a structural string (s-string) into a pointer sequence of double length so that a single prepending operation can affect at most one position, which is a different approach from that of [5]. We develop an index on this transformed pointer sequence using its space-efficient representation. As an overview, the proposed method can be described the following:

1. We represent an n-length s-string as a $(2n)$-length pointer sequence such that pointers at even (odd, resp.) positions refer to the next occurrence of the character (equal-group character, resp.).
2. We construct two suffix arrays by sorting the suffixes starting at even and odd positions separately.
3. The searching procedure is performed by navigating these two suffix arrays alternatingly; the so-called *LF-mapping* of a suffix at one suffix array is defined to be a suffix at the other suffix array.
4. The LF-mappings can be represented using four arrays L_{even}, F_{odd}, L_{odd}, and F_{even}, which can be stored in $2n \lg \sigma + 2n + o(n)$ bits in total; using these arrays, the LF-mappings can be simply computed in $\mathcal{O}(\lg \sigma)$ time.

The remainder of this paper is organized as follows. In Sect. 2, we briefly review some backgrounds needed to develop our proposed method. In Sect. 3, we present a pointer sequence representation for the structural pattern matching problem. Section 4 describes how to organize the proposed index structure, and the searching algorithm is presented in Sect. 5. Section 6 concludes the paper.

2   Preliminaries

2.1   Structural Pattern Matching

We describe the structure pattern matching introduced in [14]. In the original paper, a string consists of two types of characters: (i) static characters for the exact matching, and (ii) parameterized characters for a more sophisticated matching. In this paper, we consider only parameterized characters for brevity. Nevertheless, we emphasize that our proposed method can easily be applied to the original problem.

Let $T[0..n-1]$ be an n-length structural string (s-string) over alphabet $\Sigma = \{0, \cdots, \sigma-1\}$. We use the 0-based index. We have a one-to-one function compl : $\Sigma \to \Sigma$ that represent the association among characters in Σ. For each $x \in \Sigma$, x is associated with its complement $\mathsf{compl}(x) \in \Sigma$. And for any $x, y \in \Sigma$, if $\mathsf{compl}(x) = y$ then $\mathsf{compl}(y) = x$. For simplicity, we assume that the alphabet size $\sigma = |\Sigma|$ is a multiple of 2, and $x \neq \mathsf{compl}(x)$. To represent the relationship defined via $\mathsf{compl}(\cdot)$, we can also use a function $g : \Sigma \to \{0, \cdots, \sigma/2-1\}$ such that for $x, y \in \Sigma$ $g(x) = g(y)$ iff $x = y$ or $x = \mathsf{compl}(y)$. We say x and y such that $g(x) = g(y)$ are equal-group characters. Two s-strings X and Y are said to match if there exists a one-to-one function $f : \Sigma \to \Sigma$ such that (i) for $0 \leq i < n$, $f(X[i]) = Y[i]$ and (ii) for any $x, y \in \Sigma$, if $f(x) = y$ then $f(\mathsf{compl}(x)) = \mathsf{compl}(y)$. For example, let $\Sigma = \{\mathtt{w}, \mathtt{x}, \mathtt{y}, \mathtt{z}\}$ and $g(\mathtt{w}) = g(\mathtt{x})$ and $g(\mathtt{y}) = g(\mathtt{z})$. Then $\mathtt{wyxxwyzw}$ matches $\mathtt{zwyyzwxz}$, while it does not match $\mathtt{yxwwyxzy}$.

2.2   Pointer Sequence Matching

In this subsection, we briefly review the pointer sequence matching described in [10]. Although the description below may be slightly different from that in the original paper in detail, the basic idea is essentially the same.

In the pointer sequence matching problem, a string is a sequence of pointers. Each pointer is either a null pointer or one refers to another element among those in its right-hand side. We represent a null pointer by a symbol ∞. We represent a pointer referring to another element by its length so that the element at position i refers to the element at position $i + X[i]$ if $X[i] \neq \infty$.

Definition 1 (Pointer sequence). *A sequence $X[0..n-1]$ of length n is a pointer sequence if, for $0 \leq i < n$, $X[i] \in \{1, \cdots, n-i-1\} \cup \{\infty\}$.*

With this representation, we say two equal-length pointer sequences *match* if they are exactly the same. To define a pattern matching problem on pointer sequences, we define a substring of a pointer sequence. Taking a substring from a pointer sequence not only copies the target part but also converts pointers going to the outside of the taken part into null pointers.

Definition 2 (Substring of a pointer sequence). *A substring $Y = X[i..j]$ of X from position i to position j is defined as follows: for $0 \leq k \leq j - i$.*

$$Y[k] = \begin{cases} X[i+k] & \text{if } X[i+k] \leq j-i-k, \\ \infty & \text{otherwise.} \end{cases} \tag{1}$$

For indexing a pointer sequence, we transform it into an encoded form, which is a sequence of sets. The encoded sequence $E(X)$ of an n-length pointer sequence X is defined as follows:

$$E(X)[i] = \{1 \le j \le i \mid X[i-j] = j\} \tag{2}$$

An element of an encoded sequence represents the set of elements pointing to it. To define the lexicographical order among encoded sequences, we define the ordering on their elements, which are sets. As we will see, an element of an encoded sequence handled in this paper is either the empty set $\emptyset$ or a singleton $\{x\}$ for some integer x. We define the ordering of sets (which are elements of encoded sequences) as follows: $A < B$ iff (i) $A \neq \emptyset$ and $B = \emptyset$ or (ii) $A = \{a\}$ and $B = \{b\}$ are singletons such that $a < b$.

We can index the set of encoded suffixes in $2n \lg n + 2n + o(n)$ bits although we do not use it directly in this paper. Rather, we develop a more space-efficient representation for the structural pattern matching problem.

Proposition 1 ([10]). *For an n-length pointer sequence, there exists a data structure that uses $2n \lg n + 2n + o(n)$ bits, and can count the number of occurrences of an m-length pattern in $\mathcal{O}(m \lg n)$ time.*

2.3 Building Blocks

The proposed index uses several well-known data structures as its building blocks.

Bitvector. For an n-length bitvector $A[0..n-1]$, a data structure that supports the following operations in $\mathcal{O}(1)$ time can be represented in $n + o(n)$ bits [2,9].

1. $A.\mathrm{rank}_x(i)$: the number of occurrences of x in $A[0..i]$.
2. $A.\mathrm{select}_x(j)$: the position of the j-th occurrence of x on A.

We also define $A.\mathrm{rank}_x(i,j) = A.\mathrm{rank}_x(j) - A.\mathrm{rank}_x(i-1)$.

Wavelet Tree. A wavelet tree of an n-length string $A[0..n-1]$ over an alphabet of size σ is a data structure that supports the following operations in $\mathcal{O}(\lg \sigma)$ time using $n \lg \sigma + o(n)$ bits [3,7,11].

1. $A(i)$: accessing $A[i]$.
2. $A.\mathrm{rank}_x(i,j)$: the number of occurrences of x in $A[i..j]$.
3. $A.\mathrm{rank\_ge}_x(i,j)$: the number of occurrences of characters that are greater than or equal to x in $A[i..j]$.
4. $A.\mathrm{select}_x(j)$: the position of the j-th occurrence of x on A.

Range Maximum Query. A range maximum query (i,j) on array $A[0..n-1]$ is to ask the index of the maximum element among $A[i..j]$, which can be performed in $\mathcal{O}(1)$ time with a $2n + o(n)$-bit data structure [4]:

1. $A.\mathrm{rMq}(i,j) = \arg\max_{i \le k \le j} A[k]$.

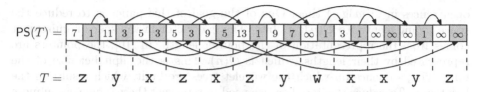

Fig. 1. Pointer sequence representation $\mathsf{PS}(T)$ of $T = $ zyxzxzywxxyz where $\Sigma = \{$w, x, y, z$\}$, $g($w$) = g($x$)$ and $g($y$) = g($z$)$. Each square represents an element of the pointer sequence. The integers inside the squares indicate the pointer lengths and ∞s indicate null pointers. White ones are pointers to its next occurrence, and shaded ones are pointers to the next occurrence of its equal-group character.

3   Pointer Sequence Representation

The basic idea of this paper is to resolve the structural pattern matching problem by solving the matching problem on pointer sequences. In this section, we present a pointer sequence representation for structural pattern matching, which will be used for developing an index structure. More specifically, we represent an n-length s-string as a $(2n)$-length pointer sequence. Each character of an s-string is corresponding to two pointers. One pointer points to the position of the nearest occurrence of the character at the current position, and the following pointer points to the position of the nearest occurrence of its equal-group character. Let $\nu(i)$ and $\mu(i)$ be the distance to the next occurrence of $T[i]$ and $T[i]$'s equal-group character, respectively. More formally, for $0 \leq i < n$,

$$\nu(i) = \min_{j > i}\{j - i : T[j] = T[i]\} \cup \{\infty\} \tag{3}$$

$$\mu(i) = \min_{j > i}\{j - i : g(T[j]) = g(T[i])\} \cup \{\infty\} \tag{4}$$

For an n-length s-string T, we define its pointer sequence representation $\mathsf{PS}(T)$ as follows: for $0 \leq i < 2n$,

$$\mathsf{PS}(T)[i] = \begin{cases} 2\nu(\frac{i}{2}) + 1 & \text{if } i = 0 \mod 2 \\ 2\mu(\frac{i-1}{2}) - 1 & \text{if } i = 1 \mod 2 \end{cases} \tag{5}$$

As an example, the pointer sequence representation of an s-string $T = $ zyxzxzywxxyz with the complement relationship $g($w$) = g($x$)$ and $g($y$) = g($z$)$ is given in Fig. 1. It is easy to see that this pointer sequence representation can be used for solving the structural pattern matching problem.

Observation 1. *For s-strings* $T, P \in \Sigma^*$*, let* $\mathsf{PS}(T)$ *and* $\mathsf{PS}(P)$ *be their pointer sequence representations. For* $0 \leq i \leq |T| - |P|$*,* P *matches* T *at position* i *if and only if* $\mathsf{PS}(P)$ *matches* $\mathsf{PS}(T)$ *at position* $2i$*.*

One can directly apply the indexing method in [10] to this representation to obtain a $4n \lg n + \mathcal{O}(n)$-bit data structure that can compute the number of occurrences in $\mathcal{O}(m \lg n)$ time. One of the goal of this paper is to reduce the space requirement into $\mathcal{O}(n \lg \sigma)$ bits. The $\lg n$ factor in the space requirement comes from the representation of the pointer length. In [10], the pointers are represented by their lengths, which is $\mathcal{O}(n)$. This is the alphabet size of the underlying sequence on which the wavelet trees are built, which results in the $\lg n$ factor. To reduce this into $\lg \sigma$, we need to represent these sequences in more compact values within a range of $\mathcal{O}(\sigma)$.

One may also notice that we do not consider occurrences of $PS(P)$ at odd positions $2i + 1$ on $PS(T)$, despite the fact that there may be (false positive) occurrences of $PS(P)$ at odd positions even if P does not match T there. When we apply the method in [10], it is inevitable to involve an additional filtering method to remove these false positives, which produces a non-negligible overhead. We will address this problem in the next section by constructing suffix arrays for suffixes at even and odd positions separately.

4   Data Structure

In this section, we present a data structure for structural pattern matching. We build two suffix arrays using the corresponding pointer sequence, one for the suffixes starting at even positions (even suffixes), the other one for those starting at odd positions (odd suffixes). Then we define integer arrays that will be used for the searching algorithm we will describe in the next section.

4.1   Suffix Arrays

For the pointer sequence $PS(T)$ of an n-length s-string T, let $\mathcal{S}_{even} = \{PS(T)[2i..] : 0 \leq i \leq n\}$ be the set of the suffixes of $PS(T)$ that start at even positions; note that $\mathcal{S}_{even}$ includes the empty string $\epsilon = PS(T)[2n..]$, which acts as the termination symbol as usually assumed in many other string indexing methods. We define the suffix array SA_{even} for the suffixes $\mathcal{S}_{even}$ using their encoded form; i.e. $SA_{even}(i) = j$ iff there are i encoded suffixes in $\mathcal{S}_{even}$ that are smaller than $E(PS(T)[2j..])$. Similarly, we define SA_{odd} from the set $\mathcal{S}_{odd} = \{PS(T)[2i + 1..] : 0 \leq i \leq n\}$ of suffixes of $PS(T)$ that start at odd positions; note that $\mathcal{S}_{odd}$ also contains the empty string as $\mathcal{S}_{even}$ does. We also define the inverse function of the two suffix arrays such that $SA_{even}^{-1}(SA_{even}(i)) = i$ and $SA_{odd}^{-1}(SA_{odd}(i)) = i$ for $0 \leq i \leq n$.

Recall that the *LF-mapping* is the one-to-one function that maps a suffix starting at position j into its previous suffix starting at position $j - 1$ in terms of their lexicographical ranks. Recall also that we defined suffix arrays separately for even and odd suffixes. The previous suffix of an even suffix is an odd suffix, and vice versa. Hence we have to define the LF-mappings to be functions from even suffixes to odd suffixes, and the other way around.

$$LF_{eo}(i) = SA_{odd}^{-1}(SA_{even}(i) + n \mod (n + 1)) \qquad (6)$$

$$LF_{oe}(i) = SA_{even}^{-1}(SA_{odd}(i)) \qquad (7)$$

4.2 F and L Arrays

In this subsection, we define four integer arrays F_{even}, F_{odd}, L_{even}, and L_{odd}, which compactly store the information used to compute the LF-mappings and to update suffix ranges in the searching algorithm.

For $0 \le i < n$, let us define $\nu(i)$ and $\mu(i)$ as Eqs. 3 and 4. We define two n-length arrays D_{even} and D_{odd} as follows. $D_{even}(i)$ indicates whether $\nu(i) = \mu(i)$ or not. D_{odd} indicates the number of distinct groups appearing between position i and $\min\{i + \mu(i), n\}$ (both exclusive). More formally, these two sequences are defined as follows.

$$D_{even}(i) = \begin{cases} 0 & \text{if } \nu(i) = \mu(i) \\ 1 & \text{otherwise.} \end{cases} \tag{8}$$

$$D_{odd}(i) = |\{g(T[j]) : i < j < \min\{i + \mu(i), n\}\}| \tag{9}$$

Now we define F_{even} and F_{odd}. Note that F_{even} represents the pointers at even positions of the pointer sequence representation and each element $F_{even}(i)$ corresponds to each entry $\mathsf{SA}_{even}(i)$ of the suffix array for even positions; similarly, F_{odd} represents the pointers at odd positions and corresponds to entries of SA_{odd}. F_{even} and F_{odd} are $(n+1)$-length arrays, which are defined as follows: $F_{even}(0) = F_{odd}(0) = -1$, and for $0 < i \le n$,

$$F_{even}(i) = D_{even}(\mathsf{SA}_{even}(i)) \tag{10}$$

$$F_{odd}(i) = D_{odd}(\mathsf{SA}_{odd}(i)) \tag{11}$$

We also define L_{even} and L_{odd} as permuted arrays of F_{odd} and F_{even}, respectively, as follows: for $0 \le i \le n$,

$$L_{even}(i) = F_{odd}(\mathsf{LF}_{eo}(i)) \tag{12}$$

$$L_{odd}(i) = F_{even}(\mathsf{LF}_{oe}(i)) \tag{13}$$

4.3   Computing $\mathsf{LF}_{eo}(i)$ and $\mathsf{LF}_{oe}(i)$

In this subsection, we present how $\mathsf{LF}_{eo}(i)$ and $\mathsf{LF}_{oe}(i)$ can be computed using the four arrays L_{even}, F_{odd}, L_{odd}, and F_{even}. The following lemma shows the key property we can use for computing the LF-mappings using the correspondence between the arrays (Fig. 2).

Lemma 1. *For* $0 \le i < j \le n$,

1. $L_{even}(i) = L_{even}(j)$ *then* $\mathsf{LF}_{eo}(i) < \mathsf{LF}_{eo}(j)$.
2. $L_{odd}(i) = L_{odd}(j)$ *then* $\mathsf{LF}_{oe}(i) < \mathsf{LF}_{oe}(j)$.

i	SA	LF	F	L	
0	12	1	−1	0	empty string
1	11	2	0	0	φ φ
2	8	3	0	0	φ φ {1} {3} φ φ {1} φ
3	10	7	1	1	φ φ {1} φ
4	7	8	1	1	φ φ {1} φ {1} {3} φ φ {1} φ
5	5	9	1	1	φ φ {1} φ φ φ {1} φ {1} {3} {7} {9} {1} {13}
6	0	0	1	−1	φ φ {1} φ φ φ {3} {7} {3} {5} {3} {5} {1} {11} {5} φ {1} {9} {1} {3} {7} {9} {1} {13}
7	9	4	0	0	φ φ φ φ {1} φ
8	6	5	0	0	φ φ φ φ {1} φ {1} {3} {7} {9} {1} φ
9	4	10	1	1	φ φ φ φ {1} φ {5} φ {1} {9} {1} {3} {7} {9} {1} {13}
10	3	11	0	1	φ φ φ φ {3} {5} {1} φ {5} φ {1} {9} {1} {3} {7} {9} {1} {13}
11	2	12	0	1	φ φ φ φ {3} {5} {3} {5} {1} φ {5} φ {1} {9} {1} {3} {7} {9} {1} {13}
12	1	6	1	0	φ φ φ φ {3} φ {3} {5} {3} {5} {1} {11} {5} φ {1} {9} {1} {3} {7} {9} {1} {13}

(a) Sorted even suffixes with SA_{even}, LF_{eo}, F_{even}, and L_{even}

i	SA	LF	F	L	
0	12	0	−1	−1	empty string
1	11	1	0	0	φ
2	10	3	0	1	φ {1} φ
3	7	4	0	1	φ {1} φ {1} {3} φ φ {1} φ
4	8	2	0	0	φ {1} φ φ φ {1} φ
5	5	5	0	1	φ {1} φ φ φ {1} φ {1} {3} {7} {9} {1} φ
6	0	6	0	1	φ {1} φ φ φ {3} φ {3} {5} {3} {5} {1} {11} {5} φ {1} {9} {1} {3} {7} {9} {1} {13}
7	9	7	1	0	φ φ φ {1} φ
8	6	8	1	0	φ φ φ {1} φ {1} {3} {7} φ {1} φ
9	4	9	1	1	φ φ φ {1} φ {5} φ {1} φ {1} {3} {7} {9} {1} {13}
10	3	10	1	0	φ φ φ {3} φ {1} φ {5} φ {1} {9} {1} {3} {7} {9} {1} {13}
11	2	11	1	0	φ φ φ {3} φ {3} {5} {1} φ {5} φ {1} {9} {1} {3} {7} {9} {1} {13}
12	1	12	1	1	φ φ φ {3} φ {3} {5} {3} {5} {1} φ {5} φ {1} {9} {1} {3} {7} {9} {1} {13}

(b) Sorted odd suffixes with SA_{odd}, LF_{oe}, F_{odd}, and L_{odd}

Fig. 2. Sorted (encoded) suffixes for $T =$ **zyxzxzywxxyz** and the related information used in the proposed data structure (white: even positions, gray: odd positions). The searching algorithm navigates two suffix arrays alternatingly by updating suffix ranges iteratively.

Proof. Observe that prepending a pointer at the beginning of a pointer sequence affects at most one position in terms of its encoded form. More specifically, consider pointer sequences X and Y such that Y can be obtained by prepending a pointer of length l at the beginning of X. Then, for $0 \leq i < |Y|$,

$$E(Y)[i] = \begin{cases} \emptyset & \text{if } i = 0, \\ E(X)[i-1] \cup \{l\} & \text{if } i = l, \\ E(X)[i-1] & \text{otherwise.} \end{cases} \tag{14}$$

We call the position on X to which a new pointer to refer *a changing position*: i.e. it refers to position $l-1$ of X in the above equation. Consider two pointer sequences, and a pointer for each sequence is to be prepended. Their relative order changes by these new pointers only if the changing position of the lexicographically greater sequence is earlier and the changing position is within their common prefix. We claim that it is impossible for two suffixes having the same L-value. Based on this observation, we prove each proposition as follows.

1. Let $k = \text{lcp}(E(\text{PS}(T)[2 \cdot \text{SA}_{even}(i)..]), E(\text{PS}(T)[2 \cdot \text{SA}_{even}(j)..]))$ be the length of the longest common prefix of the (encoded) suffixes on SA_{even} whose ranks are i and j. Let $d = |\{g(T[\text{SA}_{even}(i) + p]) : 0 \leq p < \lfloor \frac{k}{2} \rfloor\}|$ be the number of distinct groups in this longest common prefix. If $L_{even}(i) < d$ the length of the newly prepended pointers are the same, which implies the relative order does not change because the changing positions of the encoded sequences are the same. If $L_{even}(i) > d$, the changing position is out of the longest common prefix, and the relative order is determined by $E(\text{PS}(T)[2 \cdot \text{SA}_{even}(i)..])[k] < E(\text{PS}(T)[2 \cdot \text{SA}_{even}(j)..])[k]$, which do not change. Before considering the case of $L_{even}(i) = d$, note that $E(\text{PS}(T)[2 \cdot \text{SA}_{even}(i)..])[k] \leq \{k\} \neq \emptyset$ because $E(\text{PS}(T)[2 \cdot \text{SA}_{even}(i)..])[k] < E(\text{PS}(T)[2 \cdot \text{SA}_{even}(j)..])[k]$ and $\emptyset$ is considered to be the greatest. Therefore the changing position of suffix i is not k. Even if the changing position of suffix j is k, $\emptyset$ is substituted by $\{k+1\}$, which does not change the relative order.

2. Let $k = \text{lcp}(E(\text{PS}(T)[2 \cdot \text{SA}_{odd}(i) + 1..]), E(\text{PS}(T)[2 \cdot \text{SA}_{odd}(j) + 1..]))$ be the length of the longest common prefix of suffixes whose ranks are i and j, respectively. Note that, at this moment the group of the character is already determined (by the pointer at the odd position on the text pointer sequence), the pointer to be newly prepended here indicates which of the two characters in the group is actually prepended; therefore we have two candidates for the change position for each suffix. Let $c_i^{(1)}$ and $c_i^{(2)}$ be two candidate positions for suffix whose rank is i such that $c_i^{(1)} < c_i^{(2)}$. Similarly, we define $c_j^{(1)}$ and $c_j^{(2)}$ for suffix whose rank is j. Note that if $c_i^{(1)} < k$ or $c_j^{(1)} < k$, then $c_i^{(1)} = c_j^{(1)}$. Similarly, if $c_i^{(2)} < k$ or $c_j^{(2)} < k$, then $c_i^{(2)} = c_j^{(2)}$. Let l_i and l_j be the changing position of suffixes whose ranks are i and j, respectively. If $L_{odd}(i) = L_{odd}(j) = 0$, then $l_i = c_i^{(1)}$ and $l_j = c_j^{(1)}$. Similarly, if $L_{odd}(i) = L_{odd}(j) = 1$, $l_i = c_i^{(2)}$ and $l_j = c_j^{(2)}$. Therefore, if $l_i < k$ or $l_j < k$ then we have $l_i = l_j$. Let us assume that the relative order changes after prepending corresponding pointers to these suffixes. Then it must be both $l_j < k$ and $l_j < l_i$. However, if $l_j < k$, we have $l_i = l_j$. Contradiction. $\qquad \square$

From the order-preserving property described in Lemma 1, we can simply compute $LF_{eo}(i)$ and $LF_{eo}(i)$ using the rank and select operations as follows.

Corollary 1. *For $0 \le i \le n$,*

1. $LF_{eo}(i) = F_{odd}.\text{select}_x(L_{even}.\text{rank}_x(i))$ *where* $x = L_{even}(i)$.
2. $LF_{oe}(i) = F_{even}.\text{select}_x(L_{odd}.\text{rank}_x(i))$ *where* $x = L_{odd}(i)$.

4.4   Implementation

In this subsection, we describe how the proposed data structure is organized. More specifically, we show the following lemma.

Lemma 2. *There exists a data structure that uses $2n \lg \sigma + 2n + o(n)$ bits and supports the following operations in $\mathcal{O}(\lg \sigma)$ time for any $0 \le i, j \le n$, $a \in \{0, \cdots, \sigma - 1\}$, and $b \in \{0, 1\}$:*

1. *$L_{even}(i)$: access to the value $L_{even}(i)$.*
2. *$L_{odd}(i)$: access to the value $L_{odd}(i)$.*
3. *$L_{even}.\text{rank}_a(i, j)$: the number of occurrences of a in $L_{even}[i..j]$.*
4. *$L_{even}.\text{rank\_ge}_a(i, j)$: the number of elements greater than or equal to a in $L_{even}[i..j]$.*
5. *$L_{odd}.\text{rank}_b(i, j)$: the number of occurrences of b in $L_{odd}[i..j]$.*
6. *$F_{even}.\text{select}_b(i)$: the position of the i-th occurrence of b in F_{even}.*
7. *$F_{odd}.\text{select}_a(i)$: the position of the i-th occurrence of a in F_{odd}.*
8. *$LF_{eo}.\text{rMq}(i, j)$: the index of the maximum element among $LF_{eo}(i), \cdots,$ $LF_{eo}(j)$.*

Proof. In short, we build wavelet trees on L_{even} and F_{odd}, and rank/select dictionaries for bitvectors on L_{odd} and F_{even}, and a range maximum query index on LF_{eo}. More specifically,

- We build wavelet trees on L_{even} and F_{odd}, which can support the operations related on these arrays in $\mathcal{O}(\lg \sigma)$ time. Note that the alphabet size of these arrays is $\sigma/2$, thereby each wavelet tree uses $n \lg(\sigma/2) + o(n) = n \lg \sigma - n \lg 2 + o(n) = n \lg \sigma - n + o(n)$ bits.
- For L_{odd} and F_{even}, we can observe that these arrays consist of 0 and 1 except the unique -1. Thus storing the index at which -1 appears using $\mathcal{O}(\lg \sigma)$ bits, we can consider them as bitvectors, which can support rank and select queries in $\mathcal{O}(1)$ time using $n + o(n)$ bits each.
- Range maximum query on LF_{eo} requires $2n + o(n)$ bits, and can answer to a range maximum query in $\mathcal{O}(1)$ time.

As a result, the total space requirement is $2n \lg \sigma + 2n + o(n)$ bits.   □

5   Searching Algorithm

In this section, we present the searching algorithm on the proposed data structure. As other methods based on suffix arrays do, we represent the occurrences of a pattern as a contiguous interval on the suffix array, which is called a *suffix range*. Recall that we have two suffix arrays SA_{even} and SA_{odd}, and only suffix ranges on SA_{even} should be the final answer. To distinguish a suffix range on SA_{even} from one on SA_{odd}, we call a suffix range on SA_{even} a *real suffix range* and one on SA_{odd} *an imaginary suffix range*.

For a pattern P, its suffix range on SA_{even} is a pair of indices (p_s, p_e) such that for any $p_s \leq i \leq p_e$ $E(\mathsf{PS}(T)[2 \cdot \mathsf{SA}_{even}(i)..])$ has a prefix $E(\mathsf{PS}(P))$. Note that a character of an s-string is represented as two pointers. Let $x \in \Sigma$ be a character, suppose we are to prepend x to the beginning of P. Let l_1 and l_2 be the lengths of the first two pointers $\mathsf{PS}(xP)$. These are what we are about to prepend to $\mathsf{PS}(P)$ to compute the updated suffix range for xP. By prepending the latter pointer at the beginning of $\mathsf{PS}(P)$, we obtain an imaginary suffix range on SA_{odd}. Then, prepending the other pointer completes $\mathsf{PS}(xP)$, and we obtain a real suffix range on SA_{even} via a proper update procedure, which is the desired (real) suffix range for the updated pattern xP.

Our searching algorithm iteratively updates the suffix array starting from the suffix range $(p_s, p_e) = (0, n)$ on SA_{even} of the empty string, which represents all the suffixes starting at even positions. Each iteration we prepend each character of the pattern in the backward searching fashion. It is equivalent to prepend the corresponding two pointers in the pointer sequence representation. Thus the update procedure consists of two phases, in which we update the current (real) suffix range into an imaginary suffix range on SA_{odd}, followed by updating it into a real suffix range on SA_{even}. The algorithm for updating a suffix range is given in Algorithm 1.

For a suffix range (p_s, p_e) for a pointer sequence X, let l be the length of the pointer to be prepended to the beginning of X. Let (p'_s, p'_e) be the suffix range the pointer sequence after prepending the pointer. For an index $p_s \leq i \leq p_e$, we say i is a *target suffix* if $p'_s \leq \mathsf{LF}^*(i) \leq p'_e$ where $\mathsf{LF}^*(i) = \mathsf{LF}_{eo}(i)$ if $|X|$ is a multiple of 2, $\mathsf{LF}^*(i) = \mathsf{LF}_{oe}(i)$ otherwise. Now we can describe a suffix update procedure as (i) identifying the target suffixes within the current suffix range, followed by (ii) applying the LF-mapping to the identified suffixes. The remainder of the section is devoted to show the following theorem about the correctness of the algorithm.

Theorem 1. *Given a suffix range (p_s, p_e) for a pattern $P \in \Sigma^*$ and $x \in \Sigma$, Algorithm 1 correctly computes the updated suffix range (p'_s, p'_e) for pattern xP in $\mathcal{O}(\lg \sigma)$ time, provided i_g, i_c, and a can be computed in $\mathcal{O}(\lg \sigma)$ time.*

By Lemma 2, all the operations in Algorithm 1 related to the arrays take $\mathcal{O}(\lg \sigma)$ time. Considering a character $x \in \Sigma$ to be prepended to the beginning of the currently searched pattern P during the update procedure, we divide it into two cases: (i) P has x or $\mathsf{compl}(x)$, and (ii) P does not have any of them.

Algorithm 1. Update a suffix range.

```
 1: procedure UPDATE(P: current pattern s-string, (p_s, p_e): suffix range, x: character)
 2:    if either x or compl(x) appeared in P then
 3:        i_c ← min{0 ≤ j < |P| : P[i] = x} ∪ {|P|}.
 4:        i_g ← min{0 ≤ j < |P| : g(P[i]) = g(x)}.
 5:        a ← |{g(P[j]) : 0 ≤ j < i_g}|.
 6:        c ← L_even.rank_a(p_s, p_e).
 7:        p_e ← F_odd.select_a(L_even.rank_a(0, p_e)).
 8:        p_s ← p_e - c + 1.
 9:        b ← 0 if i_c = i_g, 1 otherwise.
10:        c ← L_odd.rank_b(p_s, p_e).
11:        p_e ← F_even.select_b(L_odd.rank_b(0, p_e)).
12:        p_s ← p_e - c + 1.
13:    else
14:        i* ← LF_eo.rMq(p_s, p_e).
15:        l ← L_even(i*).
16:        a ← |{g(P[j]) : 0 ≤ j < |P|}|.
17:        c ← L_even.rank_ge_a(p_s, p_e).
18:        p_e ← F_odd.select_l(L_even.rank_l(i*)).
19:        p_s ← p_e - c + 1.
20:    end if
21:    return (p_s, p_e).
22: end procedure
```

Case 1: At least one of x and compl(x) appear in P. Lines 3–12 handle this case. Let i_c be the position of the first occurrence of x on P; if x does not appear on P, then $i_c = |P|$. Let i_g be the position of first occurrence of x's equal-group character (either x or compl(x)), which must exist. Let a be the number of distinct groups in $P[0..i_g - 1]$. This value can be computed in $\mathcal{O}(\lg \sigma)$ time, if we keep a balanced binary tree keyed by positions of the first occurrences of each group, and the leftmost position of each character as similar to that described in [5]. Then the indices of the target suffixes on SA_{even} must be the suffixes having a as their L_{even}-values. The number of these suffixes can be counted by $c = L_{even}.\text{rank}_a(p_s, p_e)$. And the last index of the suffix is located by $i_e = L_{even}.\text{rank}_a(0, p_e)$. We can find the corresponding index $LF_{eo}(i_e)$ on $\text{select}_a(i_e)$, update p_e to it. Using the number c of target suffixes, we can update p_s to be $p_e - c + 1$.

Now (p_s, p_e) is an imaginary suffix range on SA_{odd}. Let b be a binary number such that $b = 0$ if $P[i_g] = x$ (i.e. $i_c = i_g$), $b = 1$ otherwise. Similarly, the target suffixes are those having b as their L_{odd}-values. We can update the suffix range correspondingly in a similar way.

Case 2: Neither x nor compl(x) appears in P. Lines 14–19 handle this case, which is a little more difficult. Let a be the number of distinct groups in P. The target suffixes are those within the current suffix range that have a L_{even} value of at least a. We can count the number c of these suffixes via $L_{even}.\text{rank\_ge}_a(p_s, p_e)$. It is easy to see, for $p_s \leq i, j \leq p_e$ such that $L_{even}(i) <$

$a \leq L_{even}(j)$, $\mathsf{LF}_{eo}(i) < \mathsf{LF}_{eo}(j)$. This is because, for such suffix i, the changing position is one of the first $|P|$ positions, which is definitely within the common prefix of the encoded sequences. Since such suffix j has a changing position beyond that of the suffix i, the suffix i becomes smaller after prepending the corresponding pointer. Therefore $\mathsf{LF}_{eo}(i) < \mathsf{LF}_{eo}(j)$. As a result, we can find i^* such that $\mathsf{LF}_{eo}(i^*)$ is the updated p_e by performing the range maximum query on LF_{eo} with the current suffix range. After updating $p_e = \mathsf{LF}_{eo}(i^*)$ by $L_{even}.\mathsf{rank}(\cdot)$ followed by $F_{odd}.\mathsf{select}(\cdot)$, p_s can also be updated using the updated p_e and the number c of the target suffixes.

To update this imaginary suffix range into a real suffix range for the update pattern xP, we observe that target suffixes are all the suffixes within the current imaginary suffix range. This is because the group corresponding to the newly prepended character x is a new group that has not been appeared in P, every suffix within the current imaginary suffix range should be considered regardless of their L_{odd}-values. And surprisingly, for $0 \leq i, j, k \leq n$ such that $i < p_s \leq k \leq p_e < j$, $\mathsf{LF}_{oe}(i) < \mathsf{LF}_{oe}(k) < \mathsf{LF}_{oe}(j)$. The lengths of the pointers to be prepended is longer than the length of the longest common prefix of (encoded) suffixes i (or j) and k, which does not change the relative order. Therefore, we do not have to update p_s and p_e anymore, and (p_s, p_e) itself is also the desired real suffix range.

6 Conclusions

In this paper, we present a space-efficient index for the structural pattern matching problem. The data structure requires $2n \lg \sigma + 2n + o(n)$ bits and it can count the number of occurrences of an m-length pattern in $\mathcal{O}(m \lg \sigma)$ time. Due to the hidden constant factor in $\mathcal{O}$ term in the previous succinct index [5], our structure can become much smaller if σ is small enough. Further, our data structure consists of building blocks that are widely used and practically implementable in many other succinct and compact data structures. Adding the sampled suffix array, we can also report each occurrence by repeatedly calling the LF-functions until reaching the sampled entry.

In the future work, the construction algorithm should be addressed. Once the suffix array of the pointer sequence for a given text s-string is given, our data structure can efficiently constructed; however, the construction of such suffix array is an open problem; besides pointer sequences, suffix sorting algorithms for a s-string have not been well-studied as well. Separating suffixes is not only for the compact representation, but it also gives many ways to generalize this problem further. For example, we may think of a problem in which a multiple number of characters are grouped together, instead of complement pairs.

Acknowledgement. The authors would like to thank anonymous reviewers for their valuable comments and suggestions.

References

1. Beal, R., Adjeroh, D.: Efficient pattern matching for RNA secondary structures. Theoret. Comput. Sci. **592**, 59–71 (2015). https://doi.org/10.1016/j.tcs.2015.05.016
2. Clark, D.: Compact pat trees. Ph.D. thesis, University of Waterloo (1996)
3. Ferragina, P., Manzini, G., Mäkinen, V., Navarro, G.: Compressed representations of sequences and full-text indexes. ACM Trans. Algorithms **3**(2), 20–es (2007). https://doi.org/10.1145/1240233.1240243
4. Fischer, J.: Optimal succinctness for range minimum queries. In: López-Ortiz, A. (ed.) LATIN 2010. LNCS, vol. 6034, pp. 158–169. Springer, Heidelberg (2010). https://doi.org/10.1007/978-3-642-12200-2_16
5. Ganguly, A., Shah, R., Thankachan, S.V.: Structural pattern matching - succinctly. In: Proceedings of the 28th International Symposium on Algorithms and Computation (ISAAC), pp. 35:1–35:13 (2017). https://doi.org/10.4230/LIPIcs.ISAAC.2017.35
6. Gog, S., Beller, T., Moffat, A., Petri, M.: From theory to practice: plug and play with succinct data structures. In: Proceedings of the 13th International Symposium on Experimental Algorithms (SEA), pp. 326–337 (2014). https://doi.org/10.1007/978-3-319-07959-2_28
7. Golynski, A., Grossi, R., Gupta, A., Raman, R., Rao, S.S.: On the size of succinct indices. In: Arge, L., Hoffmann, M., Welzl, E. (eds.) ESA 2007. LNCS, vol. 4698, pp. 371–382. Springer, Heidelberg (2007). https://doi.org/10.1007/978-3-540-75520-3_34
8. Grossi, R., Gupta, A., Vitter, J.S.: High-order entropy-compressed text indexes. In: Proceedings of the 14th Annual ACM-SIAM Symposium on Discrete Algorithms (SODA), pp. 841–850 (2003). https://doi.org/10.5555/644108.644250
9. Jacobson, G.: Space-efficient static trees and graphs. In: Proceedings of the 30th Annual Symposium on Foundations of Computer Science (FOCS), pp. 549–554 (1989). https://doi.org/10.1109/SFCS.1989.63533
10. Kim, S.H., Cho, H.G.: Indexing isodirectional pointer sequences. In: Proceedings of the 31st International Symposium on Algorithms and Computation (ISAAC), pp. 35:1–35:15 (2020). https://doi.org/10.4230/LIPIcs.ISAAC.2020.35
11. Navarro, G.: Wavelet trees for all. J. Discrete Algorithms **25**, 2–20 (2014). https://doi.org/10.1016/j.jda.2013.07.004
12. Navarro, G., Sadakane, K.: Fully functional static and dynamic succinct trees. ACM Trans. Algorithms **10**(3), 1–39 (2014). https://doi.org/10.1145/2601073
13. Sadakane, K., Navarro, G.: Fully-functional succinct trees. In: Proceedings of the 21st Annual ACM-SIAM Symposium on Discrete Algorithms, pp. 134–149 (2010). https://doi.org/10.5555/1873601.1873614
14. Shibuya, T.: Generalization of a suffix tree for RNA structural pattern matching. In: SWAT 2000. LNCS, vol. 1851, pp. 393–406. Springer, Heidelberg (2000). https://doi.org/10.1007/3-540-44985-X_34
15. Shibuya, T.: Generalization of a suffix tree for RNA structural pattern matching. Algorithmica **39**, 1–19 (2004). https://doi.org/10.1007/s00453-003-1067-9

A Shift Gray Code for Fixed-Content
Łukasiewicz Words

Paul W. Lapey and Aaron Williams[(✉)] [ID]

Williams College, Williamstown, MA 01267, USA
{pwl1,aaron.williams}@williams.edu
https://csci.williams.edu/people/faculty/aaron-williams/

Abstract. A Łukasiewicz path of length n is a lattice path from $(0,0)$ to $(n,0)$ that never goes below the x-axis, and which uses steps of the form $(1,i)$ for integers $i \geq -1$. These paths include both Dyck paths ($i \in \{-1,1\}$) and Motzkin paths ($i \in \{-1,0,1\}$). A set of fixed-content Łukasiewicz paths contains all such paths in which the frequency of each step is fixed. For example, ⛩ is the only path with one $(1,3)$ step and three $(1,-1)$ steps; equivalently, the only Łukasiewicz word with content $\{-1,-1,-1,3\}$ is $3 -1 -1 -1$ (or 4000 using 0-based values). We contribute a shift Gray code for these fixed-content sets, meaning that consecutive paths differ by moving a single line, and consecutive words differ by moving a single symbol. We also provide a successor rule for generating the next word directly from the current word, as well as loopless array-based algorithms for generating generalized fixed-content Motzkin and Schröder words. Our Gray code generalizes the cool-lex order Gray code for Dyck words.

Keywords: Łukasiewicz path · Łukasiewicz word · Dyck word · Motzkin word · Fixed-content · Gray code · Cool-lex order

1 Introduction

When the nodes of an ordered tree are labeled by their number of children, then a preorder traversal gives a Łukasiewicz word. In this paper, we efficiently order and generate Łukasiewicz words. More specifically, we consider sets of fixed-content Łukasiewicz words, which contain strings with the same multiset of symbols (see Fig. 1). These sets of strings correspond to ordered trees with the same branching sequence (see Fig. 2).

Our first result is a left-shift Gray code for fixed-content Łukasiewicz words, meaning that each string is obtained from the previous by moving one symbol to the left (see Fig. 4). There is also a relatively simple successor rule that provides the shift (see (4)) and the resulting order is a cool-lex variant of lexicographic order. Our second result is *loopless* (i.e., worst-case $O(1)$-time per string) array-based implementation for generating the special case of fixed-content Motzkin words. Both the shift Gray code and loopless algorithm generalize previous

© Springer Nature Switzerland AG 2022
C. Bazgan and H. Fernau (Eds.): IWOCA 2022, LNCS 13270, pp. 383–397, 2022.
https://doi.org/10.1007/978-3-031-06678-8_28

results for Dyck words [14]; alternate generalizations to k-ary Dyck words [4,5] and binary bubble languages [12,22] have also been considered.

To our knowledge, this paper represents the first shift Gray code for fixed-content Łukasiewicz words. Many previous investigations have focused on different orders for related sequences and special cases of these words [2,3,8,10,19,20,23]. For additional background we refer the reader to Knuth's coverage of generating combinatorial objects in Volume 4A of *The Art of Computer Programming* [7], and Mütze's recent update [9] of Savage's classic survey [16].

Section 2 introduces the relevant combinatorial objects, Sect. 3 provides the successor rule for generating our shift Gray codes, and Sect. 4 proves that the rule is correct. Section 5 provides our loopless algorithm for fixed-content Motzkin words, with Python code in the Appendix.

2   Background

In this background section, we discuss the combinatorial objects that will be generated in this paper, as well as their history and encodings.

2.1   Lattice Paths: Dyck, Motzin, Schröder, and Łukasiewicz

Lattice paths are well-studied in combinatorics, with books on the subject dating back to the 1970s (see Narayana [11]). In particular, most readers will be familiar with Dyck paths, which are paths from $(0,0)$ to $(2n,0)$ using $2n$ steps of the form $(1,1)$ (north-east) and $(1,-1)$ (south-east), and having the property that the path never goes below the x-axis. These paths can be encoded as *balanced parentheses*, or as integer strings according to several possible encoding schemes.

- North-east steps are 1 and south-east steps are 0. With this encoding, every prefix must have as many 1s as 0s.
- North-east steps are 1 and south-east steps are -1. With this encoding, every prefix must have a non-negative sum.
- North-east steps are 2 and south-east steps are 0. With this encoding, every prefix's sum must be at least as large as its length.

All of these encodings have been referred to as *Dyck words of order n*. We refer to the latter two as the -1-*based encoding* and the 0-*based encoding*, respectively. For example, the five Dyck words of order $n = 3$ are

$$\{[]\,[]\,[], []\,[[]], [[]]\,[], [[]\,[]], [[[]]]\} = \{202020, 202200, 220020, 220200, 222000\}$$

when using balanced parentheses and the 0-based encoding, respectively.

Many generalizations of Dyck paths and Dyck words have been studied under the name *generalized Dyck words*. For example, one can consider multiple types of parentheses simultaneously (e.g., '(' with ')' and '[' and ']'), or have longer inequality chains (e.g., every prefix has as many 2s as 1s as 0s).

Another approach is to vary the steps. For example, a *k-ary Dyck path of order n* is a path from $(0, 0)$ to $(kn, 0)$ using kn steps of the form $(1, k - 1)$ and $(1, -1)$ while never going below the x-axis. The corresponding *k-ary Dyck words* can again be encoded in several ways, and Dyck words are obtained when $k = 2$.

A broader step-based generalization is a *Łukasiewicz path*, which is a path from $(0, 0)$ to $(n, 0)$ that does not go below the x-axis, and which uses steps $(1, i)$ for any integer $i \geq -1$. These paths can be encoded as strings by generalizing either of the last two encodings for Dyck words discussed above.

- *−1-based encoding*: Each $(1, i)$ step is encoded as i, and every prefix must have a non-negative sum.
- *0-based encoding*: Each $(1, i)$ step is encoded as $i + 1$, and the sum of every prefix must be at least as large as its length.

We prefer the 0-based encoding, and refer to these strings as *Łukasiewicz words of order n*. Figure 1 illustrates all Łukasiewicz paths and words for $n = 4$. Although Łukasiewicz paths include Dyck paths, they differ in their use of n and the term *order*. In particular, the middle row of Fig. 1 includes all Dyck words of order $\frac{4}{2} = 2$, since the order of a Dyck word is its number of pairs.

Łukasiewicz paths include Dyck paths when the steps are $(1, i)$ for $i \in \{-1, 1\}$. They also include *Motzkin paths*, where $i \in \{-1, 0, 1\}$. A 0-based encoding is typically used for the corresponding *Motzkin words*, with $\{111, 120, 201, 210\}$ containing the four options when $n = 3$. The closely related *Schröder paths* differ from Motzkin paths in using an east step of $(2, 0)$ rather than $(1, 0)$. For example, the six *Schröder words* of order $n = 2$ are $\{11, 120, 201, 210, 2200, 2020\}$.

The Dyck, Motzkin, and Schröder paths of order n are enumerated by the nth Catalan number, Motzkin number, and big Schröder number, respectively. These sequences are illustrated below for $n \geq 0$ along with their respective entries in the Online Encyclopedia of Integer Sequences (OEIS) [17]:

$$C_n = 1, 1, 2, 5, 14, 42, 132, \ldots \qquad \text{OEISA000108} \qquad (1)$$

$$\mathcal{M}_n = 1, 1, 2, 4, 9, 21, 51, \ldots \qquad \text{OEISA001006} \qquad (2)$$

$$\mathcal{S}_n = 1, 2, 6, 22, 90, 394, 1806, \ldots \qquad \text{OEISA006318} \qquad (3)$$

The Łukasiewicz paths of order n are enumerated by C_{n+2}. Due to their connections with C_n, $\mathcal{M}_n$, and $\mathcal{S}_n$, these paths are in bijective correspondence with many interesting combinatorial objects, with Stanley's book, *Catalan Numbers*, outlining hundreds of examples [18]. In particular, Łukasiewicz paths have a particularly nice mapping to rooted ordered trees with $n + 1$ internal nodes (see Fig. 2), and for convenience, each node is labeled by its number of children. These 0-based words have also been referred to as *preorder codewords* [1].

Łukasiewicz paths are named after Jan Łukasiewicz for whom reverse Polish notation is also named. For historical notes on Łukasiewicz's life and mathematics see [6]. When considering Łukasiewicz paths for the first time, it is helpful to note that paths of order n can use steps of maximum slope $(1, n - 1)$, since otherwise there won't be enough $(1, -1)$ steps to return to the x-axis at position $(n, 0)$. This restriction also ensures that there are a finite number of such

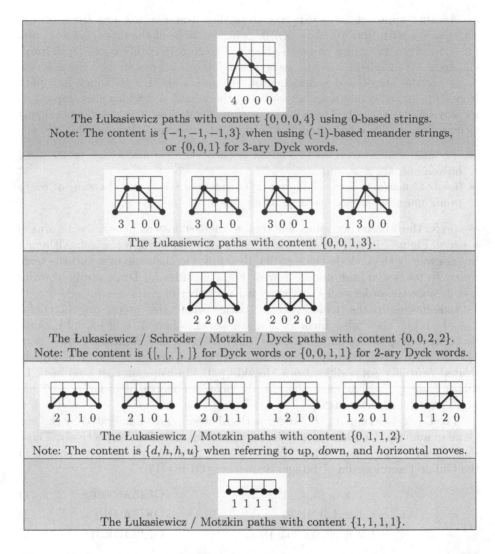

Fig. 1. All $\mathcal{C}_4 = 14$ Łukasiewicz paths of order 4 are partitioned into rows by their content (i.e., their multiset of slopes). The bottom three rows have all $\mathcal{M}_4 = 9$ Motzkin paths of order 4. The middle row has all $\mathcal{C}_2 = 2$ Dyck paths of order 2. The top row has the $\mathcal{C}_1^3 = 1$ 3-ary Dyck path of order 1. Each row is ordered lexicographically by the path's 0-based string. Other encodings are noted. For example, the second path in the middle row is encoded as 2020 (0-based), $1\,-1\,1\,-1$ ((−1)-based), *udud* (moves), [] [] (Dyck word), or 1010 (2-ary Dyck word). Our main results involve ordering and generating Łukasiewicz words (i.e., the 0-based strings) for a given content (i.e., multiset of symbols). In other words, we focus on the strings listed in the types of rows shown above.

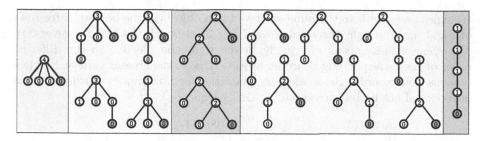

Fig. 2. The $C_4 = 14$ Łukasiewicz word of order $n = 4$ are in one-to-one correspondence with the rooted ordered trees with $n + 1 = 5$ internal nodes. Given a tree, the corresponding word is obtained by recording the number of children of each node in a preorder traversal; the last 0 (from the rightmost leaf) is omitted. For example, the two trees in the middle section correspond to 2200 (top) and 2020 (bottom). The trees are partitioned based on their branching sequence, which corresponds to the content of the associated Łukasiewicz words (see Fig. 1).

paths for all n. See [2] for a discussion of more general lattice paths using the Banderier-Flajolet model, including *excursions*, which are paths from $(0,0)$ to $(n,0)$ that do not go below the x-axis, and which use steps $(1,i)$ for any integer i.

2.2 Restriction to Fixed-Content

Lattice paths are often restricted in various ways when they are studied. We focus on *content*, which refers to the multiset of symbols used in a word, or equivalently, the multiset of steps used in a path. We use the term *fixed-content* to refer to all Łukasiewicz words, or paths, with the same content. We use $\mathcal{L}(S)$ to denote the set of (0-based) Łukasiewicz words with content S, where S is a multiset of non-negative integers whose sum is equal to its cardinality. For example, the Łukasiewicz paths in Fig. 1 are partitioned into fixed-content parts—$\mathcal{L}(\{0,0,0,4\})$; $\mathcal{L}(\{0,0,1,3\})$; $\mathcal{L}(\{0,0,2,2\})$; $\mathcal{L}(\{0,1,1,2\})$; $\mathcal{L}(\{1,1,1,1\})$—where $\{\}$ or $[]$ denotes multiset content.

The restriction to fixed-content is useful for several reasons. For example, Łukasiewicz paths generalize Dyck paths in the sense that the set of allowed steps is broadened. But it is not true that the set of Łukasiewicz paths of order n generalize the set of Dyck paths of order n; more precisely, they form a superset. On the other hand, fixed-content Łukasiewicz words do generalize Dyck words in this sense. For example, $\{202020, 202200, 220020, 220200, 222000\}$ is both the set of Dyck words of order $n = 3$ (using 0-based encoding), and the Łukasiewicz words with fixed-content $[0,0,0,2,2,2]$. Similarly, fixed-content Łukasiewicz words generalize both fixed-content Motzkin words and fixed-content Schröder words. For example, $\{120, 201, 210\}$ is the set of Motzkin, Schröder, and Łukasiewicz words with content $[0,1,2]$. Note that in this example, the Motzkin and Łukasiewicz words have order $n = 3$, while the Schröder words have order $n = 4$.

The Motzkin and Schröder numbers are partitioned by their content in OEIS A055151 and A088617, respectively. For example, the row **1**, **6**, **2** in the left triangle corresponds to the number of Motzkin objects in the bottom three rows of Fig. 1 and the right three columns of Fig. 2 (although the order is reversed). The same values appear diagonally in the right triangle due to the differing order of the corresponding Schröder object. (Due to the greater variety, it is less obvious how to order the analogous quantities for Łukasiewicz words, and the authors did not find a corresponding OEIS sequence.)

A055151				A088617					
1				1					
1				1	1				
1	1			1	3	**2**			
1	3			1	6	10	5		
1	**6**	**2**		1	10	30	35	14	
1	10	10		1	15	70	140	126	42
1	15	30	5	1	21	140	420	630	462	132

Placing a fixed-content restriction on a set of strings can also coincide with a meaningful restriction in corresponding combinatorial objects. For example, restricting Łukasiewicz words to fixed-content corresponds to restricting rooted ordered trees to a specific branching sequence. The *branching sequence* of a rooted tree is the sorted list of the number of children of each node in the tree. For example, the fourth section of Fig. 2 shows the ordered trees with branching sequence $0, 0, 1, 1, 2$, which correspond to the Łukasiewicz words with content $\{0, 1, 1, 2\}$ (as one copy of 0 is omitted).

2.3   Gray Codes for Lattice Paths and Strings

In this paper, we are not concerned with counting lattice paths, but in efficiently ordering them. More specifically, we want to create a *minimal-change order*, or *Gray code*, which means sequencing the objects so that each differs from the previous in a specific small way. Our orders are also *cyclic*, in the sense that the last object can be transformed into the first via the same type of small change.

When constructing Gray codes, it is helpful to think about the underlying graph of objects and allowable changes. For example, Fig. 3a illustrates the six Łukasiewicz words with content $\{0, 1, 1, 2\}$ as vertices, with edges connecting those that differ by a swap. A *swap*, or *adjacent-transposition*, interchanges two symbols that are immediately next to each other in the string. For example, swapping 20 with 02 changes a peak to a valley in the corresponding lattice path, and it is only valid if the path was above the x-axis at that location prior to the swap. Observe Fig. 3a does not have a Hamilton path, so $\mathcal{L}(\{0, 0, 1, 2\})$ does not have a swap Gray code. Thus, we need to broaden our notion of a minimal change in order to create a Gray code for these objects.

One generalization[1] of an adjacent-transposition is a *shift*, in which a single symbol is moved to another position. Figure 3b illustrates the associated graph, and in this case, there is a Hamilton cycle. Thus, there is a cyclic shift Gray code

---

[1] Another generalization is a *transposition*, in which two values are interchanged, without the restriction that they must be next to each other in the string.

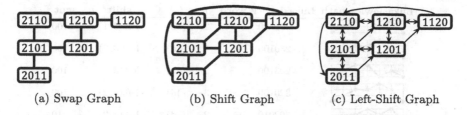

(a) Swap Graph   (b) Shift Graph   (c) Left-Shift Graph

Fig. 3. Graphs associated with Gray codes of $\mathcal{L}(S)$ for $S = \{0, 1, 1, 2\}$.

for this set of strings, and one could hope to prove that such a Gray code always exists for fixed-content Łukasiewicz words. We aim slightly higher by considering a more restrictive notion of a minimal-change. A *left-shift* moves a single symbol somewhere to the left within a string. More specifically, if $\alpha = a_1 \cdot a_2 \cdots a_n$ is a string and $i < j$, then we let

$$\text{left}_\alpha(j, i) = a_1 \cdot a_2 \cdots a_{i-1} \cdot a_j \cdot a_i \cdot a_{i+1} \cdots a_{j-1} \cdot a_{j+1} \cdot a_{j+2} \cdots a_n.$$

In other words, $\text{left}_\alpha(j, i)$ shifts a_j to the left into position i. Observe that $\text{left}_\alpha(i+1, i)$ is an adjacent-transposition or swap. We also omit α from this notation when the context is clear. The directed graph for $\mathcal{L}(\{0, 0, 1, 2\})$ with left-shifts appears in Fig. 3c. This graph has a directed Hamilton cycle, and hence, $\mathcal{L}(\{0, 0, 1, 2\})$ has a cyclic left-shift Gray code. We will establish this result for all sets of fixed-content Łukasiewicz words.

3   Successor Rule

In this section, we provide a *successor rule* that applies a left-shift to a Łukasiewicz word. The rule is given below in (4). In the statement of the rule, we assume that $\alpha = a_1 \cdot a_2 \cdots a_n \in \mathcal{L}(S)$, where S is a multiset whose sum is equal to its cardinality. We also assume that $\rho = a_1 \cdot a_2 \cdots a_m$ is α's *non-increasing prefix*. In other words, $a_1 \geq a_2 \geq \cdots \geq a_m$, and either $m = n$ (i.e., the entire string is non-increasing) or $a_m < a_{m+1}$ (i.e., there is an increase immediately following the prefix). The sum of the symbols in ρ is $\sum \rho = a_1 + a_2 + \cdots + a_m$.

$$\text{next}(\alpha) = \begin{cases} \text{left}(n, 2) & \text{if } m = n & (4a) \\ \text{left}(m+1, 1) & \text{if } m = n-1 \text{ or } a_m < a_{m+2} \text{ or} & (4b) \\ & \quad (a_{m+2} = 0 \text{ and } \sum \rho = m) \\ \text{left}(m+2, 1) & \text{if } a_{m+2} \neq 0 & (4c) \\ \text{left}(m+2, 2) & \text{otherwise} & (4d) \end{cases}$$

Figure 4 illustrates the successor rule on every string in $\mathcal{L}(S)$ for $S = \{0, 0, 0, 1, 2, 3\}$. For example, consider the top row with $\alpha = a_1 \cdot a_2 \cdot a_3 \cdot a_4 \cdot a_5 \cdot a_6 = 302100$. Here the non-increasing prefix is $a_1 \cdot a_2 = 30$, so $m = 2$, and the length of the string is $n = 6$. Thus, $m \neq n$, so (4a) is not applied. Now consider the conditions in (4b). The second condition is $a_m < a_{m+2}$, which is $a_2 = 0 < 1 = a_4$

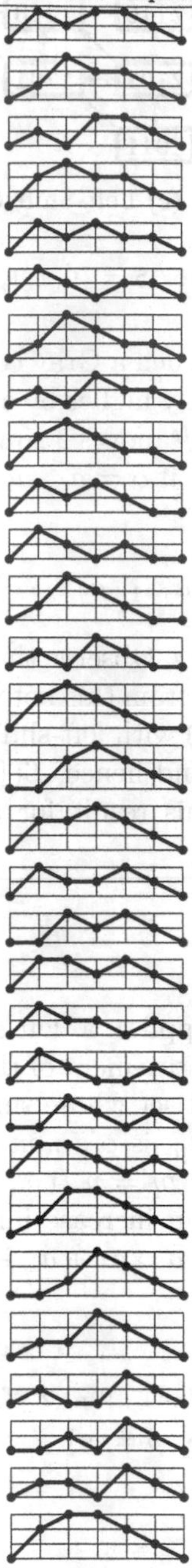

Łukasiewicz path	Łukasiewicz word	m	(4)	shift	scut
	302100	2	(4b)	left $(3, 1)$	100
	230100	1	(4d)	left $(3, 2)$	100
	203100	2	(4b)	left $(3, 1)$	100
	320100	3	(4d)	left $(5, 2)$	100
	302010	2	(4d)	left $(4, 2)$	10
	300210	3	(4b)	left $(4, 1)$	10
	230010	1	(4d)	left $(3, 2)$	10
	203010	2	(4b)	left $(3, 1)$	10
	320010	4	(4d)	left $(6, 2)$	10
	302001	2	(4d)	left $(4, 2)$	1
	300201	3	(4b)	left $(4, 1)$	1
	230001	1	(4d)	left $(3, 2)$	1
	203001	2	(4b)	left $(3, 1)$	1
	320001	5	(4b)	left $(6, 1)$	1
	132000	1	(4b)	left $(2, 1)$	2000
	312000	2	(4d)	left $(4, 2)$	2000
	301200	2	(4b)	left $(3, 1)$	200
	130200	1	(4b)	left $(2, 1)$	200
	310200	3	(4d)	left $(5, 2)$	200
	301020	2	(4d)	left $(4, 2)$	20
	300120	3	(4b)	left $(4, 1)$	20
	130020	1	(4b)	left $(2, 1)$	20
	310020	4	(4b)	left $(5, 1)$	20
	231000	1	(4c)	left $(3, 1)$	31000
	123000	1	(4b)	left $(2, 1)$	3000
	213000	2	(4d)	left $(4, 2)$	3000
	201300	2	(4b)	left $(3, 1)$	300
	120300	1	(4b)	left $(2, 1)$	300
	210300	3	(4b)	left $(4, 1)$	300
	321000	6	(4a)	left $(6, 2)$	ϵ

Fig. 4. The left-shift Gray code cool(S) for Łukasiewicz words with content $S = \{0, 0, 0, 1, 2, 3\}$. Each row gives the non-increasing prefix length m, the rule (4), and the shift that creates the next word. The right column gives the scut of each string, which illustrates the suffix-based recursive definition of cool-lex order.

for α. Since this is true, next$(\alpha) = $ left$(m + 1, 1)$ by (4b), which is left$(3, 1)$ for α. In other words, the rule left-shifts a_3 into position 1. Thus, the next string in the list is $a_3 \cdot a_1 \cdot a_2 \cdot a_4 \cdot a_5 \cdot a_6 = 230100$, as seen in the second row of Fig. 4.

3.1 Observations

Note that (4) left-shifts a symbol that is at most two symbols past the non-increasing prefix. Thus, the shifts given by (4) are usually short, and the symbols at the right side of the string are rarely changed. This implies that the order will have some similarity to co-lexicographic order, which orders strings right-to-left by increasing symbols. In fact, the order turns out to be a cool-lex order, as discussed in Sect. 4.

4 Proof of Correctness

Now we prove that the successor rule is correct. Our strategy is to define a recursive order of $\mathcal{L}(S)$, and show that (4) creates the next string in this order.

4.1 Cool-lex Order

Cool-lex order is a variation of co-lexicographic order. The order was first given for (s, t)-*combinations*, which are binary strings with s copies of 0 and t copies of 1, by Ruskey and Williams [13,15]. In this context, the order gives a *prefix-shift Gray code*, meaning that a single symbol is left-shifted into the first position. The prefix-shift Gray code was then generalized to Dyck words [14] and multiset permutations [21]. The latter result provides the recursive structure of our left-shift Gray code of fixed-content Łukasiewicz words.

Tails and Scuts. Given a multiset S of cardinality n, we define the *tail of length ℓ* to be smallest ℓ symbols arranged in a string in non-increasing order. Formally,

$$\text{tail}(\ell) = t_\ell \cdot t_{\ell-1} \cdots t_2 \cdot t_1, \tag{5}$$

where tail$(n) = t_n \cdot t_{n-1} \cdots t_1$ is the unique non-increasing string with content S.

In English, a *scut* is a short tail. We use the term for a tail that is truncated by the addition of a large first symbol. More specifically, a scut of length ℓ and a tail of length ℓ are identical, except for their first symbol, and the first symbol is larger in the scut. Formally, the *scut of length $\ell + 1$*, with respect to S is

$$\text{scut}(s, \ell) = s \cdot \text{tail}(\ell), \tag{6}$$

where $s \in S$ is greater than the first symbol tail$(\ell + 1)$. We refer to a scut of the form scut(s, ℓ) as an *s-scut*.

Recursive Order. Now we define cool(S) to be an order of $\mathcal{L}(S)$. More broadly, we define cool(S) on any multiset S with non-negative symbols whose sum is at least as large as its cardinality, and we henceforth refer to these S as *valid*. We define cool(S) recursively by grouping the strings with the same scut together. Specifically, the scuts are ordered as follows:

- The scuts are first ordered by their first symbol in increasing order. In other words, s-scuts are before $(s + 1)$-scuts.
- For a given first symbol, the scuts are ordered by decreasing length. In other words, longer s-scuts come before shorter s-scuts.
- The string tail(n) is the only string without a scut, and it is ordered last.

For example, the rightmost column of Fig. 4 illustrates this order. More specifically, the scuts appear in the following order:

$$100, 10, 1, 2000, 200, 20, 31000, 3000, 300, \tag{7}$$

with the single string tail(n) = 321000 appearing last. Note that 2, 30 and 3 are absent from (7) because there are no Łukasiewicz words with these suffixes.

In each scut group the strings are ordered recursively. In other words, the common scut is removed from the strings in a particular group, and then they are ordered according to cool(S'), where S' is the valid multiset obtained by removing the symbols of the common scut from S. For example, in Fig. 4, the strings with scut 1 are ordered according to cool(S') where $S' = \{3, 2, 1, 0, 0, 0\} - \{1\} = \{3, 2, 0, 0, 0\}$. The base case of the recursion is when $S = \emptyset$.

In the following subsection it will be helpful to know the first string that has an s-scut. By our recursive order, we know that it will have a longest s-scut. Moreover, the exact string can be obtained from the tail by a single shift. To illustrate this, consider the list in Fig. 4, and let $\alpha = \text{tail}(n) = 321000$.

- The first string with a 1-scut is $\text{left}_\alpha(4, 2) = 302100$.
- The first string with a 2-scut is $\text{left}_\alpha(3, 1) = 132000$.
- The first string with a 3-scut is $\text{left}_\alpha(2, 1) = 231000$.

In other words, the first string with a 1-scut is obtained by shifting a 0 into the second position, with the first strings with 2-scuts and 3-scuts are obtained by shifting 1 and 2 into the first position, respectively. This point is stated more generally in the following remark.

Remark 1. Let S be a valid multiset, and tail(n) = $t_n \cdot t_{n-1} \cdots t_1$ with $t_i > t_{i-1}$. The first string in cool(S) with a t_i-scut is $\text{left}_{\text{tail}(n)}(n - i + 2, 1)$ if $t_{i-1} = 0$ or $\text{left}_{\text{tail}(n)}(n - i + 2, 2)$ if $t_{i-1} > 0$.

4.2   Equivalence

Now we prove that the successor rule (4) correctly provides the next string in cool(S). This simultaneously proves that (4) is a successor rule for a left-shift Gray code of $\mathcal{L}(S)$, and that cool(S) is a recursive description of the same.

Theorem 1. *Let S be a multiset of non-negative values with cardinality n and sum $\Sigma S = n$. Also, let $\alpha \in \mathcal{L}(S)$ be a Łukasiewicz word with content S, and $\beta \in \mathcal{L}(S)$ be the next string in cool(S) taken circularly (i.e., if α is the last string in cool(S), then β is the first string in cool(S)). Then $\beta = \text{left}_\alpha(j, i)$. In other words, the successor rule in (4) transforms α into β with a left-shift.*

Proof. Let $\alpha = a_1 \cdot a_2 \cdots a_n$ and $\rho = a_1 \cdot a_2 \cdots a_m$ be α's non-increasing prefix.

- If $m = n$, then $\alpha = \text{tail}(n)$ and it is the last string in $\text{cool}(S)$. We also know that $\text{next}(\alpha) = \text{left}(n, 2)$ by (4a). This gives the first string in $\text{cool}(S)$ with a 1-scut by Remark 1, which is the first string in $\text{cool}(S)$ as expected. This is the only case where (4a) is used.
- If $m = n - 1$, then α's non-increasing prefix extends until its second-last symbol. Furthermore, we know that $a_n = 1$, since this is the only non-zero value that can appear in the rightmost position. We also know that $\text{next}(\alpha) = \text{left}(m + 1, 1) = \text{left}(n, 1)$ by (4b). Thus, Remark 1 implies that β is the first string with an x-scut, where x is the smallest symbol larger than 1 in S. This is expected since α is the last string in the order with a 1-scut.

The remaining cases are handled cumulatively (i.e., each assumes that the previous do not hold). Note that $\alpha = \rho \cdot a_{m+1} \cdot a_{m+2} \cdots a_n$ is the last string with $\text{scut}(a_{m+1}, \ell) = a_{m+1} \cdot a_{m+2} \cdots a_w$ in a sublist $\text{cool}(S - \{a_{w+1}, a_{w+2}, \ldots, a_n\})$. We also view $\text{left}_\alpha(j, i)$ in two steps: a_j is left-shifted until it joins the non-increasing prefix, then further to index i. This allows us to use Remark 1.

- If $a_m < a_{m+2}$, then the scut at this level of recursion, namely $\text{scut}(a_{m+1}, \ell)$, cannot be shortened since $\ell = 0$. So the next scut will be the longest scut with the next largest symbol, which is true by Remark 1 and $\text{next}(\alpha) = \text{left}(m + 1, 1)$ by (4b).
- If $a_{m+2} = 0$ and $\Sigma\rho = m$, then the scut cannot be shortened since the sum of the symbols before the shorter scut will be less than their cardinality. Thus, the next scut will be the longest scut with the next largest symbol, which is true by Remark 1 and $\text{next}(\alpha) = \text{left}(m + 1, 1)$ from (4b).
- If $a_{m+2} \neq 0$, then the scut at this level of recursion can be shortened to $\text{scut}(a_{m+1}, \ell - 1)$. Given this shorter scut, the order recursively adds new scuts beginning with the first x-scut, where x is the second-smallest remaining symbol. This is true by Remark 1 and $\text{next}(\alpha) = \text{left}(m + 2, 1)$ by (4c).
- Otherwise, $a_{m+2} = 0$. This is identical to the previous case, except that $a_{m+2} = 0$. Thus, Remark 1 gives $\text{next}(\alpha) = \text{left}(m + 2, 2)$ by (4d).

Therefore, (4) gives the next string in the order, which completes the proof.

5 Loopless Algorithm for Fixed-Content Motzkin Words

We now use our Gray code for fixed-content Łukasiewicz words to looplessly generate fixed-content Motzkin words[2]. More specifically, COOLMOTZKIN is an array-based algorithm, and each shift is implemented with a constant number of assignments. Pseudocode is in Fig. 5, and Python code is in the Appendix.

The algorithm follows in a similar style to previous array-based algorithms for generating (s, t)-combinations [13,15], Dyck words [14], and $1/k$-ary Dyck words in cool-lex order [4,5]. The former two are provided for the sake of comparison in Fig. 5 under the names COOLCOMBO and COOLDYCK, respectively.

A loopless cool-lex algorithm for Łukasiewicz words would require a linked list (as in [21]) since a shift can relocate an arbitrarily number of distinct symbols.

---

[2] As noted in Sect. 2.1, these strings are also fixed-content Schröder words.

(a) Combinations	(b) Dyck Words	(c) Motzkin Words
COOLCOMBO(s,t)	COOLDYCK(t)	COOLMOTZKIN(s,t)

(a) Combinations

COOLCOMBO(s,t)
$n \leftarrow s+t$
$b \leftarrow 1^t 0^s$
$x \leftarrow t$
$y \leftarrow t$
visit(b)
while $x < n$ **do**
  $b_x = 0$
  $b_y = 1$
  $x \leftarrow x+1$
  $y \leftarrow y+1$
  if $b_x = 0$ **then**
    $b_x \leftarrow 1$
    $b_1 \leftarrow 0$
    if $y > 2$ **then**
      $x \leftarrow 2$
  $y \leftarrow 1$
  visit(b)

(b) Dyck Words

COOLDYCK(t)
$n \leftarrow 2 \cdot t$
$b \leftarrow 1^t 0^t$
$x \leftarrow t$
$y \leftarrow t$
visit(b)
while $x < n$ **do**
  $b_x = 0$
  $b_y = 1$
  $x \leftarrow x+1$
  $y \leftarrow y+1$
  if $b_x = 0$ **then**
    if $x \geq 2 \cdot y - 2$ **then**
      $x \leftarrow x+1$
    else
      $b_x \leftarrow 1$
      $b_2 \leftarrow 0$
      $x \leftarrow 3$
      $y \leftarrow 2$
  visit(b)

(c) Motzkin Words

COOLMOTZKIN(s,t)
$n \leftarrow 2 \cdot s + t$
$b \leftarrow 2^s 1^t 0^s$
$x \leftarrow n - 1$
$y \leftarrow t + s + 1$
$z \leftarrow s + 1$
visit(b)
while $x < n$ **or** $b_x < 2$ **do**
  $q \leftarrow b_{x-1}$
  $r \leftarrow b_x$
  if $x + 1 \leq n$ **then**
    $p \leftarrow b_{x+1}$
  $b_x \leftarrow b_{x-1}$
  $b_y \leftarrow b_{y-1}$
  $b_z \leftarrow b_{z-1}$
  $b_1 \leftarrow r$
  $x \leftarrow x+1$
  $y \leftarrow y+1$
  $z \leftarrow y+1$
  if $p = 0$ **then**
    if $z - 2 > x - y$ **then**
      $b_1 \leftarrow 2$
      $b_2 \leftarrow 0$
      $b_x \leftarrow r$
      $x \leftarrow 3$
      $y \leftarrow 2$
      $z \leftarrow 2$
    else
      $x \leftarrow x+1$
  else if $x \leq n$ **and** $q \geq b_x$ **then**
    $b_x \leftarrow 2$
    $b_{x-1} \leftarrow 1$
    $b_1 \leftarrow 1$
    $z \leftarrow 1$
  if $b_2 > b_1$ **then**
    $z \leftarrow 1$
    $y \leftarrow 2$
    $x \leftarrow 2$
  visit(b)

Fig. 5. Algorithms for generating (a) (s,t)-combinations, (b) Dyck words, and (c) fixed-content Motzkin words in cool-lex order. The algorithms are loopless and store the current string in array $b = b_1 b_2 \cdots b_n$ (i.e., 1-based indexing). The parameters $s \geq 2$ and $t \geq 2$ give the number of 0s (and 2s) and 1s, respectively. Variables z, y, and x given the index after the 2s, 1s, and 0s in the non-increasing prefix, respectively. (Their initial values are exceptions to this pattern, and are set to make the first iteration work correctly.) The start of the **while** loop shifts the first increasing symbol to the left (i.e., (4b) in COOLMOTZKIN) and the **if** statements identify when this is not the correct shift, and adjust b accordingly. Also, COOLMOTZKIN uses q, r, p to save the symbols around the first increase.

Appendix: Python Code

Python3 functions for generating the cool-lex order of (s,t)-combinations, Dyck words of order t, and fixed-content Motzkin words with s copies of 0 and 2 and t copies of 1, are found in Fig. 6[3]. The first two are found in [13,15] and [14], respectively, and the latter is new to this article. To simulate the 1-based indexing used in Fig. 5, we store array b in a list and ignore its first entry b[0]. Lists are implemented as arrays in CPython, so each read and write is a worst-case $O(1)$-time operation. Hence, the implementations are loopless.

```
def coolCombo(t,s):
..n = s+t
..b = [-1]+[1]*t+[0]*s
..x = t
..y = t
..print(*b[1:],sep="")
..while x < n:
....b[x] = 0
....b[y] = 1
....x += 1
....y += 1
....if b[x] == 0:
......b[x] = 1
......b[1] = 0
......if y > 2:
........x = 2
......y = 1
....print(*b[1:],sep="")
```

```
def coolDyck(t):
..n = 2*t
..b = [-1]+[1]*t+[0]*t
..x = t
..y = t
..print(*b[1:],sep="")
..while x < n-1:
....b[x] = 0
....b[y] = 1
....x += 1
....y += 1
....if b[x] == 0:
......if x >= 2*y - 2:
........x += 1
......else:
........b[x] = 1
........b[2] = 0
........x = 3
........y = 2
....print(*b[1:],sep="")
```

```
def coolMotzkin(t,s):
..n = 2*s + t
..b = [-1]+[2]*s+[1]*t+[0]*s
..x = n-1
..y = t+s+1
..z = s+1
..print(*b[1:],sep="")
..while x < n-1 or b[x] < 2:
....q = b[x-1]
....r = b[x]
....if x + 1 <= n:
......p = b[x+1]
....b[x] = b[x-1]
....b[y] = b[y-1]
....b[z] = b[z-1]
....b[1] = r
....y += 1
....z += 1
....x += 1
....if p == 0:
......if z-2 > (x-y):
........b[1] = 2
........b[2] = 0
........b[x] = r
........z=2
........y=2
........x=3
......else:
........x+=1
....elif x <= n and q >= b[x]:
......b[x] = 2
......b[x-1] = 1
......b[1] = 1
......z = 1
....if b[2] > b[1]:
......z = 1
......y = 2
......x = 2
....print(*b[1:],sep="")
```

Fig. 6. Loopless generation of the cool-lex shift Gray codes of (s,t)-combinations, Dyck words, and fixed-content Motzkin words in Python 3. Each shift is achieved using a constant number of assignments to the list b.

---

[3] The leading spaces have been replaced with periods to ensure that the code can be reliably copy-and-pasted from digital versions of this document.

References

1. Balakirsky, V.B.: A new coding algorithm for trees. Comput. J. **45**(2), 237–242 (2002)
2. Banderier, C., Wallner, M.: The kernel method for lattice paths below a line of rational slope. In: Andrews, G.E., Krattenthaler, C., Krinik, A. (eds.) Lattice Path Combinatorics and Applications. DM, vol. 58, pp. 119–154. Springer, Cham (2019). https://doi.org/10.1007/978-3-030-11102-1_7
3. Dershowitz, N., Zaks, S.: Enumerations of ordered trees. Discret. Math. **31**(1), 9–28 (1980)
4. Durocher, S., Li, P.C., Mondal, D., Ruskey, F., Williams, A.: Cool-lex order and k-ary Catalan structures. J. Discrete Algorithms **16**, 287–307 (2012)
5. Durocher, S., Li, P.C., Mondal, D., Williams, A.: Ranking and loopless generation of k-ary Dyck words in cool-lex order. In: Iliopoulos, C.S., Smyth, W.F. (eds.) IWOCA 2011. LNCS, vol. 7056, pp. 182–194. Springer, Heidelberg (2011). https://doi.org/10.1007/978-3-642-25011-8_15
6. Hodgson, J.: Rediscovered: the Jan Łukasiewicz papers. https://rylandscollections.com/2018/05/16/rediscovered-the-jan-lukasiewicz-papers
7. Knuth, D.E.: Art of Computer Programming, Volume 4, Fascicle 4, The: Generating All Trees–History of Combinatorial Generation. Addison-Wesley (2013)
8. Korsh, J.F., LaFollette, P.: Loopless generation of trees with specified degrees. Comput. J. **45**(3), 364–372 (2002)
9. Mütze, T.: Combinatorial Gray codes-an updated survey. arXiv preprint arXiv:2202.01280 (2022)
10. Nakano, S.I.: Listing all trees with specified degree sequence (acceleration and visualization of computation for enumeration problems). RIMS Kôkyûroku Bessatsu 1644, 55–62 (2009)
11. Narayana, T.V.: Lattice Path Combinatorics with Statistical Applications; Mathematical Expositions 23. University of Toronto Press (1979)
12. Ruskey, F., Sawada, J., Williams, A.: Binary bubble languages and cool-lex order. J. Comb. Theor. Ser. A **119**(1), 155–169 (2012)
13. Ruskey, F., Williams, A.: Generating combinations by prefix shifts. In: Wang, L. (ed.) COCOON 2005. LNCS, vol. 3595, pp. 570–576. Springer, Heidelberg (2005). https://doi.org/10.1007/11533719_58
14. Ruskey, F., Williams, A.: Generating balanced parentheses and binary trees by prefix shifts. In: Proceedings of the Fourteenth Symposium On Computing: The Australasian Theory. Vol. 77, pp. 107–115. Citeseer (2008)
15. Ruskey, F., Williams, A.: The coolest way to generate combinations. Discret. Math. **309**(17), 5305–5320 (2009)
16. Savage, C.: A survey of combinatorial Gray codes. SIAM Rev. **39**(4), 605–629 (1997)
17. Sloane, N.J.A., The OEIS Foundation Inc.: The on-line Encyclopedia of integer sequences (2020). https://oeis.org/
18. Stanley, R.P.: Catalan numbers. Cambridge University Press (2015)
19. Van Baronaigien, D.R.: A loopless algorithm for generating binary tree sequences. Inf. Process. Lett. **39**(4), 189–194 (1991)
20. Wallner, M.: Combinatorics of Lattice Paths and Tree-Like Structures. Ph.D. thesis. Wien (2016)

21. Williams, A.: Loopless generation of multiset permutations using a constant number of variables by prefix shifts. In: Proceedings of the Twentieth Annual ACM-SIAM Symposium on Discrete Algorithms, pp. 987–996. SIAM (2009)
22. Williams, A.M.: Shift Gray Codes. Ph.D. thesis. University of Victoria (2009)
23. Zaks, S., Richards, D.: Generating trees and other combinatorial objects lexicographically. SIAM J. Comput. 8(1), 73–81 (1979)

Learning from Positive and Negative Examples: Dichotomies and Parameterized Algorithms

Jonas Lingg[1] , Mateus de Oliveira Oliveira[2] , and Petra Wolf[3(✉)]

[1] Eberhard Karls Universität Tübingen, Tübingen, Germany
[2] University of Bergen, Bergen, Norway
mateus.oliveira@uib.no
[3] Universität Trier, Trier, Germany
wolfp@informatik.uni-trier.de

Abstract. We take a closer look on the complexity landscape of one of the most fundamental and well-studied problems in computational learning theory: the problem of learning a finite automaton A consistent with a set P of positive examples and with a set N of negative examples. By consistency, we mean that A accepts all strings in P and rejects all strings in N. It is well known that this problem is NP-hard when parameterized only by the number of states of the automaton. Therefore, our analysis takes a more refined parameterization: we consider the number k of states in A, the size $|\Sigma|$ of the alphabet, the maximum size l of a string in $P \cup N$, and the number $c = |P \cup N|$ of strings in both sets. First, we prove several Pvs. NP-hard dichotomy results for these parameters when the learned automaton is drawn from different classes of finite automata. One of our dichotomy results closes a gap for the general DFA consistency problem, as here, for fixed alphabet size, the NP-hardness proofs in the literature have some issues. Interestingly, our NP-hardness results hold even for severely restricted classes of automata, such as partially-ordered automata and permutation automata. On the other hand, we provide parameterized algorithms for several combinations of parameters and show that most of them are optimal under the exponential time hypothesis.

1 Introduction

In the *DFA-consistency problem* (DFA-CON) we are given a pair of disjoint sets of strings $P, N \subseteq \Sigma^*$ and a positive integer k. The goal is to determine whether there is a deterministic finite automaton A, DFA for short, with at most k states that accepts all strings in P and rejects all strings in N. Despite the problem cannot be approximated [37], it has become one of the most central problems in computational learning theory [2,14,15,34,36], with applications that span several subfields of artificial intelligence and related areas, such as automated synthesis of controllers [38], model checking [16,26] optimization [4,7,31,45] neural networks [18,19,28,29], multi-agent systems [33], and others [20,32,39].

© Springer Nature Switzerland AG 2022
C. Bazgan and H. Fernau (Eds.): IWOCA 2022, LNCS 13270, pp. 398–411, 2022.
https://doi.org/10.1007/978-3-031-06678-8_29

Despite having been studied for at least five decades, certain questions concerning the computational complexity of DFA-CON have remained open. In this work, we analyze some of these questions using the framework of parameterized complexity theory [8,9]. Our main parameters, are the most intuitive ones associated with DFA-CON: the number of states k of the target DFA, the size $|\Sigma|$ of the alphabet, the maximum length l of a string in $P \cup N$, the number $c = |P \cup N|$ of strings, and the size T of the prefix-tree of words in $P \cup N$.

In our first result, we show that DFA-CON is NP-complete for binary alphabets (Theorem 1). It is worth noting that this problem has been claimed to be NP-hard in [20], and in [12]. Nevertheless, a careful inspection of both proofs reveals that these hardness results do not hold in the context of DFAs, but rather in the context of Mealy machines, a more concise model of computation. Indeed, both proofs are based on adaptations of a classical NP-hardness result obtained in [15] in the context of Mealy machines. Unfortunately, in the presented way, this technique does not carry over to usual DFAs, and the problem of determining if DFA-CON over binary alphabets is NP-hard has remained open [10]. In this work, we settle this problem using techniques that are completely distinct from the ones proposed in [12,20], and which do not rely on results from [15].

The parameterized complexity of DFA-CON has been studied in details in [12]. In this work, we revisit some results from [12] and analyze hardness of DFA-CON for two important restricted sub-classes of DFAs: *partially ordered automata* (PODFA-CON) and *permutation automata* (PA-CON). In particular, we establish several new dichotomy results for both problems. A summary of our results can be found in Table 1. It is worth noting that hardness for these restricted classes of automata is not implied by hardness of DFA-CON . Indeed for some of the parameters, the threshold value to ensure NP-hardness is increased by one. We further get faster FPT-algorithms for these types of automata.

2   Preliminaries

For a finite alphabet Σ, we call Σ^* the set of all words over Σ. For a regular expression r and a fixed number k, we denote k concatenations of r with r^k. A *deterministic finite automaton* (DFA) is a tuple $A = (Q, \Sigma, \delta, s_0, F)$ where Q is a finite set of states, Σ a finite alphabet, $\delta : Q \times \Sigma \to Q$ a total transition function, s_0 the initial state and $F \subseteq Q$ the set of final states. We call A a *complete* DFA if we want to highlight that δ is a *total* function. If δ is partial, we call A a *partial* DFA. We generalize δ to words by $\delta(q, aw) = \delta(\delta(q, a), w)$ for $q \in Q, a \in \Sigma, w \in \Sigma^*$. We further generalize δ to sets of input letters $\Gamma \subseteq \Sigma$ by $\delta(q, \Gamma) = \bigcup_{\gamma \in \Gamma} \{\delta(q, \gamma)\}$. A DFA A accepts a word $w \in \Sigma^*$ if and only if $\delta(s_0, w) \in F$. We let $\mathcal{L}(A)$ denote the *language of A*, i.e., the set of all words accepted by A. The length of a word w is denoted by $|w|$ and the number of occurrences of a letter $\sigma \in \Sigma$ in a word w by $|w|_\sigma$. We denote the empty word with ϵ, $|\epsilon| = 0$. The DFA-CON problem is formally defined as follows.

Definition 1 (DFA-Con)

Input: Finite set of words $P, N \subseteq \Sigma^*$ with $P \cap N = \emptyset$, and integer k.

Output: Is there a DFA $A = (Q, \Sigma, \delta, s_0, F)$ with $|Q| \leq k$, $P \subseteq \mathcal{L}(A)$, and $\mathcal{L}(A) \cap N = \emptyset$?

Table 1. New results are depicted in red. In the left half, we list values of parameters with NP-hard consistency problems. In the right half, we list algorithms for combinations of parameters. The entries in black concerning DFA-Con were listed in [12]. The parameters are: k is the number of states, $|\Sigma|$ the size of the alphabet, l the maximum length of a word in $P \cup N$, T the size of the prefix tree of $P \cup N$, and $c = |P \cup N|$. We also obtain a complete proof for the claim that DFA-Con is NP-hard for $|\Sigma| = 2$. Although some previous works had claimed this particular result, the proofs turned out to be flawed. The notation $\mathcal{O}^*$ hides factors polynomial in the input.

Param.	DFA-Con	poDFA-Con	PA-Con	Param.	DFA-Con	poDFA-Con	PA-Con
k	$k = 2$	$k = 3$	$k = 3$	$k, |\Sigma|$	$\mathcal{O}^*(k^{k|\Sigma|})$	$\mathcal{O}^*(k!^{|\Sigma|})$	$\mathcal{O}^*(k!^{|\Sigma|})$
$|\Sigma|$	$|\Sigma| = 2$	$|\Sigma| = 2$	$|\Sigma| = 2$	k, T	$-$	$\mathcal{O}^*(k^T)$	$\mathcal{O}^*(k^T)$
l	$l = 2$	$l = 2$	$l = 2$	k, l, c	$\mathcal{O}^*(k^{cl})$	$\mathcal{O}^*(k^{cl})$	$\mathcal{O}^*(k^{cl})$
k, l	$k \cdot l = 6$	$k \cdot l = 8$	$k \cdot l = 6$	l, c	$\mathcal{O}^*((cl)^{cl})$	$\mathcal{O}^*((cl)^{cl})$	$\mathcal{O}^*((cl)^{cl})$

3   Dichotomies for PO Automata

A DFA is called a *partially ordered* deterministic automaton (poDFA) if there exists an order $\leq$ on Q that fulfills the constraint: $\forall \sigma \in \Sigma, \forall p, q \in Q \colon (\delta(p, \sigma) = q) \rightarrow p \leq q$. In other words, the underlying graphs of poDFAs are directed acyclic graphs, possibly augmented by self-loops on each state. For this reason these automata are also called *acyclic automata* in the literature, i.e., in [23].

Partially ordered DFAs form a sub-class of aperiodic automata, and therefore they only recognize star-free languages [40,42]. From an algebraic perspective, the class of languages associated with poDFAs has been characterized in terms of Green's relation (see [6]) in [5]. The sub-class of confluent poDFAs characterizes the $\frac{1}{2}$ level of the Straubing-Thérien hierarchy (also known as piecewise-testable languages) [24,43] which consists of unions of languages of the form $\Sigma^* a_1 \Sigma^* a_2 \Sigma^* \ldots a_n \Sigma^*$ where a_i are single letters in Σ. If we build the Kleene closure on this class (i.e., allow a $*$ on the whole language without introducing new concatenations), we get a permutation language accepted by a deterministic permutation automaton (Sect. 4).

From a computational perspective, deterministic poDFAs are relevant because they can be converted into regular expressions without an exponential blow-up (since they accept star-free languages) [17]. They are also relevant in the context of learning theory. It was shown in [11] that with the technique of alignment-based learning, an important sub-class of poDFAs called simple-looping automata can be learned in the limit from positive examples. Additionally, partially ordered automata have strong connections with an important class of regular expressions called (extended) chain regular expressions (eCHAREs) introduced in [27]. This fragment is a super-class of the regular expressions most frequently used in schema languages for XML and a superset of the sequence motifs used in bioinformatics [30]. Further, eCHAREs are used in verification of lossy channel systems [1].

For DFA-CON restricted to binary alphabets ($|\Sigma| = 2$), two proposed proofs of NP-hardness have appeared in the literature: [20] and [12]. Nevertheless, both of them follow the same approach, and both contain inaccuracies that invalidate the proofs. They have in common that they are adaptations of the construction by Gold for the consistency problem of Mealy automata [15][1] but as the Mealy automata considered in [15] (mapping $\Sigma^* \to \Gamma$, $|\Gamma| = 2$) can be more compact (when interpreted as language acceptors) than DFAs recognizing the same language, the difference in number of states causes the adaptations to fail. For instance, we can give a counterexample to this claim. Consider the satisfiable formula $\varphi_1 = \neg x_1 \wedge x_2 \wedge x_3$ with $n = 3$ and implicit variable order $x_1 < x_2 < x_3$ as well as clause order $\{\neg x_1\} < \{x_2\} < \{x_3\}$. Following the construction in Theorem 2 in [15], one can verify that there is a three state Mealy automaton recognizing the words in the state characterization matrix correctly, but there is no three state DFA consistent with the data.

In the next theorem, we solve this issue by presenting a reduction that is essentially different from Gold's reduction. Since the presented construction allows only DFAs as separators which are already poDFAs, we also observe that for fixed alphabet size, the line between tractability and intractability of PODFA-CON lies between $|\Sigma| = 1$ and $|\Sigma| = 2$.

Theorem 1. PODFA-CON *admits a linear-time algorithm for* $|\Sigma| = 1$. *On the other hand, both* PODFA-CON *and* DFA-CON *are NP-complete for* $|\Sigma| \geq 2$.

Proof. Linear-time algorithm for $|\Sigma| = 1$. We are given P, N with $P \cap N = \emptyset$ and $k \in \mathbb{N}$. As each unary poDFA consists of a path with a self-loop on the last state, there exists a poDFA with at most k states that is consistent with P and N if and only if at least one of the longest words in P and N is shorter than k.

NP-hardness for $|\Sigma| \geq 2$. We show that there is a reduction from POS-ONE-IN-THREE 3SAT[2] to DFA-CON [and PODFA-CON] mapping each instance of POS-ONE-IN-THREE 3SAT with n variables and m clauses to an instance of DFA-CON [PODFA-CON] with $k = 4(n+1)+2$, $|\Sigma| = 2$, $|P|, |N| = \mathcal{O}(n^2 + nm)$, $l = \mathcal{O}(n)$, $T = \mathcal{O}(n^3 + n^2 m)$ in time $\mathcal{O}(n^3 + n^2 m)$.

Let ϕ be a Boolean formula in 3CNF with n variables $V = \{x_0, x_1, \ldots, x_{n-1}\}$ and m clauses $C = \{c_0, c_1, \ldots, c_{m-1}\}$, where each clause contains only positive literals. W.l.o.g., we assume that each variable in V appears in at least one clause. We construct from ϕ two sets $P, N \subseteq \{a, t\}^*$ and $k = 4(n + 1) + 2$

---

[1] Also in [2] Mealy automata instead of DFAs are considered.

[2] Given a Boolean formula in conjunctive normal form where all clauses have exactly three literals (this form is called 3CNF) and all literals are positive. Is there a variable assignment such that *exactly* one literal per clause evaluates to **true**?

such that there exists a variable assignment $\beta\colon V \to \{\texttt{false}, \texttt{true}\}$ such that under β exactly one literal is true in each clause $c \in C$ if and only if there exists a DFA with at most k states that is consistent with P and N. We first present the automaton structure determined by P and N in Fig. 1, where some transitions are still missing. The precise definitions of the sets P and N is given in Table 2 via regular expressions. We note that since n is fixed, all expressions describe finite sets. All variables are assumed to be picked from $\mathbb{N}_0$, i.e., their range starts with 0. First, we discuss why P and N determine the automaton structure in Fig. 1. Then, we argue that the realization of the missing transitions will correspond to the sought variable assignment β. We show in Lemma 1 that the only three possibilities to realize the missing t-transitions for even-numbered states are the ones depicted in Fig. 2.

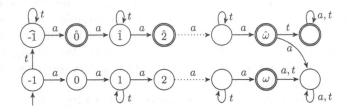

Fig. 1. Structure of a DFA enforced by P and N. The number-labels of states are only depicted for the first 3 states of both chains. We have $\omega = 2n$.

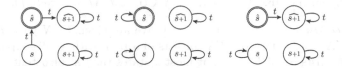

Fig. 2. Only three possibilities to realize the transition t for even states $s = 2i$ and $\hat{s} = \widehat{2i}$ in the lower and upper chain. The left (middle and right) scheme corresponds to setting the variable x_i to true (false). The right scheme can only appear for variables which are set to false and do not appear after a variable set to true in any clause.

The words in (1) demand a chain of $2n + 1$ non-accepting states followed by an accepting state. By (2), the transition t brings the first state of the lower chain to a second chain of equal length, where accepting and rejecting states alternate. Both chains end with an a in a common rejecting trap state. While the lower chain also ends with a t in the rejecting trap state, the upper chain ends with a final t in an accepting trap state. We now reached the maximal number of states k. The words in (3) state that from every even-numbered state s in the lower chain the letter t maps exactly to s or $\hat{s}$ since otherwise too many or too few alternations of accepting and rejecting states would lie to the right of

Table 2. Definition of the sets P and N in the proof of Theorem 1.

Words in P		Words in N	

(1) Words determining the states of the lower chain.

$$a^i \qquad\qquad i \leq 2n$$

$$a^{2n+1} \qquad\qquad\qquad a^{2n+1}(a|t)(\epsilon|a|t)^2$$

$$a^{2n+2+i},\ a^{2n+1}ta^i \qquad i \leq 2n+1$$

(2) Structure of the upper chain.

$$ta(aa)^i \qquad\qquad i \leq n \qquad\qquad t(aa)^i \qquad\qquad\qquad i \leq n+1$$

$$ta(aa)^n t(\epsilon|a|t)^2 \qquad\qquad\qquad t(aa)^{n+1}(\epsilon|a|t)^2 a^i \qquad i \leq 2n+1$$

(3) Relation between the two chains.

$$a(aa)^i t(aa)^{n-i} \qquad i \leq n-1 \qquad a(aa)^i ta(aa)^j \qquad i \leq n-1,\, j \leq n-i$$

$$a(aa)^i ta(aa)^{n-i}(\epsilon|a|t)^2 \qquad i \leq n-1$$

$$ta(aa)^i t(aa)^{n-i-1} atat(\epsilon|a) \qquad i \leq n \qquad ta(aa)^i t(aa)^{n-i} aa^j \qquad i \leq n,\, j \leq 2n$$

$$a^i tt \qquad\qquad\qquad i \leq 2n+2$$

(4) Loops in odd states in lower chain.

$$aa(aa)^i t(aa)^{n-i-1} a \qquad i \leq n-1 \qquad aa(aa)^i ta^j \quad i \leq n-1,\, j \leq 2(n-i-1)$$

$$aa(aa)^i t(aa)^{n-i-1} at(\epsilon|a|t)^2 \qquad i \leq n$$

(5) Loops in odd states in upper chain.

$$t(aa)^i ta(aa)^j \qquad i \leq n,\, j \leq n-i \qquad t(aa)^i t(aa)^j \qquad i \leq n,\, j \leq n-i+1$$

$$t(aa)^i t(aa)^{n-i} at(\epsilon|a|t)^2 \qquad i \leq n \qquad t(aa)^i t(aa)^{n-i+1}(\epsilon|a|t)^2 \qquad i \leq n$$

(6) Clause words. For each clause $c = \{x_i, x_j, x_k\}$ with $i < j < k < n$.

$$l \leq n-k \qquad\qquad\qquad\qquad\qquad l \leq n-k$$

$$a(aa)^i t(aa)^{j-i} t(aa)^{k-j} t(aa)^l \qquad a(aa)^i t(aa)^{j-i} t(aa)^{k-j} t(aa)^l a$$

$$a(aa)^i t(aa)^{j-i} t(aa)^{k-j} t(aa)^{n-k+1}$$

$\delta(s,t)$. Further, if $\delta(s,t)$ is an accepting state, this state must be left to its right neighbor with another t, i.e., $\delta(\delta(s,t),t) = \widehat{s+1}$. If this transition would lead anywhere else, then the second half of (3) (i.e., $ta(aa)^i t(aa)^{n-i-1} atat(\epsilon|a) \in P$, $i \leq n$, and $a^i tt \in N$, $i \leq 2n+2$) would be violated. The words in (4) and (5) demand that the letter t acts as the identity on all states $s, \hat{s}$ with an odd number > 0 and on $\widehat{-1}$ (note that for the initial state -1, t maps to the upper chain). The words in (6) demand that for every clause $c = \{x_i, x_j, x_k\}$ with $i < j < k$, we are in the accepting state $\hat{k}$ after reading the prefix $a(aa)^i t(aa)^{j-i} t(aa)^{k-j} t$. We will observe later that this is only possible if exactly one variable appearing in c has been set to true. Therefore, we observe in Lemma 1 that the only three possibilities to realize the missing transitions t for the even-numbered states, are the three depicted in Fig. 2. Due to the enforced structure, every DFA consistent with P and N has at least $4(n+1)+2$ states.

For the correctness of the construction, first assume there exists a DFA A with k states which is consistent with P and N. Due to Lemma 1 for each state s with even number, $\delta(s,t)$ is either s or $\hat{s}$. We extract from those transitions a variable assignment β. For each $x_i \in V$, we set $\beta(x_i) = \text{true}$ if for the state s

with number $2i$ $\delta(s,t) = \hat{s}$, and $\beta(x_i) = \texttt{false}$ if $\delta(s,t) = s$. For every clause word, it is only possible to switch from the lower chain to the upper chain with a letter t. Hence, from the acceptance of each clause word $a(aa)^i t(aa)^{j-i} t(aa)^{k-j} t$ for a clause $c = \{x_i, x_j, x_k\}$, we know that at least one literal in c is true. Now, assume more than one literal in c is true. Then, A switches from the lower chain to the upper chain with the first true literal. The letter t corresponding to a second true literal will then be read from a state in the upper chain with even number. But according to Lemma 1 this state is mapped to the next odd state with t. From now on A is in an odd state when reading a t and stays with t in that state according to (5). But as the odd states are rejecting, the word $a(aa)^i t(aa)^{j-i} t(aa)^{k-j} t$ is no longer accepted. If in contrast only one literal per clause is true, then A stays in the even states after all letters t in the clause words and accepts the clause words. Hence, exactly one literal per clause is true under β.

For the other direction, assume there exists a satisfying variable assignment β such that exactly one literal per clause is true under β. Realizing the missing transitions in Fig. 1, according to the left and middle scheme in Fig. 2 which corresponds to the assignment under β, gives a DFA A, which is consistent with P and N by the discussion above.

We can impose an order on the states in Fig. 1 from left to right and bottom to top such that no transition is leading backwards in the order. Further, for all completions of the automaton structure in Fig. 1 according to Lemma 1 and Fig. 2, the obtained automaton is a poDFA. Hence, the construction also works for PODFA-CON as all consistent DFAs are already poDFAs.   □

Lemma 1. *For each even states $s, \hat{s}$ in the lower and upper chain with $s, \hat{s} < 2n$, the only possible realizations of the transition t such that the resulting automaton A is consistent with P and N are the three schemes in Fig. 2.*

While DFA-CON is NP-complete for $k = 2$, we show in the following theorem that PODFA-CON is still in P for $k = 2$ and becomes NP-complete for $k \geq 3$.

Theorem 2. *PODFA-CON can be solved in linear time for $k \leq 2$, and is NP-complete for $k \geq 3$.*

Proof. Linear time algorithm for $k \leq 2$. For $k = 1$, the case is clear as every automaton with only one state can accept only the empty or universal language. For $k = 2$, every word accepted by a poDFA $A = (\{q_0, q_1\}, \Sigma, \delta, q_0, F)$ with two states can only cause a state switch once, as non-trivial loops are forbidden. As we are only considering complete poDFAs, we can partition Σ into Σ_0 and Σ_1, where $\delta(q_0, \Sigma_0) = \{q_0\}$ and $\delta(q_0, \Sigma_1) = \{q_1\}$. For A, in order to distinguish two non-empty sets P and N, one of the states q_0 and q_1 must be accepting and one must be non-accepting. If q_0 (resp., q_1) is the accepting state, then every word $w \in N$ (resp., P) must cause a transition to q_1 in contrast to the words in P (resp., N). Based on this consideration, we give in Algorithm 1 a linear time algorithm that determines the sets F, Σ_0, and Σ_1 (and hence specifies A) for a

---

Algorithm 1. Solving PODFA-CON for $k \leq 2$

Input: N, P
Output: F, partition $\Sigma = \Sigma_0 \dot{\cup} \Sigma_1$ with $\delta(q_0, \Sigma_0) = \{q_0\}$ and $\delta(q_0, \Sigma_1) = \{q_1\}$.

1: **for** $w \in N \cup P$ **do**
2:     set $\Sigma_w := \{\sigma \mid |w|_\sigma > 0\}$
3: **end for**
4: set $\Sigma_+ := \bigcup_{w \in P} \Sigma_w$
5: set $\Sigma_- := \bigcup_{w \in N} \Sigma_w$
6: **if** $\forall w \in N \colon \Sigma_w \backslash \Sigma_+ \neq \emptyset$ **then**
7:     **return** $F = \{q_0\}$, $\Sigma_0 = \Sigma_+$, $\Sigma_1 = \Sigma \backslash \Sigma_+$
8: **else if** $\forall w \in P \colon \Sigma_w \backslash \Sigma_- \neq \emptyset$ **then**
9:     **return** $F = \{q_0\}$, $\Sigma_0 = \Sigma_-$, $\Sigma_1 = \Sigma \backslash \Sigma_-$
10: **else**
11:     **return** 'no automaton possible'
12: **end if**

---

given instance of $P, N \subseteq \Sigma^*$ if and only if there exists a poDFA with two states that is consistent with P and N.

NP-hardness for $k \geq 3$. For the NP-hardness proof, we will show that there is a reduction from ONE-IN-THREE 3SAT to PODFA-CON s.t. each ONE-IN-THREE 3SAT instance with n variables and m clauses is mapped to a PODFA-CON instance with $k = 3$, $|\Sigma| = 2n$, $|P| = 2n + m$, $|N| = 4n + 1$, $l = 4$, $T = \mathcal{O}(n + m)$ in time $\mathcal{O}(n + m)$.

Given a ONE-IN-THREE 3SAT Boolean Formula ϕ with set of variables $V = \{x_1, \ldots, x_n\}$ and clauses $C = \{c_1, \ldots, c_m\}$. We construct from ϕ a PODFA-CON instance with $k = 3$ and alphabet $\Sigma = \{x_i, \overline{x_i} \mid x_i \in V\}$ where $x_i, \overline{x_i}$ are symbols representing the positive and negative literals of the variable x_i. Each clause $c_j = \{l_{j_1}, l_{j_2}, l_{j_3}\}$ is represented by a word $\sigma_{c_j} = l_{j_1} l_{j_2} l_{j_3}$. We complete the definition of the PODFA-CON instance by defining the sets

$$P = \{x_i \overline{x_i}, \ \overline{x_i} x_i \mid x_i \in V\} \cup \{\sigma_{c_j} \mid c_j \in C\},$$
$$N = \{x_i \overline{x_i} x_i \overline{x_i}, \ \overline{x_i} x_i \overline{x_i} x_i, \ x_i x_i, \ \overline{x_i} \overline{x_i} \mid x_i \in V\} \cup \{\epsilon\}.$$

First, assume A is a poDFA with states $\{q_0, q_1, q_2\}$ and alphabet Σ that accepts all words of P and rejects all of N. Observe that the initial state of A (further called q_0) cannot be an accepting state due to $\epsilon \in N$. From the remaining states q_1 and q_2 exactly one must be accepting and one must be rejecting since for every $x_i \in V$, $\delta(q_0, x_i \overline{x_i}) \in F$ and $\delta(\delta(q_0, x_i \overline{x_i}), x_i \overline{x_i}) \notin F$ while the partially ordered property of A enforces $q_0 \neq \delta(\delta(q_0, x_i \overline{x_i}), x_i \overline{x_i})$. W.l.o.g., assume $\delta(q_0, x_i \overline{x_i}) = q_1$ and $\delta(q_1, x_i \overline{x_i}) = q_2$. We can further observe that $\delta(q_0, \overline{x_i} x_i) = q_1$ and $\delta(q_1, \overline{x_i} x_i) = q_2$. Under these observations, the only possible realizations of the single transitions x_i and $\overline{x_i}$ for each $x_i \in V$ are the two depicted in Fig. 3. We show that for a consistent poDFA A, the chosen realization of the transitions for each variable x_i corresponds to a satisfying variable assignment for the corresponding formula ϕ such that exactly one literal per clause is true. We set

Fig. 3. Only possible realizations of the combined transitions $x_i\overline{x_i}$ and $\overline{x_i}x_i$. The top scheme corresponds to a variable assignment $x_i \mapsto 1$; the bottom scheme to $x_i \mapsto 0$.

each variable x_i to 1 if $\delta(q_0, x_i) = q_1$ and to 0 if $\delta(q_0, x_i) = q_0$. As A accepts all words σ_{c_j} for $c_j \in C$ in each clause at least one literal is true since q_0 has been left into q_1 with each σ_{c_j}. Further, since q_1 is also left with every true literal it follows from $\delta(q_0, \sigma_{c_j}) = q_1$ that in each clause *exactly* one literal is true.

For the other direction, we start with a satisfying variable assignment where for each clause exactly one literal is true. For each variable x_i, we pick the left (respectively right) scheme to realize the transitions x_i and $\overline{x_i}$ if x_i is set to `true` (respectively `false`). As exactly one literal is true per clause each word σ_{c_j} causes exactly one state switch and hence $\delta(q_0, \sigma_{c_j}) = q_1 \in F$. Clearly, the so obtained poDFA is also consistent with the remaining words in P and N. $\square$

Finally, we establish next a dichotomy theorem for PODFA-CON parameterized by the maximum size l of a string in the set $P \cup N$.

Theorem 3. PODFA-CON *can be solved in polynomial time for $l = 1$ and is NP-complete for $l \geq 2$.*

Proof. If $|w| \leq 1$ for every word $w \in P \cup N$, then PODFA-CON is trivially solvable in polynomial time, since in this case P and N are just subsets of letters from Σ. On the other hand, next we show that allowing arbitrary words of length 2 renders PODFA-CON NP-complete. The proof will follow by showing that there is a reduction from 3-COLORING to PODFA-CON mapping each instance of 3-COLORING with $|V| = n$ and $|E| = m$ to a PODFA-CON instance with $k = 4$, $|\Sigma| = n + 2m$, $|P| = |N| = 2m$, $l = 2$, and $T = n + 2m$ in time $\mathcal{O}(n + m)$.

Let $G = (V, E)$ be a graph with vertex set V and edge set E. W.l.o.g., we assume (i) that G does not contain isolated vertices. We construct from G an instance of PODFA-CON with $k = 4$ as: $\Sigma = V \cup \{e_{ij} \mid \{v_i, v_j\} \in E\}$, $P = \{v_i e_{ij}, v_j e_{ji} \mid \{v_i, v_j\} \in E\}$, and $N = \{v_i e_{ji}, v_j e_{ij} \mid \{v_i, v_j\} \in E\}$.

Let $A = (\{s_0, s_1, s_2, s_3\}, \Sigma, \delta, s_0, F)$ be a poDFA that is consistent with P and N. W.l.o.g. we assume that $s_0 \leq s_1 \leq s_2 \leq s_3$ and that all states are pairwise distinct. We prove that the existence of A implies that G is 3-colorable. For $0 \leq q \leq 3$, let $V_q = \{v_i \in V \mid \delta(s_0, v_i) = s_q\}$. We show that each V_q is an independent set in G and that $V = V_0 \cup V_1 \cup V_2$. Assume for an edge $\{v_a, v_b\} \in E$, that v_a and v_b are in the same set V_q: by construction, P contains $v_a e_{ab}$ and N contains $v_b e_{ab}$ where the former is accepted and the latter is rejected by A. This observation implies that $\delta(s_0, v_a) \neq \delta(s_0, v_b)$ which contradicts the assumption that v_a, v_b are in the same V_q. Further, we show by contradiction that $V_3 = \emptyset$, meaning that no $v_i \in V$ maps s_0 to s_3. Therefore, assume there is $v_i \in V$ with $\delta(s_0, v_i) = s_3$. Then, by the state order of A we have that

$\delta(s_0, v_i e_{ij}) = s_3$ and $\delta(s_0, v_i e_{ji}) = s_3$ while the former word is in P and the latter in N. Hence, the states s_0, s_1, s_2 split V into three pairwise disjunctive independent sets V_0, V_1, V_2[3].

For the other direction, assume that V_0, V_1, V_2 are independent sets in G. We construct a poDFA $A = (\{s_0, s_1, s_2, s_3\}, \Sigma, \delta, s_0, \{s_2\})$ that is consistent with P and N. We define δ as follows. All other transitions act as the identity.

$\delta(s_0, v) = s_q$ for $0 \le q \le 2$, $v \in V_q$, $\delta(s_q, e_{ij}) = s_2$ for $0 \le q \le 2$, $v_i \in V_q$,
$\delta(s_q, e_{ji}) = s_1$ for $0 \le q \le 1$, $v_i \in V_q$, $\delta(s_2, e_{ji}) = s_3$ for $v_i \in V_2$.

It can easily be verified that A is consistent with P and N.   □

4   Dichotomies for Permutation Automata

In this section, we focus on *permutation automata* (PA for short), introduced in [44]. A permutation automaton is a DFA where the transition monoid forms a group, or in other words, each letter of the alphabet induces a bijective mapping from the state stet into itself. This allows the reversal of every edge of the automaton graph while preserving the property of being deterministic, see [35]. This powerful property allows to deterministically navigate back and forth between the steps of a computation. Note that this property forbids trap states which we cannot left, hence, intuitively, PAs can never "get stuck", or in other words, the underlying graph of the automaton consists of strongly connected components (scc for short) with no arcs in between. These properties make permutation DFAs interesting for learning black-boxed systems with those nice properties. The latter property of being backward deterministic characterizes the class of zero-reversible languages considered in [3] where the accepting DFA needs to be backward-deterministic and only allows one final state. For this sub-class of DFA (which is incomparable to permutation DFAs as in general the first property of consisting only of one strongly connected component is missing), a polynomial time algorithm for both learning from positive and negative finite sets of samples as well as for learning in the limit from positive samples were obtained. Here, one wanted to learn the smallest zero-reversible language consistent with the samples. Note that this smallest language might not have the smallest representation in terms of a zero-reversible DFA (i.e., a permutation DFA with only one final state) consistent with the samples.

We now focus on permutation automata and begin by showing that for $k = 2$ PA-CON is solvable in P while it becomes NP-complete for $k \ge 3$. This stands in contrast with the results for general DFA in [12], as here the problem is already NP-complete for $k = 2$.

In the proof of the next theorem we will deal with commutative languages. A language $L \subseteq \Sigma^*$ is *commutative* if for each $w \in L$ and each permutation τ of the letters in w also $\tau(w) \in L$ holds. Hence, for a commutative language L the membership $w \in L$ is fully determined by $|w|_\sigma$ for all $\sigma \in \Sigma$.

Theorem 4. PA-CON *is decidable in polynomial time for* $k \le 2$ *and is NP-complete for any combination of* k *and* l *such that* $k \ge 3$ *and* $l \ge 2$.

---

[3] In general V_0, V_1, V_2 is not a partition of V since some of the sets might be empty if G is 1- or 2-colorable.

Proof. Polynomial time algorithm for $k \leq 2$. For one-state PAs, the claim follows trivially as those automata can only accept the empty or the universal language. Hence, for a two-state PA we may assume that one state is final and one is not. Let P and N be an instance of PODFA-CON with $k = 2$. We may assume $|P| > 0$ and $|N| > 0$. Let $A = (\{q_0, q_1\}, \Sigma, \delta, q_0, F)$ be a two-state PA. Then:

(1) Each letter $\sigma \in \Sigma$ can only implement one of two mappings: Either $\delta(q_0, \sigma) = q_0$ and $\delta(q_1, \sigma) = q_1$ or $\delta(q_0, \sigma) = q_1$ and $\delta(q_1, \sigma) = q_0$. Note that in both cases $\delta(q_0, \sigma\sigma) = q_0$ and $\delta(q_1, \sigma\sigma) = q_1$ (*).

(2) The language accepted by A is commutative as each letter behaves the same on all states, i.e., it either maps to the other state or it is the identity.

(3) Let $\Sigma = \{\sigma_1, \sigma_2, \ldots, \sigma_n\}$. It follows from (*) and (2) that we can reduce an instance $P, N \subseteq \Sigma^*$ to an instance $P', N' \subseteq \Sigma^*$ where each word $w \in P$ (and N, respectively) with $|w|_{\sigma_1} = m_1, |w|_{\sigma_2} = m_2, \ldots, |w|_{\sigma_n} = m_n$ is replaced by the word $\sigma_1^{(m_1 \bmod 2)} \sigma_2^{(m_2 \bmod 2)} \ldots \sigma_n^{(m_n \bmod 2)}$ where P, N is a yes-instance for $k = 2$ if and only if P', N' is a yes-instance for $k = 2$. Note that this reduction can be performed in poly-time.

We now present a polynomial time algorithm for solving the instance P', N'. First, check whether $P' \cap N' \neq \emptyset$, if so, return false. If $\epsilon \in P'$, we set $F = \{q_0\}$, if $\epsilon \in N'$, we set $F = \{q_1\}$. If both is not the case, we need to try both variants and start with assuming $F = \{q_0\}$. We construct a system of linear equations over $\mathbb{Z}_2$ which can be solved using Gaussian elimination in polynomial time as stated in [41]. We introduce a variable over $\mathbb{Z}_2$ for each letter in Σ. For each word $w \in P'$ with $|w| = k$, we introduce the equation $w[1] \oplus w[2] \oplus \ldots \oplus w[k] = 0$. Note that each letter appears at most once in w. For a word $w \in N'$ with $|w| = k$, we introduce the equation $w[1] \oplus w[2] \oplus \ldots \oplus w[k] = 1$. For $F = \{q_1\}$, exchange 0 and 1 on the right side of the above equations. For a variable $\sigma_i \in \Sigma$, assigning σ_i with 1 corresponds to defining the letter σ_i to cause a state change, whereas setting the variable σ_i to 0 corresponds to defining the letter σ_i as the identity. Clearly, the PA constructed from a solution of the system of equations is consistent with the input sets, and a variable assignment that solve the system of equations can be read off the transition function.

NP-Hardness for $k \geq 3, l \geq 2$ (Sketch). The proof follows by showing that there is a reduction from 3-COLORING to PA-CON mapping each 3-COLORING instance with $|V| = n$, $|E| = m$ to a PA-CON instance with $k = 3$, $|\Sigma| = n + 2m$, $|P| = |N| = 2m$, $l = 2$, and $T = n + 2m$ in time $\mathcal{O}(n + m)$. □

As for DFAs and poDFAs, we can observe NP-hardness of PA-CON for a fixed alphabet size of $|\Sigma| = 2$. The details of the construction differ, as for instance we need three chains of states, through which we switch with the letter t. Whereas, the construction for DFAs and poDFAs did not rely on every clause consisting of exactly 3 variables (here, the reduction would also work for a ONE-IN-n variant), the construction for PAs relies on this upper bound since we cannot use trap states to catch superfluous satisfied literals.

Theorem 5. PA-CON *is NP-complete for* $|\Sigma| = 2$.

5   Algorithmic Results

In this section, we summarize several rather simple parameterized algorithms, relying on iterating through all possible paths in the automaton induced by the input word, and contrast them with the conditional lower bounds obtained by the above discussed reductions. We recall that the notation O^* suppresses factors polynomial in the size of the instance. In other words, we write $O^*(f(k))$ to denote a running time of the form $f(k) \cdot n^{O(1)}$ where n is the size of the input.

Theorem 6. PODFA-CON *and* PA-CON *can be solved in time* $O^*(k^{k|\Sigma|})$.

Theorem 7. PODFA-CON *and* PA-CON *can be solved in time* $O^*(k^{c \cdot l})$.

Theorem 8. DFA-CON, PODFA-CON, *and* PA-CON *can be solved in* $O^*(k^T)$.

In the case that $k \geq T$ the problems can even be solved in linear time by computing the prefix-tree and completing it. Hence, w.l.o.g., $k < T$. Further, with the poDFA and PA restrictions we lose, while guessing the transitions, for every letter in each step one state as a potential image, leaving us with $k!$ possible mappings for each letter σ, which leads to the following improvement.

Theorem 9. PODFA-CON *and* PA-CON *are* FPT *in time* $O^*((k!)^{|\Sigma|})$ *in the combined parameter* k *and* $|\Sigma|$.

The exponential time hypothesis (ETH) is a standard conjecture in computational complexity theory which essentially states that 3SAT cannot be solved in time $2^{o(n)}$ on instances with n variables [21,22]. Using appropriate notions of fine grained reductions, ETH can be used to provide conditional lower bounds for many problems [13,25]. In our context, using the reductions from Theorems 1, 2, 3, 4 and 5, ETH can be used to establish the following lower bounds.

Corollary 1. *Under ETH,* PODFA-CON *and* PA-CON *cannot be solved: for constant parameters* k *and* l, *in time* $O^*(2^{o(|\Sigma|)})$, *nor* $O^*(2^{o(c)})$, *nor* $O^*(2^{o(T)})$; *and for constant parameter* $|\Sigma|$, *not in time* $O^*(2^{o(k)})$, *nor* $O^*(2^{o(l)})$.

Acknowledgements. Petra Wolf was supported by DFG project FE 560/9-1, and Mateus de Oliveira Oliveira by the RCN projects 288761 and 326537.

References

1. Abdulla, P.A., Collomb-Annichini, A., Bouajjani, A., Jonsson, B.: Using forward reachability analysis for verification of lossy channel systems. Formal Methods Syst. Des. **25**(1), 39–65 (2004)
2. Angluin, D.: On the complexity of minimum inference of regular sets. Inf. Control **39**(3), 337–350 (1978)
3. Angluin, D.: Inference of reversible languages. J. ACM **29**(3), 741–765 (1982)
4. Bouhmala, N.: A multilevel learning automata for MAX-SAT. Int. J. Mach. Learn. Cybern. **6**(6), 911–921 (2015)

5. Brzozowski, J.A., Fich, F.E.: Languages of R-trivial monoids. J. Comput. Syst. Sci. **20**(1), 32–49 (1980)
6. Clifford, A.H., Preston, G.B.: The Algebraic Theory of Semigroups, Volume II, vol. 2. American Mathematical Soc., Providence (1967)
7. Coste, F., Kerbellec, G.: Learning automata on protein sequences. In: JOBIM, pp. 199–210 (2006)
8. Cygan, M., et al.: Parameterized Algorithms. Springer, Cham (2015). https://doi.org/10.1007/978-3-319-21275-3_15
9. Downey, R.G., Fellows, M.R.: Fundamentals of Parameterized Complexity, vol. 4. Springer, London (2013). https://doi.org/10.1007/978-1-4471-5559-1
10. Fernau, H.: Personal communication (2021)
11. Fernau, H.: Algorithms for learning regular expressions from positive data. Inf. Comput. **207**(4), 521–541 (2009)
12. Fernau, H., Heggernes, P., Villanger, Y.: A multi-parameter analysis of hard problems on deterministic finite automata. J. Comput. Syst. Sci. **81**(4), 747–765 (2015)
13. Fernau, H., Krebs, A.: Problems on finite automata and the exponential time hypothesis. Algorithms **10**(1), 24 (2017)
14. Gold, E.M.: Language identification in the limit. Inf. Control **10**(5), 447–474 (1967)
15. Gold, E.M.: Complexity of automaton identification from given data. Inf. Control **37**(3), 302–320 (1978)
16. Groce, A., Peled, D., Yannakakis, M.: Adaptive model checking. In: Katoen, J.-P., Stevens, P. (eds.) TACAS 2002. LNCS, vol. 2280, pp. 357–370. Springer, Heidelberg (2002). https://doi.org/10.1007/3-540-46002-0_25
17. Gruber, H., Holzer, M.: Finite automata, digraph connectivity, and regular expression size. In: Aceto, L., Damgård, I., Goldberg, L.A., Halldórsson, M.M., Ingólfsdóttir, A., Walukiewicz, I. (eds.) ICALP 2008. LNCS, vol. 5126, pp. 39–50. Springer, Heidelberg (2008). https://doi.org/10.1007/978-3-540-70583-3_4
18. Guo, H., Wang, S., Fan, J., Li, S.: Learning automata based incremental learning method for deep neural networks. IEEE Access **7**, 41164–41171 (2019)
19. Hasanzadeh-Mofrad, M., Rezvanian, A.: Learning automata clustering. J. Comput. Sci. **24**, 379–388 (2018)
20. De la Higuera, C.: Grammatical Inference: Learning Automata and Grammars. Cambridge University Press, Cambridge (2010)
21. Impagliazzo, R., Paturi, R.: On the complexity of k-SAT. J. Comput. Syst. Sci. **62**(2), 367–375 (2001)
22. Impagliazzo, R., Paturi, R., Zane, F.: Which problems have strongly exponential complexity? J. Comput. Syst. Sci. **63**(4), 512–530 (2001)
23. Jirásková, G., Masopust, T.: On the state and computational complexity of the reverse of acyclic minimal DFAs. In: Moreira, N., Reis, R. (eds.) CIAA 2012. LNCS, vol. 7381, pp. 229–239. Springer, Heidelberg (2012). https://doi.org/10.1007/978-3-642-31606-7_20
24. Klíma, O., Polák, L.: Alternative automata characterization of piecewise testable languages. In: Béal, M.-P., Carton, O. (eds.) DLT 2013. LNCS, vol. 7907, pp. 289–300. Springer, Heidelberg (2013). https://doi.org/10.1007/978-3-642-38771-5_26
25. Lokshtanov, D., Marx, D., Saurabh, S., et al.: Lower bounds based on the exponential time hypothesis. Bull. EATCS **3**(105), 41–71 (2013)
26. Mao, H., Chen, Y., Jaeger, M., Nielsen, T.D., Larsen, K.G., Nielsen, B.: Learning deterministic probabilistic automata from a model checking perspective. Mach. Learn. **105**(2), 255–299 (2016). https://doi.org/10.1007/s10994-016-5565-9
27. Martens, W., Neven, F., Schwentick, T.: Complexity of decision problems for XML schemas and chain regular expressions. SIAM J. Comput. **39**(4), 1486–1530 (2009)

28. Mayr, F., Yovine, S.: Regular inference on artificial neural networks. In: Holzinger, A., Kieseberg, P., Tjoa, A.M., Weippl, E. (eds.) CD-MAKE 2018. LNCS, vol. 11015, pp. 350–369. Springer, Cham (2018). https://doi.org/10.1007/978-3-319-99740-7_25

29. Meybodi, M.R., Beigy, H.: New learning automata based algorithms for adaptation of backpropagation algorithm parameters. Int. J. Neural Syst. 12(01), 45–67 (2002)

30. Mount, D.W., Mount, D.W.: Bioinformatics: Sequence and Genome Analysis, vol. 1. Cold Spring Harbor Laboratory Press, Cold Spring Harbor, NY (2001)

31. Najim, K., Pibouleau, L., Le Lann, M.: Optimization technique based on learning automata. J. Optim. Theory Appl. 64(2), 331–347 (1990)

32. Najim, K., Poznyak, A.S.: Learning Automata: Theory and Applications. Elsevier, Amsterdam (2014)

33. Nowé, A., Verbeeck, K., Peeters, M.: Learning automata as a basis for multi agent reinforcement learning. In: Tuyls, K., Hoen, P.J., Verbeeck, K., Sen, S. (eds.) LAMAS 2005. LNCS (LNAI), vol. 3898, pp. 71–85. Springer, Heidelberg (2006). https://doi.org/10.1007/11691839_3

34. Parekh, R., Honavar, V.: Learning DFA from simple examples. Mach. Learn. 44(1), 9–35 (2001)

35. Pin, J.-E.: On reversible automata. In: Simon, I. (ed.) LATIN 1992. LNCS, vol. 583, pp. 401–416. Springer, Heidelberg (1992). https://doi.org/10.1007/BFb0023844

36. Pitt, L.: Inductive inference, DFAs, and computational complexity. In: Jantke, K.P. (ed.) AII 1989. LNCS, vol. 397, pp. 18–44. Springer, Heidelberg (1989). https://doi.org/10.1007/3-540-51734-0_50

37. Pitt, L., Warmuth, M.K.: The minimum consistent DFA problem cannot be approximated within any polynomial. J. ACM 40(1), 95–142 (1993)

38. Ramadge, P.J., Wonham, W.M.: Supervisory control of a class of discrete event processes. SIAM J. Control. Optim. 25(1), 206–230 (1987)

39. Rezvanian, A., Saghiri, A.M., Vahidipour, S.M., Esnaashari, M., Meybodi, M.R.: Recent Advances in Learning Automata. SCI, vol. 754. Springer, Cham (2018). https://doi.org/10.1007/978-3-319-72428-7

40. Ryzhikov, A.: Synchronization problems in automata without non-trivial cycles. Theoret. Comput. Sci. 787, 77–88 (2019)

41. Schaefer, T.J.: The complexity of satisfiability problems. In: Lipton, R.J., Burkhard, W.A., Savitch, W.J., Friedman, E.P., Aho, A.V. (eds.) Proceedings of the 10th Annual ACM Symposium on Theory of Computing, pp. 216–226. ACM (1978)

42. Schützenberger, M.P.: On finite monoids having only trivial subgroups. Inf. Control 8(2), 190–194 (1965)

43. Simon, I.: Piecewise testable events. In: Brakhage, H. (ed.) GI-Fachtagung 1975. LNCS, vol. 33, pp. 214–222. Springer, Heidelberg (1975). https://doi.org/10.1007/3-540-07407-4_23

44. Thierrin, G.: Permutation automata. Mathem. Syst. Theory 2(1), 83–90 (1968)

45. Yazidi, A., Bouhmala, N., Goodwin, M.: A team of pursuit learning automata for solving deterministic optimization problems. Appl. Intell. 50(9), 2916–2931 (2020)

Using Edge Contractions and Vertex Deletions to Reduce the Independence Number and the Clique Number

Felicia Lucke⬥ and Felix Mann[✉]⬥

Université de Fribourg, Boulevard de Pérolles 90, 1700 Fribourg, Switzerland
{felicia.lucke,felix.mann}@unifr.ch

Abstract. We consider the following problem: for a given graph G and two integers k and d, can we apply a fixed graph operation at most k times in order to reduce a given graph parameter π by at least d? We show that this problem is NP-hard when the parameter is the independence number and the graph operation is vertex deletion or edge contraction, even for fixed $d = 1$ and when restricted to chordal graphs. We also give a polynomial time algorithm for bipartite graphs when the operation is edge contraction, the parameter is the independence number and d is fixed. Further, we complete the complexity dichotomy on H-free graphs when the parameter is the clique number and the operation is edge contraction by showing that this problem is NP-hard in $(C_3 + P_1)$-free graphs even for fixed $d = 1$. Our results answer several open questions stated in [Diner et al., Theoretical Computer Science, 746, p. 49–72 (2012)].

Keywords: Blocker problems · Edge contraction · Vertex deletion · Independence number · Clique number

1 Introduction

Blocker problems are a type of graph modification problems which are characterised by a set $\mathcal{O}$ of graph modification operations (for example vertex deletion or edge contraction), a graph parameter π and an integer threshold $d \geq 1$. The aim of the problem is to determine, for a given graph G, the smallest sequence of operations from $\mathcal{O}$ which transforms G into a graph G' such that $\pi(G') \leq \pi(G) - d$.

As in the case of regular graph modification problems, we often consider a set of operations consisting of a single graph operation, typically vertex deletion, edge contraction, edge addition or edge deletion. Amongst the parameters which have been studied are the chromatic number χ (see [12]), the matching number μ (see [14]), the length of a longest path (see [3,10]), the (total or semitotal) domination number γ (γ_t and γ_{t2}, respectively) (see [6–8]), the clique number ω (see [11]) and the independence number α (see [2]).

In this paper, the set of allowed graph operations will always consist of only one operation, either *vertex deletion* or *edge contraction*. Given a graph G, we

© Springer Nature Switzerland AG 2022
C. Bazgan and H. Fernau (Eds.): IWOCA 2022, LNCS 13270, pp. 412–424, 2022.
https://doi.org/10.1007/978-3-031-06678-8_30

denote by $G - U$ the graph from which a subset of vertices $U \subseteq V(G)$ has been deleted. Given an edge $uv \in E(G)$, contracting the edge uv means deleting the vertices u and v and replacing them with a single new vertex which is adjacent to every neighbour of u or v. We denote by G/S the graph in which every edge from an edge set $S \subseteq E(G)$ has been contracted. We consider the following two problems, where $d \geq 1$ is a fixed integer.

---

d-DELETION BLOCKER (π)

Instance: A graph G and an integer k.
Question: Is there a set $U \subseteq V(G)$, $|U| \leq k$, such that
$\pi(G - U) \leq \pi(G) - d$?

---

d-CONTRACTION BLOCKER (π)

Instance: A graph G and an integer k.
Question: Is there a set $S \subseteq E(G)$, $|S| \leq k$, such that $\pi(G/S) \leq \pi(G) - d$?

When d is not fixed but part of the input, the problems are called DELETION BLOCKER(π) and CONTRACTION BLOCKER(π), respectively.

When $\pi = \alpha$ or $\pi = \omega$, both problems above are NP-hard on general graphs [5], so it is natural to ask if these problems remain NP-hard when the input is restricted to a special graph class.

Table 1. The table of complexities for some graph classes. Here, P means solvable in polynomial time, whereas NP-h and NP-c mean NP-hard and NP-complete, respectively. A question mark means that the case is open. Everything in **bold** are new results from this paper, all other cases are referenced in [5], where an older version of this table is given.

Class	CONTRACTION BLOCKER(π)		DELETION BLOCKER(π)	
	$\pi = \alpha$	$\pi = \omega$	$\pi = \alpha$	$\pi = \omega$
Tree	P	P	P	P
Bipartite	NP-h; **d fixed: P**	P	P	P
Cobipartite	$d = 1$: NP-c	NP-c; d fixed: P	P	P
Cograph	P	P	P	P
Split	NP-c; d fixed: P	NP-c; d fixed: P	NP-c; d fixed: P	NP-c; d fixed: P
Interval	?	P	?	P
Chordal	**d=1: NP-c**	$d = 1$: NP-c	**d=1: NP-c**	$d = 1$: NP-c
Perfect	$d = 1$: NP-h	$d = 1$: NP-h	**d=1: NP-c**	$d = 1$: NP-c

The authors of [5] show that CONTRACTION BLOCKER(α) in bipartite and chordal graphs as well as DELETION BLOCKER(α) in chordal graphs are NP-hard when the threshold d is part of the input. However, as an open question, they ask for the complexity of the problem when d is fixed. We show that CONTRACTION BLOCKER(α) in bipartite graphs is solvable in polynomial time if d is fixed and that the other problems are NP-hard even if $d = 1$. An overview of the complexities in some graph classes is given in Table 1.

A *monogenic* graph class is characterised by a single forbidden induced subgraph H. For a given graph parameter π, it is interesting to establish a *complexity dichotomy for monogenic graphs*, that is, to determine the complexity of $(d\text{-})$DELETION BLOCKER(π) or $(d\text{-})$CONTRACTION BLOCKER(π) in H-free graphs, for every graph H. For example, such a dichotomy has been established for DELETION BLOCKER(π) for all $\pi \in \{\alpha, \omega, \chi\}$ and CONTRACTION BLOCKER(π) for $\pi \in \{\alpha, \chi\}$ (all [5]), CONTRACTION BLOCKER(γ_{t2}) (for $d = k = 1$, [8]), CONTRACTION BLOCKER(γ_t) (for $d = k = 1$, [6]) and CONTRACTION BLOCKER(γ) (for $d = k = 1$, [7]). In [5], the computational complexity of CONTRACTION BLOCKER(ω) in H-free graphs has been determined for every H except $H = C_3 + P_1$. We show that this case is NP-hard even when $d = 1$ and complete hence the dichotomy.

The paper is organised as follows: In Sect. 2 we explain notation and terminology. In Sect. 3 we give the proofs of NP-hardness or NP-completeness of the aforementioned problems. Finally, in Sect. 4 we give a polynomial-time algorithm for d-CONTRACTION BLOCKER(α) in bipartite graphs.

2 Preliminaries

Throughout this paper, we assume that all graphs are connected unless stated differently.

We refer the reader to [4] for any terminology not defined here.

For a graph G we denote by $V(G)$ the vertex set of the graph and by $E(G)$ its edge set. For two graphs G and H we denote by $G+H$ the disjoint union of G and H. For two vertices $u, v \in V(G)$ we denote by $\text{dist}_G(u, v)$ the *distance* between u and v, which is the number of edges in a shortest path between u and v. For two sets of vertices $U, W \subseteq V(G)$, the *distance between U and W*, denoted by $\text{dist}_G(U, W)$, is given by $\min_{u \in U, w \in W} \text{dist}_G(u, w)$. For a set of edges $S \subseteq E(G)$ we denote by $V(S)$ the set of vertices in $V(G)$ which are endpoints of at least one edge of S. Let $v \in V(G)$, then the *closed neighbourhood of v*, denoted by $N_G[v]$, is the set $\{u \in V(G) : \text{dist}_G(u, v) \leq 1\}$. Similarly, we define for a set $U \subseteq V(G)$ the closed neighbourhood of U as $N_G[U] = \{u \in V(G) : \exists v \in U, \text{dist}_G(v, u) \leq 1\}$. For a vertex $v \in V(G)$ and a set of vertices $U \subseteq V(G)$, we say that v *is complete to U* if v is adjacent to every vertex of U. Let G be a graph and $S \subseteq E(G)$. We denote by $G\big|_S$ the graph whose vertex set is $V(G)$ and whose edge set is S. For any $U \subseteq V(G)$, we denote by $G[U]$ the subgraph of G induced by U. For any $U \subseteq V(G)$, we denote by $G - U$ the graph $G[V(G) \setminus U]$. For any vertex $v \in V(G)$, we denote by $G - v$ the graph $G - \{v\}$.

Let $S \subseteq E(G)$. We denote by G/S the graph whose vertices are in one-to-one correspondence to the connected components of $G|_S$ and two vertices $u, v \in V(G/S)$ are adjacent if and only if their corresponding connected components A, B of $G|_S$ satisfy $\text{dist}_G(V(A), V(B)) = 1$. This is equivalent to the regular notion of contracting the edges in S. However, this definition allows us to make the notation in the proofs simpler and less confusing.

We say that a set $I \subseteq V(G)$ is *independent* if the vertices contained in it are pairwise non-adjacent. We denote by $\alpha(G)$ the size of a maximum independent set in G. The decision problem INDEPENDENT SET takes as input a graph G and an integer k and outputs YES if and only if there is an independent set of size at least k in G. We say that a set $U \subseteq V(G)$ is a *clique* if every two vertices in U are adjacent. We denote by $\omega(G)$ the size of a maximum clique in G. We call a set $U \subseteq V(G)$ a *vertex cover*, if for every edge $uv \in E(G)$ we have that $u \in U$ or $v \in U$. The decision problem VERTEX COVER takes as input a graph G and an integer k and outputs YES if and only if there is a vertex cover of size at most k in G. We denote by $\tau(G)$ the size of a minimum vertex cover in G. Furthermore, we call a graph M a *matching* of a graph G, if $V(M) \subseteq V(G)$, $E(M) \subseteq E(G)$ and each vertex in M has exactly one neighbour in M. We say that a matching is a *maximum matching* if it contains the maximum possible number of edges and denote this number by $\mu(G)$. Observe that we did not use the standard definition of a matching as a set of non-adjacent edges. This was done in order to simplify the notation in the proofs. However, the edge set of a matching in our definition follows the conventional definition.

A graph without cycles is called a *forest* and a connected forest is a *tree*. It is well-known that a tree has one more vertex than it has edges. A graph is said to be *chordal* if it has no induced cycle of length at least four. A graph G is *bipartite* if we can find a partition of the vertices into two sets $V(G) = U \cup W$ such that U and W are both independent sets. For a given graph H, we say that the graph G is *H-free* if it does not contain H as an induced subgraph.

For a positive integer i we denote by P_i and C_i the path and the cycle on i vertices, respectively. We call the graph which is given in Fig. 1 a *paw*.

Fig. 1. The paw

For a given graph parameter π we say that a set $S \subseteq E(G)$ is *π-contraction-critical* if $\pi(G/S) < \pi(G)$. We say that a set $U \subseteq V(G)$ is *π-deletion-critical* if $\pi(G - U) < \pi(G)$.

We will use the following two results. The first one is due to Kőnig, the second one is well-known and easy to see.

Lemma 1 (see [4]). *Let G be a bipartite graph. Then $\mu(G) = \tau(G)$.*

Lemma 2. *Let G be a graph and let $I \subseteq V(G)$ be a maximum independent set. Then $V(G) \setminus I$ is a minimum vertex cover and hence $\tau(G) + \alpha(G) = |V(G)|$.*

In [13] it was shown that INDEPENDENT SET is NP-complete in C_3-free graphs. This and Lemma 2 imply the following corollary.

Corollary 1. VERTEX COVER *is NP-complete in C_3-free graphs.*

3 Hardness Proofs

We begin by restating VERTEX COVER as a satisfiability problem in order to simplify the notation in the proofs.

WEIGHTED POSITIVE 2-SAT

Instance: A variable set X, a clause set C in which all clauses contain exactly two literals and every literal is positive, as well as an integer k.
Question: Is there a truth assignment of the variables (that is, a mapping $f \colon X \to \{\text{true, false}\}$) such that at least one literal in each clause is true and there are at most k variables which are true.

If $\Phi = (G, k)$ is an instance of VERTEX COVER then taking $X = V(G)$ as the variable set and $C = \{(u \vee w) \colon uw \in E(G)\}$ as the set of clauses yields an instance (X, C, k) of WEIGHTED POSITIVE 2-SAT which is clearly equivalent to Φ. Since VERTEX COVER is known to be NP-hard (see Corollary 1), it follows that WEIGHTED POSITIVE 2-SAT is NP-hard, too.

Let G be a graph and $S, S' \subseteq E(G)$ such that for every connected component A of $G|_S$ there is a connected component A' of $G|_{S'}$ with $V(A) = V(A')$. Then, $G/S = G/S'$ and thus we get the following corollary.

Corollary 2. *Let G be a graph and $S \subseteq E(G)$ a minimal α-contraction-critical set of edges. Then, $G|_S$ is a forest.*

Theorem 1. *1-CONTRACTION BLOCKER(α) is NP-complete in chordal graphs.*

Proof. It was shown in [9] that INDEPENDENT SET can be solved in polynomial time for chordal graphs. Since the family of chordal graphs is closed under edge contractions, for a given chordal graph G and a set $S \subseteq E(G)$, it is possible to check in polynomial time whether S is α-contraction-critical. It follows that 1-CONTRACTION BLOCKER(α) is in NP for chordal graphs. In order to show NP-hardness, we reduce from WEIGHTED POSITIVE 2-SAT, which was shown to be NP-hard above. Let $\Phi = (X, C, k)$ be an instance of WEIGHTED POSITIVE 2-SAT. We construct a chordal graph G such that (G, k) is a YES-instance for 1-CONTRACTION BLOCKER(α) if and only if Φ is a YES-instance for WEIGHTED POSITIVE 2-SAT, as follows:

For every variable $x \in X$, we introduce a set of vertices G_x with $G_x = \{v_x\} \cup K_x$, where K_x is a set of $2k + 1$ vertices which induce a clique. We make

v_x complete to K_x. For every clause $c \in C$, we introduce a vertex v_c. We define $K_C = \bigcup_{c \in C} \{v_c\}$. We add edges so that $G[K_C]$ is a clique. For every clause $c \in C$, $c = (x \vee y)$, we make v_c complete to K_x and K_y (see Fig. 1 for an example).

Observe first that the graph G is indeed chordal: if a cycle of length at least four contains at least three vertices of K_C, it follows immediately that the cycle cannot be induced, since K_C induces a clique. Otherwise, such a cycle contains at most two vertices of K_C. If there are two vertices w and w' of the cycle which are contained in G_x and G_y, respectively, with $x, y \in X, x \neq y$, then the cycle has to contain a chord in $G[K_C]$ and is thus not induced. If all vertices of the cycle are in $K_C \cup G_x$ for some fixed $x \in X$, then there are at least two vertices w and w' contained in K_x. Hence, the cycle cannot be induced since w and w' are adjacent and have the same neighbourhood. It follows that G cannot have any induced cycle of length at least 4 and is thus chordal (Fig. 2)

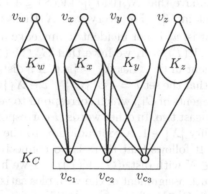

Fig. 2. This is the graph corresponding to the instance of WEIGHTED POSITIVE 2-SAT given by the variables w, x, y, z and the clauses $c_1 = w \vee x, c_2 = x \vee y$ and $c_3 = x \vee z$. The rectangular box corresponds to $G[K_C]$, the vertices contained in it induce a clique. Every set K_i induces a clique and the lines between a vertex and a set K_i mean that this vertex is complete to K_i.

Since G_x induces a clique for every $x \in X$, it can contain at most one vertex in any independent set; the same applies to K_C. Thus, $\alpha(G) \leq |X| + 1$. Let $c \in C$. Since the set $\{v_x : x \in X\} \cup \{v_c\}$ is an independent set of size $|X| + 1$, it follows that $\alpha(G) = |X| + 1$.

Let us assume that Φ is a YES-instance of WEIGHTED POSITIVE 2-SAT. Let X_p be the set of positive variables of a satisfying assignment of Φ. For each $x \in X_p$, let e_x be an edge incident to v_x and let $S = \{e_x | x \in X_p\}$. Let $G' = G/S$. We claim that $\alpha(G') < \alpha(G)$. To see this, observe first that for any $x \in X_p$, contracting e_x is equivalent to deleting the vertex v_x, since $N_G(v_x) = K_x$ induces a clique. Therefore, we have that $G' \simeq G - \{v_x : x \in X_p\}$. Suppose for a contradiction that there is an independent set I of G' of size $|X| + 1$. Since $|I \cap K_x| \leq 1$ (for $x \in X_p$) and $|I \cap G_x| \leq 1$ (for all $x \in X \setminus X_p$), it follows that there exists $c \in C$ such that $v_c \in K_C \cap I$. Furthermore, the inequalities above all have to be equalities. By the choice of X_p, it follows that there is $x \in X_p$ such that x is a literal in c. Since $|I \cap K_x| = 1$, there is a vertex $w \in I \cap K_x$ which is adjacent to v_c, contradicting the fact that I is independent. It follows that S is α-contraction-critical.

For the other direction, assume that $\Phi' = (G, k)$ is a YES-instance of 1-CON-TRACTION BLOCKER(α). Let S be a minimum α-contraction-critical set of edges such that $|S| \leq k$. By Corollary 2, the graph $G|_S$ is a forest.

For any $x \in X$, there is a vertex $u_x \in K_x \setminus V(S)$. This follows from the fact that k edges can be incident to at most $2k$ vertices and $|K_x| = 2k + 1$. Let H be the graph with vertex set $V(H) = K_C$ and edge set $E(H) = \{uv \in S : u, v \in K_C\}$.

Suppose for a contradiction that there is a connected component T of H such that for every $x \in X$ with $\text{dist}_G(G_x, V(T)) = 1$ we have $G_x \cap V(S) = \varnothing$. In other words, for every $c = (x \vee y) \in C$ with $v_c \in V(T)$ we have $G_x \cap V(S) = G_y \cap V(S) = \varnothing$. So we have that $N_G[V(T)] \cap V(S) \subseteq V(T)$, and thus T is also a connected component in $G|_S$. For every $x \in X$ the set $\{u_x\}$ is a connected component in $G|_S$, that is, u_x is not incident to any edge in S. Further, for every $x \in X$ where $\text{dist}_G(G_x, V(T)) = 1$, we have that $G_x \cap V(S) = \varnothing$ and thus $\{v_x\}$ is a connected component in $G|_S$. Let $X_1 = \{x \in X : \text{dist}_G(u_x, V(T)) = 1\}$ and $X_2 = X \setminus X_1$. Observe that the set $I = T \cup \{\{v_x\} : x \in X_1\} \cup \{\{u_x\} : x \in X_2\}$ is a set of connected components of $G|_S$ which correspond to vertices in G/S who are pairwise at distance at least two. In other words, I corresponds to an independent set in G/S of cardinality $|X| + 1$, a contradiction to the assumption that S is α-contraction-critical. It follows that there is no connected component T of H such that for every $x \in X$ with $\text{dist}_G(G_x, V(T)) = 1$ we have $G_x \cap V(S) = \varnothing$.

We can obtain a truth assignment of the variables satisfying Φ as follows: Set every x to true for which $G_x \cap V(S)$ is non-empty. For every clause $c = (x \vee y) \in C$ for which both $G_x \cap V(S)$ and $G_y \cap V(S)$ are empty, set one of its variables to true. This assignment is clearly satisfying, it remains to show that we set at most $|S| \leq k$ variables to true. Consider a connected component T of H. Recall that T is a tree, and so its number of vertices is one more than its number of edges. We have shown that there is a vertex $v_c \in V(T)$, $c = (x \vee y)$, for which $G_x \cap V(S) \neq \varnothing$. Thus, there are at most $|E(T)|$ vertices $v_c \in T$, $c = (x \vee y)$, for which both $G_x \cap V(S)$ and $G_y \cap V(S)$ are empty. This implies that for every connected component T of H we set at most $|E(T)|$ variables to true. Further, the number of variables $x \in X$ which we set to true because $G_x \cap V(S) \neq \varnothing$ is at most the number of edges of S which are not contained in $G[K_C]$. This shows that, in total, we set at most $|S|$ variables to true, which concludes the proof.

Interestingly, 1-DELETION BLOCKER(α) and 1-CONTRACTION BLOCKER(α) are equivalent on the instance Φ' constructed in the proof of Theorem 1 and thus the same construction can be used to show NP-hardness of 1-DELETION BLOCKER(α) in chordal graphs.

Theorem 2. 1-DELETION BLOCKER(α) *is* NP-*complete in chordal graphs.*

Proof. It has been shown in [9] that it is possible to determine the independence number of chordal graphs in polynomial time. Since chordal graphs are closed under vertex deletion, it is possible to check in polynomial time whether the deletion of a given set of vertices reduces the independence number. Hence 1-DELETION BLOCKER(α) is in NP for chordal graphs.

In order to show NP-hardness, we reduce from WEIGHTED POSITIVE 2-SAT. Let Φ be an instance of WEIGHTED POSITIVE 2-SAT, $\Phi = (X, C, k)$. Let $\Phi' = (G, k)$ be the instance of 1-CONTRACTION BLOCKER(α) which is described in Theorem 1 and which has been shown to be equivalent to Φ. Further, let K_x, G_x and v_x for each $x \in X$, K_C, and v_c for each $c \in C$ be as in the proof of Theorem 1. Recall that we have shown that $\alpha(G) = |X| + 1$ and that G is chordal.

We show that Φ' is a YES-instance of 1-DELETION BLOCKER(α) if and only if Φ is a YES-instance of WEIGHTED POSITIVE 2-SAT.

Assume first that Φ is a YES-instance of WEIGHTED POSITIVE 2-SAT and that X_p is the set of positive variables in a satisfying assignment of Φ. We have shown in the proof of Theorem 1 that $\alpha(G - \{v_x : x \in X_p\}) < \alpha(G)$, hence (G, k) is a YES-instance of 1-DELETION BLOCKER(α).

Conversely, assume that Φ' is a YES-instance of 1-DELETION BLOCKER(α) and let W be an α-deletion-critical set of vertices of cardinality at most k. For every $x \in X$ there is $u_x \in K_x \setminus W$, since $|W| < |K_x|$. Define a set $Z = \{x \in X : v_x \in W\}$ and initialize a set $Z' = \varnothing$. For every clause $c \in C$ with $v_c \in W$ we choose one of the variables contained in c and add it to Z'. We claim that setting the variables of $Z \cup Z'$ to true yields a satisfying assignment of Φ. Observe first that $|Z \cup Z'| \leq |W| \leq k$ by construction. Suppose for a contradiction that there is a clause $c \in C$, $c = (x \vee y)$, such that neither x nor y is contained in $Z \cup Z'$. It follows that $v_x, v_y, v_c \notin W$. But then $\{v_c, v_x, v_y\} \cup \{u_z : z \in X \setminus \{x, y\}\}$ is an independent set of size $|X| + 1$ in $G - W$, a contradiction to the α-deletion-criticalness of W. Hence the assignment is satisfying and the theorem follows.

Since perfect graphs are a superclass of chordal graphs, we obtain the following corollary.

Corollary 3. 1-DELETION BLOCKER(α) *is* NP-*complete in perfect graphs.*

Observe that Corollary 3 could also be shown as follows. Complements of perfect graphs are again perfect graphs. Further, 1-DELETION BLOCKER(α) is a YES-instance for a graph G if and only if 1-DELETION BLOCKER(ω) is a YES-instance for $\overline{G}$. Since it was shown in [5] that 1-DELETION BLOCKER(ω) is NP-hard in perfect graphs the corollary follows.

The last theorem in this section answers a question asked in [5]. Indeed, Theorem 4 settles the missing case of [5, Theorem 24] and completes the complexity dichotomy for H-free graphs, which is as follows.

Theorem 3. *Let H be a graph. If H is an induced subgraph of P_4 or of the paw, then* CONTRACTION BLOCKER(ω) *is polynomial-time solvable for H-free graphs, otherwise it is NP-hard or co-NP-hard for H-free graphs.*

Theorem 4. *The decision problem* 1-CONTRACTION BLOCKER(ω) *is NP-hard in* ($C_3 + P_1$)*-free graphs.*

Proof. We use a reduction from VERTEX COVER in C_3-free graphs which is NP-complete due to Corollary 1. Let (G, k) be an instance of VERTEX COVER where G is a C_3-free graph. Since VERTEX COVER is trivial to solve on a graph without edges, we can assume that $E(G)$ is non-empty. We construct an instance (G', k) of 1-CONTRACTION BLOCKER(ω) such that (G, k) is a YES-instance of VERTEX COVER if and only if (G', k) is a YES-instance of 1-CONTRACTION BLOCKER(ω) and G' is $(C_3 + P_1)$-free. Let G' be a graph with $V(G') = V(G) \cup \{w\}$, $w \notin V(G)$, and $E(G') = E(G) \cup \{wv, v \in V(G)\}$. In other words, we add a universal vertex w to G in order to obtain G'.

Since G is C_3-free, every copy of C_3 in G' has to contain w. Furthermore, since w is adjacent to every other vertex in $V(G')$, it follows that every vertex of G' has distance at most one to every copy of C_3. Thus, G' is $(C_3 + P_1)$-free. Also, note that $\omega(G') = 3$ and that every maximum clique in G' is a copy of C_3 which contains w and exactly two vertices of $V(G)$.

Let us assume that (G, k) is a YES-instance of VERTEX COVER. Let $\{v_1, \ldots, v_k\} \subseteq V(G)$ be a vertex cover of G. Set $S = \{v_i w : i \in \{1, \ldots, k\}\}$ and let $G^* = G'/S$. We claim that S is ω-contraction-critical. Notice that the contraction of an edge $vw \in S$ is equivalent to deleting the vertex v, since the new vertex remains adjacent to all other vertices. Thus, G^* is isomorphic to $G - (V(S) \setminus \{w\})$. Since $\{v_1, \ldots, v_k\}$ is a minimum vertex cover of G, there are no edges in $G^* - w$, meaning that G^* is C_3-free and thus $\omega(G^*) \leq 2$. Hence (G', k) is a YES-instance of 1-CONTRACTION BLOCKER(ω).

For the other direction, assume that (G', k) is a YES-instance of 1-CONTRACTION BLOCKER(ω). Let $S \subseteq E(G')$ be a minimum ω-contraction-critical set of edges with $|S| \leq k$ and let $G^* = G'/S$.

We construct a set U of vertices of G as follows: For the connected component T of $G'|_S$ that contains w, add every vertex of $V(T)$ except w to U. For every other connected component T of $G'|_S$ we add to U all vertices of $V(T)$ except one, which can be chosen arbitrarily. We claim that U is a vertex cover of G of size at most k.

To see that $|U| \leq k$, observe that for every connected component T of $G'|_S$ we have added $|V(T)| - 1$ vertices to U. Since T is a tree (see Corollary 2), we have that $|V(T)| - 1 = |E(T)|$. Thus, we have added as many vertices to U as there are edges in S and hence $|U| = |S| \leq k$.

In order to show that U is a vertex cover, suppose for a contradiction that there is an edge $uv \in E(G)$ for which neither u nor v is contained in U. Consider the connected components A_u, A_v and A_w of $G'|_S$ which contain u, v and w, respectively. It follows from the construction of U that in every connected component T of $G'|_S$ there is at most one vertex of T which is not contained in U. Hence, $A_u \neq A_v$. We have that $w \notin U$ by construction, so the same argument can be used to show that $A_u \neq A_w$ and $A_v \neq A_w$. Thus, A_u, A_v, A_w correspond to three different vertices in G^* and since the components are pairwise at distance one, their corresponding vertices induce a C_3 in G^*, a contradiction to S being ω-contraction-critical. Thus, U is a vertex cover in G and (G, k) a YES-instance of VERTEX COVER.

4   Algorithms

In this section we give a polynomial-time algorithm for d-CONTRACTION BLOCKER(α) in bipartite graphs.

Theorem 5. *Let G be a connected, bipartite graph with $|V(G)| \geq 2d + 2$ and $\alpha(G) \geq d + 1$, where $d \geq 1$ is an integer. Then $(G, 2d + 1)$ is a YES-instance of d-CONTRACTION BLOCKER(α).*

Proof. Let G be a bipartite graph with $|V(G)| \geq 2d+2$ and $\alpha(G) \geq d+1$. Let M be a maximum matching of G. Since G is connected, M is non-empty. Consider the following algorithm which constructs a tree T, which is a subgraph of G.

Algorithm 1

Input: A bipartite graph G, a maximum matching M in G, an integer $d \geq 1$
Output: A tree T
    Choose an arbitrary edge $uu' \in E(M)$.
2: Set $V(T) = \{u, u'\}, E(T) = \{uu'\}$.
    while $|E(T)| \leq 2d - 1$ **do**
4:      Choose two vertices $w \in N_G(T) \setminus V(T)$, and $w' \in N_G(w) \cap V(T)$.
      if $w \in V(M)$ **then**
6:         Let $v \in V(M)$ s.t. $vw \in E(M)$.
         $V(T) = V(T) \cup \{v, w\}, E(T) = E(T) \cup \{w'w, vw\}$
8:      **else** $V(T) = V(T) \cup \{w\}, E(T) = E(T) \cup \{w'w\}$
      end if
10: **end while**
    return T

We claim that the resulting graph T is a tree. Indeed, the initial graph is a single edge and thus a tree. Further, observe that every time there are vertices and edges added to T in lines 7 or 8, the resulting graph remains connected and the number of added vertices and added edges is the same. It follows that T is connected and has exactly one more vertex than it has edges and is thus a tree. It is easy to see that T has $2d$ or $2d + 1$ edges.

We consider the graph $G' = G - V(T)$. For every $v \in V(M) \cap V(T)$ the unique vertex $u \in V(M)$ with $uv \in E(M)$ is also contained in $V(T)$ and $uv \in E(T)$. Thus, there are at most $\left\lfloor \frac{|V(T)|}{2} \right\rfloor$ edges in $E(M)$ which have an endvertex in T. Since $M - V(T)$ is a matching in G' we have that $\mu(G') \geq \mu(G) - \left\lfloor \frac{|V(T)|}{2} \right\rfloor$.

Applying Lemma 1 and Lemma 2, we get for the independence number of G':

$$\alpha(G') = |V(G')| - \mu(G') \leq |V(G)| - |V(T)| - \mu(G) + \left\lfloor \frac{|V(T)|}{2} \right\rfloor$$

$$= \alpha(G) - \left\lceil \frac{|V(T)|}{2} \right\rceil = \alpha(G) - d - 1.$$

Let $G^* = G/E(T)$. Observe that $G\big|_{E(T)}$ contains exactly one connected component, say A, which has more than one vertex, namely the connected component corresponding to T. Let $v^* \in V(G^*)$ be the vertex which corresponds to A. Since $G^* - v^*$ is isomorphic to G', we obtain that $\alpha(G^*) \leq \alpha(G') + 1 \leq \alpha(G) - d$.

Algorithm 2

Input: A bipartite graph G, an integer k, a fixed integer d
Output: YES if (G, k) is a YES-instance of d-CONTRACTION BLOCKER(α), NO
    if not
    for every $S \subseteq E(G)$ of size at most k **do**
2:    Let $\beta = 0$.
       Let $G' = G/S$.
4:    Let $U = \{v \in V(G') : v$ *corresponds to a connected component of* $G\big|_S$
                                which contains at least 2 vertices$\}$.
       for every subset $U' \subseteq U$ **do**
6:       **if** U' is independent **then**
             $\beta = \max(\beta, \alpha(G' - (U \cup N_{G'}(U'))) + |U'|)$
8:       **end if**
       end for
10:   **if** $\beta \leq \alpha(G) - d$ **then**
       return YES
12:   **end if**
   end for
14: **return** NO

Theorem 6. *d-CONTRACTION BLOCKER(α) is solvable in polynomial time in bipartite graphs.*

Proof. Let G be a bipartite graph and k a positive integer. If $|V(G)| \leq 2d + 1$ there are at most $2^{d(d+1)}$ subsets of $E(G)$ and at most 2^{2d+1} subsets of $V(G)$. We can check for every subset $S \subseteq E(G)$ if $\alpha(G/S) \leq \alpha(G) - d$ in constant time by computing the graph G/S and checking for each subset of $V(G/S)$ if it is independent. Thus, we can check in constant time if G is a YES-instance for d-CONTRACTION BLOCKER(α).

    Since contracting edges in a non-empty graph cannot reduce the number of vertices to zero, it follows that if $\alpha(G) \leq d$ it is not possible to reduce $\alpha(G)$

by d via edge-contractions. Hence, we can assume that $|V(G)| \geq 2d + 2$ and $\alpha(G) \geq d + 1$. By Theorem 5, we know that for $k \geq 2d + 1$, it is always possible to contract at most k edges to reduce the independence number of G by at least d, so we can further assume that $k \leq 2d$.

Consider now Algorithm 2 which takes as input G, k and d and outputs YES or NO. Algorithm 2 considers every subset $S \subseteq E(G)$ of edges of cardinality at most k and computes $\alpha(G/S)$. If there is some S such that $\alpha(G/S) \leq \alpha(G) - d$ then we return YES, and NO otherwise. In order to compute $\alpha(G/S)$ for such a subset S of edges, we first set $G' = G/S$ and consider the set of vertices $U \subseteq V(G')$ which have been formed by contracting some edges in S (see line 4 of the algorithm). Observe that $G[V(G') \setminus U]$ is isomorphic to $G - V(S)$ and induces thus a bipartite graph. Every independent set of G' can be partitioned into a set $U' \subseteq U$ and a set $W \subseteq V(G') \setminus (U \cup N_{G'}(U'))$. Thus, we can find the independence number of G' by considering every independent subset U' of U and computing $\alpha(G' - (U \cup N_{G'}(U'))) + |U'|$. The largest of these values is then $\alpha(G')$. The independence number of the bipartite graph $G' - (U \cup N_{G'}(U'))$ can be computed in polynomial time, see Lemma 2 and [1].

The number of subsets of $E(G)$ of cardinality at most k is in $O(|E(G)|^k) = O(|V(G)|^{4d})$. For any such subset S, the number of subsets $U' \subseteq U$ is at most $2^k \leq 2^{2d}$. Thus, the running time of Algorithm 2 is polynomial.

References

1. Ahuja, R.K., Magnanti, T.L., Orlin, J.B.: Network Flows - Theory, Algorithms and Applications, pp. 461–509. Prentice Hall, New Jersey (1993)
2. Bentz, C., Costa, M.C., de Werra, D., Picouleau, C., Ries, B.: Minimum d-transversals of maximum-weight stable sets in trees. Electron. Notes Discrete Math. **38**, 129–134 (2011). https://doi.org/10.1016/j.endm.2011.09.022. The Sixth European Conference on Combinatorics, Graph Theory and Applications, EuroComb 2011
3. Cerioli, M.R., Lima, P.T.: Intersection of longest paths in graph classes. Discrete Appl. Math. **281**, 96–105 (2020). https://doi.org/10.1016/j.dam.2019.03.022. lAGOS 2017: IX Latin and American Algorithms, Graphs and Optimization Symposium, C.I.R.M., Marseille, France (2017)
4. Diestel, R.: Graph Theory, p. 37. Springer, Heidelberg (2016)
5. Diner, Ö.Y., Paulusma, D., Picouleau, C., Ries, B.: Contraction and deletion blockers for perfect graphs and H-free graphs. Theoret. Comput. Sci. **746**, 49–72 (2018). https://doi.org/10.1016/j.tcs.2018.06.023
6. Galby, E., Mann, F., Ries, B.: Blocking total dominating sets via edge contractions. Theoret. Comput. Sci. **877**, 18–35 (2021). https://doi.org/10.1016/j.tcs.2021.03.028
7. Galby, E., Mann, F., Ries, B.: Reducing the domination number of (p3+kp2)-free graphs via one edge contraction. Discrete Appl. Math. **305**, 205–210 (2021). https://doi.org/10.1016/j.dam.2021.09.009
8. Galby, E., Lima, P.T., Mann, F., Ries, B.: Using edge contractions to reduce the semitotal domination number (2021). 10.48550/ARXIV.2107.03755. https://arxiv.org/abs/2107.03755

9. Gavril, F.: Algorithms for minimum coloring, maximum clique, minimum covering by cliques, and maximum independent set of a chordal graph. SIAM J. Comput. 1(2), 180–187 (1972). https://doi.org/10.1137/0201013

10. Paik, D., Reddy, S., Sahni, S.: Deleting vertices to bound path length. IEEE Trans. Comput. 43(9), 1091–1096 (1994). https://doi.org/10.1109/12.312117

11. Paulusma, D., Picouleau, C., Ries, B.: Reducing the clique and chromatic number via edge contractions and vertex deletions. In: Cerulli, R., Fujishige, S., Mahjoub, A.R. (eds.) ISCO 2016. LNCS, vol. 9849, pp. 38–49. Springer, Cham (2016). https://doi.org/10.1007/978-3-319-45587-7_4

12. Picouleau, C., Paulusma, D., Ries, B.: Reducing the chromatic number by vertex or edge deletions. Electron. Notes Discrete Math. 62, 243–248 (2017). https://doi.org/10.1016/j.endm.2017.10.042. IAGOS 2017 - IX Latin and American Algorithms, Graphs and Optimization

13. Poljak, S.: A note on stable sets and colorings of graphs. Commentationes Mathematicae Universitatis Carolinae 15(2), 307–309 (1974)

14. Zenklusen, R., Ries, B., Picouleau, C., de Werra, D., Costa, M.C., Bentz, C.: Blockers and transversals. Discrete Math. 309(13), 4306–4314 (2009). https://doi.org/10.1016/j.disc.2009.01.006

Shortest Unique Palindromic Substring Queries in Semi-dynamic Settings

Takuya Mieno[1(✉)] and Mitsuru Funakoshi[2,3]

[1] Faculty of IST, Hokkaido University, Sapporo, Japan
`takuya.mieno@ist.hokudai.ac.jp`
[2] Department of Informatics, Kyushu University, Fukuoka, Japan
`mitsuru.funakoshi@inf.kyushu-u.ac.jp`
[3] Japan Society for the Promotion of Science, Tokyo, Japan

Abstract. A palindromic substring $T[i..j]$ of a string T is said to be a shortest unique palindromic substring (SUPS) in T for an interval $[p,q]$ if $T[i..j]$ is a shortest one such that $T[i..j]$ occurs only once in T, and $[i,j]$ contains $[p,q]$. The SUPS problem is, given a string T of length n, to construct a data structure that can compute all the SUPSs for any given query interval. It is known that any SUPS query can be answered in $O(\alpha)$ time after $O(n)$-time preprocessing, where α is the number of SUPSs to output [Inoue et al., 2018]. In this paper, we first show that α is at most 4, and the upper bound is tight. Also, we present an algorithm to solve the SUPS problem for a sliding window that can answer any query in $O(\log \log W)$ time and update data structures in amortized $O(\log \sigma)$ time, where W is the size of the window, and σ is the alphabet size. Furthermore, we consider the SUPS problem in the after-edit model and present an efficient algorithm. Namely, we present an algorithm that uses $O(n)$ time for preprocessing and answers any k SUPS queries in $O(\log n \log \log n + k \log \log n)$ time after single character substitution. As a by-product, we propose a fully-dynamic data structure for range minimum queries (RmQs) with a constraint where the width of each query range is limited to polylogarithmic. The constrained RmQ data structure can answer such a query in constant time and support a single-element edit operation in amortized constant time.

1 Introduction

A substring $T[i..j]$ of a string T is said to be a *shortest unique palindromic substring* (in short, *SUPS*) for an interval $[p,q]$ if $T[i..j]$ is the shortest substring such that $T[i..j]$ is a palindrome, $T[i..j]$ occurs only once in T, and the occurrence contains $[p,q]$, i.e., $[p,q] \subseteq [i,j]$. The notion of SUPS was introduced by Inoue et al. [19] in 2018[1], motivated by bioinformatics: for example, in DNA/RNA sequences, the presence of unique palindromic sequences can affect the immunostimulatory activities of oligonucleotides [21,31]. Given a string T of

---

[1] A preliminary version of [19] appeared in IWOCA 2017 [26].

© Springer Nature Switzerland AG 2022
C. Bazgan and H. Fernau (Eds.): IWOCA 2022, LNCS 13270, pp. 425–438, 2022.
https://doi.org/10.1007/978-3-031-06678-8_31

length n, the SUPS problem is to construct a data structure that can compute all SUPSs for any given query interval. The SUPS problem was formalized by Inoue et al. [19], and they showed that all SUPSs for a query interval can be enumerated in $O(\alpha)$ time after $O(n)$-time preprocessing, where α is the number of SUPSs to output. Watanabe et al. [30] considered the SUPS problem on run-length encoded strings and showed that all SUPSs for a query can be enumerated in $O(\sqrt{\log r/\log\log r} + \alpha)$ time after $O(r\log\sigma_R + r\sqrt{\log r/\log\log r})$ time preprocessing, where r is the size of the run-length encoded string and σ_R is the number of distinct runs in the input.

Both of the above results are for a static string. It is a natural question whether we can compute SUPSs efficiently in a *dynamic* string. In fact, since DNA sequences contain errors and change dynamically, it is worthwhile to consider them in a dynamic string setting. However, there is no research for solving the SUPS problem on a dynamic string to the best of our knowledge. Thus, in this paper, as a first step to designing dynamic algorithms, we consider the problem on two *semi-dynamic* models: the *sliding-window* model and the *after-edit* model. The sliding-window model aims to compute some objects (e.g., data structure, compressed string, statistics, and so on) w.r.t. the window sliding over the input string left to right. The after-edit model aims to compute some objects w.r.t. the string after applying an edit operation to the input string. Edit operations are given as queries, and they are discarded after finishing to process for the query. As related work, the set of *minimal unique palindromic substrings* (*MUPSs*) can be maintained efficiently in the sliding-window model [25]. Also, the set of MUPSs can be updated efficiently in the after-edit model [14]. Since MUPSs are strongly related to SUPSs, we utilize the above known results for MUPSs as black boxes.

In this paper, we propose an algorithm to solve the SUPS problem for a sliding window and an algorithm to solve the SUPS problem after single-character substitution. Also, we show that the number α of SUPSs for any single interval is at most 4, and the upper bound is tight. Furthermore, as a by-product, we propose a fully-dynamic data structure for the range minimum query (RmQ) in which the width of each query range is in $O(\mathsf{polylog}(n))$. The data structure can answer such a query in constant time and update in (amortized) constant time for any single-element edit operation. Note that, for the original RmQ without any additional constraint, it is known that we need $\Omega(\log n/\log\log n)$ time for answering a query when $O(\mathsf{polylog}(n))$ updating time is allowed [2].

Related Work. A typical application to the sliding-window model is string compression such as LZ77 [32] and PPM [9]. The sliding-window LZ77 compression is based on the sliding-window suffix tree [12,22,27,28]. Also, the sliding-window suffix tree can be applied to compute minimal absent words [10] and minimal unique substrings [24], which are significant concepts for bioinformatics, in the sliding-window model. Recently, the sliding-window palindromic tree was proposed, and it can be applied to compute MUPSs in the sliding-window model [25].

The after-edit model was formalized by Amir et al. [5] in 2017. They tackled the problem of computing the longest common substring (LCS) for two strings in the after-edit model, and proposed an algorithm running in polylogarithmic time. Afterward, Abedin et al. [1] improved the complexities. Also, the problems of computing the longest Lyndon substring [29], the longest palindrome [15], and the set of MUPSs [14] were considered in the after-edit model.

As for more general settings, Amir et al. [6] proposed a fully-dynamic algorithm for computing LCS for two dynamic strings. They also developed a general (probabilistic) scheme for dynamic problems on strings and applied it to the computation of the longest Lyndon substring and the longest palindrome in a dynamic string. Besides that, there are several studies for dynamic settings (e.g., [4,8,16]). In particular, a fully-dynamic and deterministic algorithm for computing the longest palindrome was shown in [3]. Very recently, a suffix array data structure for a dynamic string, where we can update and access any element in polylogarithmic time, was proposed in [20].

Paper Organization. The rest of this paper is organized as follows: In Sect. 2, we give basic notation and algorithmic tools. In Sect. 3, we show the tight bounds on the maximum number of SUPSs for any interval. In Sect. 4, we consider how to update SUPS data structures in semi-dynamic settings and propose efficient algorithms. Finally, in Sect. 5, we propose a fully-dynamic data structure for answering RmQs, in which the width of each query range is limited to $O(\text{polylog}(n))$.

2 Preliminaries

2.1 Strings

Let Σ be an alphabet. An element of Σ is called a character. An element of Σ^* is called a string. The length of a string T is denoted by $|T|$. The empty string ε is the string of length 0. For each i with $1 \leq i \leq |T|$, we denote by $T[i]$ the i-th character of T. If $T = xyz$, then x, y, and z are called a prefix, substring, and suffix of T, respectively. For each i, j with $1 \leq i \leq j \leq |T|$, we denote by $T[i..j]$ the substring of T starting at position i and ending at position j. For convenience, let $T[i'..j'] = \varepsilon$ for any i', j' with $i' > j'$. We say that string w is unique in T if w occurs only once in T. For convenience, we define that the empty string ε is not unique in any string. For a string T and a positive integer p with $p \leq |T|$, the integer p is a period of T if $T[i] = T[i + p]$ holds for every i with $1 \leq i \leq |T| - p$. We also say that T has a period p if p is a period of T.

Let T^R denote the reversal of a string T, i.e., $T[i] = T^R[n - i + 1]$ for every i with $1 \leq i \leq n$. A string P is called a palindrome if $P = P^R$ holds. A palindrome P is called an even-palindrome (resp., odd-palindrome) if $|P|$ is even (resp., odd). The length-$\lceil |P|/2 \rceil$ prefix (resp., suffix) of a palindrome P is called the left arm (resp., right arm) of P. Let $w = T[i..j]$ be a palindromic substring of T. The center of w is $(i + j)/2$ and is denoted by center(w). For a non-negative integer ℓ, $x = T[i - \ell..j + \ell]$ is said to be an expansion of w if $1 \leq i - \ell \leq j + \ell \leq n$ and x is a palindrome. Also, $T[i + \ell..j - \ell]$ is said to be a contraction of w. Further, if $i = 1$, $j = n$, or $T[i - 1] \neq T[j + 1]$, then w is said to be a maximal palindrome.

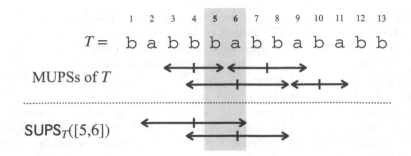

Fig. 1. MUPSs in string T = babbbabbababb are bbb, bbabb, abba, and aba. SUPSs for interval $[5, 6]$ in T are $T[2..6]$ = abbba and $T[4..8]$ = bbabb. The first SUPS $T[2..6]$ is an expansion of MUPS $T[3..5]$ = bbb, and the second SUPS $T[4..8]$ itself is a MUPS.

A palindromic substring $u = T[i..j]$ of a string T is said to be a minimal unique palindromic substring (MUPS) in T if u is unique in T and $T[i+1..j-1]$ is not unique in T. A palindromic substring $v = T[i..j]$ of a string T is said to be a shortest unique palindromic substring (SUPS) for an interval $[p, q]$ in T if v is unique in T, the occurrence contains interval $[p, q]$, and any shorter palindromic substring of T that contains $[p, q]$ is not unique in T. We denote by $\mathsf{SUPS}_T([p, q])$ the set of SUPSs for $[p, q]$. Note that all palindromes in $\mathsf{SUPS}_T([p, q])$ have equal lengths. See also Fig. 1 for examples.

In what follows, we fix a string T of arbitrary length $n > 0$ over an integer alphabet of size $\sigma = O(\mathsf{poly}(n))$. Also, our computational model is a standard word RAM model of word size $\Omega(\log n)$.

2.2   Tools

Longest Common Extension. A longest common extension (in short, LCE) query on string T is, given two integers i, j with $1 \le i \le j \le n$, to compute the length of the longest common prefix (LCP) of two suffixes $T[i..n]$ and $T[j..n]$. It is known (e.g., [18]) that any LCE query can be answered in constant time using the suffix tree of $T\$$ enhanced with a lowest common ancestor data structure, where $\$$ is a special character that is not in Σ. Once we build an LCE data structure on string $T\#T^R\$$, we can answer any LCE query in any direction on T in constant time, where $\# \notin \Sigma$ is another special character. Namely, we can compute in constant time the length of (1) the LCP length of any two suffixes of T, (2) the LCP length of the reverses of any two prefixes of T, and (3) the LCP length of any suffix of T and the reverse of any prefix of T. We call such a data structure a bidirectional LCE data structure.

RmQ, Predecessor and Successor. A range minimum query (RmQ) on integer array A is, given two indices i, j on A with $i \le j$, to compute the index of a minimum value within $A[i..j]$.

A predecessor (resp., successor) query on non-decreasing integer array B is, given an integer x, to compute the maximum (resp., minimum) value that is smaller (resp., greater) than x. We use the famous van Emde Boas tree data structure [11] to answer predecessor/successor queries. Namely, we can answer a query and update the data structure in $O(\log \log U)$ time on a dynamic array, where U is the universe size. Also, the space complexity is $O(U)$. Throughout this paper, we will apply this result to only the case of $U = n$.

2.3 Our Problems

This paper handles SUPS problems under two variants of semi-dynamic models: the sliding-window model and the after-edit model. The sliding-window SUPS problem is to support any sequence of queries that consists of the following:

- pushback(c): append a character c to the right end of the string.
- pop(): remove the first character from the string.
- sups($[p, q]$): output all SUPSs of the string for an interval $[p, q]$.

The after-edit SUPS problem on a string T is, given a substitution operation and a sequence of intervals, to compute SUPSs of T' for each interval where T' is the string after applying the substitution *to the original string* T. Note that each substitution is discarded after the corresponding SUPS queries are answered.

From the point of view of how the string changes, there are differences between the above two problems. On the one hand, in the sliding-window problem, the string can be changed dynamically under the constraints of positions to be edited. On the other hand, in the after-edit problem, any position of the string can be changed, however, the string returns to the original one after the SUPS queries are answered.

3 Tight Bounds on Maximum Number of SUPSs for Single Query

In this section, we prove the following theorem:

Theorem 1. *For any interval $[p, q]$ over T, the inequality $|\mathsf{SUPS}_T([p, q])| \leq 4$ holds. Also, this upper bound is tight.*

First, we prove Lemma 1. Roughly speaking, Lemma 1 states that there must be periodicity when two palindromes overlap enough.

Lemma 1. *Let $x = T[i..i+\ell-1]$ and $y = T[j..j+\ell-1]$ be palindromic substrings of length ℓ of string T with $i < j$. If x and y overlap, then $z = T[i..j+\ell-1]$ has period $2d$ where d is the distance between their center positions.*

Proof. Firstly, $d = (2j+\ell-1)/2 - (2i+\ell-1)/2 = j-i$ holds. Let $z = rst$ where $r = T[i..j-1]$, $s = T[j..i+\ell-1]$, and $t = T[i+\ell..j]$. Since x and y are palindromes, s^R is a prefix of x and a suffix of y. Namely, s^R is both a prefix and a suffix of z, and thus, z has period $|z| - |s^R| = (j-i+\ell) - (i-j+\ell) = 2(j-i) = 2d$. $\square$

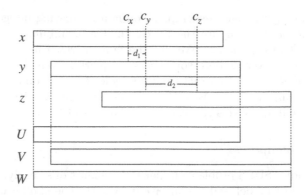

Fig. 2. Illustration for three overlapped palindromes x, y, and z.

Now we are ready to prove Theorem 1.

Proof (of Theorem 1). Let us focus on the SUPSs, each of whose center is at most p. For the sake of contradiction, we assume that there are three such SUPSs of length ℓ for a single query interval $[p, q]$. Let x, y, and z be the SUPSs from left to right, and let c_x, c_y, and c_z be their center positions (see also Fig. 2). Further let $d_1 = c_y - c_x$ and $d_2 = c_z - c_y$. Since the center positions of the three SUPSs are at most p, and they cover the position p, they overlap at least $\ell/2$ each other. Namely, $d_1 \leq \ell/2$, $d_2 \leq \ell/2$, and $d_1 + d_2 \leq \ell/2$ hold. Next, let $U = T[\lceil c_x - \ell/2 \rceil .. \lfloor c_y + \ell/2 \rfloor]$, $V = T[\lceil c_y - \ell/2 \rceil .. \lfloor c_z + \ell/2 \rfloor]$, and $W = T[\lceil c_x - \ell/2 \rceil .. \lfloor c_z + \ell/2 \rfloor]$. By Lemma 1, U has period $2d_1$ and V has period $2d_2$. Thus, y has periods both $2d_1$ and $2d_2$. Also, since $2d_1 + 2d_2 \leq \ell$ holds, y has a period $g = \gcd(2d_1, 2d_2)$ by the periodicity lemma [13] where $\gcd(a, b)$ denotes the greatest common divisor of a and b. Then W also has period g since $g < |y| = \ell$ and g divides both period $2d_1$ of U and period $2d_2$ of V. Furthermore, since $g \leq \min(2d_1, 2d_2) \leq d_1 + d_2$ and $d_1 + d_2 + \ell = |W|$, the inequality $g + \ell \leq |W|$ holds, and thus, $x = W[1..\ell] = W[g + 1..g + \ell]$ holds by the periodicity. This contradicts the uniqueness of x.

We have shown that the maximum number of SUPSs, each of whose center is at most p, is two. Symmetrically, the maximum number of SUPSs, each of whose center is at least p is also two. Thus, the maximum number of SUPSs for a single query interval is four.

Finally, we show that the upper bound is tight. Let us consider the following string $S \in \{a, b, c, A\}^*$:

$S =$cababacababacababacabacabacabacabaca   \\ length 36

   $+$ Aababacababacababa   \\ length 18

   $+$ Abacababacababacab   \\ length 18

   $+$ Abacabacabacabacab   \\ length 18.

The operator $+$ denotes the concatenation of strings. For this string and query interval $[18, 18]$ (highlighted in blue in the figure), $\mathsf{SUPS}_S([18, 18]) = \{[1, 19], [4, 22], [16, 34], [18, 36]\}$ holds. Note that palindromes $S[2..18]$, $S[5..21]$, and $S[17..33]$, which are shorter than 19 and cover the interval $[18, 18]$, are not unique since each of them has another occurrence in the artificial gadgets concatenated by $+$ operators. Also, it can be easily checked that all palindromes of length at most 18 that cover the interval $[18, 18]$ are not unique.   □

The above example having four SUPSs is of length 90, and the length of each SUPS is 19. The smallest period of the former two SUPSs is 6, and that of the latter two SUPSs is 4. We do not know if the example is the shortest one. As a side node, it is open whether the upper bound is tight for binary strings. We could find binary strings that have three SUPSs, e.g., the string baaababab has three SUPSs baaab, ababa, and babab for position 5.

4   SUPS Data Structures

In this section, we introduce SUPS data structures under two semi-dynamic models: the sliding-window model and the after-edit model. Our results are based on the static method proposed by Inoue et al. [19]. We first review their data structure for a static string.

4.1   Static Data Structure

The SUPS data structure proposed in [19] consists of the following:

- the set of MUPSs of T,
- the set of maximal palindromes of T
  (or a bidirectional LCE data structure on T, instead),
- a successor data structure on the starting positions of MUPSs,
- a predecessor data structure on the ending positions of MUPSs, and
- an RmQ data structure on the lengths of MUPSs.

Given a query interval $[p, q]$, we can compute all SUPSs for $[p, q]$ as follows: First, we determine whether the interval $[p, q]$ covers some MUPS or not by querying the predecessor of q on the ending positions of MUPSs and the successor of p on the starting positions of MUPSs. If $[p, q]$ covers only one MUPS, the shortest expansion of the MUPS that covers $[p, q]$ is the only SUPS for $[p, q]$ if such a palindrome exists, and there are no SUPSs for $[p, q]$ otherwise. If $[p, q]$ covers more than one MUPS, then there are no SUPSs for $[p, q]$ since any SUPS covers exactly one MUPS [19]. Otherwise, i.e., if $[p, q]$ covers no MUPS, all SUPSs are categorized into following three types (see also Fig. 3):

(1) an expansion of the rightmost MUPS M_l which ends before q,
(2) a MUPS covers $[p, q]$, or
(3) an expansion of the leftmost MUPS M_r which begins after p.

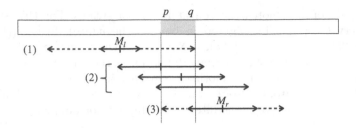

Fig. 3. Illustration for candidates for SUPSs for interval $[p, q]$. Solid arrows represent MUPSs, and dashed arrows represent expansions of MUPSs. Note that dashed arrows may not be palindromes in general.

We call M_l and M_r the left-neighbor MUPS and the right-neighbor MUPS of the interval $[p, q]$, respectively. We can find M_l by querying the predecessor of q. Also, we can determine whether there is an expansion of M_l, which covers $[p, q]$ by looking at the maximal palindrome centered at $c_l = \text{center}(M_l)$. Precisely, if the maximal palindrome centered at c_l covers $[p, q]$, its shortest contraction covering $[p, q]$ is the only candidate of type (1). Otherwise, there is no SUPS of type (1). We emphasize that we can also compute the maximal palindrome centered at c_l by querying bidirectional LCE once, without the precomputed maximal palindromes. The candidate of type (3) can be treated similarly. Finally, all SUPSs of type (2) can be computed by querying RmQ recursively on the array MUPSlen of lengths of MUPSs sorted by their starting positions[2]. Let $[b_i, e_i], \ldots, [b_j, e_j] \in \text{MUPS}(T)$ be all MUPSs covering $[p, q]$. Recall that such range $[i, j]$ of MUPSs can be detected by querying predecessor and successor (see above). We query the RmQ on MUPSlen for the range $[i, j]$, and obtain the index k that is the answer of the RmQ. Namely, $[b_k, e_k]$ is a shortest one within $[b_i, e_i], \ldots, [b_j, e_j]$. Then, we further query the RmQ on MUPSlen for the ranges $[i, k-1]$ and $[k+1, j]$, and repeat it recursively while obtained MUPS is a SUPS for $[p, q]$. The above operations can be done in time linear in the number of SUPSs to output by using linear size data structures of predecessor, successor, and RmQ.

4.2 Sliding-Window Data Structures

We adapt some modifications to the above static data structures to answer any SUPS queries for a sliding window.

It is shown in [25] that the number of changes of MUPSs is constant when we append a character or delete the first character, and we can detect the changes in amortized $O(\log \sigma)$ time. Further, predecessor and successor data structures on the MUPSs can be updated dynamically in $O(\log \log n)$ time using van Emde Boas trees [11].

---

[2] Since MUPSs cannot be nested [19], they are also sorted by their ending positions.

For a dynamic RmQ data structure, we can use the one proposed by Brodal et al. [7]. However, if we directly apply their data structure to our problem, the updating time is in $\Omega(\log n/\log\log n)$, and it becomes a bottleneck. In order to avoid such a situation, we use another dynamic data structure with some constraints which suffice for our problem.

As in the algorithm described in the previous subsection, we will use RmQ on the sequence of the lengths of MUPSs. The width of a query range of RmQ is bounded by the number of MUPSs covering query interval $[p, q]$ (see also Fig. 3). It is known that the number of MUPSs covering any interval is $O(\log n)$ [14], hence the width of a query range of RmQ is also $O(\log n)$. We call the range minimum query such that the width of any query is constrained in $O(\mathsf{polylog}(n))$ LogRmQ. Later, in Sect. 5, we show the following lemma:

Lemma 2. *There exists a linear size data structure for a dynamic array A that supports any LogRmQ on A in constant time. We can maintain the data structure in constant time when an element of A is substituted by another value. Also, we can maintain the data structure in amortized constant time when some element is inserted to (or deleted from) A.*

Finally, we show that the set of maximal palindromes for a sliding window can be maintained efficiently. We generalize Manacher's algorithm [23] to the sliding-window model.

Manacher's Algorithm for Sliding Window. Manacher's algorithm is an online algorithm that computes the set of maximal palindromes in a string. In this subsection, we apply Manacher's algorithm to the sliding-window model. The problem had been solved in [17], however, we will describe a sliding-window algorithm for completeness.

Important invariants of Manacher's algorithm before reading the i-th character are (1) we know the center position c of the longest palindromic suffix of $T[1..i-1]$, and (2) we know all the maximal palindromes, each of whose center is at most c. Note that for the SUPS query, we are interested in maximal palindromes, which are *unique* in the string. Since any palindrome whose center is greater than c is not unique, the second invariant is sufficient for our purpose.

When a character is appended to the current window, we update the set of maximal palindromes in the online manner of the original Manacher algorithm. When the first character of the window is deleted, we do not need to do anything if the window $T[b..e]$ itself is not a palindrome. Instead, when we refer to the arm-length of the maximal palindrome centered at a specified position, we need to consider that the left-end of the palindrome may exceed the left-end of the window. Namely, if the stored arm-length for center m is ℓ_m, the actual arm-length is $\min\{\ell_m, \lceil m - b\rceil\}$.

If the window $T[b..e]$ itself is a palindrome, we need to update the longest palindromic suffix to keep the first invariant. This can be done in amortized $O(1)$ time as in Manacher's algorithm. More precisely, for every (half) integers $j = 0.5, 1, 1.5\ldots$, the arm-length of the maximal palindrome of center $\frac{b+e}{2} + j$

is equal to that of center $\frac{b+e}{2} - j$. Thus, we copy them for incremental j's until we find a palindromic suffix of $T[b+1..e]$. Then, we set c to the center position of the suffix palindrome we found. Since the sequence of center positions of the longest palindromic suffixes of the windows is non-decreasing while running the algorithm, the total processing time is $O(n)$.

Therefore, we obtain the following:

Theorem 2. *There exists a data structure of size $O(W)$ for the sliding-window SUPS problem that supports* sups$([p, q])$ *in* $O(\log \log W)$ *time and* pushback(c) *and* pop$()$ *in amortized* $O(\log \sigma + \log \log W)$ *time, where W is the size of the window.*

4.3   After-Edit Data Structure

In this subsection, we design a SUPS data structure for the after-edit model. Basically, the idea is the same as the previous one. The only difference is that we do not maintain maximal palindromes in the after-edit SUPS problem. Instead, we use a bidirectional LCE on the original string T.

Theorem 3. *There exists a data structure of size $O(n)$ for the after-edit SUPS problem that can be updated in amortized $O(\log \sigma + (\log \log n)^2 + d \log \log n)$ time for a single substitution and can answer any subsequent SUPS queries in $O(k \log \log n)$ time, where d is the number of changes of MUPSs when the substitution is applied to T, and k is the number of the SUPS queries after the substitution. Also, given a string T, the data structure can be constructed in $O(n)$ time.*

Proof. Given a substitution operation, we can detect all the changes of MUPSs in $O(\log \sigma + (\log \log n)^2 + d)$ time [14]. Then, the set of MUPSs can be updated in $O(d)$ time, the predecessor/successor data structures can be updated in $O(d \log \log n)$ time, and the LogRmQ data structure can be updated in amortized $O(d)$ time by Lemma 2. Finally, we can compute the maximal palindromes in T' that are expansions of the left-neighbor and the right-neighbor MUPSs by answering a constant number of bidirectional LCE queries on T while skipping the edited position (so-called kangaroo jumps). Also, it is known that the set of MUPSs of T, the predecessor/successor data structures, and the LCE data structure can be computed in $O(n)$ time. Further, the LogRmQ data structure can be computed in $O(n)$ time by Lemma 2. □

Note that the total query time can be written as $\tilde{O}(1)$ time if k is in polylog(n) since $d \in O(\log n)$ always holds [14].

5   Dynamic LogRmQ

In this section, we give a proof of Lemma 2. We assume that the width of the query range is constrained in $O(\log^c n)$ for a fixed constant c. We first consider

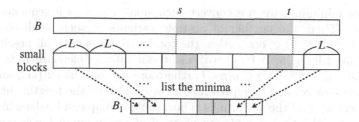

Fig. 4. Illustration for dividing a large block B into small blocks. Query $[s,t]$ on B can be reduced to queries inside the fourth and the seventh small blocks, and query $[5,6]$ on B_1.

dividing the input array A into blocks of size $\log^c n$. We call each of the blocks large block. Then, we build a linear size dynamic RmQ data structure on each large block. Given a query range of width $O(\log^c n)$, we get range minima from a constant number of large blocks and then naively compare them.

We update the RmQ data structure on the large block containing the edited position when the input array A is edited. If the size of a large block becomes far from $\log^c n$ by insertions or deletions, then we split a block or merge continuous blocks to keep the size in $\Theta(\log^c n)$. For example, we split a block into two blocks when the block size exceeds $2\log^c n$ and merge two adjacent blocks when the block size falls below $\frac{1}{2}\log^c n$. If each large block can be updated in amortized constant time, the whole data structure can also be updated in amortized constant time. In the next subsection, we consider how to treat a large block.

5.1 Recursive Structure of Large Block

In order to update large blocks efficiently, we apply the path minima data structure proposed by Brodal et al. [7]. They treated the problem of path minima queries on a tree, a generalization of range minimum queries on an array.

First, we divide a large block B of length $\Theta(\log^c n)$ into small blocks each of length $L = \Theta(\log^\varepsilon n)$ where $\varepsilon < 1$ is an arbitrary small constant.

Let B_1 be the array of length $\Theta(\log^{c-\varepsilon} n)$ that stores the minima of small blocks on B. A query on large block B can be reduced to at most two queries on small blocks and at most one query on B_1 (see Fig. 4). Similarly, for every $i \geq 2$, we divide B_{i-1} into small blocks of the fixed-length L and let B_i be the array of size $\Theta(\log^{c-i\varepsilon} n)$ that stores the minima of small blocks on B_{i-1}. A query on B_i can be reduced to at most two queries on small blocks and at most one query on B_{i+1} at the next level. We recursively apply such division until the size of B_i becomes a constant. The recursion depth is $O(c/\varepsilon)$, i.e., a constant. Notice that recursion occurs at most once at each level. Thus, if we answer RmQ inside a small block in constant time, then the total query time is also a constant.

We answer a query on each small block by using a lookup-table, where the index is a pair of a small block and a query range, and the value is the answer (the position of a minimum). Since RmQ returns the *position* corresponding to a range minimum, we can convert each small block as a sequence of their *local ranks*. Then, the number of possible variants of such small blocks is at most $O((\log^\varepsilon n)^{\log^\varepsilon n}) \subset o(n)$. Also, the total variations with all possible query intervals are still $O((\log^\varepsilon n)^2) \subset o(n)$, i.e., the number of elements in the lookup-table is $O((\log^\varepsilon n)^{\log^\varepsilon n+2}) \subset o(n)$. Furthermore, a small block (i.e., an element in the lookup-table) can be represented in $o(\log n)$ bits: the length, the pointers to each element, and the local ranks. Thus, table lookup can be done in constant time. Namely, the time complexity of an RmQ on a small block is constant. For substitutions (resp., insertions and deletions), updating small blocks can be done in worst-case (resp., amortized) constant time by combining another lookup-table and Q-heap (cf. [7]). Therefore, we have proven Lemma 2.

Acknowledgements. We would like to thank Professor Jeffrey Shallit (University of Waterloo), a PC member of IWOCA 2022, for his interest in our paper and his advice to simplify our proofs. We would also like to thank the anonymous referees for their helpful comments on the manuscript. This work was supported by the JSPS KAKENHI Grant Numbers JP20J11983 (TM) and JP20J21147 (MF).

References

1. Abedin, P., Hooshmand, S., Ganguly, A., Thankachan, S.V.: The heaviest induced ancestors problem: better data structures and applications. Algorithmica (2022). https://doi.org/10.1007/s00453-022-00955-7
2. Alstrup, S., Husfeldt, T., Rauhe, T.: Marked ancestor problems. In: 39th Annual Symposium on Foundations of Computer Science, FOCS 1998, 8–11 November 1998, Palo Alto, California, USA, pp. 534–544. IEEE Computer Society (1998). https://doi.org/10.1109/SFCS.1998.743504
3. Amir, A., Boneh, I.: Dynamic palindrome detection. CoRR abs/1906.09732 (2019). http://arxiv.org/abs/1906.09732
4. Amir, A., Boneh, I., Charalampopoulos, P., Kondratovsky, E.: Repetition detection in a dynamic string. In: Bender, M.A., Svensson, O., Herman, G. (eds.) 27th Annual European Symposium on Algorithms, ESA 2019, 9–11 September 2019, Munich/Garching, Germany. LIPIcs, vol. 144, pp. 5:1–5:18. Schloss Dagstuhl - Leibniz-Zentrum für Informatik (2019). https://doi.org/10.4230/LIPIcs.ESA.2019.5
5. Amir, A., Charalampopoulos, P., Iliopoulos, C.S., Pissis, S.P., Radoszewski, J.: Longest common factor after one edit operation. In: Fici, G., Sciortino, M., Venturini, R. (eds.) SPIRE 2017. LNCS, vol. 10508, pp. 14–26. Springer, Cham (2017). https://doi.org/10.1007/978-3-319-67428-5_2
6. Amir, A., Charalampopoulos, P., Pissis, S.P., Radoszewski, J.: Dynamic and internal longest common substring. Algorithmica **82**(12), 3707–3743 (2020). https://doi.org/10.1007/s00453-020-00744-0
7. Brodal, G.S., Davoodi, P., Srinivasa Rao, S.: Path minima queries in dynamic weighted trees. In: Dehne, F., Iacono, J., Sack, J.-R. (eds.) WADS 2011. LNCS, vol. 6844, pp. 290–301. Springer, Heidelberg (2011). https://doi.org/10.1007/978-3-642-22300-6_25

8. Charalampopoulos, P., Gawrychowski, P., Pokorski, K.: Dynamic longest common substring in polylogarithmic time. In: Czumaj, A., Dawar, A., Merelli, E. (eds.) 47th International Colloquium on Automata, Languages, and Programming, ICALP 2020, 8–11 July 2020, Saarbrücken, Germany (Virtual Conference). LIPIcs, vol. 168, pp. 27:1–27:19. Schloss Dagstuhl - Leibniz-Zentrum für Informatik (2020). https://doi.org/10.4230/LIPIcs.ICALP.2020.27

9. Cleary, J.G., Witten, I.H.: Data compression using adaptive coding and partial string matching. IEEE Trans. Commun. **32**(4), 396–402 (1984). https://doi.org/10.1109/TCOM.1984.1096090

10. Crochemore, M., Héliou, A., Kucherov, G., Mouchard, L., Pissis, S.P., Ramusat, Y.: Absent words in a sliding window with applications. Inf. Comput. **270** (2020). https://doi.org/10.1016/j.ic.2019.104461

11. van Emde Boas, P.: Preserving order in a forest in less than logarithmic time and linear space. Inf. Process. Lett. **6**(3), 80–82 (1977). https://doi.org/10.1016/0020-0190(77)90031-X

12. Fiala, E.R., Greene, D.H.: Data compression with finite windows. Commun. ACM **32**(4), 490–505 (1989). https://doi.org/10.1145/63334.63341

13. Fine, N.J., Wilf, H.S.: Uniqueness theorems for periodic functions. Proc. Am. Math. Soc. **16**(1), 109–114 (1965). https://doi.org/10.1090/S0002-9939-1965-0174934-9

14. Funakoshi, M., Mieno, T.: Minimal unique palindromic substrings after single-character substitution. In: Lecroq, T., Touzet, H. (eds.) SPIRE 2021. LNCS, vol. 12944, pp. 33–46. Springer, Cham (2021). https://doi.org/10.1007/978-3-030-86692-1_4

15. Funakoshi, M., Nakashima, Y., Inenaga, S., Bannai, H., Takeda, M.: Computing longest palindromic substring after single-character or block-wise edits. Theor. Comput. Sci. **859**, 116–133 (2021). https://doi.org/10.1016/j.tcs.2021.01.014

16. Gawrychowski, P., Karczmarz, A., Kociumaka, T., Lacki, J., Sankowski, P.: Optimal dynamic strings. In: Czumaj, A. (ed.) Proceedings of the Twenty-Ninth Annual ACM-SIAM Symposium on Discrete Algorithms, SODA 2018, New Orleans, LA, USA, 7–10 January 2018, pp. 1509–1528. SIAM (2018). https://doi.org/10.1137/1.9781611975031.99

17. Gawrychowski, P., Merkurev, O., Shur, A.M., Uznański, P.: Tight tradeoffs for real-time approximation of longest palindromes in streams. Algorithmica **81**(9), 3630–3654 (2019). https://doi.org/10.1007/s00453-019-00591-8

18. Gusfield, D.: Algorithms on Strings, Trees, and Sequences - Computer Science and Computational Biology. Cambridge University Press (1997). https://doi.org/10.1017/cbo9780511574931

19. Inoue, H., Nakashima, Y., Mieno, T., Inenaga, S., Bannai, H., Takeda, M.: Algorithms and combinatorial properties on shortest unique palindromic substrings. J. Discrete Algorithms **52–53**, 122–132 (2018). https://doi.org/10.1016/j.jda.2018.11.009

20. Kempa, D., Kociumaka, T.: Dynamic suffix array with polylogarithmic queries and updates. CoRR abs/2201.01285 (2021). https://arxiv.org/abs/2201.01285

21. Kuramoto, E., et al.: Oligonucleotide sequences required for natural killer cell activation. Jpn. J. Cancer Res. **83**(11), 1128–1131 (1992). https://doi.org/10.1111/j.1349-7006.1992.tb02734.x

22. Larsson, N.J.: Extended application of suffix trees to data compression. In: Storer, J.A., Cohn, M. (eds.) Proceedings of the 6th Data Compression Conference (DCC 1996), Snowbird, Utah, USA, 31 March–3 April 1996, pp. 190–199. IEEE Computer Society (1996). https://doi.org/10.1109/DCC.1996.488324

23. Manacher, G.K.: A new linear-time "on-line" algorithm for finding the smallest initial palindrome of a string. J. ACM **22**(3), 346–351 (1975). https://doi.org/10.1145/321892.321896

24. Mieno, T., Fujishige, Y., Nakashima, Y., Inenaga, S., Bannai, H., Takeda, M.: Computing minimal unique substrings for a sliding window. Algorithmica (9), 1–24 (2021). https://doi.org/10.1007/s00453-021-00864-1

25. Mieno, T., Watanabe, K., Nakashima, Y., Inenaga, S., Bannai, H., Takeda, M.: Palindromic trees for a sliding window and its applications. Inf. Process. Lett. **173**, 106174 (2022). https://doi.org/10.1016/j.ipl.2021.106174

26. Nakashima, Y., Inoue, H., Mieno, T., Inenaga, S., Bannai, H., Takeda, M.: Shortest unique palindromic substring queries in optimal time. In: Brankovic, L., Ryan, J., Smyth, W.F. (eds.) IWOCA 2017. LNCS, vol. 10765, pp. 397–408. Springer, Cham (2018). https://doi.org/10.1007/978-3-319-78825-8_32

27. Senft, M.: Suffix tree for a sliding window: an overview. In: WDS 2005, pp. 41–46 (2005)

28. Ukkonen, E.: On-line construction of suffix trees. Algorithmica **14**(3), 249–260 (1995). https://doi.org/10.1007/BF01206331

29. Urabe, Y., Nakashima, Y., Inenaga, S., Bannai, H., Takeda, M.: Longest Lyndon substring after edit. In: Navarro, G., Sankoff, D., Zhu, B. (eds.) Annual Symposium on Combinatorial Pattern Matching, CPM 2018, 2–4 July 2018, Qingdao, China. LIPIcs, vol. 105, pp. 19:1–19:10. Schloss Dagstuhl - Leibniz-Zentrum für Informatik (2018). https://doi.org/10.4230/LIPIcs.CPM.2018.19

30. Watanabe, K., Nakashima, Y., Inenaga, S., Bannai, H., Takeda, M.: Fast algorithms for the shortest unique palindromic substring problem on run-length encoded strings. Theory Comput. Syst. **64**(7), 1273–1291 (2020). https://doi.org/10.1007/s00224-020-09980-x

31. Yamamoto, S., Yamamoto, T., Kataoka, T., Kuramoto, E., Yano, O., Tokunaga, T.: Unique palindromic sequences in synthetic oligonucleotides are required to induce IFN [correction of INF] and augment IFN-mediated [correction of INF] natural killer activity. J. Immunol. **148**(12), 4072–4076 (1992). https://www.jimmunol.org/content/148/12/4072

32. Ziv, J., Lempel, A.: A universal algorithm for sequential data compression. IEEE Trans. Inf. Theory **23**(3), 337–343 (1977). https://doi.org/10.1109/TIT.1977.1055714

On Relative Clique Number
of Triangle-Free Planar Colored
Mixed Graphs

Soumen Nandi[1], Sagnik Sen[2], and S. Taruni[2(✉)]

[1] Institute of Engineering & Management Kolkata, Kolkata, India
[2] Indian Institute of Technology Dharwad, Dharwad, India
taruni.sridhar@gmail.com

Abstract. An (n, m)-graph is a graph with n types of arcs and m types of edges. A homomorphism of an (n, m)-graph G to another (n, m)-graph H is a vertex mapping that preserves the adjacencies along with their type and direction. An (n, m)-relative clique R of an (n, m)-graph G is a vertex subset of G for which no two distinct vertices of R get identified under any homomorphism of G to H. The (n, m)-relative clique number of G, denoted by $\omega_{r(n,m)}(G)$, is the maximum $|R|$ such that R is a relative clique of G. In this article, we prove that $\omega_{r(n,m)}(G) \leq 2(2n + m)^2 + 2$, for any triangle-free planar colored mixed graph G for all $(2n+m) \geq 10$. Moreover, we show that this bound is tight. This partially settles a recent conjecture due to Chakroborty, Das, Nandi, Roy and Sen (accepted in Discrete Applied Mathematics).

Keywords: Colored mixed graphs · Homomorphism · Relative clique number · Planar graphs

1 Introduction

In 2000, Nešetril and Raspaud [15] introduced the concept of colored homomorphisms of colored mixed graphs as a generalization of notion of graph homomorphisms.

An (n, m)-colored mixed graph, or simply, an (n, m)-graph G is a graph with n different types of arcs and m different types of edges. We denote the set of vertices, arcs, and edges of G by $V(G), A(G)$, and $E(G)$ respectively. Also we denote the underlying graph of G by $und(G)$.

In this article, we restrict ourselves to studying only those (n, m)-graphs whose underlying graphs are simple graphs. Therefore, for specific values of (n, m), we can capture the notions of simple graphs when $(n, m) = (0, 1)$ [20], oriented graphs when $(n, m) = (1, 0)$ [9,17–19], 2-edge-colored graphs or signed graphs when $(n, m) = (0, 2)$ [11,13,14,16,21], general m-edge colored graphs

S. Sen—Research partially supported by IFCAM MA/IFCAM/18/39 and SRG/2020/001575.

C. Bazgan and H. Fernau (Eds.): IWOCA 2022, LNCS 13270, pp. 439–450, 2022.
https://doi.org/10.1007/978-3-031-06678-8_32

when $(n, m) = (0, m)$ [1], etc. well-studied families of graphs in a generalized set up.

In this work though, whenever we use the term (n, m)-graphs, we mean it for all values of $(n, m) \neq (0, 1)$ unless otherwise stated. Moreover, the upcoming definitions of graph homomorphisms, chromatic numbers, clique numbers for (n, m)-graphs truly capture the existing concepts of the same for all values of (n, m), including $(0, 1)$.

A *homomorphism* of an (n, m)-graph G to another (n, m)-graph H is a vertex mapping $f : V(G) \rightarrow V(H)$ such that for any arc (resp., edge) xy in G, their images induces an arc (resp., edge) $f(x)f(y)$ of the same type in H. If there exists a homomorphism of G to H, then we denote it by $G \rightarrow H$.

Using the concept of homomorphisms of such graphs, Nešetril and Raspaud [15] further presented a generalization of the notion of chromatic number by introducing the (n, m)-chromatic number of an (n, m)-graph as

$$\chi_{n,m}(G) := \min\{|V(H)| : G \rightarrow H\}.$$

For a family $\mathcal{F}$ of simple graphs, the (n, m)-chromatic number can be defined as

$$\chi_{n,m}(\mathcal{F}) := \max\{\chi_{n,m}(G) : und(G) \in \mathcal{F}\}.$$

This parameter is studied for the family of graphs having bounded acyclic chromatic number, bounded arboricity and acyclic chromatic number, bounded maximum degree, sparse graphs, planar graphs and planar graphs with girth restrictions, partial k-trees and partial k-trees with girth restrictions, outerplanar graphs and outerplanar graphs with girth restrictions, etc. across several papers [7, 10, 12, 15].

Bensmail, Duffy, and Sen [2] contributed to this line of work by introducing and studying generalizations of the concept of clique number. The generalization ramifies into two parameters in the context of (n, m)-graphs. An (n, m)-*relative clique* $R \subseteq V(G)$ is a vertex subset satisfying $|f(R)| = |R|$ for all homomorphisms f of G to any H. The (n, m)-relative clique number of a graph G denoted by $\omega_{r(n,m)}(G)$, is,

$$\omega_{r(n,m)}(G) = \max\{|R| : R \text{ is an } (n, m)\text{-relative clique of } G\}.$$

An (n, m)-*absolute clique* $A \subseteq G$ is a subgraph of G satisfying $\chi_{n,m}(A) = |V(A)|$. The (n, m)-absolute clique number of G denoted by $\omega_{a(n,m)}(G)$, is,

$$\omega_{a(n,m)}(G) = \max\{|V(A)| : A \text{ is an } (n, m)\text{-absolute clique of } G\}.$$

For family $\mathcal{F}$ of simple graphs, both the parameters are defined similarly like (n, m)-chromatic number. That is,

$$p(\mathcal{F}) := \max\{p(G) : und(G) \in \mathcal{F}\}$$

where $p \in \{\omega_{r(n,m)}, \omega_{a(n,m)}\}$. The two parameters are primarily being studied for graphs with bounded degrees, planar, partial 2-trees and outerplanar graphs, and their subfamilies due to girth restrictions [2, 3].

Observe that the above defined three parameters trivially satisfies the following relation

$$\omega_{a(n,m)}(G) \leq \omega_{r(n,m)}(G) \leq \chi_{n,m}(G).$$

Therefore, the study of one impacts the other. We are now going to cherry pick some results to motivate our work. First of all, the best known lower and upper bounds for the (n, m)-chromatic number of planar graphs is given below. For convenience, we denote the family of planar graphs with girth at least g by the notation $\mathcal{P}_g$.

Theorem 1 ([7,15]). *For the family of planar graphs $\mathcal{P}_3$, we have:*

$$(2n + m)^3 + 2(2n + m)^2 + (2n + m) + 1 \leq \chi_{n,m}(\mathcal{P}_3) \leq 5(2n + m)^4, m > 0 \ even$$
$$(2n + m)^3 + (2n + m)^2 + (2n + m) + 1 \leq \chi_{n,m}(\mathcal{P}_3) \leq 5(2n + m)^4, \ otherwise.$$

The theorem above can be treated as an approximate analogue of the Four-Color Theorem in the context of (n, m)-graphs. For triangle-free planar graphs, no such dedicated studies have been made, that is, an approximate analogue of the Grötzsch's Theorem is open till date. On the other hand, it turns out, even though finding clique numbers for these two families, that is, the families of planar and triangle-free planar graphs, is trivial for $(n, m) = (0, 1)$, and it is quite a challenging problem for the other values of (n, m). There has been dedicated studies to explore these parameters even for $(n, m) = (1, 0)$ [4–6], and later for general values [2,3].

Let us recall the existing bounds for the two clique numbers for the families of planar and triangle-free graphs to place our work into context. For convenience, and due to space constraints, we present all these results under one theorem.

Theorem 2. *For the families $\mathcal{P}_3$ of planar and $\mathcal{P}_4$ triangle-free planar graphs, we have:*

(i) $3(2n + m)^2 + (2n + m) + 1 \leq \omega_{r(n,m)}(\mathcal{P}_3) \leq 42(2n + m)^2 - 11$ [3],
(ii) $3(2n + m)^2 + (2n + m) + 1 \leq \omega_{a(n,m)}(\mathcal{P}_3) \leq 9(2n + m)^2 + 2(2n + m) + 2$ [2],
(iii) $(2n + m)^2 + 2 \leq \omega_{r(n,m)}(\mathcal{P}_4) \leq 14(2n + m)^2 + 1$ [3],
(iv) $\omega_{a(n,m)}(\mathcal{P}_4) = (2n + m)^2 + 2$ [3].

If one notices, except the last parameter, exact general values are not known for the other parameters. In fact, there are conjectures that claim the lower bound to be tight in the cases of Theorem 2(i) and (ii) [2,3]. For Theorem 2(iii), the conjecture is the following.

Conjecture 1 ([3]). *For the family $\mathcal{P}_4$ of triangle-free planar graphs we have*

$$\omega_{r(n,m)}(\mathcal{P}_4) = 2(2n + m)^2 + 2.$$

The above conjecture has been solved for $(n, m) = (1, 0)$ [6] but, to the best of our knowledge, is open for other values of it. Our main contribution of this paper is to positively settle this conjecture for all $(2n + m) \geq 10$ by proving the following theorem.

Theorem 3. *For the family* $\mathcal{P}_4$ *of triangle-free planar graphs we have*

$$\omega_{r(n,m)}(\mathcal{P}_4) = 2(2n+m)^2 + 2$$

for all $(2n+m) \geq 10$.

2   Preliminaries

In this section, we introduce some notations to help us write the proofs. Also, we follow West [20] for standard definitions, notation and terminology.

Given an (n,m)-graph G the different types of arcs in G are distinguished by n different labels $2, 4, \cdots, 2n$. To be precise, an arc xy with label $2i$ is an arc of type i from x to y. In fact, in such a situation, y is called a $2i$-neighbor of x, or equivalently, x is called a $(2i-1)$-neighbor of y. The different types of edges in G are distinguished by m different types of labels $2n+1, 2n+2, \cdots, 2n+m$. Here also, an edge xy with label $2n+j$ is an edge of type $2n+j$ between x and y. In this case, x and y are called $(2n+j)$-neighbors of each other.

Usually, throughout the article, we will use the Greek alphabets such as α, β, γ and their variants (like α', β', γ', etc.) to denote these labels. Therefore, whenever we use such a symbol, say α, one may assume it to be an integer between 1 and $(2n+m)$. Immediately applying this nomenclature, let us present the notation of the set of all α-neighbors of x as $N^\alpha(x)$. Two vertices x, y *agree* on a third vertex z if $z \in N^\alpha(x) \cap N^\alpha(y)$ for some α, and *disagree* on z otherwise.

3   Proof of the Theorem 3

This section will be dedicated to the proof of Theorem 3. We will prove the lower bound to begin with after introducing an important characterization of (n,m)-relative cliques.

A *special 2-path* is a 2-path uwv such that u, v disagrees with each other on w. In an (n,m)-graph G, a vertex u *sees* a vertex v if they are either adjacent, or are connected by a special 2-path. If u and v are connected by a special 2-path with w as the *internal vertex*, then it is said that u sees v via w or equivalently, v sees u via w. We recall a useful characterization of an (n,m)-relative clique due to Bensmail, Duffy and Sen [2].

Lemma 1 ([2]). *Two distinct vertices of an (n,m)-graph G are part of a relative clique if and only if they are either adjacent or connected by a special 2-path in G, that is, they see each other.*

With this, we are ready to prove the lower bound.

Lemma 2. *There exists a triangle-free planar (n,m)-graph G such that*

$$\omega_{r(n,m)}(G) = 2(2n+m)^2 + 2.$$

Proof. We are going to construct an example to prove this. We start with a $K_{2,2(2n+m)^2}$. Let the vertices in the partite set consisting of only two vertices of it be x and y. The vertices from the other partite set be

$$\{g_{\alpha\beta}, g'_{\alpha\beta} : \text{ for all } \alpha, \beta \in \{1, 2, \cdots, 2n, 2n+1, 2n+2, \cdots, 2n+m\}\}.$$

Next, assign colors and directions to the edges in such a way that $g_{\alpha\beta}$ and $g'_{\alpha\beta}$ become α-neighbors of x and β-neighbors of y. Finally, add special 2-paths of the form $g_{\alpha\beta}h_{\alpha\beta}g'_{\alpha\beta}$ connecting $g_{\alpha\beta}$ and $g'_{\alpha\beta}$.

Notice that, $g_{\alpha\beta}$ (or $g'_{\alpha\beta}$) sees $g_{\alpha'\beta'}$ (or $g'_{\alpha'\beta'}$) via x if $\alpha \neq \alpha'$ or via y if $\beta \neq \beta'$. If $\alpha = \alpha'$ and $\beta = \beta'$, then they the two vertices are $g_{\alpha\beta}$ and $g'_{\alpha\beta}$ and they see each other via $h_{\alpha\beta}$. On the other hand, x and y are adjacent to each $g_{\alpha\beta}$, while they see each other via $g_{\alpha'\beta'}$ for some $\alpha' \neq \beta'$. Thus, the $g_{\alpha\beta}$'s together with x and y, form an (n, m)-relative clique of cardinality $2(2n+m)^2 + 2$. □

Next we concentrate on the upper bound. We prove the upper bound by the method of contradiction. Therefore, let us assume that $\omega_{r(n,m)}(\mathcal{P}_4) > 2(2n+m)^2 + 2$.

Also we need some basic groundwork for the proof. A *critical (n, m)-relative clique H* for the family $\mathcal{P}_4$ of triangle-free planar graphs is an (n, m)-graph H satisfying the following properties:

(i) $und(H) \in \mathcal{P}_4$,
(ii) $\omega_{r(n,m)}(H) = \omega_{r(n,m)}(\mathcal{P}_4)$,
(iii) $\omega_{r(n,m)}(H^*) < \omega_{r(n,m)}(\mathcal{P}_4)$ if, in the dictionary ordering, we find that $(|V(H^*)|, |E(und(H^*))|) < (|V(H)|, |E(und(H))|)$, where $H^* \in \mathcal{P}_4$.

Let us fix a particular critical (n, m)-relative clique H for the rest of this proof. As H is a triangle-free planar graph, we fix a particular planar embedding of it. Whenever we deal with something dependent on embedding of H or a portion of it, we are actually referring to this particular fixed embedding. We also fix a particular (n, m)-relative clique R of cardinality $\omega_{r(n,m)}(H)$. Furthermore, the vertices of R are called *good vertices* and those of $S = V(H) \setminus R$ are called the *helper vertices*. Notice that,

$$\omega_{r(n,m)}(H) = |R| = \omega_{r(n,m)}(\mathcal{P}_4) > 2(2n+m)^2 + 2$$

due to our basic assumption. Thus, proving that there are at most $2(2n+m)^2$ good vertices in H, leads to a contradiction. We recall some useful results due to Chakraborty, Das, Nandi, Roy and Sen [3].

Lemma 3. ([3]). *The (n, m)-graph H is connected and the set S of helper vertices in H is an independent set.*

We recall a modified version of another lemma due to Chakraborty, Das, Nandi, Roy and Sen [3] which we will use for our proofs.

Lemma 4. ([3]). *A vertex $x \in V(H)$ can have at most $2(2n+m)$ good α-neighbors in H for any α.*

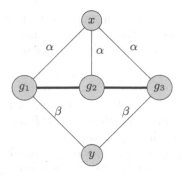

Fig. 1. The exceptional configuration of Lemma 5. The symbols α and β are arc/edge labels with respect to x, y, respectively. The thick edges g_1g_2 and g_2g_3 denote a special 2-path.

We prove a similar result for common good neighbors of two distinct vertices as well, but for an exception. Notice that, it is possible for two vertices x, y of H to agree on three good vertices as depicted in Fig. 1 where the good vertices are also able to see each other via distinct helpers. Turns out this is the only exception, otherwise, two distinct vertices of H can agree on at most 2 good vertices.

Lemma 5. *Two distinct vertices x, y of H can agree on at most 2 good vertices, except when x, y are as in Fig. 1.*

Proof. Suppose x and y have three good neighbors g_1, g_2, g_3 from $N^\alpha(x) \cap N^\beta(y)$, and at least another common good neighbor g_4. Also assume that g_1, g_2, g_3, g_4 are arranged in an anti-clockwise order around x in the fixed planar embedding of H. Notice that, as H is triangle-free, g_1 must reach g_3 via some $h \notin \{x, y, g_1, g_2, g_3, g_4\}$. This is not possible to achieve keeping the graph planar. Hence, if x and y have four common good neighbors, then it is not possible for three of the neighbors to agree on x and y.

If x and y have exactly three common neighbors g_1, g_2, g_3, and all three agree on x and y, then as H is triangle-free, g_i must see g_j via some h_{ij} for all $i < j$. These h_{ij}'s must be distinct, otherwise a $K_{3,3}$ will be created. Hence Fig. 1 is forced whenever there is an exception. $\qquad\square$

In general, two vertices in H can have at most $2(2n + m)^2$ many common good neighbors.

Lemma 6. *Two distinct vertices x, y of H can have at most $2(2n + m)^2$ good vertices in their common neighborhood.*

Proof. As $(2n+m) \geq 2$, and thus, $2(2n+m)^2 \geq 8$, the exceptional case mentioned in Lemma 5 satisfies the condition of the statement of this lemma.

In other case, Lemma 5 implies that

$$|N^\alpha(x) \cap N^\beta(y) \cap R| \leq 2.$$

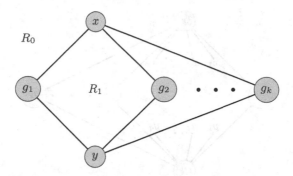

Fig. 2. The configuration $\mathbb{F}_k$.

As α, β can be chosen from a set of $(2n + m)$ integers, we can have a total of $(2n+m)^2$ combinations of (α, β) pairs. Therefore, there can be at most $2(2n+m)^2$ good vertices in $N(x) \cap N(y)$. $\qquad\square$

The configuration $\mathbb{F}_k$ consists of two vertices x, y of H which may not necessarily be good vertices and k common good neighbors of x, y, namely, $g_1, g_2, \cdots, g_k$. Also, we assume a default embedding of $\mathbb{F}_k$ by assuming that $g_1, g_2, \cdots, g_k$ are arranged in an anti-clockwise order around x. Also, without loss of generality, we may assume the embedding of $\mathbb{F}_k$, as part of the fixed embedding of H, to be such that the boundary of the unbounded region is the 4-cycle xg_1yg_kx. Thus the 4-cycle xg_1yg_kx divides the plane into two regions: a bounded region containing other g_i's called R, and an unbounded region called R_0. Moreover, the 2-paths of the type xg_iy further subdivides the region R into $(k-1)$ regions. These regions are called $R_1, R_2, \cdots, R_{k-1}$, where R_i is bounded by $xg_iyg_{i+1}x$, for $i \in \{1, 2, \cdots, k-1\}$. See Fig. 2 for a pictorial representation of this planar embedding.

If it is not possible for H to contain a $\mathbb{F}_k$ for some k, then we say that $\mathbb{F}_k$ is *forbidden* in H. Therefore, Lemma 6 essentially says that $\mathbb{F}_k$, for all $k \geq 2(2n+m)^2 + 1$ is forbidden. We are going to show that $\mathbb{F}_k$, for all $k \geq 3$ is forbidden as a key step of our proof.

Lemma 7. *The configuration $\mathbb{F}_k$, for all $k \geq 3$ is forbidden in H.*

Proof. We will prove this by induction. We suppose that the statement is true for all $k \geq t + 1$ and will prove that it is true for $k = t$, where $t \geq 3$. The base case is taken care by Lemma 6, where it is proved that $\mathbb{F}_k$ is forbidden for all $k \geq 2(2n+m)^2 + 1$ in H.

Now we come to the induction step. Notice that, if a good vertex g which is not part of $\mathbb{F}_k$, belongs to region R_0 (resp., R_1, R_2), then it must see g_2 (resp., g_3, g_1) via x or y. Observe that, it is not possible for g to be adjacent to both x and y, as otherwise, it will create a $\mathbb{F}_{t+1}$, which is forbidden due to the induction hypothesis.

Therefore, every good vertex of H, other than x, y, are either adjacent to both x and y or exactly one of them. If in case, all of them are adjacent to x, then we

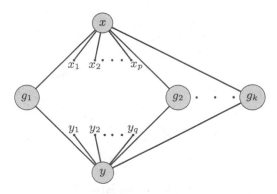

Fig. 3. Private good neighbors of x and y.

can say that there are maximum $2(2n + m)^2$ good vertices in the neighborhood of x due to Lemma 4. This implies that H has at most $2(2n + m)^2 + 2$ good vertices, the additional two vertices coming from counting x and y. Thus it is not possible for all good neighbors other than x, y to be adjacent to x. Similarly, it is not possible for all good neighbors other than x, y to be adjacent to y.

Therefore, there must be at least one good vertex which is adjacent to x but not to y and at least one good vertex that is adjacent to y but not to x. Notice that, these vertices must belong to the same region R_i for some $i \in \{0, 1, \cdots, R_{k-1}\}$, as otherwise they will not be able to see each other. Hence, without loss of generality, we may assume that all good vertices other than the ones contained in $\mathbb{F}_k$ belong to R_1. These good vertices are adjacent to exactly one of x and y. Let us assume that $x_1, x_2, \cdots, x_p$ are the good vertices adjacent to x and not to y while $y_1, y_2, \cdots, y_q$ are the good vertices adjacent to y and not to x. Also suppose that $x_1, x_2, \cdots, x_p$ are arranged in an anti-clockwise order around x and $y_1, y_2, \cdots, y_q$ are arranged in a clockwise order around y. Also for convenience, x_i's are called *private good neighbors* of x and y_j's are called private good neighbors of y. Similarly, g_i's are called common good neighbors of x and y. See Fig. 3 for a pictorial reference. Furthermore, due to the symmetry, we may assume that $p \geq q \geq 1$.

Observe that, if x has a private good α-neighbor, then it cannot have a common good α-neighbor other than, possibly, g_1, g_2. Similar statement holds for y and its private neighbors as well. Thus, if x (resp., y) has a (resp., b) different types of adjacencies with its private neighbors, then x, y can have at most $(2n + m - a)(2n + m - b)$ types (with respect to adjacencies with x and y) of common neighbors (other than, possibly, g_1, g_2). Thus, due to Lemma 5, assuming $\mathbb{F}_k$ is not the exceptional case, there are at most $2(2n+m-a)(2n+m-b)$ common good neighbors of x, y (other than, possibly, g_1, g_2). On the other hand, due to Lemma 4, there are at most $2(2n + m)a$ private good neighbors of x and $2(2n + m)b$ private good neighbors of y. Notice that, it is possible to count g_1 (resp., g_2) among the number of common good neighbors of x, y if it is neither adjacent to x with any of the a types of adjacencies, nor to y with any of the

b types of adjacencies. Otherwise, it is possible to count them along with the private neighbors. Thus the total number of good neighbors in H is at most

$$2(2n+m-a)(2n+m-b)+2(2n+m)a+2(2n+m)b+2 = 2(2n+m)^2+2ab+2$$

However, the above equation gets modified when $p \geq q \geq 2$. In that case note that if x_1 is adjacent to y_q, then it is not possible for x_p to see y_1. Hence, x_1 must see y_q via some h. This will force every x_i to see y_j via h, except for, maybe when $i = j = 1$ or when $i = p, j = q$. If x_1 agrees with y_1 on h, then it is not possible for any other x_i (resp., y_1), for $i, j \neq 1$, to agree with x_1 on h, as otherwise, they cannot see y_1 (resp., x_1). Similar statement holds if x_p agrees with y_q on h. Suppose if there are c types of adjacencies of h among private neighbors of x and d types of adjacencies among private neighbors of y, the value of $p + q$ can be at most $2ac + 2bd$. Therefore, without loss of generality assuming $a \geq b$, the revised upper bound of $p + q$ is

$$p+q \leq 2ac + 2bd \leq 2(2n+m)a.$$

Hence, the revised upper bound for the total number of good neighbors in H is

$$2(2n+m-a)(2n+m-b)+2(2n+m)a+2 \leq 2(2n+m)^2+2.$$

Thus, we have a contradiction in the case when $p \geq q \geq 2$.

Next let us concentrate on the case when $q = 1$. In this case, each x_i must see y via y_1 or via some h_i. However, if x_i sees y via some h_i and is non-adjacent to y_1, then it must see y_1 via some h'_i. Hence, in the worst case scenario, x_1 and x_p will see y and y_1 via h_1, h_p and h'_1, h'_p, respectively while the other x_i's will see y via y_1. In this case, unless $p \leq 2$, it is not possible for x_1 to agree with x_p on x. Also, x_i's (for $i \neq 1, p$) disagree with y on y_1. Therefore, $p+q \leq 2(2n+m)a+1$ in this scenario. Hence, when $q = 1$ and $p \geq 3$, the total number of good vertices in H is at most

$$2(2n+m-a)(2n+m-1)+2(2n+m)a+1+2 \leq 2(2n+m)^2+2.$$

On the other hand, if $p \leq 2$, then the total number of private neighbors are at most 3. In this scenario, the total number of good vertices in H is at most

$$2(2n+m-a)(2n+m-1)+3+2 \leq 2(2n+m)^2-2(2n+m)(a+1)+a+3+2.$$

Notice that the above bound is at most $2(2n+m)^2+2$ as $2n+m \geq 2$ and $a = 1$ or 2. This completes the proof, except when $\mathbb{F}_k$ is the exceptional case.

If $\mathbb{F}_k$ is the exceptional case, then also it is possible to prove that the total number of good vertices in H is at most $2(2n+m)^2 + 2$. Here, the counting becomes simpler as the total number of common good neighbors of x, y is 3, that is, the minimum possible value. Hence, all the above equations for the upper bound of the total number of good vertices in H will give us $2(2n+m)^2+2$ even when $\mathbb{F}_k$ is the exceptional case. Hence, we are done. $\square$

We now restrict the number of good neighbors for any vertex of H. To do so, we need some supporting results.

Lemma 8. *It is not possible for a vertex $x \in V(H)$ to have four or more good neighbors, with at least three of them being α-neighbors.*

Proof. Suppose x has at least four good neighbors g_1, g_2, g_3, g_4 arranged in an anti-clockwise order around x in the fixed embedding of H, where g_1, g_2, g_3 are α-neighbors.

Observe that, as H is a triangle-free graph, the only way for g_1 to see g_3 is via some vertex $h \notin \{x, g_1, g_2, g_3, g_4\}$. Similarly, g_2 must see g_i via some h_i, for all $i \in \{1, 3\}$. Notice that, h, h_1, h_3 are all distinct vertices, as otherwise, a $\mathbb{F}_3$ will be created which is forbidden due to Lemma 7. Also as there is no way for h_1 and h_3 to see g_3 and g_1 respectively, h_1, h_3 cannot be good vertices.

On the other hand, the 4-cycle xg_1hg_3x divides the plane into two connected regions: the bounded region containing g_2 and the unbounded region containing g_4. Let us call the unbounded region as R. Notice that, the bounded region is further subdivided into three regions , say R_1, R_2, and R_3 bounded by the cycles $xg_1h_1g_2x$, $xg_2h_3g_3x$, and $g_1h_1g_2h_3g_3hg_1$, respectively.

Notice that, any good vertex g belonging to R_3 must reach g_4 via h. However, then x and h will have three common good neighbors g_1, g_3, g_4, creating a $\mathbb{F}_3$. Therefore, R_3 cannot contain any good vertices.

Furthermore, any good vertex belonging to R_1 or R_2 must see g_4 via x, and any vertex belonging to R must see g_2 via x. Thus, every good vertex, maybe except h and x itself, is adjacent to x. As x can have at most $2(2n + m)^2$ good neighbors due to Lemma 4, there can be at most $2(2n + m)^2 + 2$ good vertices in H, counting x and h as well. $\square$

With the above lemma, we can further restrict the number of good neighbors of any vertex in H.

Lemma 9. *A vertex $x \in V(H)$ can have at most $2(2n+m)$ good neighbors in H.*

Proof. If x has at least $2(2n + m) + 1$ good neighbors, then it will have at least four neighbors with at least three of them being α-neighbors by the Pigeonhole principle. According to Lemma 8, this is not possible. Hence x can have at most $2(2n + m)$ good neighbors. $\square$

Finally, we are ready to prove Theorem 3.

Proof of Theorem 3. We know that there exists no helper vertex of degree one in H. Also, two distinct helper vertex of degree two cannot have the exact same neighborhood as H is critical. Let us delete all the helper vertices from $und(H)$ having degree two and put an edge between their neighbors. This so-obtained graph H^* will be a simple planar graph with helpers having degree at least three. Also, the set of helper vertices still remain an independent set here.

We know that in every planar graph with minimum degree three, there exists an edge xy such that the sum of the degrees of x and y is at most 13 due to

Kotzig [8]. As helper vertices are independent by Lemma 3 and have at least degree three, one of the end points of such an edge must be a good vertex of degree at most 10. Let x be such a good vertex.

Notice that, all good neighbors of x must be contained in its neighborhood or its second neighborhood. As any vertex can have at most $2(2n + m)$ good neighbors, each of the neighbors of x can have at most $2(2n + m) - 1$ many good neighbors, other than x, in their neighborhood. Thus, x will have at most $2(2n+m) - 10$ good second neighbors, as it has at most 10 neighbors. Therefore, the total number of good neighbors in H is at most $2(2n+m)10+1$. Therefore, it is a contradiction to the assumed number of good vertices in H for all $(2n+m) \geq 10$. This concludes the proof.   □

4 Conclusions

In this paper, we partially prove a conjecture that claims $\omega_{r(n,m)}(\mathcal{P}_4) = 2(2n + m)^2 + 2$ by showing its true for all values of $(2n + m) \geq 10$. During the proof, we have only used this relation $(2n + m) \geq 10$ in the last part. We propose to settle the conjecture for all values of $2n + m$ as a future work.

References

1. Alon, N., Marshall, T.H.: Homomorphisms of edge-colored graphs and coxeter groups. J. Algebr. Comb. **8**(1), 5–13 (1998). https://doi.org/10.1023/A: 1008647514949
2. Bensmail, J., Duffy, C., Sen, S.: Analogues of cliques for (m, n)-colored mixed graphs. Graphs Comb. **33**(4), 735–750 (2017)
3. Chakraborty, D., Das, S., Nandi, S., Roy, D., Sen, S.: On clique numbers of colored mixed graphs. Discrete Appl. Math. (2021, accepted)
4. Das, S., Mj, S., Sen, S.: On oriented relative clique number. Electron. Notes Discrete Math. **50**, 95–101 (2015)
5. Das, S., Prabhu, S., Sen, S.: A study on oriented relative clique number. Discrete Math. **341**(7), 2049–2057 (2018)
6. Das, S.S., Nandi, S., Sen, S.: The relative oriented clique number of triangle-free planar graphs is 10. In: Changat, M., Das, S. (eds.) CALDAM 2020. LNCS, vol. 12016, pp. 260–266. Springer, Cham (2020). https://doi.org/10.1007/978-3-030-39219-2_22
7. Fabila-Monroy, R., Flores, D., Huemer, C., Montejano, A.: Lower bounds for the colored mixed chromatic number of some classes of graphs. Commentationes Mathematicae Universitatis Carolinae **49**(4), 637–645 (2008)
8. Jendrol, S., Voss, H.-J.: Light subgraphs of graphs embedded in the plane-a survey. Discrete Math. **313**(4), 406–421 (2013)
9. Klostermeyer, W.F., MacGillivray, G.: Homomorphisms and oriented colorings of equivalence classes of oriented graphs. Discrete Math. **274**(1–3), 161–172 (2004)
10. Lahiri, A., Nandi, S., Taruni, S., Sen, S.: On chromatic number of (n, m)-graphs. In: Nešetřil, J., Perarnau, G., Rué, J., Serra, O. (eds.) Extended Abstracts EuroComb 2021. TM, vol. 14, pp. 745–751. Springer, Cham (2021). https://doi.org/10.1007/978-3-030-83823-2_119

11. Montejano, A., Ochem, P., Pinlou, A., Raspaud, A., Sopena, É.: Homomorphisms of 2-edge-colored graphs. Discrete Appl. Math. **158**(12), 1365–1379 (2010)
12. Montejano, A., Pinlou, A., Raspaud, A., Sopena, É.: Chromatic number of sparse colored mixed planar graphs. Electron. Notes Discrete Math. **34**, 363–367 (2009)
13. Naserasr, R., Rollová, E., Sopena, É.: Homomorphisms of signed graphs. J. Graph Theory **79**(3), 178–212 (2015)
14. Naserasr, R., Sopena, E., Zaslavsky, T.: Homomorphisms of signed graphs: an update. Eur. J. Comb. **91**, 103222 (2021)
15. Nešetřil, J., Raspaud, A.: Colored homomorphisms of colored mixed graphs. J. Comb. Theory Ser. B **80**(1), 147–155 (2000)
16. Ochem, P., Pinlou, A., Sen, S.: Homomorphisms of 2-edge-colored triangle-free planar graphs. J. Graph Theory **85**(1), 258–277 (2017)
17. Sopena, E.: The chromatic number of oriented graphs. J. Graph Theory **25**(3), 191–205 (1997)
18. Sopena, E.: Oriented graph coloring. Discrete Math. **229**(1–3), 359–369 (2001)
19. Sopena, É.: Homomorphisms and colourings of oriented graphs: an updated survey. Discrete Math. **339**(7), 1993–2005 (2016)
20. West, D.B.: Introduction to Graph Theory, vol. 2. Prentice Hall, Upper Saddle River (2001)
21. Zaslavsky, T.: Signed graphs. Discrete Appl. Math. **4**(1), 47–74 (1982)

Exact Polynomial Time Algorithm
for the Response Time Analysis
of Harmonic Tasks

Thi Huyen Chau Nguyen[1], Werner Grass[2], and Klaus Jansen[3(✉)]

[1] Department of Information Technology, Thang Long University (TLU),
Dai Kim, Hoang Mai, Hanoi, Vietnam
chaunth@thanglong.edu.vn
[2] Faculty of Computer Science and Mathematics, University of Passau,
Passau, Germany
grass@fim.uni-passau.de
[3] Department of Computer Science, Christian-Albrechts-University Kiel,
Kiel, Germany
kj@informatik.uni-kiel.de

Abstract. In some important application areas of hard real-time systems, e.g., avionics, automotive, industrial controls, and robotics, preemptive sporadic tasks with harmonic periods and constrained deadlines running on a uni-processor platform play an important role. For such applications we have to check the system task set for guaranteed compliance with deadlines. For this purpose, we present a new algorithm that has a lower computational complexity than known algorithms for the same system class. For this we determine the *worst-case response time* for each task with a linear computational complexity in the number of tasks, if the task priorities are defined according to their periodic request rates. Otherwise we have to add the time for task ordering.

Keywords: Real-time systems · Response-time analysis · Polynomial time algorithm · Workload function · Fixed point iteration · Harmonic tasks

1 Introduction

Hard real-time embedded systems must deliver functional correct results related to their initiating events within specified time limits. Such systems are typically modeled as a composition of a finite number of recurring tasks, each of which releases a potentially infinite sequence of jobs. In the sporadic task model, the jobs arrive at a time distance that is greater than or equal to the inter-arrival time (called period), which thus represents an important task parameter. The processing of a job must be completed at the latest with the relative deadline of the associated task. An important step in the design of such a system is therefore the schedulability analysis, with which compliance with the time conditions is checked.

In this paper we consider task executions by a single processor, tasks having fixed priorities and we allow a task being preempted in order to perform a higher priority task. A common method of schedulability analysis for these characteristics is response time analysis (RTA) [1,8,16].

Real-time systems with harmonic tasks (the periods are integer multiples of each other) have several advantages over systems with arbitrary periods, for example, the processor utilization can be larger than in the general case and the *worst-case response times* for the different tasks can be determined in polynomial time [3], while in the general case RTA is NP-hard [6].

In the following we introduce an iterative method to determine the exact *worst-case response times* of tasks in a harmonic task system.

1.1 Related Work

In 1973, Liu and Layland [12] had generalized the priority assignment result of [7] to the optimality of Rate Monotonic (RM) Scheduling, where the smaller the assigned periods, the larger the priorities of the tasks to be scheduled. They also presented a simple sufficient schedulability test for periodic tasks with fixed priorities under RM and the assumption that the deadline of a task is equal to its period. Kuo and Mok [9] have shown that for harmonic periods keeping the total processor utilization ≤ 1 is sufficient to schedule the task system with the scheduling policy considered in [12]. After the pioneering paper by Liu and Layland, much work has been done in the area of analysis for fixed-priority preemptive scheduling FPPS, e.g. for any static prioritisations and deadlines lower than or equal to the task periods.

The exact Response Time Analysis (RTA) for the FPPS-model was first introduced in 1986 by Joseph and Pandya [8], then in 1993 Audsley et al. [1] showed how the response time could be calculated by solving a non-linear equation (sum of ceiling terms) by fixed point iteration starting from a suitable initial time, thus providing an exact schedulability test. Some work has been done [4,15] increasing the initial value, some [13,14] have increased the step size of the iterations. The goal of this work is to keep the number of iterations as small as possible, while keeping it unknown in advance.

A different exact schedulability test has been proposed by Lehoczky et al. [11] and is based on the *time demand function* and called time demand analysis (TDA). It must be tested if the time demand function meets the timeline for all multiples of the periods of the tasks with higher priority, which in the case of harmonic tasks means that all multiples of the smallest period up to the maximum period have to be examined.

Bonifaci et al. [3] allow any fixed priorities that are not dependent on any other task parameters. The basic task and scheduling model is the same as in our approach but the schedulability test is TDA based. Their algorithm computes the response time R_n of task τ_n in time $\mathcal{O}(n \cdot \log n + n \cdot \log T_{max})$, where $T_{max} = \max_{1 \leq i \leq n}(T_i)$. In [17] Xu et al. have adapted the TDA method to harmonic task systems and RM schedules. Caused by the stronger assumptions their algorithm uses a slightly smaller number of iterations than [3] which results in a total running time $\mathcal{O}(n \cdot \log T_{max}/T_{min})$, where $T_{min} = \min_{1 \leq i \leq n}(T_i)$.

1.2   This Research

We present an algorithm that determines the exact worst-case response time for fixed priority preemptive sporadic harmonic tasks with *constrained deadlines* running on an uni-processor platform. Our method is based on the standard RTA approach which performs a fixed point iteration on the basis of the *processor demand function* which takes into account the worst-case execution time of the examined task as well as the time duration in which it is preempted by higher-priority tasks (*total worst-case interference*).

In contrast to the standard approach we present a parametric approximation of this *total worst-case interference* that contributes to the response time of the task considered. This approximation proceeds in n phases of fine-tuning to get the exact *total worst-case interference* hence arriving to the exact response time. Our main result improving the previous results in [3,17] is:

Theorem 1. *There is an algorithm A which, given a list of n tasks $\tau_1, \ldots, \tau_n$ with harmonic periods computes the response time R_n of task τ_n in time $\mathcal{O}(n \cdot \log n)$. If the priorities are rate monotonic, the running time is $\mathcal{O}(n)$.*

1.3   Organization

We formally define the terminology, notation and task model in Sect. 2. In Sect. 3, we present our new algorithm for getting the *worst-case response time* for a task in a time that is linear in n assuming that the higher priority tasks are ordered by non-increasing periods. The correctness of the algorithm is proved in Sect. 4. Since this algorithm operates with floating point numbers, errors can accumulate. Therefore, in Sect. 5 we modify our algorithm so that only integer arithmetic is needed.

2   System Model and Background

In this work, we analyze a list $\Gamma_n = \tau_1, \tau_2, \ldots, \tau_n$ of n hard real-time sporadic *tasks,* each one releasing a sequence of *jobs* at time 0. The tasks are scheduled over a single processor by Preemptive Fixed Priorities (FPP). Task τ_i is characterized by:

- a minimum time T_i (that we call *period*) between the arrival of two consecutive jobs,
- a worst-case *execution time* C_i, and
- a *relative deadline* D_i which is the time interval between the arrival time of a job and the time at which the job should be completed

− a *fixed priority* which is implicitly given by the order in the task list: τ_i has a higher priority than τ_j if and only if $i < j$. At a given time, the processor executes the task with the highest priority of all tasks currently ready for execution. A running task at this time with a lower priority is preempted.

The task periods are assumed to be harmonic that is T_i divides T_j or vice versa $(T_i|T_j)$ or $(T_j|T_i)$. All task parameters are positive integer numbers.

We assume *constrained deadlines* i.e., $C_i \leq D_i \leq T_i$. The ratio $U_i = C_i/T_i$ denotes the *utilization* of task τ_i, that is, the fraction of time required by τ_i to execute.

At the time instants denoted by $a_{i,j}$ (arrival time), the i-th task demands the processor for executing its j-th job [2]. Two consecutive requests of the same task cannot be separated by less than T_i, that is,

$$\forall ij, \qquad a_{i,j+1} \geq a_{i,j} + T_i.$$

We denote the *finishing time* of the j-th job of the i-th task by $f_{i,j}$.

We use abbreviated notations for the cumulative utilization of tasks with successive indexes $U_{\iota...\kappa} = \sum_{i=\iota...\kappa} U_i$. Notice that the utilization $U_{1..n}$ of entire task set is equal to $\sum_{j=1}^{n} C_i/T_i \leq 1$.

Also, we recall some basic notions related to fixed-priority preemptive scheduling. In 1990, Lehoczky [10] introduced the notion of *level-i busy period*.

Definition 1. *An interval* $[f(i,j) - a(i,j))$ *is called* level-i busy period *[10] if (a) no task of priority i or higher becomes ready strictly before* $a(i,j)$. *(b) task* τ_i *becomes ready at time* $a(i,j)$ *and its executions ends strictly before time* $f(i,j)$ *possibly suffering preemption by higher priority tasks within the interval.*

Different *level-i busy periods* correspond to response times of different jobs of a task and can have different lengths, which is why the notion of critical instants is important.

Definition 2. *A critical instant* $a(i,k)$ *[12] of a task* τ_i *leads to the worst-case response time for any job of that task (where k needs not to be unique):*

$$f(i,k) - a(i,k) = max_{j=1,2,...} \{f(i,j) - a(i,j)\}$$

Under FPPS this is a time instant task τ_i is released simultaneously with all higher priority tasks. Furthermore it is assumed that the subsequent jobs arrive as soon as possible. Without loss of generality such an instant is set equal to zero since the first jobs of all tasks arrive at this time instant (for all $1 \leq i \leq n, a(i,1) = 0$).

The *worst-case response time* R_i of a task τ_i is therefore

$$R_i = f_{i,1} - a_{i,1} \tag{1}$$

In order to check the schedulability of a task system, it is sufficient to test the response times for the first jobs on compliance with the condition $R_i \leq D_i$ [10].

A task set is said to be *schedulable* when for all tasks τ_i the *worst-case response time* R_i is lower than the relative deadline [5]:

$$\forall i, \quad R_i \leq D_i.$$

For simplification in the notation, from now on we consider the *worst-case response time* of task τ_n but our results could easily be applicable for any $i < n$.

The *total worst-case interference* $I_{n-1}(t)$ describes the amount of time that is taken for executing the tasks with a higher priority than τ_n during the time interval $[0, t)$.

$$I_{n-1}(t) = \sum_{i=1}^{n-1} C_i \cdot \left\lceil \frac{t}{T_i} \right\rceil \tag{2}$$

The total time demand for a complete execution of the n-th task is given by the *processor demand function*:

$$W_n(t) = C_n + I_{n-1}(t) = C_n + \sum_{i=1}^{n-1} C_i \cdot \left\lceil \frac{t}{T_i} \right\rceil \tag{3}$$

The *worst-case response time* is the first point in time $(t > 0)$ at which $t - I_{n-1}(t) = C_n$. We therefore determine the *worst-case response time* R_n as the least fixed point [1]:

$$R_n \overset{\text{def}}{=} \min \{t | W_n(t) = t > 0\}$$

At time $t = R_n$ we have therefore for the first time the equality:

$$R_n = C_n + \sum_{i=1}^{n-1} C_i \cdot \left\lceil \frac{R_n}{T_i} \right\rceil \tag{4}$$

3   Main Results

According to (4) and as proven in [15] R_n may be determined by an iterative technique starting with $R_n^{(0)}$ and producing the values $R_n^{(1)}, R_n^{(2)}, R_n^{(3)}, \ldots$ by applying the recurrence:

$$R_n^{(0)} = \frac{C_n}{1 - U_{1\ldots n-1}} \qquad R_n^{(k)} = C_n + \sum_{i=1}^{n-1} C_i \cdot \left\lceil \frac{R_n^{(k-1)}}{T_i} \right\rceil \tag{5}$$

The iteration stops when $R_n^{(k)} = R_n^{(k-1)}$. Then we get $R_n = R_n^{(k)}$. Although the iteration converges for $U_{1\ldots n} \leq 1$ the number of iteration steps can be high.

One of the main results of our paper is the introduction of a completely different sequence of exactly n approximations to the true value of R_n. It is presented in Theorem 2.

In preparation of the theorem, we introduce a lemma that justifies the admissibility of reordering tasks $\tau_1, \ldots, \tau_{n-1}$ when calculating the response time R_n of task τ_n. Such rearrangements were also made in [3] and [2].

Lemma 1. *The task order* $\tau_1, \ldots, \tau_{n-1}$ *is immaterial for the cumulative worst-case response time of task* τ_n.

Proof. By (4) the worst-case response time of task τ_n is computed by summing up the terms $C_i \lceil R_n/T_i \rceil$. The result is independent of the task order. $\square$

Now in order to keep the calculation of the indices simple in the various processing steps described below, we choose an inverse rate-monotonic order. To formally describe this reordering we introduce a bijective mapping

$$\pi : 1 \ldots n-1 \rightarrow 1 \ldots n-1, \tag{6}$$

in which $\pi(i) = k$ signifies that task τ_k with priority k is at position i in the new order. The reverse rate monotonic order satisfies the condition that for all $i < j$ period $T_{\pi(j)}$ divides the period $T_{\pi(i)}$ having the priorities $\pi(i)$ and $\pi(j)$, respectively.

Theorem 2. *Let a set of n tasks be given, where the first $n-1$ tasks are ordered such that $T_{\pi(n-1)} | T_{\pi(n-2)} | \ldots | T_{\pi(2)} | T_{\pi(1)}$ and $U_{\pi(1)\ldots\pi(n-1)} < 1$ applies. Then the least fixed point*

$$R_n = C_n + \sum_{i=1}^{n-1} C_{\pi(i)} \left\lceil R_n/T_{\pi(i)} \right\rceil \tag{7}$$

can be obtained in $\mathcal{O}(n)$ time by applying the iterative formula:

$$\widetilde{R}_n^{(0)} = \frac{C_n}{1 - U_{\pi(1)\ldots\pi(n-1)}} \tag{8}$$

$$1 \leq i \leq n-1, \quad \widetilde{R}_n^{(i)} = \widetilde{R}_n^{(i-1)} + \frac{C_{\pi(i)} \left(\left\lceil \frac{\widetilde{R}_n^{(i-1)}}{T_{\pi(i)}} \right\rceil - \frac{\widetilde{R}_n^{(i-1)}}{T_{\pi(i)}} \right)}{1 - U_{\pi(i+1)\ldots\pi(n-1)}} \tag{9}$$

we finally get $R_n = \widetilde{R}_n^{(n-1)}$.

The algorithm of Theorem 2 uses floating point operations. In Corollary 1, we will modify the algorithm so that only integer operations are required.

4   Proof of Theorem 2

To obtain the result of Theorem 2, in this section we introduce a set of functions $C_n + \widetilde{I}_{n-1}^{(i)}(t)$ with growing i that represent closer approximations of the function $C_n + I_{n-1}(t)$. We start with the approximation where all ceiling terms $\lceil t/T_{\pi(i)} \rceil$ in Eq. (2) are replaced with their respective lower bounds $t/T_{\pi(i)}$ and $C_{\pi(i)}/T_{\pi(i)}$ with $U_{\pi(i)}$. In each subsequent step, we replace one lower bound again with a corresponding ceiling term, starting with the largest and ending with the smallest period.

$$0 \leq i \leq n-1, \quad \widetilde{I}_{n-1}^{(i)}(t) \overset{\text{def}}{=} \sum_{i+1 \leq j \leq n-1} U_{\pi(j)} \cdot t + \sum_{1 \leq j \leq i} C_{\pi(j)} \left\lceil \frac{t}{T_{\pi(j)}} \right\rceil \tag{10}$$

Notice that by this construction, $I_{n-1}(t) = \widetilde{I}_{n-1}^{(n-1)}(t)$ i.e. at the end there is no subexpression that is linear in t.

In the following we prove that the values $\widetilde{R}_n^{(i)}$ calculated with Theorem 2 define the lowest points of intersections $(\widetilde{R}_n^{(i)}, \widetilde{R}_n^{(i)})$ between the identity function Id(t) and the functions $C_n + \widetilde{I}_{n-1}^{(i)}(t)$. With the equality $R_n = \widetilde{R}_n^{(n-1)}$ we then get the fixed point for the actual problem.

We start with finding the sequence $t_0, \ldots t_{n-1}$ of least fixed points of the equations

$$t = C_n + \widetilde{I}_{n-1}^{(i)}(t) \tag{11}$$

for all i. Since for all t and $i < j$: $\widetilde{I}_{n-1}^{(i)}(t) \leq \widetilde{I}_{n-1}^{(j)}(t)$ it is $t_i \leq t_j$. It follows that one can determine the fixed points in the order $t_0, t_1, \ldots, t_{n-1}$ and thereby start the iteration for t_i with the initial value t_{i-1}.

First, we reshape the Eqs. (10) for $i < n-1$ by summing up the terms that are linear in t. For $i = 0$ we get the fixed point equation:

$$t = C_n + \widetilde{I}_{n-1}^{(0)} = C_n + \sum_{1 \leq j \leq n-1} U_{\pi(j)} \cdot t \tag{12}$$

We set $U_{\pi(1)\ldots\pi(n-1)}$ for $\sum_{1 \leq j \leq n-1} U_{\pi(j)}$ and get the fixed point t_0:

$$t_0 = \frac{C_n}{1 - U_{\pi(1)\ldots\pi(n-1)}} \tag{13}$$

We now consider some i and Eq. (10) and transform this equation into a fixed point equation that is better suited for fixed point iteration since the RHS does not contain terms linear in t.

$$t = C_n + \sum_{i+1 \leq j \leq n-1} U_{\pi(j)} \cdot t + \sum_{1 \leq j \leq i} C_{\pi(j)} \left\lceil \frac{t}{T_{\pi(j)}} \right\rceil \tag{14}$$

We replace $\sum_{i+1 \leq j \leq n-1} U_{\pi(j)}$ by $U_{\pi(i+1)\cdots\pi(n-1)}$ and summarize the linear subexpressions on the left side of the following equation.

$$t \cdot (1 - U_{\pi(i+1)\cdots\pi(n-1)}) = C_n + \sum_{1 \leq j \leq i} C_{\pi(j)} \left\lceil \frac{t}{T_{\pi(j)}} \right\rceil$$

We divide the equation by $(1 - U_{\pi(i+1)\cdots\pi(n-1)})$ and get the fixed point equations:

$$0 \leq i \leq n-1, \quad t = \frac{C_n + \sum_{1 \leq j \leq i} C_{\pi(j)} \left\lceil \frac{t}{T_{\pi(j)}} \right\rceil}{1 - U_{\pi(i+1)\ldots\pi(n-1)}} \tag{15}$$

The solutions of these equations can be interpreted as intersections of the identity function $Id(t) = t$ and a function we denote by $K_i(t)$.

$$0 \leq i \leq n-1, \ K_i(t) \stackrel{\text{def}}{=} \frac{C_n + \sum_{1 \leq j \leq i} C_{\pi(j)} \left\lceil \frac{t}{T_{\pi(j)}} \right\rceil}{1 - U_{\pi(i+1)...\pi(n-1)}} \tag{16}$$

so that for the fixed points t_i the equations $t_i = K_i(t_i)$ apply.

Notice, that we get for $i = n-1$ $K_{n-1}(t) = C_n + \sum_{1 \leq j \leq n-1} C_{\pi(j)} \left\lceil \frac{t}{T_{\pi(j)}} \right\rceil$ such that the fixed point $t_{n-1} = K_{n-1}(t_{n-1})$ is the response time R_n.

In the following we propose not to use the usual iteration to solve the fixed point Eq. (5), but to determine the sequence $t_0, t_1, t_2, ..., t_{n-1}$ and we will show that t_i can be determined from t_{i-1} by applying exactly one iteration step, so that a total of n steps are required.

First of all we have to show that we can solve the fixed point equations $t = K_i(t)$ by iteration.

Lemma 1. *For any i with $0 \leq i \leq n-1$, the fixed point equation $t = K_i(t)$ can be solved by iteration starting with a value $t \leq t_i$.*

Proof. We define a different task system with $i+1$ tasks $\tau_1', \tau_2', \ldots, \tau_i', \tau_n'$ having the periods $T_{\pi(1)}, T_{\pi(2)}, \ldots T_{\pi(i)}, T_n$ and the execution times

$$C_{\pi(1)}/(1-U_{\pi(i+1)...\pi(n-1)}) \ldots C_{\pi(i)}/(1-U_{\pi(i+1)...\pi(n-1)}), C_n/(1-U_{\pi(i+1)...\pi(n-1)})$$

To determine the response time for this task system, we need to find the fixed point $t_i = K_i(t_i)$. This can be done by iteration if the total utilization for this new task system is ≤ 1 [15]. Notice that in [15] the convergence of the fixed point iteration is shown also for positive rational values of the *execution times*. The total utilization is:

$$\sum_{1 \leq j \leq i} \frac{C_{\pi(j)}}{T_{\pi(j)} \cdot (1 - U_{\pi(i+1)...\pi(n-1)})} + \frac{C_n}{T_n \cdot (1 - U_{\pi(i+1)...\pi(n-1)})}$$

By introducing $U_{\pi(j)} = C_{\pi(j)}/T_{\pi(j)}$ and using the abbreviation $U_{\pi(1)...\pi(i)} = \sum_{1 \leq j \leq i} C_{\pi(j)}/T_{\pi(j)}$ this can also be written as

$$\frac{U_{\pi(1)...\pi(i)} + U_n}{1 - U_{\pi(i+1)...\pi(n-1)}}$$

The value of this fraction must be lower than or equal to 1. Since according to our basic assumption $U_{\pi(1)...\pi(i)} + U_n + U_{\pi(i+1)...\pi(n-1)} = U_{\pi(1)...\pi(n-1)} + U_n \leq 1$, the proof is given. $\qquad \square$

The functions (16) are also used in [14] to reduce the number of iterations applying the RTA method.

The solution defined by any of the equations $t = K_i(t)$ is equal to that of the corresponding equation $t = C_n + \tilde{I}_i(t)$ as shown in the following lemma.

Lemma 2. $t_i = C_n + \widetilde{I}_i(t_i) \Leftrightarrow t_i = K_i(t_i)$

Proof. We start with the definition of t_i:

$$t_i = C_n + \widetilde{I}_i(t_i) = C_n + U_{\pi(i+1)...\pi(n-1)} \cdot t_i + \sum_{1 \leq j \leq i} C_{\pi(j)} \left\lceil \frac{t_i}{T_{\pi(j)}} \right\rceil \qquad (*)$$

Using the transformation steps from (14) to (15) and the definition of $K_i(t)$ in (16) we get

$$t_i = \frac{C_n + \sum_{1 \leq j \leq i} C_{\pi(j)} \left\lceil \frac{t_i}{T_{\pi(j)}} \right\rceil}{1 - U_{\pi(i+1)...\pi(n-1)}} = K_i(t_i) \qquad (**)$$

which proves the $\Rightarrow$ direction. We can also start with Eq. (**) and make the reverse conversion to Eq. (*). This proves the $\Leftarrow$ direction. $\qquad \square$

In Eq. (16) we have introduced the functions $K_i(t)$ which we also can represent in terms of $K_{i-1}(t)$. This is advantageous because we plan an iterative determination of the fixed point $t_i = K_i(t_i)$ with known fixed point $t_{i-1} = K_{i-1}(t_{i-1})$. We start with Eq. (16):

$$K_{i-1}(t) = \frac{C_n + \sum_{1 \leq j \leq i-1} C_{\pi(j)} \left\lceil t/T_{\pi(j)} \right\rceil}{1 - U_{\pi(i)...\pi(n-1)}}$$

Multiplying this equation by $1 - U_{\pi(i)...\pi(n-1)}$ results in the equality

$$K_{i-1}(t) \cdot (1 - U_{\pi(i)...\pi(n-1)}) = C_n + \sum_{1 \leq j \leq i-1} C_{\pi(j)} \left\lceil t/T_{\pi(j)} \right\rceil$$

We add $C_{\pi(i)} \left\lceil \frac{t}{T_{\pi(i)}} \right\rceil$ and get

$$K_{i-1}(t) \cdot (1 - U_{\pi(i)...\pi(n-1)}) + C_{\pi(i)} \left\lceil t/T_{\pi(i)} \right\rceil = C_n + \sum_{1 \leq j \leq i} C_{\pi(j)} \left\lceil t/T_{\pi(j)} \right\rceil$$

The RHS of this equality is the numerator of $K_i(t)$ as defined in (16) and we replace the RHS by the LHS in terms of $K_{i-1}(t)$

$$K_i(t) = \frac{C_n + \sum_{1 \leq j \leq i} C_{\pi(j)} \left\lceil t/T_{\pi(j)} \right\rceil}{1 - U_{\pi(i+1)...\pi(n-1)}} = \frac{K_{i-1}(t)(1 - U_{\pi(i)...\pi(n-1)}) + C_{\pi(i)} \left\lceil t/T_{\pi(i)} \right\rceil}{1 - U_{\pi(i+1)...\pi(n-1)}}$$

$$(17)$$

Using the equality $U_{\pi(i)...\pi(n-1)} = U_{\pi(i)} + U_{\pi(i+1)...\pi(n-1)}$ we get

$$K_i(t) = K_{i-1}(t) + \frac{-K_{i-1}(t) \cdot U_{\pi(i)} + C_{\pi(i)} \left\lceil t/T_{\pi(i)} \right\rceil}{1 - U_{\pi(i+1)...\pi(n-1)}} \qquad (18)$$

We now use this result to determine the smallest fixed point $t_i = K_i(t_i)$ given a known fixed point $t_{i-1} = K_{i-1}(t_{i-1})$.

Figure 1 illustrates the situation formally dealt with in the following lemma with an example. It shows two functions $C_n + \tilde{I}_i(t)$ and $C_n + \tilde{I}_{i-1}(t)$. We are interested in the points of intersection with the identify function $Id(t) = t$ and want to construct the solution of $t_i = C_n + \tilde{I}_i(t_i) = K_i(t_i)$ knowing the solution of $t_{i-1} = C_n + \tilde{I}_{i-1}(t_{i-1}) = K_{i-1}(t_{i-1})$. So if we start at point (t_{i-1}, t_{i-1}) and go upward to curve $K_i(t)$, then $K_i(t_{i-1}) = t_i$, if we now go to the right we reach the point (t_i, t_i).

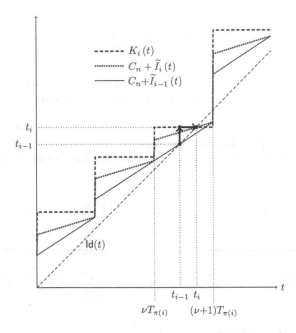

Fig. 1. The figure shows an example of functions $C_n + \tilde{I}_i(t)$, $C_n + \tilde{I}_{i-1}(t)$, $K_i(t)$, and $Id(t)$ as well as the solutions of $t_{i-1} = C_n + \tilde{I}_{i-1}(t)$ and $t_i = K_i(t_i)$

Lemma 3. *For a harmonic task system with n tasks where the first $n-1$ tasks are ordered such that $T_{\pi(n-1)}|T_{\pi(n-2)}|\dots|T_{\pi(2)}|T_{\pi(1)}$ and $U_{\pi(1)\dots\pi(n-1)} < 1$, given a known fixed point $t_{i-1} = K_{i-1}(t_{i-1})$ we obtain the fixed point $t_i = K_i(t_i)$ where*

$$t_i = \frac{t_{i-1} \cdot \left(1 - U_{\pi(i)\dots\pi(n-1)}\right) + C_{\pi(i)} \cdot \left\lceil \frac{t_{i-1}}{T_{\pi(i)}} \right\rceil}{1 - U_{\pi(i+1)\dots\pi(n-1)}} \tag{19}$$

Proof. By (18) we get for $t = t_{i-1}$ and $K_{i-1}(t_{i-1}) = t_{i-1}$:

$$K_i(t_{i-1}) = t_{i-1} + \frac{-t_{i-1}U_{\pi(i)} + C_{\pi(i)}\left\lceil t_{i-1}/T_{\pi(i)} \right\rceil}{1 - U_{\pi(i+1)\dots\pi(n-1)}} \geq t_{i-1}.$$

In the case that t_{i-1} is a multiple of $T_{\pi(i)}$, the equality holds, i.e. $K_i(t_{i-1}) = t_{i-1}$, so that t_{i-1} is already the fixed point t_i we are looking for.

Otherwise t_{i-1} lies in an open interval $(vT_{\pi(i)}, (v+1)T_{\pi(i)})$ with some integer v in which $K_i(t)$ is constant, since the values of the ceiling terms in $K_i(t)$ can change only for integer multiples of $T_{\pi(i)}$. Due to the assumption that $T_{\pi(i)}$ does not divide t_{i-1} we can write

$$t \in (vT_{\pi(i)}, (v+1)T_{\pi(i)}) : \ K_i(t) = K_i(t_{i-1}) =$$
$$t_{i-1} + \frac{-t_{i-1}U_{\pi(i)} + C_{\pi(i)}\lceil t_{i-1}/T_{\pi(i)}\rceil}{1 - U_{\pi(i+1)...\pi(n-1)}} > t_{i-1}$$

This property is obtained formally because of the divisibility $T_{\pi(i)}|T_{\pi(j)}$ for $j < i$ or $T_{\pi(j)} = aT_{\pi(i)}$ with a integer. With t being in $(vT_{\pi(i)}, (v+1)T_{\pi(i)})$ and $T_{\pi(j)} = aT_{\pi(i)}$ we get $t \in ((v/a)T_{\pi(j)}, ((v+1)/a)T_{\pi(j)})$. The ceiling terms in $K_i(t)$ only assume new values at times that are multiples of the respective periods. We therefore use $\lfloor v/a\rfloor \le v/a$ and since $v+1$ and a are positive integers we can use the equality $\lceil((v+1)/a)\rceil = \lfloor((v+1-1)/a)\rfloor + 1 = \lfloor((v)/a)\rfloor + 1$ and make the factors of $T_{\pi(j)}$ integer: $t \in (\lfloor v/a\rfloor T_{\pi(j)}, (\lfloor v/a\rfloor + 1)T_{\pi(j)})$. This also means that $(vT_{\pi(i)}, (v+1)T_{\pi(i)}) \subseteq (\lfloor v/a\rfloor T_{\pi(j)}, (\lfloor v/a\rfloor + 1)T_{\pi(j)})$ and $t_{i-1} \in (\lfloor v/a\rfloor T_{\pi(j)}, (\lfloor v/a\rfloor + 1)T_{\pi(j)})$. The values of the ceil-terms $\lceil t/T_{\pi(j)}\rceil$ are therefore constant for all t in the interval. Since t_{i-1} lies in the interval $(vT_{\pi(i)}, (v+1)T_{\pi(i)})$, we can state as an interim result:

$$j \le i, t \in (vT_{\pi(i)}, (v+1)T_{\pi(i)}) : \lceil t/T_{\pi(j)}\rceil = \lceil t_{i-1}/T_{\pi(j)}\rceil$$

i.e. the value of the ceiling terms does not change in these intervals and by (16)

$$t \in (vT_{\pi(i)}, (v+1)T_{\pi(i)}) : K_i(t) = K_i(t_{i-1}).$$

We now consider the point in time

$$y_i = t_{i-1} + \frac{-t_{i-1}U_{\pi(i)} + C_{\pi(i)}\lceil t_{i-1}/T_{\pi(i)}\rceil}{1 - U_{\pi(i+1)...\pi(n-1)}} \tag{20}$$

and show that y_i is a fixed point, i.e. $K_i(y_i) = y_i$. We use the equality $U_{\pi(i)} = C_{\pi(i)}/T_{\pi(i)}$ and factor out $C_{\pi(i)}/T_{\pi(i)}$.

$$y_i = t_{i-1} + \frac{(\lceil t_{i-1}/T_{\pi(i)}\rceil T_{\pi(i)} - t_{i-1})C_{\pi(i)}/T_{\pi(i)}}{1 - U_{\pi(i+1)...\pi(n-1)}}$$

To show that this point also lies within the interval $(vT_{\pi(i)}, (v+1)T_{\pi(i)})$, we first show that it lies below the upper interval boundary.

$$y_i < t_{i-1} + \lceil t_{i-1}/T_{\pi(i)}\rceil T_{\pi(i)} - t_{i-1} = \lceil t_{i-1}/T_{\pi(i)}\rceil T_{\pi(i)} = (v+1)T_{\pi(i)}$$

Here we use the inequality $1 - U_{\pi(i+1)...\pi(n-1)} > U_{\pi(i)} = C_{\pi(i)}/T_{\pi(i)}$, which is derived from the total utilization bound 1.

Furthermore with Eq. (20) we get $y_i > t_{i-1} > vT_{\pi(i)}$. Since $K_i(t)$ is constant in this interval, it follows that

$$K_i(y_i) = K_i(t_{i-1}) = t_{i-1} + \frac{(\lceil t_{i-1}/T_{\pi(i)}\rceil T_{\pi(i)} - t_{i-1})C_{\pi(i)}/T_{\pi(i)}}{1 - U_{\pi(i+1)...\pi(n-1)}} = y_i.$$

Thus y_i is a point where the functions $K_i(t)$ and $Id(t)$ intersect.

Because of the case distinction here t_{i-1} is not a multiple of $T_{\pi(i)}$ and we therefore have $y_i = K_i(t_{i-1}) > t_{i-1}$, so the iteration to determine the least fixed point can start with t_{i-1}. After this time $K_i(t)$ remains constant until time $(v+1)T_{\pi(i)}$, so that y_i is the least fixed point.   □

We now summarize the proven statements in a proof of Theorem 2:

Proof. of Theorem 2 We carry out an induction proof.

Base case: In (13) we have defined the time t_0 with $t_0 = K_0(t_0)$. This value is equal with $R_n^{(0)}$

Induction step: Assume we have found the time $t_{i-1} = K_{i-1}(t_{i-1})$ equal to $\widetilde{R}_n^{(i-1)}$ in (9), then by Lemma 3 we have shown that the fixed point $t_i = K_i(t_i)$ can be determined (19) as

$$t_i = t_{i-1} + \frac{-t_{i-1} \cdot U_{\pi(i)} + C_{\pi(i)} \cdot \left\lceil \frac{t_{i-1}}{T_{\pi(i)}} \right\rceil}{1 - U_{\pi(i+1)...\pi(n-1)}} \tag{21}$$

The Eq. (9) is equivalent to (21).if we substitute t_i by $\widetilde{R}_n^{(i)}$ and t_{i-1} by $\widetilde{R}_n^{(i-1)}$.   □

5   Improving the Algorithm

The algorithm in Theorem 2 uses floating-point arithmetic to determine the response time, which means that rounding errors may accumulate and lead to a wrong value. We therefore perform simple transformations to calculate R_n with integer operations. We start the transformation of the algorithm with a definition.

Definition 3. *The worst-case i-idle time $H_{\pi(i)...\pi(n-1)}$ is the time within a period $\left(vT_{\pi(i)}, (v+1)T_{\pi(i)}\right]$ the processor is not busy with any of the tasks $\tau_{\pi(i)}, \ldots, \tau_{\pi(n-1)}$.*

One could determine this amount of time by applying

$$H_{\pi(i)...\pi(n-1)} \stackrel{\text{def}}{=} T_{\pi(i)} \cdot (1 - U_{\pi(i)...\pi(n-1)}) \tag{22}$$

Notice that

$$H_{\pi(n-1)} = T_{\pi(n-1)} - C_{\pi(n-1)}$$

In order to determine the other values of the H-functions we consider (22) for i and $i + 1$: We divide the equations by the respective period:

$$H_{\pi(i)...\pi(n-1)}/T_{\pi(i)} = (1 - U_{\pi(i)...\pi(n-1)}) \tag{23}$$

and

$$H_{\pi(i+1)...\pi(n-1)}/T_{\pi(i+1)} = (1 - U_{\pi(i+1)...\pi(n-1)}) \tag{24}$$

From these equalities follows with $U_{\pi(i)...\pi(n-1)} - U_{\pi(i)} = U_{\pi(i+1)...\pi(n-1)}$:

$$H_{\pi(i)...\pi(n-1)}/T_{\pi(i)} + U_{\pi(i)} = H_{\pi(i+1)...\pi(n-1)}/T_{\pi(i+1)} \tag{25}$$

We solve this equation by $H_{\pi(i)}$ using $U_{\pi(i)} = C_{\pi(i)}/T_{\pi(i)}$ and get:

$$H_{\pi(i)...\pi(n-1)} = H_{\pi(i+1)...\pi(n-1)} \cdot T_{\pi(i)}/T_{\pi(i+1)} - C_{\pi(i)} \tag{26}$$

Notice that $T_{\pi(i+1)} | T_{\pi(i)}$ such that $H_{\pi(i)...\pi(n-1)}$ is an integer.

We now can introduce the algorithm that uses only integer arithmetic:

Corollary 1. *We define for* $0 \le i \le n - 1$, $\widetilde{Q}_n^{(i)} \overset{\text{def}}{=} \widetilde{R}_n^{(i)} \cdot (1 - U_{\pi(i+1)...\pi(n-1)})$ *and get*

$$\widetilde{Q}_n^{(0)} = C_n \tag{27}$$

and for $1 \le i \le n - 1$

$$\widetilde{Q}_n^{(i)} = \widetilde{Q}_n^{(i-1)} + C_{\pi(i)} \left\lceil \frac{\widetilde{R}_n^{(i-1)}}{H_{\pi(i)...\pi(n-1)}} \right\rceil \tag{28}$$

Finally, we get the response time $R_n = \widetilde{Q}_n^{(n-1)}$

Proof. By Theorem 2 we have:

$$\widetilde{R}_n^{(0)} = \frac{C_n}{1 - U_{\pi(1)...\pi(n-1)}}$$

By definition of $\widetilde{Q}_n^{(0)}$ it follows $\widetilde{Q}_n^{(0)} = C_n$

For the general case Theorem 2 Eq. (9) gives the solution

$$1 \le i \le n - 1, \quad \widetilde{R}_n^{(i)} = \widetilde{R}_n^{(i-1)} + \frac{C_{\pi(i)} \cdot \left(\left\lceil \frac{\widetilde{R}_n^{(i-1)}}{T_{\pi(i)}} \right\rceil - \frac{\widetilde{R}_n^{(i-1)}}{T_{\pi(i)}} \right)}{1 - U_{\pi(i+1)...\pi(n-1)}}$$

After multiplying the equation by $1 - U_{\pi(i+1)...\pi(n-1)}$, combining the terms linear in $\widetilde{R}_n^{(i-1)}$, and observing the equality $1 - U_{\pi(i+1)...\pi(n-1)} - C_{\pi(i)}/T_{\pi(i)} = 1 - U_{\pi(i+1)...\pi(n-1)} - U_{\pi(i)} = 1 - U_{\pi(i)...\pi(n-1)}$ we get:

$$\widetilde{R}_n^{(i)}(1 - U_{\pi(i+1)...\pi(n-1)}) = \widetilde{R}_n^{(i-1)} \cdot (1 - U_{\pi(i)...\pi(n-1)}) + C_{\pi(i)} \cdot \left\lceil \frac{\widetilde{R}_n^{(i-1)}}{T_{\pi(i)}} \right\rceil$$

By definition of $\widetilde{Q}_n^{(i)} = \widetilde{R}_n^{(i)} \cdot (1 - U_{\pi(i+1)...\pi(n-1)})$ and $\widetilde{Q}_n^{(i-1)} = \widetilde{R}_n^{(i-1)} \cdot (1 - U_{\pi(i)...\pi(n-1)})$ it follows

$$\widetilde{Q}_n^{(i)} = \widetilde{Q}_n^{(i-1)} + C_{\pi(i)} \cdot \left\lceil \frac{\widetilde{R}_n^{(i-1)}}{T_{\pi(i)}} \right\rceil$$

In Eq. (22) we have defined $H_{\pi(i)...\pi(n-1)} = T_{\pi(i)} \cdot (1 - U_{\pi(i)...\pi(n-1)})$ such that we can expand the fraction that is argument of the ceiling operation by $1 - U_{i...n-1}$:

$$\widetilde{Q}_n^{(i)} = \widetilde{Q}_n^{(i-1)} + C_{\pi(i)} \cdot \left\lceil \frac{\widetilde{Q}_n^{(i-1)}}{H_{\pi(i)...\pi(n-1)}} \right\rceil$$

Finally, for $i = n - 1$ we have $\widetilde{Q}_n^{(n-1)} = \widetilde{R}_n^{(n-1)} \cdot (1 - U_{n...\pi(n-1)}) = \widetilde{R}_n^{(n-1)} = R_n$ (by Theorem 2).

We can take advantage of the fact that $\widetilde{Q}_n^{(i-1)}$ and $H_{\pi(i)...\pi(n-1)}$ are positive integers. Therefore, we can use the equivalence $\left\lceil \frac{\widetilde{Q}_n^{(i-1)}}{H_{\pi(i)...\pi(n-1)}} \right\rceil = \left\lfloor \frac{\widetilde{Q}_n^{(i-1)} - 1}{H_{\pi(i)...\pi(n-1)}} \right\rfloor + 1$ and perform an integer division instead of applying the floor function.

6   Conclusions

Because of the manifold practical applications of task systems with harmonic tasks it is important to take advantage of the special features resulting from the divisibility of periods by all smaller periods. For example, response time analysis is possible in strongly polynomial time with harmonic tasks, while in the general case it has pseudo-polynomial complexity and is known to be NP-hard [6]. We have introduced a new algorithm that calculates the exact *worst-case response time* of a task in linear time $\mathcal{O}(n)$ when the higher-priority tasks are ordered by non-increasing periods and in $\mathcal{O}(n \cdot \log n)$ in general.

References

1. Audsley, N., Burns, A., Richardson, M., Tindell, K., Wellings, A.J.: Applying new scheduling theory to static priority pre-emptive scheduling. Softw. Eng. J. **8**, 284–292 (1993)
2. Bini, E., Parri, A., Dossena, G.: A quadratic-time response time upper bound with a tightness property. In: Proceedings of the IEEE International Real-Time Systems Symposium (RTSS 2015), San Antonio, TX, USA, December 2015
3. Bonifaci, V., Marchetti-Spaccamela, A., Megow, N., Wiese, A.: Polynomial-time exact schedulability tests for harmonic real-time tasks. In: RTSS 2013, pp. 236–245 (2013)
4. Davis, R.I., Zabos, A., Burns, A.: Efficient exact schedulability tests for fixed priority real-time systems. IEEE Trans. Comput. **57**(9), 1261–1276 (2008). http://doi.ieeecomputersociety.org/10.1109/TC.2008.66

5. Davis, R., Burns, A.: Response time upper bounds for fixed priority real-time systems. In: Proceedings of the IEEE International Real-Time Systems Symposium (RTSS 2008), pp. 407–418 (2008). https://doi.org/10.1109/RTSS.2008.18
6. Eisenbrand, F., Rothvoss, T.: Static-priority real-time scheduling: response time computation is NP-hard. In: Proceedings of the 29th IEEE Real-Time Systems Symposium, RTSS 2008, Barcelona, Spain, 30 November–3 December 2008, pp. 397–406. IEEE Computer Society (2008). https://doi.org/10.1109/RTSS.2008.25
7. Fineberg, M., Serlin, O.: Multiprogramming for hybrid computation. In: Proceedings of the AFIPS Fall Joint Computing Conference, pp. 1–13 (1967)
8. Joseph, M., Pandya, P.: Finding response times in a real-time system. Comput. J. **29**(5), 390–395 (1986). https://doi.org/10.1093/comjnl/29.5.390. http://comjnl. oxfordjournals.org/content/29/5/390.abstract
9. Kuo, T., Mok, A.: Load adjustment in adaptive real-time systems. In: Proceedings of the IEEE International Real-Time Systems Symposium (RTSS 1991), pp. 160–170, December 1991. https://doi.org/10.1109/REAL.1991.160369
10. Lehoczky, J.: Fixed priority scheduling of periodic task sets with arbitrary deadlines. In: Proceedings of the IEEE International Real-Time Systems Symposium (RTSS 1990), pp. 201–209 (1990). https://doi.org/10.1109/REAL.1990.128748
11. Lehoczky, J., Sha, L., Ding, Y.: The rate monotonic scheduling algorithm: exact characterization and average case behavior. In: Proceedings of the IEEE International Real-Time System Symposium (RTSS 1989) pp. 166–171 (1989)
12. Liu, C., Layland, J.: Scheduling algorithms for multiprogramming in a hard-real-time environment. J. ACM **20**(1), 46–61 (1973). https://doi.org/10.1145/321738. 321743. http://doi.acm.org/10.1145/321738.321743
13. Lu, W., Hsieh, J., Shih, W., Kuo, T.: A faster exact schedulability analysis for fixed-priority scheduling. J. Syst. Softw. **79**(12), 1744–1753 (2006). https://doi. org/10.1016/j.jss.2006.03.023
14. Lu, W., Hsieh, J., Shih, W.K.: A precise schedulability test algorithm for scheduling periodic tasks in real-time systems. In: Proceedings of the 2006 ACM Symposium on Applied Computing, SAC 2006, pp. 1451–1455. ACM, New York (2006). https://doi.org/10.1145/1141277.1141616. http://doi.acm.org/ 10.1145/1141277.1141616
15. Sjodin, M., Hansson, H.: Improved response-time analysis calculations. In: Proceedings of the IEEE International Real-Time Systems Symposium (RTSS 1998), pp. 399–408 (1998). https://doi.org/10.1109/REAL.1998.739773
16. Tindell, K., Burns, A., Wellings, A.J.: An extendible approach for analysing fixed priority hard real-time tasks. Real-Time Syst. **6**(2), 133–151 (1994). https://doi. org/10.1007/BF01088593
17. Xu, Y., Cervin, A., Arzén, K.E.: LQG-based scheduling and control co-design using harmonic task periods. Technical Reports TFRT-7646, Department of Automatic Control, Lund Institute of Technology, Lund University (2016)

Computing a Minimum Subset Feedback Vertex Set on Chordal Graphs Parameterized by Leafage

Charis Papadopoulos and Spyridon Tzimas[(✉)]

Department of Mathematics, University of Ioannina, Ioannina, Greece
charis@uoi.gr, roytzimas@hotmail.com

Abstract. Given a vertex-weighted graph $G = (V, E)$ and a set $S \subseteq V$, the SUBSET FEEDBACK VERTEX SET (SFVS) problem asks for a vertex set of minimum weight that intersects all cycles containing a vertex of S. SFVS is known to be polynomial-time solvable on interval graphs, whereas SFVS remains NP-complete on split graphs and, consequently, on chordal graphs. Towards a better understanding of the complexity of SFVS on subclasses of chordal graphs, we exploit structural properties of a tree model in order to cope with the hardness of SFVS. Here we consider variants of the *leafage* that measures the minimum number of leaves in a tree model. We show that SFVS can be solved in polynomial time for every chordal graph with bounded leafage. In particular, given a chordal graph on n vertices with leafage ℓ, we provide an algorithm for SFVS with running time $n^{O(\ell)}$, thus improving upon $n^{O(\ell^2)}$, the running time of the previously known algorithm obtained for graphs with bounded mim-width. We complement our result by showing that SFVS is W[1]-hard parameterized by ℓ. Pushing further our positive result, it is natural to consider a slight generalization of leafage, the *vertex leafage*, which measures the minimum upper bound on the number of leaves of every subtree in a tree model. However, we show that it is unlikely to obtain a similar result, as we prove that SFVS remains NP-complete on undirected path graphs, i.e., chordal graphs having vertex leafage at most two. Lastly, we provide a polynomial-time algorithm for SFVS on rooted path graphs, a proper subclass of undirected path graphs and graphs of mim-width one, which is faster than the previously known algorithm obtained for graphs with bounded mim-width.

Keywords: Subset feedback vertex set · Leafage · W-hardness

1 Introduction

Several fundamental optimization problems are known to be intractable on chordal graphs, however they admit polynomial time algorithms when restricted

Research supported by the Hellenic Foundation for Research and Innovation (H.F.R.I.) under the "First Call for H.F.R.I. Research Projects to support Faculty members and Researchers and the procurement of high-cost research grant", Project FANTA (eFficient Algorithms for NeTwork Analysis), number HFRI-FM17-431.

C. Bazgan and H. Fernau (Eds.): IWOCA 2022, LNCS 13270, pp. 466–479, 2022.
https://doi.org/10.1007/978-3-031-06678-8_34

to a proper subclass of chordal graphs such as interval graphs. Typical examples of this type of problems are domination or induced path problems [2,5,12,23,25,32]. Towards a better understanding of why many intractable problems on chordal graphs admit polynomial time algorithms on interval graphs, we consider the algorithmic usage of the structural parameter named leafage. Leafage, introduced by Lin et al. [30], is a graph parameter that captures how close is a chordal graph of being an interval graph. As it concerns chordal graphs, leafage essentially measures the smallest number of leaves in a clique tree, an intersection representation of the given graph [19]. Here we are concerned with the SUBSET FEEDBACK VERTEX SET problem, SFVS for short: given a vertex-weighted graph and a set S of its vertices, compute a vertex set of minimum weight that intersects all cycles containing a vertex of S. Although SUBSET FEEDBACK VERTEX SET does not fall to the themes of domination or induced path problems, it is known to be NP-complete on chordal graphs [16], whereas it becomes polynomial-time solvable on interval graphs [34]. Thus our research study concerns to what extent the structure of the underlying tree representation influences the computational complexity of SUBSET FEEDBACK VERTEX SET.

An interesting remark concerning SUBSET FEEDBACK VERTEX SET, is the fact that its unweighted and weighted variants behave computationally different on hereditary graph classes. For example, SUBSET FEEDBACK VERTEX SET is NP-complete on H-free graphs for some fixed graphs H, while its unweighted variant admits a polynomial time algorithm on the same class of graphs [7,35]. Thus the unweighted and weighted variants of SUBSET FEEDBACK VERTEX SET do not align. SUBSET FEEDBACK VERTEX SET remains NP-complete on bipartite graphs [39] and planar graphs [18], as a generalization of FEEDBACK VERTEX SET. Notable differences between the two latter problems regarding their complexity status is the class of split graphs and $4P_1$-free graphs for which SUBSET FEEDBACK VERTEX SET is NP-complete [16,35], as opposed to the FEEDBACK VERTEX SET problem [7,11,38]. Inspired by the NP-completeness on chordal graphs, SUBSET FEEDBACK VERTEX SET restricted on (subclasses of) chordal graphs has attracted several researchers to obtain fast, still exponential-time, algorithms [21,37].

On the positive side, SUBSET FEEDBACK VERTEX SET can be solved in polynomial time on restricted graph classes [6,7,34,35]. Cygan et al. [14] and Kawarabayashi and Kobayashi [29] independently showed that SUBSET FEEDBACK VERTEX SET is fixed-parameter tractable (FPT) parameterized by the solution size, while Hols and Kratsch provided a randomized polynomial kernel for the problem [24]. Related to the structural parameter mim-width, Bergougnoux et al. [1] recently proposed an $n^{O(w^2)}$-time algorithm that solves SUBSET FEEDBACK VERTEX SET given a decomposition of the input graph of mim-width w. As leaf power graphs admit a decomposition of mim-width one [26], from the later algorithm SUBSET FEEDBACK VERTEX SET can be solved in polynomial time on leaf power graphs if an intersection model is given as input. However, to the best of our knowledge, it is not known whether the intersection model of a leaf power graph can be constructed in polynomial time. Moreover, even for graphs of mim-width one that do admit an efficient construction of the corresponding decomposition, the exponent of the running time given in [1] is relatively high.

Habib and Stacho [22] showed that the leafage of a connected chordal graph can be computed in polynomial time. Their described algorithm also constructs a corresponding clique tree with the minimum number of leaves. Here we show that SUBSET FEEDBACK VERTEX SET is polynomial-time solvable for every chordal graph with bounded leafage. In particular, given a chordal graph with a tree model having ℓ leaves, our algorithm runs in $O(\ell n^{2\ell+1})$ time. Thus, by combining the algorithm of Habib and Stacho [22], we deduce that SUBSET FEEDBACK VERTEX SET is in XP, parameterized by the leafage.

One advantage of leafage over mim-width is that we can compute the leafage of a chordal graph in polynomial time, whereas we do not know how to compute in polynomial time the mim-width of a chordal graph. However we note that a graph of bounded leafage implies a graph of bounded mim-width and, further, a decomposition of bounded mim-width can be computed in polynomial time [17]. This can be seen through the notion of H-graphs which are exactly the intersection graphs of connected subgraphs of some subdivision of a fixed graph H. The intersection model of subtrees of a tree T having ℓ leaves is a T'-graph where T' is obtained from T by contracting nodes of degree two. Thus the size of T' is at most 2ℓ, since T has ℓ leaves. Moreover, given an H-graph and its intersection model, a (linear) decomposition of mim-width at most $2|E(H)|$ can be computed in polynomial time [17]. Therefore, given a graph of leafage ℓ, there is a polynomial-time algorithm that computes a decomposition of mim-width $O(\ell)$. Combined with the algorithm via mim-width [1], one can solve SUBSET FEEDBACK VERTEX SET in time $n^{O(\ell^2)}$ on graphs having leafage ℓ. Notably, our $n^{O(\ell)}$-time algorithm is a non-trivial improvement on the running time obtained from the mim-width approach.

We complement our algorithmic result by showing that SUBSET FEEDBACK VERTEX SET is W[1]-hard parameterized by the leafage of a chordal graph. Thus we can hardly avoid the dependence of the exponent in the stated running time. Our reduction is inspired by the W[1]-hardness of FEEDBACK VERTEX SET parameterized by the mim-width given in [27]. However we note that our result holds on graphs with arbitrary vertex weights and we are not aware if the unweighted variant of SUBSET FEEDBACK VERTEX SET admits the same complexity behavior.

Our algorithm works on an expanded tree model that is obtained from the given tree model and maintains all intersecting information without increasing the number of leaves. Then in a bottom-up dynamic programming fashion, we visit every node of the expanded tree model in order to compute partial solutions. At each intermediate step, we store all necessary information of subsets of vertices that are of size $O(\ell)$. As a byproduct of our dynamic programming scheme and the expanded tree model, we show how our approach can be extended in order to handle rooted path graphs. Rooted path graphs are the intersection graphs of rooted paths in a rooted tree. They form a subclass of leaf powers and have unbounded leafage (through their underlying tree model). Although rooted path graphs admit a decomposition of mim-width one [26] and such a decomposition can be constructed in polynomial time [15,20], the running time obtained through the bounded mim-width approach is rather unpractical, as it requires to store a table of size $O(n^{13})$ even in this particular case [1]. By analyzing further subsets of vertices at each intermediate step, we manage to derive an algorithm for SUBSET FEEDBACK VERTEX SET on rooted path graphs that runs in $O(n^2 m)$ time. Observe that the stated running time is comparable to the $O(nm)$-time algorithm on interval graphs [34] and interval graphs form a proper subclass of rooted path graphs.

Moreover, inspired by the algorithm on bounded leafage graphs we consider its natural generalization concerning the *vertex leafage* of a graph. Chaplick and Stacho [10] introduced the vertex leafage of a graph G as the smallest number k such that there exists a tree model for G in which every subtree corresponding to a vertex of G has at most k leaves. As leafage measures the closeness to interval graphs (graphs with leafage at most two), vertex leafage measures the closeness to undirected path graphs which are the intersection graphs of paths in a tree (graphs with vertex leafage at most two). We prove that the unweighted variant of SUBSET FEEDBACK VERTEX SET is NP-complete on undirected path graphs and, thus, the problem is para-NP-complete parameterized by the vertex leafage. An interesting remark of our NP-completeness proof is that our reduction comes from the MAX CUT problem as opposed to known reductions for SUBSET FEEDBACK VERTEX SET which are usually based on, more natural, covering problems [16,35].

2   Preliminaries

All graphs considered here are finite undirected graphs without loops and multiple edges. We refer to the textbook by Bondy and Murty [4] for any undefined graph terminology and to the recent book of [13] for the introduction to Parameterized Complexity. For a positive integer p, we use $[p]$ and $-[p]$ to denote the sets of integers $\{1, \dots, p\}$ and $\{-1, \dots, -p\}$, respectively. For a graph $G = (V_G, E_G)$, we use V_G and E_G to denote the set of vertices and edges, respectively. We use n to denote the number of vertices of a graph and use m for the number of edges. Given $x \in V_G$, we denote by $N_G(x)$ the neighborhood of x. The *degree* of x is the number of edges incident to x. Given $X \subseteq V_G$, we define the neighborhood $N_G(X)$ of X to be $(\cup\{N_G(x) : x \in X\}) \setminus X$. We denote by $G - X$ the graph obtained from G by the removal of the vertices of X. If $X = \{u\}$, we also write $G - u$. The *subgraph induced by* X is denoted by $G[X]$, and has X as its vertex set and $\{uv \mid u, v \in X$ and $uv \in E_G\}$ as its edge set. A *clique* is a set $K \subseteq V_G$ such that $G[K]$ is a complete graph.

Given a collection $\mathcal{C}$ of sets, the graph $G = (\mathcal{C}, \{\{X, Y\} : X, Y \in \mathcal{C}$ and $X \cap Y \neq \emptyset\})$ is called *the intersection graph of* $\mathcal{C}$. Structural properties and recognition algorithms are known for intersection graphs of (directed) paths in (rooted) trees [9,31,33]. Depending on the collection $\mathcal{C}$, we say that a graph is *chordal* if $\mathcal{C}$ is a collection of subtrees of a tree, *undirected path* if $\mathcal{C}$ is a collection of paths of a tree, *rooted path* if $\mathcal{C}$ is a collection of directed paths of a rooted tree, and *interval* if $\mathcal{C}$ is a collection of subpaths of a path. For any undirected tree T, we use $L(T)$ to denote the set of its leaves, i.e., the set of nodes of T having degree at most one. If T contains only one node then we let $L(T) = \emptyset$. Let T be a rooted tree. We assume that the edges of T are directed away from the root. We denote the unique directed path from a node w to a node v by $w \rightarrow v$. If $w \rightarrow v$ exists in T, we say that v is a *descendant* of w and that w is an *ancestor* of v. The leaves of a rooted tree T are exactly the nodes of T having in-degree one and out-degree zero. Observe that for an undirected tree T with at least one edge we have $|L(T)| \geq 2$, whereas in a rooted tree T with at least one edge $|L(T)| \geq 1$ holds.

A binary relation, denoted by $\leq$, on a set V is called *partial order* if it is transitive and anti-symmetric. For a partial order $\leq$ on a set V, we say that two elements x and y of V are *comparable* if $x \leq y$ or $y \leq x$; otherwise, x and y are called *incomparable*. If $x \leq y$ and $x \neq y$ then we simply write $x < y$. Given $X, Y \subseteq V$, we write $X \leq Y$ if for any $x \in X$ and $y \in Y$, we have $x \leq y$; if X and Y are disjoint then $X \leq Y$ is denoted by $X < Y$. Given a rooted tree T, we define a partial order on the nodes of T as follows: $x \leq_T y \Leftrightarrow x$ is a descendant of y. It is not difficult to see that if $x \leq_T y$ and $x \leq_T z$ then y and z are comparable, as T is a rooted tree.

A *tree model* of a graph $G = (V_G, E_G)$ is a pair $(T, \{T_v\}_{v \in V_G})$ where T is a tree, called a *host tree*[1], each T_v is a subtree of T, and $uv \in E_G$ if and only if $V(T_u) \cap V(T_v) \neq \emptyset$. We say that a tree model $(T, \{T_v\}_{v \in V_G})$ *realizes* a graph H if its corresponding graph G is isomorphic to H. It is known that a graph is chordal if and only if it admits a tree model [8,19]. The tree model of a chordal graph is not necessarily unique. The *leafage* of a chordal graph G, denoted by $\ell(G)$, is the minimum number of leaves of the host tree among all tree models that realize G, that is, $\ell(G)$ is the smallest integer ℓ such that there exists a tree model $(T, \{T_v\}_{v \in V_G})$ of G with $\ell = |L(T)|$ [30]. Moreover, every chordal graph G admits a tree model for which its host tree T has the minimum $|L(T)|$ and $|V(T)| \leq n$ [10,22]; such a tree model can be constructed in $O(n^3)$ time [22]. Thus the leafage $\ell(G)$ of a chordal graph G is computable in polynomial time.

A generalization of leafage is the *vertex leafage* introduced by Chaplick and Stacho [10]. The vertex leafage of a chordal graph G, denoted by $v\ell(G)$, is the smallest integer k such that there exists a tree model $(T, \{T_v\}_{v \in V_G})$ of G where $|L(T_v)| \leq k$ for all $v \in V_G$. Clearly, we have $v\ell(G) \leq \ell(G)$.

We will only consider tree models of chordal graphs where the host tree is a rooted tree. Under these terms, observe that $\ell(G) = 0$ iff G is a clique, $\ell(G) \leq 1$ iff G is an interval graph, $v\ell(G) \leq 1$ iff G is a rooted path graph, and $v\ell(G) \leq 2$ if G is an undirected path graph.

---

[1] The host tree is also known as a *clique tree*, usually when we are concerned with the maximal cliques of a chordal graph [19].

By an induced cycle of G we mean a chordless cycle. A *triangle* is a cycle on 3 vertices. Hereafter, we consider subclasses of chordal graphs, that is graphs that do not contain induced cycles on more than 3 vertices.

Given a graph G and $S \subseteq V(G)$, we say that a cycle of G is an *S-cycle* if it contains a vertex in S. Moreover, we say that an induced subgraph F of G is an *S-forest* if F does not contain an S-cycle. Thus an induced subgraph F of a chordal graph is an S-forest if and only if F does not contain any S-triangle. In these terms, the SUBSET FEEDBACK VERTEX SET problem asks for a vertex set of minimum (weight) size such that its removal results in an S-forest. The set of vertices that do not belong to an S-forest is referred to as *subset feedback vertex set*. In our dynamic programming algorithms, we focus on the equivalent formulation of computing a maximum weighted S-forest.

For a collection $\mathcal{C}$ of sets of vertices, we write $\max\limits_{\text{weight}} \{C \in \mathcal{C}\}$ to denote $\arg\max\limits_{C \in \mathcal{C}} \{weight(C)\}$, where $weight(C)$ is the sum of weights of the vertices in C. The collection of S-forests of a graph G, is denoted by $\mathcal{F}_S$. For any $X, Y \subseteq V_G$ such that $X \cap Y = \emptyset$ and $G[Y] \in \mathcal{F}_S$, we write A_X^Y to denote an arbitrary element of the collection $\max\limits_{\text{weight}} \{U \subseteq X : G[U \cup Y] \in \mathcal{F}_S\}$. We use the operator $\leftrightarrow$ between any two expressions involving such sets to denote that for any particular evaluation of one there exists an evaluation of the other such that both yield the same result. Our desired optimal solution is any element $A_{V_G}^{\emptyset}$ of $\max\limits_{\text{weight}} \{U \subseteq V_G : G[U] \in \mathcal{F}_S\}$. We will subsequently show that in order to compute $A_{V_G}^{\emptyset}$ it is sufficient to compute A_X^Y for a polynomial number of sets X and Y.

Let $G = (V_G, E_G)$ be a chordal graph and let $X, Y \subseteq V_G$ such that $X \cap Y = \emptyset$ and $G[Y] \in \mathcal{F}_S$. A partition $\mathcal{P}$ of X is called *nice* if for any S-triangle S_t of $G[X \cup Y]$, there is a partition class $P_i \in \mathcal{P}$ such that $V(S_t) \cap X \subseteq P_i$. In other words, any S-triangle of $G[X \cup Y]$ is involved with at most one partition class of a nice partition $\mathcal{P}$ of X. With respect to the optimal defined solutions A_X^Y, we observe the following[2]:

Observation 2.1. Let $G = (V_G, E_G)$ be a chordal graph and let $X, Y \subseteq V_G$ such that $X \cap Y = \emptyset$ and $G[Y] \in \mathcal{F}_S$. Then, the following hold:

(1) $A_X^Y \leftrightarrow A_{X'}^{Y'}$ for any $Y \supseteq Y' \supseteq Y \cap N(X')$ where $X' = X \setminus \{u \in X \setminus S : Y \cap N(u) \subseteq Y \setminus S\}$.
(2) $A_X^Y \leftrightarrow \bigcup_{X' \in \mathcal{P}} A_{X'}^Y$ for any nice partition $\mathcal{P}$ of X.

By Observation 2.1, we search for nice partitions of the vertex set X in order to consider smaller instances of A_X^Y. More precisely, Observation 2.1 (2) suggests how to consider the sets X' that form a nice partition of X, whereas Observation 2.1 (1) indicates which vertices of Y are relative to each set X'.

---

[2] In this extended abstract all proofs are omitted due to space constraints. See a preliminary full version [36] for all the details.

3   Expanded Tree Model and Related Vertex Subsets

Given a tree model of a chordal graph, we are interested in defining a partial order on the vertices of the graph that takes advantage of the underlying tree structure. For this reason, it is more convenient to consider the tree model as a natural rooted tree and each of its subtrees to correspond to at most one vertex of the graph. Here we show how a tree model can be altered in order to capture the appropriate properties in a formal way. We assume that G is a chordal graph that admits a tree model $(T, \{T_v\}_{v \in V_G})$ such that $|L(T)| = \ell(G)$. We will concentrate on the case in which $|L(T)| \geq 2$ and T contains a non-leaf node. The rest of the cases (i.e., $|V(T)| \leq 2$) are handled by the algorithm on interval graphs [34] in a separate way. For this purpose we say that a chordal graph G is *non-trivial* if $|V(T)| > 2$.

A tree model $(T, \{T_v\}_{v \in V_G})$ of G is called *expanded tree model* if

- the host tree T is rooted (and, consequently, all of its subtrees are rooted),
- for every $v \in V_G$, $L(T_v) \neq \emptyset$ holds, and
- every node of T is either the root or a leaf of at most one subtree T_v that corresponds to a vertex v of G.

We show that any non-trivial chordal graph admits an expanded tree model that is *close* to its tree model. In fact, we provide an algorithm that, given a tree model of a non-trivial chordal graph G, constructs an expanded tree model that realizes G.

Lemma 3.1. *For any tree model $(T, \{T_v\}_{v \in V_G})$ of G with $|L(T)| = \ell \geq 2$ and $|L(T_v)| \leq v\ell \leq \ell$ for all $v \in V_G$, there is an expanded tree model $(T', \{T'_v\}_{v \in V_G})$ of G such that:*

- $|L(T')| = \ell$,
- $|L(T_v)| - 1 \leq |L(T'_v)| \leq |L(T_v)|$ *for every $v \in V_G$, and*
- $|V(T')| \leq |V(T)| + (1 + v\ell)(n - 1)$.

Moreover, given $(T, \{T_v\}_{v \in V_G})$, the expanded tree model can be constructed in time $O(n^2)$.

Hereafter we assume that $(T, \{T_v\}_{v \in V_G})$ is an expanded tree model of a non-trivial chordal graph G. For any vertex u of G, we denote the root of its corresponding rooted tree T_u in T by $r(u)$. We define the following partial order on the vertices of G: for all $u, v \in V_G$, $u \leq v \Leftrightarrow r(u) \leq_T r(v)$. In other words, two vertices of G are comparable (with respect to $\leq$) if and only if there is a directed path between their corresponding roots in T. For all $u \in V_G$, we define $V_u = \{u' \in V_G : u' \leq u\}$.

Observation 3.1. Let $u, v, w, z \in V_G$. Then, the following hold:

(1) If $uv \in E_G$, then u and v are comparable.
(2) If $u \leq v$, $z \leq w$, and u and z are comparable, then v and w are comparable.
(3) If $u < v < w$ and $uw \in E_G$, then $vw \in E_G$.

Lemma 3.2. *For every $u \in V_G$, we have $N(V_u) \subseteq N(u)$.*

For all $u \in V_G$, we denote the set of all maximal proper predecessors of u by $\lhd u$. Notice that such vertices correspond to the children of $r(u)$. We extend the previous case of a single vertex, on subsets of vertices with respect to an edge. For all $u, v \in V_G$ such that $uv \in E_G$, we denote by $\lhd uv$ the set of all maximal vertices of V_G that are proper predecessors of both u and v but are not adjacent to both, so $\lhd uv = \max_G((V_u \cap V_v) \setminus (N[u] \cap N[v]))$. Recall that for any edge $uv \in E_G$, either $u < v$ or $v < u$ by Observation 3.1 (1). If $u < v$ holds, then $\lhd uv = \max_G(V_u \setminus (N[u] \cap N(v)))$. For all $U \subseteq V_G$, we define $\mathcal{V}_U = \{V_u : u \in U\}$. The following two lemmas are crucial for our algorithms, as they provide natural partitions into smaller instances.

Lemma 3.3. *For every $u \in V_G$, the collection $\mathcal{V}_{\lhd u}$ is a partition of $V_u \setminus \{u\}$ into pairwise disconnected sets. For every $u, v \in V_G$ such that $u < v$ and $uv \in E_G$, $\mathcal{V}_{\lhd uv}$ is a partition of $V_u \setminus (N[u] \cap N(v))$ into pairwise disconnected sets.*

Lemma 3.4. *For every $u \in V_G$, the collection $\mathcal{V}_{\lhd u}$ is a nice partition of $V_u \setminus \{u\}$. For every $u, v \in V_G$ such that $u < v$ and $uv \in E_G$, the collection $\mathcal{V}_{\lhd uv}$ is a nice partition of $V_u \setminus (N[u] \cap N(v))$.*

Having defined the necessary predecessors of u, we next analyze specific solutions described in $A_{V_u}^Y$ with respect to the vertices of $\lhd u$. Both statements follow by carefully applying Lemma 3.2 and Lemma 3.4.

Lemma 3.5. *Let $Y \subseteq V_G \setminus V_u$. (i) If $u \notin A_{V_u}^Y$ then $A_{V_u}^Y \leftrightarrow \bigcup\limits_{u' \in \lhd u} A_{V_{u'}}^{Y \cap N(u')}$.*

(ii) *Moreover,* $A_{V_u}^{\emptyset} \leftrightarrow \max\limits_{weight} \left\{ \bigcup\limits_{u' \in \lhd u} A_{V_{u'}}^{\emptyset}, \{u\} \cup \bigcup\limits_{u' \in \lhd u} A_{V_{u'}}^{\{u\} \cap N(u')} \right\}.$

4   SFVS on Graphs with Bounded Leafage

In this section our goal is to show that SFVS can be solved in polynomial time on chordal graphs with bounded leafage. In particular, we concern ourselves with chordal graphs that have an intersection model tree with at most ℓ leaves and we show that SFVS can be solved in $n^{O(\ell)}$ time on such graphs. In the case of $\ell \leq 2$, the input graph is an interval graph, so SFVS can be solved in $O(nm)$ time [34]. We subsequently assume that we are given a chordal graph G that admits an expanded tree model $(T, \{T_v\}_{v \in V_G})$ with $\ell = L(T) \geq 2$, due to Lemma 3.1.

Given a subset of vertices of G, we collect the leaves of their corresponding subtrees: for every $U \subseteq V_G$, we define $L(U) = \cup_{u \in U} L(T_u)$. Notice that for any non-empty $U \subseteq V_G$, we have $L(U) \neq \emptyset$, since $(T, \{T_v\}_{v \in V_G})$ is an expanded tree model. Moreover, we associate the nodes of T with the vertices of G for which the nodes appear as leaves in their corresponding subtrees: for every $V \subseteq V_T$, we define $L^{-1}(V)$ to be the set $\{u \in V_G : L(T_u) \cap V \neq \emptyset\}$. For $V \subseteq V_T$, we denote by $\min_T V$ the subset of minimal nodes of V with respect to $\leq_T$. Observe that $\min_T V$ is a set of pairwise incomparable nodes, so $|\min_T V| \leq |\min_T V_T| \leq \ell$.

Lemma 4.1. *Let $U \subseteq V_G$ and $V \subseteq L(U)$. Then $L^{-1}(V) \subseteq U$.*

Instead of manipulating with the actual vertices of U, our algorithm deals with the *representatives* of U which contain the vertices of $L^{-1}(\min_T L(U))$. In particular, we are interested in the set of vertices $F_{\leq 2}(U) = F_1(U) \cup F_2(U)$, where $F_1(U) = L^{-1}(\min_T\{L(U)\})$ and $F_2(U) = L^{-1}(\min_T\{L(U \setminus F_1(U))\})$. We show that the representatives hold all the necessary information needed from their actual vertices.

Lemma 4.2. *Let $u \in V_G$ and $W \subseteq V_G \setminus V_u$ such that $W \neq \emptyset$, $G[\{u\} \cup W]$ is a clique, and $G[W] \in \mathcal{F}_S$, and let $u \in A_{V_u}^W$.*

- *If $(\{u\} \cup W) \cap S \neq \emptyset$ then $W = \{w\}$ and no vertex of $V_u \cap N(u) \cap N(w)$ belongs to $A_{V_u}^{\{w\}}$.*
- *If $(\{u\} \cup W) \cap S = \emptyset$ then $A_{V_{u'}}^{W \cap N(u')} \leftrightarrow A_{V_{u'}}^{F_{\leq 2}((\{u\} \cup W) \cap N(u'))}$, for any vertex $u' \in \lhd\, u$.*

We next show that Lemma 3.5 (ii) and Lemma 4.2 are enough to develop a dynamic programming scheme. As the size of the representatives is bounded with respect to ℓ by Lemma 4.1, we are able to store a bounded number of partial subsolutions. In particular we show that we only need to compute A_X^Y such that $|X| = O(n)$ and $|Y| \leq 2\ell + 1$.

Theorem 4.1. *There is an algorithm that, given a connected chordal graph G with leafage $\ell \geq 2$ and an expanded tree model of G, solves the weighted SUBSET FEEDBACK VERTEX SET problem in $O(n^{2\ell+1})$ time.*

If we let the leafage of a chordal graph to be the maximum over all of its connected components then we reach to the following result.

Corollary 4.1. *The weighted SUBSET FEEDBACK VERTEX SET problem can be solved in time $n^{O(\ell)}$ for chordal graphs with leafage at most ℓ.*

We next prove that we can hardly avoid the dependence of the exponent in the stated running time, since we show that SUBSET FEEDBACK VERTEX SET is W[1]-hard parameterized by the leafage of a chordal graph. Our reduction is inspired by the W[1]-hardness of FEEDBACK VERTEX SET parameterized by the mim-width given by Jaffke et al. in [27].

Theorem 4.2. *The weighted SUBSET FEEDBACK VERTEX SET decision problem on chordal graphs is W[1]-hard when parameterized by its leafage.*

5 SFVS on Rooted Path Graphs

Here we show how to extend our previous approach for SFVS on rooted path graphs. Rooted path graphs are exactly the intersection graphs of rooted paths on a rooted tree. Notice that rooted path graphs have unbounded leafage. Our main goal is to derive a recursive formulation for A_X^Y, similar to Lemma 4.2. In particular, we show that it is sufficient to consider sets Y containing at most one vertex.

For any vertex u of G, we denote the leaf of its corresponding rooted path in T by $l(u)$. We need to define further special vertices and subsets. Let $u, v \in V_G$ such that $u < v$. The (unique) maximal predecessor u' of v such that $l(u') < r(u) \le r(u')$ is denoted by $u \lhd v$. Moreover, for every $V_1, V_2, V_3 \subseteq V_G$, we define the following sets:

- $V_{\langle V_1|V_2|V_3\rangle} = \{u \in V_G : r(v_1) < l(u) < r(v_2) < r(u) \le r(v_3) \text{ for some } v_i \in V_i, i \in \{1,2,3\}\}$
- $V_{\langle |V_2|V_3\rangle} = \{u \in V_G : l(u) < r(v_2) < r(u) \le r(v_3) \text{ for some } v_i \in V_i, i \in \{2,3\}\}$
- $V_{\langle V_1||V_3\rangle} = \{u \in V_G : r(v_1) < l(u) < r(u) \le r(v_3) \text{ for some } v_i \in V_i, i \in \{1,3\}\}$

Vertical bars indicate the placements of $l(u), r(u)$ with respect to V_1, V_2, V_3.

Lemma 5.1. *Let $u, w \in V_G \setminus S$ such that $u < w$ and $uw \in E_G$. Then the collection $\{V_{\langle \lhd uw||\lhd u\rangle} \setminus S\} \cup \{V_{u'} \cup (V_{\langle |\{u'\}|\{u'\lhd u\}\rangle} \setminus S)\}_{u' \in \lhd uw}$ is a nice partition of $X = V_u \setminus (\{u\} \cup (N(u) \cap N(w) \cap S))$ with respect to any $Y \subseteq V_G \setminus X$ such that $Y \cap S = \emptyset$.*

For every appropriate u, v, we denote the set $V_u \cup (V_{\langle |\{u\}|\{v\}\rangle} \setminus S)$ by $V_{u,v}$. Observe that $V_{u,u}$ is simply V_u. First we consider the set $A_{V_u}^{\{w\}}$ for which $u < w$ and $uw \in E_G$. If $u \notin A_{V_u}^{\{w\}}$ then $A_{V_u}^{\{w\}} = \bigcup_{u' \in \lhd u} A_{V_{u'}}^{\{w\} \cap N(u')}$ by Lemma 3.5 (i). Also, recall that $A_{V_u}^\emptyset$ is described by the formula given in Lemma 3.5 (ii).

Lemma 5.2. *Let $u, w \in V_G$ such that $u < w$ and $uw \in E_G$, and let $u \in A_{V_u}^{\{w\}}$.*

- *If $u \in S$ or $w \in S$ then $A_{V_u}^{\{w\}} \leftrightarrow \{u\} \cup \bigcup_{u' \in \lhd uw} A_{V_{u'}}^{\{u,w\} \cap N(u')}$.*

- *If $u, w \notin S$ then $A_{V_u}^{\{w\}} \leftrightarrow \{u\} \cup (V_{\langle \lhd uw||\lhd u\rangle} \setminus S) \cup \bigcup_{u' \in \lhd uw} A_{V_{u'}, u' \lhd u}^{\{u,w\} \cap N(u')}$.*

We next deal with the sets $A_{V_{u,v}}^Y$ for which $u < v$, $|Y| \le 1$ and no vertex of $\{v\} \cup Y$ belongs to S. Observe that $V_{u,v}$ is not necessarily described by a set V_w for some $w \in V_G$. Thus we need appropriate formulas that handle such sets. For doing so, notice that $V_{u,v} \setminus \{v\} = V_{u,u \lhd v}$, since $V_{u,v} = V_u \cup (V_{\langle |\{u\}|\{v\}\rangle} \setminus S)$ and $u \le u \lhd v < v$. This means that if $v \notin A_{V_{u,v}}^Y$, we have $A_{V_u}^Y = A_{V_u \setminus \{v\}}^Y = A_{V_{u,u \lhd v}}^Y$. We subsequently assume that $v \in A_{V_{u,v}}^{\{w\}}$.

Notice that given a partition $\mathcal{P}$ of a set X and a set $X' \subseteq X$, the collection $\mathcal{P}' = \{P \cap X'\}_{P \in \mathcal{P}}$ is a partition of X'. Furthermore, observe that if $\mathcal{P}$ is a nice partition of X with respect to a set $Y \subseteq V_G \setminus X$ such that $Y \cap S = \emptyset$, then $\mathcal{P}'$ is a nice partition of X' with respect to Y.

Lemma 5.3. *Let $u \in V_G$ and $v, w \in V_G \setminus S$ such that $u < v < w$ and $\{u, v, w\}$ induce a clique and let $v \in A_{V_{u,v}}^{\{w\}}$. Then, $A_{V_{u,v}}^{\{w\}} \leftrightarrow \{v\} \cup (V_{\langle \lhd vw|\{u\}|\{u\lhd v\}\rangle} \setminus S) \cup \bigcup_{u' \in V_u \cap \lhd vw} A_{V_{u'}, u' \lhd v}^{\{v,w\} \cap N(u')}$.*

Now we are in position to state our claimed result, which is obtained in a similar fashion with the algorithm given in Theorem 4.1.

Theorem 5.1. *The weighted* Subset Feedback Vertex Set *problem can be solved on rooted path graphs in $O(n^2 m)$ time.*

6   Vertex Leafage to Cope with SFVS

Due to Theorem 4.1 and Corollary 4.1, it is interesting to ask whether our results can be further extended on larger classes of chordal graphs. Here we consider graphs of bounded vertex leafage as a natural candidate towards such an approach. However we show that Subset Feedback Vertex Set is NP-complete on undirected path graphs which are exactly the graphs of vertex leafage at most two. In particular, we provide a polynomial reduction from the NP-complete Max Cut problem. Given a graph G, the Max Cut problem is concerned with finding a partition of $V(G)$ into two sets A and $\overline{A}$ such that the number of edges with one endpoint in A and the other one in $\overline{A}$ is maximum among all partitions. For two disjoint sets of vertices X and Y, we denote by $E(X, Y)$ the set $\{\{x, y\} \mid x \in X, y \in Y\}$. In such terminology, Max Cut aims at finding a set $A \subseteq V(G)$ such that $|E(A, V(G) \setminus A) \cap E(G)|$ is maximum. The *cut-set* of a set of vertices A is the set of edges of G with exactly one endpoint in A, which is $E(A, V(G) \setminus A) \cap E(G)$. The Max Cut problem is known to be NP-hard for general graphs [28] and remains NP-hard even when the input graph is restricted to be a split or 3-colorable or undirected path graph [3]. We mention that our reduction is based on Max Cut on general graphs.

Towards the claimed reduction, for any graph G on n vertices and m edges, we will associate a graph H_G on $12n^2 + 4n + 2m$ vertices. First we describe the vertex set of H_G. For every vertex $v \in V(G)$ we have the following sets of vertices:

- $X(v) = \{x_v^1, \ldots, x_v^{2n}\}$ and $\overline{X}(v) = \{\overline{x}_v^1, \ldots, \overline{x}_v^{2n}\}$,
- $Y(v) = \{y_v^1, \ldots, y_v^{2n+1}\}$ and $\overline{Y}(v) = \{\overline{y}_v^1, \ldots, \overline{y}_v^{2n+1}\}$,
- $Z(v) = \{z_v^1, \overline{z}_v^1, \ldots, z_v^{2n+1}, \overline{z}_v^{2n+1}\}$, and
- $E(v) = \{(v, x) \mid \{v, x\} \in E(G)\}$.

Observe that for every edge $\{u, v\} \in E(G)$ there are two vertices in H_G that correspond to the ordered pairs (u, v) and (v, u). We denote by $\overline{E}(v)$ the set $\{(x, v) \mid \{x, v\} \in E(G)\}$. The edge set of H_G contains precisely the following:

- all edges required for the set $\bigcup_{v \in V(G)}(Y(v) \cup \overline{Y}(v) \cup E(v))$ to form a clique
- for every vertex $v \in V(G)$,
  - all elements of the sets $E(X(v), Y(v))$, $E(\overline{X}(v), \overline{Y}(v))$, $E(X(v), E(v))$, and $E(\overline{X}(v), \overline{E}(v))$;
  - $\{x_v^i, x_v^{n+i}\}, \{\overline{x}_v^i, \overline{x}_v^{n+i}\}$ for each $i \in [n]$;
  - $\{y_v^j, z_v^j\}, \{y_v^j, \overline{z}_v^j\}, \{\overline{y}_v^j, z_v^j\}, \{\overline{y}_v^j, \overline{z}_v^j\}$ for each $j \in [2n + 1]$.

This completes the construction of H_G. An example of H_G is given in Fig. 1.
In the following two lemmas, we show our main result of this section.

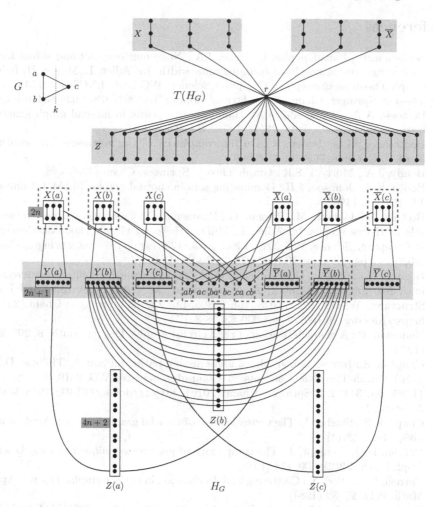

Fig. 1. Illustrating the undirected path graph H_G. On top left we show a graph G on three vertices and on the bottom part we illustrate the corresponding graph H_G. A tree model $T(H_G)$ for H_G, is given on the top right part. The vertices of H_G that lie on the grey area form a clique.

Lemma 6.1. *For any graph G, H_G is an undirected path graph.*

Lemma 6.2. *Let G be a graph with $A \subseteq V(G)$, H_G be the undirected path graph of G, and let $X = \bigcup X(v)$, $\overline{X} = \bigcup \overline{X}(v)$, and $Z = \bigcup Z(v)$. For the set of vertices $S = X \cup \overline{X} \cup Z$ of H_G, there is a subset feedback vertex set U of (H_G, S) such that $|U| = 4n^2 + n + 2m - k$, where k is the size of the cut-set of A in G.*

Theorem 6.1. *The unweighted* SUBSET FEEDBACK VERTEX SET *decision problem is NP-complete on undirected path graphs.*

References

1. Benjamin, B., Papadopoulos, C., Telle, J.A.: Node multiway cut and subset feedback vertex set on graphs of bounded mim-width. In: Adler, I., Müller, H. (eds.) Graph-Theoretic Concepts in Computer Science, WG 2020. LNCS, vol. 12301, pp. 388–400. Springer, Cham (2020). https://doi.org/10.1007/978-3-030-60440-0_31
2. Bertossi, A.A., Bonuccelli, M.A.: Hamiltonian circuits in interval graph generalizations. Inf. Process. Lett. **23**(4), 195–200 (1986)
3. Bodlaender, H.L., Jansen, K.: On the complexity of the maximum cut problem. Nord. J. Comput. **7**(1), 14–31 (2000)
4. Bondy, J.A., Murty, U.S.R.: Graph Theory. Springer, Cham (2008)
5. Booth, K.S., Johnson, J.H.: Dominating sets in chordal graphs. SIAM J. Comput. **11**, 191–199 (1982)
6. Brettell, N., Johnson, M., Paesani, G., Paulusma, D.: Computing subset transversals in H-free graphs. In: Adler, I., Müller, H. (eds.) Graph-Theoretic Concepts in Computer Science, WG 2020. LNCS, vol. 12301, pp. 187–199. Springer, Cham (2020). https://doi.org/10.1007/978-3-030-60440-0_15
7. Brettell, N., Johnson, M., Paulusma, D.: Computing weighted subset transversals in H-free graphs. In: Lubiw, A., Salavatipour, M. (eds.) Algorithms and Data Structures, WADS 2021. LNCS, vol. 12808, pp. 229–242. Springer, Cham (2021). https://doi.org/10.1007/978-3-030-83508-8_17
8. Buneman, P.: A characterization of rigid circuit graphs. Discret. Math. **9**, 205–212 (1974)
9. Chaplick, S.: Intersection graphs of non-crossing paths. In: Sau, I., Thilikos, D.M. (eds.) Graph-Theoretic Concepts in Computer Science, WG 2019. LNCS, vol. 11789, pp. 311–324. Springer, Cham (2019). https://doi.org/10.1007/978-3-030-30786-8_24
10. Chaplick, S., Stacho, J.: The vertex leafage of chordal graphs. Discret. Appl. Math. **168**, 14–25 (2014)
11. Corneil, D.G., Fonlupt, J.: The complexity of generalized clique covering. Discret Appl. Math. **22**(2), 109–118 (1988)
12. Corneil, D.G., Perl, Y.: Clustering and domination in perfect graphs. Discret. Appl. Math. **9**(1), 27–39 (1984)
13. Cygan, M., et al.: Parameterized Algorithms. Springer, Cham (2015). https://doi.org/10.1007/978-3-319-21275-3
14. Cygan, M., Pilipczuk, M., Pilipczuk, M., Wojtaszczyk, J.O.: Subset feedback vertex set is fixed-parameter tractable. SIAM J. Discret. Math. **27**(1), 290–309 (2013)
15. Dietz, P.: Intersection graph algorithms. Ph.D. Thesis, Cornell University (1984)
16. Fomin, F.V., Heggernes, P., Kratsch, D., Papadopoulos, C., Villanger, Y.: Enumerating minimal subset feedback vertex sets. Algorithmica **69**(1), 216–231 (2014). https://doi.org/10.1007/s00453-012-9731-6
17. Fomin, F.V., Golovach, P.A., Raymond, J.: On the tractability of optimization problems on h-graphs. Algorithmica **82**(9), 2432–2473 (2020). https://doi.org/10.1007/s00453-020-00692-9
18. Garey, M.R., Johnson, D.S.: Computers and Intractability. W.H. Freeman and Co, New York (1978)
19. Gavril, F.: The intersection graphs of subtrees of trees are exactly the chordal graphs. J. Comb. Theory Ser. B **16**, 47–56 (1974)
20. Gavril, F.: A recognition algorithm for the intersection graphs of directed paths in directed trees. Discret. Math. **13**(3), 237–249 (1975)

21. Golovach, P.A., Heggernes, P., Kratsch, D., Saei, R.: Subset feedback vertex sets in chordal graphs. J. Discret. Algorithms **26**, 7–15 (2014)
22. Habib, M., Stacho, J.: Polynomial-time algorithm for the leafage of chordal graphs. In: Fiat, A., Sanders, P. (eds.) Algorithms - ESA 2009. LNCS, vol. 5757, pp. 290–300. Springer, Heidelberg (2009). https://doi.org/10.1007/978-3-642-04128-0_27
23. Heggernes, P., van't Hof, P., van Leeuwen, E.J., Saei, R.: Finding disjoint paths in split graphs. Theory Comput. Syst. **57**(1), 140–159 (2015)
24. Hols, E.C., Kratsch, S.: A randomized polynomial kernel for subset feedback vertex set. Theory Comput. Syst. **62**, 54–65 (2018). https://doi.org/10.1007/s00224-017-9805-6
25. Ioannidou, K., Mertzios, G.B., Nikolopoulos, S.D.: The longest path problem has a polynomial solution on interval graphs. Algorithmica **61**(2), 320–341 (2011). https://doi.org/10.1007/s00453-010-9411-3
26. Jaffke, L., Kwon, O., Strømme, T.J.F., Telle, J.A.: Mim-width III. Graph powers and generalized distance domination problems. Theor. Comput. Sci. **796**, 216–236 (2019)
27. Jaffke, L., Kwon, O., Telle, J.A.: Mim-width II. The feedback vertex set problem. Algorithmica **82**(1), 118–145 (2020)
28. Karp, R.M.: Reducibility among combinatorial problems. In: Miller, R.E., Thatcher, J.W. (eds.) Complexity of Computer Computations, pp. 85–103. Plenum Press (1972)
29. Kawarabayashi, K., Kobayashi, Y.: Fixed-parameter tractability for the subset feedback set problem and the s-cycle packing problem. J. Comb. Theory Ser. B **102**(4), 1020–1034 (2012)
30. Lin, I., McKee, T.A., West, D.B.: The leafage of a chordal graph. Discuss. Math. Graph Theory **18**(1), 23–48 (1998)
31. Monma, C.L., Wei, V.K.: Intersection graphs of paths in a tree. J. Comb. Theory Ser. B **41**(2), 141–181 (1986)
32. Natarajan, S., Sprague, A.P.: Disjoint paths in circular arc graphs. Nord. J. Comput. **3**(3), 256–270 (1996)
33. Panda, B.S.: The separator theorem for rooted directed vertex graphs. J. Comb. Theory Ser. B **81**(1), 156–162 (2001)
34. Papadopoulos, C., Tzimas, S.: Polynomial-time algorithms for the subset feedback vertex set problem on interval graphs and permutation graphs. Discret. Appl. Math. **258**, 204–221 (2019)
35. Papadopoulos, C., Tzimas, S.: Subset feedback vertex set on graphs of bounded independent set size. Theor. Comput. Sci. **814**, 177–188 (2020)
36. Papadopoulos, C., Tzimas, S.: Computing subset feedback vertex set via leafage. CoRR abs/2103.03035 arXiv:2103.03035 (2021)
37. Philip, G., Rajan, V., Saurabh, S., Tale, P.: Subset feedback vertex set in chordal and split graphs. Algorithmica **81**(9), 3586–3629 (2019). https://doi.org/10.1007/s00453-019-00590-9
38. Spinrad, J.P.: Efficient Graph Representations. American Mathematical Society, Fields Institute Monograph Series 19 (2003)
39. Yannakakis, M.: Node-deletion problems on bipartite graphs. SIAM J. Comput. **10**(2), 310–327 (1981)

Linear Time Construction of Indexable Elastic Founder Graphs

Nicola Rizzo$^{(\boxtimes)}$ ⓘ and Veli Mäkinen ⓘ

Department of Computer Science, University of Helsinki, Helsinki, Finland
{nicola.rizzo,veli.makinen}@helsinki.fi

Abstract. The pattern matching of strings in labeled graphs has been widely studied lately due to its importance in genomics applications. Unfortunately, even the simplest problem of deciding if a string appears as a subpath of a graph admits a quadratic lower bound under the Orthogonal Vectors Hypothesis (Equi et al. ICALP 2019, SOFSEM 2021). To avoid this bottleneck, the research has shifted towards more specific graph classes, e.g. those induced from multiple sequence alignments (MSAs). Consider segmenting $\mathrm{MSA}[1..m, 1..n]$ into b blocks $\mathrm{MSA}[1..m, 1..j_1]$, $\mathrm{MSA}[1..m, j_1 + 1..j_2]$, ..., $\mathrm{MSA}[1..m, j_{b-1} + 1..n]$. The distinct strings in the rows of the blocks, after the removal of gap symbols, form the nodes of an *elastic founder graph* (EFG) where the edges represent the original connections observed in the MSA. An EFG is called *indexable* if a node label occurs as a prefix of only those paths that start from a node of the same block. Equi et al. (ISAAC 2021) showed that such EFGs support fast pattern matching and gave an $O(mn \log m)$-time algorithm for preprocessing the MSA in a way that allows the construction of indexable EFGs maximizing the number of blocks and, alternatively, minimizing the maximum length of a block, in $O(n)$ and $O(n \log \log n)$ time respectively. Using the suffix tree and solving a novel ancestor problem on trees, we improve the preprocessing to $O(mn)$ time and the $O(n \log \log n)$-time EFG construction to $O(n)$ time, thus showing that both types of indexable EFGs can be constructed in time linear in the input size.

Keywords: Multiple sequence alignment · Pattern matching · Data structures · Segmentation algorithms · Dynamic programming · Suffix tree

1 Introduction

Searching strings in a graph has become a central problem along with the development of high-throughput sequencing techniques. Namely, thousands of human genomes are now available, forming a so-called *pangenome* of a species [20]. Such pangenome can be used to enhance various analysis tasks that have previously been conducted with a single reference genome [3,8,11,13,14,18,19]. The most popular representation for a pangenome is a graph, whose paths spell the input genomes. The basic primitive required on such pangenome graphs is to be able

© Springer Nature Switzerland AG 2022
C. Bazgan and H. Fernau (Eds.): IWOCA 2022, LNCS 13270, pp. 480–493, 2022.
https://doi.org/10.1007/978-3-031-06678-8_35

to search occurrences of query strings (short reads) as subpaths of the graph. Unfortunately, even finding exact matches of a query string of length q in a graph with e edges cannot be done significantly faster than $O(qe)$ time, and no index built in polynomial time allows for subquadratic-time string matching, unless the Orthogonal Vectors Hypothesis (OVH) is false [4,5]. Therefore, practical tools deploy various heuristics or use other pangenome representations as a basis.

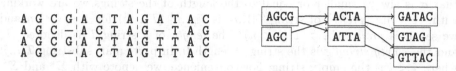

Fig. 1. An indexable elastic founder graph induced from a segmentation of an MSA. The example is adapted from Equi et al. [6].

Due to the difficulty of string search in general graphs, Mäkinen et al. [12] and Equi et al. [6] studied graphs induced from multiple sequence alignments (MSAs), as we describe in Sect. 2. Any segmentation of an MSA naturally induces a graph consisting of nodes partitioned into blocks with edges connecting consecutive blocks. Such *elastic founder graph* (EFG) is illustrated in Fig. 1. The key observation is that if the resulting node labels do not appear as a prefix of any other path than those starting at the same block, then there is an index structure for the graph that supports fast pattern matching [6,12]. Equi et al. [6] also showed that such indexability property is required, as the OVH-based lower bound holds for EFGs derived from MSAs. Mäkinen et al. [12] gave an $O(mn)$ time algorithm to construct an indexable EFG with minimum maximum block length, given a gapless MSA[1..m, 1..n]. Equi et al. [6] extended the result to general MSAs. They obtained an $O(mn \log m)$-time preprocessing algorithm which allows the construction of indexable EFGs maximizing the number of blocks and, alternatively, minimizing the maximum length of a block, in $O(n)$ and in $O(n \log \log n)$ time, respectively. We recall these results in Sect. 3.

In this paper, we improve the preprocessing algorithm of Equi et al. to $O(mn)$ by performing an in-depth analysis of their solution based on the generalized suffix tree GST_{MSA} built from the gaps-removed rows of the MSA (Sect. 4). Although removing gaps constitutes a loss of essential information, this information can be fed back into the structure by considering the right subsets of its nodes or leaves. Then, the main step in preprocessing the MSA is solving a novel ancestor problem on the tree structure of GST_{MSA} that we call the *exclusive ancestor set problem*, and as our main contribution, we identify such problem and provide a linear-time solution. This directly improves the solution by Equi et al. for constructing indexable EFGs maximizing the number of blocks from $O(mn \log m)$ to $O(mn)$ time. Moreover, in Sect. 5 we give a new algorithm that after the $O(mn)$-time preprocessing can construct indexable EFGs minimizing the maximum block length in $O(n)$ time. In our subsequent work [16], we extend these techniques to minimize the maximum block height.

2   Definitions

We follow the notation of Equi et al. [6].

Strings. We denote integer intervals by $[x..y]$. Let $\Sigma = [1..\sigma]$ be an alphabet of size $|\Sigma| = \sigma$. A *string* $T[1..n]$ is a sequence of symbols from Σ, i.e. $T \in \Sigma^n$, where Σ^n denotes the set of strings of length n over Σ. In this paper, we assume that σ is always smaller or equal to the length of the strings we are working with. A *suffix* (*prefix*) of string $T[1..n]$ is $T[i..n]$ ($T[1..i]$) for $1 \leq i \leq n$ and we say it is *proper* if $i > 1$ ($i < n$). The *length* of a string T is denoted $|T|$ and the *empty string* ε is the string of length 0. In particular, substring $T[i..j]$ where $j < i$ is the empty string. For convenience, we denote with Σ^* and Σ^+ the set of finite strings and finite non-empty strings over Σ, respectively. The *lexicographic order* of two strings A and B is naturally defined by the order of the alphabet: $A < B$ iff $A[1..i] = B[1..i]$ and $A[i+1] < B[i+1]$ for some $i \geq 0$. If $i + 1 > \min(|A|, |B|)$, then the shorter one is regarded as smaller. However, we usually avoid this implicit comparison by adding an *end marker* \$ to the strings and we consider \$ to be the smallest character lexicographically. The concatenation of strings A and B is denoted as $A \cdot B$, or just AB.

Elastic Founder Graphs. MSAs can be compactly represented by elastic founder graphs, the vertex-labeled graphs that we formalize in this section.

A *multiple sequence alignment* $\mathsf{MSA}[1..m, 1..n]$ is a matrix with m strings drawn from $\Sigma \cup \{-\}$, each of length n, as its rows. Here, $- \notin \Sigma$ is the *gap* symbol. For a string $X \in (\Sigma \cup \{-\})^*$, we denote $\mathsf{spell}(X)$ the string resulting from removing the gap symbols from X. If an MSA does not contain gaps then we say it is *gapless*, otherwise we say that it is a *general* MSA. Let $\mathcal{P}$ be a *partitioning* of $[1..n]$, that is, a sequence of subintervals $\mathcal{P} = [x_1..y_1], [x_2..y_2], \ldots, [x_b..y_b]$ where $x_1 = 1$, $y_b = n$, and for all $j > 2$, $x_j = y_{j-1} + 1$. A *segmentation* S of $\mathsf{MSA}[1..m, 1..n]$ based on partitioning $\mathcal{P}$ is the sequence of b sets $S^k = \{\mathsf{spell}(\mathsf{MSA}[i, x_k..y_k]) \mid 1 \leq i \leq m\}$ for $1 \leq k \leq b$; in addition, we require for a (proper) segmentation that $\mathsf{spell}(\mathsf{MSA}[i, x_k..y_k]) \neq \varepsilon$ for any i and k. We call set S^k a *block*, while $\mathsf{MSA}[1..m, x_k..y_k]$ or just $[x_k..y_k]$ is called a *segment*. The *length* of block S^k or its segment $[x_k..y_k]$ is $L(S^k) = L([x_k..y_k]) = y_k - x_k + 1$.

Definition 1 (Block graph). *A* block graph *is a graph* $G = (V, E, \ell)$ *where* $\ell : V \to \Sigma^+$ *is a function that assigns a string label to every node and for which the following properties hold:*

1. *set V can be partitioned into a sequence of b blocks $V^1, V^2, \ldots, V^b$, that is, $V = V^1 \cup V^2 \cup \cdots \cup V^b$ and $V^i \cap V^j = \emptyset$ for all $i \neq j$;*
2. *if $(v, w) \in E$ then $v \in V^i$ and $w \in V^{i+1}$ for some $1 \leq i \leq b - 1$; and*
3. *if $v, w \in V^i$ then $|\ell(v)| = |\ell(w)|$ for each $1 \leq i \leq b$ and if $v \neq w$, $\ell(v) \neq \ell(w)$.*

Definition 2 (Elastic block and founder graphs). *We call a block graph* elastic *if its third condition is relaxed in the sense that each V^i can contain non-empty variable-length strings. An* elastic founder graph *(EFG) is an elastic* *block graph* $G(S) = (V, E, \ell)$ induced *by a segmentation* S *as follows: for each* $1 \leq k \leq b$ *we have* $S^k = \{\text{spell}(\text{MSA}[i, x_k..y_k]) \mid 1 \leq i \leq m\} = \{\ell(v) : v \in V^k\}$. *It holds that* $(v, w) \in E$ *if and only if there exist* $k \in [1..b-1]$, $i \in [1..m]$ *such that* $v \in V^k$, $w \in V^{k+1}$, *and* $\text{spell}(\text{MSA}[i, x_k..y_{k+1}]) = \ell(v)\ell(w)$.

For example, in the general MSA$[1..4, 1..13]$ of Fig. 1, the segmentation based on partitioning $[1..4], [5..8], [9..13]$ induces an EFG $G(S) = (V^1 \cup V^2 \cup V^3, E, \ell)$ where the nodes in V^1 and V^3 have labels of variable length.

By definition, (elastic) founder and block graphs are acyclic. For convention, we interpret the direction of the edges as going from left to right. Consider a path P in $G(S)$ between any two nodes. The label $\ell(P)$ of P is the concatenation of the labels of the nodes in the path. Let Q be a query string. We say that Q *occurs* in $G(S)$ if Q is a substring of $\ell(P)$ for any path P of $G(S)$.

Definition 3 ([12])**.** *EFG* $G(S)$ *is* repeat-free *if each* $\ell(v)$ *for* $v \in V$ *occurs in* $G(S)$ *only as a prefix of paths starting with* v.

Definition 4 ([12])**.** *EFG* $G(S)$ *is* semi-repeat-free *if each* $\ell(v)$ *for* $v \in V$ *occurs in* $G(S)$ *only as a prefix of paths starting with* $w \in V$, *where* w *is from the same block as* v.

For example, the EFG of Figure 1 is not repeat-free, since AGC occurs as a prefix of two distinct labels of nodes in the same block, but it is semi-repeat-free since all node labels $\ell(v)$ with $v \in V^k$ occur in $G(S)$ only starting from block V^k, or they do not occur at all elsewhere in the graph. We will discuss these two indexability properties together as the (semi-)repeat-free property, when applicable.

Basic Tools. A *trie* [2] of a set of strings is a rooted directed tree with outgoing edges of each node labeled by distinct symbols such that there is a root-to-leaf path spelling each string in the set; the shared part of the root-to-leaf paths of two different leaves spell the common prefix of the corresponding strings. In a *compact trie*, the maximal non-branching paths of a trie become edges labeled with the concatenation of labels on the path. The *suffix tree* of $T \in \Sigma^*$ is the compact trie of all suffixes of string $T\$$. In this case, the edge labels are substrings of T and can be represented in constant space as an interval. Such tree takes linear space and can be constructed in linear time, assuming that $\sigma \leq |T|$, so that when reading the leaves from left to right the suffixes are listed in their lexicographic order. [7,21] We say that two or more leaves of the suffix tree are *adjacent* if they succeed one another when reading them left to right. A *generalized suffix tree* is one built on a set of strings. In this case, string T above is the concatenation of the strings with symbol $\$$ between each.

Let $Q[1..m]$ be a query string. If Q occurs in T, then the *locus* or *implicit node* of Q in the suffix tree of T is (v, k) such that $Q = XY$, where X is the path spelled from the root to the parent of v and Y is the prefix of length k of the edge from the parent of v to v. The leaves in the subtree rooted at v, or *the leaves covered by* v, are then all the suffixes sharing the common prefix Q. Let

aX and X be the paths spelled from the root of a suffix tree to nodes v and w, respectively. Then one can store a *suffix link* from v to w.

String $B[1..n]$ from a binary alphabet is called a *bitvector*. The operation $\text{rank}(B, i)$ returns the number of 1s in $B[1..i]$, whereas the operation $\text{select}(B, j)$ returns the index i containing the j-th 1 in B. Both queries can be answered in constant time using an index requiring $o(n)$ bits in addition to the bitvector itself and computable in linear time [9,10].

3   Overview of EFG Construction Algorithms

Equi et al. have shown that (semi-)repeat-free EFGs are easy to index for fast pattern matching [6], and as we describe in Sect. 3.1 they extended the previous research for the gapless and repeat-free setting showing that finding (semi-)repeat-free elastic founder graphs is equivalent to finding (semi-)repeat-free MSA segmentations. Moreover, to show that the (semi-)repeat-free property does not hinder the flexibility in choosing the resulting EFGs, they considered the following score functions for MSA segmentations: *i.* maximizing the number of blocks, and *ii.* minimizing the maximum length of a block.

In the gapless and repeat-free setting, scores *i.* and *ii.* admit the construction of indexable founder graphs in $O(mn)$ time, thanks to previous research on founder graphs and MSA segmentations [1,12,15]. In the general and semi-repeat-free setting, Equi et al. have given $O(mn \log m)$ and $O(mn \log m + n \log \log n)$-time algorithms for scores *i.* and *ii.*, respectively, based on a common preprocessing of the MSA that we review in Sect. 3.2.

3.1   Segmentation Characterization for Indexable EFGs

Consider a segmentation $S = S^1, S^2, \ldots, S^b$ that induces a (semi-)repeat-free EFG $G(S) = (V, E, \ell)$, as per Definition 2. The strings occurring in graph $G(S)$ are a superset of the strings occurring in the original MSA rows because each node label can represent *multiple* rows and each edge $(v, w) \in E$ means the existence of *some* row spelling $\ell(v)\ell(w)$ in the corresponding consecutive segments. For example, string GACTAGT occurs in the EFG of Fig. 1 but it does not occur in any row of the original MSA.

The (semi-)repeat-free property involves graph $G(S)$, but luckily it does not depend on the new strings added in the founder graph and can be checked only against the MSA and segmentation S. This simplifies choosing a segmentation resulting in an indexable founder graph and it was initially proven by Mäkinen et al. in the gapless and repeat-free setting.

Lemma 1 (Characterization, gapless setting [12]). *We say that a segment $[x..y]$ of a gapless MSA$[1..m, 1..n]$ is repeat-free if string MSA$[i, x..y]$ occurs in the MSA only at position x of some row, for all $1 \leq i \leq m$. Then $G(S)$ is repeat-free if and only if all segments defining S are repeat-free.*

Equi et al. in [6] refined this property for MSAs with gaps, but did not provide an explicit proof. Since it is essential for the correctness of the construction algorithms, we provide such a proof in the full version of this paper [17].

Lemma 2 (Characterization [6]). *We say that segment* $[x..y]$ *of a general* MSA$[1..m, 1..n]$ *is semi-repeat-free if for any* $i, i' \in [1..m]$ *string* spell(MSA$[i, x..y]$) *occurs in gaps-removed row* spell(MSA$[i', 1..n]$) *only at position* $g(i', x)$, *where* $g(i', x)$ *is equal to* x *minus the number of gaps in* MSA$[i', 1..x]$. *Similarly,* $[x..y]$ *is repeat-free if the eventual occurrence of* spell(MSA$[i, x..y]$) *at position* $g(i', x)$ *in row* i' *also ends at position* $g(i', y)$. *Then* $G(S)$ *is (semi-)repeat-free if and only if all segments of* S *are (semi-)repeat-free.*

3.2 EFG Construction Algorithms

Just as in the gapless and repeat-free setting, Lemma 2 implies that the optimal score $s(j)$ of a (semi-)repeat-free segmentation of the general MSA prefix MSA$[1..m, 1..j]$ can be computed recursively for a variety of scoring schemes:

$$s(j) = \bigoplus_{\substack{j' : 0 \leq j' < j \text{ s.t.} \\ \text{MSA}[1..m, j'+1..j] \text{ is} \\ \text{(semi-)repeat-free}}} E(s(j'), j', j) \tag{1}$$

where operator $\bigoplus$ and function E depend on the desired scoring scheme. Indeed: *i.* for $s(j)$ to be equal to the optimal score of a segmentation maximizing the number of blocks, set $\bigoplus = \max$ and $E(s(j'), j', j) = s(j') + 1$; for a correct initialization set $s(0) = 0$ and if there is no (semi-)repeat-free segmentation set $s(j) = -\infty$; *ii.* for minimizing the maximum block length, set $\bigoplus = \min$ and $E(s(j'), j', j) = \max(s(j'), L([j' + 1, j])) = \max(s(j'), j - j')$; set $s(0) = 0$ and if there is no (semi-)repeat-free segmentation set $s(j) = +\infty$.

Equi et al. studied the computation of semi-repeat-free segmentations optimizing for these two scores [6]. The algorithms they developed—and that we will improve in Sects. 4 and 5—are based on a common preprocessing of the valid semi-repeat-free segmentation ranges, based on the following observation.

Observation 1 (Semi-repeat-free right extensions [6]). *Given a general* MSA$[1..m, 1..n]$, *for any* $x < y$ *we say that segment* $[x + 1..y]$ *is an extension of prefix* MSA$[1..m, 1..x]$. *If extension* $[x + 1..y]$ *is semi-repeat-free, then extension* $[x + 1..y']$ *is semi-repeat-free for all* $y < y' \leq n$.

Note that in the presence of gaps Observation 1 does not hold if we swap the semi-repeat-free notion with the repeat-free one, or if we swap the right extensions with the symmetrically defined left extensions.

To compute $s(j)$, Eq. (1) considers all semi-repeat-free right extensions $[j' + 1..j]$ ending at column j. Equi et al. discovered that the computation of values $s(j)$ can be done efficiently by considering that each semi-repeat-free right extension $[j' + 1..j]$ has as prefix a minimal (semi-repeat-free) right extension $[j' + 1..f(j')]$, with function f defined as follows.

Definition 5 (Minimal right extensions [6]**).** *Given* MSA$[1..m, 1..n]$, *for each* $0 \leq x \leq n - 1$ *we define value* $f(x)$ *as the smallest integer greater than* x *such that segment* $[x+1..f(x)]$ *is semi-repeat-free, or, in other words,* $[x+1..f(x)]$ *is the minimal (semi-repeat-free) right extension of prefix* MSA$[1..m, 1..x]$. *If there is no semi-repeat-free extension, we define* $f(x) = \infty$.

Indeed, Equi et al. in [6] developed an algorithm computing values $f(x)$ in time $O(mn \log m)$. Using only these values, described by a list of pairs $(x, f(x))$ sorted in increasing order by the second component, they developed two algorithms computing the score of an optimal semi-repeat-free segmentation: in time $O(n)$ for the maximum number of blocks score and in time $O(n \log \log n)$ for the maximum block length score. We will explain in detail how the latter works in Sect. 5, as we will improve its run time to $O(n)$.

4   Preprocessing the MSA in Linear Time

In this section, we study the computation of the minimal right extensions $f(x)$, for $0 \leq x \leq n - 1$ (Definition 5). Equi et al. in [6] proposed an $O(nm \log m)$-time solution using the following structure, built from the gaps-removed MSA rows.

Definition 6. *Given* MSA$[1..m, 1..n]$ *from alphabet* $\Sigma \cup \{-\}$, *we define* GST$_{MSA}$ *as the generalized suffix tree of the set of strings* $\{$spell$($MSA$[i, 1..n]) \cdot \$_i : 1 \leq i \leq m\}$, *with* $\$_1, \ldots, \$_m$ m *new distinct terminator symbols not in* Σ.[1]

An example of GST$_{MSA}$ is given in Fig. 2. From the suffix tree properties, it follows that for any gaps-removed row $\alpha_i := $ spell$($MSA$[i, 1..n])\$_i$, with $1 \leq i \leq m$: each suffix $\alpha_i[x..|\alpha_i|]$ corresponds to a unique leaf $\ell_{i,x}$ of GST$_{MSA}$ and vice versa, with $1 \leq x \leq |\alpha_i|$; each substring $\alpha_i[x..y]$ corresponds to an explicit or implicit node of GST$_{MSA}$ in the root-to-$\ell_{i,x}$ path; and each explicit or implicit node corresponds to one or more such substrings, uniquely identifiable thanks to the leaves covered by the node. Also, note that GST$_{MSA}$ does not contain any information about the gap symbols of the MSA, as this information will be added back into the structure thanks to the set of leaves and nodes considered.

In Sect. 4.1 we perform an analysis of GST$_{MSA}$ similar to that of Equi et al., showing that semi-repeat-free segments of the MSA correspond to a specific set of nodes of GST$_{MSA}$ covering exactly m leaves. Then, in Sect. 4.2, we show that the novel resulting problem on the tree structure of GST$_{MSA}$, that we call the *exclusive ancestor set problem*, can be solved efficiently, resulting in an algorithm computing the minimal right extensions in linear time, described in Sect. 4.3.

---

[1] We added the m new distinct terminators for simplicity, whereas Equi et al. used the suffix tree of the concatenation of all gaps-removed rows with a single new symbol $\$$ between each. The suffix tree of this string, if a second unique terminator $\#$ is concatenated to this string, is equivalent to GST$_{MSA}$ for our purposes.

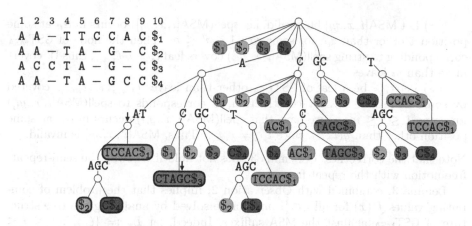

Fig. 2. Example of an MSA[1..4, 1..10] and its GST$_{MSA}$, where the label to each leaf has been moved inside the leaf itself. We have also highlighted the leaves corresponding to suffixes spell(MSA[i, 1..n]) (black outline) and its exclusive ancestors (arrows).

4.1   Semi-repeat-Free Segments in the Generalized Suffix Tree

The following has been implicitly stated and exploited in [6].

Definition 7 (Semi-repeat-free substrings). *Recall the definition of semi-repeat-free segment (Lemma 2). Given substring MSA[i, $x..y$] of MSA[1..m, 1..n] such that* spell(MSA[i, $x..y$]) $\in \Sigma^+$, *we say that MSA[i, $x..y$] is semi-repeat-free if, for all $1 \le i' \le m$, string* spell(MSA[i, $x..y$]) *occurs in gaps-removed row i' only at position $g(i', x)$ (or it does not occur at all).*

Observation 2. *Segment [$x..y$] is semi-repeat-free if and only if all substrings MSA[i, $x..y$] are semi-repeat-free, for $1 \le i \le m$. If MSA[i, $x..y$] is semi-repeat-free, then MSA[i, $x..y'$] is semi-repeat-free for all $y < y' \le n$. Let $f^i(x)$ be the smallest integer greater than x such that substring MSA[i, $x + 1..f^i(x)$] is semi-repeat-free: it is easy to see that $f(x) = \max_{i=1}^{m} f^i(x)$.*

This translates into a specific set of implicit or explicit nodes of GST$_{MSA}$. The fact that we added a unique terminator symbol to each row is equivalent to the addition of an MSA column spelling $\$_1 \cdots \$_m$ at position $n + 1$, which means that [$x + 1..n + 1$] is always semi-repeat-free and the minimal right extensions such that $f(x) = \infty$ become $f(x) = n + 1$.

Lemma 3. *Given m row substrings MSA[i, $x..y_i$] of MSA[1..m, 1..n] such that* spell(MSA[i, $x..y_i$]) $\in \Sigma^+$ *for $1 \le i \le m$, let $W = \{w_1, \ldots, w_k\}$ be the set of implicit or explicit nodes of GST$_{MSA}$ corresponding to strings* {spell(MSA[i, $x..y_i$]) : $1 \le i \le m$}. *Then MSA[i, $x..y_i$] is semi-repeat-free for all $1 \le i \le m$ if and only if W covers exactly m leaves in GST$_{MSA}$.*

Proof. By construction of GST$_{MSA}$, W covers the m leaves $\ell_{1,z_1}, \ldots, \ell_{m,z_m}$, with $z_i = g(i, x)$, so we only need to prove that if some MSA[i, $x..y_i$] is not semi-repeat-free, or *invalid*, then W covers more than m leaves, and vice versa.

($\Leftarrow$) Let $\mathsf{MSA}[i, x..y_i]$ be invalid, i.e. $\mathsf{spell}(\mathsf{MSA}[i, x..y_i])$ occurs in $\alpha_{i'}$ at some position $\hat{z}$ other than $z_{i'}$, for some row $1 \le i' \le m$. Then the node of $\mathsf{GST_{MSA}}$ corresponding to string $\mathsf{spell}(\mathsf{MSA}[i, x..y_i])$ covers leaf $\ell_{i',\hat{z}} \ne \ell_{i',z_{i'}}$, thus W covers more than m leaves.

($\Rightarrow$) Let $\ell_{i',\hat{z}}$ be a leaf of $\mathsf{GST_{MSA}}$ other than leaves $\ell_{1,z_1}, \dots, \ell_{m,z_m}$ covered by some node $w \in W$. By construction, w corresponds to $\mathsf{spell}(\mathsf{MSA}[i, x..y_i])$ for some $1 \le i \le m$, so we have that $\mathsf{spell}(\mathsf{MSA}[i, x..y_i])$ occurs in $\alpha_{i'}$ at some position other than $g(i', x)$, since $\ell_{i',\hat{z}} \ne \ell_{i',z_{i'}}$. Thus, $\mathsf{MSA}[i', x..y_i]$ is invalid.

Note that the correctness of Lemma 3 does not hold if we swap the semi-repeat-free notion with the repeat-free one.

Lemma 3, combined with Observation 2, implies that the problem of computing values $f^i(x)$ for all $i \in [1..m]$ can be solved by analyzing the tree structure of $\mathsf{GST_{MSA}}$ against the MSA suffixes. Indeed, let $L_x := \{\ell_{i,z_i} : 1 \le i \le m, z_i = g(i, x+1)\}$ be the leaves of $\mathsf{GST_{MSA}}$ corresponding to the suffixes $\mathsf{spell}(\mathsf{MSA}[i, x+1..n])$. For each row $1 \le i \le m$, the first semi-repeat-free prefix of $\mathsf{spell}(\mathsf{MSA}[i, x+1..n])$ corresponds to the first implicit or explicit node v of $\mathsf{GST_{MSA}}$ in the root-to-ℓ_{i,z_i} path such that v covers only leaves in L_x. The fact that $\mathsf{GST_{MSA}}$ is a compacted trie is not an issue: the parent of v in the suffix trie is branching, since it covers more leaves than v, so the first explicit node of $\mathsf{GST_{MSA}}$ in the root-to-ℓ_{i,z_i} path covering only leaves in L_x is the first explicit descendant w of v, thus we can identify v by finding w. Finally, $f^i(x)$ is computed by retrieving the smallest column index y such that $\mathsf{spell}(\mathsf{MSA}[i, x+1..y]) = \mathsf{sstring}(\mathsf{parent}(w)) \cdot \mathsf{char}(w)$, where $\mathsf{sstring}(u)$ is the concatenation of edge labels of the root-to-u path, and $\mathsf{char}(u)$ is the first symbol of the edge label from $\mathsf{parent}(u)$ to u. In other words, y corresponds to the k-th non-gap symbol of MSA row i, with $k = \mathsf{rank}(\mathsf{MSA}[i, 1..n], x) + \mathsf{stringdepth}(\mathsf{parent}(w)) + 1$, where $\mathsf{rank}(\mathsf{MSA}[i, 1..n], x)$ is the number of non-gap symbols in $\mathsf{MSA}[i, 1..x]$ and $\mathsf{stringdepth}(u) = |\mathsf{sstring}(u)|$. For example, in Fig. 2 the leaves of L_0 have been marked and so have the shallowest ancestors covering only leaves in L_0.

4.2   Exclusive Ancestor Set

The results of the previous section show that we can compute the minimal right extensions by solving multiple instances of the following problem on the tree structure of $\mathsf{GST_{MSA}}$.

Problem 1 (Exclusive ancestor set). Let $T = (V, E, \mathrm{root})$ be a rooted ordered tree, with $L^T \subseteq V$ the set of its leaves. Given T and a subset of leaves $L \subseteq L^T$, find the minimal set W of exclusive ancestors of L in T, i.e. the minimal set $W \subseteq V$ such that W covers all leaves in L and only leaves in L. Can T be preprocessed to support the efficient solving of multiple instances of the problem?

As is the case for $\mathsf{GST_{MSA}}$, we can assume that each internal node of T has at least two children, otherwise, a linear-time processing of T can be employed to compact its unary paths. Indeed, after a linear-time preprocessing of T, any instance of exclusive ancestor set can be solved in time $O(|L|)$ by a careful traversal of the tree with the following procedure, that we describe informally:

1. partition L in k maximal sets $L_1, \ldots, L_k$ of leaves contiguous in the ordered traversal of T, to be processed independently (if two leaves belong to different contiguous sets, any common ancestor cannot be part of the solution);
2. for each L_i, with $1 \leq i \leq k$, start from the leftmost leaf ℓ_i and ascend in the tree until the closest ancestor of ℓ_i that covers some leaf not in L_i;
3. upon failure in step 2., add the last safe ancestor to the solution W and if there are still uncovered leaves in L_i repeat steps 2. and 3. starting from the leftmost uncovered leaf.

An example of the procedure is shown in Fig. 3. The failure condition of step 2. can be evaluated by checking if both the leftmost leaf and the rightmost leaf in the subtree of the candidate replacement are still in set L_i, and step 2. always terminates if we assume that L is a nontrivial instance: if $L \subset L^T$, then the root of T is not the solution to the problem.

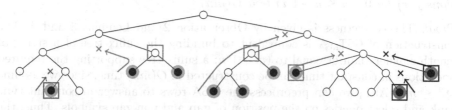

Fig. 3. Example of an instance of exclusive ancestor set, where the set of leaves L corresponds to the black leaves: the algorithm partitions L into sets of contiguous leaves (shown as brown, blue, and purple leaves), and for each set it finds the exclusive ancestors (marked with rectangles). Each arrow shows the ascent of step 2. up the tree until the node corresponding to the failure condition, marked with a cross. (Color figure online)

Assuming the leaves of T are sorted, step 1. can be implemented efficiently: we can partition L into sets of contiguous leaves by coloring leaves in L and finding all the leaves with the preceding leaf not in L. We can easily preprocess T to support the required operations in constant time, leading to a time complexity of $O(|L|)$, since any forest built on top of leaves L has $O(|L|)$ nodes.

Lemma 4. *The exclusive ancestor set problem on a rooted ordered tree $T = (V, E, \text{root})$ and a subset L of its leaves can be solved in time $O(|L|)$, after a $O(|V|)$-time preprocessing to support operations v.leftmostleaf, v.rightmostleaf on any node $v \in V$ and operations ℓ.prevleaf, ℓ.nextleaf, and the binary coloring of any leaf $\ell \in L^T$ in constant time.*

4.3   Computing the Minimal Right Extensions

Returning to the problem of computing values $f(x)$, the representation of GST_{MSA} needs to support the operations on its tree structure described by Lemma 4 plus operations v.stringdepth, returning the length of the string corresponding to the root-to-v path in GST_{MSA} of an explicit node v, and ℓ.suffixlink, implementing the suffix links of the leaves. The final algorithm, described in the full version of this paper [17], computes leaf sets $L_0, L_1, \ldots, L_{n-1}$ corresponding to the MSA suffixes starting at column $1, 2, \ldots, n$, respectively, and for each L_x with $0 \leq x < n$:

1. it marks the leaves in L_x and partitions them in sets of contiguous leaves, by finding all their left boundaries ℓ such that ℓ.prevleaf is not marked;
2. it solves the exclusive ancestor set problem on each set of contiguous leaves and whenever it finds an exclusive ancestor, covering leaves $\ell_{i_1}, \ldots, \ell_{i_k}$, it computes values $f^i(x)$ for $i \in \{i_1, \ldots, i_k\}$ (see the conclusion of Sect. 4.1);
3. after processing all leaves, it finally computes $f(x) = \max_{i=1}^{m} f^i(x)$ and transforms L_x into L_{x+1} by taking the suffix links[2] of only leaves ℓ_i such that MSA$[i, x+1] \neq -$.

Theorem 1. *Given* MSA$[1..m, 1..n]$, *we can compute the minimal right extensions* $f(x)$ *for* $0 \leq x \leq n - 1$ *in time* $O(mn)$.

Proof. The correctness is given by Observation 2 and Lemmas 3 and 4. The construction of GST$_{\mathsf{MSA}}$ is equivalent to building the suffix tree of a string of length smaller than or equal to $(m + 1)n$: a suffix tree supporting the required operations in constant time can be constructed in $O(mn)$ time, since we assume $|\Sigma| \leq mn$. Also, we can preprocess the MSA rows to answer in constant time rank and select queries on the position of gap and non-gap symbols. Thus, the computation of each $f(x)$ takes time $O(|L_x| + m) = O(m)$, so $O(mn)$ time in total.

Corollary 1. *Given* MSA$[1..m, 1..n]$ *from* $\Sigma \cup \{-\}$, *with* $\Sigma = [1..\sigma]$ *and* $\sigma \leq mn$, *the construction of an optimal semi-repeat-free segmentation minimizing the maximum number of blocks can be done in time* $O(mn)$.

Proof. Algorithm [6, Algorithm 1] by Equi et al. solves the problem in $O(n)$ time, assuming it is given the minimal right extensions $(x, f(x))$ sorted in increasing order by the second component, which we can now compute and sort in time $O(mn)$ thanks to Theorem 1.

5 Minimizing the Maximum Block Length

The improvement on the computation of the minimal right extensions in the case of general MSAs from $O(nm \log m)$ to $O(nm)$ gives us the motivation to improve the $O(n \log \log n)$-time algorithm of Equi et al. [6, Algorithm 2] for an optimal semi-repeat-free segmentation minimizing the maximum block length. As mentioned in Sect. 3.2, we can compute $s(j)$ by processing the recursive solutions corresponding to all right extensions $(x, f(x))$ with $f(x) \leq j$. For the maximum block length there are two types of recursion for an optimal solution of MSA$[1..m, 1..j']$ using semi-repeat-free $[x + 1..j']$ as its last segment (Fig. 4):

---

[2] As noted by an anonymous reviewer, the support for suffix links is not strictly necessary, since we are exploring leaves only. Indeed, a traversal of the tree can easily fill an $m \times n$ table containing $L_0, \ldots, L_{n-1}$, that we then have to store.

non-leader recursion: if $j' \leq x + s(x)$ then the score of $s(j')$ is equal to $s(x)$, because the length of segment $[x+1..j']$ is less than or equal to $s(x)$; in this case, we say that $[x+1..j']$ is a *non-leader segment*;

leader recursion: otherwise, if $j' > x + s(x)$, we say that $[x+1..j']$ is a *leader segment*, since it gives score $j' - x$ to an optimal solution constrained to use it as its last segment.

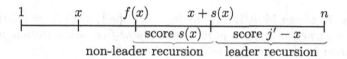

Fig. 4. Scheme for the score of an optimal semi-repeat-free segmentation of MSA$[1..m, 1..j']$ constrained to use $[x+1..j']$ as its last segment.

Note that if $x + s(x) < f(x)$ then the non-leader recursion does not occur for $(x, f(x))$. Then, it is easy to see that

$$s(j) = \min \left(\min_{\substack{(x,f(x)): \\ f(x) \leq j \leq x+s(x)}} s(x), \quad \min_{\substack{(x,f(x)): \\ j > f(x) \,\wedge\, j > x+s(x)}} j - x \right) \quad (2)$$

so Equi et al. correctly solve the problem by keeping track of the two types of recursions with two one-dimensional search trees: the first keeps track of ranges $[f(x)..x + s(x)]$ with score $s(x)$, the second tracks ranges $[x + s(x) + 1..n]$ where the leader recursion must be used, saving only the $-x$ part of score $j' - x$. With two semi-infinite range minimum queries, for ranges $[j + 1.. + \infty]$ and $[-\infty..j]$ respectively, we can compute $s(j)$ and solve the problem in time $O(n \log \log n)$.

Instead, we can reach a linear time complexity using simpler data structures, thanks to the following observations: the data structure for the leader recursion can be replaced by a single variable S holding value $\min\{j - x : j > f(x) \wedge j > x + s(x)\}$, so that S is the best score of a segmentation ending with a leader segment $[x + 1..j]$; for the non-leader recursion, we can swap the structure of Equi et al. with an equivalent array $C[1..n]$ such that $C[k]$ counts the number of available solutions with score k using the non-leader recursion so that a variable $K = \min\{k : C[k] > 0\}$ is equal to the best score of a segmentation ending with a non-leader segment $[x + 1..j]$. The final and crucial observation is that the two types of recursion are closely related: when $[x + 1..j]$ goes from being a non-leader segment to a leader segment, that is, $j = x + s(x) + 1$, we decrease $C[s(x)]$ by one and update S with value $s(x) + 1 = j - x$ if needed. Therefore, when the best score of $C[1..n]$ is removed in this way, we do not need to update K to $\min\{k : C[i] > 0\}$, but it is sufficient to increment K by 1 to ensure that $s(j) = \min(K, S)$, unless other updates of C and S result in a better score.

Theorem 2. *Given the minimal right extensions* $(x, f(x))$ *of* MSA$[1..m, 1..n]$, *we can compute in time* $O(n)$ *the score of an optimal semi-repeat-free segmentation minimizing the maximum block length.*

Proof. The correctness of the algorithm, described in the full version of this paper [17], follows from that of [6, Algorithm 2] and from the fact that when $C[K] = 0$ we have that $C[j'] = 0$ for $1 \leq j' \leq K$ and $S \leq K + 1$. Similarly, the processing of minimal right extensions $(x, f(x))$ and the dynamic management of intervals $[f(x)..s(x) + j']$ takes time $O(n)$ in total, thus the algorithm takes linear time.

Combined with Theorem 1, we get our second main result.

Corollary 2. *Given* $\mathsf{MSA}[1..m, 1..n]$ *from* $\Sigma \cup \{-\}$, *with* $\Sigma = [1..\sigma]$ *and* $\sigma \leq mn$, *the construction of an optimal semi-repeat-free segmentation minimizing the maximum block length can be done in time* $O(mn)$.

Acknowledgements. This project has received funding from the European Union's Horizon 2020 research and innovation programme under the Marie Skłodowska-Curie grant agreement No 956229.

References

1. Cazaux, B., Kosolobov, D., Mäkinen, V., Norri, T.: Linear time maximum segmentation problems in column stream model. In: Brisaboa, N.R., Puglisi, S.J. (eds.) String Processing and Information Retrieval, SPIRE 2019. LNCS, vol. 11811, pp. 322–336. Springer, Cham (2019). https://doi.org/10.1007/978-3-030-32686-9_23
2. De La Briandais, R.: File searching using variable length keys. In: Western Joint Computer Conference, IRE-AIEE-ACM 1959 (Western), 3–5 March 1959, pp. 295–298. Association for Computing Machinery, New York (1959). https://doi.org/10.1145/1457838.1457895
3. Eggertsson, H.P., et al.: Graphtyper2 enables population-scale genotyping of structural variation using pangenome graphs. Nat. Commun. **10**(1), 5402 (2019). https://doi.org/10.1038/s41467-019-13341-9
4. Equi, M., Grossi, R., Mäkinen, V., Tomescu, A.I.: On the complexity of string matching for graphs. In: Baier, C., Chatzigiannakis, I., Flocchini, P., Leonardi, S. (eds.) 46th International Colloquium on Automata, Languages, and Programming, ICALP 2019, LIPIcs, 9–12 July 2019, Patras, Greece, vol. 132, pp. 55:1–55:15. Schloss Dagstuhl - Leibniz-Zentrum für Informatik (2019)
5. Equi, M., Mäkinen, V., Tomescu, A.I.: Graphs cannot be indexed in polynomial time for sub-quadratic time string matching, unless SETH fails. In: Bureš, T., et al. (eds.) SOFSEM 2021: Theory and Practice of Computer Science. LNCS, vol. 12607, pp. 608–622. Springer, Cham (2021). https://doi.org/10.1007/978-3-030-67731-2_44
6. Equi, M., Norri, T., Alanko, J., Cazaux, B., Tomescu, A.I., Mäkinen, V.: Algorithms and complexity on indexing elastic founder graphs. In: Ahn, H., Sadakane, K. (eds.) 32nd International Symposium on Algorithms and Computation, ISAAC 2021, LIPIcs, 6–8 December 2021, Fukuoka, Japan, vol. 212, pp. 20:1–20:18. Schloss Dagstuhl - Leibniz-Zentrum für Informatik (2021). https://doi.org/10.4230/LIPIcs.ISAAC.2021.20
7. Farach, M.: Optimal suffix tree construction with large alphabets. In: Proceedings 38th Annual Symposium on Foundations of Computer Science, pp. 137–143. IEEE (1997)

8. Garrison, E., et al.: Variation graph toolkit improves read mapping by representing genetic variation in the reference. Nat. Biotechnol. **36** (2018). https://doi.org/10. 1038/nbt.4227
9. Jacobson, G.: Space-efficient static trees and graphs. In: Proceedings of FOCS, pp. 549–554 (1989)
10. Jacobson, G.J.: Succinct static data structures. Carnegie Mellon University (1988)
11. Kim, D., Paggi, J., Park, C., Bennett, C., Salzberg, S.: Graph-based genome alignment and genotyping with HISAT2 and HISAT-genotype. Nat. Biotechnol. **37** (2019). https://doi.org/10.1038/s41587-019-0201-4
12. Mäkinen, V., Cazaux, B., Equi, M., Norri, T., Tomescu, A.I.: Linear time construction of indexable founder block graphs. In: Kingsford, C., Pisanti, N. (eds.) 20th International Workshop on Algorithms in Bioinformatics, WABI 2020, LIPIcs, 7–9 September 2020, Pisa, Italy (Virtual Conference), vol. 172, pp. 7:1–7:18. Schloss Dagstuhl - Leibniz-Zentrum für Informatik (2020). https://doi.org/10. 4230/LIPIcs.WABI.2020.7
13. Mäkinen, V., Navarro, G., Sirén, J., Välimäki, N.: Storage and retrieval of highly repetitive sequence collections. J. Comput. Biol. **17**(3), 281–308 (2010)
14. Norri, T., Cazaux, B., Dönges, S., Valenzuela, D., Mäkinen, V.: Founder reconstruction enables scalable and seamless pangenomic analysis. Bioinformatics **37**(24), 4611–4619 (2021). https://doi.org/10.1093/bioinformatics/btab516
15. Norri, T., Cazaux, B., Kosolobov, D., Mäkinen, V.: Linear time minimum segmentation enables scalable founder reconstruction. Algorithms Mol. Biol. **14**(1), 12:1-12:15 (2019). https://doi.org/10.1186/s13015-019-0147-6
16. Rizzo, N., Mäkinen, V.: Indexable elastic founder graphs of minimum height. In: Proceedings of 33rd Annual Symposium on Combinatorial Pattern Matching (CPM 2022) (2022). To appear
17. Rizzo, N., Mäkinen, V.: Linear time construction of indexable elastic founder graphs. CoRR abs/2201.06492 arXiv:2201.06492 (2022)
18. Schneeberger, K., Hagmann, J., Ossowski, S., Warthmann, N., Gesing, S., Kohlbacher, O., Weigel, D.: Simultaneous alignment of short reads against multiple genomes. Genome Biol. **10**, R98 (2009)
19. Sirén, J., Välimäki, N., Mäkinen, V.: Indexing graphs for path queries with applications in genome research. IEEE/ACM Trans. Comput. Biol. Bioinform. **11**(2), 375–388 (2014)
20. The computational pan-genomics consortium: computational pan-genomics: status, promises and challenges. Brief. Bioinform. **19**(1), 118–135 (2018). https://doi.org/ 10.1093/bib/bbw089
21. Ukkonen, E.: On-line construction of suffix trees. Algorithmica **14**(3), 249–260 (1995). https://doi.org/10.1007/BF01206331

On Critical Node Problems
with Vulnerable Vertices

Jannik Schestag, Niels Grüttemeier[(✉)] [iD], Christian Komusiewicz[iD],
and Frank Sommer[iD]

Fachbereich für Mathematik und Informatik, Philipps-Universität Marburg,
Marburg, Germany
{jschestag,niegru,komusiewicz,fsommer}@informatik.uni-marburg.de

Abstract. A vertex pair in an undirected graph is called *connected* if
the two vertices are in the same connected component. In the NP-hard
CRITICAL NODE PROBLEM (CNP), the input is an undirected graph G
with integers k and x, and the question is whether we can transform G
via at most k vertex deletions into a graph whose total number of connected vertex pairs is at most x. In this work, we introduce and study
two NP-hard variants of CNP where a subset of the vertices is marked as
vulnerable and we aim to obtain a graph with at most x connected vertex
pairs where at least one vertex is vulnerable. In the first variant, which
generalizes CNP, we may delete vulnerable and non-vulnerable vertices.
In the second variant, we may only delete non-vulnerable vertices.

We perform a parameterized complexity study of both problems. For
example, we show that both problems are FPT with respect to $k + x$.
Furthermore, in case of deletable vulnerable vertices we provide a polynomial kernel for the parameter vc $+k$, where vc is the vertex cover number.
In case of non-deletable vulnerable vertices, we prove NP-hardness even
when there is only one vulnerable vertex.

1 Introduction

Detecting important vertices in graphs is a central task in network analysis.
There is an abundance of different formalizations of this natural task, many
of which adopt the view that a vertex set is important if its removal severely
affects the connectivity of the remaining graph [10]. One concrete formulation,
known as the CRITICAL NODE PROBLEM, measures connectivity by the number
of connected pairs of vertices, that is, the number of pairs of vertices that are in
the same connected component. The aim is to look for a set of vertices whose
deletion decreases this number as much as possible.

CRITICAL NODE PROBLEM (CNP)
Input: A graph $G = (V, E)$, and two integers $k, x \in \mathbb{N}$.
Question: Is there a vertex set $C \subseteq V$ of size at most k such that $G - C$
has at most x connected pairs of vertices?

Most of the results of this work are also contained in the first author's Master's
thesis [12].
F. Sommer—Supported by the DFG, project EAGR KO 3669/6-1.

C. Bazgan and H. Fernau (Eds.): IWOCA 2022, LNCS 13270, pp. 494–508, 2022.
https://doi.org/10.1007/978-3-031-06678-8_36

One application of this formulation is to model the influence of vertices in the spreading of viruses in computer networks or social networks [10]. Taking the latter view, the entities represented by a set C that minimizes the number of connected pairs in $G - C$ would be good candidates for being vaccinated or removed from the network via other interventions. The number x of connected pairs would be a rough measure for the amount of virus spreading in the remaining network, as vertices that are connected to many other vertices are more likely to contract the virus. For some vertices in the network, however, it may be irrelevant whether they contract the virus, for example because they are not prone to develop a severe disease in case of infection. Conversely, it may be critical that some vertices in the network are protected from the virus, because they belong to a high risk group. This aspect is missing from the CNP problem. One way to model this aspect is to label some vertices as *vulnerable* and to consider only the number of connected pairs for the vulnerable vertices. In other words, we only count those vertex pairs that contain at least one vulnerable vertex.

Definition 1. *Let* $G = (V, E)$ *be a graph and let* A *be a set of* vulnerable *vertices. A* vertex pair $\{u, v\}$ *is a* vulnerable connection *(with respect to* A) *in* G *if* $\{u, v\} \cap A \neq \emptyset$ *and* u *and* v *are in the same connected component of* G. *The* A-vulnerability *of* G *is the number of vulnerable connections of* G.

Note that it is not required that the set A of vulnerable vertices is a subset of the vertex set V. Thus, given a graph $G = (V, E)$ with $A \subseteq V$ and a subgraph $G' = (V', E')$ of G, we may refer to the A-vulnerability of G' even if $A \not\subseteq V'$. Replacing the number of connected pairs by A-vulnerability leads to the following problem definition.

CRITICAL NODE PROBLEM WITH VULNERABLE NODES (CNP-V)
Input: A graph $G = (V, E)$, $A \subseteq V$, and two integers $k, x \in \mathbb{N}$.
Question: Is there a vertex set $C \subseteq V$ of size at most k such that the A-vulnerability of $G - C$ is at most x?

A further complication may be that, for several reasons, vulnerable vertices may not be removed. This is modelled by the following problem.

CRITICAL NODE PROBLEM WITH NON-DELETABLE VULNERABLE NODES (CNP-NDV)
Input: A graph $G = (V, E)$, $A \subseteq V$, and two integers $k, x \in \mathbb{N}$.
Question: Is there a vertex set $C \subseteq V \setminus A$ of size at most k such that the A-vulnerability of $G - C$ is at most x?

The set C is called a *critical node cut*. We study the parameterized complexity of these two problems.

Related Work. Arulselvan et al. [3] showed that CNP is NP-complete; the NP-hardness follows also directly from the fact that CNP is a generalization of VERTEX COVER ($x = 0$). As a consequence, CNP is NP-hard even on subcubic graphs. CNP is also NP-hard on split and bipartite graphs [1] and on power

Table 1. Overview of our results.

Parameter	CNP-V	CNP-NDV
x	NP-hard for $x = 0$ [9]	W[1]-hard (Theorem 2)
		XP (Proposition 2)
y	FPT (Theorem 6)	W[1]-hard (Theorem 7)
	No poly kernel [9]	XP (Proposition 3)
k	W[1]-hard [9]	W[1]-hard (Theorem 7)
	XP (Proposition 1)	XP (Proposition 1)
$k + x$	FPT (Corollary 2, Theorem 5)	FPT (Corollary 1)
$k + y$	FPT (Theorem 6)	W[1]-hard (Theorem 7)
$|A|$	XP (Proposition 3)	NP-hard for $|A| = 1$ (Theorem 2)
$|A| + x$	FPT (Corollary 3)	
vc	FPT (Theorem 8)	FPT (Theorem 8)
vc $+x$	poly kernel (Corollary 5)	FPT (Theorem 8)
vc $+ k + x$		poly kernel (Corollary 6)

law graphs [13]. In contrast, CNP can be solved in polynomial time on trees [5] and, more generally, on graphs with constant treewidth [1]. The parameterized complexity of CNP has been studied with respect to the parameters k, x, and the treewidth tw of G [9]: On the negative side, CNP is W[1]-hard with respect to k [9] or tw [9], and even with respect to $k +$ tw [2]. On the positive side, the problem is FPT with respect to $k + x$ and with respect to the parameter y which is defined as $\ell - x$, where ℓ is the number of connected pairs in G. In other words, y is the number of connected pairs that we want to remove at least by deleting the k vertices.

Other formulations of graph modifications for limiting disease spreading consider for example edge deletions and limiting the size of the largest remaining connected component [7]. For an overview of different formulations of critical vertex detection, refer to the survey of Lalou et al. [10].

Our Results. We study the parameterized complexity of the problems CNP-V and CNP-NDV with respect to a number of natural parameters. Our main findings are as follows (an overview is given in Table 1). We transfer the FPT algorithm for $k + x$ from CNP to the two new problems. We then show that, while being solvable in polynomial time for constant values of x, CNP-NDV is W[1]-hard with respect to x even when $|A| = 1$. In contrast, CNP-V is solvable in polynomial time for constant $|A|$ and NP-hard already for $x = 0$. Thus, the complexity of the two problems differs quite drastically with respect to very natural parameters. This can be also observed for the parameter y for which CNP-V has a subexponential FPT algorithm while CNP-NDV is W[1]-hard even with respect to $k + y$. We remark that the algorithm for CNP-V with subexponential running time for parameter y improves on a previous algorithm for CNP with exponential running time in y [9].

Finally, we consider parameterizations using the vertex cover number vc of G. This is motivated by the fact that CNP is W[1]-hard with respect to the treewidth tw [2,9] and thus larger structural parameters need to be considered. We show that both problems are FPT with respect to vc, and provide polynomial kernels for both problems parameterized by vc $+x$ and vc $+k+x$, respectively.

Further FPT results for parameters such as the neighborhood diversity of G or $|V \setminus A|$ have been obtained in the first author's Master thesis [12]. Due to lack of space, the proofs of several results (marked with ($\star$)) are deferred to a full version.

Preliminaries. For two integers p and q, $p \leq q$, we denote $[p, q] := \{p, \dots, q\}$. We consider undirected simple graphs G and let $V(G)$ denote the vertex set and $E(G)$ the edge set of a graph G. We use n to denote the number of vertices of G and m to denote the number of edges. For a vertex set S, we let $N(S) = \{u \mid \{u, v\} \in E(G), v \in S\} \setminus S$ and $N[S] := S \cup N(S)$ denote the open and closed neighborhood of S, respectively. For a vertex v, we denote $N(v) := N(\{v\})$ and $N[v] := N[\{v\}]$. For a vertex set S, we let $G[S] := (S, \{\{u, v\} \in E(G) \mid u, v \in S\})$ denote the subgraph induced by S, and $G - S := G[V(G) \setminus S]$ denote the subgraph of G obtained by deleting S and its incident edges. For the relevant definitions of parameterized complexity refer to the standard monographs [4,6].

2   Basic Observations

Vulnerability. First, observe that the A-vulnerability of a graph can be computed in linear time via depth-first search.

Lemma 1 ($\star$). *Let $G = (V, E)$ and let $A \subseteq V$. The A-vulnerability of G can be computed in $\mathcal{O}(n + m)$ time.*

For constant k, CNP-V and CNP-NDV can thus be solved in polynomial time by trying all $\mathcal{O}(n^k)$ possibilities of deleting k vertices (in the case of CNP-NDV only deletions in $V \setminus A$ are considered).

Proposition 1. *CNP-V and CNP-NDV can be solved in $\mathcal{O}(n^k \cdot (n+m))$ time.*

Moreover, for CNP-NDV at most x non-vulnerable vertices can be connected to vulnerable vertices in $G - C$. Thus, one may find a critical node cut by considering all $\mathcal{O}(n^x)$ possible sets B for these vertices, deleting all neighbors of $A \cup B$, and checking whether the number of deletions is at most k and the A-vulnerability of the resulting graph is at most x.

Proposition 2. *CNP-NDV can be solved in $\mathcal{O}(n^x \cdot (n + m))$ time.*

Reduction Rules. We provide a collection of simple reduction rules for CNP-V and CNP-NDV. The first rule removes trivial components from the input.

Rule 1. *Let* $I := (G, A, k, x)$ *be an instance of* CNP-V *or* CNP-NDV *and let* C *be a connected component of* G. *If* C *contains no vulnerable vertex or* C *is an isolated vulnerable vertex, then delete* C *from* G.

Rule 1 is correct since no vertex of C is part of a vulnerable connection. For the rest of this work, we assume that all instances of CNP-V and CNP-NDV are reduced with respect to Rule 1. The next rule identifies instances of CNP-V and CNP-NDV that are trivial because k is sufficiently large.

Rule 2. *a) Let* (G, A, k, x) *be an instance of* CNP-V. *If* $y \leq k$, *then return yes.*
b) Let (G, A, k, x) *be an instance of* CNP-NDV *such that* $y \leq k$. *If* $|V \setminus A| \geq y$, *then return yes. If* $|V \setminus A| < y$, *check if the number of vulnerable connections in* $G - (V \setminus A)$ *is at most* x. *If this is the case, return yes. Otherwise, return no.*

The correctness of Rule 2 can be seen as follows: Since the instance is reduced with respect to Rule 1, every vertex of the graph is in at least one vulnerable connection. If we remove y vertices, we remove at least y vulnerable connections and therefore, the instance is a yes-instance. In case of CNP-NDV, we might not be able to remove y vertices if $|V \setminus A|$ is too small. In this case we can trivially solve the instance by checking if $G - (V \setminus A)$ contains at most x vulnerable connections. Hence, we may assume $y > k$ throughout the rest of this work.

In case of CNP-V, we can identify a further class of yes-instances. An instance of CNP-V with $|A| \leq k$ is a trivial yes-instance, since adding all vulnerable vertices to a critical node cut destroys all vulnerable connections.

Rule 3. *Let* (G, A, k, x) *be an instance of* CNP-V. *If* $|A| \leq k$, *then return yes.*

The final rule deals with the case where one vertex has too many vulnerable neighbors. The idea behind the rule is that a vertex that causes too many vulnerable connections in his neighborhood belongs to every possible solution.

Rule 4. *a) If in an instance* (G, A, k, x) *of* CNP-V *a vertex* $v \in V$ *exists with* $|N(v) \cap A| > k + \sqrt{2x}$, *then remove* v *from* G *and decrease* k *by 1.*
b) If in an instance (G, A, k, x) *of* CNP-NDV *a vertex* $v \in V \setminus A$ *exists with* $|N(v) \cap A| > \sqrt{2x}$, *then remove* v *from* G *and decrease* k *by one.*

Recall that CNP-V and CNP-NDV can be solved in $\mathcal{O}(n^k \cdot (n + m))$ time due to Proposition 1. Since we can assume $y > k$ due to Rule 2 and, for CNP-V, $|A| > k$ due to Rule 3, we obtain the following.

Proposition 3. CNP-V *and* CNP-NDV *can be solved in* $\mathcal{O}(n^y \cdot (n+m))$ *time;* CNP-V *can be solved in* $\mathcal{O}(n^{|A|} \cdot (n + m))$ *time.*

Component Information. We next show that CNP-V is solvable in polynomial time if we have additional information about the connected components of the input graph. We apply this fact to obtain efficient algorithms for CNP-V when the connected components are small. Let $I := (G, A, k, x)$ be an instance of CNP-V, and let $C_1, \ldots, C_t \subseteq V$ be the connected components of the input graph G. The *component information* $T[i, k']$ of some integers $i \in [1, t]$ and $k' \in [0, \min(k, |C_i|)]$ is defined as the minimal number of vulnerable connections in $G[C_i] - S$ among all subsets $S \subseteq C_i$ of size exactly k'. A table T containing all component information $T[i, k']$ is called a *component table of the instance* I. We now show that CNP-V can be solved in polynomial time if we have a component table of the input instance, the algorithm was also described by Hermelin et al. [9] for CNP.

Lemma 2 ($\star$). *Given an instance* $I := (G, A, k, x)$ *of* CNP-V *and a component table* T *of* I, *we can compute in* $\mathcal{O}(n \cdot k^2)$ *time, whether* I *is a yes-instance of* CNP-V.

Observe that, for an instance where the input graph has maximum component size c for some constant c, a component table can be computed in $\mathcal{O}(2^c \cdot (n + m)) = \mathcal{O}(n)$ time by iterating over every subset of each connected component.

Proposition 4. CNP-V *can be solved in* $\mathcal{O}(n \cdot k^2)$ *time if the input graph has maximum component size* c *for some constant* c.

NP-Hardness of CNP-NDV. In contrast to CNP-V, the problem CNP-NDV is not an obvious generalization of VERTEX COVER. We show the following by a simple reduction.

Theorem 1 ($\star$). CNP-NDV *is* NP-*hard on planar graphs, even if the input graph has maximum degree* 4.

3 Parameterization by the Targeted Vulnerability

First, we consider parameterization by x alone. CNP-V is NP-hard for $x = 0$ since it is a generalization of CNP. We now show that, in contrast, CNP-NDV is W[1]-hard with respect to x, even if G contains only one vulnerable vertex.

Theorem 2 ($\star$). CNP-NDV *is* W[1]-*hard with respect to the parameter* x, *even if* $|A| = 1$ *and* diam $= 2$.

We now show an FPT algorithm for CNP-V and CNP-NDV parameterized by $k + x$. To this end, we consider the following more general problem.

CNP-VNDV
Input: A graph $G = (V, E)$, two sets A, N, and two integers $k, x \in \mathbb{N}$.
Question: Is there a vertex set $C \subseteq V \setminus N$ of size at most k such that the A-vulnerability of $G - C$ is at most x?

Hermelin et al. [9] showed that CNP can be solved in $\mathcal{O}(3^{k+x} \cdot (x^{k+2} + n))$ time. The idea of this algorithm is to branch for each edge $\{u, v\}$ whether one of u and v is deleted or whether this is one of the x remaining connections. In the following, we use similar ideas to provide two search tree algorithms for the more general CNP-VNDV. The first algorithm solves instances of CNP-VNDV with $A \subseteq N$ in $\mathcal{O}(2^{k+x} \cdot (n+m))$ time. This implies that CNP-NDV can be solved within the same running time. The second algorithm solves arbitrary instances of CNP-VNDV in $\mathcal{O}(3^{k+x} \cdot (n+m))$ time, which implies that CNP-V can be solved in $\mathcal{O}(3^{k+x} \cdot (n+m))$ time. Moreover, since CNP is a special case of CNP-V this improves over the algorithm for CNP by Hermelin et al. [9]. The next lemma describes the mechanism of the branching rule.

Lemma 3 ($\star$). *Let* $I = (G = (V, E), A, N, k, x)$ *be an instance of* CNP-VNDV *and let* $v \in V \setminus N$. I *is a yes-instance of* CNP-VNDV *if and only if* $I_1 = (G - \{v\}, A, N, k - 1, x)$ *or* $I_2 = (G, A, N \cup \{v\}, k, x)$ *is a yes-instance of* CNP-VNDV.

Theorem 3. *An instance* $I := (G, A, N, k, x)$ *of* CNP-VNDV *with* $A \subseteq N$ *can be solved in* $\mathcal{O}(2^{k+x} \cdot (n+m))$ *time.*

Proof. Intuition: In the algorithm we pick a neighbor v of N and branch into removing v from the graph or making v non-deletable.

Algorithm: **Step 0.** If $k < 0$ or the A-vulnerability of $G[N]$ is greater than x, return no. If the A-vulnerability of G is at most x, return yes.

Step 1. Compute the set $N' \subseteq N$ such that N' contains all vulnerable vertices A and also all vertices that are connected to a vulnerable vertex in $G[N]$.

Step 2. If the neighborhood of N' is empty, return yes. Otherwise, pick a neighbor v of N' and branch into the following instances: $I_1 := (G - \{v\}, A, N, k - 1, x)$ and $I_2 := (G, A, N \cup \{v\}, k, x)$ of CNP-VNDV.

The correctness of the algorithm follows rather directly from Lemma 3. The running time can be seen as follows: The depth of the search tree is bounded by $k + x$ since in each branch, we either add a vertex to N (which increases the A-vulnerability of $G[N]$ by at least one) or delete a vertex (which decreases k). Thus, the search tree has size $\mathcal{O}(2^{k+x})$; the steps at each search tree node can be clearly performed in linear time. □

Corollary 1. CNP-NDV *can be solved in* $\mathcal{O}(2^{k+x} \cdot (n+m))$ *time.*

Theorem 4 ($\star$). CNP-VNDV *can be solved in* $\mathcal{O}(3^{k+x} \cdot (n+m))$ *time.*

Corollary 2. CNP-V *can be solved in* $\mathcal{O}\left(3^{k+x} \cdot (n+m)\right)$ *time.*

In the following, we provide an algorithm that solves CNP-V in $\mathcal{O}((\frac{4}{3}x + 2)^k \cdot m \cdot x)$ time. This running time is preferable, when x is much larger than k. The idea of the algorithm is that we search a set B of at most $\frac{4}{3}x + 2$ vertices of G such that the A-vulnerability of $G[B]$ is larger than x. Then, if there exists a critical node cut C, at least one vertex of B is in C.

Theorem 5 ($\star$). *An instance* $I := (G, A, k, x)$ *of* CNP-V *can be solved in* $\mathcal{O}((\frac{4}{3}x + 2)^k \cdot m \cdot x)$ *time.*

After Rule 3 is applied, we can assume $|A| > k$ for instances of CNP-V. Hence, we also obtain the following.

Corollary 3. CNP-V *has an FPT-algorithm for the parameter* $|A| + x$.

4   Parameterization by the Decrease in Vulnerability

In this section, we consider the parametrization by $y := \ell - x$, where ℓ is the A-vulnerability of the input graph. In other words, y counts how many vulnerable connections shall be removed.

An FPT Algorithm for Deletable Vulnerable Vertices. CNP is fixed-parameter tractable with respect to y [9], based on the following observations: If some connected component has at least y vertices, then we have a yes-instance. Afterwards, we may compute the component information in $\mathcal{O}(2^y \cdot y^2 \cdot (n+m))$ time and combine it using the dynamic programming algorithm presented also in Sect. 2. We now extend the FPT result to the more general CNP-V problem. Moreover, we improve the running time to a subexponential running time in y.

Theorem 6. CNP-V *can be solved in* $2^{\mathcal{O}(\sqrt{y}\log y)} \cdot n^{\mathcal{O}(1)}$ *time.*

Proof. Let $I := (G, A, k, x)$ be an instance of CNP-V and let $C_1, \ldots, C_t$ be the connected components of G. Recall that we assume that I is reduced regarding Rule 1 and therefore each connected component has a non-empty intersection with A. Moreover, we assume that $k \geq 1$ since otherwise we can solve I in polynomial time by computing the number of vulnerable connections of G.

We first assume that there exists a connected component C_i of size at least y. Since we assume that every connected component of G contains some vertices from A, let $v \in C_i \cap A$. Since $|C_i| \geq y$, we can remove at least y vulnerable connections by deleting v. Together with the fact that $k \geq 1$ we conclude that the instance I is a yes-instance. Throughout the rest of the proof, we assume that $|C_i| < y$ for every connected component of G.

In the remainder of the proof, we show that a component table T of I can be computed in $2^{\mathcal{O}(\sqrt{y}\log y)} \cdot n^{\mathcal{O}(1)}$ time. With a component table at hand, we can then solve CNP-V in polynomial time due to Lemma 2. Recall that a component table T of I has entries of type $T[i, k']$ with $i \in [1, t]$ and $k' \in [0, k]$ such that $T[i, k']$ is the minimum number of vulnerable connections in $G[C_i]$ that remain after deleting exactly k' vertices in C_i.

Let C_i be a connected component. We now describe how to compute all component information $T[i, k']$ with $k' \in [0, k]$ in $2^{\mathcal{O}(\sqrt{y}\log y)} \cdot n^{\mathcal{O}(1)}$ time. Then, since there are at most n connected components, the statement follows. We first consider the case where $k < \sqrt{y}$. Note that for each $k' \in [0, k]$, there are at most $\binom{|C_i|}{k'} \leq |C_i|^{k'}$ subsets $S \subseteq C_i$ of size k'. Since $|C_i| \leq y$ and $k' \leq k < \sqrt{y}$, we can compute all component information $T[i, k']$ in $y^{\sqrt{y}} \cdot n^{\mathcal{O}(1)} = 2^{\mathcal{O}(\sqrt{y}\log y)} \cdot n^{\mathcal{O}(1)}$ time. Next, let $k \geq \sqrt{y}$. For this, we first identify a further case, where I is a yes-instance.

Claim. If $k \geq \sqrt{y}$ and there exists a connected component C_i such that $|C_i| \geq \frac{3\sqrt{y}+1}{2}$ and $|C_i \cap A| \geq \sqrt{y}$, then I is a yes-instance.

Proof. Since $|C_i \cap A| \geq \sqrt{y}$ and $k \geq \sqrt{y}$, we may delete $\sqrt{y}$ vulnerable vertices from C_i. This decreases the number of vulnerable connections by at least

$$\underbrace{\binom{\sqrt{y}}{2}}_{=:c_1} + \underbrace{\sqrt{y} \cdot (|C_i| - \sqrt{y})}_{=:c_2},$$

where c_1 corresponds to the vulnerable connections between the deleted vertices and c_2 corresponds to vulnerable connections between the deleted vertices and the remaining vertices in C_i. Then, since $|C_i| \geq \frac{3\sqrt{y}+1}{2}$, the number of vulnerable connections is decreased by at least $\binom{\sqrt{y}}{2} + \sqrt{y} \cdot \left(\frac{3\sqrt{y}+1}{2} - \sqrt{y} \right) = y$. Therefore, I is a yes-instance.   ◇

Due to the previous case distinction, we may immediately return *yes* if C_i satisfies the two constraints stated in the claim. For the rest of the proof we may assume that this is not the case. Consequently, we have $|C_i| < \frac{3\sqrt{y}+1}{2}$ or $|C_i \cap A| < \sqrt{y}$. Consider the following cases.

Case 1: $|C_i| < \frac{3\sqrt{y}+1}{2}$. We can then compute the component information of the connected component C_i by iterating over all subsets $S \subseteq C_i$ and computing the number of vulnerable connections in $G[C_i] - S$. Since $|C_i| < \frac{3\sqrt{y}+1}{2}$, there are at most $2^{\frac{1}{2} \cdot (3\sqrt{y}+1)} \in 2^{\mathcal{O}(\sqrt{y})}$ subsets. Therefore, all component information $T[i, k']$ can be computed in $2^{\mathcal{O}(\sqrt{y} \log y)} \cdot n^{\mathcal{O}(1)}$ time.

Case 2: $|C_i \cap A| < \sqrt{y}$. Then, since $k \geq \sqrt{y}$, we have $T[i, k'] = 0$ for all $k' \geq |C_i \cap A|$ since one may remove all vulnerable vertices in C_i and afterwards, no vertex of C_i is part of a vulnerable connection anymore. It remains to compute component information $T[i, k']$ with $k' < \sqrt{y}$ by iterating over every $S \subseteq C_i$ of size k'. Since there are at most $|C_i|^{k'}$ such subsets, this can be done in $y^{\sqrt{y}} \cdot n^{\mathcal{O}(1)} = 2^{\sqrt{y} \log y} \cdot n^{\mathcal{O}(1)}$ time.

By the above argumentation, we can compute the component table T of I in $2^{\sqrt{y} \log y} \cdot n^{\mathcal{O}(1)}$. Together with Lemma 2, we conclude that CNP-V can be solved within the claimed running time.   □

Hardness for Non-Deletable Vulnerable Vertices. Now, we show that, in contrast to CNP-V, the CNP-NDV problem is W[1]-hard with respect to the parameter $k+y$, even if the input graph only contains one vulnerable vertex. We reduce from CLIQUE which has as input graph G and an integer ℓ, and asks whether G contains a set of ℓ vertices that are pairwise adjacent. It is well-known that CLIQUE is W[1]-hard with respect to ℓ [4,6].

The reduction follows the spirit of a reduction of Fomin et al. [8] that shows W[1]-hardness of the CUTTING AT MOST k VERTICES WITH TERMINAL problem. The reduction of Fomin et al. [8] already shows W[1]-hardness of CNP-NDV with respect to the parameter k, even if $|A| = 1$. We adapt the reduction to show hardness with respect to the larger parameter $k + y$.

Theorem 7 (⋆). *CNP-NDV is W[1]-hard with respect to the parameter $k+y$, even if $|A| = 1$ and the input graph has diameter 2.*

5   Parameters Related to the Vertex Cover Number

First, we obtain an FPT algorithm for the vertex cover number vc for the generalization CNP-VNDV of both problems via a combination of branching and dynamic programming.

Theorem 8. CNP-VNDV *can be solved in* $4^{\mathrm{vc}} \cdot n^{\mathcal{O}(1)}$ *time.*

Proof. Let (G, A, N, k, x) be an instance of CNP-VNDV. The first step of the algorithm is to compute a minimum vertex cover S of G. Then, we branch into all possible cases for $D := C \cap (S \setminus N)$. In other words, we consider all possible cases for vertex deletions in the vertex cover S. Consider one such possibility. Let $G' := G - D$ and let $k' := k - |D|$. Observe that $S' := S \setminus D$ is a vertex cover of G'. The question is now whether there is a set C' of at most k' vertices such that $G' - C'$ has A-vulnerability at most x and such that C' contains no vertices of N. To answer this question, we use dynamic programming over subsets of S'. More precisely, we fill a dynamic programming table T with entries of the type $T[S^*, k^*]$ where S^* is a subset of S' and $k^* \in \{1, \ldots, k'\}$. To define the meaning of a table entry, let $N_p(S^*)$, for $S^* \subseteq S'$ denote the neighbors of S^* that are not neighbors of $S' \setminus S^*$. That is, $N_p(S^*) := N(S^*) \setminus N(S' \setminus S^*)$.

A table entry $T[S^*, k^*]$ contains the minimum A-vulnerability of any graph that is obtained from $G'[S^* \cup N_p(S^*)]$ by deleting at most k^* vertices of $N_p(S^*) \setminus N$. The value of $T[S', k']$ then is the minimum A-vulnerability of any graph that can be obtained from G' by deleting at most k' vertices from $N(S') \setminus N$. If this number is smaller than x, then we have a yes-instance; otherwise, the CNP-VNDV instance has no critical node cut that contains D.

Informally, the recurrence to compute the value of $T[S^*, k^*]$ is to consider the possibilities of how one connected component created by the critical node cut may intersect with S^*. To simplify the description somewhat, we will define $T[S^*, k^*] = +\infty$ for all $k^* < 0$. The base cases of the recurrence are the A-vulnerability values that we get when $S^* \cup N_p(S^*)$ remains connected after the deletion of k^* vertices. More precisely, let $Q[S^*, k^*]$ contain the minimum A-vulnerability of any connected graph that is obtained from $G'[S^* \cup N_p(S^*)]$ by deleting at most k^* vertices of $N_p(S^*) \setminus N$. This value can be computed greedily by first deleting as many vertices of $(N_p(S^*) \setminus N) \cap A$ as possible and then deleting up to $k^* - |(N_p(S^*) \setminus N) \cap A|$ vertices of $(N_p(S^*) \setminus N) \setminus A$. Assuming that the values of $Q[S^*, k^*]$ have been precomputed, we may now compute $T[S^*, k^*]$ by the recurrence

$$T[S^*, k^*] = \min_{\tilde{S} \subseteq S^*} \min_{\tilde{k} \leq k^*} Q[\tilde{S}, \tilde{k}] + T[S^* \setminus \tilde{S}, k^* - \tilde{k} - \delta(\tilde{S}, S^* \setminus \tilde{S})]$$

where $\delta(\tilde{S}, S^* \setminus \tilde{S}) = |N(\tilde{S}) \cap N(S^* \setminus \tilde{S}) \cap N_p(S^*)|$ if

- $|N(\tilde{S}) \cap N(S^* \setminus \tilde{S}) \cap N_p(S^*)|$ contains no vertices of N, and
- there are no edges with one endpoint in $\tilde{S}$ and one endpoint in $S^* \setminus \tilde{S}$,

and $\delta(\tilde{S}, S^* \setminus \tilde{S}) = k + 1$, otherwise. That is, δ counts the number of vertex deletions that are necessary to disconnect $\tilde{S}^*$ and $S^* \setminus \tilde{S}$ in $G[S^* \cup N_p(S^*)]$ if it is possible to disconnect the two sets without deleting vertices in $N \cup S^*$. Otherwise, the value of δ is sufficiently large to ensure that the equation evaluates to ∞. . We omit a formal proof of the correctness and now bound the running time of the algorithm. A minimum vertex cover S can be computed in $\mathcal{O}(2^{\mathrm{vc}}(n + m))$ time using the standard search tree algorithm. Afterwards, we consider every subset D of S and fill the table T for the possibility where we delete exactly the vertex set D from S. Filling the table needs $3^{\mathrm{vc} - |D|} \cdot n^{\mathcal{O}(1)}$ time since each evaluated term corresponds to a 3-partition of $S \setminus D$. Thus, the overall running time is $\sum_{i=0}^{\mathrm{vc}} \binom{\mathrm{vc}}{i} \cdot 3^{\mathrm{vc} - i} \cdot n^{\mathcal{O}(1)}$. Using the binomial theorem, the overall running time for all possibilites of D is thus $4^{\mathrm{vc}} \cdot n^{\mathcal{O}(1)}$ time.   □

Next, we show that CNP-V has a polynomial-size kernel for the parameter vc $+ x$ and that CNP-NDV has a polynomial-size kernel for the parameter vc $+ k + x$. To this end, we first make a simple observation on k and the vertex cover number of the input graph. Let (G, A, k, x) be an instance of CNP-V or CNP-NDV. For the rest of the section, we fix a vertex set Z which is a 2-approximation of the minimum vertex cover of G, that is, $|Z| \leq 2 \cdot$ vc. Note that Z can be computed in linear time.

Consider CNP-V. Removing S from G results in an edgeless graph and therefore, there are no vulnerable connections in $G - S$. Thus, we may immediately return yes if k is at least as big as the size of Z.

Rule 5. *Let (G, A, k, x) be an instance of* CNP-V*. Return yes, if $k \geq 2 \cdot$ vc.*

Recall that we assume that the input instance of CNP-V is reduced with respect to Rules 1 and 4 and therefore we might assume that there are no isolated vertices and that $|N(v) \cap A| \leq k + \sqrt{2x}$ for every vertex v. In the following, we show that we can use these assumptions to bound the size of A in vc $+x$.

Lemma 4 ($\star$). *After Rules 1, 4, and 5 have been applied exhaustively, in an instance (G, A, k, x) of* CNP-V*, the set A contains less than $(2\,\mathrm{vc}) \cdot ((2\,\mathrm{vc}) + \sqrt{2x} + 1)$ vertices.*

Next, we define a subset B of the vertices. We provide two different definitions for CNP-V or CNP-NDV: For CNP-V, we define $B := A \cup Z$. For CNP-NDV, we define $B := Z$. We call B the *base*. We then have $|B| \leq 2 \cdot$ vc when we deal with an instance of CNP-NDV and by Lemma 4 we have $|B| \leq (2\,\mathrm{vc}) \cdot ((2\,\mathrm{vc}) + \sqrt{2x} + 2)$ when we deal with an instance of CNP-V. It remains to bound the size of the set $Y := V \setminus B$. Note that Y is an independent set because B contains a vertex cover. Moreover, Y does not contain isolated vertices since the instance is reduced with respect to Rule 1. In the following, we provide a reduction rule that in instances of CNP-NDV helps us to handle vulnerable vertices in the set Y. After the reduction rule has been applied exhaustively, if a vertex v has a neighborhood of size at least $k + x + 1$, all neighbors of v are non-vulnerable. This rule should only be applied on instances of CNP-NDV.

Rule 6. *Let* (G, A, k, x) *be an instance of* CNP-NDV *with base B. If a vertex* $v \in B$ *has more than* $k + x$ *neighbors of which one is vulnerable, then do the following*

1. If $v \notin A$, *then remove* v *from the graph and decrease* k *by one.*
2. If $v \in A$, *then return no.*

Lemma 5 (⋆). *For an instance of* CNP-NDV, *Rule 6 is safe and can be applied exhaustively in* $\mathcal{O}(n^2)$ *time.*

This reduction rule can only be applied on instances of CNP-NDV, because, if $v \notin A$, we know that we have to add v to a critical node cut. However, in CNP-V there remain three options: we can add the vulnerable vertex d, or the vertex v, or both to a critical node cut. Thus, in order to avoid such a decision for instances of CNP-V, we added all vulnerable vertices to the base B.

In the last reduction rule, we use the Expansion Lemma. The Expansion Lemma was introduced by Prieto-Rodríguez [11]. We use the formulation by Cygan et al. [4].

Lemma 6 (Expansion Lemma *[4]*)**.** *Let* H *be a bipartite graph with vertex bipartition* (R, T). *For a positive integer* q, *a set of edges* $M \subseteq E(H)$ *is called a* q-*expansion of* C *into* T, *if every vertex of* R *is incident with exactly* q *edges of* M *and the edges in* M *are incident with exactly* $q \cdot |R|$ *vertices in* T.

Let $q \geq 1$ *be a positive integer and* H *be a bipartite graph with vertex bipartition* (R, T) *such that* $|T| \geq q \cdot |R|$ *and there are no isolated vertices in* T. *Then, there exist nonempty vertex sets* $P \subseteq R$ *and* $Q \subseteq T$ *such that there is a* q-*expansion of* P *into* Q *and* $N_H(Q) \subseteq P$. *Furthermore, the sets* P *and* Q *can be found in time polynomial in the size of* H.

Since the Expansion Lemma can only be applied to bipartite graphs, in the next reduction rule we define a bipartite graph that is an induced subgraph of G. We apply the Expansion Lemma on the graph G' which contains the vertices $V' := V(G)$ and the set of edges $E' := E(G) \backslash E(G[B])$. This is a bipartite graph, because we do not consider the edges within B and, by definition, Y is an independent set. Thus, G' is a bipartite graph with vertex bipartition (B, Y).

Now, we assume that Rules 1 and 6 are exhaustively applied.

Rule 7. *If the set* Y *contains at least* $(k + x + 2) \cdot |B|$ *vertices, then, in the graph* G' *that we defined before this reduction rule, compute non-empty vertex sets* $P \subseteq B$ *and* $Q \subseteq Y$ *such that there is a* $k + x + 2$-*expansion of* P *into* Q. *Remove an arbitrary vertex* $v \in Q$ *from* G.

Lemma 7. *For an instance of* CNP-V *or* CNP-NDV, *Rule 7 is safe and can be applied exhaustively in polynomial time.*

Proof. Safeness: Let (G, A, k, x) be an instance of CNP-V or CNP-NDV with base B for which the inequality $|Y| \geq (k + x + 2) \cdot |B|$ is correct. Let G' be the graph defined before this reduction rule.

We start by showing that we can apply the Expansion Lemma. After Rule 1 has been applied exhaustively, all vertices in Y are adjacent to at least one vertex in B. Thus, all conditions for the Expansion Lemma are fulfilled. From the Expansion Lemma, we know that we can then find non-empty vertex sets $P \subseteq B$ and $Q \subseteq Y$ such that there is a $k + x + 2$-expansion of P into Q in polynomial time. Also, the sets fulfill $N_G(Q) \subseteq P$.

For the rest of the proof, let v be an arbitrary but fixed vertex of Q. We show that (G, A, k, x) is a yes-instance of CNP-V or CNP-NDV, if and only if $(G - \{v\}, A, k, x)$ is a yes-instance of the same problem. Observe that v is non-vulnerable: In an instance of CNP-V we defined $A \subseteq B$ and thus $A \cap Y = \emptyset$ and in particular $A \cap Q = \emptyset$. In an instance of CNP-NDV, after Rule 6 has been applied exhaustively, a vertex of B with a neighbor in $A \cap Y$ has at most $k + x$ neighbors. Thus, a described $k + x + 2$-expansion of P into Q cannot exist if $A \cap Q \neq \emptyset$.

Because $G - \{v\}$ is an induced subgraph of G, $C \setminus \{v\}$ is a critical node cut for $(G - \{v\}, A, k, x)$ if C is a critical node cut for (G, A, k, x).

Conversely, let $(G - \{v\}, A, k, x)$ be a yes-instance of CNP-V or CNP-NDV and let C be a corresponding critical node cut. From the Expansion Lemma we know $N(Q) \subseteq P$. In $(G - \{v\}) - C$ there is no vulnerable connection $\{d, u\}$ with $d \in A$ and $u \in P$: Otherwise, for all $w \in (N_G(u) \cap Q) \setminus (\{v\} \cup C)$ also $\{d, w\}$ is a vulnerable connection in $(G - \{v\}) - C$. By the definition of P and Q, the size of $(N_G(u) \cap Q)$ is at least $k + x + 1$ and thus $\{u\} \cup ((N_G(u) \cap Q) \setminus (\{v\} \cup C))$ contains more than x vertices. This is a contradiction to C being a critical node cut. By the same argument, the sets A and $P \setminus C$ are not connected in $(G - \{v\}) - C$. It follows that in $(G - \{v\}) - C$ the sets $P \setminus C$ and $Q \setminus C$ are in connected components that do not contain a vulnerable vertex. Since $N_G(v) \subseteq P$, the A-vulnerability of $(G - \{v\}) - C$ is the A-vulnerability of $G - C$ and C is also a critical node cut for (G, A, k, x).

Clearly, the rule can be performed in polynomial time.   □

It remains to give a bound on the size of the computed kernel.

Theorem 9 ($\star$). *An instance (G, A, k, x) of CNP-V or CNP-NDV contains less than $|B| \cdot (k + x + 3)$ vertices after Rules 1, 6, and 7 have been applied exhaustively.*

Since $|B| \leq |A| + 2 \cdot \text{vc}$ for CNP-V and due to Lemma 4, we obtain the following.

Corollary 4. *For an instance (G, A, k, x) of CNP-V, we can compute a kernelization with less than $((2\,\text{vc})((2\,\text{vc}) + \sqrt{2x} + 1) \cdot (k + x + 3)$ vertices in polynomial time.*

Since the instance is reduced regarding Rule 5, we obtain the following.

Corollary 5. *For an instance (G, A, k, x) of CNP-V, we can compute a kernelization with less than $((2\,\text{vc})((2\,\text{vc}) + \sqrt{2x} + 1) \cdot (2\,\text{vc} + x + 3)$ vertices in polynomial time.*

Since $|B| \leq 2 \cdot vc$ for CNP-NDV, we obtain the following.

Corollary 6. *For an instance* (G, A, k, x) *of* CNP-NDV, *we can compute a kernelization with less than* $2 \cdot vc \cdot (k + x + 3)$ *vertices in polynomial time.*

6   Conclusion

We introduced two new critical node detection problems problems CRITICAL NODE PROBLEM WITH VULNERABLE NODES (CNP-V) and CRITICAL NODE PROBLEM WITH NON-DELETABLE VULNERABLE NODES (CNP-NDV), that take into account that we are only interested in the number of connected pairs for a specified set of vulnerable vertices. We performed a parameterized complexity analysis for some of the most natural parameters and their combinations. We left open, however, the complexity of a number of natural parameterizations. For example, is CNP-V FPT with respect to $|A|$? At the moment we only have an XP-algorithm for A and an FPT algorithm for $|A| + x$. Moreover, does either problem admit a polynomial kernel for the vertex cover number vc?

References

1. Addis, B., Di Summa, M., Grosso, A.: Identifying critical nodes in undirected graphs: complexity results and polynomial algorithms for the case of bounded treewidth. Discret. Appl. Math. **161**(16–17), 2349–2360 (2013)
2. Agrawal, A., Lokshtanov, D., Mouawad, A.E.: Critical node cut parameterized by treewidth and solution size is W[1]-hard. In: Bodlaender, H.L., Woeginger, G.J. (eds.) Graph-Theoretic Concepts in Computer Science, WG 2017. LNCS, vol. 10520, pp. 32–44. Springer, Cham (2017). https://doi.org/10.1007/978-3-319-68705-6_3
3. Arulselvan, A., Commander, C.W., Elefteriadou, L., Pardalos, P.M.: Detecting critical nodes in sparse graphs. Comput. Oper. Res. **36**(7), 2193–2200 (2009)
4. Cygan, M., et al.: Parameterized Algorithms. Springer, Cham (2015). https://doi.org/10.1007/978-3-319-21275-3
5. Di Summa, M., Grosso, A., Locatelli, M.: Complexity of the critical node problem over trees. Comput. Oper. Res. **38**(12), 1766–1774 (2011)
6. Downey, R.G., Fellows, M.R.: Fundamentals of Parameterized Complexity. Texts in Computer Science, Springer, Cham (2013). https://doi.org/10.1007/978-1-4471-5559-1
7. Enright, J.A., Meeks, K.: Deleting edges to restrict the size of an epidemic: a new application for treewidth. Algorithmica **80**(6), 1857–1889 (2018). https://doi.org/10.1007/s00453-017-0311-7
8. Fomin, F.V., Golovach, P.A., Korhonen, J.H.: On the parameterized complexity of cutting a few vertices from a graph. In: Chatterjee, K., Sgall, J. (eds.) Mathematical Foundations of Computer Science 2013, MFCS 2013. LNCS, vol. 8087, pp. 421–432. Springer, Heidelberg (2013). https://doi.org/10.1007/978-3-642-40313-2_38
9. Hermelin, D., Kaspi, M., Komusiewicz, C., Navon, B.: Parameterized complexity of critical node cuts. Theor. Comput. Sci. **651**, 62–75 (2016)
10. Lalou, M., Tahraoui, M.A., Kheddouci, H.: The critical node detection problem in networks: a survey. Comput. Sci. Rev. **28**, 92–117 (2018)

11. Prieto-Rodríguez, E.: Systematic kernelization in FPT algorithm design. Ph.D. Thesis, The University of Newcastle (2005)
12. Schestag, J.: Critical Node Problem with Vulnerable Vertices. Master's Thesis, Philipps-Universität Marburg (2021)
13. Shen, Y., Nguyen, N.P., Xuan, Y., Thai, M.T.: On the discovery of critical links and nodes for assessing network vulnerability. IEEE/ACM Trans. Netw. **21**(3), 963–973 (2013)

Winner Determination Algorithms for Graph Games with Matching Structures

Kanae Yoshiwatari[1]([⊠]) [iD], Hironori Kiya[2]([⊠]) [iD], Tesshu Hanaka[2]([⊠]) [iD], and Hirotaka Ono[1]([⊠]) [iD]

[1] Nagoya University, Nagoya, Japan
yoshiwatari.kanae.w1@s.mail.nagoya-u.ac.jp, ono@nagoya-u.ac.jp
[2] Kyushu University, Fukuoka, Japan
h-kiya@econ.kyushu-u.ac.jp, hanaka@inf.kyushu-u.ac.jp

Abstract. Cram, Domineering, and Arc Kayles are well-studied combinatorial games. They are interpreted as edge-selecting-type games on graphs, and the selected edges during a game form a matching. In this paper, we define a generalized game called Colored Arc Kayles, which includes these games. Colored Arc Kayles is played on a graph whose edges are colored in black, white, or gray, and black (resp., white) edges can be selected only by the black (resp., white) player, although gray edges can be selected by both black and white players. We first observe that the winner determination for Colored Arc Kayles can be done in $O^*(2^n)$ time by a simple algorithm, where n is the order of a graph. We then focus on the vertex cover number, which is linearly related to the number of turns, and show that Colored Arc Kayles, BW-Arc Kayles, and Arc Kayles are solved in time $O^*(1.4143^{\tau^2+3.17\tau})$, $O^*(1.3161^{\tau^2+4\tau})$, and $O^*(1.1893^{\tau^2+6.34\tau})$, respectively, where τ is the vertex cover number. Furthermore, we present an $O^*((n/\nu+1)^\nu)$-time algorithm for Arc Kayles, where ν is neighborhood diversity. We finally show that Arc Kayles on trees can be solved in $O^*(2^{n/2})(= O(1.4143^n))$ time, which improves $O^*(3^{n/3})(= O(1.4423^n))$ by a direct adjustment of the analysis of Bodlaender et al.'s $O^*(3^{n/3})$-time algorithm for Node Kayles.

Keywords: Arc Kayles · Combinatorial game theory · Exact exponential-time algorithm · Vertex cover · Neighborhood diversity

1 Introduction

1.1 Background and Motivation

Cram, Domineering, and Arc Kayles are well-studied two-player mathematical games and interpreted as combinatorial games on graphs. Domineering (also

This work is partially supported by JSPS KAKENHI JP17H01698, JP17K19960, JP19K21537, JP20H05967, JP21H05852, JP21K17707, JP21K19765, JP21K21283, JP22H00513.

C. Bazgan and H. Fernau (Eds.): IWOCA 2022, LNCS 13270, pp. 509–522, 2022.
https://doi.org/10.1007/978-3-031-06678-8_37

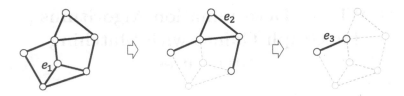

Fig. 1. A play example of Arc Kayles

called Stop-Gate) was introduced by Göran Andersson around 1973 under the name of Crosscram [6,8]. Domineering is usually played on a checkerboard. The two players are denoted by Vertical and Horizontal. Vertical (resp., Horizontal) player is only allowed to place its dominoes vertically (resp., horizontally) on the board. Note that placed dominoes are not allowed to overlap. If no place is left to place a domino, the player in the turn loses the game. Domineering is a partisan game, where players use different pieces. The impartial version of the game is Cram, where two players can place dominoes both vertically and horizontally.

An analogous game played on an undirected graph G is Arc Kayles. In Arc Kayles, the action of a player in a turn is to select an edge of G, and then the selected edge and its neighboring edges are removed from G. If no edge remains in the resulting graph, the player in the turn loses the game. Figure 1 is a play example of Arc Kayles. In this example, the first player selects edge e_1, and then the second player selects edge e_2. By the first player selecting edge e_3, no edge is left; the second player loses. Note that the edges selected throughout a play form a maximal matching on the graph.

Similarly, we can define BW-Arc Kayles, which is played on an undirected graph with black and white edges. The rule is the same as the ordinary Arc Kayles except that the black (resp., white) player can select only black (resp., white) edges. Note that Cram and Domineering are respectively interpreted as Arc Kayles and BW-Arc Kayles on a two-dimensional grid graph, which is the graph Cartesian product of two path graphs.

To focus on the common nature of such games with matching structures, we newly define Colored Arc Kayles. Colored Arc Kayles is played on a graph whose edges are colored in black, white, or gray, and black (resp., white) edges can be selected only by the black (resp., white) player, though grey edges can be selected by both black and white players. BW-Arc Kayles and ordinary Arc Kayles are special cases of Colored Arc Kayles. In this paper, we investigate Colored Arc Kayles from the algorithmic point of view.

1.2   Related Work

Cram and Domineering. Cram and Domineering are well studied in the field of combinatorial game theory. In [8], Gardner gives winning strategies for some simple cases. For Cram on $a \times b$ board, the second player can always win if both a and b are even, and the first player can always win if one of a and b is even and the other is odd. This can be easily shown by the so-called Tweedledum and Tweedledee strategy. For specific sizes of boards, computational studies have been conducted [17]. In [16], Cram's endgame databases for all board sizes with at most 30 squares are constructed. As far as the authors know, the complexity to determine the winner for Cram on general boards still remains open.

Finding the winning strategies of Domineering for specific sizes of boards by using computer programs is well studied. For example, the cases of 8×8 and 10×10 are solved in 2000 [3] and 2002 [4], respectively. The first player wins in both cases. Currently, the status of boards up to 11×11 is known [15]. In [18], endgame databases for all single-component positions up to 15 squares for Domineering are constructed. The complexity of Domineering on general boards also remains open. Lachmann, Moore, and Rapaport show that the winner and a winning strategy Domineering on $m \times n$ board can be computed in polynomial time for $m \in \{1, 2, 3, 4, 5, 7, 9, 11\}$ and all n [11].

Kayles, Node Kayles, and Arc Kayles. Kayles is a simple impartial game, introduced by Henry Dudeney in 1908 [7]. The name "Kayles" derives from French word "quilles", meaning "bowling". The rule of Kayles is as follows. Given bowling pins equally spaced in a line, players take turns to knock out either one pin or two adjacent pins, until all the pins are gone. As graph generalizations, Node Kayles and Arc Kayles are introduced by Schaefer [14]. Node Kayles is the vertex version of Arc Kayles. Namely, the action of a player is to select a vertex instead of an edge, and then the selected vertex and its neighboring vertices are removed. Note that both generalizations can describe the original Kayles; Kayles is represented as Node Kayles on sequentially linked triangles or as Arc Kayles on a caterpillar graph.

Node Kayles is known to be PSPACE-complete [14], whereas the winner determination is solvable in polynomial time on graphs of bounded asteroidal numbers such as cocomparability graphs and cographs by using Sprague-Grundy theory [1]. For general graphs, Bodlaender et al. propose an $O(1.6031^n)$-time algorithm [2]. Furthermore, they show that the winner of Node Kayles can be determined in time $O(1.4423^n)$ on trees. In [10], Kobayashi sophisticates the analysis of the algorithm in [2] from the perspective of the parameterized complexity and shows that it can be solved in time $O^*(1.6031^\mu)$, where μ is the modular width of an input graph[1]. He also gives an $O^*(3^\tau)$-time algorithm, where τ is the vertex cover number, and a linear kernel when parameterized by neighborhood diversity.

Different from Node Kayles, the complexity of Arc Kayles has remained open for more than 30 years. Even for subclasses of trees, not much is known. For example, Huggans and Stevens study Arc-Kayles on subdivided stars with three paths [9]. To our best knowledge, no exponential-time algorithm for Arc Kayles is presented except for an $O^*(4^{\tau^2})$-time algorithm proposed in [13].

---

[1] The $O^*(\cdot)$ notation suppresses polynomial factors in the input size.

1.3 Our Contribution

In this paper, we address winner determination algorithms for Colored Arc Kayles. We first propose an $O^*(2^n)$-time algorithm for Colored Arc Kayles. Note that this is generally faster than applying the Node Kayles algorithm to the line graph of an instance of Arc Kayles; it takes time $O(1.6031^m)$, where m is the number of the original edges. We then focus on algorithms based on graph parameters. We present an $O^*(1.4143^{\tau^2 + 3.17\tau})$-time algorithm for Colored Arc Kayles, where τ is the vertex cover number. The algorithm runs in time $O^*(1.3161^{\tau^2 + 4\tau})$ and $O^*(1.1893^{\tau^2 + 6.34\tau})$ for BW-Arc Kayles, and Arc Kayles, respectively. This is faster than the previously known time complexity $O^*(4^{\tau^2})$ in [13].

On the other hand, we give a bad instance for the proposed algorithm, which implies the running time analysis is asymptotically tight. Furthermore, we show that the winner of Arc Kayles can be determined in time $O^*((n/\nu + 1)^\nu)$, where ν is the neighborhood diversity of an input graph. This analysis is also asymptotically tight, because there is an instance having $(n/\nu + 1 - o(1))^{\nu(1-o(1))}$. We finally show that the winner determination of Arc Kayles on trees can be solved in $O^*(2^{n/2}) = O(1.4143^n)$ time, which improves $O^*(3^{n/3})(= O(1.4423^n))$ by a direct adjustment of the analysis of Bodlaender et al.'s $O^*(3^{n/3})$-time algorithm for Node Kayles.

2 Preliminaries

2.1 Notations and Terminology

Let $G = (V, E)$ be an undirected graph. We denote $n = |V|$ and $m = |E|$, respectively. For an edge $e = \{u, v\} \in E$, we define $\Gamma(e) = \{e' \mid e \cap e' \neq \emptyset\}$. For a graph $G = (V, E)$ and a vertex subset $V' \subseteq V$, we denote by $G[V']$ the subgraph induced by V'. For simplicity, we denote $G - v$ instead of $G[V \setminus \{v\}]$. For an edge subset E', we also denote by $G - E'$ the subgraph obtained from G by removing all edges in E' from G. A vertex set S is called a *vertex cover* if $e \cap S \neq \emptyset$ for every edge $e \in E$. We denote by τ the size of a minimum vertex cover of G. Two vertices $u, v \in V$ are called *twins* if $N(u) \setminus \{v\} = N(v) \setminus \{u\}$.

Definition 1. *The* neighborhood diversity $\nu(G)$ *of* $G = (V, E)$ *is defined as the minimum number w such that V can be partitioned into w vertex sets of twins.*

In the following, we simply write ν instead of $\nu(G)$ if no confusion arises. We can compute the neighborhood diversity of G and the corresponding partition in polynomial time [12]. For any graph G, $\nu \leq 2^\tau + \tau$ holds.

2.2 Colored Arc Kayles

Colored Arc Kayles is played on a graph $G = (V, E_G \cup E_B \cup E_W)$, where E_G, E_B, E_W are mutually disjoint. The subscripts G, B, and W of E_G, E_B, E_W respectively, stand for gray, black, and white. For every edge $e \in E_G \cup E_B \cup E_W$, let $c(e)$ be the color of e, that is, $c(e) = $ G if $e \in E_G$, B if $e \in E_B$, and W if $e \in E_W$. If $\{u, v\} \notin E_G \cup E_B \cup E_W$, we set $c(\{u, v\}) = \emptyset$ for convenience. As explained below, the first (black or B) player can choose only gray or black edges, and the second (white or W) player can choose only gray or white edges.

Two players alternatively choose an edge of G. Player B can choose an edge in $E_G \cup E_B$ and player W can choose an edge in $E_G \cup E_W$. That is, there are three types of edges; E_B is the set of edges that only the first player can choose, E_W is the set of edges that only the second player can choose, and E_G is the set of edges that both the first and second players can choose. Once an edge e is selected, the edge and its neighboring edges (i.e., $\Gamma(e)$) are removed from the graph, and the next player chooses an edge of $G - \Gamma(e)$. The player that can take no edge loses the game. Since (Colored) Arc Kayles is a two-person zero-sum perfect information game and ties are impossible, one of the players always has a winning strategy. We call the player having a winning strategy the *definite winner*, or simply *winner*.

The problem that we consider in this paper is defined as follows:

Input: $G = (V, E_G \cup E_B \cup E_W)$, active player in $\{B, W\}$.
Question: Suppose that players B and W play Colored Arc Kayles on G from the active player's turn. Which player is the winner?

Remark that if $E_B = E_W = \emptyset$, Colored Arc Kayles is equivalent to Arc Kayles and if $E_G = \emptyset$, it is equivalent to BW-Arc Kayles.

To simply represent the definite winner of Colored Arc Kayles, we introduce two Boolean functions f_B and f_W. The $f_B(G)$ is defined such that $f_B(G) = 1$ if and only if the winner of Colored Arc Kayles on G from player B's turn is player B. Similarly, $f_W(G)$ is the function such that $f_W(G) = 1$ if and only if the winner of Colored Arc Kayles on G from player W's turn is the player W. If two graphs G and G' satisfy that $f_B(G) = f_B(G')$ and $f_W(G) = f_W(G')$, we say that G and G' have the same game value on Colored Arc Kayles.

3   Basic Algorithm

In this section, we show that the winner of *Colored* Arc Kayles on G can be determined in time $O^*(2^n)$. We first observe that the following lemma holds by the definition of the game.

Lemma 1. *Suppose that Colored Arc Kayles is played on $G = (V, E_G \cup E_W \cup E_B)$. Then, player B (resp., W) wins on G with player B's (resp., W's) turn if and only if there is an edge $\{u, v\} \in E_G \cup E_B$ (resp., $\{u, v\} \in E_G \cup E_W$) such that player W (resp., B) loses on $G - u - v$ with player B's (resp., W's) turn.*

This lemma is interpreted by the following two recursive formulas:

$$f_B(G) = \bigvee_{\{u,v\} \in E_G \cup E_B} \neg \left(f_W(G - u - v) \right), \tag{1}$$

$$f_W(G) = \bigvee_{\{u,v\} \in E_G \cup E_W} \neg \left(f_B(G - u - v) \right). \tag{2}$$

By these formulas, we can determine the winner of G with either first or second player's turn by computing $f_B(G)$ and $f_W(G)$ for all induced subgraphs of G. Since the number of all induced subgraphs of G is 2^n, it can be done in time $O^*(2^n)$ by a standard dynamic programming algorithm.

Theorem 1. *The winner of Colored Arc Kayles can be determined in time* $O^*(2^n)$.

4   FPT Algorithm Parameterized by Vertex Cover

In this section, we propose winner determination algorithms for Colored Arc Kayles parameterized by the vertex cover number. As mentioned in Introduction, the selected edges in a play of Colored Arc Kayles form a matching. This implies that the number of turns is bounded above by the maximum matching size of G and thus by the vertex cover number. Furthermore, the vertex cover number of the input graph is bounded by twice of the number r of turns of Arc Kayles. Intuitively, we may consider that a game taking longer turns is harder to analyze than games taking shorter turns. In that sense, the parameterization by the vertex cover number is quite natural.

In this section, we propose an $O^*(1.4143^{\tau^2+3.17\tau})$-time algorithm for Colored Arc Kayles, where τ is the vertex cover number of the input graph. It utilizes similar recursive relations shown in the previous section, but we avoid to enumerate all possible positions by utilizing equivalence classification.

Before explaining the equivalence classification, we give a simple observation based on isomorphism. The isomorphism on edge-colored graphs is defined as follows.

Definition 2. *Let* $G = (V, E)$ *and* $G' = (V', E')$ *be edge-colored graphs where* $E = \bigcup_{i=1}^r E_i$ *and* $E' = \bigcup_{i=1}^r E_i'$. *Then* G *and* G' *are called* isomorphic *if for any pair of* $u, v \in V$ *there is a bijection* $f : V \to V'$ *such that* $\{u, v\} \in E_i$ *if and only if* $\{f(u), f(v)\} \in E_i'$.

The following proposition is obvious.

Proposition 1. *If edge-colored graphs* G *and* G' *are isomorphic,* G *and* G' *have the same game value for Colored Arc Kayles.*

Let S be a vertex cover of $G = (V, E_G \cup E_W \cup E_B)$, that is, any $e = \{u, v\} \in E_G \cup E_W \cup E_B$ satisfies that $\{u, v\} \cap S \neq \emptyset$. Note that for $v \in V \setminus S$, $N(v) \subseteq S$ holds. We say that two vertices $v, v' \in V \setminus S$ are *equivalent with respect to* S *in* G if $N(v) = N(v')$ and $c(\{u, v\}) = c(\{u, v'\})$ holds for $\forall u \in N(v)$. If two vertices $v, v' \in V \setminus S$ are equivalent with respect to S in G, $G - u - v$ and $G - u - v'$ are isomorphic because the bijective function swapping only v and v' satisfies the isomorphic condition. Thus, we have the following lemma.

Lemma 2. *Suppose that two vertices* $v, v' \in V \setminus S$ *are equivalent with respect to* S *in* G. *Then, for any* $u \in N(v)$, $G - u - v$ *and* $G - u - v'$ *have the same game value.*

By the equivalence with respect to S, we can split $V \setminus S$ into equivalence classes. Note here that the number of equivalence classes is at most $4^{|S|}$, because for each $u \in S$ and $v \in V \setminus S$, edge $\{u, v\}$ does not exist, or it can be colored with one of three colors if exists; we can identify an equivalent class with $x \in \{\emptyset, G, B, W\}^S$, a 4-ary vector with length $|S|$. For $S' \subseteq S$, let $x[S']$ denotes the vector by dropping the components of x except the ones corresponding to S'. Also for $u \in S$, $x[u]$ denotes the component corresponding to u in x. Then, V is partitioned into $V_S^{(x)}$'s, where $V_S^{(x)} = \{v \in V \setminus S \mid \forall u \in S : c(\{v, u\}) = x[u]\}$. We arbitrarily define the representative of non-empty $V_S^{(x)}$ (e.g., the vertex with the smallest ID), which is denoted by $\rho(V_S^{(x)})$. By using ρ, we also define the representative edge set by

$$E^R(S) = \bigcup_{x \in \{\emptyset, G, B, W\}^S} \{\{u, \rho(V_S^{(x)})\} \in E_G \cup E_B \cup E_W \mid u \in S\}.$$

By Lemma 2, we can assume that both players choose an edge only in $E^R(S)$, which enables to modify the recursive equations (1) and (2) as follows: For a vertex cover S of G, we have

$$f_B(G) = \bigvee_{\{u,v\} \in (E_G \cup E_B) \cap (E^R(S) \cup S \times S)} \neg (f_W(G - u - v))), \tag{3}$$

$$f_W(G) = \bigvee_{\{u,v\} \in (E_G \cup E_W) \cap (E^R(S) \cup S \times S)} \neg (f_B(G - u - v))). \tag{4}$$

Note that this recursive formulas imply that the winner of Colored Arc Kayles can be determined in time $O^*((\tau^2 + \tau \cdot 4^\tau)^\tau) = O^*((4^{\tau + \log_4 \tau})^\tau) = O^*(4^{\tau^2 + \tau \log_4 \tau}) = O^*(5.6569^{\tau^2})$, because the recursions are called at most $|S|$ times and $\tau + \log_4 \tau \le 1.25\tau$ for $\tau \ge 1$.

In the following, we give a better estimation of the number of induced subgraphs appearing in the recursion. Once such subgraphs are listed up, we can apply a standard dynamic programming to decide the necessary function values, or we can compute f_B and f_W according to the recursive formulas with memorization, by which we can skip redundant recursive calls. In order to estimate the number of induced subgraphs appearing in the recursion, we focus on the fact that the position of a play in progress corresponds to the subgraph induced by a matching.

Lemma 3. *The number of nodes in recursion trees of equations (3) and (4) for Colored Arc Kayles is $O((r + 1)^{|S|^2/4} 3^{|S|} |S|^2)$, where r is the used colors.*

Proof. Suppose that S is a vertex cover of G and players play Colored Arc Kayles on G. At some point, some edges selected by players together with their neighboring edges are removed, and the left subgraph represents a game position. Note that at least one endpoint of such a selected edge is in S, and selected edges form a matching. To define such a subgraph, let us imagine that some M is the set of edges that have been selected until the point. Although we do not specify M, the M defines a partition (X, Y, Z) of S; $X = \{\{u, v\} \in M \mid |\{u, v\} \cap S| = 2\}$, $Y = \{\{u, v\} \in M \mid |\{u, v\} \cap S| = 1\}$, and $Z = \{\{u, v\} \in M \mid |\{u, v\} \cap S| = 0\}$. We now count the number of positions having a common (X, Y, Z). Since X and Y are removed, the remaining vertices in $V \setminus S$ are classified into $V_Z^{(x)}$'s $x \in \{\emptyset, G, B, W\}^{|Z|}$. This is the common structure defined by (X, Y, Z), and the positions vary as $|Y|$ vertices in $\bigcup_{x \in \{\emptyset, G, B, W\}^{|Z|}} V_Z^{(x)}$ are matched with Y. Thus, we estimate the number of positions by counting the number of choices of $|Y|$ vertices in $\bigcup_{x \in \{\emptyset, G, B, W\}^{|Z|}} V_Z^{(x)}$. Here, let r be the number of used colors of edges. For example, BW-Arc Kayles and ordinary Arc Kayles use $r = 2$ and $r = 1$ colors, respectively, and which may reduce the numbers of x's for smaller r. Then, it is above bounded by the multiset coefficient of $(r+1)^{|Z|}$ multichoose $|Y|$,

$$\left(\!\!\binom{(r+1)^{|Z|}}{|Y|}\!\!\right) = \binom{(r+1)^{|Z|} + |Y| - 1}{|Y|} \le (r+1)^{|Z||Y|}.$$

This is an upper bound of the number of subgraphs to consider with respect to (X, Y, Z). By considering all possible (X, Y, Z), the total number of subgraphs is bounded by

$$\sum_{X, Y, Z \subseteq S} (r+1)^{|Z||Y|} \le 3^{|S|} \cdot |S| \sum_{y=0}^{|S|} (r+1)^{y(|S|-y)} \le 3^{|S|} \cdot |S|^2 \cdot (r+1)^{\frac{|S|^2}{4}},$$

where the first inequality comes from the choices of X, Y, and Z from S.   □

The following theorem immediately holds by Lemma 3 and the fact that a minimum vertex cover of G can be found in time $O^*(1.2738^\tau)$, where τ is the vertex cover number of G [5].

Theorem 2. *The winners of Colored Arc Kayles, BW-Arc Kayles, and Arc Kayles can be determined in time* $O^*(1.4143^{\tau^2+3.17\tau})$, $O^*(1.3161^{\tau^2+4\tau})$, *and* $O^*(1.1893^{\tau^2+6.34\tau})$, *respectively, where* τ *is the vertex cover number of a graph.*

We have shown that the winner of Arc Kayles can be determined in time $O^*(1.1893^{\tau^2+6.34\tau})$. The following theorem shows that the analysis is asymptotically tight, which implies that for further improvement, we need additional techniques apart from ignoring vertex-cover-based isomorphic positions. We here give such an example in Fig. 2.

Theorem 3. *There is a graph for which the algorithm requires* $2^{\tau^2/2}$ *recursive calls for Colored Arc Kayles.*

Proof. We explain how we systematically construct such a graph G (see Fig. 2). Let k be an even number. We first define $U = \{u_1, \ldots, u_{k/2}\}$ and $V = \{v_1, \ldots, v_{k/2}\}$ as vertex sets. The union $U \cup V$ will form a vertex cover after the graph G are constructed. For every 4-ary vector $x \in \{\emptyset, G, B, W\}^U$ and every $i \in \{1, \ldots, k/2\}$, we define $x_{i,x}$ as a vertex, and let X be the collections of $x_{i,x}$'s, i.e.,

$X = \{x_{i,x} \mid x \in \{\emptyset, G, B, W\}^U, i \in \{1, \ldots, k/2\}\}$. These are the vertices of G. We next define the set of edges of G. We connect v_i and $x_{i,x}$'s for each i by edges with no-color, i.e., $E_G = \{\{v_i, x_{i,S(x)}\} \mid x \in \{\emptyset, G, B, W\}^U, i \in \{1, \ldots, k/2\}\}$. Furthermore, we connect $x_{i,x}$ and $u \in U$ for each x so that $\{u, x_{i,x}\}$ has color $x[u]$; if $x[u] = \emptyset$, no edge exists between $x_{i,x}$ and $u \in U$. Figure 2 shows an example how we connect $x_{1,x}$ and u_i's, where $x[u_1] = \emptyset$, $x[u_2] = G$, $x[u_i] = W$, and $x[u_{k/2}] = B$. Notice that in Fig. 2, $\{x_{1,x}, u_1\}$ is connected with edge $\emptyset$ for explanation, which means that there is no edge between $x_{1,x}$ and u_1. Note that the number of vertices in G is $|U| + |V| + |X| = k + 4^{k/2}k/2$. Moreover, U, V, and X form independent sets and X separates U and V. Thus, $U \cup V$ forms a vertex cover of size k in G.

We are ready to explain that G has different $2^{\tau^2/2}$ subgraphs called by the algorithm. Starting from G, we call the recursive formulas (3) and (4) $k/2$ times by selecting edges incident to only v_i's. Then, all the vertices in V are removed from G, and the neighbors of remaining $x_{i,x}$'s are in U. That is, each $x_{i,x}$ has its inherent set of neighbors, and thus $x_{i,x}$'s are not equivalent each other for vertex cover $U \cup V$. This implies that if the set of removed edges are different, the resulting subgraphs are also different.

In a step before $k/2 + 1$, an edge connecting some v_i is removed, and such an edge is chosen from $\{\{v_i, x_{i,x}\} \mid x \in \{\emptyset, G, B, W\}^U\}$, that is, the number of candidates is $4^{k/2}$ for each i. Thus, the total way to choose edges is $(4^{k/2})^{k/2} = 4^{k^2/4} = 2^{k^2/2}$; at least $2^{k^2/2}$ recursion calls occur. □

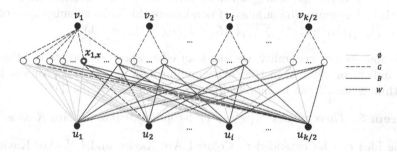

Fig. 2. The constructed graph $G = (U \cup V \cup X, E)$.

By the similar construction, we can show the following theorem.

Theorem 4. *There is a graph for which the algorithm requires* 1.3161^{τ^2} *and* 1.1893^{τ^2} *recursive calls for BW-Arc Kayles and Arc Kayles, respectively.*

Remark 1. Although Theorems 3 and 4 give lower bounds on the running time of the vertex cover-based algorithms, the proof implies a stronger result. In the proof of Theorem 3, we use ID's of the vertices in U. By connecting $2i$ pendant vertices to u_i, we can regard them as ID of u_i. Furthermore, we make U a clique by adding edges. These make the graphs not automorphic, which implies that the time complexity of an algorithm utilizing only isomorphism is at least the value shown in Theorems 3 or 4.

5   XP Algorithm Parameterized by Neighborhood Diversity

In this section, we deal with neighborhood diversity ν, which is a more general parameter than vertex cover number. We first give an $O^*((n/\nu+1)^\nu)$-time algorithm for Arc Kayles. This is an XP algorithm parameterized by neighborhood diversity. On the other hand, we show that there is a graph having at least $(n/\nu + 1 - o(1))^{\nu(1-o(1))}$ non-isomorphic induced subgraphs, which implies the analysis of the proposed algorithm is asymptotically tight.

By Proposition 1, if we list up all non-isomorphic induced subgraphs, the winner of Arc Kayles can be determined by using recursive formulas (1) and (1). Let $\mathcal{M} = \{M_1, M_2, \ldots, M_\nu\}$ be a partition such that $\bigcup_i M_i = V$ and vertices of M_i are twins each other. We call each M_i a *module*. We can see that non-isomorphic induced subgraphs of G are identified by how many vertices are selected from which module.

Lemma 4. *The number of non-isomorphic induced subgraphs of a graph of neighborhood diversity ν is at most $(n/\nu + 1)^\nu$.*

Proof. By the definition of neighborhood diversity, vertices in a module are twins each other. Therefore, the number of non-isomorphic induced subgraphs of G is at most $\prod_{i=1}^{\nu}(|M_i| + 1) \leq (\sum_{i=1}^{\nu}(|M_i| + 1)/\nu)^\nu \leq (n/\nu + 1)^\nu$.   □

Without loss of generality, we select an edge whose endpoints are the minimum indices of vertices in the corresponding module. By Proposition 1, the algorithm in Sect. 3 can be modified to run in time $O^*((n/\nu + 1)^\nu)$.

Theorem 5. *There is an $O^*((n/\nu + 1)^\nu)$-time algorithm for Arc Kayles.*

The idea can be extended to Colored Arc Kayles and BW-Arc Kayles. In $G = (V, E_G \cup E_B \cup E_W)$, two vertices $u, v \in V$ are called *colored twins* if $c(\{u, w\}) = c(\{v, w\})$ holds $\forall w \in V \setminus \{u, v\}$. We then define the notion of colored neighborhood diversity.

Definition 3. *The* colored neighborhood diversity *of $G = (V, E)$ is defined as minimum ν' such that V can be partitioned into ν' vertex sets of colored twins.*

In Colored Arc Kayles or BW-Arc Kayles, we can utilize a partition of V into modules each of which consists of colored twins. If we are given a partition of the vertices into colored modules, we can decide the winner of Colored Arc Kayles or BW-Arc Kayles like Theorem 5. Different from ordinary neighborhood diversity, it might be hard to compute colored neighborhood diversity in polynomial time.

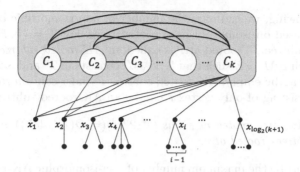

Fig. 3. The constructed graph G with neighborhood diversity $\nu = k + 2\log_2(k + 1)$.

Theorem 6. *Given a graph $G = (V, E_G \cup E_B \cup E_W)$ with a partition of V into ν' modules of colored twins, we can compute the winner of Colored Arc Kayles on G in time $O^*((n/\nu' + 1)^{\nu'})$.*

In the rest of this section, we give a bad instance for the proposed algorithm as shown in Fig. 3, although the detailed proof is omitted. The result implies that the analysis of Theorem 5 is asymptotically tight.

Theorem 7. *There is a graph having at least $(n/\nu + 1 - o(1))^{\nu(1-o(1))}$ non-isomorphic positions of Arc Kayles.*

6   Arc Kayles for Trees

In [2], Bodlaender et al. show that the winner of Node Kayles on trees can be determined in time $O^*(3^{n/3}) = O(1.4423^n)$. It is easy to show by a similar argument that the winner of Arc Kayles can also be determined in time $O(1.4423^n)$. It is also mentioned that the analysis is sharp apart from a polynomial factor because there is a tree for which the algorithm takes $\Omega(3^{n/3})$ time. The example is also available for Arc Kayles; namely, as long as we use the same algorithm, the running time cannot be improved.

In this section, we present that the winners of Arc Kayles on trees can be determined in time $O^*(2^{n/2}) = O(1.4143^n)$, which is attained by considering a tree (so-called) unordered. Since a similar analysis can be applied to Node Kayles on trees, the winner of Node Kayles on trees can be determined in time $O^*(2^{n/2})$. We omit the proof for Node Kayles to avoid repetition.

Let us consider a tree $T = (V, E)$. By Sprague-Grundy theory, if all connected subtrees of T are enumerated, one can determine the winner of Arc Kayles. Furthermore, by Proposition 1, once a connected subtree T' is listed, we can ignore subtrees isomorphic to T'. Here we adopt isomorphism of rooted trees.

Definition 4. *Let $T = (V, E, r)$ and $T' = (V', E', r')$ be trees rooted at r and r', respectively. Then, T and T' are called isomorphic if for any pair of $u, v \in V$ there is a bijection $f : V \to V'$ such that $\{u, v\} \in E_i$ if and only if $\{f(u), f(v)\} \in E'_i$ and $f(r) = f(r')$.*

In the following, we estimate the number of non-isomorphic connected subgraphs of T based on isomorphism of rooted trees. For $T = (V, E)$ rooted at r, a connected subtree T' rooted at r is called an *AK-rooted subtree* of T, if there exists a matching $M \subseteq E$ such that $T[V \setminus M]$ consists of T' and isolated vertices. Note that M can be empty, AK-rooted subtree T' must contain root r of T, and the graph consisting of only vertex r can be an AK-rooted subtree.

Lemma 5. *Any tree rooted at r has $O^*(2^{n/2})(= O(1.4143^n))$ non-isomorphic AK-rooted subtrees rooted at r.*

Proof. Let $R(n)$ be the maximum number of non-isomorphic AK-rooted subtrees of any tree rooted at some r with n vertices. We claim that $R(n) \leq 2^{n/2} - 1$ for all $n \geq 4$, which proves the lemma.

We will prove the claim by induction. For $n \leq 4$, the values of $R(n)$'s are as follows: $R(1) = 1, R(2) = 1, R(3) = 2$, and $R(4) = 3$. These can be shown by concretely enumerating trees. For example, for $n = 2$, a tree T with 2 vertices is unique, and an AK-rooted subtree of T containing r is also unique, which is T itself. For $n = 3$, the candidates of T are shown in Fig. 4. For Type A in Fig. 4, AK-rooted subtrees are the tree itself and isolated r, and for Type B, an AK-rooted subtree is only the tree itself; thus we have $R(3) = 2$. Similarly, we can show $R(4) = 3$ as seen in Fig. 5. Note that $R(1) > 2^{1/2} - 1$, $R(2) = 1 \leq 2^{2/2} - 1 = 1$, $R(3) = 2 > 2^{3/2} - 1$, and $R(4) = 3 \leq 2^{4/2} - 1 = 3$. This $R(4)$ is used as the base case of induction.

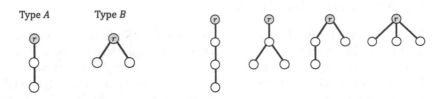

Fig. 4. Trees with 3 vertices rooted at r **Fig. 5.** Trees with 4 vertices rooted at r

As the induction hypothesis, let us assume that the claim is true for all $n' < n$ except 1 and 3, and consider a tree T rooted at r on n vertices. Let $u_1, u_2, \ldots, u_p$ be the children of root r, and T_i be the subtree of T rooted at u_i with n_i vertices for $i = 1, 2, \ldots, p$. Note that for an AK-rooted subtree T' of T, the intersection of T' and T_i for each i is either empty or an AK-rooted subtree of T_i rooted at u_i. Based on this observation, we take a combination of the number of AK-rooted subtrees of T_i's, which gives an upper bound on the number of AK-rooted subtrees of T. We consider two cases: (1) for any i, $n_i \neq 3$, (2) otherwise. For case (1), the number of AK-rooted subtrees of T is at most

$$\prod_{i:n_i>1} (R(n_i) + 1) \cdot \prod_{i:n_i=1} 1 \leq \prod_{i:n_i>1} 2^{n_i/2} = 2^{\sum_{i:n_i>1} n_i/2} \leq 2^{(n-1)/2} \leq 2^{n/2} - 1.$$

That is, the claim holds in this case. Here, in the left hand of the first inequality, $R(n_i)+1$ represents the choice of AK-rooted subtree of T_i rooted at u_i or empty, and "1" for i with $n_i = 1$ represents that u_i needs to be left as is, because otherwise edge $\{r, u_i\}$ must be removed, which violates the condition "rooted at r". The first inequality holds since any n_i is not 3 and thus the induction hypothesis can be applied. The last inequality holds by $n \geq 5$.

For case (2), we further divide into two cases: (2.i), for every i such that $n_i = 3$, T_i is Type B, and (2.ii) otherwise. For case (2.i), since a AK-rooted subgraph of T_i of Type B in Fig. 4 is only T_i itself, the number is $1 \leq 2^{3/2} - 1$. Thus, the similar analysis of Case (1) can be applied as follows:

$$\prod_{i:n_i \neq 1,3} (R(n_i) + 1) \cdot \prod_{i:\text{Type B} T_i} (2^{3/2} - 1 + 1) \leq \prod_{i:n_i > 1} 2^{n_i/2} \leq 2^{n/2} - 1,$$

that is, the claim holds also in case (2.i).

Finally, we consider case (2.ii). By the assumption, at least one T_i is Type A in Fig. 4. Suppose that T has q children of r forming Type A, which are renamed $T_1, \ldots, T_q$ as canonicalization. Such renaming is allowed because we count non-isomorphic subtrees. Furthermore, we can sort AK-rooted subtrees of $T_1, \ldots, T_q$ as canonicalization. Since each Type A tree can form in T' empty, a single vertex, or Type A tree itself, $T_1, \ldots, T_q$ of T, the number of possible forms of subforests of $T_1, \ldots, T_q$ of T is

$$\left(\!\!\binom{q}{3}\!\!\right) = \binom{q+2}{2}.$$

Since the number of subforests of T_i's other than $T_1, \ldots, T_q$ are similar evaluated as above, we can bound the number of AK-rooted subtrees by

$$\binom{q+2}{2} \prod_{i:i>q} 2^{n_i/2} \leq \frac{(q+2)(q+1)}{2} 2^{\sum_{i:i>q} n_i/2} \leq \frac{(q+2)(q+1)}{2} 2^{(n-3q-1)/2}.$$

Thus, to prove the claim, it is sufficient to show that $(q+2)(q+1)2^{(n-3q-3)/2} \leq 2^{n/2} - 1$ for any pair of integers n and q satisfying $n \geq 5$ and $1 \leq q \leq (n-1)/3$. This inequality is transformed to the following

$$\frac{(q+1)(q+2)}{2^{\frac{3(q+1)}{2}}} \leq 1 - \frac{1}{2^{\frac{n}{2}}}.$$

Since the left hand and right hand of the inequality are monotonically decreasing with respect to q and monotonically increasing with respect to n, respectively, the inequality always holds if it is true for $n = 5$ and $q = 1$. In fact, we have

$$\frac{(1+1)(1+2)}{2^{\frac{3(1+1)}{2}}} = \frac{3}{4} = 1 - \frac{1}{2^2} \leq 1 - \frac{1}{2^{\frac{5}{2}}},$$

which completes the proof.   □

Theorem 8. *The winner of Arc Kayles on a tree with n vertices can be determined in time $O^*(2^{n/2}) = O(1.4143^n)$.*

References

1. Bodlaender, H.L., Kratsch, D.: Kayles and nimbers. J. Algorithms **43**(1), 106–119 (2002)
2. Bodlaender, H.L., Kratsch, D., Timmer, S.T.: Exact algorithms for Kayles. Theor. Comput. Sci. **562**, 165–176 (2015)
3. Breuker, D., Uiterwijk, J., van den Herik, H.: Solving 8 × 8 domineering. Theor. Comput. Sci. **230**(1), 195–206 (2000)
4. Bullock, N.: Domineering: solving large combinatorial search spaces. ICGA J. **25**(2), 67–84 (2002)
5. Chen, J., Kanj, I., Xia, G.: Improved upper bounds for vertex cover. Theor. Comput. Sci. **411**(40–42), 3736–3756 (2010)
6. Conway, J.H.: On Numbers and Games. CRC Press, Boca Raton (2000)
7. Dudeney, H.: The Canterbury Puzzles. Courier Corporation, North Chelmsford (2002)
8. Gardner, M.: Mathematical games: cram, crosscram and quadraphage: new games having elusive winning strategies. Sci. Am. **230**(2), 106–108 (1974)
9. Huggan, M.A., Stevens, B.: Polynomial time graph families for Arc Kayles. Integers **16**, A86 (2016)
10. Kobayashi, Y.: On structural parameterizations of node Kayles. In: Akiyama, J., Marcelo, R.M., Ruiz, M.-J.P., Uno, Y. (eds.) Discrete and Computational Geometry, Graphs, and Games, JCDCGGG 2018. LNCS, vol. 13034, pp. 96–105. Springer, Cham (2021). https://doi.org/10.1007/978-3-030-90048-9_8
11. Lachmann, M., Moore, C., Rapaport, I.: Who wins domineering on rectangular boards? arXiv preprint math/0006066 (2000)
12. Lampis, M.: Algorithmic meta-theorems for restrictions of treewidth. Algorithmica **64**(1), 19–37 (2012). https://doi.org/10.1007/s00453-011-9554-x
13. Lampis, M., Mitsou, V.: The computational complexity of the game of set and its theoretical applications. In: Pardo, A., Viola, A. (eds.) LATIN 2014: Theoretical Informatics, LATIN 2014. LNCS, vol. 8392, pp. 24–34. Springer, Heidelberg (2014). https://doi.org/10.1007/978-3-642-54423-1_3
14. Schaefer, T.J.: On the complexity of some two-person perfect-information games. J. Comput. Syst. Sci. **16**(2), 185–225 (1978)
15. Uiterwijk, J.W.: 11 ×11 domineering is solved: the first player wins. In: Plaat, A., Kosters, W., van den Herik, J. (eds.) Computers and Games, vol. 10068, pp. 129–136. Springer, Cham (2016). https://doi.org/10.1007/978-3-319-50935-8_12
16. Uiterwijk, J.W.: Construction and investigation of cram endgame databases. ICGA J. **40**(4), 425–437 (2018)
17. Uiterwijk, J.W.: Solving cram using combinatorial game theory. In: Cazenave, T., van den Herik, J., Saffidine, A., Wu, I.C. (eds.) Advances in Computer Games, vol. 12516, pp. 91–105. Springer, Cham (2019). https://doi.org/10.1007/978-3-030-65883-0_8
18. Uiterwijk, J.W., Barton, M.: New results for domineering from combinatorial game theory endgame databases. Theor. Comput. Sci. **592**, 72–86 (2015)

Author Index

Printed in the United States
by Baker & Taylor Publisher Services

Printed in the United States
by Baker & Taylor Publisher Services